SPECTRAL DATA OF NATURAL PRODUCTS

SPECTRAL DATA
OF
NATURAL PRODUCTS

VOLUME I

K. YAMAGUCHI

Torii & Co., Ltd.,
Nihonbashi, Tokyo, Japan

ELSEVIER PUBLISHING COMPANY

Amsterdam London New York

1970

ELSEVIER PUBLISHING COMPANY

335 Jan van Galenstraat

P. O. Box 211, Amsterdam, The Netherlands

ELSEVIER PUBLISHING CO. LTD.

Barking, Essex, England

AMERICAN ELSEVIER PUBLISHING COMPANY, INC.

52 Vanderbilt Avenue

New York, New York 10017

Library of Congress Card Number: 72–105623

Standard Book Number: 444–40841–X

Preface

In the late 1950's, the chemistry of natural products underwent a dramatic change as regards the methods of structure determination. This change has as its direct basis the recent developments in electronics, and completely upset the classical chemical methods of recognizing unsaturation, functional groups and even the carbon skeleton. The techniques which played the most important role in this revolution were various spectroscopic methods: UV, IR, NMR, ORD, CD and MS, and more recently NMDR and NOE as well as X-ray analysis. Applying these techniques, organic chemists have elucidated the structures of various natural products with the most extraordinary and fascinating structures, which had never been dreamed of by any classical chemist.

The usefulness of these methods naturally prompted the publication of a number of instructive books and collections of spectra, large and small, on all of these methods. To the author's knowledge, however, no collective book has been published which classifies natural products according to structural type, and presents them with the spectral data used for their structure determination. The reason for attempting the present series is that such a systematic collection should be of great help for chemists working in this field.

It consists of two volumes; the first one deals with the compounds whose structures were elucidated before 1963, and the second those structures which were determined in 1964 and 1965. In each volume the natural products are classified into 20 groups according to their structure types. The structure, available physical constants and absorption spectral data (UV, IR, NMR, MS, ORD and CD) of each compound are presented, together with their natural origin and the relevant literature references. The second volume carries, in addition, an Appendix listing the natural origins and references for the compounds whose structures were reported during 1966-1968.

Throughout the preparation, the author felt hopelessly aware of the limitation of what one individual can achieve for this kind of classification and collection, in spite of his volition and effort. Now he, a Don Quixote in 1969, has to confess to himself that, if this kind of collection is needed in the future for the further development of the chemistry of natural products, it can and should be done only by the co-operation of several specialists, using an electronic computer.

The major part of these books was completed in 1966 during the author's previous position at the National Institute of Hygienic Sciences, and the

Appendix was collected at the Research Laboratories, Torii & Co., Ltd., where the author is currently working.

Many colleagues helped in all directions, and the author particularly acknowledges the encouragement of Emeritus Professor Tatsuo Kariyone, the previous Director of the National Institute of Hygienic Sciences, and Professor Shô Itô, of Tohoku University. He is indebted to Professor Hiroshi Ageta, Showa College of Pharmacy; Dr. Shinsaku Natori, Head of the Pharmacognosy Division, National Institute of Hygienic Sciences; and to Dr. Mitsuaki Kodama, Department of Chemistry, Tohoku University, for their helpful advice and meticulous reading and correction of the manuscript. Thanks are also due to Mr. Setsuo Hirokawa, the Managing Director of Hirokawa Publishing Co., and Mr. Takashi Torii, the President of Torii & Co., Ltd., for their help which enabled this book to appear.

JUNE 1969   K. YAMAGUCHI

Abbreviations for the names of periodicales

Acta Chem. Scand.	Acta chemica Scandinavica
Agr. Biol. Chem. Japan	Agricultural and Biological Chemistry (Japan)
Angew. Chem.	Angewandte Chemie
Ann.	Justus Liebigs Annalen der Chemie
Arch. d. Pharm.	Archiev der Pharmazie und Berichte der deutschen pharmazeutischen Gesellschaft
Austr. J. Chem.	Australian Journal of Chemistry
Ber.	Chemische Berichte
Biochem. J.	The Biochemical Journal
Bull. Chem. Japan	Bulletin of the Chemical Society of Japan
Bull. soc. chim. France	Bulletin de la Sociētē Chimique de France
C.A.	Chemical Abstracts
Canad. J. Chem.	Canadian Journal of Chemistry
Chem. & Ind.	Chemistry and Industry (London)
Chem. Pharm. Bull. Japan	Chemical and Pharmaceutical Bulletin (Tokyo)
Collection Czech. Chem. Comm.	Collection of Czechosolovak Chemical Communication
Experientia	Experientia
Helv.	Helvetica Chimica Acta
J.A.C.S.	The Journal of American Chemical Society
J. Biol. Chem.	The Journal of Biological Chemistry
J.C.S.	Journal of Chemical Society (London)
J. Indian Chem. Soc.	Journal of the Indian Chemical Society
J. Org. Chem.	The Journal of the Organic Chemistry
J. Pharm. Sci.	Journal of Pharmaceutical Sciences
Kashi	Nippon Kagaku Zasshi (日本化学雑誌)=Journal of the Chemical Society of Japan
Nature	Nature
Naturwissenschaften	Die Naturwissenschaften
Nôgeishi	Nippon Nôgei-Kagaku Kaishi (日本農芸化学会誌)=Journal of the Agricultural Chemical Society of Japan
Phytochem.	Phytochemistry
Proc. Chem. Soc.	Proceedings of the Chemical Society (London)
Tetrahedron	Tetrahedron
Tetrahedron Letters	Tetrahedron Letters
Y.Z.	Yakugaku Zasshi (薬学雑誌)=Journal of the Pharmaceutical Society of Japan

Other periodicals are referred to Chemical Abstracts and Nihon Kagaku Sôran (日本化学総覧) and abbreviated by the system of Chemical Abstracts.

CONTENTS

1 Hydrocarbons and the Derivatives

[A] NMR Spectra

(1) **Chemical shifts of protons** are indicated in Table 1[1]

Table 1 (Jungnickel *et al.*)[1]

Sample in 30% CCl_4 solv., sweep rate to 10 cps/s, δ=ppm to TMS (internal reference)

Compound	Proton of	Chem. shift δ
CH_3 $H_2C=CH-\overset{\mid}{\underset{\mid}{C}}-CH_3$ CH_3	=CH— =CH_2 —CH_3	5.75 4.83 1.00
$CH_3-CH=CH-CH_2-CH_2-CH_2-CH_3$	=CH— —CH_3, —CH_2—	5.3 0.9~1.9
$HC\equiv C-CH_2-CH_3$	$\equiv C-CH_2-$ $\equiv CH$ —CH_3	2.17 1.78 1.14
$H_3C\diagup\!\!\overset{\overset{H_2}{C}}{\triangle}\!\!\diagdown CH_3$	—CH_3 ring—CH_2—	1.03 0.25
CH_3-CH_2-OH	—OH —CH_3 —CH_2—	4.90 1.17 3.60
$CH_3-CH_2-CH_2-CH_2-OH$	—OH —O-CH_2— —CH_3, —CH_2—	4.7 3.5 0.8~1.7

1) Jungnickel, J.Z *et al*: *Anal. Chem.*, **35**, 938 (1963).

Compound	Proton of	Chem. shift δ
$(CH_3-CH_2-CH_2-)_2O$	$-CH_2-O-$ $\beta-CH_2-$ CH_3-	3.30 1.50 0.90
$(CH_3-CH_2-CH_2-CH_2-)_2O$	$-CH_2-O-$ CH_3-, $-CH_2-$	3.3 0.9~1.4
$H_2C\overset{O}{\diagup}CH-CH_3$	$-CH_2-O-CH-$ $-CH_3$	2.2~2.8 1.22
$CH_3-CH-CH_2-CH_3$ $\overset{|}{SH}$	$-S-\overset{|}{C}H$ $-SH$, $-CH_3$, $-CH_2-$	2.83 0.9~1.6
$CH_3-S-CH_2-CH_2-CH_3$	$S-CH_2-$ CH_3-S, $-CH_2-CH_3$	2.4 0.9~2.0
$\overset{CH_3CH_2}{\underset{CH_3CH_2}{>}}C=O$	$-CO-CH_2-$ $-CH_3$	2.35 1.0
$CH_3(CH_2)_5-\overset{O}{\overset{\|}{C}}-O-(CH_2)_3-CH_3$	$-CO-O-CH_2-$ CH_3-, $-CH_2-$	4.00 0.9~2.2
cyclopentene-CH_3	$=CH-$ CH_3-, $-CH_2-$	5.20 1.7~2.2
cyclopentane-$CH=CH_2$	$=CH-$ $=CH_2$ $-CH_2-$, $>CH-$	5.66 4.83 1.5~2.4
cyclohexane$=CH-CH_3$	$=CH-$ CH_3-, $-CH_2-$	5.05 1.5~2.1
benzene-CH_2-CH_3	Arom. H Ar. $-CH_2-$ $-CH_3$	7.05 2.53 1.15
benzene-$CH_2-CH=CH_2$	Arom. H $=CH-$ $=CH_2$	7.05 5.8 5.0
benzene-NH_2	Arom. H Ar. $-NH_2$	6.3~7.2 3.3
tetralin	Arom. H $\alpha-CH_2-$ $\beta-CH_2-$	6.85 2.60 1.66
pyrrolidine $\overset{}{\underset{N}{}}\overset{}{H}$	$\alpha-CH_2-$ $\beta-CH_2-$, $>NH$	2.84 1.5~1.8

(2) **Approximate τ values of various protons** are indicated in Table 2

Table 2   (Jones, R. A. Y *et al.*[1a])

Atomic Group	τ	Atomic Group	τ
CH₃–C	9. 1	CH₃–O	6. 7
C–CH₂–C	} 8.5~8.8	Ar–NH₂	6. 5
C–CH–C ⎮ C		CH₃–O–CO	6. 2
C–NH₂	8. 4	CH₃–O–Ar	6. 2
CH₃–C=C	8. 2	H₂C=C	5. 3
CH₃–C=O	8. 0	H–C=C ⎮	4. 7
CH₃–S	8. 0	C–OH	4. 7
CH₃–N	7. 8	CO–NH₂	3. 0
H–C≡C	7. 7	H–Ar	2~3
CH₃–Ar	7. 6	Ar–OH	2. 3
CH₃–N–CO ⎮	7. 2	CHO	0. 2
		COOH	−0. 8

(3) **Typical coupling constants** are indicated in Table 3

Table 3   (Jones, R. A. Y *et al.*)

Group	Jab (c/s)	Group	Jab (c/s)
C⟨ Ha / Hb	12~15	Ha⟩C=C⟨Hb	11~18
Ha⟩C–C⟨Hb	8~12	Ha⟩C=C⟨Hb	6~14
Ha⟩C–C⟨Hb	2~4	Ha⟩C=C⟨CHb	0.5~2
Ha–C–C–C–Hb	0	Ha⟩C=C⟨CHb	0

1a) Jones, R. A. Y *et al* : *Chem. & Ind.*, **1962**, 522.

[B]   Saturated Hydrocarbons and Plant Waxes

(1)   Mass Spectra of Saturated Hydrocarbons

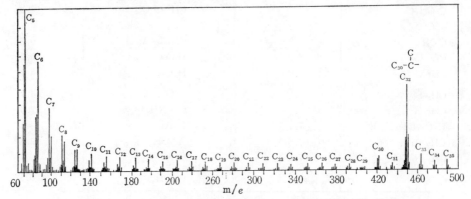

Fig. 1   Phthiocerane (3–Methyltetratriacontane)[1b] (Biemann, K)

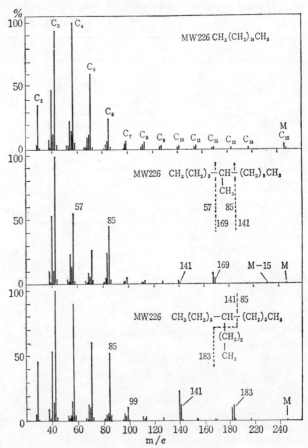

Fig. 2   Three isomeric Hydrocarbons
$C_{16}H_{34}$[1c] (Schumacher, E)

1b) Biemann, K: *Angew. Chem.*, **74**, 102 (1962).

1c) Schumacher, E: *Helv. Chim. Acta*, **46**, 1295 (1963).

(2) Mass spectra of vegetable paraffins and primary alcohols[1d]

(3) Eicosan–1–ol, docosan–1–ol, octacosan–1–ol, tricontan–1–ol and, octadecanoic, eicosanoic, docosanic, tetracosanoic, hexacosanoic, octacosanoic, triacontanoic and dotriacontanoic acids from the heart wood of *Vitex divaricata*[2].

(4) *n*–Alkanes from tridecane to tetracosane, palmitic and stearic acids, β–amyrin, diterpenoids and α–spinasterol from the heart wood of *Manikara bidentata*[3].

[C] Naturally Occurring Polyyne Compounds

(1) Hydrocarbons

(a) **Capillene** $CH_3-C\equiv C-C\equiv C-CH_2-C_6H_5$ (I)

≪Occurrence≫   *Artemisia capillaris*[4], *A. dracunculus* L. and *Chrysanthemum frutescens* L.[5]

UV λ_{max} mμ (logε): 239, 253 (2.73, 2.63)[4]

IR V_{max}cm^{-1}: 2270, 2210, 2160, 1960, 1600, 1500, 1380, 1075, 730, 695[4]

(b) **Benzyldiacetylene** $HC\equiv C-C\equiv C-CH_2-C_6H_5$ (II)

≪Occurrence≫   *Artemisia frutescens* L.[5] bp. $_{0.01}$ 45~50°, n^{22}1.5726

UV λ_{max}^{Hexane}mμ(ε): 267, 263.5, 257, 252, 247, 238.5, 226, 206.5, 196 (115, 184, 270, 400, 310, 560, 660, 12000, 19000)[5]

(c) **Aethusin** $H_3C-CH=CH-(C\equiv C)_2-(CH=CH)_2-CH_2-CH_3$ (III)

≪Occurrence≫   *Aethusa cynapium* L.[6],   mp. 23°, Colorless oil

UV (see Fig. 3), IR (see Fig. 4).

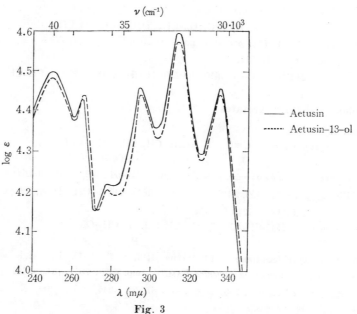

Fig. 3

1d) Waldon, J. D *et al*: *Biochem. J.*, **78**, 435 (1961).

2) Cocker, W *et al*: *Soc.*, **1962**, 5194.   3) Cocker, W *et al*: *ibid.*, **1963**, 677.

4) Harada, R: *J. Chem. Soc. Japan*, **78**, 1031 (1957).   5) Bohlmann, F *et al*: *Ber.*, **95**, 39 (1962).

6) Bohlmann, F *et al*: *ibid.*, **93**, 981 (1960).

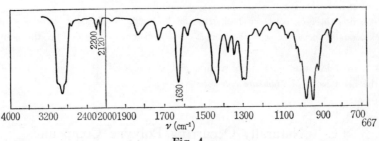

Fig. 4

(d) Tetraynediene[7] $H_3C-CH=CH-(C\equiv C)_4-CH=CH_2$ (IV)

≪Occurrence≫   *Centaurea cyanus* L. and other *Centaurea* spp.,   mp. 87°,

UV $\lambda_{max}^{Et2O}m\mu$: 389.5, 360.5, 336, 314.5, 287, 271, 258

IR $\nu_{max}^{CCl_4}cm^{-1}$: 2190, 2120 (—C≡C—), 1850, 965, 925 (—CH=CH₂), 1615, 945 (—CH=CH—)

(e) Hydrocarbon[7] $H_3C-CH=CH-(C\equiv C)_2-(CH=CH)_2-(CH_2)_4-CH=CH_2$ (V)

≪Occurrence≫   *Centaurea ruthenica* LAM.

UV, IR.

(f) Centaur X₄[8] $H_3C-CH=CH-(C\equiv C)_2-(CH=CH)_2-(CH_2)_4-CH=CH_2$ (VI)

≪Occurrence≫   *Centaurea ruthenica* L., (*trans* compound) mp. 28.5~29.5°

UV $\lambda_{max}^{Et2O} m\mu(\varepsilon)$: 336, 314, 296, 279.5, 266, 250.5(36300, 43400, 32050, 19650, 32800, 41000)

IR $\nu_{max}^{CCl_4}$ cm⁻¹: 1820, 1630, 912(–CH=CH₂), 1625, 945(–CH=CH–), 980(—(CH=CH)₂—*trans*), 2190, 2115

            (—C≡C—)

(g) Tridecatriyne–(7,9,11)–triene–(1,3,5)[9]   (VII)

            $H_3C-(C\equiv C)_3-CH=CH-CH=CH-CH=CH_2$

≪Occurrence≫   *Achillea ptarmica* and other *Achillea* spp.; *Anthemis* spp., mp.88~93°

UV $\lambda_{max}^{Et2O} m\mu$: 356, 331, 311, 283

IR $\nu_{max}cm^{-1}$: 2240 (—C≡C—), 1610, 1007 ((CH=CH)₂), 1830, 910 (—CH=CH₂), 1385 (≡C–CH₃)

(2) Alcohols and their Acetates

(a) Centaur X₂[7]

            $H_3C-CH=CH-(C\equiv C)_2-(CH=CH)_2-CH_2-CH-(CH_2)_2-OCOCH_3$

                                    $\overset{|}{O}COCH_3$

≪Occurrence≫ *Centaurea macrocephala*,   mp. 54~56°, $[\alpha]_D$ —5° (MeOH)

UV[7], IR ν_{max}^{CHCl3} cm⁻¹: 2190, 2125 (—C≡C—), 1615, 945 (—CH=CH—), 980 ((CH=CH)₂),

            1725, 1375, 1240 (—OAc)

[Compounds] $H_3C-CH=CH-(C\equiv C)_3-CH=CH-CH-CH_2-OR_1$ (IX)

                                    $\overset{|}{R_2}$

(b) 12, 13–Diacetoxytridecadiene (2, 10)–triyne–(4, 6, 8)[7][8]   (IX) R_1=Ac, R_2=OAc

≪Occurrence≫ *Centaurea ruthenica* LAM.,   mp. 64°[7], 76°[8], $[\alpha]_D$ +102.2° (CHCl₃)[7]

UV[7] $\lambda_{max}^{Et2O} m\mu$ (ε)[8]: 354.5, 330, 309, 290, 267, 255, 246, 235 (20400, 27800, 20900, 12400, 60400,

            75800, 72900, 52000)

---

7) Bohlmann, F: *Ber.*, **91**, 1631, 1642 (1958).   8) Bohlmann, F *et al*: *ibid.*, **92**, 1319 (1959).
9) Bohlmann, F *et al*: *ibid*, **95**, 1315 (1962).

IR $\nu_{max}^{CCl_4}$ cm$^{-1\ 8)}$: 2180, 2160 (—C≡C—), 1735, 1225 (—OCOCH$_3$), 1610, 945 (—CH=CH—)

(c) 12-Chlorotridecadiene-(2, 10)-triyne (4, 6, 8)-ol (13)[7)8)] (IX) R$_1$=OH, R$_2$=Cl

≪Occurrence≫ *Centaurea ruthenica* LAM., mp. 65.5~66.5° 8), $[\alpha]_D$ —54.5° (CHCl$_3$)

UV λ_{max}^{Et2O} mμ (ε)[8)]: 356, 331, 309, 291, 256, 247, 236 (18000, 25000, 18900, 11500, 69500, **67400,**

47700)

IR $\nu_{max}^{CHCl_4}$ cm$^{-1\ 8)}$: 3530 (OH), 2185, 2170 (—C≡C—), 1610, 945 (—CH=CH—)

(d) 13-Acetoxy-12-chlorotridecadiene-(2, 10)-triyne (4, 6, 8)[8)] (IX) R$_1$=Ac, R$_2$=Cl

≪Occurrence≫ *Centaurea ruthenica* LAM., mp. 40°

UV λ_{max}^{Et2O} mμ (ε)[8)]: 356, 331, 310, 291, 256, 248, 236 (16400, 22200, 15800, 8700, 67100, **64600,**

44700)

(e) Aethusanol A[6)]

$$H_3C-CH=CH-(C≡C)_2-CH=CH-CH-CH_2-CH_2-CH_3 \qquad (X)$$
$$OH$$

≪Occurrence≫ *Aethusa cynapium* L. $[\alpha]_D$ +14.40° (MeOH)

UV: See Fig. 5, IR $\nu_{max}^{CCl_4}$cm^{-1}: 2200, 2120(—C≡C—), 1630, 950(—CH=CH—), 3600, 3350(OH)

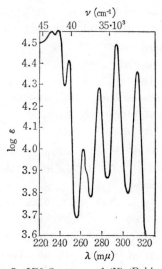

Fig. 5 UV Spectrum of (X) (Bohlmann)

(f) Aethusanol B[6)]

$$HOCH_2-CH=CH-(C≡C)_2-(CH=CH)_2-CH_2-CH_3 \qquad (XI)$$

≪Occurrence≫ *Aethusa cynapium* L., mp. 69~72°

UV: See Fig. 3, IR $\nu_{max}^{CCl_4}$ cm^{-1}: 2200, 2120 (—C≡C—), 1630, 950 (CH=CH), 1630, 1580, **980**

((CH=CH)$_2$), 3600, 3350 (OH)

[Compounds] $H_3C-(C≡C)_2-(CH=CH)_3-CH_2-CH_2-OR$ (XII)

(g) Chamomilla alcohol[10)] (XII) R=H

≪Occurrence≫ *Matricaria chamomilla* L. and *M. discoidea* DC., mp. 86.5°

UV λ_{max}^{Et2O} mμ (ε): 332.5, 316, (302), 251, 242 (62500, 60200, 34500, 20000, 11200)

(h) Chamomilla ester[10)] (XII) R=Ac

≪Occurrence≫ *Matricaria chamomilla* L. and *M. discoidea* DC., mp. 65~66.5°

UV λ_{max}^{Et2O} mμ (ε): 331.5, 315, (300), 250, 241 (62800, 60600, 36000, 21500, 11700)

IR ν_{max}^{CCl4}cm^{-1}: 2220, 2130 (C≡C), 1740, 1230 (OAc), 1600, 993 ((CH=CH)$_n$), 1460, 1423 (δ_{CH2})

(i) Triyne diene[11]   H$_3$C–(C≡C)$_3$–(CH=CH)$_2$–CH$_2$–CH$_2$–CH$_2$–OCOCH$_3$ (XIII)

≪Occurrence≫   *Tanacetum vulgare* L.,   mp. 36~37° (Synthetic sample)[12]

UV λ_{max} mμ (ε): 348, 325, 305.5, 288, 268.5, 258 (33900, 41600, 27500, 14600, 114500, 56300)

(j) Matricaricanol[13]   H$_3$C–CH=CH–(C≡C)$_2$–CH=CH–CH$_2$OH   (XIV)

≪Occurrence≫   *Centaurea ruthenica* LAM.,   mp. 95~99°

UV λ_{max}^{Et2O} mμ (ε): 310, 291, 274, 263, 235 (16900, 21000, 14400, ·7600, 27500)

IR ν_{max}^{CHCl3} cm^{-1}: 3630, 3450, 1010 (OH), 2220 (—C≡C—), 952 (CH=CH)

(k) Compound[13]   H$_3$C–CH=CH–(C≡C)$_2$–(CH=CH)$_2$–CH–CH$_2$–OAc   (XV)
                                                                          |
                                                                  OCOCH$_3$

≪Occurrence≫   *Centaurea ruthenica* LAM.,   mp. 57°, $[\alpha]_D$ +94.5° (Acetone)

UV λ_{max}^{Et2O} mμ (ε): 335, 314, 295, 277, 265, 246 (28600, 37900, 27800, 16600, 34600, 38000)

IR ν_{max}^{CCl4} cm^{-1}: 2180, 2150, 2120 (—C≡C—), 1760 (C=O), 1620, 987, 950 (—CH=CH—)

(l) Compound   (IX) R$_1$=H, R$_2$=OAc[13]

≪Occurrence≫   *Centaurea ruthenica* LAM.,   mp. 86°, $[\alpha]_D$ +39° (CHCl$_3$)

UV λ_{max}^{Et2O} mμ (ε): 352, 329, 307.5, 289, 267.5, 254, 245, 234 (16100, 22800, 17100, 9500, 52000,
                                              62900, 61200, 41200)

IR ν_{max}^{CHCl3} cm^{-1}: 3600, 3460, 1043 (OH), 2190, 2170 (C≡C), 1750 (C=O), 1630w, 995 (—CH=CH—)

(m) Compound   (IX) R$_1$=H, R$_2$=OH[13]

≪Occurrence≫   *Centaurea ruthenica* LAM.,   mp. 127~128°, $[\alpha]_D$ +30.8° (MeOH)

UV λ_{max}^{Et2O} mμ (ε): 353, 328.5, 307.5, 289, 267, 254, 245, 235 (19100, 26800, 20000, 10700, 61600,
                                              76000, 70500, 46400)

IR ν_{max}^{CHCl3} cm^{-1}: 3610, 3290, 1048 (O–H), 2190, 2170w (—C≡C—), 1630, 953 (—CH=CH)

(n) Centaur Z[14]   AcO–CH=CH–C≡C–C≡C–CH=CH–CH=CH–OAc   (XVI)

≪Occurrence≫   *Centaurea montana* L.

UV $\lambda_{max}^{cyclohexane}$ mμ (ε): 338.5, 316.5, 297.5, 280.4, 267.5, 249.5, 215.0, 208.6 (34000, 44900, 30800,
                                              16700, 30500, 35100, 15200, 14700)

IR ν_{max}^{CCl4} cm^{-1}: 2119, 2183 (—C≡C–C≡C—), 1364 (ν_{C-H} of AcO), 1742 ($\nu_{C=O}$ of AcO), 624, 657,
                                    1016, 1041, 1222 (AcO), 946, 979 (all *trans*-CH=CH–), no *cis* absorption at 650~
                                    820

(o) Tridecatetrayne (4,6,8,10)–diene (3,12)–ol (1)[9]   H$_2$C=CH–(C≡C)$_4$–CH=CH–CH$_2$OH   (XVII)

≪Occurrence≫   *Bidens leucanthus* L.

UV: see Fig. 6

(3) **Aldehydes** and **Ketones**

(a) Tridecatetrayne (4,6,8,10)–diene (3,12)–al (1)[9]   H$_2$C=CH–(C≡C)$_4$–CH=CH–CHO   (XVIII)

≪Occurrence≫   *Bidens leucanthus* L.

UV: see Fig. 6,   IR: see Fig. 7

---

10) Bohlmann, F *et al*: *Ber.*, **93**, 1931 (1960).   11) Bohlmann, F *et al*: *Ber.*, **93**, 1937 (1960).
12) Bohlmann, F *et al*: *ibid.*, **94**, 3189 (1961).   13) Bohlmann, F *et al*: *ibid.*, **94**, 3179 (1961).
14) Löfrgen, N *et al*: *Acta Chem. Scand.*, **17**, 1065 (1963).

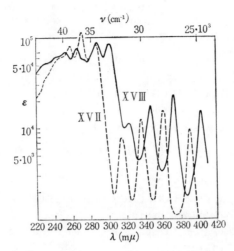

Fig. 6 UV Spectra of (XVII) and (XVIII)

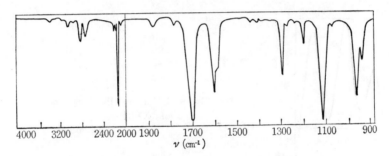

Fig. 7 IR Spectrum of (XVIII) (Bohlmann)

(b) Dehydrofalcariunone $H_2C=CH-(CH_2)_5-CH=CH-CH_2-C\equiv C-C\equiv C-CO-CH=CH_2$ (XIX)

《Occurrence》   *Artemisia campestris*

UV λ_{max}^{Et2O} mμ (ε): 283, 268, 253.5, 240.5, 227 (6700, 8700, 6100, 3200, 2300)

IR ν_{max} cm^{-1}: 2240, 2150 ($-C\equiv C-$), 1680 (C=O), 1840, 916 ($-CH=CH_2$)

(c) Capillin[15] $C_6H_5-CO-C\equiv C-C\equiv C-CH_3$ (XX)

UV λ_{max}^{Et2O} mμ[5]: 297.5, 293.5, 265

IR λ_{max} μ: 5~6, 6.24, 6.32 (arom.), 14.33 (〈benzene〉$-$), 4.45, 4.64 ($-C\equiv C-C\equiv C-$),

7.25 w ($-C\equiv C-CH_3$), 6.09 ($-CO-C\equiv C-$)

(d) Phenylpentadiyne−(2, 4)−one−(1)[18] $C_6H_5CO-C\equiv C-C\equiv CH$ (XX–a)

《Occurrence》   *Chrysanthemum segetum* L.,   mp. 106°

UV λ_{max} mμ (ε): 287, 273, 262.5 (13500, 18200, 14100)

IR ν_{max} cm^{-1}: 3300, 2250, 2205, 2075 ($-C\equiv C-C\equiv C-H$), 1665 (C=O)

(e) Falcarindione[12] $n-C_7H_{15}-CH=CH-CO-(C\equiv C)_2-CO-CH=CH_2$ (XXI)

《Occurrence》   *Carum carvi* L.[16]

---

15) Imai: *Y. Z.*, **76**, 405 (1956).   16) Bohlmann, F *et al*: *Ber.*, **94**, 958 (1961).

(Synthetic sample)[12)]

   UV λ_{max} mμ (ε): 303, 285, 268, 253, 242 (6780, 9850, 11400, 12780, 14300)

   IR ν_{max} cm^{-1}: 2200, 2130 (—C≡C—), 1660 (C=O), 980, 964 (—CH=CH$_2$);

(Natural sample)[16)] yellowish oil

   UV λ_{max}^{CS2} cm^{-1}: 2194, 2134 (—C≡C—), 1659 (CO), 1613, 980, 964 (—CH=CH$_2$) 675 (*cis*–CH=CH—)

(f) Falcarinolone[16)]

$$n\text{–}C_7H_{15}\text{–}CH{=}CH\text{–}CH\text{–}C{\equiv}C\text{–}C{\equiv}C\text{–}CO\text{–}CH{=}CH_2 \qquad\qquad (XXII)$$
$$\underset{\displaystyle OH}{\big|}$$

≪Occurrence≫   *Carum carvi* L., yellowish oil, $[\alpha]_D^{22}+255°$ (Et$_2$O)

   IR ν_{max}^{CS2} cm^{-1}: 2236, 2146 (—C≡C—), 1660 (C=O), 1610, 980, 966, (–CH=CH$_2$), 3600, 3440 (OH),
           675 (*cis*–CH=CH—)

(g) Falcarinone[16)]

$$n\text{–}C_7H_{15}\text{–}CH{=}CH\text{–}CH_2\text{–}C{\equiv}C\text{–}C{\equiv}C\text{–}CO\text{–}CH{=}CH_2 \qquad\qquad (XXIII)$$

≪Occurrence≫   *Falcaria vulgaris* BENTH.

   IR ν_{max}^{CS2} cm^{-1}: 2231, 2160, 2086 (—C≡C—), 1652 (C=O), 3020, 1615, 980, 964 (—CH=CH$_2$), 675
           (*cis*–CH=CH—)

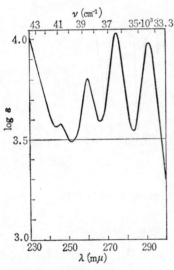

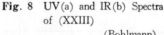

Fig. 8   UV(a) and IR(b) Spectra
            of (XXIII)
                        (Bohlmann)

Fig. 8a

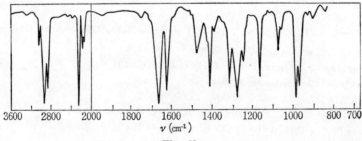

Fig. 8b

(4)　Ethers and Epoxides

(a)　Acetylenic ether　$C_{12}H_{22}O^{17)}$

$$CH_3-CH-(CH_2)_7-C\equiv CH \hspace{3cm} \text{(XXIV)}$$
$$\overset{|}{O}CH_3$$

≪Occurrence≫　　*Litsea odorifera* VAL.,　　　d^{20} 0.8424,　n_D^{20} 1.4400

IR　ν_{max} cm^{-1}: 1090 s(-O-),　3300 m,　2150 w(ν_{C-H}, $\nu_{C\equiv C}$)

NMR:　8.23τ(C≡C-H) multiplet due to hydrogen bond formation with the solvent, methyl doublet
　　　　($CH_3-\underset{|}{CH}-CH_2-$)

(b)　Pontica epoxide[19)]

$$H_3C-(C\equiv C)_3-CH=CH-CH-CH-CH=CH_2 \hspace{2cm} \text{(XXV)}$$
$$\underset{O}{\diagdown\diagup}$$

≪Occurrence≫　　*Artemisia pontica* L.　　mp. 66°,　$[\alpha]_D^{23}$ +201° (acetone)

UV　λ_{max} mμ (ε): 333.5, 311.5, 292, 275.5, 250.5, (243.5) (19800, 28300, 21000, 11100, 75000, 64500)

IR　ν_{max} cm^{-1}: 2210 (-C≡C-), 1640, 945 (-CH=CH-), 985, 915 (-CH=CH$_2$)

(c)　Compound[13) 24)]

$$H_3C-(C\equiv C)_4-CH=CH-\overset{O}{\overset{\diagup\diagdown}{CH-CH}}_2 \hspace{2cm} \text{(XXVI)}$$

≪Occurrence≫　　*Centaurea ruthenica* LAM.,　　mp. 95.5~97°,　$[\alpha]_D$ +75.1° (CHCl$_3$)

UV　λ_{max}^{Et2O} mμ (ε): 374, 347, 323.5, 303, 285, 272, 258, 239, 229, 219 (12000, 18700, 16000, 9800,
　　　　　6650, 155100, 107600, 81400, 77100, 62100)

IR　$\nu_{max}^{CCl_4}$ cm^{-1}: 2230, 2150(-C≡C-), 1295, 1245, 1170, 1135, 850(-CH-CH-), 950, 925(-CH=CH-)
　　　　　　　　　　　　　　　　　　　　　　　$\underset{O}{\diagdown\diagup}$

(d)　Chlorohydrin[13) 24)]

$$H_3C-(C\equiv C)_4-CH=CH-CH-CH_2-OH \hspace{2cm} \text{(XXVII)}$$
$$\overset{|}{C}l$$

≪Occurrence≫　　*Centaurea ruthenica* LAM.,　　mp. 111~112° (decomp.),　$[\alpha]_D$ -88.5° (CHCl$_3$)

UV　λ_{max}^{Et2O} mμ (ε): 372, 347, 323, 302, 284, 270, 256, 238~239(9400, 15800, 13200, 7800, 5050,
　　　　　152100, 106100, 84700)

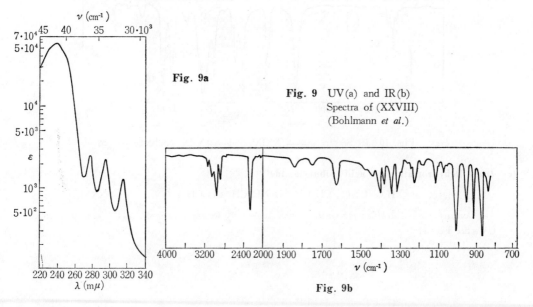

Fig. 9a

Fig. 9　UV(a) and IR(b)
Spectra of (XXVIII)
(Bohlmann *et al.*)

Fig. 9b

---

17)　Matthews, W. S *et al*: *Chem. & Ind.*, **1963**, 122.　　18)　Bohlmann, F *et al*: *Ber.*, **96**, 226 (1963).

18a)　Bohlmann, F *et al*: *Ann.*, **668**, 51 (1963).　　　19)　Bohlmann, F *et al*: *Ber.*, **93**, 1937 (1960).

IR ν_{max}^{CHCl3} cm^{-1}: 3600, 3400, 1070 (O–H), 2250, 2200, 2170, 2100 (—C≡C—), 1630, 950 (—CH=CH—)

(e) Compound[9]

$$H_3C-(C≡C)_3-HC\overset{\displaystyle CH-CH=CH-CH=CH_2}{\underset{O}{\diagdown\diagup}}$$ (XXVIII)

《Occurrence》   *Artemisia cota* L.

(5) Acidesters and Acidamides

(a) Frutescin[20]

$$\underset{OCH_3}{\overset{CH_2-C≡C-C≡C-CH_3}{\diagup}}COOCH_3$$ (XXIX)

《Occurrence》   *Chrysanthemum frutescens* L.,   mp. 67°,

UV λ_{max}^{Et2O} mμ (ε): 280 (2900),   IR: See Fig. 10

(b) Desmethylfrutescin[20]

$$\underset{OCH_3}{\overset{CH_2-C≡C-C≡CH}{\diagup}}COOCH_3$$ (XXX)

《Occurrence》   *Chrysanthemum frutescens* L.,   mp. 68.5°

UV λ_{max}^{Et2O} mμ (ε): 280(2700),

IR: See Fig. 10

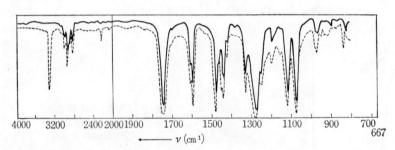

Fig. 10 IR Spectra of——(XXIX) and······(XXX) (Bohlmann *et al.*)

(c) *trans*-Dehydromatricaric acid isobutylamide[20a]

$$H_3C-(C≡C)_3-CH=CH-CO-NH-CH_2-CH(CH_3)_2$$ (XXXI)

《Occurrence》   *Achillea ptarmica*,   mp. 133～139°(decomp.) (natural), 144.5～145.5° (Synth.)

UV: See Fig. 11

---

20) Bohlmann, F *et al*: *Ber.*, **95**, 602 (1962).

20a) Bohlmann, F *et al*: *ibid.*, **95**, 1742 (1962).

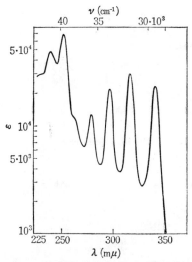

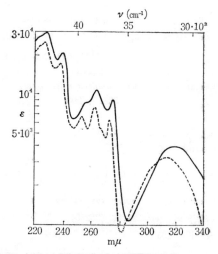

Fig. 11 UV Spectrum of (XXXI)
(Bohlmann *et al.*)

Fig. 12 UV Spectra of (XXXII) ———
and Isocoumarin (Bohlmann) ·······

(6) Coumarins

(a) Butin–(2)–yl–isocoumarin[5][20]

$$-CH_2-C\equiv C-CH_3 \qquad (XXXII)$$

≪Occurrence≫   *Artemisia dracunculus* L.,   mp. 124°

UV: See Fig. 12[5],   IR: See Fig. 13[5]

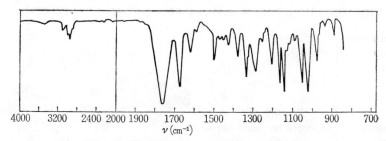

Fig. 13 IR Spectrum of (XXXII) (Bohlmann)

(7) Furano Derivatives

(a) Atractylodin   (1–(2–Furyl)–*trans*–1, *trans*–7–nonadiene–3, 5–diyne)[21]

$$\overset{t}{-}CH=CH-C\equiv C-C\equiv C-\overset{t}{-}CH=CH-CH_3 \qquad (XXXIII)$$

≪Occurrence≫   *Atractylodes* spp.

UV λ_{max} mμ (ε): 258, 272, 336, 354(12700, 14300, 40800, 32500)

---

21) Yoshioka, I *et al*: *Chem. Pharm. Bull.*, **8**, 949 (1960).

IR ν_{max}^{nujol} cm^{-1}: 1014, 882, 735 (furan); $\nu_{max}^{CCl_4}$ cm^{-1}: 3135, 2193w (—C≡C—), 2120w (—C≡C–C≡C—), 1550w, 1484, 1014, 883 (—C=C—)

(b) 2-[Nonatriyne (3, 5, 7)-ene-(1)-yl-(1)]-2, 3-dihydrofuran[22]

$$\text{⟨O⟩-CH=CH-(C≡C)}_3\text{-CH}_3 \qquad (XXXIX)$$

≪Occurrence≫ *Chrysanthemum leucanthemum* L., mp. 58°

UV λ_{max} mμ (ε): 330, 308.5, 290, 273, 258, 241, 230 (10900, 15600, 11500, 6300, 4150, 107500, 72000)

IR ν_{max} cm^{-1}: 2230 (—C≡C—), 1610, 945 (—CH=CH—) 1635, 1058 (enol ether)

(c) Compound $C_{14}H_{14}O_2$[22a]

$$H_3C-(C≡C)_2-CH=⟨O⟩=CH-CO-CH_2-CH_3 \qquad (XXXIX-a)$$

≪Occurrence≫ *Anacyclus radiatus* Lois., mp. 78.5°

UV λ_{max}^{Et2O} mμ (ε): 291.5, (282.5), 229.5 (27900, (25200), 11800)

IR: See Fig. 13a

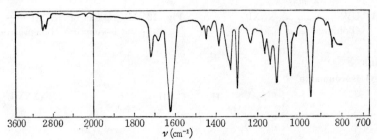

Fig. 13a IR Spectrum of (XXXIX-a) (Bohlmann *et al.*)

NMR 60Mc, in CCl$_4$

TMS τ: 8.97t (J=7cps) (-CO-CH$_2$-CH$_3$), 7.6q (J=7) (-CO-CH$_2$-CH$_3$), 7.99d (J=1.2) (≡C-CH$_3$), 7.02m (-CH$_2$-CH$_2$-), 5.32q (J=1.2) (≡C-CH), 3.99t (J=1.5) (=CH-CO-)

(8) Miscellaneous

(a) Compound[22]

$$\underset{CH_3-(C≡C)_2}{\overset{H}{\underset{}{C=}}} \qquad (XL)$$

≪Occurrence≫ *Matricaria matricarioides* L.

UV: See Fig. 14 IR: See Fig. 15

(b) Compound[22]

$$\underset{CH_3-(C≡C)_2}{\overset{H}{\underset{}{C=}}} \qquad (XLI)$$

≪Occurrence≫ *Matricaria matricarioides* L.

UV: See Fig. 14 IR: See Fig. 15

22) Bohlmann, F *et al*: *Ber.*, **94**, 3193 (1961).
22a) Bohlmann, F *et al*: *ibid.*, **96**, 588 (1963).

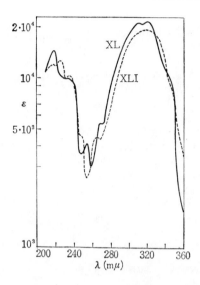

Fig. 14 UV Spectra of (XL) and (XLI)

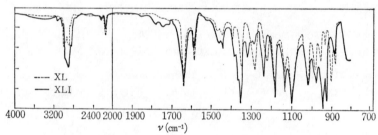

Fig. 15 IR Spectra of (XL) and (XLI) (Bohlmann *et al.*)

(c) Compounds[22]

(XLII) ⇌ (XLIII)

《Occurrence》 *Chrysanthemum leucanthemum* L.

(XLII) mp. 91°, UV λ_{max} mμ (ε): (326), 313.5, 265, (250~248), (238~233), (229~226) (17850, 20600, 5800 (5400), (13100), (13400))

(XLIII) mp. 123°, UV λ_{max} mμ (ε): 337, 317.5, 306, 255, 240.5, 233 (14300, 24500, 22700, 4600, 12100, 12500)

(d) Compound[22]

(XLIV)

《Occurrence》 *Chrysanthemum leucanthemum* L. mp. 103°

UV λ_{max} mμ (ε): 291.5, 276, 263, 222, 214 (13300, 15800, 11100, 33500, 30600)

(e) Compound $C_{15}H_{12}O_5$[18a] (XLV–a)

≪Occurrence≫ *Chrysanthemum leucanthemum* L. mp. 153.5°, $[\alpha]_{546} -78°(Et_2O)$

UV λ_{max}^{Et2O} mμ (ε): 292, 277, 263.5(251), 222.5 (15500, 18900, 13400, (7200), 3200)

IR $\nu_{max}^{CHCl3(CCl4)}$ cm^{-1}: 2160 (—C≡C—), 1750, 1240 (OAc), 1670, 1630 (—C≡C—), 875 (epoxide)

NMR 60 Mc, in CDCl$_3$, internal standard TMS τ:

(f) Compounds[18]

(XLV) R=H

≪Occurrence≫ *Chrysanthemum arcuticum* L., mp. 78°

UV λ_{max}^{Et2O} mμ (ε): 317, 263, 248, 235.5(18200, 4400, 4700, 10200)

IR ν_{max}^{CCl4} cm^{-1}: 2190, 2150(—C≡C—), 1640, 1590(enol ether)

NMR 60 Mc, in CCl$_4$, τ: 8.10 (6) (—(CH$_2$)$_3$—) ; 7.93 (3) (≡C–CH$_3$) ; 6.02(2) (—CH$_2$O—) ; 4.88 (1) (—C=) ; 3.6 d, 3.49 d (1), 3.14, 3.06(1) $\left(\begin{smallmatrix}H\\\end{smallmatrix}\rangle C=C \langle\begin{smallmatrix}H\\\end{smallmatrix}\right)$

(XLV) R=—OCOCH$_3$

≪Occurrence≫ *Chrysanthemum arcuticum* L., mp. 125~126°

UV λ_{max}^{Et2O} mμ (ε): 316, 264.5, 249, 236.5, 223.5, 215.5 (24950, 6800, 7080, 16000, 17700, 18400)

IR ν_{max}^{CCl4} cm^{-1}: 2190, 2150 (—C≡C—), 1750, 1250 (—O–COCH$_3$), 1640, 1590 (enolether)

(g) S–Compound[22] C_6H_5–CH$_2$–C≡C–CH=CH–S–CH$_3$ (XLVI)

≪Occurrence≫ *Chrysanthemum segetum* L., bp. 110°/$_{0.1}$

UV λ_{max}^{Et2O} mμ (ε): 281, 274, (18400, 18400)

IR ν_{max}^{CCl4} cm^{-1}: 2270, 2230 (—C≡C—), 1610, 1580 (—CH=CH–S—), 3070, 3235, 1970, 1870, 1820 1505 (arom.)

NMR: ~2.7 m (5) (C$_6$H$_5$), 3.7 d (1) (—CH=CH–S—) ~4.6 two triplets, ~6.3 d (2) (—CH$_2$—), 7.7 (3) (—S–CH$_3$)

(h) S–Compound[22] C_6H_5–CO–C≡C–CH=CH–S–CH$_3$ *cis* (XLVII)

≪Occurrence≫ *Chrysanthemum segetum* L., mp. 42.5°

UV λ_{max}^{Et2O} mμ (ε): 350, 259(13600, 15200)

IR ν_{max}^{CCl4} cm^{-1}: 2230, 2180 ; 1650 (C=O)

---

23) Bohlmann, F *et al*: *Ber.*, **95** 1733 (1962).   24) Bohlmann, F *et al*: *ibid.*, **95**, 2939 (1962).

(i)　S–Compound[23)]

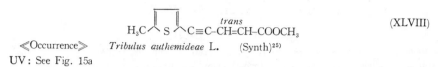

$$H_3C \diagdown S \diagdown C \equiv C\text{-}CH\text{=}CH\text{-}COOCH_3 \qquad \text{(XLVIII)}$$

trans

≪Occurrence≫　*Tribulus authemideae* L.　(Synth)[25)]

UV: See Fig. 15a

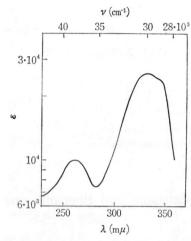

Fig. 15a　UV Spectrum of (XLVIII) (Bohlmann *et al.*)

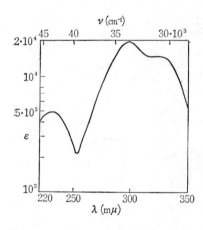

(a)　in ether

Fig. 16　UV (a) and IR (b) Spectra of (XLVIII–a)
(Bohlmann *et al.*)

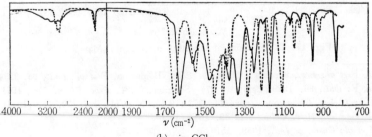

(b)　in CCl₄

(j) S–Compound[25) 25a)]

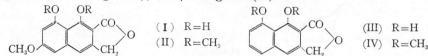

$$H_3C-C\equiv C \overset{OH}{\underset{S}{\diagdown}} CO\cdot CH_3 \qquad \text{(XLVIII–a)}$$

≪Occurrence≫   *Artemisia arborescens* L.[25a)],   (Synth.)[25)] mp. 99.5~100°

UV λ_{max}^{Et2O} $m\mu$ (ε): 323, 299, 232 (14700, 19500, 4480), See Fig. 16a

IR ν_{max}^{CCl4} cm⁻¹: 2250 (—C≡C—), 1630 (—COCH₃), 1570 ( ), See Fig. 16b

NMR in CDCl₃, τ: 7.9 (3) (≡C–CH₃), 7.65 (3) (—CO–CH₃), 3.3 (1)

(k) S–Compound[26)]

$$OHC \overset{}{\underset{S}{\diagdown}} \overset{}{\underset{S}{\diagdown}} C\equiv C-CH=CH_2 \qquad \text{(XLIX)}$$

≪Occurrence≫   *Flavaria repanda* Lag.

Naturally occurring acetylenic compounds[27)] (Review)

［D］ Naphthalene and Phenanthrene Derivatives

Naturally occurring non–basic naphthalene and phenanthrene derivatives are described here. No hydrocarbon of such type appears in vegetable kingdom.

(1) Naphthalides α–**Sorigenin** (I) and β–**Sorigenin** (III)[27a~d)]

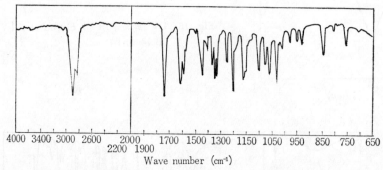

(I) R=H
(II) R=CH₃

(III) R=H
(IV) R=CH₃

Fig. 17   IR Spectrum of Dimethyl–α–sorigenin (Matsui *et al.*)

25) Bohlmann, F *et al*: *Ber.*, **96**, 584 (1963).   25a) Bohlmann, F *et al*: *ibid.*, **95**, 2934 (1962).
26) Bohlmann, F: *ibid.*, **96**, 1229 (1963).   27) Sörensen, N.A: *Proc. Chem. Soc.*, **1961**, 98.
27a) Nikuni, Z: *J. Agr. Chem. Soc. Japan* **14**, 352 (1938), N *et al*: *ibid.*, **20**, 283 (1944).
27b) Haber, R.G *et al*: *Helv. Chim. Acta* **39**, 1654 (1956).
27c) Horü, Z *et al*: *Chem. & Ind.*, **1959**, 1576.
27d) Matsui, M *et al*: *Agr. Biol Chem.* (*Tokyo*) **27**, 40 (1963).

≪Occurrence≫ *Rhamnus japonica*

α–Sorinin $C_{24}H_{28}O_{14}$＝Primverose＋(I), β–Sorinin $C_{23}H_{26}O_{13}$＝Primverose＋(III)

IR ν_{max}^{nujol} cm⁻¹ of Dimethyl–α–sorigenin (II)： 1756 (lactone), 1631, 1620, 1595, 1515, (naphthalenering), See Fig. 17

(2) **Aristolochic acid** and the **Derivative**

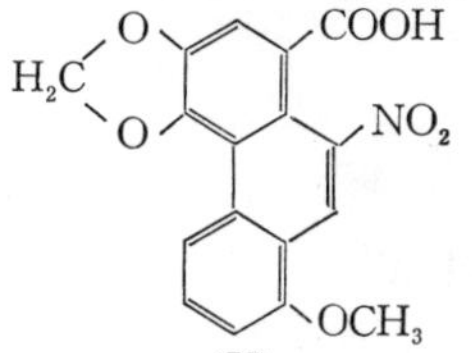

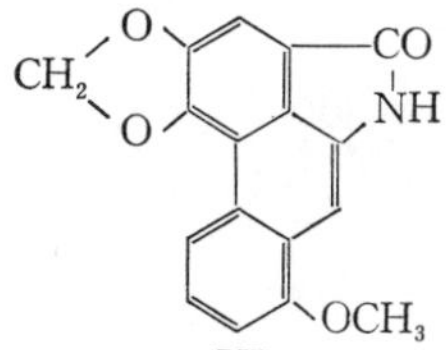

≪Occurrence≫ *Aristolochia reticulata* L., *A. serpentaria* L., *A. longa* L., *A. indica* L.[27e], and *A. funghi* Wu[27f]

(a) **Aristolochic acid** (V)　mp. 275°

UV λ_{max} mμ (ε)： 223, 250, 318, 390 (30000, 30600, 11500, 5700)[27e]；

IR λ_{max}^{nujol} μ： 5.93 (C=O), 6.56, 7.43 (NO₂)[27f]

(b) **Aristololactam**[27f] **or Aristo-red** (VI)[27e]　mp. 305°

UV λ_{max} mμ (log ε)： 241, 250, 259, 291, 300 (4.50, 4.47, 4.56, 4.16, 4.15)

IR λ_{max}^{nujol} μ： 3.15 (N–H), 5.91 (γ–lactam)

(3) 5,6,7,8-**Tetrahydronaphthalene**-1,2,4-**triol** (VII) [Note]

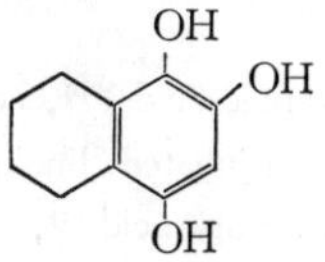

≪Occurrence≫ Spruce (*Picea* spp.),　mp. 140～141°, $C_{10}H_{12}O_3$,

Trimethyl ether　mp. 55～56°

NMR in CCl₄, τ： 7.33 (α–CH₂), 8.27 (β–CH₂), 2.5～3.4 (arom. and vinyl), 6.15, 6.21, 6.38 (OCH₃)

27e) Coutts, R.T *et al*: *J. Rh. Rh.*, **11**, 607, (1959).

27f) Tomita, M *et al*: *Y. Z.* **79**, 973 (1959).

[Note] Existing with piceatannol：

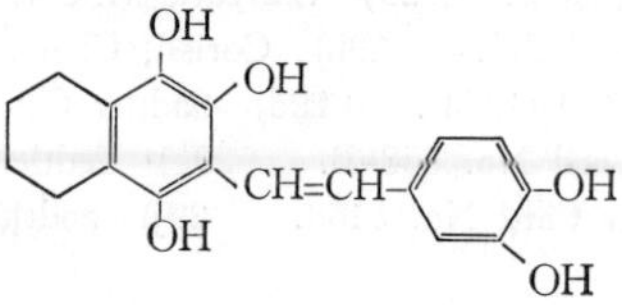

2 Carboxylic Acids

[A] UV Spectra[28]

Formic, acetic and, *n*–butylic acids, ethyl acetate[28a] palumitic and, myristic acids[28b)28c], oxalic, succinic and pyruvic acids[28)28c], ascorbic acid[28c]. Acrylic[28c], crotonic, sorbic, octatrienic[28d], decatetraenic[28e] and 2,3–oleic acids[28e]. Monoenic or nonconjugated dienic acids and conjugated trienic acids[28f]. Oleic, elaidic, ricinelaidic acids and their oxides[28g]. Linolenic, elaidolin olenic, pseudo elaeostearic and α–elaeostearic acids[28h]. 9, 10– Linolic, α and β–elaeostearic[28i] and parinaric acids[28j]. Ethyl linolate and it's oxidated products. *Cis* and *trans* cinnamic acids.

[B] IR Spectra

C_{10}, C_{14}, C_{18}, C_{24}–*n*–Aliphatic acids[29a] and decanoic acids[29b], acetic, propionic and butyric acids[29b], $C_2 \sim C_{12}$–*n*–saturated fatty acids[29b], elaidic, linoelaidic, conjugated linoleic, α and β–elaeostearic acids and their methylesters[29c]. Oleic acid[29d], methyl linolate[29e], tartaric acid[29f], and citric acid[29g]. Ethyl acetoacetate[29h], ethyl levulate[29i], cinnamic acid[29j] and 2–benzofuroic acid[29k].

[C] Miscellaneous

(1) Caffeic acid (I) and chicoric acid (II)[29]

---

28) Yamaguchi, K: Shokubutsu Seibun Bunseki Ho, Vol. II (Nankodo Pub. Inc., Tokyo, 1961)
28a) Ley *et al*: *Z. physik. Chem.*, **B 17**, 177 (1932).　　28b) Bielecki: *Ber.*, **47**, 1690 (1914).
28d) Hausser *et al*: *Z. Physik. Chem.*, **B 29**, 371 (1935).
28e) Hulst *et al*: *Rec. Trav. Chim. Pays. Bas.*, **54**, 639, 644 (1935).
28f) Wendland, Wheeler: *Anal. Chem.*, **26**, 1469 (1954).　　28g) King: *Soc.*, **1950**, 2897.
28h) Holman *et al*: *J. A. C. S.*, **67**, 1392 (1945).　　28i) Mannecke *et al*: *Farben. Ztg.*, **32**, 2887 (1927).
28j) Kaufmann *et al*: *Fett. u Seifen.*, **45**, 302 (1938).　　29) Scarpati, M. C *et al*: *Tetrahedron* **4**, 43 (1958).
29a) Meikelejohn *et al*: *Anal. Chem.*, **29**, 329 (1957).　　29b) Corish, Chapman: *Soc.*, **1957**, 1746.
29c) Wendland, Wheeler: *Anal. Chem.*, **26**, 1469 (1954).　　29d) Sadtler Card No. 915–B.
29e) Sadtler Card No. 913.　　29f) Sadtler Card No. 658–B.　　29g) Sadtler Card No. 765–B.
29h) Sadtler Card No. 101–B.　　29i) Sadtler Card No. 5168.　　29j) Sadtler Card No. 29–B.
29k) Sadtler Card No. 1463.

HOOC–CH=CH– (OH, OH benzene ring)

CH–O–CO–CH=CH– (OH, OH benzene ring)
CH–O–CO–CH=CH– (OH, OH benzene ring)
COOH / COOH structure

(I)   (II)

≪Occurrence≫ *Chicorium intybus*,   (II) mp. 206°, $[\alpha]_D$ +383.5 (MeOH)   UV: See **Fig. 18**,
IR (I), (II): See **Fig. 19**.

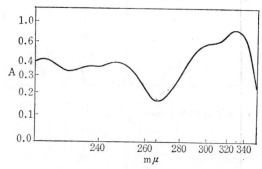

Fig. 18 UV Spectrum of Chicoric acid.

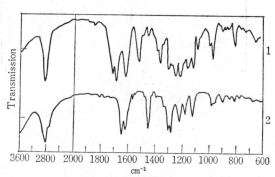

Fig. 19 IR Spectra of *d*–Chicolic acid (1) and Caffeic acid (2)
(KBr)   (Scarpati *et al.*)

(2) Chaulmoogric acid[30]

$(CH_2)_{12}$–COOH   (III)

≪Occurrence≫ *Taraketogenos Kurzii* etc, mp. 66°, $[\alpha]_D$ +60.24°
IR ν_{max} cm^{-1}: 724, 921, 1111, 1190, 1205, 1232, 1266, 1282, 1316, 1379, 1418, 1429, **1471**, **1724**

(3) Densipolic acid (IV, R=H)[31]

$$CH_3–CH_2–\overset{16}{C}H=\overset{15}{C}H–(CH_2)_2–CH–CH_2–\overset{10}{C}H=\overset{9}{C}H–(CH_2)_7–COOR \quad (IV)$$
$$\underset{OH}{|}$$

30) Bhattacharyya, S. C: *Tetrahedron*, **19**, 1189 (1963).
31) Smith, C. R *et al*: *J. Org. Chem.*, **27**, 3112 (1962).

Methylester (IV), $R=CH_3$, $C_{19}H_{34}O_3$

UV λ_{max}: No absorption below 220 mμ;　　IR $\lambda_{max}^{CCl_4}$ μ: 5.73 (ester), 2.75 (OH), no absorption at 10.0~11.5

NMR　60 Mc, in CCl_4, TMS, τ: 4.5~4.8 (4H of C_9, C_{10}, C_{15}, C_{16}), 6.4~6.8 (4H on C_{12} and OCH_3), 7.5~8.2 (11H of C_2, C_8, C_{11}, C_{14}, C_{17} and —OH), 8.3~8.9 (12H on C_3~C_7 and C_{13}), 8.9~9.3 (centered at 9.05 t ($J=7$)) (3H on C_{18}) (shows a $-\overset{15}{C}=\overset{16}{C}-$), see Fig. 20.

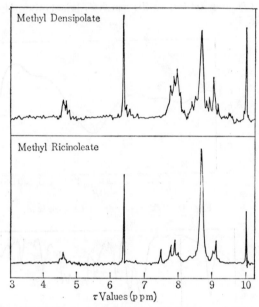

Fig. 20　NMR Spectra of Methyl densipolate and Methyl
ricinolate　　　　　　　　(Smith *et al.*)

(4)　Sterculic acid[32)]

$$CH_3-(CH_2)_7-\overset{\overset{\displaystyle CH_2}{\frown}}{C=C}-(CH_2)_7-COOH \qquad\qquad (V)$$

≪Occurrence≫　*Sterculia foetida*,　　mp. 19.3~19.9°

IR　λ_{max} μ: 5.35, 9.92 (cyclopropene) [Note], 5.86 (—COOH), 7.78, 10.7 s (charact. in fatty acid spectra), 7.27 (CH_2)

NMR　40 Mc, pure liquid, zero to H_2O, δ: +6.88 (—COOH), −3.50 (CH_2), −3.95, −3.85 d (CH_3), −2.53 (—CH_2-C=), −4.03 ($-\overset{\overset{\displaystyle CH_2}{\frown}}{C=C}-$)

[Note]　Dihydrosterculic acid: 9.79 μ (cyclopropane).

32)　Rinehart, K L *et al*: *J.A.C.S.*, **80**, 503 (1958).

(5)　NMR of α,β-Unsaturated Fatty Acids[33]

Table 4　NMR of α,β-Unsaturated Fatty Acids (in CCl_4, 25°)

(Fujiwara *et al.*)

Compounds $\begin{smallmatrix}(3)&&(2)\\&C=C\\(4)&&COOH\end{smallmatrix}$	Chem. shift (ppm)	H–H $J_{2\cdot3}$	H–H $J_{2\cdot4}$	H–H $J_{3\cdot4}$	H–CH$_3$ $J_{2\cdot3}$	H–CH$_3$ $J_{2\cdot4}$	H–CH$_3$ $J_{3\cdot4}$
$\begin{smallmatrix}CH_3&&H\\&C=C\\H&&COOH\end{smallmatrix}$	$\delta_{H_2}-\delta_{H_4}=1.28$		15.5		1.2		6.1
$\begin{smallmatrix}H&&H\\&C=C\\CH_3&&COOH\end{smallmatrix}$	$\delta_{H_2}-\delta_{H_3}=0.19$	9.0				≦1.0	7.0
$\begin{smallmatrix}H&&CH_3\\&C=C\\H&&COOH\end{smallmatrix}$	$\delta_{H_3}-\delta_{H_4}=0.62$			1.9	~0	1.7	
$\begin{smallmatrix}CH_3&&H\\&C=C\\CH_3&&COOH\end{smallmatrix}$	$\delta_{Me_3}-\delta_{Me_4}=0.27$				≦1.0	≦1.0	
$\begin{smallmatrix}H&&CH_3\\&C=C\\CH_3&&COOH\end{smallmatrix}$	$\delta_{Me_2}-\delta_{Me_4}=0.2$					~0	6.5
$\begin{smallmatrix}CH_3&&CH_3\\&C=C\\H&&COOH\end{smallmatrix}$	$\delta_{Me_2}-\delta_{Me_3}=0$					≦1.0	6.0

(6)　Malvalic acid[34]

$$CH_3\text{–}(CH_2)_x\text{–}\underset{\underset{\displaystyle CH_2}{\diagdown\diagup}}{C{=}C}\text{–}(CH_2)_y\text{–}COOH$$

$$x+y=13 \qquad\qquad (VI)$$

《Occurrence》　*Malva verticillata, M. parviflora*

IR : See[34]

(7)　Lactobacillic acid[35]

$$H_3C\text{–}(CH_2)_x\text{–}\underset{\displaystyle H\ H}{\overset{\displaystyle \overset{H\ \ H}{\diagdown C\diagup}}{C\text{–}C}}\text{–}(CH_2)_y\text{–}COOH \qquad (VII)$$

《Occurrence》　A lipid of various microorganismus.

(8)　Coronaric acid[36]

$$CH_3\text{–}(CH_2)_4\text{–}\underset{cis}{CH=\overset{12}{C}H}\text{–}CH_2\text{–}\underset{\diagdown O\diagup}{\overset{10}{C}H\text{–}\overset{9}{C}H}\text{–}(CH_2)_7\text{–}COOR$$

$$R=H \qquad\qquad (VIII)$$

《Occurrence》　*Chrysanthemum coronarium*

33)　Fujiwara, *S. et al* : *Bull. Chem. Soc. Japan*, **33**, 428 (1960).
34)　Macfarlane *et al* : *Nature*, **179**, 830 (1957).　　35)　Hofmann, K *et al* : *J.A.C.S.*, **80**, 5717 (1958).
36)　Smith, C.R *et al* : *J. Org. Chem.*, **25**, 218 (1960).

(9) 15, 16-Epoxy-9, 12-octadecadienoic acid (IX)[37]
≪Occurrence≫   *Camelina sativa*

(10) Vernalic acid[38]   (*cis*-12, 13-Epoxy-*cis*-9-octadecanoic acid)   (X)
≪Occurrence≫   *Vernonia anthelmintica, Clarkia elegans*

(11) Stereospecific Synthesis of *dl*-Quinic acid (XI)[39] **and D-(−)-Shikimic acid** (XII)[40]

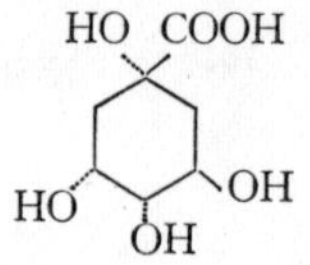

HO COOH

HO   OH
    OH   (XI)

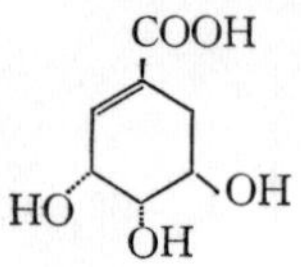

COOH

HO   OH
    OH   (XII)

(12) Neochlorogenic acid (XIII) **and " Band 510 "** (XIV)[41]

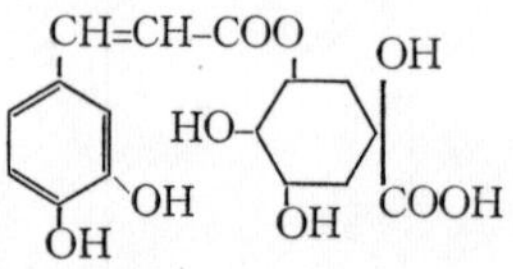

CH=CH-COO   OH

HO-

OH   OH   COOH
OH

CH=CH-COO OH   OH

OH   OH   COOH
OH

(XIII) $C_{16}H_{18}O_9$
mp. 204∼206°, $[\alpha]_D + 3.1°$

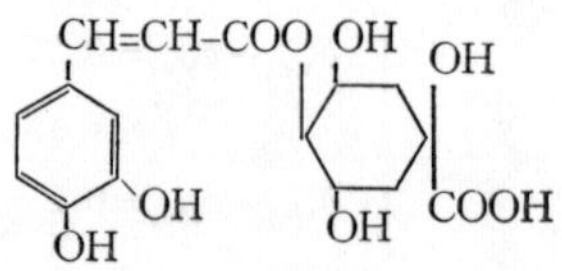

(XIV) $C_{16}H_{18}O_9$
$[\alpha]_D - 69°$

≪Occurrence≫   Coffee beans, artichoke leaves

37) Gunstone *et al*: *Soc.*, **1959**, 2127.   38) Gunstone: *Soc.*, **1954**, 1161.
39) Snissman, E.E *et al*: *J.A.C.S*, **85**, 2184 (1963).   40) McCrindle *et al*: *Soc.*, **1960**, 1560.
41) Scarpati, M.L *et al*: *Tetr. Letters*, **1963**, No. 18, 1147.

3 Amino Acids

[A] UV Spectra[28]

Glycocol, kreatin, alanine, valine, leucine and aspartic acid[28a)42]. Asparagin, glutamic acid, arginine, cystine and cysteine[42a]. Phenylalanine and tyrosine[42b], oxyproline[28a], tryptophane[42b] and histidine[42a], domoic acid.

[B] IR Spectra

(1) Leucine and isoleucine[43], β–alanine[43a], DL–serine[43b], DL–valine[43c], DL–leucine[43d], DL–isoleucine[43e], DL–lysine monohydrochloride[43f], DL–threonine[43g], arginine[43h], DL–aspartic acid[43i], glutamine[43j], D–glutamicacid[43k], L–cystine[43l], L(+)cysteine[43m], (−)hydrochloride[43n], D–methionine[43o], DL–phenylalanine[43p], L–tyrosine[43q], 4–hydroxy–DL–proline and DL–tryptophane[43r]. α–Kainic acid, β–kainic acid, α–kainic lactone, α–isokainic acid and α–allokainic lactone, domoic acid[43s].

(2) IR Spectra of Amino acids in an aqueous solution[44)45)46]

IR spectra of the following amino acids in an aqueous solution using BaF_2 cell are shown in Fig. 21~26.[46]

DL–valine, L–leucine, L–isoleucine, DL–serine, L–threonine, L–proline, L–hydroxyproline, L–aspartic acid, L–alanine, β–alanine, L–glutamic acid, DL–α–aminobutyric acid, α–aminobutyric acid and tyramine–HCl.

42) Castille *et al*: *Bull. Soc. Chim. Biol.*, **10**, 623 (1928).
42a) Mackinney: *J. Biol. Chem.*, **140**, 315 (1941). 42b) Holiday: *Biochem. J.*, **30**, 1795, 1799 (1936).
43) Darmon, Sutherland: *Biochem. J.*, **42**, 508 (1948). 43a) Sadtler Card No. 795.
43b) Sadtler Card No. 454–B. 43c) Sadtler Card No. 453–B. 43d) Sadtler Card No. 789.
43e) Sadtler Card No. 787. 43f) Sadtler Card No. 435. 43g) Sadtler Card No. 784.
43h) Sadtler Card No. 785. 43i) Sadtler Card No. 7919. 43j) Sadtler Card No. 2125.
43k) Sadtler Card No. 5561. 43l) Sadtler Card No. 323. 43m) Sadtler Card No. 326.
43n) Sadtler Card No. 5564. 43o) Sadtler Card No. 436–B. 43p) Sadtler Card No. 5569.
43q) Sadtler Card No. 786–B. 43r) Sadtler Card No. 432–B.
43s) Morimoto: *Y. Z.*, **75**, 901, 916 (1955).
44) Gore, R.C *et al*: *Anal. Chem.*, **21**, 382 (1949).
45) Blout, E.R *et al*: *Ann. N.Y. Acad. Sci.*, **69**, 84 (1957).
46) Parker, F.S *et al*: *Nature*, **187**, 386 (1960).

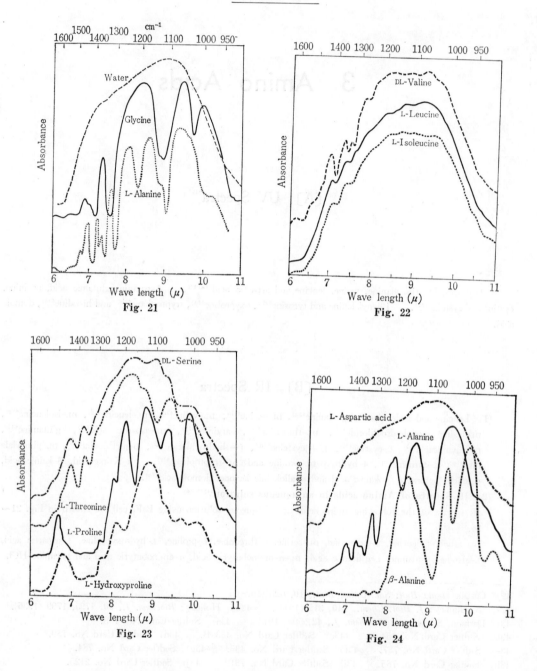

Fig. 21

Fig. 22

Fig. 23

Fig. 24

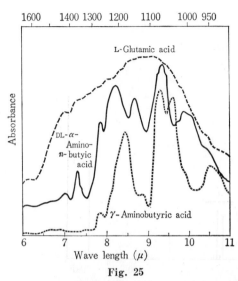

Fig. 25

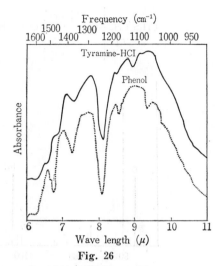

Fig. 26

Fig. 21~26 IR Spectra of Amino acids in an aquous solution (Parker *et al.*)

(3) **Mass spectra of Amino acids**

Mass spectra of the ethylesters of 24 amino acids have been determined and interpretated in terms of their structures[47].

(a) **Isoleucine**   *m/e* 131

$$CH_3-CH_2-CH-CH\vdots COOH$$
$$\qquad\quad CH_3\ \ NH_2\vdots$$
$$\qquad\qquad\qquad\quad 86\vdots$$

See Fig. 27

(b) **Isoleucine ethylester**   *m/e* 159

$$\qquad\qquad 57\vdots\ \ 102\vdots$$
$$CH_3-CH_2\vdots CH\vdots CH\vdots COOC_2H_5$$
$$\qquad\qquad\ CH_3\vdots\ \ NH_2\vdots$$
$$\qquad\quad 130\ \ \vdots\qquad 86\vdots$$

See Fig. 28

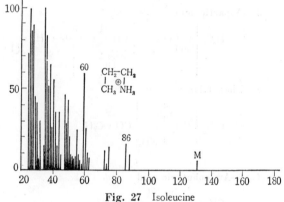

Fig. 27 Isoleucine

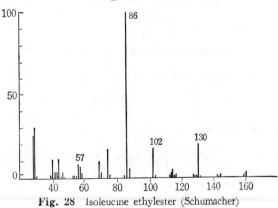

Fig. 28 Isoleucine ethylester (Schumacher)

47) Biemann, K *et al* : *J. A. C. S.*, **83**, 3795 (1961).

(c) Methionine ethylester   m/e 177

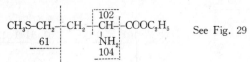

See Fig. 29

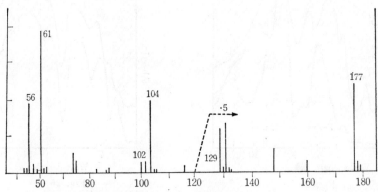

Fig. 29 Methionine ethylester (Biemann)

(d) Aspartic acid   m/e 189

$$H_5C_2 \!-\! O\!-\!CO\!-\!CH_2 \underset{NH_2}{\overset{102\ 144}{-CH-}} \!-\! CO \!-\! O\!-\!C_2H_5$$
160   116   (I)

189w, 160m, 144m, 116s, 102m, 88w, 74m, 70s, 44m, 29s

(e) Phenylalanine   m/e 193

$$C_6H_5 \underset{77}{\overset{91\ 102}{-CH_2-}} \underset{NH_2}{\overset{}{CH}} \!-\! CO\!-\!OC_2H_5$$
120   (II)

193w, 131w, 120s, 103m, 102s, 91m, 71w, 74m, 29m

(f) α–Aminobutyric acid   m/e 131

$$CH_3\!-\!CH_2 \underset{NH_2}{\overset{102}{-CH-}} \!-\! CO\!-\!OC_2H_5$$
58   (III)

131w, 102w, 74w, 58s, 30w

(g) β–Aminobutyric acid   m/e 131

$$CH_3 \underset{\underset{116}{NH_2}}{\overset{44}{-CH-}} \!-\! CH_2\!-\!COO \!-\! C_2H_5$$
102   (IV)

131w, 116w, 102w, 44s

(h) Glycine   m/e 103

$$H_2N\!-\!CH_2 \overset{30}{-} COOC_2H_5 \quad (V)$$

103w, 30s

(i) Alanine   m/e 117

$$CH_3\!-\!\underset{NH_2}{\overset{44}{CH}}\!-\!COOC_2H_5 \quad (VI)$$

117w, 44s

(j) Valine   m/e 145

$$CH_3-\underset{\underset{130}{\overset{|}{CH_3}}}{\overset{43}{\overset{|}{CH}}}-\underset{\underset{72}{\overset{|}{NH_2}}}{\overset{102}{\overset{|}{CH}}}-COOC_2H_5 \quad (VII)$$

145w, 130w, 102w, 74w, 72s, 55m, 30w

(k) Leucine   m/e 159

$$CH_3-\underset{\underset{43}{\overset{|}{CH_3}}}{\overset{}{\overset{|}{CH}}}-\underset{116}{CH_2}-\underset{\underset{86}{\overset{|}{NH_2}}}{\overset{102}{\overset{|}{CH}}}-COOC_2H_5 \quad (VIII)$$

159w, 144w, 127w, 116w, 102w, 86s, 74w, 44s, 30s

(l) Proline   m/e 143

$$\underset{H}{\overset{}{N}}\underset{70}{>}-COOC_2H_5 \quad (IX)$$

143m, 70s

(m) Tyrosine   m/e 209

$$HO-\left\langle\bigcirc\right\rangle-\underset{107}{CH_2}-\underset{\underset{136}{\overset{|}{NH_2}}}{\overset{102}{\overset{|}{CH}}}-COOC_2H_5 \quad (X)$$

209s, 136s, 107s, 102s, 74s

(n) Tryptophane   m/e 232

$$\underset{H}{\overset{}{N}}\underset{130}{CH_2}-\underset{\underset{159}{\overset{|}{NH_2}}}{\overset{102}{\overset{|}{CH}}}-COOC_2H_5 \quad (XI)$$

232s, 159w, 130s

(o) Serine   m/e 133

$$\underset{\overset{|}{OH}}{\overset{31}{\overset{|}{CH_2}}}-\underset{\underset{60}{\overset{|}{NH_2}}}{\overset{102}{\overset{|}{CH}}}-COOC_2H_5 \quad (XII)$$

133w, 115m, 102m, 74m, 60s, 42m, 30m

(p) Threonine   m/e 147

$$CH_3-\underset{\underset{132}{\overset{|}{OH}}}{\overset{45}{\overset{|}{CH}}}-\underset{\underset{130}{\overset{|}{NH_2}}}{\overset{102}{\overset{|}{CH}}}-COOC_2H_5 \quad (XIII)$$

147w, 132s, 103s, 74s, 56s, 45m, 30s

(q) Glutamic acid   m/e 203

$$H_5C_2O\underset{174}{OC-CH_2-CH_2}-\underset{\underset{130}{\overset{|}{NH_2}}}{\overset{102}{\overset{|}{CH}}}-COOC_2H_5 \quad (XIV)$$

204w, 174s, 157m, 130s, 102w, 84s, 74w, 56s, 29s

(r) Lysine   m/e 174

$$H_2N-\underset{30}{CH_2}-CH_2-CH_2-CH_2-\underset{\underset{101}{\overset{|}{NH_2}}}{\overset{102}{\overset{|}{CH}}}-COOC_2H_5 \quad (XV)$$

174m, 167m, 128s, 101m, 84s, 56s, 44m, 30s

(s) Cysteine   m/e 149

$$HS-\underset{47}{CH_2}-\underset{\underset{76}{\overset{|}{NH_2}}}{\overset{102}{\overset{|}{CH}}}-COOC_2H_5 \quad (XVI)$$

149w, 132w, 102s, 76s, 74m, 59m

(t) Methionine m/e 177

$$CH_3\text{-}S\text{-}CH_2\text{-}CH_2\text{-}CH\text{-}COOC_2H_5 \quad (XVII)$$

with fragment markers 75, 102; NH$_2$; 61, 116, 104

177s, 160w, 148m, 131s, 129s, 116w, 104s, 102w

(u) Histidine m/e 183

$$\text{N}\diagup\diagdown\text{NH}\text{-}CH_2\text{-}CH\text{-}COOC_2H_5 \quad (XVIII)$$

with fragment markers 81, 102; NH$_2$; 110

183w, 138w, 110s, 102w, 82s, 74w

Mass Fragmentation

$$\underset{44}{\overset{CH_2=CH}{\underset{^+NH_3}{}}} \quad \underset{30}{\overset{CH_2^+}{\underset{NH_2}{}}} \quad \underset{56}{H_2N\text{-}\overset{+}{C}=C=O} \quad \underset{74}{\overset{^+CH_2\text{-}C\diagup^{OH}_{\diagdown O}}{\underset{NH_2}{}}} \quad \underset{56}{\overset{CH_2=CH\text{-}CH^+}{\underset{NH_2}{}}}$$

(4) Relationship between the Charge Distributions[48]

(Q_H and Q_C) and the chemical shift (δ_H)

$$\delta_H = -AQ_C - BQ_H + C$$

where: A=9.92, B=133.93 and C=9.67

(5) Hypoglycine A

$$CH_2=C\text{—}CH\text{-}CH_2\text{-}CH(NH_2)\text{-}COOH \quad (XIX)$$
$$\diagdown CH_2$$

≪Occurrence≫   *Blighia sapida*[49]~[53]

(6) 1-Aminocyclopropane-1-carboxylic acid   (XX)

≪Occurrence≫   Perry pears[54], *Vaccinium vitis-idaea*[55]

(7) α-Methylenecyclopropylglycine

$$CH_2=C\text{—}CH\text{-}CH(NH_2)\text{-}COOH \quad (XXI)$$
$$\diagdown CH_2$$

≪Occurrence≫   *Litchi chinensis*[56]

(8) Pichrostachic acid[57]

$$HOOC\text{-}CH(OH)\text{-}CH_2\text{-}SO_2\text{-}CH_2\text{-}S\text{-}CH_2\text{-}CH\text{-}COOH \quad (XXII)$$
$$NH_2$$

≪Occurrence≫   *Pichrostachys glomerata*

(9) (−)-S-Propenyl-L-cysteine[57a]

$$CH_3\text{-}CH=CH\text{-}S\text{-}CH_2\text{-}CH(NH_2)\text{-}COOH \quad (XXII\text{-}a)$$

≪Occurrence≫   *Allium sativum*

48) Del Re. G *et al*: *Biochim. et Biophys. Acta*, **75**, 153 (1963).
49) Hassall, C. H *et al*: *Nature*, **173**, 356 (1954).   50) H. *et al*: *Biochem. J.*, **60**, 334 (1956).
51) Anderson, H. V *et al*: *Chem. & Ind.*, **1958**, 330.
52) Renner, V *et al*: *Helv. Chim. Acta*, **41**, 588 (1958).   53) Hassall *et al*: *Chem. & Ind.*, **1958**, 329.
54) Burroughs, L. F: *Nature*, **179**, 360 (1957).
55) Vähätalo, M. L *et al*: *Acta Chem. Scand.*, **11**, 741 (1957).
56) Gray, D. O *et al*: *Biochem. J.*, **82**, 385 (1962).   57) Gmelin, R: *Hoppe Seyler*, **327**, 186 (1962).
57a) Sugii, M *et al*: *Chem. Pharm. Bull.*, **11**, 548 (1963).

(10) Zizyphin   Peptide $C_{33}H_{49}O_6N_5$ (XXII–b)[57b]

≪Occurrence≫   *Zizyphus oenoplia* MILL., mp. 121°d, $[\alpha]_D$ −464.7° (CHCl$_3$), pKs 5.8, 6.23.

UV λ_{max}^{EtOH} mμ (ε): 206, 267, 319 (38000, 10800, 8270)

IR μ; 2.67, 2.87, 3.32, 3.42, 3.47, 3.53, 5.87, 6.05, 6.25, 6.61, 7.05, 7.55, 8.16, 8.43, 9.5, 12.3.

NMR 60Mc, TMS, cps: 52s, 58s, 120b, 138s, 157s, 163s (total 32H); 204b, 214b, 229s (total 5H); 253b, 262s, 268s, 276s, 282s (total 5H); 313q (1); 357d, J=9.5 (1), 412s, 415q (total 4H); 502d (1).

Mass m/e 611, 568, 554, 496, 411, 384, 372, 357, 356, 277, 235, 216, 193, 192, 191, 100, 70, 57, 44, 28[57b]

(11) Zizyphinin   Peptide $C_{32}H_{47}O_6N_5$ (XXII–c)

≪Occurrence≫   *Zizyphus oenoplia* MILL., Amorphous, $[\alpha]_D$ −457° (CHCl$_3$)

UV λ_{max} mμ (ε): 267, 320 (10220, 7760)

NMR: 52s, 55s, 58s, 61s, 121b, 141s, 144s (total 27H); 175d, J=4.5, 201b, 213b (total 3H), 228s (3); 265b (5), 318b (1), 357d, J=9 (1), 406s, 411s, 417s, 434s (total 5H), 457d (1), 504d (1).

Mass m/e: 597, 554, 498, 455, 426, 385, 358, 216, 202, 191, 114.

57b) Menard, E. L *et al*: *Helv. Chim. Acta*, **46**, 1801 (1963).   58) Crombie, L: *Soc.*, **1955**, 995.

4 Acidamides

[A] UV Spectra

(1) *N*-**Isobutylpolyeneamides** and **monoeneamides** (see Table 5)[58]

Table 5 (Crombie)

	Compounds	λ_{max}^{EtOH} m$\mu(\varepsilon)$			
(I)	Me–CH=CH–CH=CH–CH=CH–(CH$_2$)$_2$–CH=CH–CONH–Bu Neoherculin	227 (13000)	259 (31500)	269 (43500)	280 (36000)
(II)	(α)Me–(CH$_2$)$_3$–CH=ĊH–CH=ĊH–CH=ĊH–(CH$_2$)$_7$–COOH	226 (4000)	261 (36000)	271 (47000)	281 (38000)
(III)	(β)Me–(CH$_2$)$_3$–CH=ĊH–CH=ĊH–CH=ĊH–(CH$_2$)$_7$–COOH	226 (5500)	259 (47000)	268 (61000)	279 (49000)
(IV)	Me–(CH$_2$)$_6$–CH=ĊH–CO–NH–Bu	226 (10300)			
(V)	Me–(CH$_2$)$_6$–CH=ĊH–CO–NH–Bu	226 (10000)			

(2) *N*-**Isobutylamides of stereoisomeric deca-2, 4-dienoic acids**[59] (see Table 6)

H$_3$C–(CH$_2$)$_4$–CH=CH–CH=CH–CONH–*iso*-Bu (VI)

Table 6 (Crombie)

Configuration	bp/mm.	n_D	λ_{max}^{EtOH}	ε
trans–2 : *trans*–4	(mp 90°)	—	258	29500
trans–2 : *cis*–4	142°/0.1	1.5062(20°)	258	26000
cis–2 : *trans*–4	130°/0.07	1.5088(19°)	259	23500
cis–2 : *cis*–4	140°/0.3	1.4985(13°)	258	17500

(3) *N*-**Isobutylamides of decadienoic** (VI) and **dodecatrienoic acids** (VII)[60] (see Table 7)

CH$_3$–CH$_2$–CH$_2$–CH=CH–CH$_2$–CH$_2$–CH=CH–CH=CH–CONH–*iso* Bu (VII)

59) Crombie, L: *Soc.*, **1955**, 1007. 60) Crombie, L: *ibid.*, **1955**, 4244.

〔B〕 IR Spectra

Table 7 (Crombie)

Configuration			Acid λ_{max}^{EtOH} m$\mu(\varepsilon)$	Amide λ_{max}^{EtOH} m$\mu(\varepsilon)$
2:3	4:5	8:9		
trans	*trans*	*trans*	259(27000)	259(32000)
trans	*trans*	*cis*	259(25500)	259(29000)
trans	*trans*	—	257(28500)	258(29500)

(4) **N-Isobutylamides of undeca-1,7-diene-1-carboxylic acids**[61] (see Table 8)

$$CH_3-CH_2-CH=CH-(CH_2)_4-CH=CH-CO-NH-iso\text{-}Bu \quad (VIII)$$

Table 8 (Crombie)

Configuration	λ_{max}^{EtOH} mμ	ε
trans–1 : *trans*–7	226	11500
cis–1 : *trans*–7	226	10700
trans–1 : *cis*–1	227	10500
cis–1 : *cis*–1	227	8700

〔B〕 IR Spectra

(1) **Assignments for stereoisomeric N-isobutyldeca-2,4-dienamides**[59] (see Table 9)

$$H_3C-(CH_2)_4-CH=CH-CH=CH-CO-NH-iso \text{ Bu} \quad (VI)$$

Table 9 (Crombie)

Configuration		ν_{NH}		Amide A			Mean		Amide B	δ'	
2	4			$\nu_{C=O}$	$\nu_{C=C\,I}$	$\nu_{C=C\,II}$	$\nu_{C=C}$	$\Delta_{C=C}$	$\nu_{C=O}$	[CH=CH]$_2$	
trans	*trans*	3295	3075	1625	1654	1614	1634	40	1550	994	—
trans	*cis*	3285	3075	1623	1653	1609	1631	44	1549	993	963
cis	*trans*	3285	3090	1632	1650	1605	1627	45	1544	998	961
cis	*cis*	3285	3075	1628	1649	1595	1622	54	1546	994	962

(Liquid film or paraffin mulls)

(2) **Assignments for stereoisomeric Dodecatrienoic acids** and their **Isobutylamides**[50]

(see Table 10)

$$CH_3-CH_2-CH_2-CH=CH-CH_2-CH_2-CH=CH-CH=CH-CONH-iso\text{-}Bu \quad (VII)$$

Table 10 (Crombie)

Compounds	2:3	4:5	8:9	ν_{NH}		$\nu_{C=O}$	$\nu_{C=C\,I}$	$\nu_{C=C\,II}$	$\Delta_{C=C}$	δ'(CH=CH)$_2$	
iso-Bu–amides	*trans*	*trans*	*trans*	3265	3060	1623	1655	1613	42	993	967
	trans	*trans*	*cis*	3295	3060	1626	1657	1618	39	998	—
Acids	*trans*	*trans*	*trans*	—	—	1684	1634	1610	24	1005	968
	trans	*trans*	*cis*	—	—	1684	1636	1613	23	1000	—

(Paraffin mulls)

61) Crombie, L: *Soc.*, **1952**, 3000.
62) Crombie, L: *ibid.*, **1955**, 995.

[C] Naturally Occurring N-Isobutylpolyene amides

(1) Affinin[63] = Spiranthol (Stereoisomer of N-Isobutyl-all-*trans*-2, 6, 8-decatrieneamide[64])

$$CH_3-\overset{t}{CH}=CH-\overset{t}{CH}=CH-(CH_2)_2-\overset{t}{CH}=CH-CO-NH-iso\text{-}Bu \quad (VIII)$$

Affinin (natural)

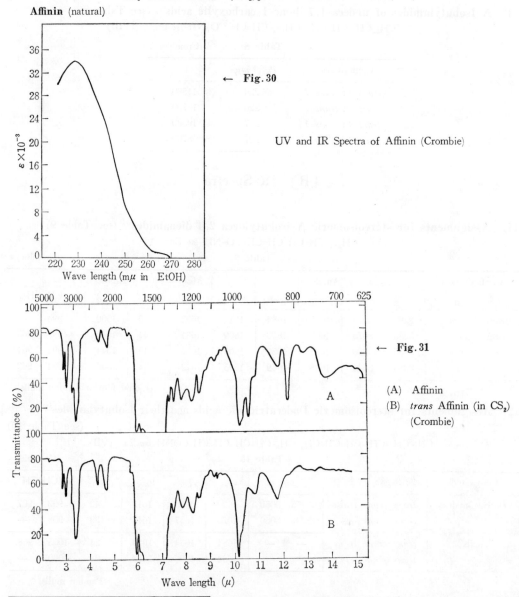

← Fig. 30

UV and IR Spectra of Affinin (Crombie)

← Fig. 31

(A) Affinin
(B) *trans* Affinin (in CS₂)
(Crombie)

63) Jacobson, M *et al*: *J. Org. Chem.*, **10**, 236, 449 (1945), **12**, 731 (1947).
64) Jacobson, M: *J.A.C.S.*, **76**, 4606 (1954).

≪Occurrence≫   *Heliopsis longipes*   mp. 23°, bp. 157°/0.26, [n]$_D$ 1.5134, $C_{14}H_{23}ON$

UV λ_{max}^{EtOH} m$\mu(\varepsilon)$: 228.5(33700)   Isomerize to all–*trans*–affinin.

all-*trans*-Affinin (VIII) mp. 91.5°.   UV λ_{max}^{EtOH} m$\mu(\varepsilon)$: 228.5(37100), see Fig. 30, IR: see Fig. 31.

Spiranthol[65]

≪Occurrence≫   *Spilanthes acmella* L. *f. fusca* Makino (キバナオランダセンニチ).

Spiranthol had previously been regarded as N–isobutyl–4,6–decadienamide[66], which structure was revised to (VIII)[65].

(2) **Neoherculin**=α-Sanshoöl (*trans*-2: *cis*-6: *trans*-8: *trans*-10-N-Isobutyldodecatetraenamide)

$$CH_3-\overset{10}{CH}=CH-CH=\overset{8}{CH}-CH=\overset{6}{CH}-(CH_2)_2-\overset{2}{CH}=CH-CO-NH-\textit{iso}\text{-Bu} \qquad (IX)$$

Neoherculin (IX)[61][62]

≪Occurrence≫   *Zanthoxylum clava-herculis* L.,   mp. 69~70°, $C_{16}H_{25}ON$

UV λ_{max}^{EtOH} m$\mu(\varepsilon)$: 227, 259, 269, 280 (13000, 31500, 43500, 36000)

IR: See Fig. 32[61]

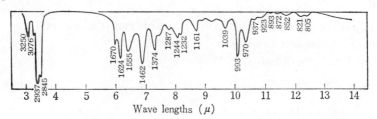

Fig. 32 IR Spectrum of Neoherculin in Paraffin mulls (Crombie)

α–Sanshoöl[67]

≪Occurrence≫   *Zanthoxylum piperitum* DC.,   mp. 69~70°, $C_{16}H_{25}ON$

UV λ_{max}^{EtOH} m$\mu(\varepsilon)$: 260, 269, 278.5 (36000, 47000, 37500).

IR ν_{max}^{nujol} cm^{-1}: 3200 m, 3060 w(NH), 1668 m(α-C=C—), 1626 S (—CONH—), 1560 (amide B), 994, 972 (*trans–trans–cis*-triene and *trans* α–CH=CH).

(3) **β-Sanshoöl**[67]

(All-*trans*-N-Iso butyldodecatetraenamide) (IX)

≪Occurrence≫   *Zanthoxylum piperitum* DC.,   mp. 112°, $C_{16}H_{25}ON$

UV λ_{max}^{EtOH} m$\mu(\varepsilon)$: 259, 267.5, 278 (38500, 48500, 39000), IR ν_{max}^{nujol} cm^{-1} 3250 m, 3060 w (NH), 1667 m (α-C=C), 1623 s (CO–NH), 1549 s (amide B), 996 s (all-*trans*-triene) 978 (*trans* α–CH=CH).

(4) **Pellitorine**[68] is a mixture of at least three amides. A major component is regarded as the amide (X: *trans*-2, *trans*-4).

$$CH_3-(CH_2)_4-\overset{t}{CH}=CH-CH=\overset{t}{CH}-CO-NH-\textit{iso}\text{-Bu} \qquad (X)$$

≪Occurrence≫   *Anacyclus pyrethrum* DC.,   mp. (pellitorine) 72°, (X: *t*–2, *t*–4) 90°.

UV λ_{max}^{EtOH} mμ ($E_{1cm}^{1\%}$): (X: *t*–2, *t*–4) 258, 264 inf. (1330, 1200)

65) Jacobson, M: *Chem. & Ind.*, **1957**, 50.   66) Asano, M *et al*: *J. Pharm. Soc. Japan*, **47**, 521 (1927).
67) Crombie, L *et al*: *Soc.*, **1957**, 2760.   68) Crombie, L: *Soc.*, **1955**, 999.

IR: of pellitorine: See Fig. 33[69].

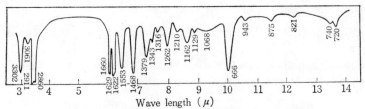

Fig. 33 IR Spectrum of Pellitorine in paraffin mulls (Crombie)

Table 11 IR spectra of synthetic (X: t-2, t-4) and pellitorine are shown as:

ν_{max}^{nujol}		ν_{NH}	$\nu_{C=O}$	$\nu_{C=C\ I}$	$\nu_{C=C\ II}$	δ-[CH=CH]$_2$
X: t-2, t-4	3295	3075	1625	1654	1614	994
Pellitorine	3300	3070	1625	1655	1615	995

(5) **Anacyclin**[68]

$$CH_3\text{-}(CH_2)_2\text{-}C\equiv C\text{-}C\equiv C\text{-}CH_2\text{-}CH_2\text{-}\overset{t}{CH}=CH\text{-}\overset{t}{CH}=CH\text{-}CONH\text{-}iso\text{-}Bu \quad (XI)$$

≪Occurrence≫　*Anacyclus pyrethrum* DC.,　mp. 121°, $C_{18}H_{25}ON$

UV　λ_{max}^{EtOH} mμ($E_{1cm}^{1\%}$): 259, 265 inf.(1240, 1100).

IR　ν_{max}^{nujol} cm^{-1}: 3300, 3060 (NH), 2233 w, 2206 vw (C≡C), 1625 (amide A), 1544 (amide B), 1655, 1612

$$(\text{-}\overset{t}{C}=C\text{-}CO),\ 996\ (C=C)$$

(6) **Herclavin**

$$CH_3\text{-}\hexagon\text{-}CH_2\text{-}CH_2\text{-}N(CH_3)\text{-}CO\text{-}CH=CH\text{-}\hexagon \quad (XII)$$

≪Occurrence≫　*Zanthoxylum clava-herculis* L.

UV　λ_{max} mμ (ε): 223, 280 (21500, 22700).

IR　ν_{max} cm^{-1}: 1655 s, 1610 s, 1576 w, 1509 m, 1036 i, 1027 s, 991 m, 982 s.

[D] Miscellaneous

(1) **Capsaicin** has been regarded as a mixture which was separated to (XIII) and (XIV)[70]

$$HO\text{-}\underset{OCH_3}{\hexagon}\text{-}CH_2\text{-}NH\text{-}CO(CH_2)_4\text{-}CH=CH\text{-}\underset{CH_3}{\overset{CH_3}{CH}} \quad (XIII)$$

N-(3-Methoxy-4-hydroxybenzyl)-8-methylnon-*trans*-6-enamide

$$HO\text{-}\underset{OCH_3}{\hexagon}\text{-}CH_2\text{-}NH\text{-}CO(CH_2)_6\text{-}\underset{CH_3}{\overset{CH_3}{CH}} \quad (XIV)$$

N-(3-Methoxy-4-hydroxybenzyl)-8-methylnonamide

69) Crombie L: *Soc.*, **1952**, 4342.　　70) Kosuge, S *et al*: *J. agricul. Chem. Japan*, **32**, 720 (1960).

IR: See Fig. 33a

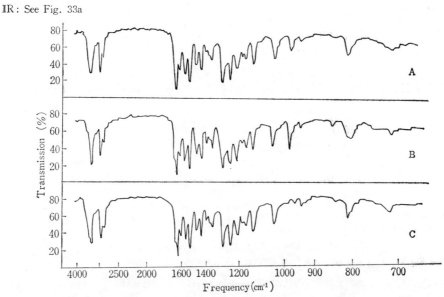

Fig. 33a IR Spectra of (A) Capsaicin, (B) (XIII), (C) (XIV) (Kosuge *et al.*)

≪Occurrence≫　　*Capsicum* spp.,　　mp. (capsaicin) 63.5～65°, (XIII) 65.9～66.3°, (XIV) 65.6～
65.8°[70], (natural capsaicin) 64°[71]

UV　λ_{max}^{EtOH} mμ(ε): (Natural capsaicin)[71]: 227, 281 (7000, 2500).

IR　ν_{max}^{nujol} cm^{-1}: (Natural capsaicin)[71]: 3320 broad (NH and OH), 3090, 1650 (C=C), 1629 (amide A,
C=O), 1598 (Ph$_I$), 1553 (amide B), 1511 (Ph$_{II}$), 1454, 1429, 1372, 1349, 1282, 1241, 1201,
1172, 1155, 1121, 1032, 968.5, 937.5, 843, 809, 718

(2) Aegeline[72]

$$CH_3O-\langle\!\!\!\bigcirc\!\!\!\rangle-\underset{\underset{OH}{|}}{CH}-CH_2-NH-CO-CH=CH-C_6H_5 \qquad (XV)$$

≪Occurrence≫　　*Aegle marmelos* CRREA,　　mp. 176°, C$_{18}$H$_{19}$O$_3$N
UV　λ_{max} mμ (log ε): 275, 223, 217 (4.605, 4.52, 4.53).
IR　ν_{max} cm^{-1}: 3250, 3060, 2830, 1665, 1627, 1580, 1520, 1062, 1040, 990, 982.

(3) Leonurine[73]

$$\begin{array}{c}CH_3O\\ HO-\langle\!\!\!\bigcirc\!\!\!\rangle-CO-N-CH_2-CH_2-CH_2-CH_2OH\\ CH_3O \quad\quad HN^{\diagup C}{}_{\diagdown}NH_2\end{array} \qquad (XVI)$$

≪Occurrence≫　　*Leonurus sibricus* L.,　　HCl salt　　mp. 193～194°, C$_{14}$H$_{21}$O$_5$N$_3$·HCl·H$_2$O
UV　λ_{max}^{MeOH} mμ (log ε): 276 (4.0).

71) Crombie, L: *Soc.*, **1955**, 1025.　　72) Chatterjee, A *et al*: *J. Org. Chem.*, **24**, 687 (1959).
73) Hayashi *et al*: *Y. Z.*, **82**, 1020 (1962), **83**, 271 (1963).
74) Goto, T *et al*: *Tetr. Letters*, **1962**, 545.

$\lambda_{max}^{NaOH-MeOH}$ mμ (log ε): 239, 326 (4.3, 4.7).

IR ν_{max}^{KBr} cm^{-1}: 3300, 3120, 1704, 1667, 1617, 1335, 1264, 1222, 1197, 1102.

NMR τ: 8.2 (C–H), 6.2 (N–C–H or O–C–H)

(4) Casimiroedine (*N*-D-glucoside of *N-trans* or *cis*-cinnamoyl-*N*-methylhistamine) (XVII)[75][76][77]

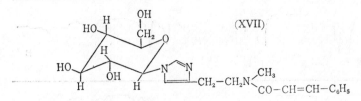

《Occurrence》   *Casimiroa edulis*

mp. 226.5~228°[77], [α]$_D$ $-27°$ (1% HCl)[76]

(XVII) $\longrightarrow$ Casimidine $C_{12}H_{21}O_5N_3$+Cinnamic acid

UV: See Fig.34, λ_{max} mμ (ε): 218, 281 (20000, 21500)

IR: See Fig.35, ν_{max} cm^{-1}: 3360 (OH), 3100 (bonding NH), 1645 $\left(-CON\langle^R_R\right)$, 1583 (arom.), 1498, 740

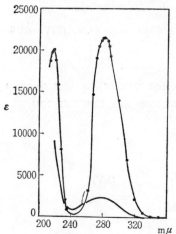

← **Fig.34**
Upper: Casimiroedin
Under: Tetrahydrocasimiroedin

Fig.34 UV and Fig.35 IR Spectra of Casimiroedin (Aebi)

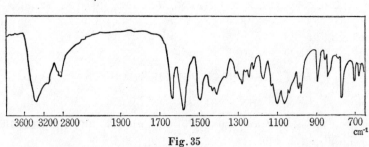

Fig.35

75) Raman, S *et al*: *Tetr. Letters*, **1962**, 357.
76) Djerrassi, C *et al*: *J. Org. Chem.*, **21**, 1510 (1956).
77) Aebi *et al*: *Helv. Chim. Acta.*, **39**, 1495 (1956).

(5) Lunarine

$C_{25}H_{31}O_4N_3$ (XVIII)[78]

≪Occurrence≫ *Lunaria biennis, L. redivava.*

UV λ_{max}^{EtOH} $m\mu$ (ε): 208, 224, 296, 315 (29200, 26000, 22100, 17300); $\lambda_{max}^{0\cdot1M-NaOH}$: 224, 355 (28800, 27500);

$\lambda_{max}^{10M-HCl}$: 236, 354 (24200, 24400).

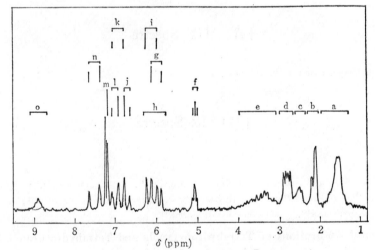

(XVIII)

NMR 60 Mc, in CDCl₃, TMS, δ: see Fig. 36.

Fig. 36 NMR Spectrum of Lunarine (Potier *et al.*)

78) Potier, P *et al*: *Tetr. Letters,* **1963**, 293.

5 Carbohydrates, Uronic Acids and Related Compounds

[A] UV Spectra

Glucose[79], arabinose, rhamnose[80], galactose, and maltose[79].

[B] IR Spectra

(1)

α–and β–Monosaccharides[81], sucrose[82], sorbitol[83], mannitol[84], L–(+)–arabinose[85], D–glucose[86], D–fructose[87], mannose[88], D–galactose[89], L–(+)–rhamnose[90], D–lactose[91], D–gluconic acid[92], gluconic acid–δ–lactone[93], D–glucuronic acid[94], glucosamine hydrochloride[95].

(2) Fundamental Absorptions of Tetrahydropyranols and Tetrahydrofuranols[96]

Table 12 shows the absorption of tetrahydropyran (I) and tetrahydropyranols (II)~(V).

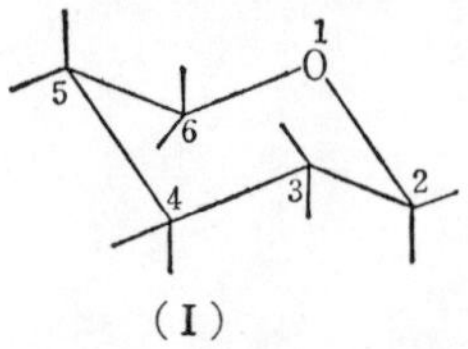

(I)

2–ol (II)
3–ol (III)
4–ol (IV)
2–hydroxymethyl (V)

79) Gabryelski *et al*: *Biochem. Zeitschr.*, **261**, 394 (1933).
80) Marchlewski *et al*: *Biochem. Zeitschr.*, **262** 250 (1933).
81) Whistler, House: *Anal. Chem.*, **25**, 1463 (1953).   82) *J.A.O.A.C.*, **34**, 756 (1951).
83) Sadtler Card No. 992   84) Sadtler Card No. 992–B   85) Sadtler card No. 1043–B
86) Sadtler Card No. 1040–B   87) Sadtler Card No. 2755   88) Sadtler Card No. 993–B
89) Sadtler Card No. 989   90) Sadtler Card No. 1018–B   91) Sadtler Card No. 1049–B
92) Sadtler Card No. 8335   93) Sadtler Card No. 5145   94) Sadtler Card No. 1037–B
95) Sadtler Card No. 5122   96) Baggett, N *et al*: *Soc.*, **1960**, 4565.

Table 13 shows the absorption of tetrahydrofuran (**VI**) and tetrahydrofuranols (**VII**)~(**IX**).

2–ol (**VII**)
3–ol (**VIII**)
2–hydroxymethyl (**IX**)

(**II**)

Table 12 Assignments for tetrahydropyran (**I**), –2–ol (**II**), –3–ol (**III**),–4–ol (**IV**) and tetrahydro–2–hydroxymethylpyran (**V**)

$\nu_{max}^{liq\ film}$ cm^{-1}

(Baggett *et al.*)

Assignment Compounds	OH	CH$_2$ antisym	CH$_2$ sym	CH$_2$ scissors		δ_{C-H}
I	—	2910vs	2735m 2700w	1470m 1445vs	1455s 1386s	—
II	3395vs	2940vs 2850s	2730w 2670w	1475w 1449s,	1463m 1389m	1425w
III	3395vs	2940vs 2810vs	2730w 2700w	1475m 1448s	1387w	1420w
IV	3350vs	2910vs 2825vs	2740w 2690w 2675w	1474m 1435w	1452s 1385m	1425m
V	3400vs	2910vs 2827vs	2710w	1467w 1446s	1457m 1383m	1415m

Assignment Compounds	CH$_2$ wagging		δ_{C-H}	CH$_2$ twisting		CH$_2$(rock?)	Ring freq.	
I	1363w 1303s	1352m 1275s	—	1260m 1197vs		1160m	1093vw 1030s	1047vs 1013m
II	1361s 1280m	1300m	1334w	1263m 1201s	1247vw 1175s	1141s	1080vs 1030s	1057m 1018s
III	1370s 1318wv 1280s	1354vw 1300m	1338w	1245w 1185w	1228vs 1168m	1130vs	1082vs 1035m	1065m 1010s
IV	1361w 1276m	1306w	—	1240w 1165vw		1132m	1072s	1008m
V	1356m 1255vw	1274s	1325vw	1268w 1208s	1224m 1183m	1160w	1095vs 1050vs	1075m 1005m

Assignment Compounds	δ_{C-OH}	CH$_2$ rock	ring	CH$_2$ rock	ring	CH$_2$ rock
I	—	970s	873vs	856m	817s	—
II	981vs	—	869s	842m	820m	760vw
III	983w	965s	874m 863m	846m	803m	—
IV	993s 987m	963m	863m	863vs	815s	750m
V	992m	962w	872m	858s	800m	—

Table 13 Assignments for tetrahydrofuran (VI), –2–ol (VII), –3–ol (VIII)
and tetrahydro–2–hydroxymethylfuran (IX)

$\nu_{max}^{liq\ film}$, cm^{-1} (Baggett)

Assignment / Compounds	OH	CH$_2$ antisym	CH$_2$ sym	CH$_2$ scissors	δ_{C-H}
VI	—	2981vs 2947w 2933w	2878s 2868s	1462vs	—
VII	3400vs	2955vs	2890vs 2720w	1495vw 1469s 1451m	1432m
VIII	3380vs	2950vs	2870vs 2720w 2680w	1490w 1470vw 1450s	1431m
IX	3400vs	2930vs	2855vs	1495vw 1490vw 1470s, 1462w 1455m	1308w

Assignment / Compounds	CH$_2$ wagging	δ_{C-H}	CH$_2$ twisting	ν_{C-OH}	ring
VI	1366s 1332w 1286s	—	1233s 1181s	—	1069vs 1031s
VII	1373s 1345m 1275s	1330m	1245m 1190s 1123s	1123s	1068vs 1040vs
VIII	1375w 1338s 1295m	1350vw	1258w 1235m 1208w	1125s	1075vs 1052m
IX	1368m 1338w 1292w	1308w	1250m 1230m 1190s	1110m	1075vs 1055vs

Assignment / Compounds	δ_{C-OH}	ring	CH$_2$ rock	ring
VI	—	911vs	761vw	—
VII	990vs	923vs	852s 760m	—
VIII	973s	903vs	860m 773s	—
IX	990s	928s	857m 773s	—

[C] Relationship between the stereochemical Structures and the NMR Spectra of α–and β–Acetylpyranoses[97]

Fig 37. and Table 14 show the difference of NMR spectra of β–acetylpyranoses, having different stereochemical structure.

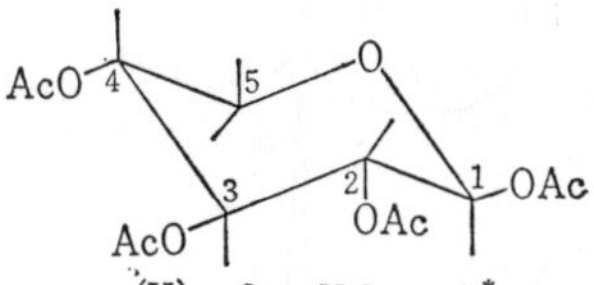

(X) β-D-Xylose-A* 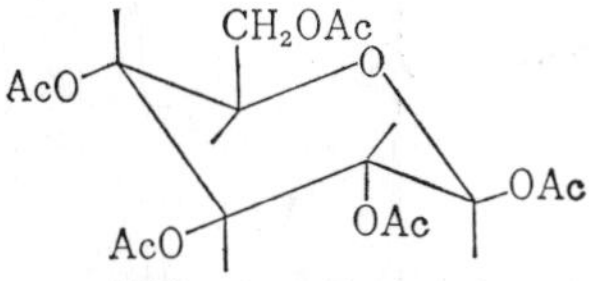   (XI) β-D-Glucose-A 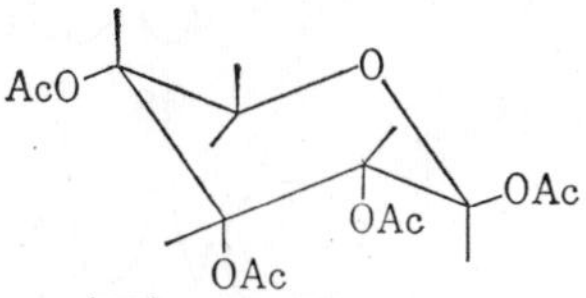   (XII) β-D-Ribose-A

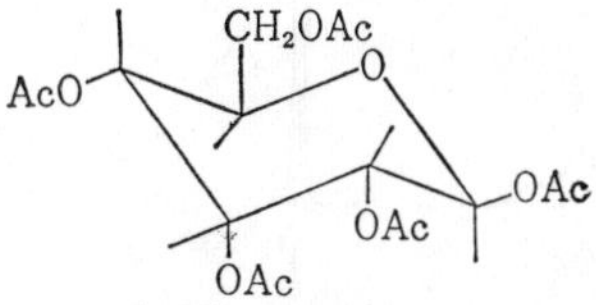

(XIII) β-D-Allose-A 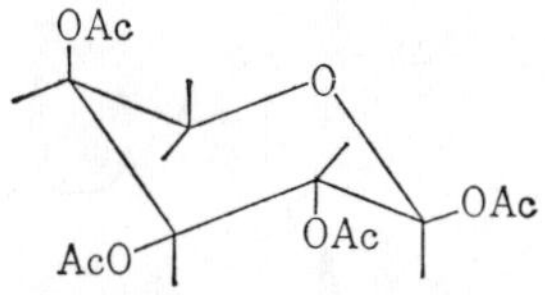   (XIV) α-L-Arabinose-A 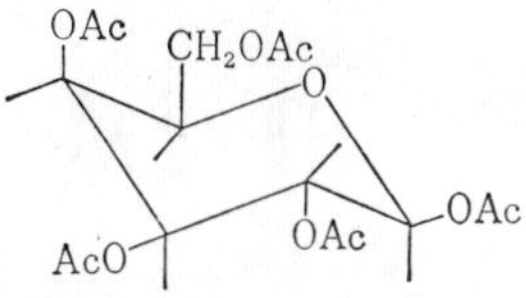   (XV) β-D-Galactose-A

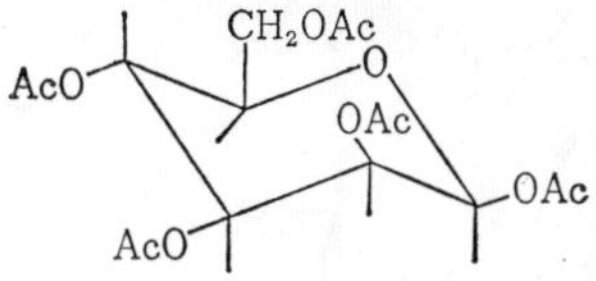

(XVI) β-D-Mannose-A 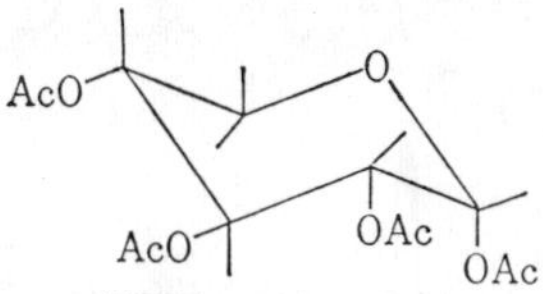   (XVII) α-D-Xylose-A 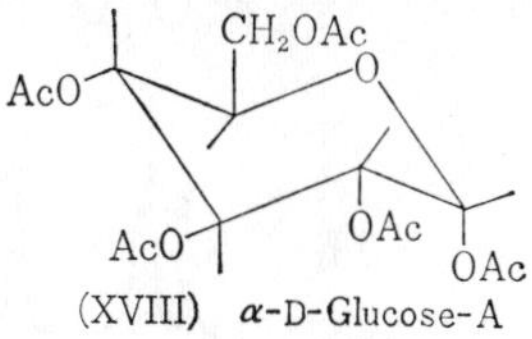   (XVIII) α-D-Glucose-A

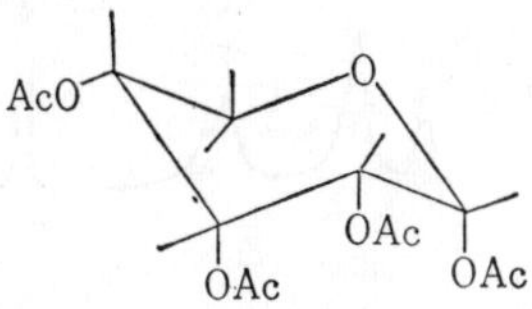

(XIX) α-D-Ribose-A 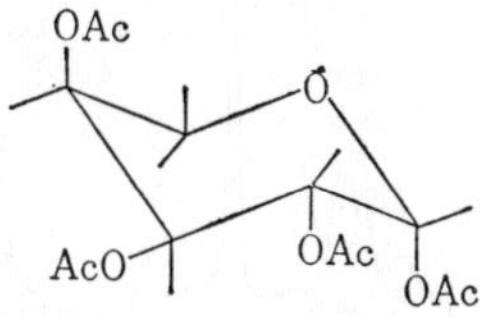   (XX) β-L-Arabinose-A

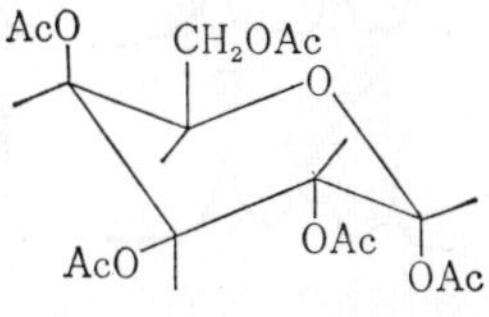

(XXI) α-D-Galactose-A 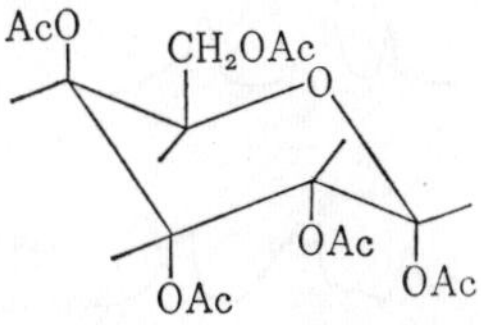   (XXII) β-L-Gulose-A

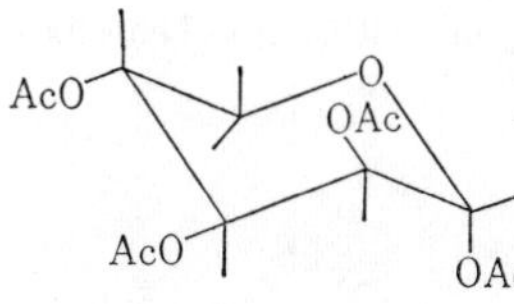

(XXIII) α-D-Lyxose-A 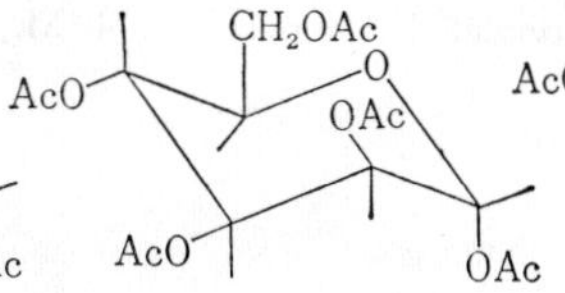   (XXIV) α-D-Mannose-A 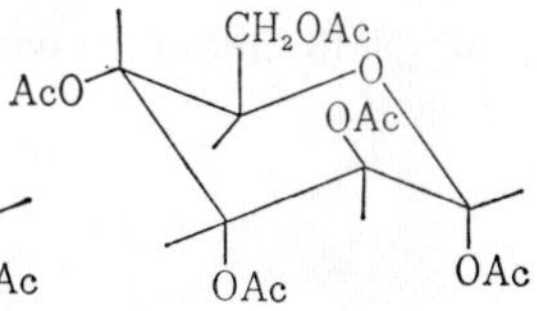   (XXV) α-D-Altrose

---

* –A means fully acetylated aldopyranoses.

97) Lemieux, R. U et al : J.A.C.S., **80**, 6098 (1958).

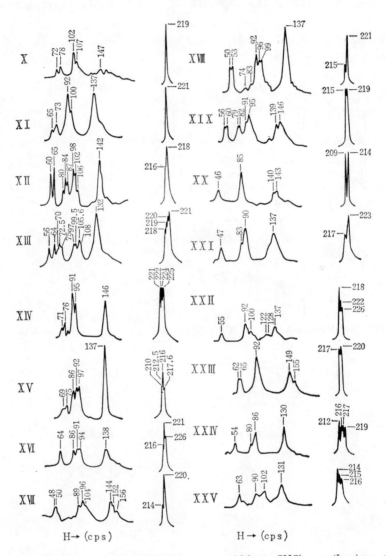

Fig. 37 NMR Spectra of acetylated Aldopyranoses 40 Mc, in CHCl₃, cps (Lemieux *et al.*)

Table 14 NMR Spectra for the anomeric Hydrogen of acetylated Aldopyranoses

(Lemieux *et al.*)

Fully acetylated aldopyranose		Position in the magnetic field cps	Spin-spin coupling constant J
1. Axial anomeric hydrogen			
(a) Axial 2–hydrogen			
β–D–Xylose	(X)	75	6
β–D–Ribose	(XII)	62	5
α–L–Arabinose	(XV)	73	~8
β–D–Glucose	(XI)	69	8
β–D–Allose	(XIII)	60	8
β–D–Galactose	(XV)	72	6
(b) Equatorial 2–hydrogen			
β–D–Mannose	(XVI)	63.5	3
2. Equatorial anomeric hydrogen			
(a) Axial 2–hydrogen			
α–D–Xylose	(XVII)	49.5	3
α–D–Ribose	(XIX)	58	2
β–D–Arabinose	(XX–L–)	46	3
α–D–Glucose	(XVIII)	51	3.2
α–D–Galactose	(XXI)	47	3
α–D–Gulose		55	6.2
(b) Equatorial 2–hydrogen			
α–D–Lyxose	(XVII)	63	3
α–D–Mannose	(XXIV)	52	3
α–D–Altrose	(XXV)	63	3

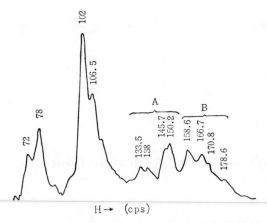

Fig. 38 High resolution Spectrum of β–D–Xylosetetraacetate (Lemieux)

Fig 38 shows fein structure of the chemical shifts due to methylene protons at 5 position in the molecule of β–D–xylosetetraacetate (X).

It's NMR spectrum shows ABX spin-spin coupling due to the protons at 4 and 5 positions, $J_{AB}=12$, $J_{BX}=8$, $J_{AX}=3.2$ (cps). Chemical shifts (cps) of the protons of acetylaldopyranoses (X)∼(XXV) (Fig. 37) are summarized as follows: 45∼75 (C_1 proton), 70∼105 (three protons, attaching to secondary carbons of C_2, C_3, C_4, respectively), 120∼150 (two methylene protons at C_5 or C_6), 210∼230 (twelve protons of four AcO)

[D] Mass Spectra of Acetates of methylated Pentoses and Hexoses[98]

As an example, mass spectrum (Fig. 39) and the fragmentation (1) of methyl–β–D–xylopyranoside triacetate are shown.

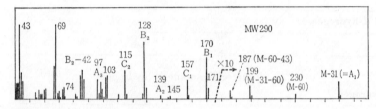

43 | 69 | 128 B_2 | MW290 | 170 B_1 | B_2−42 97 | 115 C_2 | 157 C_1 | ×10 187 (M-60-43) | 74 | A_3 103 | 139 A_2 145 | 171 | 199 (M-31-60) 230 (M-60) | M-31 (=A_1)

Fig. 39 Mass spectrum of Methyl β–D–xylopyranoside triacetate (Biemann *et al.*)

(1) Methyl–β–D–xylopyranoside triacetate

(XXVI) mw 290   →(−CH₃O) A_1, M−31 (m/e 259)   →(−CH₃COOH)   m/e 199 A_1−60

→(−CH₃COOH) m/e 139 A_2 →(−CH₂CO) m/e 97 A_3

(XXVI) → [AcO— ...—OAc]⁺ B_1 (m/e 170) →(−CH₂CO) B_2 (m/e 128) →(−CH₂CO) B_2−42 (m/e 86)

R_1\C≡C≡C/R_2 or (triangle structure)

C_1: R_1, R_2=CH₃CO₂ (m/e 157), C_2: R_1+R_2=CH₃CO₂+OH (m/e 115)
C_1' m/e 129 (C_1−28), C_2' m/e 87 (C_2−28)

98) Biemann, K *et al*: *J.A.C.S.*, **85**, 2289 (1963).

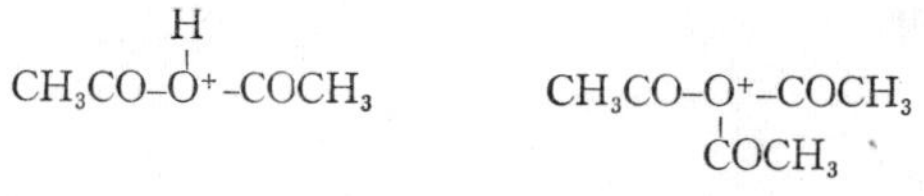
H
CH_3CO-O$^+$-$COCH_3$
CH_3CO-O$^+$-$COCH_3$
$COCH_3$

m/e 103 m/e 145

Mass fragmentation of methylpyranose acetates (XXVII)~(XXXIII) are described in (2)~(8).

(2) 2-*O*-Methyl-*β*-D-xylopyranose triacetate

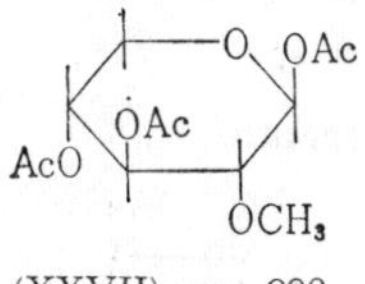
O
OAc
OAc
AcO
OCH_3

(XXVII) mw 290

247 w (M−43), 231 w (M−59), 187 w (M−43−60), 142 m (B_1'), 129 m (C_1'), 111 (A_2'), 100 s (B_2'), 87 s (C_2'), 74 w, 43 s.

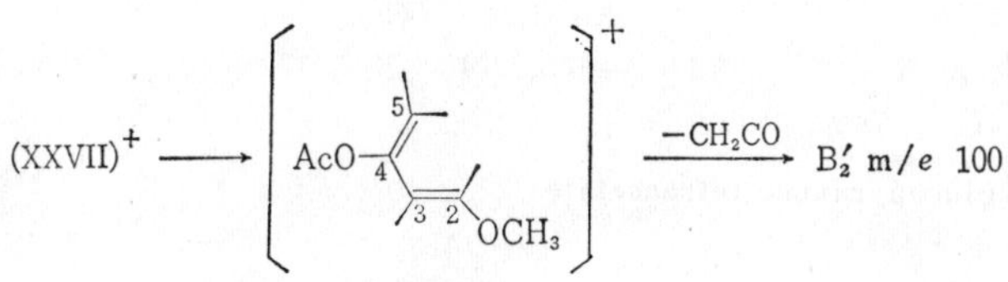
(XXVII)$^+$
AcO
5
4
3 2
OCH_3
+
$-CH_2CO$
B_2' m/e 100

B_1' m/e 142

A_2' m/e 111 (A_2−28)

(3) 3-*O*-Methyl-*β*-D-xylopyranose triacetate

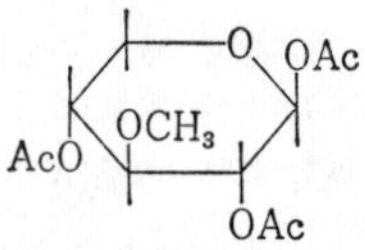
O
OAc
OCH_3
AcO
OAc

(XXVIII) mw 290

247 w (M−43), 231 m (M−59=A'_1), 199 s (A_1'−32), 188 (M−60−42), 170 m (B_1), 157 m (C_1), 145, 139 (A_2), 129 m (C_1'), 128 s (B_2), 115 m (C_2), 111 w (A_2'), 103 m, 100 m (B_2'), 97 (A_3), 87 s (C_2'), 74 m, 71 s, 69 s, 43 s.

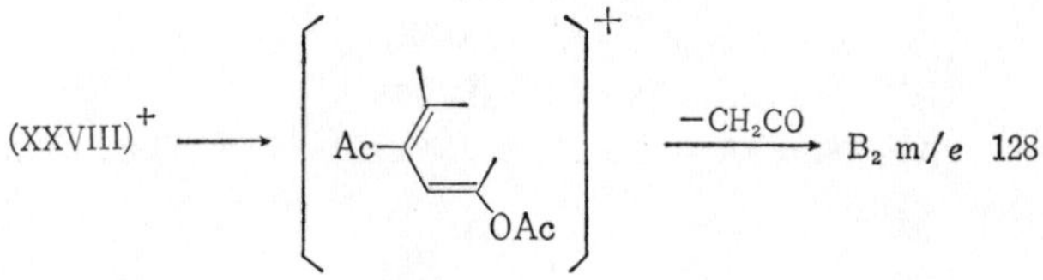
(XXVIII)$^+$
Ac
OAc
+
$-CH_2CO$
B_2 m/e 128

B_1 m/e 170

(4) 5-*O*-Methyl-*β*-D-xylofuranose triacetate

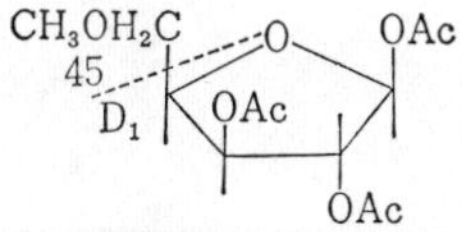
CH_3OH_2C
O
OAc
45
D_1
OAc
OAc

(XXIX) Mw 290

245 w (M−45=D_1), 231 w (M−59), 203 w (D_2), 157 w (C_1), 145 w, 143 s (D_3), 129 s (M−59−102), 115 m (C_2), 103 m, 101 s (D_4), 87 m (C_2'), 45 s, 43 s.

(5) **Methyl-α-D-mannopyranoside tetraacetate**

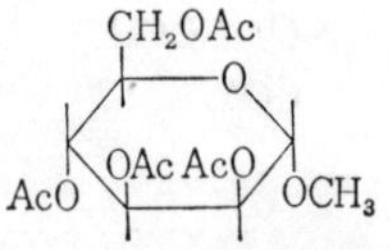

$$CH_2OAc$$

(XXX) Mw 362

331 s (M−31=A_1), 302 w (M−60), 289 w (M−73), 242 m (B_1), 200 m (B_2), 169 m (A_3), 157 s (C_1), 145 s, 140 s (B_3), 115 s (C_2), 103 s, 98 s (B_4), 87 s, 85 s, 81 s, 74 s, 43 s.

(6) **2-O-Methyl-β-D-glucopyranose tetraacetate**

$$CH_2OAc$$

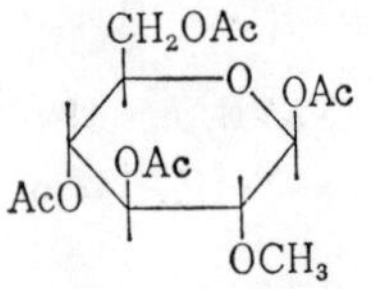

(XXXI) Mw 362

319 w (M−43), 303 w (A_1'), 259 w (M−43−60), 214 w (B_1'), 172 w (B_2'), 140 w (B_3), 129 s (C_1'), 115 w (C_2), 112 w (B_3'), 87 s (C_2''), 74 m, 43 s.

(7) **3-O-Methyl-β-D-glucopyranose tetraacetate**

$$CH_2OAc$$

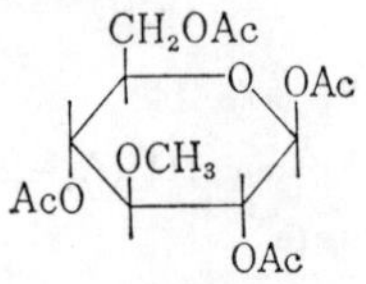

(XXXII) Mw 362

319 w (M−43), 303 m (A_1'), 247 s (M−73−42), 242 s (B_1), 211 w (A_2), 200 m (B_2), 183 w (A_2'), 169 m (A_3), 157 s (C_1), 145 s, 141 w (A_3'), 140 m (B_3), 129 w (C_1'), 115 s (C_2), 109 m (A_4), 103 s, 98 s (B_4), 87 s (C_2'), 74 s, 43 s.

(8) **2, 3, 4, 6-Tetra-O-methyl-β-D-glucopyranose acetate**

$$CH_2OCH_3$$

(XXXIII) Mw 278

233 w (M−45), 219 w (A_1), 187 w (A_1−32), 101 vs (C_1''), 88 s, 74 w, 71 m, 45 s, 43 s.

〔E〕 Five-membered Sugar-lactones[98a)]

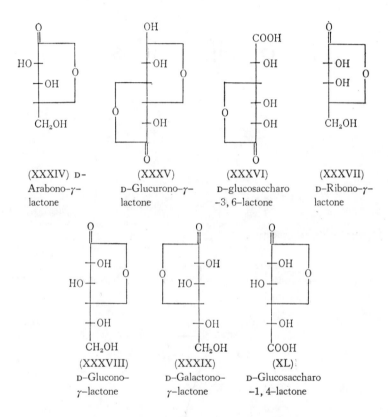

(XXXIV) D-Arabono-γ-lactone

(XXXV) D-Glucurono-γ-lactone

(XXXVI) D-glucosaccharo-3, 6-lactone

(XXXVII) D-Ribono-γ-lactone

(XXXVIII) D-Glucono-γ-lactone

(XXXIX) D-Galactono-γ-lactone

(XL) D-Glucosaccharo-1, 4-lactone

Table 14 a UV, IR and ORD of Five-membered Sugar-lactones

(Okuda *et al.*)

Lactone	Str. No.	$\lambda_{max}^{MeOH}m\mu$	$\nu_{max}^{C=O}cm^{-1}$	$[ORD]_{MeOH}$	
D-Arabono-γ-lactone	XXXIV	222	1760	Cotton Eff.	+
D-Glucurono-γ-lactone	XXXV	220	1760	"	+
D-Glucosaccharo-3, 6-lactone	XXXVI	219	1770	"	+
D-Ribono-γ-lactone	XXXVII	222	1760	"	−
D-Glucono-γ-lactone	XXXVIII	226	1770	"	−
D-Galactono-γ-lactone	XXXIX	221	1790	"	−
D-Glucosaccharo-1, 4-lactone	XL	220sh	1770	"	−

98a) Okuda, T *et al*: *Chem. Pharm. Bull.*, **12**, 504 (1964).

6 Glycosides

[A] Cycas Glycosides

(1) **Cycasin** (Glucosyloxyazoxymethane)

$$CH_3-\overset{\overset{O}{\uparrow}}{N}=N-CH_2O-C_6H_{11}O_5 \quad (I)$$

≪Occurrence≫   *Cycas revoluta* Thunb.[99)100)101)] (ソテツ), *C. circinalis* L.[102)], mp. 144～145° (decomp.)

UV $\lambda_{max}^{0.4N-H_2SO_4}$: 217 m$\mu$, see Fig. 40.

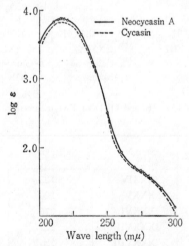

Fig. 40 UV Spectra of Cycas glycosides (Nishida *et al.*)

(2) **Neocycasin A** (β–Laminaribiosylazoxymethane)[101)].

99) Nishida, K *et al* : *Bull. Agr. Chem. Soc. Japan*, **19**, 77, 172 (1955) ; *Seikagaku. (Tokyo)*, **28**, 70 (1956).
100) Nishida, K *et al* : *Agr. Biol. Chem. (Tokyo)*, **23**, 460 (1959).
101) Kobayashi, A : *Agr. Biol. Chem. (Tokyo)*, **26**, 208 (1962).
102) Matsumoto, H *et al* : *Arch. Biochem. & Biophys.*, **101**, 299 (1963).

[C] Iridoids

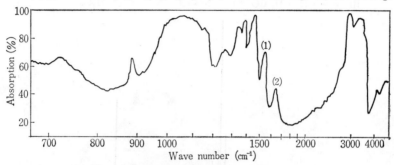

(II)

≪Occurrence≫   *Cycas revoluta* Thunb.

mp. 162~163°, $[\alpha]_D$ −35.1° (H_2O), $C_{14}H_{26}O_{12}N_2 \cdot H_2O$,   UV: See Fig. 40, IR: See Fig. 41.

Fig. 41   IR Spectrum of Neocycasin A (Nishida *et al.*)

[B]   Thioglycosides

(1)   **Allylisothiocyanate** $CH_2=CH-CH_2-N=C=S$ (I)

≪Occurrence≫   Allyl mustard oil from *Brassica* spp., *Eutrema Wasabi* Maxim.(ワサビ)
UV[103] and IR[104]

(2)   **Alliin**[104a]

$$CH_2=CH-CH_2-\overset{O}{\underset{\,}{S}}-CH_2-CH-COOH \quad (II)$$
$$\underset{NH_2}{|}$$

≪Occurrence≫   *Allium cepa* (タマネギ),   mp. 146~148 (decomp.), $C_6H_{11}O_3NS$
IR: See Fig. 42.   Mass: See Fig. 43.

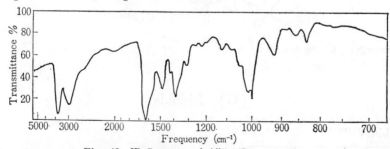

Fig. 42   IR Spectrum of Alliin (Spåre *et al.*)

103)   Nagashima, Z. Nakagawa M: *Nogeikagaku Kaishi (Tokyo)*, **31**, 416 (1957).
104)   Carol, Ramsey: *J. A. O. A. C.*, **36**, 967 (1953).
104a)   Spåre, C. G *et al*: *Acta Chem. Scand.*, **17**, 641 (1963).

(3) **Gluconorcappasalin**[105]

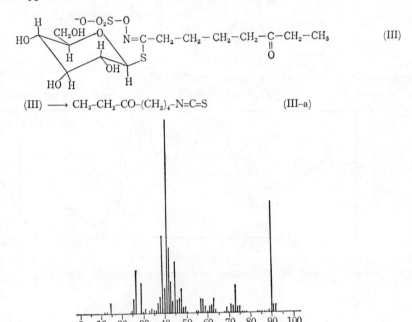

$$(III) \longrightarrow CH_3-CH_2-CO-(CH_2)_4-N=C=S \qquad (III-a)$$

Fig. 43 Mass Spectrum of Alliin (in H_2O, 1 min) (Spåre *et al.*)

≪Occurrence≫ *Capparis salicifolia* GRISEB.

(III–a) IR ν_{max}: 1720 cm^{-1} (C=O)

Mass MW171, m/e 142, 114, 72, 57, 29.

(4) **Glucocleomin**[106]

≪Occurrence≫ *Cleome Spinosa* JACQ.

[C] Iridoids

(1) **Aucubin**(V)[107)~111)]

105) Kjaer, A *et al*: *Acta Chem. Scand*, **17**, 561 (1963).
106) Christensen, B. O. W *et al*: *ibid.*, **17**, 278 (1963).
107) Fujise, S *et al*: *J. Chem. Soc. Japan*, **81**, 1716, 1720, 1723 (1960).
108) Fujise, S *et al*: *4th Symposium on the Chemistry of Natural Products* p. 50. (Tokyo, 1960)
109) Fujise, S *et al*: *Chem. & Ind.*, **1963**, 1688.   110) Birch, A. J *et al*: *Soc.*, **1961**, 5194.
111) Schmid, H *et al*: *Helv. Chim. Acta*, **43**, 1440 (1960).

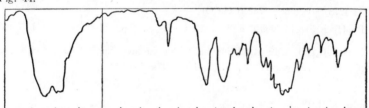

[C] Iridoids

OH
HO–CH$_2$ O–C$_6$H$_{11}$O$_5$
(β–D–glucose)
(V)

$\xrightarrow{\text{H}_2,\ \text{PtO}_2,\ \text{HCl}}$

OH
O
O
(VI) Tetrahydroanhydro
–aucubigenin

《Occurrence》 *Aucuba japonica* (アオキ)[107], *Plantago lanceolata* (オオバコ)[112].

mp. 182°, [α]$_D$ −162.0°(H$_2$O), C$_{15}$H$_{22}$O$_9$·H$_2$O

UV and IR: UV and IR absorptions of 2,4–dinitrophenylhydrazones of tetrahydroanhydroaucubigenin (VI) and the model compounds (VII) and (VIII) are indicated in Table 15[107].

(CH$_2$)$_{2\sim3}$–CHO
OH
(VII)

(CH$_2$)$_{2\sim4}$–OH
O

(CH$_2$)$_3$–OH
O

(VIII)

Table 15   (Fujise *et al.*)

2,4–DNP of	Aucubigenin Derivs. (VI) etc:	(VII)	(VIII)
UV λ_{max} mμ	347～ 355	348～ 357	360～ 365
IR ν_{max} (N–H)cm^{-1}	3280～3299	3280～3292	3310～3320

IR: See Fig. 44.

Fig. 44 IR Spectrum of Aucubin in nujol (Fujise *et al.*)

《2》 **Catalposide (IX) and Catalpol (X)**

RO H
O
O
HO·H$_2$C O–C$_6$H$_7$O (OH)$_4$–β
(a) Bobbitt[113]

or

H–
R–
O
O
O
HO·H$_2$C O–C$_6$H$_7$O (OH)$_4$
(b) Lunn[114]

Catalposide (a) or (b), R=HO–⟨⟩–CO   (IX)

《Occurrence》 *Catalpa ovata* G. Don[115] (キササゲ), *C. bignonioides* Walt.[113)114]

mp. 215～216.5°(decomp.), [α]$_D$ −184°(MeOH), C$_{22}$H$_{26}$O$_{12}$

Catalpol (a) or (b), R=H   (X)

《Occurrence》 *Catalpa ovata* G. Don[115], *C. bignonioides* Walt[114]

mp. 199～202° (decomp.), C$_{15}$H$_{22}$O$_{10}$

IR λ_{max}^{nujol} cm$^{-1\ 115)}$: 3350, 1665, 1460, 1375, 1310, 1235, 1103, 1036, 910, 833, 750

112) Karrer, P *et al*: *Helv. Chim. Acta*, **29**, 525 (1946).
113) Bobbitt, J. M *et al*: *J. Org. Chem.*, **26**, 3090 (1961), *Tetr. Letters*, **1962**, 321.
114) Lunn, W. H *et al*: *Canad. J. Chem.*, **40**, 104 (1962).
115) Kimura, K *et al*: *Y. Z.*, **83**, 635 (1963).

(3)   Asperuloside   (XI: R=H)[116]

(XI)

<Occurrence>   *Coprosma robusta* (ユズリハ)[117], *Pyrola monotropa* (イチヤクソウ), *Monotropa strum*. *Chimaphila* spp.[115)118)]

mp. 131.5~132°, $[\alpha]_D$ −200.4°(H$_2$O)[117], C$_{18}$H$_{22}$O$_{11}$·H$_2$O

UV λ_{max}^{EtOH} mμ (log ε): 234.5 (3.83)

IR ν_{max}^{nujol} cm^{-1}: 3497w, 3300m, 3165m, 1786m, 1748s, 1701s, 1661s, 1534w, 1508w, 1330w, 1282s, 1217m, 1185m, 1081s, 1059s, 1025s, 990s, 913m, 862m, 815m, 765w, 745m, 727w[117].

ν_{max}^{KBr} cm^{-1}: 1773 (γ–lactone), 1742 (CH$_3$CO)

NMR   (XI: R=H) in CDCl$_3$, 100Mc, δ[116]: 7.20 (C$_3$–H), 5.75 (C$_7$–H), 5.68d (C$_1$–H) 5.49d (J=6) (C$_6$–H).

3.48td (C$_5$–H), 5.49~4.65 (four axial H on the sugar ring), 4.65 (—$\overset{10}{C}H_2$—), 4.24m ABXtype (C$_6$′–H), 3.79b (C$_5$′–H), 3.23d (C$_9$–H), 2.17s (CH$_3$CO)

(4)   Monotropein   (XII)[118]

cis-Nepetaic acid

<Occurrence>   *Pyrola monotropa*, *Monotropastrum*, and *Chimaphla*[118]. *Monotropa hypopithys* L. (欧州産シャクジョウソウ)[120]

mp. 175°(decomp.), $[\alpha]_D$ −130.4°(H$_2$O)

UV λ_{max} mμ: 235

IR ν_{max} cm^{-1}: 2800~2500, 1700, 1675, 1645, 1615

116)   Briggs, L. H *et al*: *Tetr. Letters*, **1963**, 69.   117)   Briggs, L. H *et al*: *Soc.*, **1954**, 4182.

118)   Inoue, H *et al*: *7th Symposium on the Chemistry of Natural Products* (Fukuoka, 1963) p. 90.

119)   Grimshaw, J: *Chem. & Ind.* **1961**, 403.

120)   Bridel, M: *Compt. rend.*, **176**, 1742 (1923); *Bull. Soc. Chim. Biol.*, **5**, 722 (1923).

《5》 Plumieride[121]

COOCH$_3$

O—(β-D-glucose)

C—OH
    ''H

(XIII)

≪Occurrence≫　*Plumiera acutifolia* and *P. rubra.*
mp. 156~158° (monohydrate), 224~225°, $[\alpha]_D$ −114°(H$_2$O), −80°(MeOH) C$_{21}$H$_{26}$O$_{12}$
UV　λ_{max}^{EtOH} mμ: 218; IR　λ_{max}^{nujol} μ: 5.72, 5.91, 6.06, 6.13.

《6》 Gentiopicrin (Gentiopicroside)

OC$_6$H$_{11}$O$_5$

glucose-O

(Canonica)　　　　　　(Former Str. by Korte)[124]
(XIV)[122)123]

≪Occurrence≫　*Gentiana* spp., *Swertia* spp.
mp. 176~177°, 191° (EtOH or EtAc), 122° (H$_2$O), $[\alpha]_D$ −200°, C$_{16}$H$_{20}$O$_9$
UV: See Fig. 45[124)~126]
IR: See Fig. 46[124]

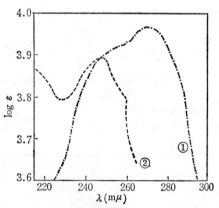

Fig. 45　UV Spectra of Gentiopicrin ① and Tetrahydrogentiopicrin
　　② in MeOH　　　　　　　　　　　　(Korte)

121)　Schmid, H *et al*: *Helv. Chim. Acta*, **41**, 1109 (1958).
122)　Canonica, L *et al*: *Tetrahedron*, **16**, 192 (1962): *Tetr. Letters*, **1960**, (5), **7**.
123)　Kubota, T *et al*: *ibid.*, **1960**, 176; *Bull. Chem. Soc. Japan*, **34**, 1345 (1961).
124)　Korte, F *et al*: *Ber.*, **87**, 512, 769, 880 (1954).
125)　Korte, F *et al*: *ibid.*, **87**, 512 (1954), **89**, 2405 (1956).
126)　Personen, S, Ramstad, E: *J. Am. Pharm. Assoc. Sci. Ed.* **45**, 522 (1956).

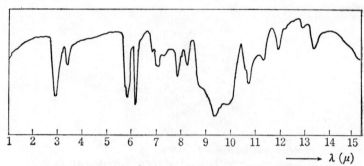

Fig. 46 IR Spectrum of Gentiopicrin (Korte)

(XIV) ⟶

Tetraacetylgentiopicroside

Ac$_4$–glucose–O

| 2 H$_2$

Ac$_4$–glucose–O

(XV) λ_{max} 247mμ

Ac$_4$–glucose–O

(XVI) λ_{max} 232mμ

See next item (7) Swertiamarin

(7) Swertiamarin (XVII)[123]

glucose–O

(XVII)

Ac$_2$O ⟶

Ac–glucose–O

(XVIII)

BF$_3$ in Ac$_2$O ⟶

Ac–glucose–O

(XIX)

(XIX)=Gentiopicrin acetate ← Gentiopicrin (XIV)

≪Occurrence≫ *Swertia japonica* MAKINO. (センブリ), mp. (XVII) 112～114°, C$_{16}$H$_{22}$O$_{10}$ Tetraacetates[127].

	Swertiamarin–Ac$_4$	Gentiopicrin–Ac$_4$
mp.	190～191°	134°
Formula	C$_{22}$H$_{30}$O$_{14}$	C$_{24}$H$_{28}$O$_{13}$
[α]$_D$	−100.8°	−164°
λ_{max} mμ	234	272

(XVIII)[123] mp. 151～152°. UV λ_{max} mμ (log ε): 245 (3.65), IR λ_{max} μ: 12.2 (>=<^{H})
(XIX)[123] mp. 137～138°. UV λ_{max} mμ (log ε): 270 (3.8). See (8) Erythrocentaullin.

127) Tomita, Y: *J. Chem. Soc. Japan*, **81**, 1726 (1960).
128) Tomita, Y *et al*: *ibid.*, **81**, 1729 (1960).
129) Bandow: *Biochem. Zeitschr.*, **296**, 112 (1938).

(8) Erythrocentaullin[128]

$$(XX)$$

≪Occurrence≫   *Erythrea centaurium* PERSON, *Swertia japonica*[128].

Swertiamarin $\xrightarrow{\text{hydrolysis by emulsion}}$ (XX)

(XX) mp. 140∼141°, $C_{10}H_8O_3$

UV λ_{max} mμ (log ε): 223 (4.30), IR λ_{max} μ: 5.91 (–C=C–CO), 5.82 (α, β–unsatur. lactone).

[D] Phenolic Glycosides

(1) UV Spectra[129) 130)]

Arbutin, phenyl–β–D–glucoside (G), *o, m, p*–tolyl–D–G., guaiacol–D–G., tymol–D–G., α, β–naphtyl–D–G., resorcinol–D–G., helicin, *p*–hydroxy benzaldehyde–D–G., *o,p*–hydroxyacetophenone–D–G., paeonoside and paeonolide.

(2) IR Spectra   IR Spectra of salicin (I), populin (II), tremuloidin (III), 6–benzohelicin and 2–benzohelicin are indicated in Fig. 47[131].

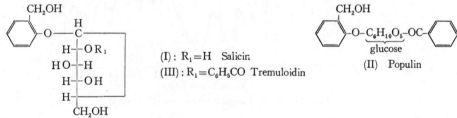

(I): R_1=H Salicin

(III): R_1=C_6H_5CO Tremuloidin

(II) Populin

Tremuloidin   (III)[131]

≪Occurrence≫   *Populus tremuloides*, mp. 207∼208°, $[\alpha]_D$ +17.1° → +19.5° (in pyridine after 72 hrs.)

(3) Homoarbutin (IV)[132] and Isohomoarbutin (IV–a)[132a]

(IV)

(IV–a)

(a) **Homoarbutin** (IV)

≪Occurrence≫   *Pirola incarnata* FISCH. (イチヤクソウ),   mp. 192∼193°,$[\alpha]_D$ −79.2°, $FeCl_3$ (−),
Gibb's R (+),   UV λ_{max} mμ (log ε): 287 (3.43)

(b) **Isohomoarbutin** (IV–a)

≪Occurrence≫   *Chimaphila japonica* MIQ. (ウメガサソウ), mp. 175∼176°, $[\alpha]_D$ −64° (EtOH),
$C_{13}H_{18}O_7$

130) Kariyone, T *et al*: *Y. Z.*, **78**, 939 (1958).

131) Pearl, I. A *et al*: *J. Org. Chem.*, **24**, 731 (1959).

132) Inoue, H: *Chem. Pharm. Bull.*, **4**, 281 (1956).   132a) Inoue, H *et al*: *Y. Z.*, **84**, 337 (1964).

Frequency cm⁻¹

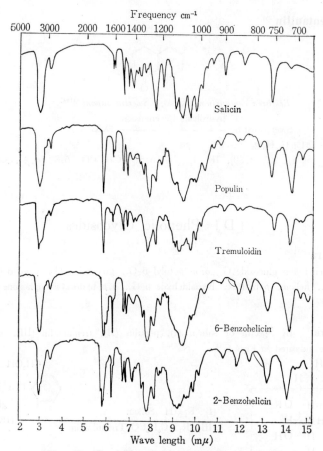

Fig. 47   IR of Phenolic glycosides (Pearl *et al.*)

UV λ_{max}^{EtOH} mμ (log ε): 287 (3.43)

(4)   Periplanetin[133)]

$$C_6H_5-CO-O-\underset{H\ \ H}{C}-\underset{H}{\overset{OH}{C}}-\underset{OH\ H}{\overset{OH}{C}}-\underset{H}{C}-\underset{H}{\overset{O}{C}}-CH_2OH$$ (V)

≪Occurrence≫   *Periplaneta americana* L. and *Blatta orientallis* L.,   mp. 120~121°

(5)   Pungenoside[134)]

COCH₃

O-C₆H₁₁O₅

OH   β-D-glucose   (VI)

≪Occurrence≫   *Picea pungens* ENGLM (=オイモミ).,   mp. 190~191°, C₁₄H₁₈O₈

133)   Quilico, A *et al*: *Tetrahedron*, **5**, 10 (1959).
134)   Takahashi, M: *Y. Z.*, **80**, 782 (1960).

《6》 Pirolatin[135]

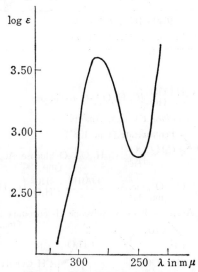

(VII)

《Occurrence》 *Pirola incarnata* FISCH (イチヤクソウ), *P. japonica* SIEB. and other Pirolaceous plants.
mp. 163～164°
UV: See Fig. 48.   IR: See Fig. 49.

Fig. 48 UV of Pirolatin in H_2O

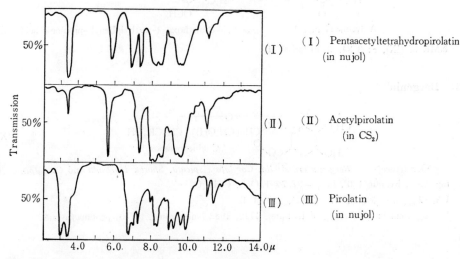

(I) Pentaacetyltetrahydropirolatin (in nujol)

(II) Acetylpirolatin (in CS_2)

(III) Pirolatin (in nujol)

Fig. 49 IR of Pirolatin and related compounds (Inoue *et al.*)

IR (see Fig. 49): λ_{max} μ: (Pirolatin) 11.93, 6.00, (Acetylpirolatin) 12.3, 5.95 (>CH=CH–)

---

135) Inoue, H *et al*: *Y. Z.*, **72**, 223, 228, 731 (1952); *Chem. Pharm. Bull.*, **1**, 401 (1953); **2**, 359 (1954), **6**. 655 (1958).

Related Compounds :[135a]

(VIIa) $R=CH_2-CH=C-CH_2-CH_2-CH=C-CH_3$
                    $|$ $|$
                   CH_3 CH_3

(VIIb) $R=CH_2-CH=C-CH_2-CH_2-CH_2-C=CH_2$
                    $|$ $|$
                   CH_3 CH_3

(VIIc) $R=CH_2-CH_2-C=CH-CH_2-CH_2-C=CH_2$
                       $|$ $|$
                      CH_3 CH_3

	$>C=CH_2$	$>C=CH-$
cm^{-1}	1640~1651, 890~905	1665~1679, 800~840
μ	6.10~6.06, 11.24~11.05	6.00~5.95, 12.50~11.89

(7) Paeoniflorin[136]

$$C_6H_5COO-\begin{Bmatrix}C_{10}H_{11}O_2\end{Bmatrix}-O-\text{D-glucose} \qquad\qquad (VIII)$$
$$\quad HO-$$

≪Occurrence≫   *Paeonia albiflora* PALLAS, Amorphous.

Pentaacetate [mp. 158°]

(VIII) < $C_6H_5COO-\begin{Bmatrix}C_{10}H_{11}O_2\end{Bmatrix}-O-\text{glucose Ac}_4 \xrightarrow{MgI,Ag_2O}$
           $HO-$ [mp 196°]

$Bz.CO-O-\begin{Bmatrix}C_{10}H_{11}O_2\end{Bmatrix}-O-\text{gluc.Ac}_4 \xrightarrow{LiAlH_4} \begin{matrix}HO-\\CH_3O-\end{matrix}\begin{Bmatrix}C_{10}H_{11}O_2\end{Bmatrix}-O-C_6H_{11}O_5$
CH_3O- [mp. 123°]   [mp. 194°]

⟶ Aglycone F ⟶ Aglycone H–leucodimethylether
                                    [mp. 114°]
      ↘ CrO₃   ↗ ‖

Aglycone H mp. 133°   2-(2,5-Dimethoxy-4-methylphenyl)-propionic acid (IX)

NMR of (IX), τ: 3.34, 3.40 (arom H)

(8) Bergenin[136a]

$$(X)$$

≪Occurrence≫   *Bergenia crassifolia, Corylopsis spicata, Shorea leprosula* and *Caesalpinia digyna*
mp. 238°, hydrate 130°, $[\alpha]_D$ −37.7° (EtOH), $C_{14}H_{16}O_9$
UV λ_{max} mμ (log ε): 275 (3.92), 220 (4.42)
IR ν_{max} cm^{-1}: 1966 ($\nu_{C=O}$ of hydrate), 1712, 1682 (anhydrous), (1740) pentamethylate

135a) Barnard, D *et al*: *Soc.*, **1950**, 915.
136) Shibata, S *et al*: *Chem. Pharm. Bull.*, **11**, 372, 379 (1963).
136a) Evelyn Hay, J *et al*: *Soc.*, **1958**, 2231.

(9) 1, 2, 4-Trihydroxy-5, 6, 7, 8-tetrahydronaphthalene (XI)[136b]

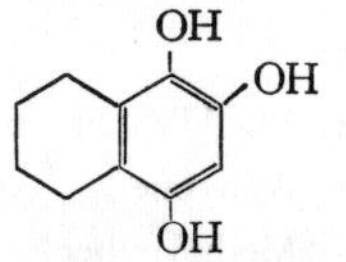

(XI)

≪Occurrence≫   As a glucoside in " Fichte bast ".
mp. 120° (decomp.), $C_{10}H_{10}O_3$
UV λ_{max}^{EtOH} mμ (log ε) : 205 (3. 97), 272. 5 (4. 18)

(10) Renifolin (XII)[136c]

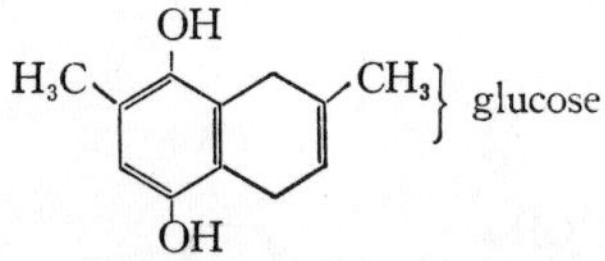

(XII)

≪Occurrence≫   *Pyrola renifolia* MAXIM.
mp. 236〜238°, $[\alpha]_D$ −3. 75° (EtOH), $C_{18}H_{24}O_7$
UV λ_{max}^{EtOH} mμ (log ε) : 283 (3. 30)

〔E〕 Stilbene Glycosides

(1) Piceid   (I, R=glucose)[137]

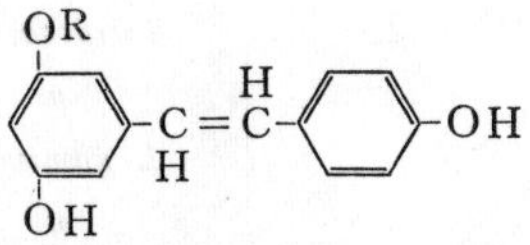

(I)

≪Occurrence≫   *Picea glehnii* MASTERS (アカエゾマツ)
mp. 225〜226°, $[\alpha]_D$ −75. 7° (EtOH), $C_{20}H_{22}O_8 \cdot H_2O$
  Piceid aglycone   (3, 4′, 5-Trihydroxystilbene (I, R=H)),   mp. 256〜257°, $C_{14}H_{12}O_3$
UV : See Fig. 50, λ_{max}^{EtOH} mμ : 307, 315

IR : ν_{max} cm^{-1} : 965, 830, 675 ($trans$-CH=CH—, —◯—, ◯—)

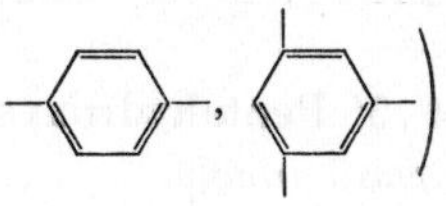

136b)   Endres, H *et al* : *Ber.*, **94**, 438 (1961).
136c)   Inoue, H *et al* : *Chem. Pharm. Bull.*, **12**, 533 (1964).
137)   Kariyone, T. *et al* : *Y. Z.*, **79**, 219, 394 (1959).

6. Glycosides
6. Glycosides

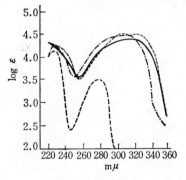

Fig. 50 UV Spectra of Piceid-aglycone and the Derivatives

— Aglycone
····· Methyl ether
–·–· Acetate
—— Dihydroaglycone in 95% EtOH

(Kariyone *et al.*)

(2) **Piceatannol Glucoside** (II)[138]

(II)

≪Occurrence≫ " Fichtenrindenbast " (German)

(3) **Naturally occurring Hydroxystilbens**

Table 16 (Hathway *et al.*)[139]

4 $\underset{5\ 6}{\overset{3\ 2}{\bigcirc}}$-CH=CH-$\underset{6'\ 5'}{\overset{2'\ 3'}{\bigcirc}}$4'	Family	Reference
(III) 3,5,4′-tri–OH–stilben	*Liliaceae*	140)
(IV) 3,5,2′,4′-tetra–OH–stilben	"	140)
(V) 3,5-diCH$_3$O–4′–OH–stilben	*Leguminosae*	141),142)
(VI) 3,5,3′,4′-tetra–OH–stilben	"	143)
(VII) 3,5,3′,4′,5′-penta–OH–stilben	"	143)
(VIII) 3,5,2′,4′-tetra–OH–stilben	*Moraceae*	144),145)
(IX) 2-Homogeranyl–3,5,2′,4′-tetra–OH–stilben	"	146)
(X) 4-CH$_3$O–3,5,3′-tri–OH–stilben	*Polygonaceae*	147)

(4) *cis*- and *trans*-3,5,3′,4′,5′-**Pentahydroxystilben Glycosides** (XI)[148]

≪Occurrence≫ *Eucalyptus astringens*

138) Endres, H : *Ber.*, **91**, 636(1958). 139) Hathway, D.E *et al* : *Biochem. J.*, **72**, 369 (1959).
140) Takaoka : *J. Chem. Soc. Japan*, **61**, 30 (1940) 141) Späth, E *et al* : *Ber.*, **73**, 881 (1940).
142) King, F.E *et al* : *Soc.*, **1953**, 3693. 143) King, F.E *et al* : *ibid.*, **1956**, 4477.
144) Barnes, P.A *et al* : *J.A.C.S.*, **77**, 3259 (1955).
145) Mongolsuk, S *et al* : *Soc.*, **1957**, 2231. 146) King, F.E *et al* : *Soc.*, **1949**, 3348.
147) Kawamura, J : *Y. Z.*, **58**, 405 (1938).
148) Hillis, W.E *et al* : *Biochem. J.*, **82**, 435 (1962).

〔F〕　Coumarins, Furocoumarins and their Glycosides

(1)　UV Spectra

Coumarin[149], acetylcoumarin[150], angelicone[151], dihydroangelicone, 6-Ac-5,7-di-CH$_3$O-coumarin, 5,7-di-CH$_3$O-coumarin[151] coumarin, 5-AcO-coumarin, 5-Me-coumarin, 5-CH$_3$O-coumarin, 5-OH-coumarin, 6-AcO-coumarin, 6-Me-coumarin, 6-CH$_3$O-coumarin, 6-OH-coumarin, 7-AcO-coumarin, 7-Me-coumarin, 7-CH$_3$O-coumarin, 7-OH-coumarin, 5,7-di AcO-coumarin, 5,7-di-CH$_3$O-coumarin, 5,7-di-OH-coumarin, 6,7-di-Ac-coumarin, 6,7-di-CH$_3$O-coumarin, 6,7di-OH-coumarin, 7,8-di-AcO-coumarin, 7,8-di-CH$_3$O-coumarin, 7,8-di-OH-coumarin[152].

Pimpinellin, isopimpinellin, sphondin, isobergaptens and bergapten[153].

(2)　7-Hydroxycoumarins

UV　Spectra of compounds (I)~(IV) are shown in Fig. 51[154].

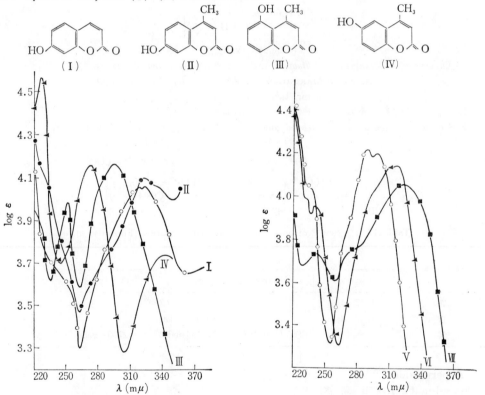

Fig. 51　UV Spectra of 7-Hydroxy-
coumarins (Sen *et al.*)

Fig. 52　UV Spectra of 4-and 3-
Hydroxycoumarins (Sen *et al.*)

149)　Harle, Lyons : *Soc.*, **1950**, 1567.　　150)　Bodforess : *Ann.*, **534**, 236 (1938).

151)　Fujita, M. Furuya, T : *Y.Z.*, **76**, 538 (1956).

152)　Nakabayashi *et al* : *Y.Z.*, **73**, 669 (1953).

153)　Fujita, Furuya : *Y.Z.*, **76**, 535 (1956).

154)　Kalyanmay Sen *et al* : *J. Org. Chem.*, **24**, 316 (1959).

Umbelliferone (I)

UV : $\lambda_{max}^{EtOH}m\mu$ (log ε) : 326, 320 sh, 256, 244 (4.27, 4.24, 3.44, 3.57)

IR : $\nu_{max}^{KBr}cm^{-1}$: 1708, 1610, 905

(3) 4-and 3-**Hydroxycoumarins**

UV spectra of the compounds (V)~(VII) are indicated in Fig. 52.

(V)   (VI)   (VII)

(4) **Glabra-lactone** (VIII)[156] (Angelicone[155])

(VIII)[156]

≪Occurrence≫   *Angelica shishiudo* KOIDZ (シシウド), *A. ursina* MAXIM (エゾニウ)[155],
A. glabra MAKINO (ウドモドキ)[156],   mp. 130°, $C_{16}H_{16}O_5$

UV $\lambda_{max}\,m\mu$: 249, 320 ($\alpha\beta$–unsatur. lactone)

IR $\nu_{max}\,cm^{-1}$: 1737 (δ–lactone), 1667 ($\alpha\beta$–conjtd. ketone)

IR Spectra of angelicone and the related coumarins (IX) and (X) are indicated in Table 17.

Angelicone

(VIII)   (IX)   (X)

Table 17 $\nu_{max}^{nujol}cm^{-1}$ of angelicone and related coumarins (Furuya *et al.*)

Compoumd	$\nu_{C=O}$			ν_{OH}	
	Lactone	Carboxyl	Ketone	Carboxyl	Phenol
(VIII)	1737 (δ)		1667(–C=C–CO–)		
(IX)	1737 (δ)		1710(nonchelated)		
(X)		1686(–C=C–CO–)	1625(chelated)	2750~2550	(–)

(5) **Asperuloside**[117) 157]

Revised to [C], Iridoids (3), (XI)[116]

≪Occurrence≫   *Galium trifolium*

155) Fujita, M *et al*: *Y.Z.*, **76**, 538 (1956) ; Furuya, T *et al*: *ibid.*, **81**, 800 (1961).
156) Hata, K *et al*: *ibid.*, **76**, 649, 666 (1956).   157) Knott, R.P *et al*: *J.P.S.*, **50**, 963 (1961).

(6) **Daphnoretin**[158]

(XII)

mp. 244~247°

UV λ_{max}^{MeOH} mμ (log ε) : 228 sh, 265, 325, 343 (4.18, 3.86, 4.28, 4.31)

IR ν_{max} cm^{-1} : ca 3400 (broad) (OH), 1715 (unsatur. δ–lactone), 1500~1600 (arom. ring), 1280 (C–O–C=O)

(7) **Avicennin** (XIII), (XIV) or (XV)[159]

(XIII)   (XIV)   (XV)

$$R = \overset{\backslash}{\underset{H}{C}} = \overset{H}{\underset{\backslash C}{/}} - C \overset{CH_3}{\underset{CH_2}{<}}$$

《Occurrence》 *Zanthoxylum avicennae*,   mp. 141~142°, [α]$_D$ ±0°, $C_{20}H_{22}O_4$

IR ν_{max} cm^{-1} : 1745 (δ–lactone)

NMR 60Mc, in CDCl$_3$, TMS, τ : 8.52s (2 $\underline{CH_3}$O), 6.24 (1 CH_3O), 4.85 ($>$C=$\underline{CH_2}$), 3.77d, 2.0d,

$J=10$ ($\overset{3}{\underset{\underline{H}}{C}} = \overset{4}{\underset{\underline{H}}{C}}$ –of coumarin), 4.37d, 3.40d, $J=10$, AB system (two *cis* H of–$\overset{3'}{\underset{\underline{H}}{C}} = \overset{4'}{\underset{\underline{H}}{C}}$–), 2.58d, 3.30d,

$J=16$, AB system (two *trans* H of R)

(8) **Bergenin**[160]

(XVI)

《Occurrence》 *Bergenia crassifolia*,   mp. 140°, [α]$_D$ −45.3° (EtOH)

UV λ_{max}^{EtOH} mμ (log ε) : 272, 215 (4.47, 5.02) ; λ_{min} : 241 (3.90)

158) Tschesche, R *et al* : *Ann.*, **662**, 113 (1963).
159) Arthur, H. R *et al* : *Soc.*, **1960**, 4654, **1963**, 3910.
160) Posternak, Th *et al* : *Helv. Chim. Acta*, **41**, 1159 (1958).

(9) (+) *cis*-**Pteryxin** (XVII) and (+) *cis*-**Suksdorfin** (XVIII)[161]

(XVII)   (XVIII)

≪Occurrence≫   *Pteryxia terebinthina* COULTER & BOSE var. *terebinthina*, *Lomatium suksdorfii* COUNTER & ROSE

(a) (+) *cis*-**Pteryxin** (XVII)

mp. 81.5~82.5°, $[\alpha]_D$ +10° (EtOH)

UV λ_{max} mμ (log ε) : 263, 218 (3.20, 4.31) ; λ_{min} 252 (3.57)

IR ν_{max}^{nujol} cm^{-1} : 1757~1745 (α-pyrone and $\alpha\beta$-satur. ester CO), 1637 (α-pyrone ring C=C), 1621, 1502 (arom. C=C), 1587 (arom. ring with conj. C=C), 1321, 1107 ($\alpha\beta$-unsatur. ester C–O), 1122 (Ar.–O–CR$_3$), 908 (CH$_2$C–O), 846 (1,2,3,4 substd. arom.), 839 (trisubstd. ethylene C–H), 1339, 1377 ((CH$_3$)$_2$C=)

(b) (+) *cis*-**Suksdorfin** (XVIII)   mp. 140.5~141°, $[\alpha]_D$ +4° (EtOH)

UV λ_{max} mμ (log ε) : 219, 234 (4.05, 3.52) ; λ_{min} : 240, 251, 263 (3.47, 3.30, 3.07)

IR ν_{max}^{nujol} cm^{-1} : 1754 (ester C=O), 1736 (α-pyrone C=O), 1631 (α-pyrone ring C=C), 1610, 1493, 1445 (arom. C=C), 1572 (arom. ring with conj. C=C), 1250, 1120 (Ar–O–CR$_3$), 902 ((CH$_3$)$_2$C–δ), 856 (1,2,3,4-substd. arom.), 1395, 1376 (C(CH$_3$)$_2$)

(10) **Coumestrol**[168]

(XIX)

≪Occurrence≫   Alfalfa, clover spp.,   mp. >350°, C$_{15}$H$_8$O$_5$

UV λ_{max}^{EtOH} mμ : 343, 304, 244 ; $\lambda_{max}^{EtOH-NaOAc}$ mμ : 387, 320, 281, 260

(11) **Wedelolactone** (5′,5,6-Trihydroxy-7′-methoxycoumarino-(3′,4′ : 3,2) coumarone) (XX, R=Me, R′=H)[162]

(XX)

≪Occurrence≫   *Wedelia calendulaceae*,   mp. 327~330° (decomp.), C$_{16}$H$_{10}$O$_7$

---

161) Soine, T.O *et al* : *J.P.S.*, **51**, 149 (1962).

162) Govindachari, T.R *et al* : *Soc.*, **1957**, 545, 548 ; *Tetrahedron*, **15**, 129 (1961).

UV of tri–*O*–methylwedelolactone (XX, R=Me, R′=Me)
See Fig. 52a and the item (20)[175]

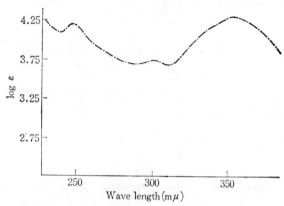

Fig. 52a UV Spectrum of Tri–*O*–methylwedelolactone
(Govindachari)

(12) Psoralidin[163]

(XX–a)

≪Occurrence≫ *Psoralea corylifolia* L., mp. 290〜292°, $C_{20}H_{16}O_5$

UV $\lambda_{max}m\mu$ (ε): 208, 244, 305, 347 (40800, 20300, 7000, 25300)

IR $\nu_{max}^{KBr}cm^{-1}$: 3350 (phenolic OH), 1710 (conj. δ–lactone), 1625 (conj. C=C), 1600, 1578, 1498 (arom.),
1261 (arom. C–O)

(13) Mammein[164]

(XXI)

≪Occurrence≫ *Mammea americana*, mp. 128.5〜129.5°, $C_{22}H_{28}O_5$

UV $\lambda_{max}^{EtOH-HCl}m\mu$ (log ε): 295 (4.34); $\lambda_{max}^{EtOH-HCl}$: 250 (3.39); $\lambda_{max}^{EtOH-KOH}m\mu$ (log ε): 232, 265, 332

(3.97, 4.04, 4.60), $\lambda_{min}^{EtOH-KOH}$: 240, 274 (3.84, 3.30)

163) Khastgir, H. N *et al*: *Tetrahedron*, **14**, 275 (1961).
164) Djerassi, C *et al*: *J.A.C.S.*, **80**, 3686 (1958); *J. Org. Chem.*, **25**, 2164 (1960).

IR $\lambda^{CHCl3}_{max}\mu$: 3.03 m, 5.78 s, 6.20 s, 6.37 m, $\lambda^{CS2}_{max}\mu$: 3.02 m, 5.71 s, 5.82 sh, 6.25 s,

assigned for : phenolic OH, lactone, conj. C=O, $(CH_3)_2$ C=CH–$\overset{\mid}{\underset{\mid}{C}}$–

(14) Novobiocin[165]~[167]

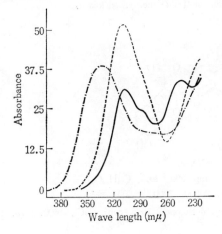

(XXII)

≪Occurrence≫   *Streptomyces niveus*,   mp. 174~178°, 152~156° (dimorphoric)

UV : See Fig. 53

Fig. 53 UV Spectra of Novobiocin
—— 70%, EtOH, pH 7.5
······ alkaline EtOH
–·–· acid EtOH
(Hinman *et al.*)

(15) Farnesiferol A[169]

(XXIII)

≪Occurrence≫   *Asa foetida*

Farnesiferol A $C_{24}H_{30}O_4$,   mp. 155~155.5°, $[\alpha]_D$ −55° $(CHCl_3)$

〃   B $C_{24}H_{30}O_4$,   113.5~114°,   +10° (〃)

〃   C $C_{24}H_{30}O_4$,   84~85°   −29° (〃)

Those compounds are described in (12) Sesquiterpenes, [D], (2), (3), [E], (3).

165) Hinman *et al* : *J.A.C.S.*, **79**, 3789 (1957).   166) Stammer *et al* : *ibid.*, **80**, 137 (1958).
167) Chambers, K *et al* : *Proc. Chem. Soc.*, **1960**, 291.   168) Jurd, L : *Tetr. Letters*, **1963**, 1151.
169) Jeger, O *et al* : *Helv. Chim. Acta*, **41**, 2278 (1958).

(16) Furocoumarins *of Angelica* spp.[170)~172)]

(XXX–c) Angelicin

		R_1	R_2
(XXIV)	Oxypeucedanin	$-O-CH_2-CH-C\langle^{CH_3}_{CH_3}$ (epoxide O)	H
(XXV)	Isoimperatorin	$-O-CH_2-CH=C(CH_3)_2$	H
(XXVI)	Imperatorin	H	$-O-CH_2-CH=C(CH_3)_2$
(XXVII)	Phellopterin	$-OCH_3$	$-O-CH_2-CH=C(CH_3)_2$
(XXVIII)	Byak-angelicin	$-OCH_3$	$-O-CH_2CH-C(CH_3)_2$ $\overset{|}{O}H\ \overset{|}{O}H$
(XXIX)	Byak-angelicol	$-OCH_3$	$-O-CH_2-CH-C(CH_3)_2$ $\diagdown O \diagup$
(XXX)	Alloimperatorin	$-OCH_2-CH=C(CH_3)_2$	$-OH$
(XXX–a)	Psoralen	H	H
(XXX–b)	Xanthotoxin	H	OCH_3

(a) *Angelica dahurica* BENTH. et HOOK var. *dahurica* (日本産ビャクシン), *A. dahurica* var. *pai-chi* KIMURA HATA et VEN (中国産ビャクシン)[170)] (XXIV)～(XXX)

(b) *Angelica formosana* BOISS (タイワンシシウド), *A. anomala* LALL. (エゾヨロイグサ)[170)] (XXIV), (XXV), (XXVI), (XXVII), bergapten, (XXXII), oxypencedanin-hydrate, umbelliferone (I)

(c) *Angelica japonica* A. GRAV. (ハマウド) Isobyak-angelicolic acid (XXXI), byak-angelicin (XXVIII), osthol.

(XXXI)

(d) *Angelica glabra* MAKINO (ウドモドキ)[172)], oxypencedanin (XXIV), ferulin, umbelliferone (I).

(e) *Angelica keiskei* KOIDZ. (アシタバ)[173)], psoralen (XXX–a), angelicin (XXX–c), bergapten (XXXII), xanthotoxin (XXX–b)

(17) IR Spectra of Furocoumarins[171)]

≪Occurrence≫ *Helacleum lanatum* var. *asiaticum* HARA (オオハナウド)

(XXXII) Bergapten (XXXIII) Isobergapten

170) Hata, K *et al*: *Y. Z.*, **83**, 606, 611 (1963). 171) Mitsuhashi, M *et al*: *ibid*., **81**, 464 (1961).
172) Hata, K *et al*: *ibid*., **80**, 742, 892 (1960). 173) Hata, K *et al* : *ibid*., **81**, 1647 (1961).

(XXXIV) Pimpinellin   (XXXV) Isopimpinellin

(XXXIII) Isobergapten   mp. 218~220°, $C_{12}H_8O_4$.   IR $\nu_{max}^{KBr}cm^{-1}$: 1730, 1605, 885, 750

(XXXII) Bergapten   mp. 191°, $C_{12}H_8O_4$.   IR $\nu_{max}^{KBr}cm^{-1}$: 1720, 1615, 1602, 890, 743

(XXXV) Isopimpinellin   mp. 116~118°, $C_{13}H_{10}O_5$.   IR $\nu_{max}^{KBr}cm^{-1}$: 1735, 1610, 880, 750

(XXXIV) Pimpinellin   mp. 116~118°, $C_{13}H_{10}O_5$.   IR $\nu_{max}^{KBr}cm^{-1}$: 1735, 1610, 880, 750

(XXV) Imperatorin   mp. 99.5~100°, $C_{16}H_{14}O_4$.   IR $\nu_{max}^{KBr}cm^{-1}$: 1700, 1610, 1595, 875, 775, 745.

(XXIX) Byak-angelicol   See Fig. 54

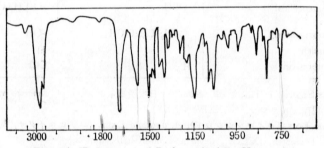

Fig. 54 IR Spectrum of Byak-angelicol[172)] (Hata *et al.*)

IR spectra of the coumarins of *Angelica edulis* MIYABE are described in the item (23)

(18)   NMR of angular (XXXVI) and **linear** (XXXVII) **Furocoumarins**[173a)]   See Fig. 55

(XXXVI)   (XXXVII)

60Mc, in $CDCl_3$, TMS, cps:

(XXXVI)   475 dd $J=1, 9$ (C_4-H), 435 d $J=9$ (C_5-H)

$\left(_H>C^4=C^5<_H\right)$, assigned the angular structure for (XXXVI)

407 (1H), 377 (C_8-H), 156, 151 (9-, 2-CH$_3$)

(XXXVII)   447 (C_4-H), 435 (C_9-H), 384 (C_3-H), 371 (7H), 149, 144 (2-, 8-CH$_3$)

173a)   Kaufmann *et al*: *J. Org. Chem.*, **27**, 2567 (1962).

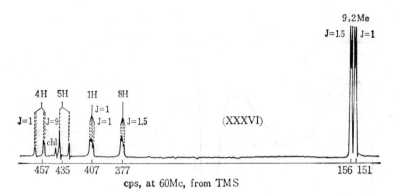

cps, at 60Mc, from TMS

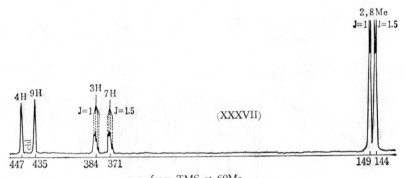

cps, from TMS at 60Mc

Fig. 55 NMR of Furocoumarins (Kaufmann)

(19) Prangolarin[174]

(XXXVIII)

《Occurrence》 *Prangos pabularia* LINDL., mp. 104~105°, $[\alpha]_D$ +20.1° (CHCl₃), C₁₆H₁₄O₅

UV λ_{max}^{EtOH} mμ (log ε) : 222, 249, 313 (4.48, 4.31, 4.25); IR λ_{max} μ : 5.75 (lactone), 7.94 (epoxide), 9.3 (benzofuran)

(20) Erosnin (XXXIX) and Pachyrrhizin (XL)[175]

(XXXIX)   (XL)

《Occurrence》 *Pachyrrhizus erosus* (Yaw beans)

The plant contains also tri–(O)–methylwedelolactone (XX, R=Me, M′=Me) in this item (11) and rotenone.

174) Chatterjee, A : *Chem. & Ind.* **1963.**, 1430.
175) Schmid, H *et al* : *Helv. Chim. Acta,* **42,** 61 (1959).

Erosnin   mp. >350° (decomp.), $C_{18}H_8O_6$, UV: See Fig. 56, IR: See Fig. 57

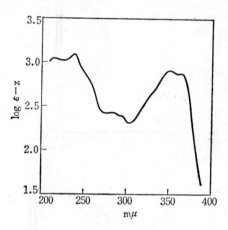

Fig. 56 UV Spectrum of Erosnin (in 96% EtOH) (Schmid)

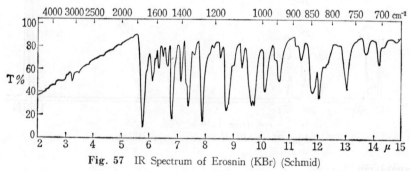

Fig. 57 IR Spectrum of Erosnin (KBr) (Schmid)

IR ν_{max}^{KBr}cm^{-1}: 1733 (C=O of pyrone), 1639 (conj C=C), 1582, 1506 (arom.) 1266 (arom. C–O)$_{\nu}$ 937, 706 (—O–CH$_2$–O—)

(21) Marmesin (XLI) and **Marmesinin** (XLII)[176)177)]

R=H Marmesin (XLI)
R=C$_6$H$_{11}$O$_5$ Marmesinin (XLII)

≪Occurrence≫   *Ammi majus* L.

(a) Marmesin   mp. 189~190°, $[\alpha]_D$ +25° (CHCl$_3$)

UV λ_{max}^{EtOH} $m\mu$ (log ε) : 212, 248, 335 (3.89, 3.50, 4.09)

(b) Marmesinin   mp. 215~216°, $[\alpha]_D$ −29° (CHCl$_3$), $C_{14}H_{14}O_4$

---

176) Chatterjee, A *et al* : *J.A.C.S.*, **71**, 606 (1949).
177) Abu-Mustafa, E.A *et al* : *J. Org. Chem.*, **26**, 161 (1961).

(22) Athamantin (XLIII)[178]

$$R=COCH_2-CH \big\langle\begin{smallmatrix}CH_3\\CH_3\end{smallmatrix}$$

(XLIII)

(XLIV) Oroselone   (XLV) Oroselol

+2 mol Isovaleric acid

≪Occurrence≫   *Athamanta oreoselinum* L. syn. *Peucedanum oreoselinum*

UV λ_{max}^{EtOH}mμ (log ε) : 217 sh, 322 (4.18, 4.17)

IR ν_{max}^{CH2Cl2}cm^{-1} : 1748 (ester CO, pyrone ring), 1629 (α–pyrone C=C, arom.), 1587 (C=C conj. arom.)

1388, 1370 (gem di–Me), 835

IR of (XLIV) and (XLV) are described in next item (23).

(23) Edultin (XLVI)[179]

(XLVI)

≪Occurrence≫   *Angelica edulis* MIYABE (マルバエゾニウ)

mp. 136～142°, [α]$_D$ +41.5° (pyridine), $C_{21}H_{22}O_7$

UV λ_{max}^{EtOH}mμ (ε) : 219, 248, 259, 299 sh, 323 (31430, 6770,

5710, 9350, 21000)

[Degradated products]

(XLVI) —

7.5%NaOH·MeOH → Oroselol–Me–ether (XLV–Me)
+ Oroselone (XLIV) + Angelicine (XXX–c)
+ HOAc + Angelic acid

0.1N–NaOH → Oroselol (XLV) + Oroselolacetate
(XLV–Ac) + HOAc + Angelic acid

IR ν_{max}^{nujol} cm^{-1} Angelicine (XXX–c) : 1740, 1620, 1590, 1460, 885, Oroselone (LIV) : 1720, 1620, 1450,

Oroselolmethylether (XLV–Me) : 1730, 1720 (δ–lactone), 1620 (C=C), 1580,

1450 (arom.)

(24) Rengasin (XLVII)[180]

(XLVII)

≪Occurrence≫   *Melanorrhea* spp. (Rengas)

mp. 220° (decomp.) $C_{16}H_{12}O_6 \cdot 2H_2O$

UV λ_{max}^{EtOH}mμ (ε) : 254, 403 (9900, 40000)

178) Schmid, H *et al* : *Helv. Chim. Acta*, **40**, 758 (1957).
179) Mitsuhashi, H *et al* : *Chem. Pharm. Bull.*, **10**, 511, 514 (1962).
180) King, F.E *et al* : *Soc.*, **1962**, 1192.

(25) **Maesopsin** (2-Benzyl-2, 4, 6, 4′-tetrahydroxycoumaranone)[181]

(XLVIII)

≪Occurrence≫   *Maesopsis eminii*,   mp. 218~220° (decomp.), $[\alpha]_D$ ±0°, $C_{15}H_{12}O_6$

UV $\lambda_{max}^{EtOH} m\mu$ (log ε) : 211, 290 (4.38, 4.28)

IR ν_{max} cm^{-1} : 1670

NMR of tetramethylether (Table 18)

Table 18   (Jones *et al.*)

Chemical shift τ	Coupling const. J cps	Relative Intensity	Assignment
2.77, 2.98, 3.15, 3.35	8.4	4	AB pattern. Coupled pairs of arom. H atoms
3.32, 3.47, 3.59	2.3	2	AB pattern. Pair of *meta* coupled arom. H atoms
5.86		2	H atoms on C ajacent to O or -CH$_2$-
6.15		9	O–Me H atoms
6.19		3	"

(26) **Hydrangenol 8-glucoside**[181a]

(XLIX)

≪Occurrence≫   *Hydrangea hortensia* DC. var. *otakusa* Maxim. (アジサイ)
mp. (180~182°), $C_{21}H_{22}O_9$

(27) **Furocoumarins of *Ficus carica* L.**[182]

(28) **Fluorometry of Coumarins**[183)184]

〔G〕 Chalcone and Flavanone Glycosides

(1) UV Spectra

Hesperidin, hesperetin[185)188], apigenidin, hesperetidin, sakuranetidin,[186)187] poncirin and isosakurarin[189].

181) Jones, N.F *et al* : *Soc.*, **1963**, 1356.
181a) Sakamura *et al* : *6th Symposium on the Chemistry of Natural Products*, (Sapporo, 1962) p. 64.
182) Fukushi *et al* : *J. Agr. Chem. Soc. Japan*, **33**, 376, 564, 1025 (1959).
183) Ichimura, Y : *Y. Z.*, **80**, 771 (1960).
184) Wheelock, C.E : *J. A. C. S.*, **81**, 1348 (1959).
185) Asahina, Shinoda, Inubuse, : *Y. Z.*, **553**, 207 (1928).   186) Asahina, Inubuse : *ibid.*, **561**, 1081 (1928).
187) Asahina, Inubuse, : *ibid.*, **559**, 868 (1928).   188) Lajos, Gerendas : *Biochem. Zeitschr.*, **291**, 229 (1937).
189) Shimokomiya : *J. A. C. S.*, **79**, 4199 (1957).

Stillopsin octaacetate, leptosin hexaacetate, butein tetraacetate, leptosidin triacetate, butein, leptosidin and $3', 4',$ ·6, 7–tetramethoxyflavanone[190]. 　Chalcone derivatives, acebogenin[191]. 　Desmethoxymatteucinol, naringin, naringenin, isosakuranetin, matteucinol, sakuranetin, hesperidin, hesperetin[192]~[217] and $2', 3, 4$–trihydroxychal·cone[218].

(2)　IR Spectra

Flavanone, 7–methoxy——, 7, 4'–dimethoxy——, 5–hydroxy——,[219] $3', 4'$–dihydroxy——, $3', 4'$–diacetoxy ——, $3', 4', 5, 7$–tetrahydroxy——, $3', 4', 5, 7$–tetraacetoxy——, 5–hydroxy–$3', 4', 7$–trimethoxy——, $3, 3', 4', 5, 7$–pentahydroxy——, $3, 3', 4', 5, 7$–pentaacetoxy——, $3, 3', 4', 5, 7$–pentamethoxy——, chalcone, $2', 3, 4$–triacetoxy ——, $2', 3, 4$–trihydroxy——, $2', 3, 3', 4, 4'$–pentahydroxy——, $2', 3, 3', 4, 4'$–pentabenzoxy——, $3, 3', 4', 5, 7$–pentahydroxy——[220].

Flavanone, 7–methoxy——, 7, 4'–dimethoxy——, 5–hydroxy——, $3', 4'$–dihydroxy——, $3', 4', 7$–trimethoxy –5–hydroxy——, chalcone, trihydroxy——, triacetoxy——,[221] naringenin[222].

(3)　UV and IR Spectra of Hydroxychalcones[223]

Chalcone＝Benzalacetophenone (Ⅰ)

Table 19　UV and IR Spectra of Hydroxychalcones　　　(Klinke et al.)

No.	Compounds	mp.	IR CO Band μ	UV $\lambda_{max}m\mu$ (ϵ) 1st Band	2nd Band	3rd Band	pKa
Ⅰ	Chalcone	57~59	6.04	205 (15800)	226 (10400)	308 (25400)	9.0
Ⅱ	4–OH–chalcone	182.5	6.09	204 (17400)	246 (11900)	347 (24300)	9.0
Ⅱ–a	4–OAc–chalcone	129	6.00	207 (19000)	241 (13700) 248 (13900)	308 (11600)	
Ⅲ	4'–OH–chalcone	174.5	6.07	207 (20200)	226 (16100)	309 (16600)	8.5
Ⅲ–a	4'–OAc–chalcone	95	6.00	206 (20300) 208 (20200)	231 (12500) 261 (13300)		

190)　Ferguson *J. A. C. S.*, **70**, 3907 (1948).　　　191)　Murakami: *Y. Z.*, **75**, 573 (1955).
192)　Klein: Handbuch der Pflanzen Analyse Ⅲ/2., 924 (Julius Springer, 1932).
193)　Shibata K *et al*: *Acta Phytochim.*, **1**, 91 (1923).　　194)　Hattori: *ibid.*, **5**, 1 (1930).
195)　Nierenstein: *J. Indian Chem. Soc.*, **8**, 144 (1931).　　196)　Hattori: *Acta Phytochim.*, **4**, 63 (1928).
197)　Hattori: *ibid.*, **6**, 131 (1932).　　198)　Tazaki: *ibid.*, **2**, 129 (1925).
199)　Tazaki: *ibid.*, **3**, 1 (1927).　　200)　Robertson *et al*: *Soc.*, **1929**, 2241.
201)　Hattori: *Acta Phytochim.*, **5**, 99 (1930).　　202)　Hattori: *ibid.*, **5**, 219 (1931).
203)　Shibata K: *Bot. Mag.*, **29**, 130 (1915).　　204)　Shibata K *et al*: *J. Biol. Chem.*, **28**, 108, 1916/17.
205)　Tazaki: *Acta Phytochim.*, **2**, 119 (1925).
206)　Asahina Y., Inubuse M: *Ber.*, **61**, 1646 (1928) ; *Y. Z.*, **48**, 1086 (1923).
207)　Hattori: *Acta Phytochim.*, **2**, 109 (1925).　　208)　Shibata K., *et al*: *Acta Phytochim.*, **1**, 134 (1923).
209)　Kondo *et al*: *Y. Z.*, **49**, 182 (1929).　　210)　Asahina Y., Inubuse M.: *Ber.*, **64**, 1256 (1931).
211)　Fuzise: *Scient. Papers. Inst. Physic. Chem. Res.*, **11**, 111 (1929).
212)　Shibata K *et al*: *Acta Phytochim.*, **2**, 37 (1924).
213)　Hattori: *Acta Phytochim.*,, **4**, 219 (1929)　　214)　Asahina Y *et al*: *Y. Z.*, **47**, 1011 (1927).
215)　Asahina Y *et al*: *ibid.*, **48**, 29 (1928).　　216)　Asahina Y *et al*: *ibid.*, **48**, 150 (1928).
217)　Asahina Y *et al*: *ibid.*, **48**, 150 (1928).　　218)　Jurd, Geissman: *J. Org. Chem.*, **21**, 1395 (1956).
219)　Shaw, Simpson: *Soc.*, **1955**, 655.　　220)　Hergert, Kurth: *J. A. C. S.*, **75**, 1622 (1953).
221)　Yamaguchi K: Syokubutuseibun Bunseki-ho Vol, Ⅱ, p. 212 (1961) (Nankodo Pub. Inc., Tokyo).
222)　Sadtler Card No. 5015.　　223)　Klinke, P *et al*: *Ber.*, **94**, 26 (1961).

No.	Compounds	mp.	IR CO Band μ	UV $\lambda_{max}m\mu$ (ε) 1st Band	2nd Band	3rd Band	pKa
IV	4, 4'-Di-OH-chalcone	200	6.07	207(16200)	233(14200)	310(15700) 345(31200)	9.2
IV-a	4, 4'-Di-OAc-chalcone	126	5.98	208(25400)	258(13000)	308(12700) 321(10500)	
V	3, 4-Di-OH-chalcone	202~203	6.06	210(21800)	263(14700)	322(10600) 363(21100)	8.7
V-a	3, 4-Di-OAc-chalcone	125	6.00	207(24900) 208(25100)	235(14300) 242(14100)	301(12700) 321(10000)	
VI	3', 4'-Di-OH-chalcone	187	6.04	205(26900)	223(13200)	305(10600) 336(16500)	8.2
VI-a	3', 4'-Di-OAc-chalcone	131	6.00	208(25400)	258(13000)	308(12700) 321(10500)	
VII	3, 4, 4'-Tri-OH-chalcone	203~204	6.07	208(24400)	230(11800)	262(9000) 326(17100)	9.0
VII-a	3, 4, 4'-Tri-OAc-chalcone	150	5.98	208(27000)	225(15600) 263(14200)	302(14600) 321(11800)	
VIII	4, 3', 4'-Tri-OH-chalcone	200	6.08	208(24200)	240(13400)	355(29600)	9.1
VIII-a	4, 3', 4'-Tri-OAc-chalcone	114	6.00	207(24900) 208(24800)	249(14200)	312(14200)	
IX	3, 4, 3', 4'-Tetra-OH-chalcone	220~221	6.08	210(31800)	253(12000)	373(27000)	8.8
IX-a	3, 4, 3', 4'-Tetra-OAc-chalcone	156~158	6.00	209(26700)	260(13700)	310(15200)	
X	3, 4, 3', 4'-Tetra-OH-hydrochalcone	188	6.07	208(25400)	229(19200)	279(12900) 307(8200)	8.5
X-a	3, 4, 3', 4'-Tetra-OAc-hydrochalcone	94~95	5.92	209(33900)	248(15800)	—	

(4) *cis, trans*-**Isomerization of Chalcone**[422]

(5) **Xanthohumol** (4, 2', 6'-Trihydroxy-4'-methoxy-3'-[γ, γ-dimethylallyl]-chalcone) (XI)[225]

≪Occurrence≫ *Humulus lupulus* (ホップ), mp. 159~161°

(6) **Dihydrochalcones of *Malus* spp.**[226]

(XII) R=β-D-glucosyl, R'=H (Phloridzin)
(XIII) R=H, R'=β-D-glucosyl
(XIV) R=β-D-glucosyl

≪Occurrence≫ (XIII) *Malus trilobata*, (XIV) *M. sieboldii*

224) Noyce, D.S *et al*: *J.A.C.S.*, **83**, 2525 (1961)   225) Riedl, W. *et al*: *Ber.*, **93**, 312 (1960).
226) Williams, A.H: *Soc*, **1961**, 4133.

(7) Bathochromic Shift on the Spectra of Flavanones[227]

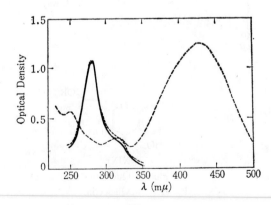

		$\lambda_{max}m\mu$		
Compounds	Flavanone	EtOH	+NaOAc	+NaOH
	7-OH-flavanone	277	338	338
Liquiritigenin	7,4'-Di-OH-flavanone	276	338	338
Butin	7,3',4'-Tri-OH-flavanone	278	338	338
Pinocembrin	5,7-Di-OH-flavanone	291	329	329
Naringenin	5,7,4'-Tri-OH-flavanone	290	328	328
Eriodictyol	5,7,3',4'-Tetra-OH-flavanone	289	328	328
Taxifolin	3,5,7,3',4'-Penta-OH-flavanone	291	330	329
Isosakuranetin	5,7-Di-OH-4'-OCH₃-flavanone	292	328	328
	5,3',4'-Tri-OH-7-OCH₃-flavanone	287	287	289
Homoeriodictyol	5,7,4'-Tri-OH-3'-OCH₃-flavanone	289	328	328
Hesperetin	5,7,3'-Tri-OH-4'-OCH₃-flavanone	288	328	328
	5-OH-7,3',4'-tri-OAc-flavanone	274		
Poncirin	Isosakuranetin-7-rhamnogluc.	283	283	285
Eriocitrin	Eriodictyol-7-rhamnogluc.	285	285	285
Hesperidin	Hesperetin-7-rutinoside	285	285	287
Neohesperidin	Hesperetin-7-neohesperidoside	285	285	287
Sakuranetin	5,4'-Di-OH-7-OCH₃-flavanone	287	287	424
Sakuranin	Sakuranetin-5-gluc.	281	281	428
Prunin	Naringenin-7-gluc.	284	284	425
	Naringenin-7-rhamnoside.	284	284	428

Table 20 (Horowitz *et al.*)

Fig. 58 Sakuranin

—— in EtOH

······ +AlCl₃

- - - - +1% NaOH, after 10 min

(Horowitz *et al.*)

227) Horowitz, R.M *et al* : *J. Org. Chem.*, **26**, 2446 (1961); *J.A.C.S.*, **82**, 2803 (1960).

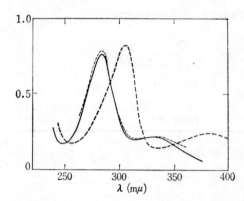

Fig. 59 Eriocitrin

—— in EtOH

······ +NaOAc

— — +AlCl₃

(Horowitz *et al.*)

(8) (−)-**Hesperetin** (XV), (−)-**Liquiritigenin** (XVI) and (+)-**Sakuranetin** (XVII)[228]

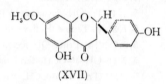

(XV) R₁=OH, R₂=OCH₃   (XVII)

(XVI) R₁=H, R₂=OH

UV: See Fig. 60

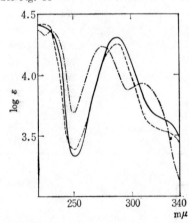

Fig. 60 UV Spectra of Flavanones in MeOH

—— (−)-Hesperetin (XV)

-·-· (−)-Liquiritigenin (XVI)

······ (+)-Sakuranetin (XVII)

(Arakawa)

(9) **Taxifolin glucoside**[229]

$$\xrightarrow{\text{NaHSO}_3}$$

(XVIII)   (XIX)

R=glucose Taxifolin–3′–glucoside   R=glucose Quercetin–3′–glucoside

R=H   Taxifolin   R=H   Quercetin

228)   Arakawa, H *et al*: *Ann.*, **636**, 110 (1960).

229)   Hergert. H. U *et al*: *J. Org. Chem.*, **27**, 700 (1958).

<Occurrence> Douglas-fir wood

Taxifolin–3′–glucoside   mp. 203～205°, $[\alpha]_D$ −23°, $C_{21}H_{22}O_{12}\cdot 2H_2O$

UV See Fig. 61 and Tab. 21

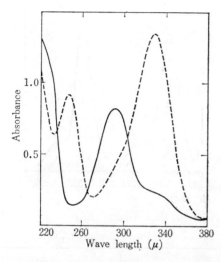

Fig. 61 UV Spectra of Taxifolin–3′–glucoside

—— $5.68\times 10^{-5}\ M$, in EtOH

······ in 0.006 N KOH in EtOH

(Hergert *et al.*)

Table 21 UV Absorption of Taxifolin and related Flavonoids (Hergert *et al.*)

Compounds	$\lambda_{max}m\mu$ (log ε) in 95% EtOH	$\lambda_{max}m\mu$ (log ε) in alkaline EtOH
Taxifolin–3′–gluc.	227(4.35), 292(4.19), 329(3.65)	249(4.24), 329(4.42)
Taxifolin	229(4.39), 291(4.27), 326(3.35)	246(4.04), 329(4.38)
Isosakuranin	227(4.40), 284(4.20), 332(3.46)	242(4.16), 336(3.96), 350(4.13)
Quercetin–3′–gluc.	253(4.29), 325(4.03), 370(4.34)	240(4.19), 329(4.35), 432(4.03)
Kaempferol	268(4.19), 324(3.99), 368(4.26)	246(4.11), 325(4.27), 430(3.93)
Quercetin	256(4.32), 301(3.89), 373(4.32)	247(4.08), 334(4.27), 420(3.86)

IR See the item [I] Flavones and Flavonoid–Glycosides (5), Fig. 68

(10) UV and IR Spectra of Flavanones and their Acetates[230]

See Fig. 62 and Table 22.

Dihydroquercetin
(XX)

Eriodictyol
(XXI)

Astilbin
(XXII)

Sakuranetin
(XXIII)

230) Aft, H: *J. Org. Chem.*, **26**, 1958 (1961).

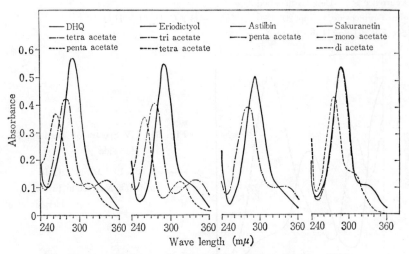

Fig. 62 UV Spectra of the Flavanoids and their Acetates (Aft)

Table 22 UV and IR Spectra of the Flavanoids and their Acetates (Aft)

Flavanoid	+Dihydroquercetin (XX)	Eriodictyol (XXI)	Astilbin (XXII)	Sakuranetin (XXIII)
mp.	240~242°	269~270°d	222~224°	153~155°
λ_{max}^{EtOH} mμ (log ε)	290(4.30)	290(4.28)	292(4.28)	289(4.26)
λ_{min}^{EtOH} mμ (log ε)	249(3.52)	250(3.53)	250(3.15)	245(3.30)
$\nu_{C=O}$cm^{-1}	1642	1635	1645	1645
Acetate	penta–	tetra–		mono–
mp.	88~90°	140~141°		140~141°
λ_{max}^{EtOH}mμ (log ε)	263(4.11) 312(3.67)	260(4.06) 315(3.59)		289(4.28) 320(3.58)
λ_{min}^{EtOH}mμ (log ε)	295(3.59)	288(3.28)		248(3.07)
$\nu_{C=O}$cm^{-1}	1705	1673		1640
Acetate	tetra–	tri–	penta–	di–
mp.	153~154°	126~127°	116~117°	116~117°
λ_{max}^{EtOH} mμ (log ε)	277(4.13) 340(3.57)	275(4.12) 340(3.60)	281(4.36) 335(3.76)	275(4.18) 305(3.74)
λ_{min}^{EtOH} mμ (log ε)	245(3.39) 307(3.35)	245(3.46) 305(3.38)	248(3.61) 320(3.74)	248(3.46)
$\nu_{C=O}$cm^{-1}	1680	1640	1634	1675

(11) Sophorol

(XXIV)

≪Occurrence≫ *Sophora japonica* L., mp. 215°, $[\alpha]_D$ −13.6° (EtOH), +9.5° (acetone)

UV See Fig. 63

$\lambda_{max}^{EtOH}m\mu$ (ε): 230 (16950), 277 (15950), 307.5 (12090); $\lambda_{max}^{0 \cdot 1N-NaOH}m\mu$ (ε): 246 (11950), 332 (27570)

IR　See Fig. 64

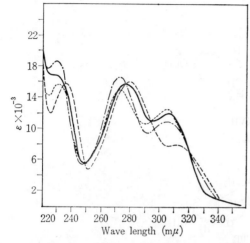

Fig. 63　UV Spectra of Sophorol and
related Compounds

——　Sophorol
-·-·　O–Dimethyl Sophorol
······　2′, 7–Dimethoxy–3′, 4′–methylenedi-
oxyisoflavanone
——　7–Hydroxy–3′, 4′–methylenedioxy-
flavanone

(in EtOH) (Suginome)

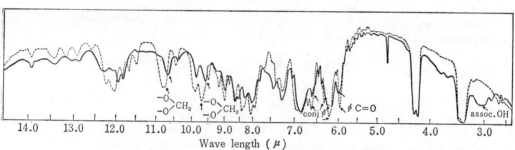

Fig. 64　IR Spectra of Sophorol (—) and O–Dimethylsophorol (······) in nujol (Suginome)

(12)　NMR of Flavanoids

(a)　(+)–**Mollisacacidin** ((2R: 3S: 4R)–3′, 4′, 7–Trihydroxyflavan–3, 4–diol) (XXV)[231]

<Occurrence>　*Acacia mearnsii*,　$[\alpha]_D$ +32° (acetone+H_2O)

3, 4–O–Diacetyl–3′, 4′, 7–trimethyl—(XXV), mp. 102°

NMR: $J_{2,3}=8.9$, $J_{3,4}=7.0$ cps (H^2, H^3, H^4) (natural and synthetic *trans–trans*), $J_{2,3}=10.3$, $J_{3,4}=3.3$
(synthetic *trans–cis*)

(b)　(+)–**Leucorobinetinidin** ([(+)–3′, 4′, 5′, 7–Tetrahydroxyflavan–3, 4–diol]) (XXVI)[231]

231) Lillya, C.P: *Chem. & Ind.*, **1963**, 783.

《Occurrence》 *Robinia pseudacacia*, $[\alpha]_D$ +35° (acetone+H_2O)

3,4–*O*–Diacetyl–3′,4′,5′,7–*O*–tetramethyl–(XXVI), mp. 121°

NMR $J_{2,3}$=8.9, $J_{3,4}$=6.9 cps (natural and synthetic *trans-trans*)

$J_{2,3}$=10.0, $J_{3,4}$=3.5 cps (synthetic *trans-cis*)

(c) Methyl matteucinol (XXVII)[232]

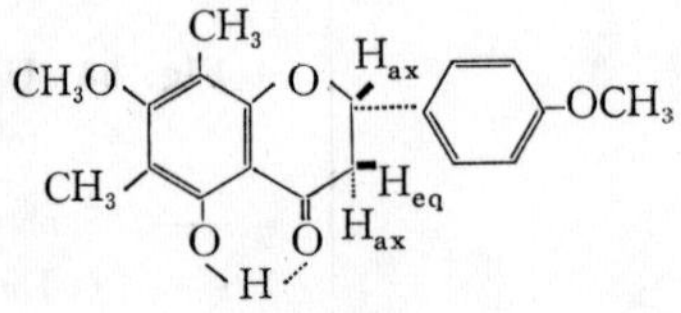

(XXVII)

NMR in $CDCl_3$: J_{AX}=12.9 cps (2(ax) 3 (ax)), J_{BX}=3.1 cps (2 (ax) 3 (eq))

(d) Cyanomaclurin[233] (XXVIII)

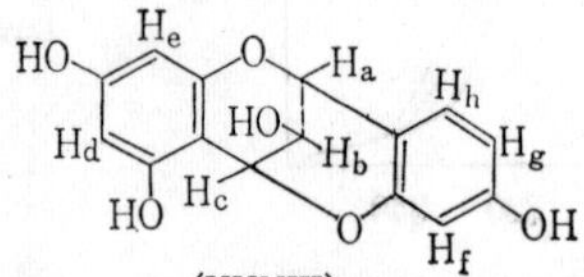

(XXVIII)

Trimethyl ether of (XXVIII) H_b/H_a *trans*, H_B/H_C *trans* structure was assigned by NMR.

(e) Nepseudin (Trimethoxy furanoisoflavanone) (XXIX)[234]

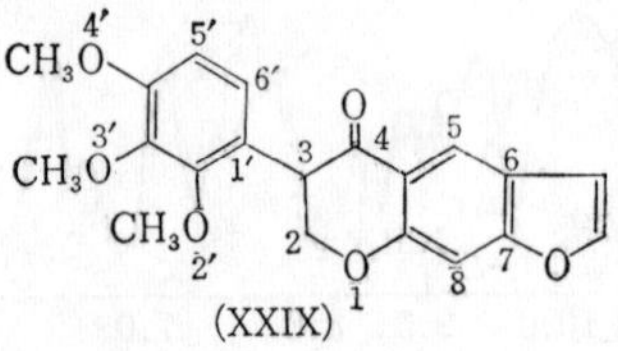

(XXIX)

《Occurrence》 *Neorautanenia pseudopachyrrhiza*, mp. 116°, $C_{20}H_{18}O_6$

UV, IR, NMR See the item 〔K〕 Rotenoids (5), (b), (X)

(f) Stereochemistry and NMR of Flavan–4β–**ols**[234a]

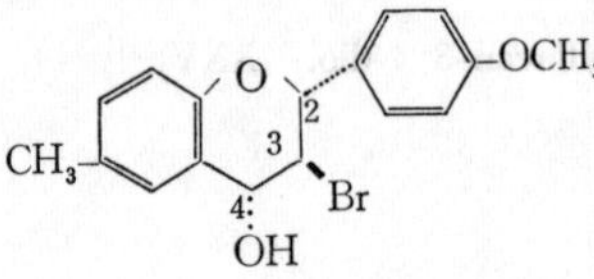

trans-trans
(XXIX–a)

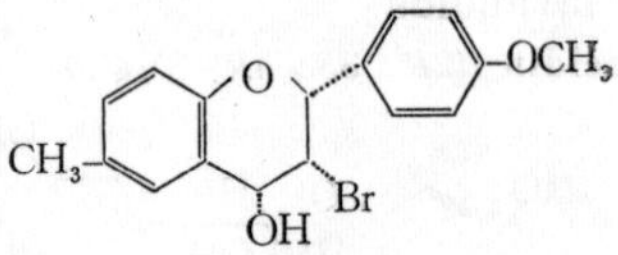

cis-cis
(XXIX–b)

NMR (XXIX–a) $J_{2,3}$=10.4, $J_{3,4}$=9.4 c/s

(XXIX–b) $J_{2,3}$=1.0, $J_{3,4}$=4.3 c/s

(13) (−) Farrerol and **(±) Angophorol**[235]

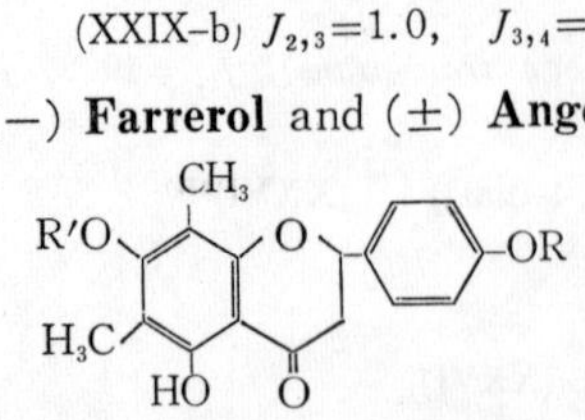

R=R′=H: (−) Farrerol (XXX)

R=H, R′=CH_3: (±) Angophorol (XXXI)

232) Horn, D.H.S *et al*: *Chem. & Ind.*, **1963**, 691.   233) Venkataraman, K: *Tetr. Letters*, **1963**, 317.

234) Crombie, L *et al*: *Chem. & Ind.*, **1962**, 1946.

234a) Clark-Lewis, J.W *et al*: *Proc. Chem. Soc.*, **1963**, 20.   235) Birch, A.J *et al*: *Soc.*, **1960**, 2063.

≪Occurrence≫   *Angophora lanceolata*

(XXX)   mp. 212~220°, $[\alpha]_D$ −20° (EtOH), $C_{17}H_{16}O_5$

UV $\lambda_{max}m\mu$ (log ε): 298 (4.25), 254 (3.28)

(XXXI)   mp. 150°, $C_{18}H_{18}O_5$; $\lambda_{max}m\mu$ (log ε): 216 (4.72), 283 (4.49), 353 (3.96)

(14) Citromitin (5,6,7,8,3′,4′-Hexamethoxyflavanone) and 5-*O*-Demethylcitromitin ("Compound B")[236]

R=CH₃: Citromitin   (XXXII)
R=H: 5-*O*-Demethylcitromitin   (XXXIII)

≪Occurrence≫   *Citrus mitis* BLANCO

(XXXII)   mp. 134~136°, $C_{21}H_{24}O_8\cdot1/2\ H_2O$,   UV $\lambda_{max}^{EtOH}m\mu$ (log ε): 335 (4.38), 272 inf (4.23)

(XXXIII)   mp. 146~147°, $C_{20}H_{22}O_3$,   UV $\lambda_{max}^{EtOH}m\mu$ (log ε): 344 (4.37), 283 (4.28)

(15) Other Flavanoids and Leucoanthocyanidins

(a) Citronin (2′-Methoxy-5,7-dihydroxyflavanone-7-rhamnoglucoside) (XXXIV)[237)238]

(XXXIV)

≪Occurrence≫   *Citrus limon* BURM. *f. ponderosa* HORT.

(b) Padmatin (XXXV)[239]

(XXXV)

≪Occurrence≫   *Prunus puddum*,   mp. 170~171°

(c) (+) *O*-Pentamethyldihydromelanoxetin (XXXVI)[240]

Gummy solid $\xrightarrow{\text{Methylation}}$

(XXXVI)

≪Occurrence≫   *Albizzia odoratissima* BENTH.

236) Sastry, G.P *et al*: *Tetrahedron*, **15**, 111 (1961).
237) Yamamoto *et al*: *J. Agr. Chem. Soc. Japan*, **7**, 312 (1931).
238) Horowitz, R.M *et al*: *Nature*, **185**, 319 (1960).
239) Seshadri, T.R *et al*: *Tetrahedron*, **5**, 91 (1959).
240) Ramachandra Row, L. *et al*: *Tetrahedron*, **19**, 1371 (1963).

(d) Neohesperidin (XXXVII)[241]

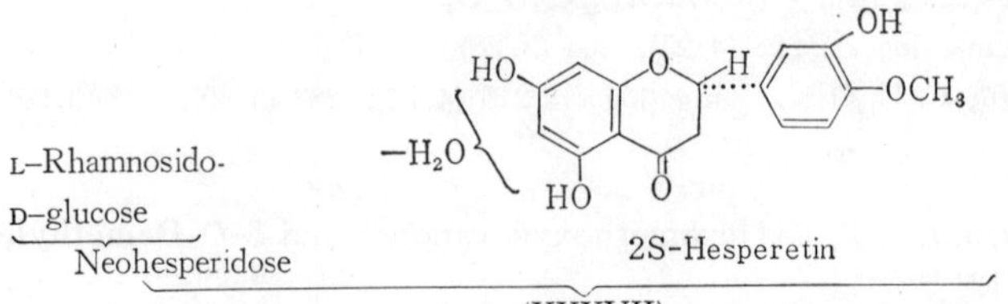

L-Rhamnosido-
D-glucose
$\underbrace{\hspace{3cm}}$
Neohesperidose

$-H_2O$

2S-Hesperetin

(XXXVII)

≪Occurrence≫   *Citrus bigaradia* RISSO (Pomeranzen),   mp. 234~235°, $[\alpha]_D$ −94.5° (pyridine), $C_{28}H_{34}O_{16}$

(e) Relation between bitter taste and stereochemical structure of **Neohesperidin** and **Hesperidin**[241a]

(i) Bitter

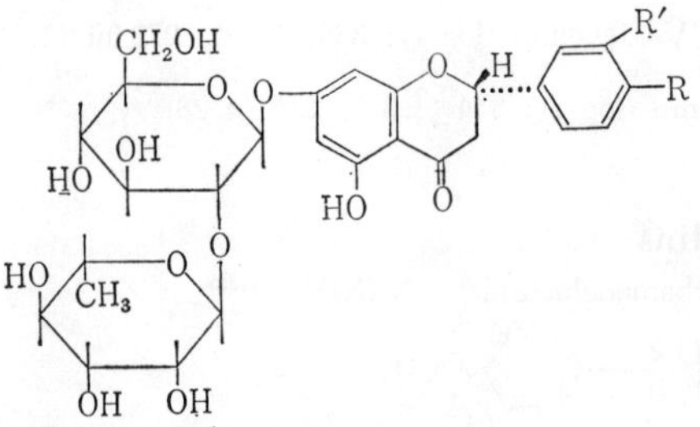

R=OCH$_3$, R'=OH : Neohesperidin
(XXXVII)
R=OH, R'=H : Naringin (XXXVII−a)
R=OCH$_3$, R'=H : Poncirin (XXXVII−b)

2−O−α−L−Rhamnopyranosyl−
D−glucopyranose

(ii) Not bitter

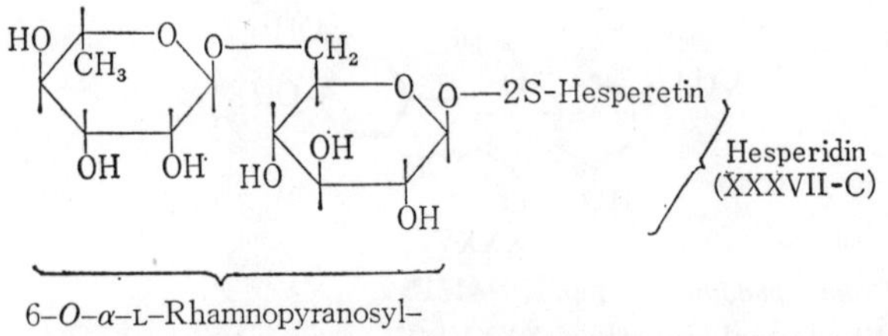

Hesperidin
(XXXVII−C)

6−O−α−L−Rhamnopyranosyl−
D−glucopyranose

(f) Teracacidin ((−)−7,8,4′−Trihydroxy−2,3−*cis*−flavan−3,4−*cis*−diol)   (XXXVIII)

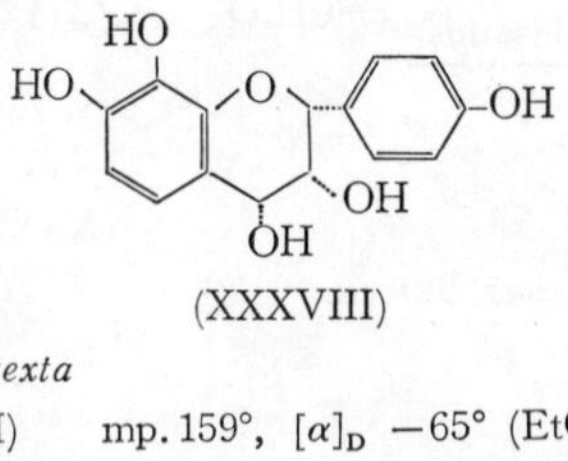

(XXXVIII)

≪Occurrence≫   *Acacia intertexta*
7,8,4′−Trimethylether of (XXXIII)   mp. 159°, $[\alpha]_D$ −65° (EtOH), $C_{18}H_{20}O_6$

---

241) Hardegger, E *et al*: *Helv. Chim. Acta*, **44**, 1413 (1961).
241a) Horowitz, R.M *et al*: *Tetrahedron*, **19**, 773 (1963).

(g) **Melacacidin** (XXXIX) and **Isomelacacidin** (XL)[242]

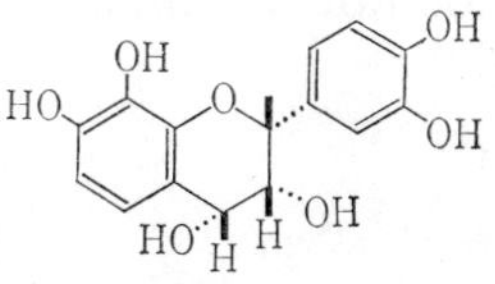

(XXXIX) 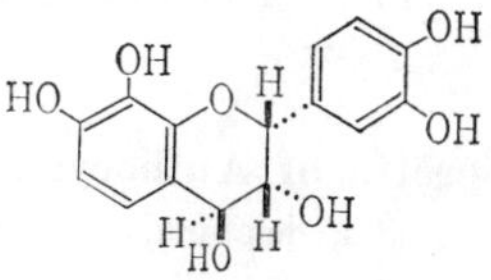   (XL)

《Occurrence》   *Acacia excelsa, A. heterophylla, A. melanoxylon*
(XXXIX)   mp. 229° (decomp.), $[\alpha]_D$ −75° (EtOH)

UV $\lambda_{max}^{EtOH}m\mu$ (log ε): 280 (3.5); $\lambda_{min}^{EtOH}m\mu$ (log ε): 254 (2.77)

(h) (+)-**Mollisacacidin** (XLI) and (−) **Leucofisetinidin** (XLII)[243]

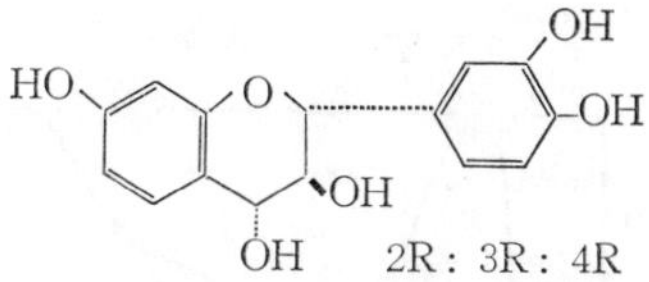

2R: 3R: 4R   2S: 3R: 4S
(XLI)=(XXV)   (XLII)

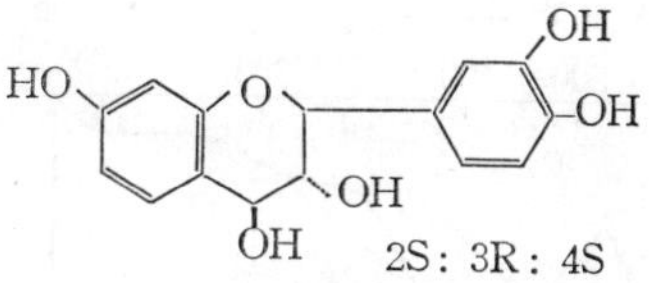

《Occurrence》   (XLI) *Acacia mearnsii*
                (XLII) *Schimnopsis* spp.

(i) **Mundulone** (XLIII)[244]

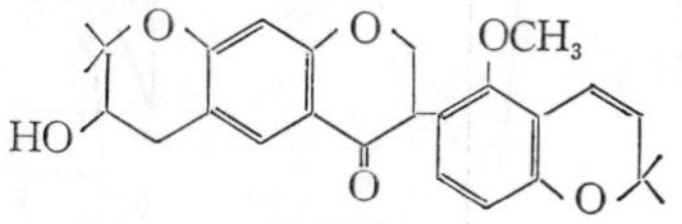

(XLIII)

《Occurrence》   *Mundulea sericea,*   mp. 180°, $[\alpha]_D$ −11.5° (CHCl₃)
IR ν_{max}^{CHCl3}cm⁻¹: 3600, 3395 (OH), 1631 (CO)
Dihydro-(XLIII)   mp. 200°
UV $\lambda_{max}^{EtOH}m\mu$ (ε): 244 (32000), 252 (32400), 309 (18000)

(j) **Peltogynol** (XLIV)[245]

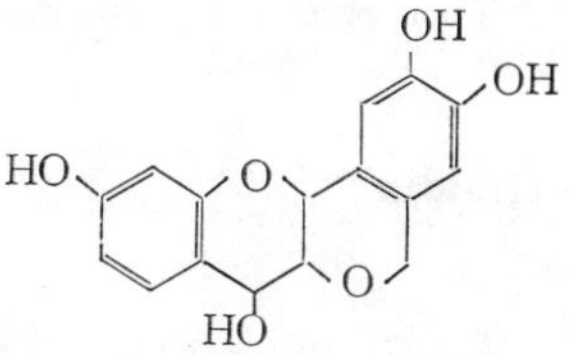

(XLIV)

《Occurrence》   *Peltogyne porphyrocardia,* $[\alpha]_D$ +273° (EtOAc), $C_{16}H_{14}O_6$
Tri-*O*-methyl—(XLIV)   mp. 203~205°, $[\alpha]_D$ +250° (CHCl₃)
UV $\lambda_{max}^{EtOH}m\mu$ (log ε): 280 (3.84), 286 (3.87)

IR ν_{max}^{KBr}cm⁻¹: 3484

242) Clark-Lewis, J. W *et al: Soc.,* **1960**, 4106.
243) Drewes, S. E *et al: Chem. & Ind.,* **1963**, 532.
244) Burrows, B. F: *Proc. Chem. Soc.* **1959**, 150.   245) Hassall, C. H *et al: Soc.,* **1958**, 3174.

[H] Anthocyan and Anthochlor Pigments

(1) Absorption Spectra of Anthocyan Pigments

(a)[192] Pelargonidin, cyanidin, delphinidin, päonidin, syringidin and malvin.

Pelargonin chloride, cyanin chloride, chrysanthemin chloride, hirusutin chloride, peonin chloride, morindin chloride, malvidin chloride galactoside, malvidin chloride xyloside, delphinidinglucoside chloride, petunidin monoglucoside chloride, mecocyanin chloride, 3-galactosidopelargonidin chloride, and bougainvillaeidin chloride.

Pelargonidin, cyanidin, delphinidin, päonidin, syringidin and malvin chloride.

(b) Fig. 65 shows examples of absorption spectra of anthocyan pigments[246].

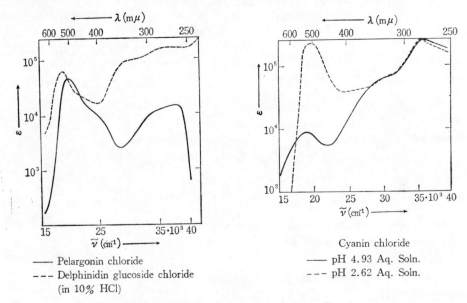

— Pelargonin chloride
--- Delphinidin glucoside chloride
(in 10% HCl)

Cyanin chloride
—— pH 4.93 Aq. Soln.
--- pH 2.62 Aq. Soln.

Fig. 65 Absorption Spectra of Anthocyans (Bayer)

(c) Uliginosin chloride[247], idaein chloride[248]

(2) Guibourtacacidin and Guibourtinidin[249]

Leucoanthocyanidin
R=OH: Leucofisetinidin
Mollisacacidin[250]=Gleditsin[251] (XLV)
R=H: Guibourtacacidin (XLVII)

Anthocyanidin
R=OH: Fisetinidin chloride (XLVI)
R=H: Guibourtinidin (XLVIII) $\lambda_{max}=490$ mμ

≪Occurrence≫   *Guibourta coleosperma* & other spp.

---

246) Bayer, E: *Ber.*, **92**, 1062(1959).   247) Sadtler Card No. 4516.   248) Sadtler Card No. 4511.
249) Roux, D. G: *Nature*, **183**, 891 (1959).   250) Keppler, H. H: *Soc.*, **1957**, 2721.
251) Mitsuno, M. *et al*: *Y. Z.*, **77**, 557, 1208 (1957).

(3) Sulfuretin and Sulfurein[252]

R=H: Sulfuretin
R=β-D-glucose: Sulfurein

(XLIX)

≪Occurrence≫　*Cosmos sulphureus*

UV λ^{EtOH}mμ

	max.			Infl.	min.		
Sulfurein	257	277	405	340	240	264	305
Sulfuretin	257	270	400	240	265	295	

(4) Alphitonin (L)[253]

(L)

≪Occurrence≫　*Alphitonia excelsa* (Australian " red ash "),　mp.　225～226°, $C_{15}H_{12}O_7$
Pentamethyl ether of (L)　mp.　119～120°,
UV　$\lambda_{\max}$ mμ (log ε): 290 (4.31); $\lambda_{\min}$ mμ (log ε): 248 (3.08); IR $\nu_{c=o}$ cm^{-1}: 1710

[I]　Flavones and Flavone Glycosides

(1) UV Spectra[192]

Flavone, primetin, toringin, chrysin, baicalin, baicalein, wogonin, apiin, apigenin, acaciin, acacetin, scutellarin, scutellarein, luteolin, galangin, kaempferitrin, robinin, kaempferol, kaempferid, morin, quercitrin, rutin, quercetin, isorhamnetin, myricitrin, myricetin, tectoridin, tectorigenin, iridin, and irigenin.

Flavone, 3-hydroxy——, 7-hydroxy——, 4'-hydroxy——.

Daitsetin, morin, quercetin, rutin, xanthorhamnin, isoquercetin, avicularin, astragalin, apigenin, 3',4'-dihydroxyflavone, *d*-catechin, 3',4'-dihydroxyflavone, 5-hydroxy——, apigenin, 3-hydroxyflavone, luteolin, kaempferol and quercetin.

(2) IR Spectra

Flavone, 7-CH$_3$O-flavone, 3'-CH$_3$O——, 4'-CH$_3$O——, 7,3'-di-CH$_3$O——, 7,4'-di-CH$_3$O——, 3',4'-di-CH$_3$O——, 3-OH——, 5-OH——, 3,5-di-OH——, 3-OH-7-CH$_3$O——, 3-OH-3',4'-di-CH$_3$O——, 3-OH-7, 3'-di-CH$_3$O——, 3-OH-7,4'-di-CH$_3$O——, 5-OH-3'CH$_3$O——, 5-OH-7,3'-di-CH$_3$O——, 7,3',4'-tri-CH$_3$O ——[253a], 3,3',4',5,7-pentahydroxyflavone, 3,3',4',5,7-penta AcO——, 3,3',4',5,7-penta CH$_3$O——, 3',4', 5,7-tetra-OH-flavone-3-rutinoside, 3,3',4',5,8-penta-OH-flavone, and 3,3',4',5,8-penta-AcO——, Quercetin, rutin, 3,3',4',5,7-pentamethoxyflavone, and 3',4',5,7-tetrahydroxy-flavonol-3-rhamnoside[253b].

252)　Farkas, L *et al*: *Ber.*, **92**, 2847 (1959).
253)　Birch, A. J *et al*: *Soc.*, **1960**, 3593.
253a)　Shaw, Simpson: *ibid.*, **1955**, 655.　　253b)　Hergert, Kurth: *J. A. C. S.*, **75**, 1622 (1953).

(3) Bathochromic Effect of NaOEt Reagent for the UV Absorption of Hydroxyflavonols[254]

(I)

Table 23.   (Horowitz)

Flavone	EtOH		NaOEt–EtOH			$\Delta\lambda$
	λ_{max}	log ε	λ_{max}	log ε after 0.1 hr	1.0 hr	
3–OH–Favone	343	4.22	407	4.20	4.13	64
	239	4.26	237		4.27	
3–OCH₃——	299	4.21	299	4.21	4.21	0
	246	4.25	246	4.25	4.25	
3,5,7,4′–Tetra–OH—— (Kaempferol)	367.5	4.32	315	4.13		
	266	4.22				
3,3′,4′–Tri–OH——	368	4.36	234	4.13		
	250	4.26	315	4.13		
Robinin	352	4.14	399	4.30	4.30	47
	268	4.18	272	4.15	4.15	
Kaempferol–4′,5–dimethylether	357	4.35	389	4.25	4.24	32
	259	4.24	275	4.37	4.37	
3,5,7,3′,4′–Penta–OH—— (Quercetin)	370	4.32	325	4.19		
	257	4.31	242	4.13		
Quercetin–3–methylether	360	4.31	412	4.33	4.30	52
	258	4.31	272	4.35	4.34	
Quercetin–3–rhamnoside (Quercitrin)	350	4.18	402	4.30	4.30	52
	258	4.30	272	4.39	4.39	
Rutin	361	4.29	415	4.36	4.36	54
	258	4.37	273	4.35	4.35	
3,5,3′,4′–Tetra–OH–7–CH₃O –flavone (Rhamnetin)	371	4.41	358	3.98		
	256	4.40	294	4.11		
Quercetin–7–glucoside (Quercimeritrin)	372	4.33	361	4.00		
	257	4.38	294	4.16		
Xanthorhamnin	362	4.22	411	4.38	4.37	49
	258	4.33	270	4.35	4.34	
Quercetin–3′,4′,5,7–tetra–Me–ether	360	4.31	403	4.23	4.23	43
	252	4.30	263	4.31	4.31	
Q–3′,4′,5,7–tetra–benzoylether	359	4.19	401	4.08		
	252	4.22	264	4.08		
Q–3,3′,4′,7–tetra–Me–ether	352	4.34	372	4.00	4.00	20
	254	4.37	284	4.40	4.42	
Gossypin	380	4.15	325	3.80		
	26₂	4.12				

254) Horowitz, R. M et al: J. Org. Chem., **22**, 1618 (1957).

(4) IR Spectra of Monohydroxyflavones and their Derivatives[225]

(a) Carbonyl Bands

Table 24   Carbonyl Absorption Bands of Monohydroxyflavones and their Derivatives

(KBr–Disk)   (Looker *et al.*)

Radicals / Position	–OH C=O	–OCH$_3$ C=O	–OCOCH$_3$ Ester C=O	–OCOCH$_3$ Flavone C=O
3	1610	1637	1757	1646
5	1653	1645	1755	1647
6	1639	1637	1750	1645
7	1628	1650	1761	1637
8	1630	1637	1759	1655
2'	1632	1642	1761	1637
3'	1620	1649	1751	1640
4'	1633	1642	1759	1643

(b) Aromatic in-plane Skeletal Vibrations

Table 25   (KBr–Disk) (Looker *et al.*)

Radicals position of	1st Band	2nd Band	3rd Band	4th Band
Flavone	1627sh, 1610s	1573m	1502m, 1474s	1457m
3-OH	1610s?	1564s, 1548sh	1486s, 1475s	1450s, 1418s?
5-OH	1617s	1590s, 1550sh	1473s	1457s, 1417s?
6-OH	1622s, 1602s	1590s, 1573sh	1478s	1403m?
7-OH	1613s	1583s–br, 1552s	1513s, 1479sh	1460m
8-OH	1603sh	1590sh, 1565s–br	1515sh, 1494s	1457s, 1419s?
2'-OH	1615s, 1608sh	1593m, 1567s	1481m	1455s, 1407m?
3'-OH	1600s	1568s	1485s	?
4'-OH	1604s	1582s, 1569s	1518m–br, 1487s	1460m–br
3-OCH$_3$	1606m	1565m	1494m, 1469s	1442s
5-OCH$_3$	1600s	1573w	1495m, 1477s	1456m, 1442s
6-OCH$_3$	1617s, 1606s	1583s, 1571s	1490s, 1473s	1437s
7-OCH$_3$	1624s, 1603s	1580sh	1494w–br, 1470w	1450sh, 1442s
8-OCH$_3$	1601s	1583s, 1573s	1496s	1447s
2'-OCH$_3$	1610s, 1600s	1583s, 1570s	1500s, 1475s	1442s
3'-OCH$_3$	1607s	1572s, 1565sh	1474s	1450s, 1438s
4'-OCH$_3$	1620m, 1607s	1573m, 1562sh	1480sh, 1470s	1445sh, 1432m

(c) IR Data for Flavone Derivatives in 1400~1000 cm^{-1} Region

Table 26   (KBr–Disk) (Looker *et al.*)

Position of Radicals	IR Bands (cm^{-1})
Flavone	1382s, 1313m, 1285w, 1264m, 1231m, 1196w, 1133m, 1105w, 1080w, 1047m, 1033m, 1013w
3-OH	1351s, 1305s, 1286s, 1248s, 1215s, 1195h, 1187sh, 1155m, 1130s, 1077s, 1035m

---

255)   Looker, J. H. *et al*: *J. Org. Chem.*, **27**, 381 (1962).

Position of Radicals	IR Bands (cm^{-1})
5–OH	1368sh, 1361m, 1328sh, 1317m, 1298m, 1258s, 1228s, 1161m, 1121m, 1105m, 1077m, 1057m, 1034m
7–OH	1387s, 1360m, 1318m, 1313m, 1287m, 1263s, 1191m, 1173m, 1140m, 1095m, 1043m, 1030m, 1004w
8–OH	1382s, 1306sh, 1300s, 1265sh, 1257m, 1220m, 1180s, 1151m, 1077m, 1051s, 1028m, 1002w
3′–OH	1391s, 1353m, 1337m, 1317m, 1308sh, 1284m, 1249m, 1229m, 1200m, 1161w, 1136m, 1089w, 1052w, 1022w, 1000w
4′–OH	1393s, 1337w, 1300m, 1265s, 1224w, 1183m, 1138w, 1117w, 1047w, 1015w
3–OCH$_3$	1382s, 1332m, 1293m, 1282sh, 1241s, 1212s, 1182m, 1167m, 1145s, 1110m, 1076m, 1034sh, 1028m, 1000m
4′–OCH$_3$	1382s, 1335m, 1316m, 1268s, 1257m, 1224sh, 1197s, 1136m, 1126m, 1058w, 1043w, 1027m, 1023sh

(d) Aromatic and Pyrone C–H out-of-plane Deformation Frequencies
Table 27 (KBr–Disk) (Looker *et al.*)

Position of Radicals	C^5–H		C^4–H	C^3–H		C^2–H	Single H	Pyrone H	Unassigned strong bands
	1st	2nd		1st	2nd				
Flavone	771s	687m	758s	—	—	—	—	855s	—
3–OH	777s	703s	758s	—	—	—	—	—	898, 663
3–OCH$_3$	796m?	687s	761s	—	—	—	—	—	893
5–OH	752s	687m	—	801s	?	—	—	847s	
5–OCH$_3$	774s	688s	—	798s	?	—	—	850m or 856m	763
6–OH	773s	683m	—	—	—	808m	872w	841s	
6–OCH$_3$	777s	693s	—	—	—	817s	868m	846s or 855m	
7–OH	776~769m	684m	—	—	—	821m	912m?	843m or 853m	
7–OCH$_3$	768m	678m	—	—	—	818m	906w?	850m	
8–OH	778m	698m	—	806m	772m	—	—	843m or 856m	
8–OCH$_3$	773s	678s or 716m	—	805s	716m	—	—	844m	749
2′–OH	—	—	756m 743m	—	—	—	—	865sh	
2′–OCH$_3$	—	—	758s 743s	—	—	—	—	857s	
3′–OH	—	—	757m	780s	706m	—	871m	849m	
3′–OCH$_3$	—	—	757s	782s or 772s	694m	—	875s	?	
4′–OH	—	—	753m or 774m	—	—	834m	—	?	
4′–OCH$_3$	—	—	750m or 770s	—	—	828s	—	845sh	

(5) **UV and IR Spectra of Quercetin Glucosides**[229]

Spectra are shown in Fig. 66 and 67. See also Table 21.

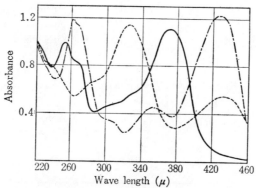

Fig. 66 UV Spectra of
———— Quercetin–3′–glucoside in EtOH
·············· in 0.002 *M* NaOEt Soln.
—·—·— in 0.04 *M* AlCl₃ in EtOH
(Hergert *et al.*)

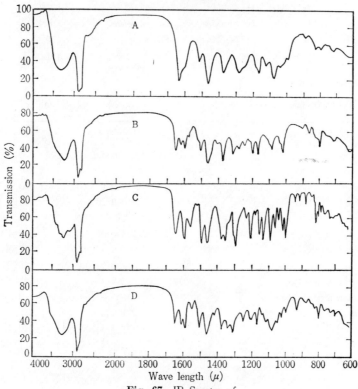

Fig. 67 IR Spectra of
A. Taxifolin–3′–glucoside
B. Quercetin–3′–glucoside
C. Quercetin–3–glucoside
D. Quercetin–7–glucoside
(Hergert *et al.*)

(6) UV Absorption of Biflavonyls and their Components[256]

	R	R'	
	H	H	Apigenin (IIa)
	H	Me	Acacetin (IIb)
	Me	H	Genkwanin(IIc)

(II)

(III)

R	R'	R''	R'''	
H	H	H	H	Amentoflavone (IIIa)
H	H	Me	H	Sotetsuflavone (IIIb)
H	Me	H	H	Bilobetin (IIIc)
Me	Me	H	H	Ginkgetin (IIId)
H	Me	H	Me	Isoginkgetin (IIIe)
Me	Me	H	Me	Sciadopitysin (IIIf)
H	Me	Me	Me	Kayaflavone (IIIg)

UV Spectra   See Table 28.

Table 28.   UV Spectra of Flavones and Biflavonyls   (λ_{max}^{EtOH} mμ (ε))   (Baker *et al.*)

No	Compound	Band I	Band II
IIId	Ginkgetin	212(76000) 271.5(42200)	335(40000)
IIIe	Isoginkgetin	213(90000) 271.5(42000)	330(36500)
IIa	Apigenin	269(18800)	300(13500) 340(20900)
IIb	Acacetin	~210 269(20300)	298(16400) 330(20800)
IIc	Genkwanin	269(17000)	300(12000) 337(19600)
	Ginkgetintetraacetate	211(63500) 248~258(34500)	317(47000)
	Isoginkgetintetraacetate	220(74000) 250(47000)	324(50500)
	Acacetindiacetate	222(21200) 258(13300)	325(26300)
	Ginkgetintetramethylether	212(63000) 267(48000)	328(45500)
	Apigenintrimethylether	212(37000) 265(21200)	325(24000)
IIIf	Sciadopitysin	271.5(37600)	330(35000)
IIIg	Kayaflavone	271.5(44100)	329(41000)
IIIc	Bilobetin	272(44500)	337(35200)

256) Baker, W *et al*: *Soc.*, **1963**, 1477.

(7) Bathochromic Shift on the Spectra of Isoflavones[227)]

See Fig. 68, 69 and Table 29.

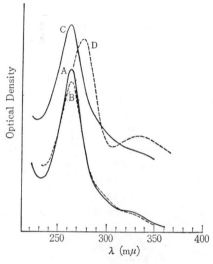

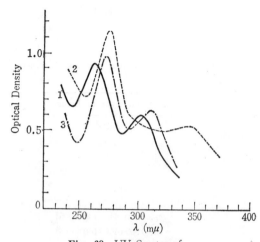

Fig. 68 UV Spectra of
(A) Genistin in EtOH
(B) Genistin in EtOH–NaOAc
(C) Sophoricoside in EtOH
(D) Sophoricoside in EtOH–NaOAc

Fig. 69 UV Spectra of
(1) Podospicatin in EtOH
(2) Podospicatin in EtOH–NaOAc
(3) Podospicatin in EtOH–AlCl₃

(Horowitz *et al.*)

Table 29　　　　　　　　　　(Horowitz *et al.*)

Compounds	Isoflavone	λ_{max} mμ		
		EtOH	+NaOAc	+AlCl₃
Formonetin	7–OH—4′–CH₃O–isoflavone	250	260	
Osajin		274	274	
Genistein	5, 7, 4′–Tri–OH–isoflavone	262	271	274
Genistin	5, 4′–Di–OH–7–gluc.–isoflavone	262	262	273
Biochanin–A	5, 7–Di–OH–4′–CH₃O–isoflavone	261	271	
Pomiferin		276	276	
Sophoricoside	5, 7–Di–OH–4′–gluc.–isoflavone	262	275	276
Santal	5, 3′, 4′–Tri–OH–7–CH₃O–isoflavone	263		274
7–O–Methylpodospicatin	5, 2′–Di–OH–6, 7, 5′–tri–CH₃O–isoflavone	265		277
Irigenin	5, 7, 3′–Tri–OH–6, 4′, 5′–tri–CH₃O–isoflavone	267	277	
Iridin	5, 3′–Di–OH–6, 4′, 5′–tri–CH₃O–7–gluc.–isoflavone	268	268	
Podospicatin	5, 7, 2′–Tri–OH–6, 5′–di–CH₃O–isoflavone	263	275	273
	5–OH–6, 7, 2′, 5′–tetra–CH₃O–isoflavone	262		275

(8) NMR of Flavones
(a) Eucalyptin (IV)[232]

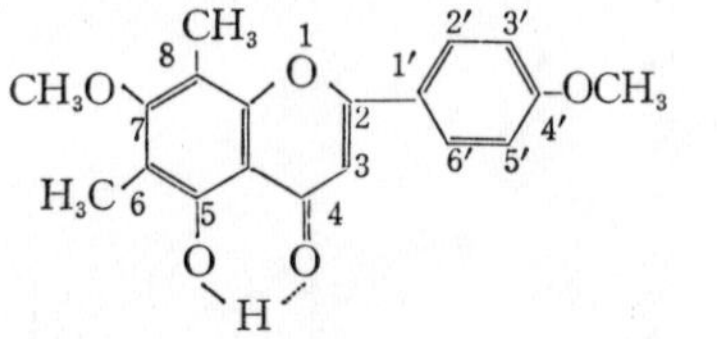

(IV)

《Occurrence》 *Eucalyptus globulus, E. cinerea, E. risdoni*

NMR TMS, δ:

Position and Type of Band	Relative Intensity	Assignment
2.18s	3	arom. Me
2.36s	3	arom. Me
3.82s	3	arom. OCH_3
3.90s	3	arom. OCH_3
6.60s	1	3 positon H
ca. 7.05 (410–430 c/s) d	2	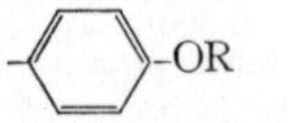
ca. 7.95 (460–480 c/s) d	2	—〈 〉—OR
12.96s	1	5 position O-H······

(Horn, *et al.*)

(b) Sericetin Osajin and Pomiferin[257]

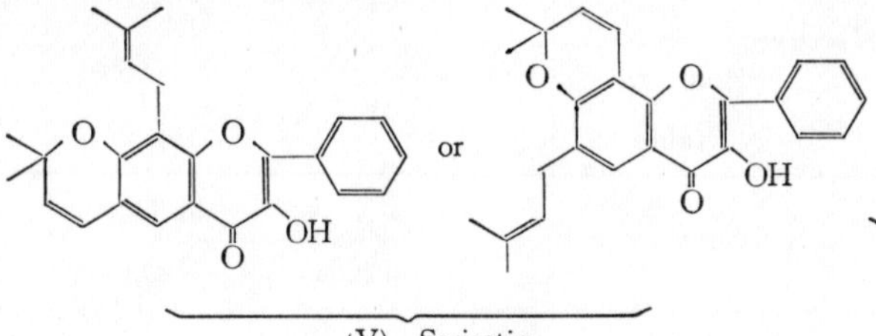

(V) Sericetin

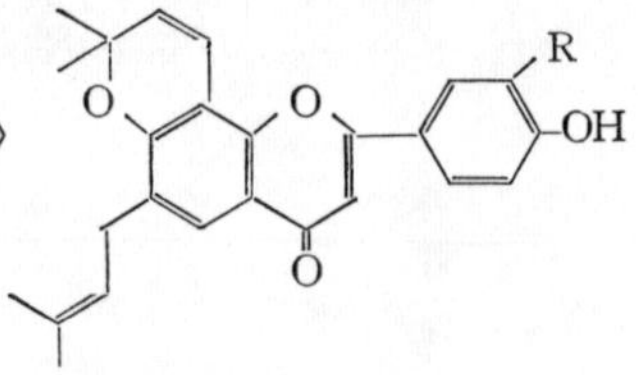

or

(VI) R=H Osajin
      R=OH Pomiferin

Sericetin (V)[257]

《Occurrence》 *Mundalea sericea*, mp. 157°, $C_{25}H_{24}O_5$,

UV λ_{max} mμ: 225~230 (2,2-dimethylchromene)

IR ν_{max}^{CHCl3} cm^{-1}: 1630 (C=C), 1598, 1563, 1493 (arom. C=C), 1375, 1361 (Me$_2$C), 1123 (ϕ-O-C$\stackrel{C}{<}_{C}$)

1375, 896

Osajin (VI, R=H)[258]

《Occurrence》 *Maclura pomifera*, mp. 189°, $C_{25}H_{24}O_5$

Pomiferin (VI, R=OH)[259]

《Occurrence》 *Maclura pomifera*, mp. 200.5°, $C_{25}H_{24}O_6$

NMR [257] in CDCl$_3$, τ value, see Table 30.

---

257) Burrows, B. F *et al: Proc. Chem. Soc.* **1960**, 177.

258) Wolfrom, M. L *et al: J. A. C. S.*, **65**, 1434 (1943), **68**, 406 (1946).

259) Wolfrom, M. L *et al: ibid.*, **63**, 1253 (1941), **64**, 308 (1942), **65**, 1434 (1943), **68**, 406 (1946).

Table 30 NMR of Sericetin, Pomiferin and Osajin (Burrows *et al.*)

Compounds	$>C(CH_3)_2$	$C=C(CH_3)_2$	$-CH_2-$[b]	$H>C=C<H_c$
Sericetin	8.55	8.33, 8.18	6.52	4.40, 3.29
Pomiferin	8.51	8.32, 8.19	6.66	? ?
Osajin	8.51	8.33, 8.18	6.66	4.40, 3.32

b: doublet, $J\sim7.0$ c/s, c: doublet, $J\sim10.0$ c/s

(c) Munetone (VII)[260)261)]

(VII)[261)]

《Occurrence》 *Mundulea suberosa,* mp. 192~193°, $C_{21}H_{18}O_4$, $C_{25}H_{24}O_5$ [261)]

UV λ_{max} mμ (log ε): 263 (4.61), 330 (4.03)[260)]

NMR[262)] Total 24 H, τ value: 8.51, 8.55 (2 gem di Me), absence of iso Pr. 2.05 s, 2.10 s (H^3, H^5), 3.18 (H^4), 3.28 (H^7, H^8), 3.43 (H^7, H^8), 3.60 (H^8, H^3) (1: 1.5: 0.5 resp. with $J=9\sim10$ c/s). Center 4.25d, 4.23d ($J=10$ c/s) H^1, H^9 center 3.35, 3.50d ($J=10$ c/s) (chromene α-protons H^8, H^2), 2.84d ($J=9$ c/s) H^6 3.35d ($J=9$ c/s) (H^7)

Mass[192)] m/e 416, 401=416−Me, 386, 193 (bisdimethylchromone), 406, 203, 188=376/2.

(9) Mass Spectra of Flavones and Isoflavones[263)]

Apigenin (VIII)   m/e=152   m/e=118

Acacetin (IX)   Leaserone (X)   m/e=234

(a) **Apigenin** (VIII) 270 (100), 269s, 242s, 241m, 153s, 152s, 133w, 128w, 129w, 124s, 123s, 121s, 118s,

(b) **Acacetin** (IX) 285s, 284 (100), 283m, 256w, 241m, 213w, 152m, 132s, 128w, 124m, 117w, 89w,

260) Dutta: *J. Indian Chem. Soc.* **33**, 716 (1956), **36**, 165 (1959), **39**, 475 (1962).
261) Dyke, S. F: *Proc. Chem. Soc.*, **1963**, 179.
262) Barnes, C. S: *Tetr. Letters*, **1963**, 281.
263) Reed, R. I. *et al*: *Soc.*, **1963**, 5949.

(c) **Leaserone** (X)   355s, 354 (100), 340m, 339s, 326m, 325s, 322m, 321s, 299m, 295m, 293m, 285s, 257s, 235s, 233s, 220s, 219s, 217m, 206s, 179s, 176s, 167s, 163m, 153m, 149s, 145m, 140m, 121m, 120s, 115m, 107m, 105s, 103m

Relative intensity for 284 (100%) s>10%, m 10~5%, w 3~5%, omitted <3%

(10) Flavones and Flavonols

(a) **Centaurein** (X-a)[264]

R=glucose : Centaurein (X-a)
R=H :   Centaureidin

≪Occurrence≫   *Centaurea jacea* L.,   mp. 205~207°, $C_{24}H_{26}O_{13}\cdot 2H_2O$

UV $\lambda_{max}\,m\mu$ (log ε): 354 (5.32), 258 (5.30)

Centaureidin mp. 130~133°

(b) **Xanthomicrol** (XI)[265]

(XI)

≪Occurrence≫   *Satureia douglassii* (*Labiatae*),   mp. 227~230°, $C_{15}H_7O_3(OCH_3)_3$

UV $\lambda_{max}\,m\mu$: 282, 296, 336; IR $\lambda_{max}^{nujol}\,\mu$: 3.08, 6.07, 6.26, 6.32, 6.43

(c) **Digitalis Flavonoids**

  (i) **5,7,4′-Trihydroxy-6,3′-dimethoxyflavone** (XII) R=H, R′=MeO
        5,7,4′-Trihydroxy-6-methoxyflavone   (XII) R=R′=H

(XII)[266]

≪Occurrence≫   *Digitalis lanata*

  (ii) **Digicitrin** (5,3′-Dihydroxy-3,5,7,8 4′,5′-hexamethoxyflavone) (XIII)[266a]

(XIII)

≪Occurrence≫   *Digitalis purpurea*,   mp. 178~179°, $[\alpha]_D \approx 0°$ (CHCl₃)

---

264) Farkas, L., Hörhammer, L: *Tetr. Letters*, **1963**, 727.
265) Stout, G. H *et al*: *Tetrahedron*, **14**, 296 (1961).
266) Whalley, W. C *et al*: *Soc.*, **1963**, 3780, 5577.
266a) Meier, W *et al*: *Helv. Chim. Acta*, **45**, 232 (1962).

(d) **Sinensetin** (5,6,8,3′,4′-Pentamethoxyflavone) (XIV)[267]

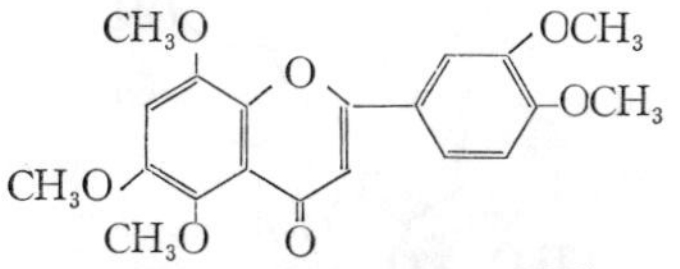

(XIV)

≪Occurrence≫ Orange peel

(e) **Pedaliin** (5,3′,4′-Trihydroxy-6-methoxyflavone-7-glucoside) (XV)[268]

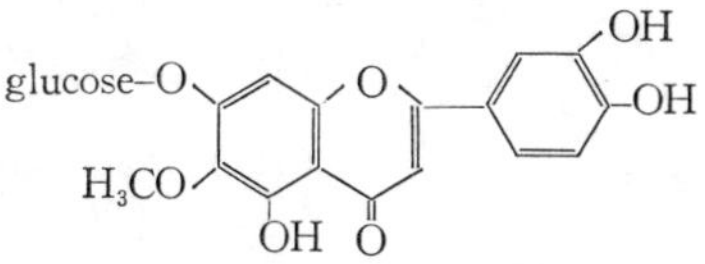

(XV)

≪Occurrence≫ *Sesamum indicum* L. (ゴマ), mp. 254°, $C_{22}H_{22}O_{12} \cdot 2^1/_2 H_2O$
Aglycone: Pedaltin $C_{16}H_{12}O_7$, mp. 300~301°, pedaltin trimethylether mp. 189~190°

UV λ_{max}^{EtOH} mμ (log ε): 282 (4.62), 340 (4.87)

(f) **Cirsimarin** (XVI)[269]

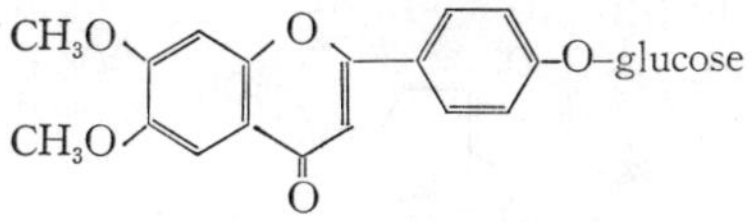

(XVI)

≪Occurrence≫ *Cirsium martimum* MAKINO (ハマアザミ)
mp. 178~179°, 243°, $[\alpha]_D$ −70.8°(pyridine–EtOH), $C_{23}H_{24}O_{11} \cdot 2H_2O$
Aglycone: Cirsimaritin (6,7-dimethyl–scutellarein), mp. 257°, $C_{15}H_8O_4(OCH_3)_2$,
UV λ_{max}^{EtOH} mμ (log ε): 277 (4.57), 337 (4.75); λ_{min}^{EtOH} mμ (log ε): 248 (4.21), 304 (4.38)

(g) **Limocitrin** (XVI-a)[269a]

(XVI-a)

≪Occurrence≫ *Citrus lemon*, mp. 274~275°, $C_{17}H_{14}O_8$
UV λ_{max}^{EtOH} mμ: 259, 273, 378 inf.; $\lambda_{max}^{EtOH+NaOAc}$ mμ: 282; $\lambda_{max}^{EtOH+NaOAc+H_3BO_3}$ mμ: 379,

$\lambda_{max}^{EtOH+AlCl_3}$ mμ: 442

267) Seshadri, T. R. *et al*: *J. Indian Chem. Soc.*, **39**, 515 (1962) ; *Index Chem.*, **8**, 24625 (1963).
268) Morita, N: *Chem. Pharm. Bull.* **8**, 59 (1960).
269) Marita, N. *et al*: *Y. Z.*, **83**, 615 (1963).
269a) Horowitz, R. M *et al*: *J. Org. Chem.*, **26**, 2899 (1961).

(h)   Azalein (5–Methylquercetin–3–rhamnoside) (XVII)[270]

(XVII)

≪Occurrence≫   *Plumbago Capensis, Rhododendron* spp.

(i)   Flavonoids of *Citrus aurantium*[271]

(XVIII)   (XIX)

(XVIII)   R=H: Auranetin[272]   (XIX)   R=CH₃: Nobiletin
(XVIII)   R=OH: 5-Hydroxyauranetin   (XIX)   R=H: Desmethylnobiletin

(j)   Vitexin (XX)[273)274]

(XX)

≪Occurrence≫   *Vitex lucens,*   mp. 263° (decomp.), $[\alpha]_D$ −14.3° (pyridine).
UV λ_{max} mμ (log ε): 270 (4.32), 335 (4.33)
IR ν_{max}^{KBr} cm⁻¹: 3425, 3392, 3284, 2889, 1650

(k)   Orientin (XXI) and **Homoorientin** (XXII)[275)276]

(XXI)   (XXII)

≪Occurrence≫   *Aspalathus acuminatus* (Rooibos Tea)
Orientin (XXI) mp. 265~267° (decomp.), $[\alpha]_D$ +18.4° (pyridine), $C_{21}H_{20}O_{11}$
UV λ_{max}^{EtOH} mμ (log ε): 221 (4.36), 258 (4.15), 272 (4.21), 355 (4.19); $\lambda_{max}^{AlCl_3-EtOH}$ mμ (log ε): 222
(4.48), 278 (4.22), 360 (4.18), 390 (4.17)
Octa–O–acetyl–(XXI) mp. 197°

270)   Harborne, J. B: *Arch. Biochem & Biophys.* **96**, 171 (1962).
271)   Seshadri, T. R *et al*: *Tetrahedron,* **6**, 64 (1960).
272)   Venkataraman, K *et al*: *J. Indian Chem. Soc.,* **1942**, 135.
273)   Briggs, L. H: *Tetrahedron,* 3, 269 (1958).   274)   Evans *et al*: *Soc.,* **1957**, 3510.
275)   Koeppen, B. H: *Chem. & Ind.* **1962**, 2145.
276)   Koeppen, B. H *et al*: *Biochem. J.* **83**, 507 (1962).

UV λ_{max}^{EtOH} mμ (log ε): 221 (4.44), 264 (4.32), 278 (4.33)

IR ν_{max} cm^{-1}: 1752 (CO–CH$_3$), 1650 (conj. CO), 1620, 1515, 1435 (arom. C–C), 1365, 1215 (ester C–O)

Homoorientin (Lutonaretin) (XXII)   mp. 235°, [α]$_D$ +30.8° (pyridine), C$_{21}$H$_{22}$O$_{12}$

UV λ_{max}^{EtOH} mμ (log ε): 220 (4.40), 259 (4.22), 272 (4.25), 357 (4.27); $\lambda_{max}^{AlCl3-EtOH}$ mμ (log ε): 221
(4.47), 279 (4.28), 365 (4.26), 390 (4.27)

IR ν_{max} cm^{-1}: 3450 (OH), 1652 (conj. CO), 1580, 1500, 1460 (arom. C–C)

(l) **Sudachitin** (XXIII)[277]

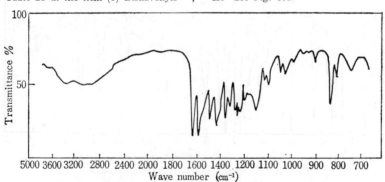

(XXIII)

《Occurrence》   *Citrus sudachi* HORT et SHIRAI,   mp. 239.5~240.5°, C$_{18}$H$_{16}$O$_8$

(m) **Ginkgetin** (XXIV)[278][279]

(XXIV)

《Occurence》   *Ginkgo biloba* (イチ ョ ウ)   mp. 336°, acetate 259°, C$_{32}$H$_{22}$O$_{10}$[279]

UV See Table 28 in the item (6) Bisflavonyls[278],   IR See Fig. 69a

Fig. 69a IR Spectrum of Ginkgetin (natural) (Nakazawa *et al.*)

(n) **Pongaglabrone**   (3′, 4′–Methylenedioxyfurano–(2″, 3″–7, 8)–flavone) (XXIV–a)[279a]

(XXIV–a)

277) Horie, T *et al*: *Bull. Chem. Soc. Japan*, **34**, 1547 (1961).
278) Koguri, A.: *J. Chem. Soc. Japan*. **80**, 1352, 1355, 1462, 1467 (1959).
279) Nakazawa, K *et al*: *Chem. Pharm. Bull.* **11**, 283 (1963).
279a) Seshadri, T. R *et al* : *Tetrahedron*, **19**, 219 (1963).

≪Occurrence≫　　*Pongamia glabra*,　　mp. 233°,　$C_{18}H_{10}O_5$

UV　See Fig. 70, λ_{max} mμ (log ε): 249 (4.44), 331 (4.32)

IR　See Fig. 71, ν_{max}^{nujol} cm^{-1}: 1035s, 935w (—O–CH$_2$–O—)

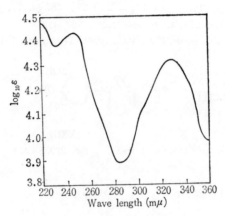

Fig. 70　UV Spectra of Pongaglabrone
(Sechadri *et al.*)

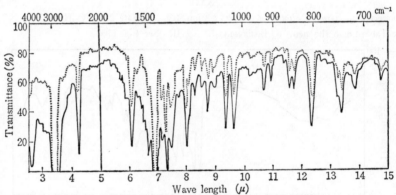

Fig. 71　IR Spectra of Pongaglabrone in nujol
Upper curve: Synth.
Lower curve: Natural
(Seshadri *et al.*)

(11)　Isoflavones

(a)　**Wistin**[280] **and Afromosin**[281] **(XXV)**

$$\text{(structure)}$$

R=D-glucose: Wistin
R=H: Afromosin

(XXV)

280)　Shibata, S *et al*: *Chem. Pharm. Bull.*, **11**, 382 (1963).

281)　Mc Murry, T. B. H *et al*: *Soc.*, **1960**, 1491.

Wistin

  ≪Occurrence≫   *Wistaria floribunda* DC.,   mp. 209~210°, $[\alpha]_D$ −67.15° (AcOH), $C_{23}H_{24}O_{10} \cdot H_2O$

UV λ_{max}^{EtOH} mμ (log ε): 261 (4.38), 320 (3.84)

IR ν_{max}^{nujol} cm^{-1}: 3400, 1636, 1612, 1585

Afromosin

  ≪Occurrence≫   *Afromosia elata* HARMS[281],   mp. 228~229°,

UV λ_{max}^{EtOH} mμ (log ε): 259 (4.42), 322 (4.02)

(b) Irisolone (XXVI)[282]

(XXVI)

  ≪Occurrence≫   *Iris nepalensis*,   mp. 269~270°, $C_{17}H_{12}O_6$
UV λ_{max} mμ (log ε): 270 (4.62), 330 inf. (3.97)

(c) Thatlancuayin (XXVII)[283]

(XXVII)

  ≪Occurrence≫   *Iresine celosioides* L.[284]   synth. mp. 147~148° [283],
UV λ_{max} mμ (log ε): 246 (4.40), 285 (4.13), 310 (3.88); λ_{min} mμ (log ε): 236 (4.36), 276 (4.11),
        307 (3.83)[283];

IR λ_{max}^{nujol} μ: 6.04 (C=O), 8.94, 9.28, 9.52 (–O–CH$_2$–O–)[283]; λ_{max}^{CHCl3} μ: 6.02, 6.08, 6.15, 6.20, 6.26[281]

(d) Maxima Isoflavones A (XXVIII) and **B** (XXIX)[285]

(XXVIII)[285]   (XXIX)

Maxima Isoflavone A   a) R= H: Pseudobaptigenin[286]
                            b) R= (CH$_3$)$_2$=C=CH–CH$_2$–: Maxima Isoflavone B

  ≪Occurrence≫   *Tephrosia maxima*

282) Gopinath, K. W *et al*: *Tetrahedron*, **16**, 201 (1961).
283) Seshadri, T. R *et al*: *Tetrahedron*, **18**, 559 (1962).
284) Crabbe, P *et al*: *J. A. C. S.*, **80**, 5258 (1958).
285) Seshadri, T. R *et al*: *Tetrahedron*, **18**, 1443 (1962).
286) Rangaswami, S *et al*: *Curr. Sci.*, **24**, 337 (1955).

(e) Maxima Isoflavone C (XXX)[287]

$(CH_3)_2=C=CH-CH_2-O-$

(XXX)

≪Occurrence≫　　*Tephrosia maxima*,　　synth.　mp. 146°, $C_{22}H_{20}O_6$

(f) Jamaicin (XXXI)[288]

(XXXI)

≪Occurrence≫　　*Piscidia erythrina* L.,　　mp. 193~195° (stable modification), 160~163° (unstable modification), $C_{22}H_{18}O_6$, $[\alpha]_D$ 0°

UV　λ_{max}^{EtOH} mμ (log ε) : 266 (4.43)

IR　ν_{max}^{nujol} cm^{-1} : OH (no) ;　　ν_{max}^{KBr} cm^{-1} : 1647 (uncherated CO of benzopyrone), 1634 (conj. C=C), 1597

1575, 1508 (arom. system), 1398, 1362 $\left(\begin{smallmatrix}Me\\Me\end{smallmatrix}>C<\begin{smallmatrix}C\\R\end{smallmatrix}\right)$, 1266 (C–O), 1117 (aryl–O–CR₃),

1038 (OCH₃), 933s, 719m (–O–CH₂–O–), 908 (CH₂=C<$\begin{smallmatrix}R\\R\end{smallmatrix}$)

(g) Podospicatin (XXXII)[289]

(XXXII)

≪Occurrence≫　　*Podocarpus spicatus*

UV　See Fig. 70 and Table 29.

Dimethyl ether　　mp. 212° $C_{17}H_{14}O_7$

(h) Caviunin (XXXIII) and Isocaviunin (XXXIII–a)[290]

(XXXIII)　　　　　　　　　　　(XXXIII–a)

≪Occurrence≫　　*Dalbergia nigra*[291],　　(XXXIII) mp. 191~193°, $C_{19}H_{18}O_8$, $[\alpha]_D$ 0°(CHCl₃)

UV　λ_{max}^{EtOH} mμ (log ε) : 263 (4.37), 297 (4.25)

IR　λ_{max}^{nujol} μ : 2.95, 6.00, 6.15, 6.30, 6.56, 8.25, 10.35, 12.08.

287) Anirudhan, C. A *et al* : *Soc.*, **1963**, 6049.
288) Büchi, J *et al* : *Helv. Chim. Acta*, **41**, 2006 (1958).
289) Briggs, L. H *et al* : *Tetrahedron*, **6**, 143 (1959).
290) Farkas, L *et al* : *Ber.*, **94**, 2501 (1963).
291) Gottlieb, O. R *et al* : *J. Org. Chem.*, **26**, 2449 (1961).

(i) **Daidzin** and **Daidzein** (XXXIV)[292]

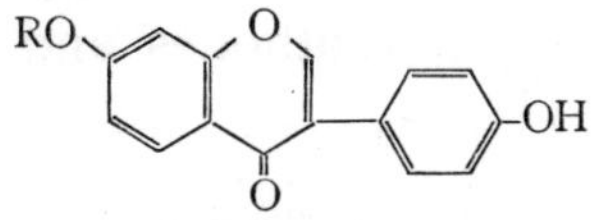

R=D-glucose: Daidzin
R=H   : Daidzein

(XXXIV)

≪Occurrence≫   *Pueraria Thunbergiana* BENTH. and other *Pueraria* spp. (カッコン)

Daidzin   mp. 215~217° (decomp.), $C_{21}H_{20}O_9 \cdot H_2O$

IR ν_{max}^{nujol} cm^{-1}; 3382 (br), 1635 (sh), 1620, 1602, 1572

Pentaacetate   UV λ_{max}^{EtOH} mμ (log ε): 252 (4.48), 294 (3.97), 305 (3.94)

Daidzein   mp. 320° (decomp.), $C_{15}H_{10}O_4$

IR ν_{max}^{nujol} cm^{-1}: 3257 (br), 1630, 1611 (sh), 1601, 1585 (sh)

Diacetate λ_{max}^{EtOH} mμ (log ε): 252 (4.26), 304 (3.73)

(j) **Puerarin** (XXXV)[293]

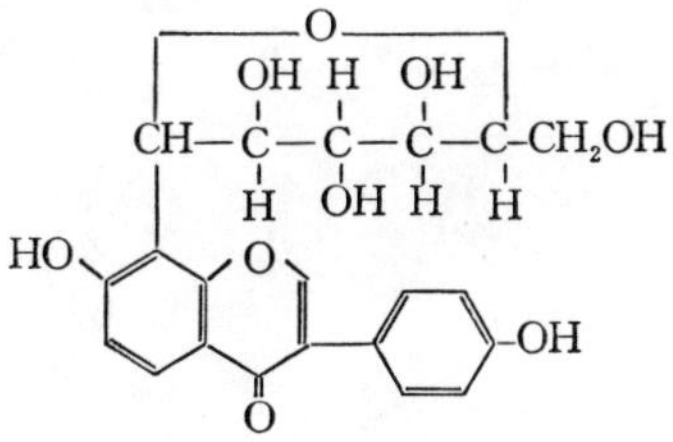

(XXXV)

≪Occurrence≫   *Pueraria Thunbergiana* BENTH. and other *Pueraria* spp.

mp. 187° (decomp.), $C_{21}H_{20}O_9$

IR ν_{max}^{nujol} cm^{-1}: 3372 (br), 1634, 1615, 1584

Acetate   mp. 129~130° (decomp.)

UV λ_{max} mμ (log ε): 249 (4.47), 307 (3.94)

292) Shibata, S *et al*: *Y. Z.*, **79**, 757 (1959) ; *Chem. Pharm. Bull.*, **7**, 134 (1959).
293) Murakami, T *et al*: *Chem. Pharm. Bull.*, **8**, 687 (1960).

[J] Chromones, Chromanes, Xanthones and their Glycosides

(1) UV Spectra

γ-Pyrone[294], α-pyrone carboxylic acid[295], γ-chromanone[296], α-tocopherol[297], xanthone[294], chromone[294], 2-methylchromone, 3-methylchromone, 2, 3-dimethylchromone[298], khellin, visnagin, khellol glucoside[299) 300), dihydrokhellin, dihydrovisnagin, visnamminol, anhydrovisnamminol, 5-norkhellin (khellinol)[301) 302], cannabinol, tetrahydrocannabinol and its synthetic isomers, cannabinol acetate, 3-acetoxy-2 N-amyl-6, 6, 9-trimethyl-6-benzopyran, cannabidiol dimethylether and cannabidiol[303) 304]

(2) IR Spectra

Khellin, visnagin, khellol glucoside[305] and xanthone[305a]

(3) UV Spectra of Hydroxymethylchromones See Fig. 72[306]

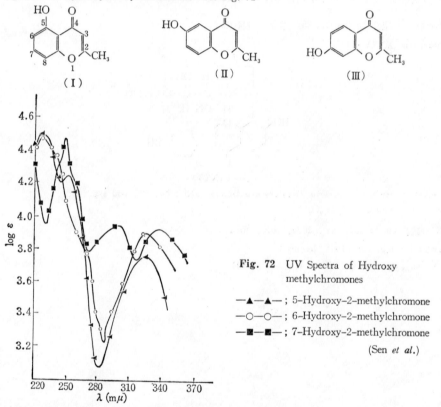

Fig. 72 UV Spectra of Hydroxy methylchromones

—▲—▲—; 5-Hydroxy-2-methylchromone

—○—○—; 6-Hydroxy-2-methylchromone

—■—■—; 7-Hydroxy-2-methylchromone

(Sen *et al.*)

294) Gibbs: *J. A. C. S.*, **52**, 4895 (1930).   295) Haworth: *Soc.*, **1938**, 711.
296) Ramart-Lucas *et al*: *Soc. Chim. France* (5) **2**, 1383 (1935).
297) Emerson: *J. Biol. Chem.*, **122**, 99 (1937).   298) Ganguly, Bagchi: *J. Org. Chem.*, **21**, 1415 (1956).
299) Bailey, Geary, de Wald: *J. Am. pharm. Assoc. Sci. Ed.*, **40**, 280 (1951).
300) Satoda, I *et al*: *Yakugaku Kenkyu*, (*Osaka*), **29**, 287 (1957).
301) Davies, Norris: *Soc.*, **1950**, 3194.
302) Bencze, Eisenbeiss, Schmid: *Helv. Chim. Acta*, **39**, 923 (1956).
303) Adams, R: *J. A. C. S.*, **62**, 2201 (1940).   304) Adams, R: *ibid.*, **62**, 732 (1940).
305) Bailey, Geary, de Ward: *J. Am. Pharm. Assoc. Sci. Ed.*, **40**, 280 (1951).
305a) Sadtler Card No. 5608.   306) Sen, K *et al*: *J. Org. Chem.* **24**, 316 (1959).

(4) IR Spectra of 4-Pyrone and Chromones[277]

(Ⅳ) 1, 4-Pyrone

(Ⅴ) R=H : Chromone
(Ⅵ) R=CH₃ : 2-Methylchromone

(a) 1, 4-Pyrone (Ⅳ)

IR ν_{max}^{KBr} cm⁻¹: 1659 (C=O), 1640 (C=C, asymm.), 1609 (C=C, symm.), 1468, 1421 (in plane skeletal),

1319 (C–O–C), 925, 855 (δ_{C-H} out of plane)

(b) Chromone (Ⅴ)

IR ν_{max}^{KBr} cm⁻¹: 1655 (C=O), 1600~1615, 1563, 1466, 1407 (arom. in plane).

(c) 2-Methylchromone (Ⅵ)

IR ν_{max}^{KBr} cm⁻¹: 1643 (C=O), 1610, 1575, 1467 or 1433, 1440 or 1426 (arom. in plane)

(d) Chromonol

IR ν_{max}^{KBr} cm⁻¹: 1638 (C=O), 1607 br, 1568 br, 1495 or 1474, 1430 (arom. in plane)

(5) Yangonine (Ⅶ) and Pseudoyangonine (Ⅷ)[307]

CH₃O– ⟨ ⟩ –CH=CH– O =O   OCH₃

CH₃O– ⟨ ⟩ –CH=CH– O –OCH₃   O

(Ⅶ)   (Ⅷ)

≪Occurrence≫   *Macropiper methysticum* (Kawa kawa roots)

(a) Yangonine   mp. 153~154°, C₁₅H₁₄O₄

UV λ_{max}^{EtOH} mμ (log ε): 360 (4.33); λ_{min}^{EtOH} mμ (log ε): 280 (3.75)

IR $\lambda_{max}^{CCl_4}$ μ: 5.80, 6.06, 6.20, 6.45, 6.64, 6.90, 7.15, 8.00, 8.53, 8.70, 9.63, 10.40, 10.45

(b) Pseudoyangonine   mp. 138~140°, C₁₅H₁₄O₄

UV λ_{max}^{EtOH} mμ (log ε): 345 (3.81); λ_{min}^{EtOH} mμ (log ε): 290 (3.63)

IR $\lambda_{max}^{CCl_4}$ μ: 6.00, 6.20, 6.40, 6.60, 7.10, 7.95, 8.48, 8.60, 9.75, 10.40, 10.85, 11.85

(6) Benzochromones
(a) UV Spectra of Linear- and Angular-Benzochromones[308][308a]

307) Chmielewska, I *et al*. *Tetrahedron*, **4**, 36 (1958).
308) Bycroft. B. W *et al*: *Soc.*, **1962**, 40.
308a) Fukushima, S *et al*: *Chem. Pharm. Bull.*, **12**, 316 (1964).

(IX) (X) (XI)
Eleuterinol

(IX–a) R=R′=R″=H: Norrubrofusarin
(IX–b) R=CH₃, R′=R″=H: Rubrofusarin
(IX–c) R=R″=CH₃, R′=H: Rubrofusarin monomethylether
(IX–d) R=R′=R″=CH₃: Rubrofusarin dimethylether
(X–a) R=CH₃: Flavasperone (Asperxanthone)

Table 31 UV Absorption Data of Linear and Angular Benzochromones (Bycroft, B. W et al.)

No.	Compound	mp.	λ_{max}^{EtOH} mμ (log ε)
IX–a	Norrubrofusarin		225(4.45), 280(4.64), 330(3.48), 414(3.74)
IX–b	Rubrofusarin	215.5°	225(4.45), 278(4.68), 326(3.51), 406(3.74)
IX–c	——monomethylether	206°	226(4.47), 275(4.73), 329(3.45), 345(3.46), 387(3.67)
IX–d	——dimethylether[309]	178°	225.5(4.44), 272(4.61), 327(3.50), 343(3.50), 379(3.72)
X–a	Flavasperone	203°	241(38.4), 282(22.4), 370(4.64)
XI	Eleutherenol[310]	>310°	240(41.7), 273(30.2), 362(10.0)

(b) UV Spectra of Benzofurochromones[308a]

(XI–a) (XI–b)

Structure	Compound	λ_{max}^{EtOH} mμ (log ε)
XI–a, R₁=OMe, R₂=OMe	Khellin	248(4.54), 282(3.66), 332(3.67)
X–a, R₁=H, R₂=OMe	Visnagin	250(4.60), 290sh(3.60), 340(3.70)
XI–a, R₁=OMe, R₂=OH	Khellinol	250(4.50), 260(4.60), 300sh(3.60), 360(3.40)
XI–b, R=OH	Khellol	220(4.2), 250(4.6), 290sh(3.7), 330(3.7)
XI–b, R=O-glucose	Khellol glucoside	220sh(4.1), 250(4.6), 290sh(3.7), 330(3.7)

(c) Rubrofusarin (IX b) and **Flavasperone** (Asperxanthone) (X–a)

Rubrofusarin[308)309)311)~314]

≪Occurrence≫   *Fusarium culmarum* Sacc.,   mp. 214.5~215.5°,

UV See Table 31[308]

IR ν_{max}^{KBr} cm⁻¹: 3389 (br), 1662s, 1627s, 1590s, 1517, 1483s, 1459, 1406s, 1376s

See Fig. 73[309]

309) Tanaka, H et al: *Agr. Biol. Chem.*, **27**, 249 (1963).
310) Schmid, H et al: *Helv. Chim. Acta*, **35**, 910 (1952).
311) Raistrick, H et al: *Biochem. J.*, **31**, 385 (1937).
312) Stout et al: *Chem. & Ind.*, **1961**, 289.
313) Tanaka, H et al: *Tetr. Letters*, **1961**, 151; *Agr. Biol. Chem.*, **27**, 48 (1963).
314) Shibata, S et al: *Chem. Pharm. Bull.*, **11**, 821 (1963).

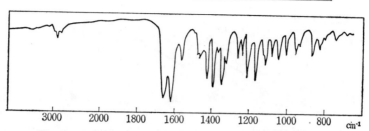

Fig. 73 IR Spectrum of Rubrofusarin Dimethylether (Tanaka)

Flavasperone[308]

≪Occurrence≫   *Aspergillus niger,*   mp. 203°

UV   See Table 31

IR ν_{max}^{KBr} cm^{-1}: 3060w, 2925w, 2855w, 2351w, 2331w, 1677s, 1622s, 1588s, 1530, 1465s, 1442s, 1397s, 1383s, 1357

(d) Ustilaginaidin (XII)[315]

≪Occurrence≫   *Claviceps virens* (イネコウジ),   mp. >320°, $[\alpha]_D$ −309° (dioxane), $C_{28}H_{18}O_{10}$

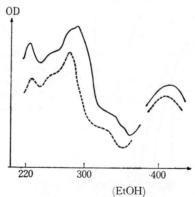

UV   See Fig. 74   (XII)

IR[314]

		ν_{CO}	$\nu_{C=C}$
Ustilaginoidin	ν_{max}^{KBr}	1645	1615, 1585
Rubrofusarin	ν_{max}^{CHCl3}	1655	1615, 1565

NMR   in CHCl$_3$, τ (hexamethyl ether) 7.83 (2 CH$_3$ of γ-pyrone), 6.19, 5.94, 5.89 –OCH$_3$), 4.18, 3.38, 3.1 (ring protons)

Fig. 74 UV Spectra of
—— Ustilaginoidin
------ Norrubrofusarin

315) Shibata, S *et al*: *6th Symposium on the Chemistry of Natural Products* (Sapporo, 1962), p 47.

(e) Ergoflavin (XIII) [316)~318)]

(XIII)

《Occurrence》   *Claviceps purpurea* (バッカク),   mp. 350° (decomp.), $[\alpha]_D$ +37.5° (acetone.), $C_{30}H_{26}O_{14}$

UV λ_{max}^{EtOH} mμ, $E_{1cm}^{1\%}$: 240 (350), 260 (346), 381 (130)

NMR in CDCl$_3$, τ: 8.84d (J=6c/s) ($>$CH–Me), 2.48d (J=8), 3.02d (J=8) (two pairs of *ortho* arom. protons)

(f) Secalonic acid and **Chrysergonic acid** (XIV–diastereoisomers)[319) 320)]

(XIV)

《Occurrence》   *Claviceps purpurea*

Secalonic acid

   mp. 268~270°, $[\alpha]_D$ +49.5° (dioxane) $C_{32}H_{34}O_{14}$

UV See Fig. 75, λ_{max}^{MeOH} mμ (α): 338 (54.3), 242 (33.5), 220 (34.8)

IR ν_{max}^{KBr} cm^{-1}: 3430m, 2900w, 1738s, 1607s, 1588m, 1562m, 1435s, 1317w, 1230m, 1158w, 1126w, 1082w, 1061m, 1039m, 998w, 883w, 820m.

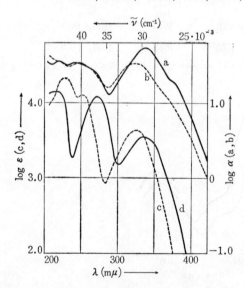

Fig. 75 UV Spectra of

a: Secalonic acid and Chrysergonic acid
b: Dimethylsecalonic acid
c: 5–Hydroxy–2–methylchromone
d: 2–Hydroxy–6–methoxyacetophenone
(Franck *et al.*)

316) Whalley, W. B *et al*: *Soc.*, **1958** 1833.
317) Apsimon, J. W *et al*: *Proc. Chem. Soc.*, **1963**, 209.
318) Asher, J. D. M *et al*: *ibid*, **1963**, 210.   319) Franck, B *et al*: *Ber.*, **95**, 1328 (1962).
320) Apsimon, J. W *et al*: *Proc. Chem. Soc.*, **1963**, 313.

NMR in n-NaOD, 60Mc, TMS 10.0 ppm, H_2O 4.8 ppm: 2.58d, 3.57d (two pairs of arom. protons)

Chrysergonic acid

mp. 244~266°, $C_{32}H_{34}O_{14}$

UV & IR identical to those of secalonic acid (see Fig. 75)

(g) Ergochrysin (XV, R=H) [320]

(XV)

≪Occurrence≫ *Claviceps purpurea,* mp. 285° (decomp.), $[\alpha]_D$ −68°, $C_{29}H_{25}O_{12}$ (COOCH₃)

IR ν_{max} cm⁻¹: 1802 (γ-lactone), 1761 ($-O-\overset{|}{\underset{|}{C}}-CO_2R$), 1639 (chelated arom. C=O).

Tri-O-methylether (VX, R=Me) mp. 179°/260°, $[\alpha]_D$ 19.1°; IR ν_{max} cm⁻¹: 1802, 1752.

NMR in CDCl₃, τ: 8.80d (J=6c/s) (6H of >CH Me), 2.36d, 3.04d (J=8), 2.51d, 2.88d (J=9) (two pairs of *ortho* arom. protons)

(h) Morellin (XVI)[321][322]

CHO

(XVI)

≪Occurrence≫ *Garcinia morella,* $C_{33}H_{36}O_7$

UV λ_{max} mμ (ε): 360 (15130) (cinnamoyl conj. w chromone)

IR ν_{max} cm⁻¹: 1730 (unconj. CO), 1668 ($\alpha\beta$-unsatur. CO), 1628 (chelated CO), 1580 (arom.)

NMR in CDCl₃, 60Mc, τ: 36H, 7CH₃ of 8.47 8.47, 8.5, 8.67 (four $-\overset{|}{C}-CH_3$), 8.2, 8.25, 8.3

(three $\underset{}{C=C}\overset{CH_3}{\diagup}$), −2.77s (chelated OH), 0.37 (−CHO), 2.4d (J=7), 3.35d (J=10), 4.43d (J=10), 3.9t (J=8), 4.75t (J=7.5) (vinyl protones)

(7) UV Spectra of Xanthones[323]

(a) λ_{max}^{MeOH} mμ (ε);[323a] Xanthone 238 (44200), 1-OAc-X. 238 (44100), 1-OCH₃-X. 236 (40300), 1-OH-X. 230 (30600), 2-OAc-1-OH-X. 238 (35500), 2-formyl-1-OH-X. 260 (34400), 4′: 5′-di-OH-4′-oxofurano (2′: 3′-1: 2)-X. 255 (38200)

(XVII) Xanthone

321) Venkataraman, K *et al*: *J. Sci. Ind. Res.*, **14B**, 135 (1955), **15B**, 128 (1956).
322) Kartha, G: *Tetr. Letters*, **1963**, 459.
323) Roberts, J. C: *Chem. Rev.* **61**, 591 (1961).
323a) Scheinmann, F *et al*: *Tetrahedron.*, **7**, 31 (1960).

(**b**) Table 32 shows the UV absorption of naturaly occurring xanthones (XVII–a~XXVIII)[323]

Trihydroxydimethoxy–xanthone
(XVII–a) Corymbiferin
$C_{15}H_{12}O_7$, mp. 268~269°
from *Gentiana corymbifera*

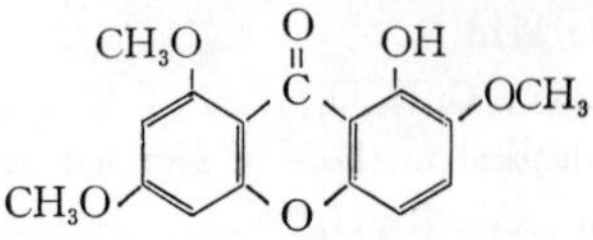

(XVIII) Decussatin
$C_{16}H_{14}O_6$, mp. 149~150°
from *Swertia decussata*

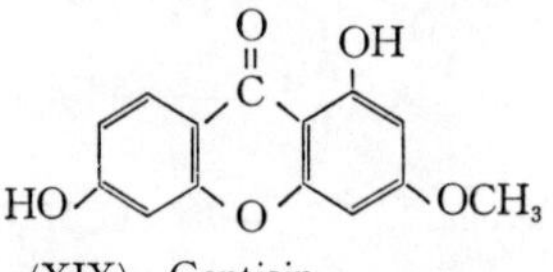

(XIX) Gentisin
$C_{14}H_{10}O_5$, mp. 274°
from *Gentiana lutea*

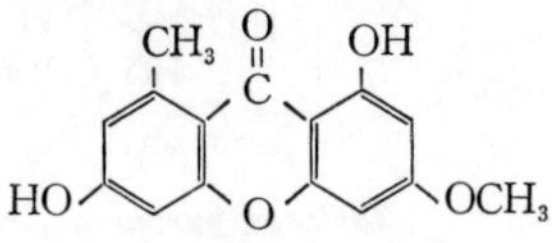

(XX) Griseoxanthone
$C_{15}H_{12}O_5$, mp. 253~255°
from *Penicillium patulum*

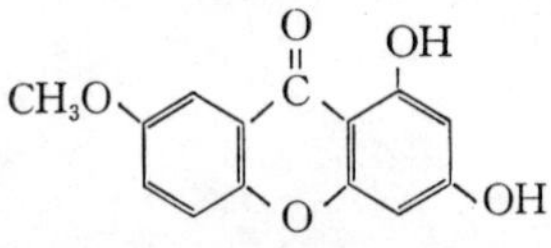

(XXI) Isogentisin
$C_{14}H_{10}O_5$, mp. 241°
from *Gentiana lutea*

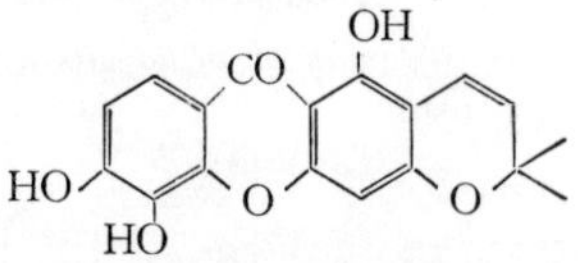

(XXII) Jacareubin
$C_{18}H_{14}O_6$, mp. 256~257°
from *Calophyllum brasiliense*

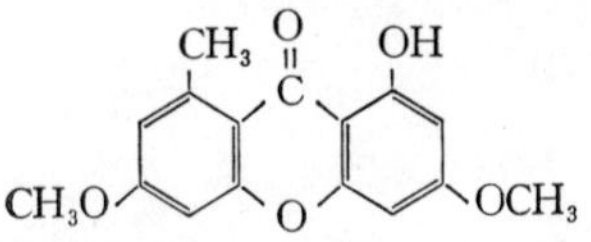

(XXIII) Lichexanthone
$C_{16}H_{14}O_5$, mp. 187°
from *Palmeria formosana*

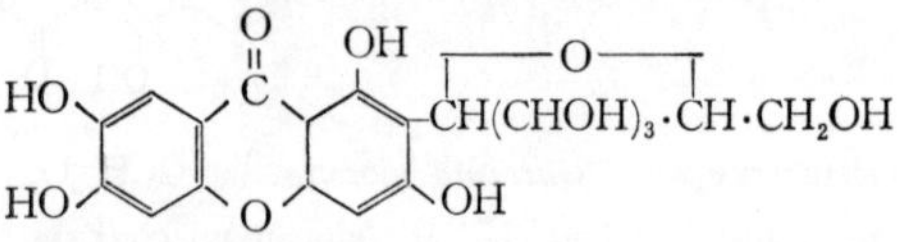

(XXIV) Mangiferin
$C_{19}H_{18}O_{11}$, mp. 271°
from *Mangifera indica*

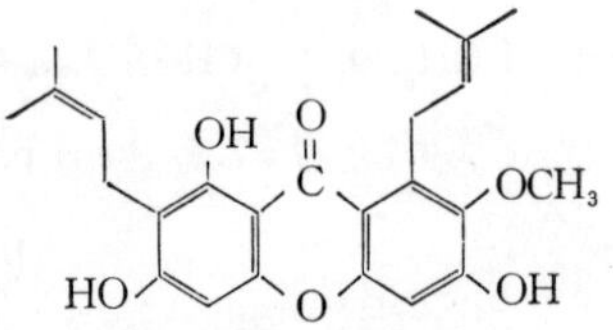

(XXV) Mangostin
$C_{24}H_{26}O_6$, mp. 182°
from *Garcinia mangostana*

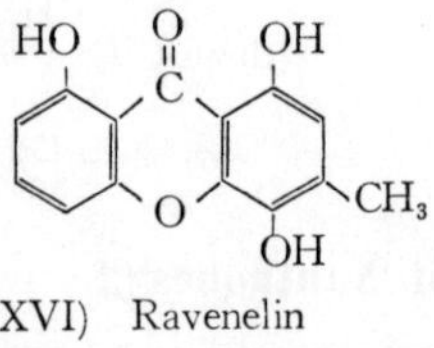

(XXVI) Ravenelin
$C_{14}H_{10}O_5$, mp 267~268°
from *Helminthosporium ravenelii*

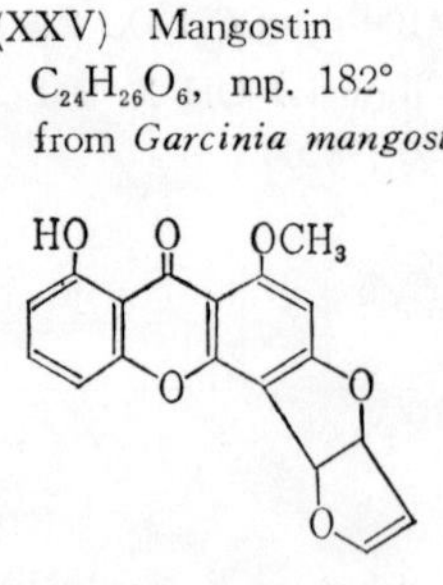

(XXVII) Sterigmatocystin
$C_{18}H_{12}O_6$, mp. 246°
from *Aspergillus versicolor*

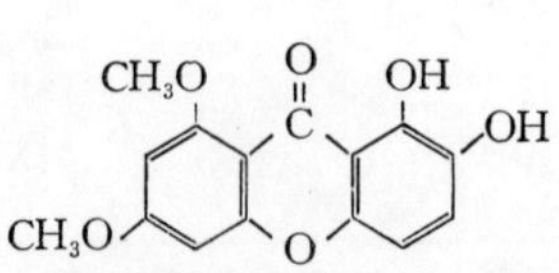

(XXVIII) Swertinin
$C_{15}H_{12}O_6$, mp. 217°
from *Swertia decussata*

Table 32 UV Absorption Spectra of naturally occurring Xanthones (Roberts)

No	Compound	Solvent	λ_{max} mμ (log ε)
XVII–a	Corymbiferin	EtOH	228(4.28), 256(4.32), 278(4.22), 352(4.10)
XVIII	Decussatin	〃	240(4.6), 260(4.6), 315(4.1), 380(3.7)
XIX	Gentisin	MeOH	232(4.43), 260(4.32), 272(4.3), 410(3.75)
XX	Griseoxanthone C	EtOH	242(4.57), 269(3.95), 309(4.36), 340(3.85)
XXI	Isogentisin	—	234(4.78), 260(4.62), 315(4.17)
XXII	Jacareubin	EtOH	240(4.09), 279(4.60), 334(4.26)
XXIII	Lichexanthone	〃	242(4.55), 254(4.43), 306(4.35), 340(3.75)
XXIV	Mangiferin	70%EtOH	242(4.44), 260(4.50), 320(4.22), 368(4.10)
XXV	Mangostin	EtOH	243(4.54), 259(4.44), 318(4.38), 351(3.86)
XXVI	Ravenelin	〃	234(4.1), 266(4.3), 345(3.8), >400(∼3.2)
XXVII	Sterigmatocystin	〃	233(4.49), 246(4.53), 325(4.21)
XXVIII	Swertinin	〃	240(4.4), 267(4.5), 329(4.1), 390(3.5)

(8) IR Spectra of Xanthones[323)324)]

(XXIX)   (XXX)   (XXXI)

Table 33[324)] IR Absorption Spectra of Xanthone Derivatives (λ_{max}^{nujol} μ) (Scheinmann)

Compound	Characteristic bands in the 6μ region
Xanthone	6.02vs, 6.19sh, 6.22vs, 6.38w
1–OH–xanthone (XXIX)	6.08vs, 6.21vs, 6.37s
2–OH–xanthone	6.04sh, 6.10vs, 6.18vs, 6.25s
3–OH–xanthone	6.10s, 6.21vs, 6.40vs
1–OCH$_3$–xanthone	6.00vs, 6.18vs, 6.25vs, 6.38s
1–OAc–xanthone	6.00vs, 6.17s, 6.22vs, 6.40m
4′: 5′–Dihydro–5′methylfurano–(2′: 3′–1: 2)–xanthone (XXX)	6.03vs, 6.17vs, 6.22vs, 6.34m
5′: 6′–Dihydropyrano–(2′: 3′–1:2) xanthone (XXXI)	6.03vs, 6.21vs, 6.28vs, 6.36m

	in the 7.5∼10.0μ region
Xanthone	7.50s, 8.07m, 8.22w, 8.46w, 8.71m, 8.97w, 9.10w, 9.75w
1–OH–xanthone (XXIX)	7.77vs, 8.11vs, 8.24s, 8.49w, 8.62m, 8.96m, 9.10w, 9.75w
2–OH–xanthone	7.63vs, 8.08s, 8.17s, 8.64w, 8.71m, 8.98w, 9.75w
3–OH–xanthone	7.64vs, 7.98s, 8.20vs, 8.47w, 8.62s, 9.02vs, 9.74w
1–OCH$_3$–xanthone	7.71vs, 8.08vs, 8.20v, 8.50m, 8.68s, 8.95s, 9.22vs, 9.75w
1–OAc–X	7.64s, 8.09s, 8.23vs, 8.53s, 8.66s, 8.99w, 9.78m
(XXX)	7.64s, 8.11vs, 8.30m, 8.35m, 8.54s, 8.67m, 8.98w, 9.08w, 9.75s
(XXXI)	7.69vs, 8.11s, 8.24m, 8.40s, 8.72m, 9.72m

---

324) Scheinmann, F: *Tetrahedron*, **18**, 853 (1962).

	in the $10.0 \sim 13.5\mu$ Region					
Xanthone	10.75m,	12.40w,	13.21vs			
1–OH–xanthone (XIXX)	10.74m,	11.60m,	12.25s,	12.77s,	13.08vs	
2–OH–xanthone	11.60w,	12.42m,	12.71m,	13.10s,	13.30m	
3–OH–xanthone	10.79w,	11.76s,	12.15m,	12.94m,	13.32vs	
1–OCH₃–xanthone	10.88vs,	11.55m,	12.40s,	12.95vs,	13.10s,	13.25vs
1–OAc–xanthone	10.80s,	11.60w,	12.22m,	12.61s,	12.72m,	13.18s
(XXX)	11.44m,	12.28m,	12.51s,	13.01vs		
(XXXI)	11.41m,	12.36m,	12.61m,	12.96s,	13.16sh	

(9)　Athyriol and Isoathyriol (XXXII)[325]

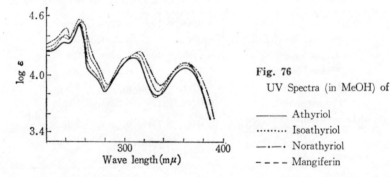

a: $R_1 = R_2 = H$, $R_3 = CH_3$　Athyriol

b: $R_1 = R_3 = H$, $R_2 = CH_3$　Isoathyriol

(XXXII)

≪Occurrence≫　*Athyrium mesosorum* MAKINO（ヌリワラビ）

Athyriol　mp. 300°, $C_{14}H_{10}O_6$

IR　ν_{max}^{KBr} cm⁻¹: 1645 (C=O)

Isoathyriol　mp. 325°, $C_{14}H_{10}O_6$

IR　ν_{max}^{KBr} cm⁻¹: 1645 (C=O)

UV　See Eig. 76

Fig. 76

UV Spectra (in MeOH) of

―――　Athyriol

‥‥‥　Isoathyriol

―・―・　Norathyriol

――――　Mangiferin

(10)　Mangiferin (XXIV)

≪Occurrence≫　*Mangifera indica* L.,[326] *Hedysarum obscurum* L.,[327] *Madhuca utilis* LAM.[328] *Iris* spp.[329]

mp.　271°, $C_{19}H_{18}O_{11}$

325)　Ueno, A: *Y. Z.*, **92**, 1482, 1486 (1962).

326)　Seshadri, T. R *et al*: *Current Sci.*, **29**, 131 (1960).

327)　Hörhammer, L *et al*: Recent Development in the Chemistry of Natural Phenolic Compounds, by Ollis, W. D., p 185 (Pergamon Press, 1961).

328)　Hawthorne, B. J *et al*: Chemistry of Natural and Synthetic Colouring Matters by Gore T. S *et al*, p 331 (Academic Press, 1962).

329)　Bate-Smith, E. C *et al*: *Nature*, **198**, 1307 (1963).

See Table 32, (XXIV)[330)]

UV λ_{max} mμ: 242, 258, 316, 364

Bathochromic shift occurs by addition of alkali, AlCl$_3$, NaOAc or NaOAc + boric acid

(11) Mangostin (XXV)[331)]

　　≪Occurrence≫　　*Garcinia mangostana (Guttiferae)*,　　mp. 181.6~182.6°, $C_{24}H_{26}O_6$

UV　See (7), Table 32, (XXV)

IR　$\lambda_{max}^{CCl_4}$ μ: 2.85, 2.93, 6.08, 6.14sh, 6.20, 6.32

O, O–Dimethylmangostin　　mp. 123.3~123.8°

NMR　in CCl$_4$. Aromatic protons and methylprotons peaks of C_6H_5·CH_3 are assigned to 100 and 1197
　　　cycles resp.: 740 (chelated OH), 1115 (OCH$_3$), 1205 (=CH-CH$_3$). Intensities ratio of the signals
　　　1205／1115=1.33~1.40≒4: 3

(12) Celebixanthone (XXXIII)[332)]

(XXXIII)

　　≪Occurrence≫　　*Cratoxylon celebicum (Guttiferae)*,　　mp. 164~165°, $C_{18}H_{15}O_5$ (OCH$_3$)

Triacetate　　mp. 164~165°

UV　λ_{max} mμ (ε): 247 (38000), 272 (12000), 345 (7800)

NMR: $\dfrac{CH_3}{CH_3}$>C=CH-CH$_2$-Ar, 1 OCH$_3$, 3 CH$_3$COO—

(13) Jacareubin (XXXIV)[333)333a)]

(XXXIV)

　　≪Occurrence≫　　*Calophyllum brasiliense* CAMB.,　　mp. 256~257°(decomp.), $C_{18}H_{14}O_6$

UV　λ_{max}^{EtOH} mμ (log ε)[333a)]: 240 (4.09), 279 (4.61), 334 (4.26)

Trimethylether　　IR: See Fig. 76a.

330) Iseda, S.: *Bull. Chem. Soc. Japan*, **30**, 625, 629 (1957).
331) Yates, P, Stout, G. H: *J. A. C. S.*, **80**, 1961 (1958).
332) Stout, G. H *et al*: *Tetr. Letters*, **1962**, 541.
333) King, F. E *et al*: *Soc.*, **1953**, 3931, **1957**, 563.
333a) Wolfrom, M. L *et al*: *J. Org. Chem.* **29**, 689, 692 (1964).

(14) Xanthones of Osage Orange[333a)]

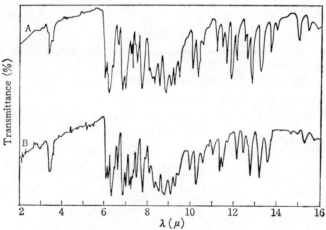

(XXXV)

≪Occurrence≫ *Maclura pomifera* RAF.

(a) **Macluraxanthone** (XXXV)[333a)]

mp. 181~182°/205~206°, $C_{23}H_{22}O_6$

UV λ_{max}^{EtOH} mμ (log ε): 242 (4.31), 283 (4.64), 338 (4.28)

Trimethylether   mp. 98°, IR See Fig. 76a

NMR in $CDCl_3$, 60Mc, τ: 1.99d, 3.04d (AB system of arom. ring, *ortho* to the xanthone carbonyl), 8.59s, 3.21d, 4.30d (J=10.0c/s) (2, 2–dimethylchromenone), 6.01s, 6.06s, 6.07s (3 CH_3O)

(b) **Osajaxanthone** (XXXVI)

mp. 264~265°, $C_{18}H_{14}O_5$

(c) **Alvaxanthone** (XXXVII)[333a) b)]

Fig. 76a   IR Spectra of the Trimethylethers of
(A) Jacareubin and (B) Macluraxanthone
(Wolfrom, *et al.*)

(15) β-**Guttilactone** (XXXVIII)[334)]

≪Occurrence≫   Rubber gum

(XXXVIII)

333b) Wolfrom, M. L *et al*: *Tetr. Letters*, **1963**, 749; *Index Chem.*, 8, 24583 (1963).
334) Antherhoff, H *et al*: *Arch. Pharm.*, **295**, 833 (1962); *Index Chem.*, 8, 24583 (1963).

〔K〕 Rotenoids

(1) UV Spectra

Rotenone, dehydrorotenone and rotenol[335]

(2) IR Spectra

Rotenone and dihydrorotenone and rotenol[336)337].

(3) (−)-Rotenone (I)

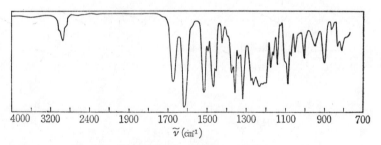

R=H　　Rotenone (I)[338)~340)]
R=OH　Sumatrol

《Occurrence》　　*Derris elliptica, D. malaccensis*,　　mp. 163°, $[\alpha]_D$ −225° (C_6H_6), $C_{23}H_{22}O_6$
IR See Fig. 77

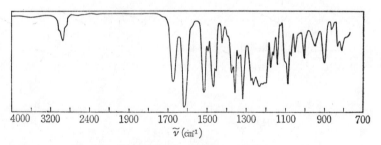

Fig. 77 IR Spectrum of Rotenone (Miyano *et al.*)

(4) α-Malaccol (IV) and β-Malaccol (V)[341]

335) Kapoor, A. L. Aebi, A. Büchi, J: *Helv. Chim. Acta*, **40** 1574 (1957).
336) Cupples, Hornstein: *J. A. C. S.*, **73** 4023 (1951).
337) Kapoor, Aebi, Büchi: *Helv. Chim. Acta*, **40** 1574 (1957).
338) Nakazaki, M *et al*: *Bull. Chem. Soc. Japan*, **34** 1247 (1961).
339) Djerassi, C *et al*: *Soc.*, **1961**, 1448.　　340) Büchi, G. *et al*: *ibid.*, **1961**, 2843.
341) Crombie, L *et al*: *ibid.*, **1961**, 5445.

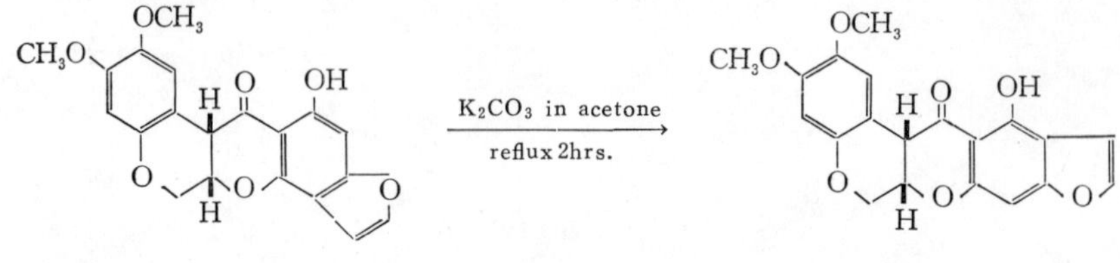

(IV)=(IId) ; R=H, R′=OH   (V)

≪Occurrence≫   *Derris elliptica, D. malaccensis*

+α-**Maiaccol** (natural)   mp. 244~246°, [α]$_D$ +190° (CHCl$_3$)

UV λ$_{max}^{EtOH}$ mμ (ε): 246(29200), 253.5(34100), 264 i(13600), 287(12000), 362(2500)

IR ν$_{max}^{CHCl3}$ cm^{-1}: 1645, 1623 i, 1605

±β-**Malaccol** mp. 249°

UV λ$_{max}^{EtOH}$ mμ (ε); 248 i(30300), 253(36100), 288(15100), 357(1800)

IR ν$_{max}^{CHCl3}$ cm^{-1}: 1653, 1629, 1590

(5) NMR of Rotenoids

(a) **Rotenoids of *Derris elliptica*** BENTH[342)]

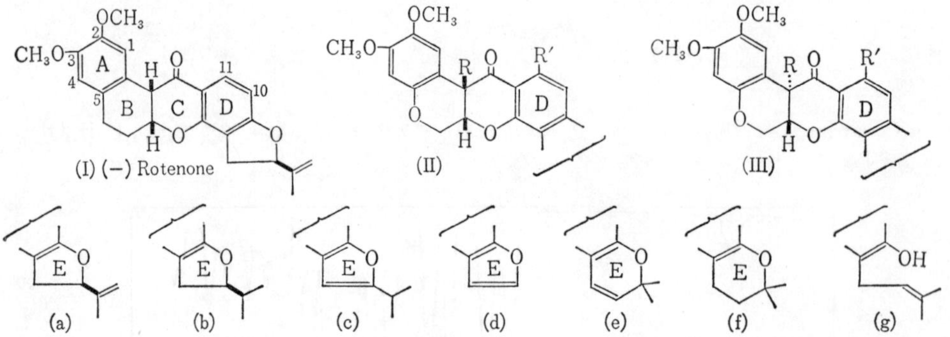

(I) (−) Rotenone   (II)   (III)

(a)   (b)   (c)   (d)   (e)   (f)   (g)

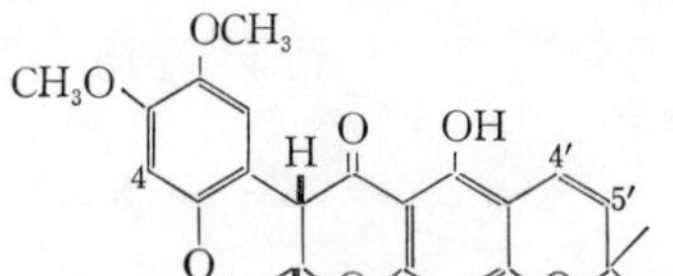

(VI) (±) β-Toxicarol

Table 34   NMR Spectra of Rotenoids (56, 445 Mc, in CDCl$_3$, τ, J in c/s) (Crombie *et al.*)

Compound	Ring A O–Me	Ring A Protons	Ring D Protons	Ring E Protons	Ring E C–Me	Misc.
(−) Rotenone (I)	6.24 6.24	(1) 3.32 (4) 3.59	(10) 3.53 (11) 2.26 } J 11	—	8.25	—

342) Crombie, L *et al*: *Soc.*, **1962**, 775.

Compound	Ring A O–Me	Ring A Protons	Ring D Protons	Ring E Protons	Ring E C–Me	Misc.
(−) Isorotenone (II–c ; R=R′=H)	6.26 6.26	(1) 3.21 (4) 3.55	(10) 2.92 ⎱ J 9.5 (11) 2.20 ⎰	(4′) 3.45	8.66 8.81	—
(±) Degnelin (II–e: R=R′=H)	6.16 6.16	(1) 3.22 (4) 3.59	(10) 3.58 ⎱ J 9 (11) 2.29 ⎰	(4′) 3.39 ⎱ J10.5 (5′) 4.47 ⎰	8.54 8.60	— —
(−) Elliptone (II–d: R=R′=H)	6.17 6.10	(1) 3.21 (4) 3.57	(10) 2.85 ⎱ J 8.5 (11) 2.10 ⎰	(4′) 3.10 ⎱ J 3 (5′) 2.40 ⎰	—	—
(−) Sumatrol (II–a: R=H, R′=OH)	6.28 6.28	(1) 3.23 (4) 3.63	(10) 4.10	—	8.22	—
(−) α-Toxicarol (II–e: R=H, R′=OH)	6.22 6.22	(1) 3.16 (4) 3.55	(10) 4.10	(4′) 3.44 ⎱ J10.8 (5′) 4:32 ⎰	8.51 8.60	OH −2.49
(±) β-Toxicarol (VI)	6.37 6.37	(1) 3.16 (4) 3.55	(8) 4.12	(4′) 2.97 ⎱ J 10 (5′) 4.55 ⎰	8.65 8.65	OH −2.22
(±) Tephrosin (II–e: R=OH, R′=H)	—	(1) 3.34 (4) 3.54	(10) 3.43 ⎱ J 9 (11) 2.34 ⎰	(4′) 3.56 ⎱ J 10 (5′) 4.40 ⎰	8.68 8.68	—

Number in parenthesis shows the position in the structure (I)

(b) Rotenoids of *Neorautanenia pseudopachyrrhiza* HARMS[343].

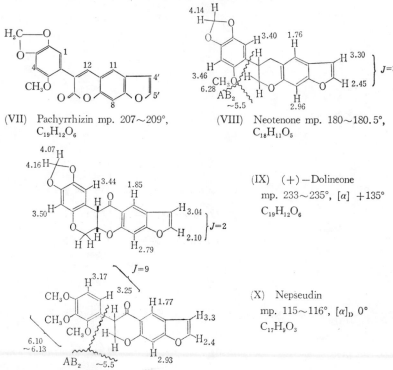

(VII) Pachyrrhizin mp. 207~209°,
  $C_{19}H_{12}O_6$

(VIII) Neotenone mp. 180~180.5°,
  $C_{18}H_{11}O_5$

(IX) (+)−Dolineone
  mp. 233~235°, [α] +135°
  $C_{19}H_{12}O_6$

(X) Nepseudin
  mp. 115~116°, [α]$_D$ 0°
  $C_{17}H_9O_3$

343) Crombie, L *et al*: *Soc.*, **1963**, 1569.

Pachyrrhizin (VII)[342] :

NMR (τ) : 6.24 (ring A,OCH$_3$), 2.23 (ring A,C$_1$–H), 3.11 (ring A, C$_4$–H), 2.53 (ring D, C$_8$–H), 2.35 (ring D, C$_{11}$–H), 2.35, 3.15 ($J \not> 3$) (ring E, C$_5'$–H : C$_4'$–H), 3.37 (C$_{12}$–H), 4.03 (CH$_2$O$_2$)

Neotenone (VIII)

UV λ_{max}^{EtOH} mμ (log ε): 235(4.68), 275(3.83), 300(3.87), 335(3.61)

IR ν_{max}^{CHCl3} cm^{-1}: 1684(C=O), 1623, 1582, 1546, (aryl.)

(+)−**Dolineone** (IX)

UV λ_{max}^{EtOH} mμ (log ε): 237(4.56), 275(3.84), 305(3.71), 335(3.50)

IR ν_{max}^{CHCl3} cm^{-1}: 1684 (C=O), 1629, 1587, 1546, 1504 (aryl.) ; ν_{max}^{mull} cm^{-1}: 1686, 1631, 1587, 1546, 1504, 937.

Nepseudin (X)

UV λ_{max}^{EtOH} mμ (log ε): 235(4.74), 257 i(4.08), 272(3.94), 335(3.61)

IR ν_{max}^{CHCl3} cm^{-1}: 1686(C=O), 1631, 1605, 1585, 1541, 1490 (aryl.) ; ν_{max}^{mull} cm^{-1}: 1686 i, 1680(C=O), 1629, 1605, 1580, 1490

(6) Mass Spectra of Rotenoids [343a]

(a) Cracking Patterns

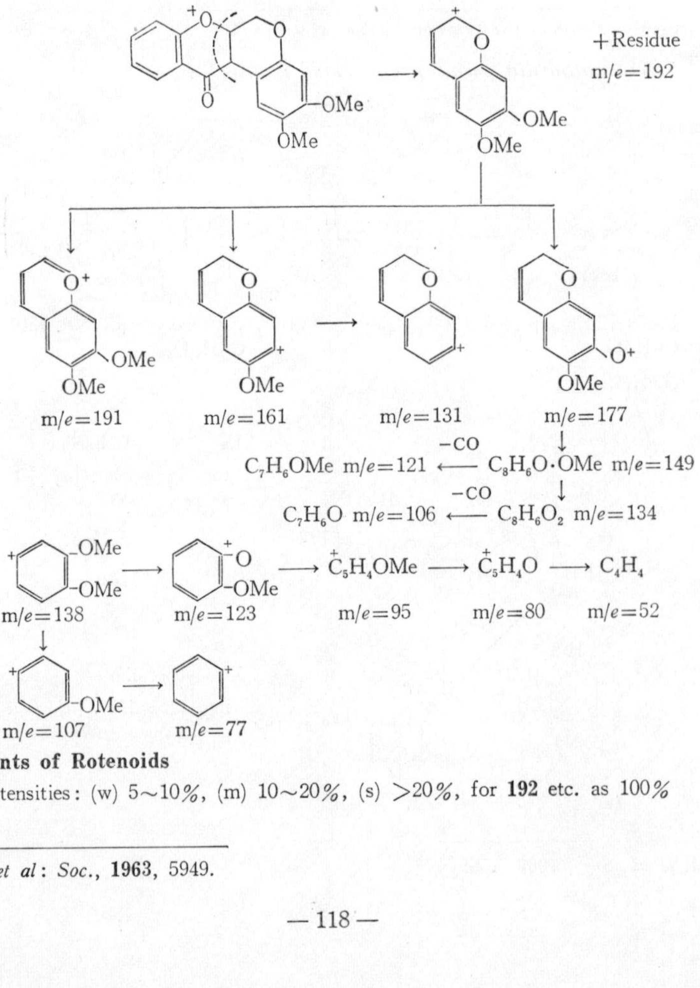

(b) Mass Fragments of Rotenoids

m/e, relative intensities: (w) 5~10%, (m) 10~20%, (s) >20%, for **192** etc. as 100%

343a) Reed, R. I et al : Soc., **1963**, 5949.

(X–a) Pachyrrhizone   (XII) Munduserone   (X–b) Sermundone

(X–c) Isoelliptone

(X–d) Toxicarol [Note]

Pachyrrhizone (X–a)   367s, 366s, 364w, 191m, 190m, 177s, **176**, 175s, 163m, 162m, 147m, 138w, 133w, 119w

Munduserone (XII)   343, 342s, 193, **184**, 191s, 177s, 150m, 149m, 134w, 121m, 106m, 93m, 91w,

Sermundone (X–b)   359s, 358s, 356m, 193s, **192**, 191s, 190w, 178w, 177s, 150m, 134w, 121w, 106m

Elliptone (II–d: R=R′=H)   353s, 352s, 193s, **192**, 191s, 178w, 177s, 160s, 149w, 121w, 106w, 93w

Isoelliptone (X–c)   353s, 352s, 193s, **192**, 191s, 177s, 163s, 161w, 149s, 147m, 134w, 132s, 121m

Rotenone (I–a)   395s, 394s, 193s, **192**, 191s, 177s

β–Dihydrorotenone (I–f)   397m, 396s, 193s, **192**, 191s, 177s, 149m, 121w

Isorotenone (I–c)   395m, 394s, 193s, **192**, 191s, 177s, 163w, 149w, 121w

Toxicarol (X–d)   441m, 410s, 396s, 395s, 219m, 204m, 203s, 198m, 193s, **192**, 191s, 190m, 179s, 177s, 163w, 161w, 151w, 149w, 133w, 121w, 106w, 103w

Rotenonic acid? (I–g)   396s, 193s, **192**, 191s, 190m, 178m, 177s, 163w, 149m, 147w, 121w, 106w

(7) Amorphigenin (XI)[344]

Amorphin ⟶ Amorphigenin+Glucose+Arabinose

(XI)

≪Occurrence≫   *Amorpha fruiticosa*, $C_{23}H_{22}O_7$

UV and IR are closely resemble to those of rotenone except the absorption $\nu_{max}^{CCl_4}$ cm^{-1}: 3498(OH)

NMR (τ): 2CH₃O, 3.19, 3.52 (1 and 4 protons), 3.47q (J=9), 2.12d (J=9) (10 and 11 protons) 5.73 (8′–protons), 4.73 (2H, vinylic H)

[Note] The different structure was assigned from that of (−)α-Toxicarol (II–e: R=H, R′=OH)[342]

344) Crombie, L *et al*: *Proc. Chem. Soc.*, **1963**, 246

(8) **Munduserone** (XII)[345]

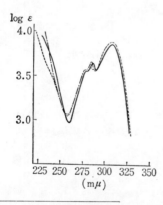

(XII)

≪Occurrence≫   *Mundulea sericea,*   mp. 162°, $[\alpha]_D$ +103° (CHCl₃)
IR ν_{max}^{CHCl3} cm⁻¹: 1666.

[L]   Coumaranochromane Derivatives

(1) **Pterocarpin** (I) and **Maackiain** (II)[345a]

R=CH₃   Pterocarpin   (I)

R=H   Maackiain   (II)

(Revised Str.)[345b]

(a) Pterocarpin (Maackiain methylether) (I)[345a)b)]

≪Occurrence≫   *Pterocarpus santalinus, P. dalbergoides, P. macrocarpus, P. osun, P. indicus, Baphia*
nitida, Sophora sabprostrata (小豆根), and *S. japonica.* (クララ)

l– mp. 164~165°, $[\alpha]_{546}$ −207.5° (CHCl₃) C₁₇H₁₄O₅

dl– mp. 185~186°, $[\alpha]_D$ ±0° C₁₇H₁₄O₅

d– mp.   159~160°, $[\alpha]_D$ +232° (CHCl₃) C₁₇H₁₄O₅

UV See Fig. 77–a, IR See Fig. 77–b (*d*–Pterocarpin) λ_{max}^{EtOH} mμ (log ε): 280(3.58), 2.86(3.65), 310

(3.87); ν_{max}^{CHCl3} cm⁻¹: 2998, 2919, 2864, 2812 (C–H), 1620, 1589 (arom.), 1037, 942 (–O–CH₂–O)

Fig. 77a UV Spectra of
——·· *l*, *d* and *dl*–Maackiain,
–·–·– *dl*–Pterocarpin,
······ *dl*–Maackiain acetate
(in EtOH) (Shibata *et al.*)

345) Finch, N. *et al*: *Proc. Chem. Soc.* **1960**, 176, 177.
345a) Shibata, S. *et al*: *Chem. Pharm. Bull.*, **11**, 167 (1963).
345b) Son Bredenberg, J. B. *et al*: *Tetr. Letters* **1961**, (9), 285.

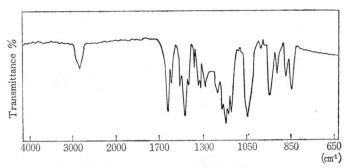

Fig. 77b IR Spectrum of *l*–Pterocarpin in $CHCl_3$
(almost superimposable to those of *d*–and *dl*–isomers)

NMR[345b)] (60 Mc, TMS, δ): 7.41d, *J*=9 c/s (5–H) 6.72 (6′–H), 6.43 (3′–H, overlap to ABX system of 5, 6, 8–protons), 5.92 (–O–CH₂–O–) 5.48d (4–H)

(b) *dl*–**Maackiain** (II)[345a)]

 ≪Occurrence≫ *Sophora japonica* (*dl*–), *Maackia nitida* (*l*–) mp. 195~196°, [α]$_D$ ±0°, $C_{16}H_{12}O_5 \cdot {}^1/_2 H_2O$

UV λ$_{max}^{EtOH}$ mμ (log ε): 281(3.58), 287(3.65), 310(3.86) (See Fig. 77–a)

IR ν$_{max}^{nujol}$ cm⁻¹: 3460~3540 (OH), 1617, 1600, 1509 (arom.), 1033, 928 (–O–CH₂–O–) (See Fig. 77–b)

Trifolirhizin (*l*–Maackiain–β–D–glucoside)

 ≪Occurrence≫ *Trifolium pratense, Sophora subprostrata.*

Sophojaponicin (*d*–Maackiain–β–D–glucoside)

 ≪Occurrence≫ *Sophora japonica,* mp. 202~204°, (decomp.) [α]$_D$ −104° (AcOH), $C_{22}H_{22}O_{10} \cdot CH_3OH$

Inermin (Desmethylpterocarpin (II))[345c)]

 ≪Occurrence≫ *Andira inermis,* mp. 100~105°/174~175°, [α]$_D$ −221° (MeOH), $C_{16}H_{12}O_5 \cdot {}^1/_2 H_2O$

UV λ$_{max}$ mμ (log ε): 286 (3.72), 309 (3.93).

IR ν$_{max}$ cm⁻¹: 3500 (OH), 1626, 1592, 1504 (arom.), 847, 709 (–O–CH₂–O–)

(2) Pisatin (III)[345d) e)]

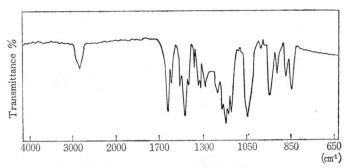

(III) [345e)]

 ≪Occurrence≫ *Pisum sativum,* mp. 72°, [α]$_{436}$ +570° (EtOH), $C_{17}H_{14}O_6$

UV λ$_{max}^{EtOH}$ mμ (log ε): 213 (4.75), 280 (3.62), 286 (3.68), 309 (3.86)

NMR (in CCl_4, 56.4 Mc external toluene reference. benzene scale, $C_6H_5 \cdot CH_3$ 5.00 ppm, C_6H_6 0.09 ppm.)
 (See Fig. 77–c, d).

345c) Cocker, W. *et al*: *Soc.,* **1962**, 4906: *Chem. & Ind.,* **1962**, 216.
345d) Perrin, D. R. *et al*: *J. A. C. S.,* **84**, 1919, 1922 (1962).
345e) Perrin, D. R. *et al*: *Nature,* **191**, 76 (1961).

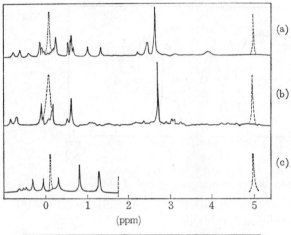

Fig. 77c NMR Spectra of
Pisatin (a), Pterocarpin (b)
(in CCl$_4$) and Anhydro-
pisatin (c) (in dioxane)
(Perrin *et al.*)

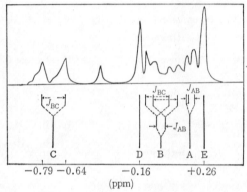

Fig. 77d NMR Spectrum of
Pisatin in CCl$_4$
(aromatic region)

2.77 ppm (3 H of CH$_3$O), $-0.79 \sim +0.26$ m (5 H of arom.), -0.79, -0.64 d, $J=9$ cps (*ortho* 2 H of
arom.), 0.55, 0.64 (–O–CH$_2$–O–), 3.90 br. (–OH), 2.54, 2.32 (2 H at C$_2$), 1.03$\sim$1.40 d, $J=21$
(4–H? [Note])

In Fig. 77–d: A (8–H), B (6–H), C (5–H), D (6′–H), E (3′–H), ABC–system due to three protons.
H$_A$, H$_B$, H$_C$ of 1, 2, 4–unsubstituted aromatic ring. ($J_{XY} \ll \delta_X - \delta_Y$), $J_{AB}=3.1$, $J_{BC}=8.7$ cps.

(IV) Edulin   (V) Neorautone

《Occurrence》   *Neorautanenia edulis* C. A. Sм.

(a) Edulin (IV)   mp. 225°, $[\alpha]_D$ $-265.3°$ (CHCl$_3$), C$_{18}$H$_{12}$O$_5$

UV λ_{max}^{CHCl3} $m\mu$ (ε): 250 (13680), 257 (12500), 305 (12600)

IR ν_{max}^{KBr} cm^{-1}: 1645, 1639, 1626, 1608, 1548, 1511 (arom.), 1166, 1040, 945 (–O–CH$_2$–O–)

NMR (in CDCl$_3$, 60 Mc, TMS, cps): (See Fig. 77–e)

[Note] Coupling constant $J=21$ is anormalous, and is intrepreted as due to hindered internal rotation.
345f) van Duuren, L.: *J. Org. Chem.*, **26**, 5013 (1961).

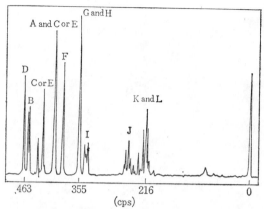

Fig. 77e NMR Spectrum of Edulin (van Duuren)

(b) **Neorautone** (V)　mp. 207°

UV　$\lambda_{max}^{CHCl_3}$ $m\mu$ (ε): 246 (41500), 291 (20000), 350 (16750).

IR　ν_{max}^{KBr} cm^{-1}: 1730 (C=O), 1620, 1580, 1540, 1500 (arom.), 1190, 1036, 922 (–O–CH$_2$–O–)

(4) **Homopterocarpin** (Baphiniton) (VI)[345g)h)]

(VI)

《Occurrence》　*Pterocarpus santalinus, P. dalbergoides, P. macrocarpus, P. osun,* and *Baphia nitida.*
mp. 87~88°, $[\alpha]_D$ −207° (CHCl$_3$), C$_{17}$H$_{16}$O$_4$

〔M〕 Pyrane Derivatives

(1) **Brasilin**(I)[345i)j)]

(I)

《Occurrence》　*Caesalpinia braziliensis,* white paly yellow crystal　C$_{16}$H$_{14}$O$_5$
Triacetate　mp. 105~106°, Trimethylether　mp. 139~140°, Tetraacetate　mp. 149~151°

345g) King, F. E. *et al*: *Soc.,* **1953**, 3693.　　345h) Akisanya, A. *et al*: *ibid.,* **1959**, 2679.
345i) Perkin *et al*: *Soc.,* **1928**, 1504.　　345j) Pfeifer *et al*: *Ber.,* **63**, 1301 (1930).

(2) **Haematoxylin** (II)[345k]

(II)

≪Occurrence≫   *Haematoxylon campechianum,*   mp. 100～120°, $C_{16}H_{14}O_6 \cdot 3H_2O$
Pentaacetate mp. 165～166°, Tetramethylether mp. 139～140°, Antibacterial activity[3451]

[N]   Tannins

(1) **UV Spectra**

Catechol, pyrogallol, gallic acid, 5,7-dioxy-2,2-dimethylchromane, (−) *epi*-catechin, (−)-gallocatechin, (±) gallocatechin, (−)-*epi*-catechin gallate, (−)-gallocatechin gallate[346]. Extracts of the bark of Western hemlock (*Tsuga heterophylla*) and other plants[347]. Dihydroquercetin, β-conidendrol, catechol and resorcinol.

(2) **Hydroxyflavan Derivatives**

UV and NMR spectra of the following hydroxyflavanes are described in the item [G] Chalcone and Flavanone glycosides.   Number in parenthesis indicates sub-item of each compound.
(+)-Mollisacacidin ((12), (a)), (+)-leuco-robinetinidin ((12), (b)), cyanomaclurin ((12), (d)), teracacidin ((15), (f)), melacacidin and isomelacacidin ((15), (g)), (+) mollisacacidin and (−) leucofisetinidin ((15), (h)), peltogynol ((15), (j)).

(3) **Kakitannin** (Leucodelphinidin-3-glucoside) (I)[348]

(I)

≪Occurrence≫   *Diospyros Kaki* L. (カキ)   Octaacetate mp. 245°, $C_{21}H_{14}O_{13}$ $(COCH_3)_{10}$

345k) Perkin *et al*: *Soc.*, **1908**, 496, 1121.   345l) Pratt, R. *et al*: *J. A. P. A.*, **48**, 69 (1959).
346) Bradfield: *Soc.*, **1948**, 2249.   347) Maranville, Goldschmid: *Anal. Chem.*, **26**, 1423 (1954).
348) Oshima, Y. *et al*: *Agr. Biol. Chem.*, **26**, 156 (1962).

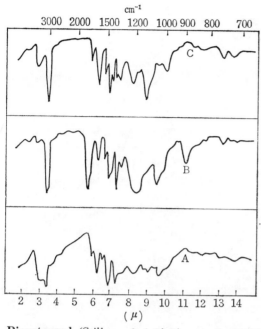

Fig. 78. IR Spectra of
A : Kakitannin,
B : Acetate,
C : Methylether
(in nujol) (Oshima *et al.*)

(4) **Piceatannol** (Stilbene-3, 4, 3', 5'-tetraol) (II)[349]

≪Occurrence≫ Spruce (*Pisca* spp.)
mp. 229° (decomp.), $C_{14}H_{12}O_4$ [Note]

IR ν_{max}^{nujol} cm^{-1}: 1611, 1522

(II)
Tetramethylether   mp. 64～66°

NMR in CCl$_4$, τ: 7.4 (α–CH$_2$), 8.29 (β–CH$_2$), 3.77 (arom. and vinyl), 6.18, 6.24, 6.29 (OCH$_3$)

(5) **Ellagic acid Derivatives**[350]~[352]

(III)   (IV)   (V)

(III) X=X'=Y=Y'=H Ellagic acid
(IV) R=H   Ellagorubin
(V)   X=X'=Y=Y'=H 5, 5'–Di–*C*–benzylellagic acid

≪Occurrence≫ Gallotannins and ellagotannins.

[Note]   5, 6, 7, 8-Tetrahydronaphthalene–1, 2, 4–triol was also found. See the item (**1**),
[D] Naphthalene and Phenanthrene Derivatives (3).

349) Cunningham, J. *et al*: *Soc.*, **1963**, 2875.
350) Jurd, L: *J. A. C. S.*, **81**, 4610 (1959).   351) Stitt, F. *et al*: *ibid.*, **81**, 4615 (1959).
352) Jurd, L. *et al*: *ibid.*, **81**, 4620 (1959).

(a)   **Ellagic acid** (III)

Sublimable, $C_{14}H_6O_8$,

UV: See Fig. 79. λ_{max}^{EtOH} mμ: 255, 352, 366.

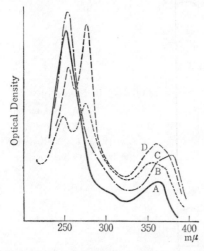

Fig. 79 UV Spectra of Ellagic acid

(A)   in EtOH

(B)   in EtOH + NaOAc

5, 5′-Di-C-benzylellagic acid

(C)   in EtOH

(D)   in EtOH+NaOAc

(Stitt *et al.*)

IR λ_{max}^{KBr} 5.00~6.15 μ: 5.90 ($\alpha\beta$-unsatur. lactone CO)

(b)   **Ellagorubin**   mp. 220°, $C_{42}H_{30}O_8$

IR λ_{max}^{KBr} 5.00~6.15 μ: 5.83 ($\alpha\beta$-unsatur lactone), 5.97 ($\alpha\beta$-γ, δ-unsatur. ketone)

(c)   **5, 5′-Di-C-benzylellagic acid**

UV See Fig 79, λ_{max}^{EtOH} mμ: 256, 363, 378

IR λ_{max}^{KBr} μ: 5.00~6.15μ: 5.84 ($\alpha\beta$-unsatur. lactone C=O)

NMR 60 Mc, in CDCl$_3$, shift (cps) from C_6H_6:

(A) O-Tetramethyl-5, 5′-di-C-benzylellagic acid (V) X=X′=Y=Y′=CH$_3$, (B) O-Tetrabenzyl-5, 5′-di-C-benzylellagic acid (V) X=X′=Y=Y′=CH$_2$C$_6$H$_5$, (C) Ellagorubin diacetate (IV) R=COCH$_3$, (D) Ellagorubin dimethylether (IV) R=CH$_3$, (E) Ellagorubin (IV) R=H.

(A)   −47s (arom. H), 97 (two C−CH$_2$−C), 131, 154 (two OCH$_3$ resp.)

(B)   center −47d (arom. H), 56, 85 (two−O−CH$_2$−resp.), 98 (two C−CH$_2$−C)

(C)   center −33 br (arom. H), 167 (four C−CH$_2$−ϕ), 238 (two COCH$_3$) ($^{167}I_{238}$=8/6)

(D)   center −39.5 br (arom. H), 142 (two−OCH$_3$), 167 d (two C−CH$_2$−ϕ resp.)

(E)   (in pyridine) center 167 q (J=14 cps) (two non equiv. methylene protons of $\rangle$C$\langle\substack{CH_2-\phi \\ CH_2-\phi}$)

(d) Trimethylellagic acid[353]

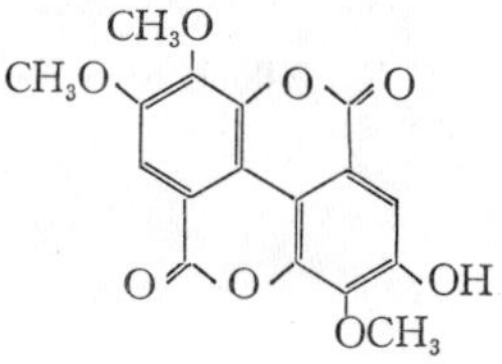

(III) X′=H, X=Y=Y′=CH₃

《Occurrence》   *Eugenia maire* A. CUNN.,   mp. 293~294°, $C_{17}H_{12}O_8$

UV λ_{max}^{EtOH} mμ (log ε): 249 (4.64), 355 (sh), 370 (4.07) (unchanged in the presence of NaOAc)

IR ν_{max}^{KBr} cm⁻¹: 3486 (OH), 2985, 2899, 1462, 1359, 1245, 1175, 1129 (OCH₃), 1754 (lactone C=O)

(e) 3, 3′–Di–*O*–methylellagic acid–4–glucoside (VI)[354]

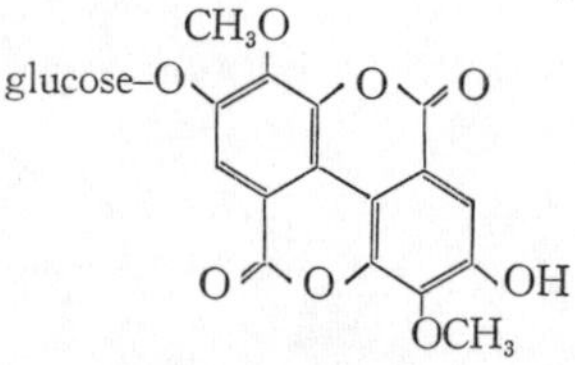

(VI)

《Occurrence》   *Terminalia paniculata* ROTH.,   mp. 214~215°, [α]_D +79° (EtOH)
Genin mp. 338~340°, $C_{16}H_{18}O_8$
UV λ_{max} mμ (log ε): (glucoside) 249 (4.67), 373 (4.07), (genin) 249 (4.75), 370 (4.15).

(6) Gallotannins

(a) Phenolic acid, composing Hydrolysable Tannins

《Occurrence》   Chinese tannin (galls, produced by *Aphis chinensis* on the leaves of *Rhus semialata*, Turkish tannin (galls, produced by *Cynips tinctoria* on the twigs of *Quercus infectoria*), sumach (leaves of *Rhus coriaria*, *R. typhina* and other spp.), tara (pods of *Caesalpinia chebula*), myrobalans (fruit of *Terminalia chebula*), valonea (acorn cups of *Quercus aegilops*), algarobilla (pods of *Caesalpinia brevifolia*).

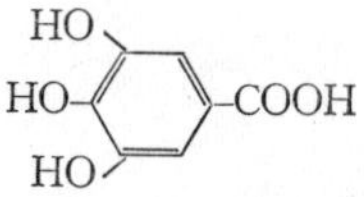

(VII) Gallic acid 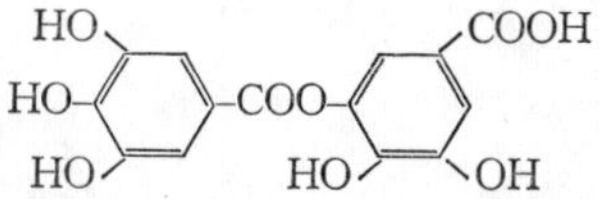   (VII–a) *m*–Digallic acid 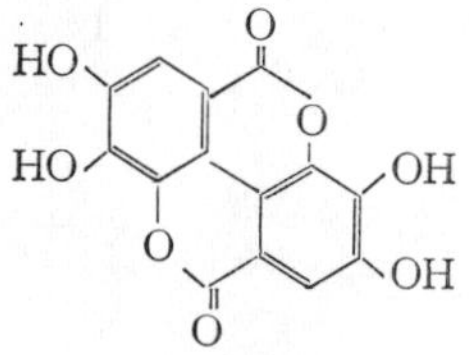   Ellagic acid=(III) X=X′=Y=Y′=H

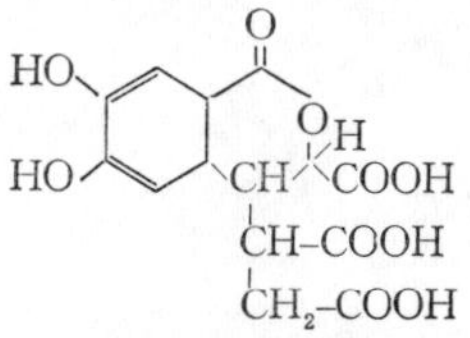

(VIII) Chebulic acid 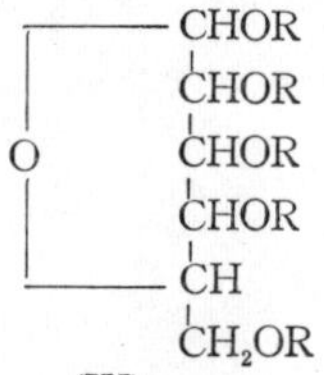   (IX)

353) Briggs, L. H *et al* : *Soc.*, **1961**, 642.
354) Ramachandra Row, L *et al* : *Tetrahedron*, **18**, 357 (1962).

(b) Chinese gallotannin

(i) β–Penta–O–m–digalloylglucose (IX: R=m–digalloyl)

(ii) 1, 2, 3, 4–Tetra–O–galloyl–6–O–m–hexagalloyl–glucose (X) R=galloyl, $n=5$ precipitated by $AlCl_3$.

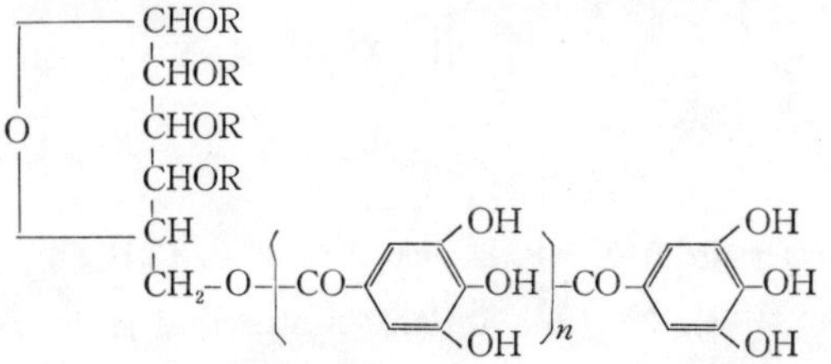

(X)

(iii) 1, 2, 3, 4–Tetra–O–galloyl–6–O–m–trigalloyl–glucose (X), R=galloyl, $n=2$
Obtained also from sumach tannin.

(c) Tara tannin (XI)

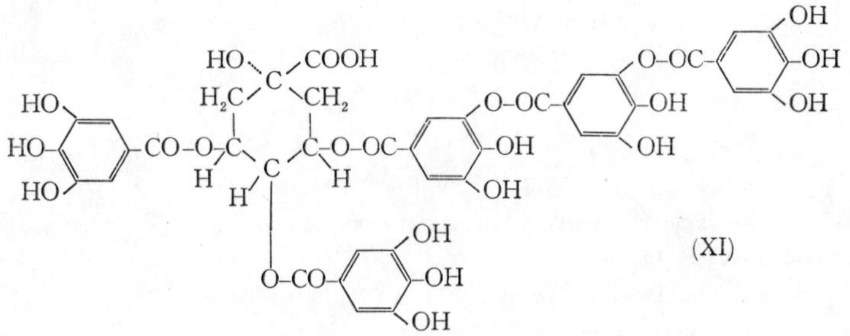

(XI)

(d) Myrobalan tannin

(i) Chebulinic acid (XII)[355]

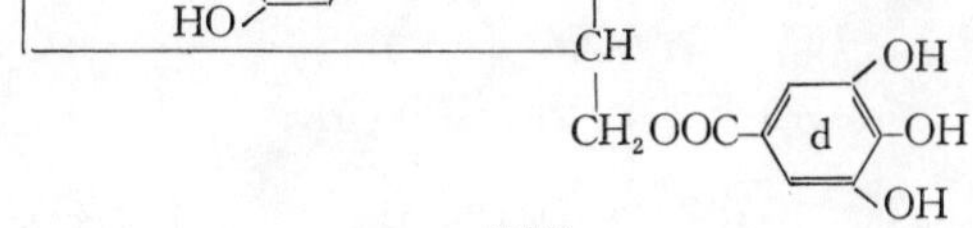

(XII)

---

355) Schmidt *et al*: *Ann.*, **569**, 149 (1950); **571**, 232 (1951); **578**, 31 (1952).

(ii) Chebulic acid (XIII)[356]

(iii) Corilagin (XIV)[355]

(iv) Chebulagic acid (XV)[355]

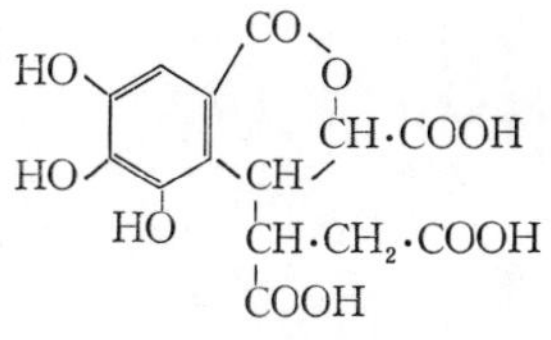

(XIII)

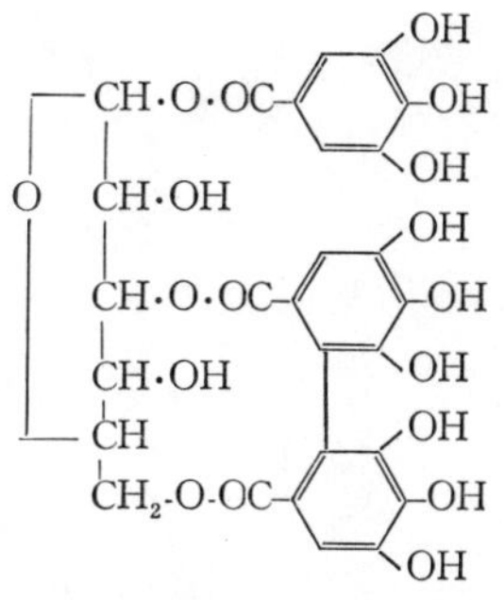

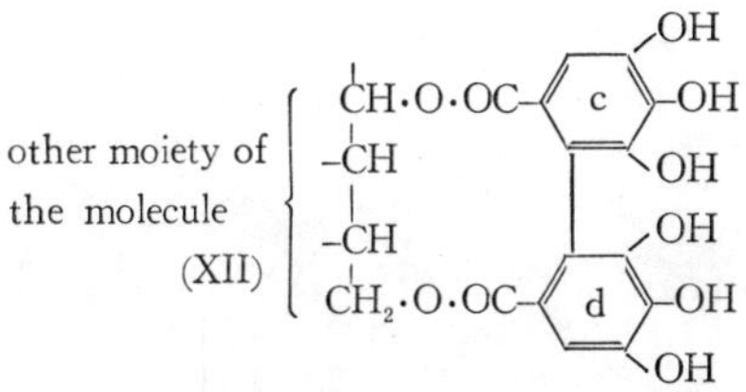

other moiety of
the molecule
(XII)

(XV)

(XIV)

(e) **Tannins of Spanish chestnut (*Castania vesea*) and Valonea**[357][358]

(i) Dehydrodigallic acid (XVI)

(ii) Valoneic acid dilactone (XVII)

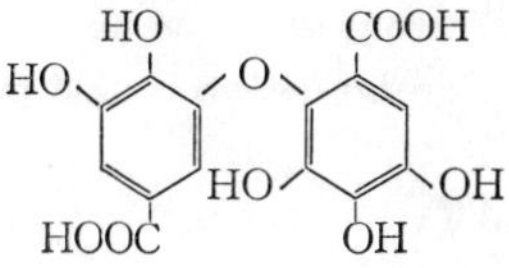

(XVI)

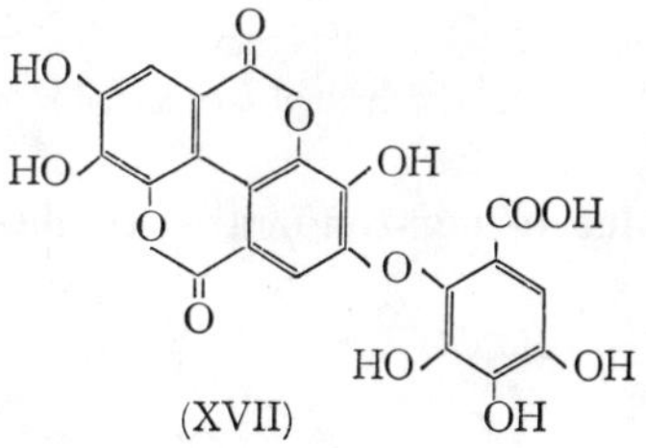

(XVII)

356) Schmidt *et al*: *Ann.*, **571**, 1 (1951).

357) Mayer: *Ann.*, **578**, 34 (1952).   358) Schmidt *et al*: *ibid.*, **591**, 156 (1955).

7 Triterpenoids

[A] UV Spectra

(1) Triterpenoids and the dehydrogenated products

Monomethyl naphthalenes, dimethylnaphtalenes and trimethylnaphthalenes[359]. Glycyrrhetic acid[360], 2-acetoxy -7-oxo-α-amyran, ——enoldiacetate, ——enolmonoacetate[361], Oleanolic acid derivatives[362][363]. 2-Acetoxy-12-oxo-ursan-28-carboxylic acid methylester, ——acetoxylactone, $\Delta^{10,11}$-2-acetoxy-12-oxo-ursene-28-carboxylic acid methylester, ——enoldiacetate[364]. β-Amyran trienol acetate and methyl acetyldehydroursolate [365]. Acetyl-α-elemolic acid methylester and diketoacetyldihydro-α-elemolic acid methylester[366].

(2) Absorption due to non-conjugated Doublebond[367]

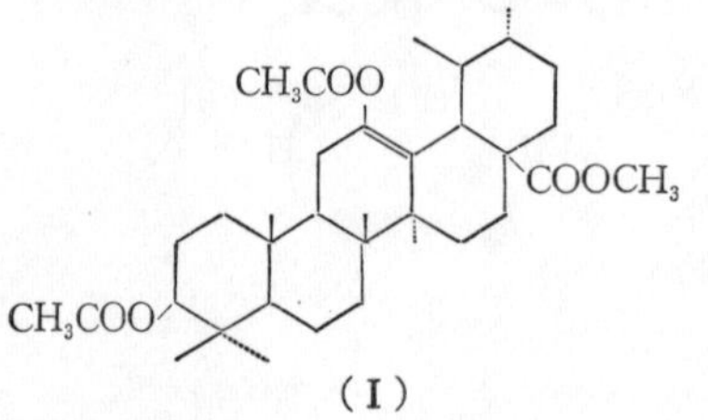

(I) 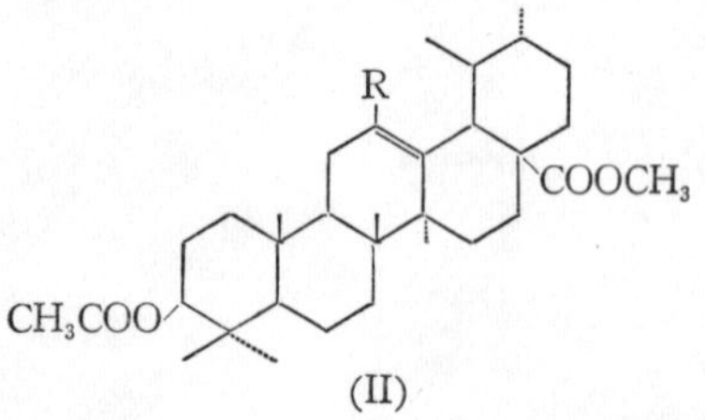 (II)

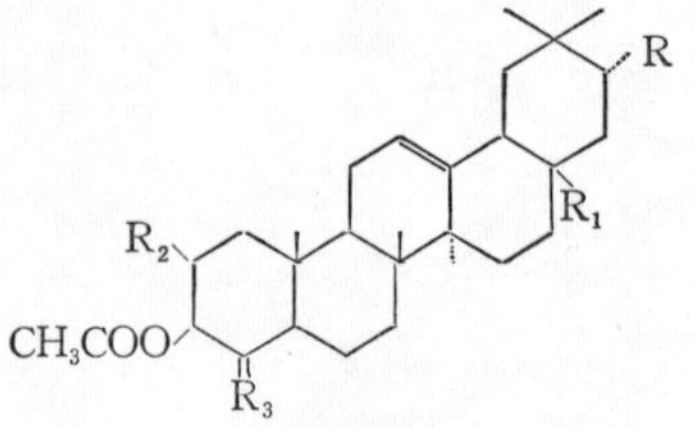

(III)

359) Heilbronner, E. *et al*: *Helv. Chim. Acta*, **32**, 2479 (1949).
360) Van Katwijk *et al*: *Rec. trav. Chim.*, **74**, 889 (1955).
361) Ruzicka: *Helv. Chim. Acta*, **28**, 199 (1945).
362) Ruzicka *et al*: *Helv. Chim. Acta*, **26**, 265 (1943).
363) Ruzicka *et al*: *ibid.*, **33**, 1835 (1950).   364) Ruzicka *et al*: *Helv. Chim. Acta*, **33** 1325 (1950).
365) Ruzicka *et al*: *ibid.*, **26**, 1235 (1943).   366) Ruzicka *et al*: *ibid.*, **25**, 1380 (1942).
367) Micheli, R.A. *et al*: *J. Org. Chem.*, **27**, 345 (1962).

Table 34′ UV Absorption due to non-conjugated Doublebond of Triterpenoids

$\lambda_{max}^{Cyclohexane}$ m$\mu(\varepsilon)$ (Micheli *et al.*)

Triterpenoid	No.	Radical	λ_{max}	ε	mp.
Me–Δ^{11} ursene–3β–12–diol–28–oate diacetate	(I)		198	9600	216~218°
Me–Δ^{12}–ursene–3β–12–diol–28–oate diacetate	(II)	R=OAc	197.5	7600	173~177°
Acetylmethyl ursolate	(II)	R=H	193	7180	246~247.5°
Soyasapogenol–B·triacetate	(III)	R=OAc, R$_1$=Me, R$_2$=H R$_3$=AcOCH$_2$ Me	194.5	9920	
Diacetoxydimethyl medicagenate	(III)	R=H, R=CO$_2$Me, R$_2$=OAc, R$_3$=Me CO$_2$Me	194.5	7570	248~250°
β–Amyrin acetate	(III)	R=R$_2$=H, R=Me R$_3$=Me Me	194.5	8870	239~240.5°

(3) **Absorption of Dehydrogenated Triterpenoids**[367a]

UV Spectra of dehydrogenated triterpenoids, produced by SeO$_2$ reveal three absorption maxima at 242, 251 and 260 mμ. (See Table 34″).

Δ^{12}, 2, 3–di OAc–28–CO$_2$Me,
Crataegolic acid methylesterdiacetate

reflux with SeO$_2$+CH$_3$COOH →

Dehydrocrataegolic acid methylesterdiacetate

Table 34″ UV Absorption of dehydrogenated Triterpenoids

(Tschesche *et al.*)

Dehydrogenated triterpenoid	λ_{max}^{MeOH} mμ (log ε)		
Isodehydro–β–amyrin	243	251(4.49)	260(4.26)
Dehydrooleanolic acid		251(4.45)	260(4.33)
Dehydrosumaresinolic acid		250(4.5)	

367a) Tschesche, R. *et al*: *Ber.*, **92**, 320 (1959).

Dehydrogenated triterpenoid	λ_{max}^{MeOH} mμ (log ε)		
Dehydrodesoxoglycyrrhetinic acid	242(4.45)	250(4.5)	260(4.25)
Dehydromyrtillogenic acid	240(4.38)	248(4.42)	258(4.23)
Dehydromedicagenic acid	243(4.17)	251(4.19)	260(4.15)
Dehydrocrataegolic acid	242(4.46)	251(4.52)	260(4.30)

［B］　IR Spectra

(1)　Triterpenoids[28]

(a)　Ethylenic Absorption——(▭) 12:13, 2:3, 18:19, 9:11, 20:29, 13, 18:17:18, monoene, $\varDelta^{12,18}$ diene, $\varDelta^{9(10):12}$ diene, $\varDelta^{11:13(18)}$diene, $\varDelta^{12}$–C$_{11}$–ketone, $\varDelta^{13(18)}$–C$_{12}$: C$_{19}$–diketone, methyl oleanolate, methyl ursolate, α–amyrin, β–amyrin, lupeol, methyl betulate, betulin and oleanol[368)369].

(b)　δ_{C-H} of ketoterpenoids——urs–12–ene–11–one, phyllanth–3–one, 3β–hydroxyoleanan–12–one, methyl ursolate, methyl 3–oxo–A–trisnorlupan–28–oate and 3β–hydroxy–30–norlupan–20–one[368)369].

(c)　Carbonyl absorption——γ–lactone, A–trisnor–C$_{(3)}$–ketone, acetate, aldehyde, methylester, formate, benzoate, C$_{19}$–ketone (oleanane series), C$_{20}$–ketone (30–norlupane series), C$_3$–ketone, C$_{11}$–ketone, C$_{12}$–ketone, carboxylic acid, conjugated aldehyde, $\varDelta^{13(18)}$–C$_{(12)}$: C$_{(19)}$–diketone(oleanane series), $\varDelta^{13(18)}$–C$_{(14)}$–ketone(〃), $\varDelta^{12}$–C$_{(11)}$–ketone, $\varDelta^{12:18}$–diene–C$_{(11)}$–ketone, δ_{C-H}——C$_{(11)}$–ketone, C$_{(3)}$–ketone, C$_{(12)}$–ketone, A–nor–C$_{(3)}$–ketone and 30–nor–C$_{(20)}$–ketone. Ursolic acidlactone, oleanolic acidlactone, α–amyrin acetate, lupenylacetate, methyl ursolate, methyl siaresinolate, methyl oleanolate, methyl betulate, α–amyrin benzoate, β–amyrin benzoate, ursolic acid and oleanolic acid[368)369].

(d)　3–Hydroxyl absorption——(*Equatorial*) α–Amyrin, β–amyrin, lupeol, methyl betulate, dihydrolanosterol, (*Axial*) epi–α–amyrin, epi–β–amyrin, epi–lupeol, methyl–3α–hydroxy–lup–20(29)–en–28–oate and lanost–8–en–3α–ol[370].

(2)　Characteristic IR Absorptions at the Region of 1392～1245 cm[371]

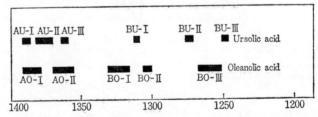

AO–I (1379～1392), AO–II (1355～1370), BO–I (1315～1330), BO–II (1299～1306), BO–III (1250～1267), AU–I (1386～1392), AU–II(1370～1383)　AU–III (1359～1364), BU–I(1308～1312), BU–II (1270～1276), BU–III (1245～1250)

Fig. 80　Characteristic IR Absorptions of Oleanolic and Ursolic acid Derivatives (1392～1245 cm⁻¹, in pyridine) (Tschesche *et al.*)

Table 35～38 indicate the characteristic absorptions of naturally occurring triterpenoids having followed skeletons (IV)～(XVII).

368)　Cole, Thornton: *Soc.*, **1957**, 1322.　369)　Cole, Thornton: *ibid.*, **1956**, 1007.

370)　Allsop, I. L. *et al*: *Soc.*, **1956**, 4868.

371)　Tschesche, R. *et al*: *Tetrahedron*, **18**, 1417 (1962).

Table 35 IR Spectra of Oleanane Series (IV) (in pyridine)　　　(Tschesche *et al.*)

Sapogenin	No.	Δ	–OH	=O	–CO₂R	glycoside with	AO–I	AO–II	Interval Region	BO–I	BO–II	BO–III	Ref.
Oleanolic acid	IV	12	3β		28		1392 S	1367 m	1348 w (1333)	1325 w	1306 m	1267 S (1258)	372)
Me-ester							1390 S	1368 m	1348 w	1324 w	1304 m	1262 S (1255)	
Crataegolic acid	IV	12	2,3		28		1387 m	1364 m	1348 w	1321 w	1302 m	1264 S 1241 w	373)

372) Karrer, W: Konstitution und Vorkommen der organischen Pflanzenstoffe, p. 796~843 (Birkhäuser, 1958).
373) Tschesche, R. *et al*: *Ber.*, **92**, 320 (1959).

Sapogenin	No.	Δ	–OH	=O	–CO₂R	glycoside with	AO–I	AO–II	Interval Region	BO–I	BO–II	BO–III	Ref.
Echinocystic acid	IV	12	3β,16α		28		1387s	1363m	1341w	1330w	1305m	1279m 1250s	372)
Cochalic acid	IV	12	3β,16β		28		1386s	1360m	1343w	1322w	1302m	(1278) 1266s	372)
Siaresinolic acid	IV	12	3β,19α		28		1387s	1360m	(1345) 1337w	1320w	1300m	1363s 1251s	374)
Hederagenin	IV	12	3β,23		28		1389s	1365m	1347w	1321w	1302m	1264s 1255s	372)
Hederacoside A	IV	12	3β,23		28	Arab. +Gluc.	1386s	1362m		1320w	1300m	1262s	372)
Bayogenin	IV	12	2β,3β, 23		28		1388m	1365s		1321w	1303m	(1376) 1264s	375)
Arjunolic acid Me-ester	IV	12	2α,3β, 23		28		1388s	1366s	(1345)	1320w	1302m	1261s	372)
Bassiaic acid Me-ester	IV	5, 12	2β,3β, 23		28		1384s	1365s	(1346) 1332w	1321m	1303m	(1275) 1259s	376)
Tomentosic acid	IV	12	2,3β, x, 23or 24		28		1392s	1367m	1348w	1323w	1305m	1265s 1249w	377)
Rehmannic acid	IV	12	22β	3	28	angelic acid ester	1388s	1365m	1352w	1325w	1309w (1305)	1256s	372)
Icterogenin	IV	12	22β,24	3	28		1385s	1365m	1352w	1325w	1306w	(1270) 1256s	372)
Terminolic acid	IV	12	2,3β,6β 23		28		(1384) 1379s	1363m	1346w 1331w	1320w	1302m	1263s	372)
Quillajaic acid	IV	12	3β,6α	23	28		1387s	1362m	1342w	1330w 1323w	1304w	1277m 1250s	372)
Gypsogenic acid	IV	12	3β		23, 28		1388s	1364m		1322w	1301m	1265s	378)
Medicagenic acid Me-ester	IV	12	2β,3β		23, 28		1386m	1364s		1320w	1300m	1252s 1239s	379)
Glycyrrhetinic acid	IV	12	3β	11	29		1390s	1360m		(1316)	1281m	1260s	372)
Katonic acid	IV	12	3α		29		1386s	1366m	1351w	1334w	1281s	1261s	374)
Morolic acid	IV	18	3β		28		1390m	1378m	1361s	1325s	1298s	1275m	372)

374) King, F. E. et al : Soc., **1960**, 4738.   375) unpublished.
376) King, T. J. et al : Soc., **1961**, 4308.
377) Row, L. R. et al : Tetr. Letters, **1960** No. 27, 12.
378) Ruzicka, L. et al : Helv. Chim. Acta, **21**, 83 (1938).
379) Djerassi, C. et al : J. A. C. S., **79**, 5292 (1957).

Sapogenin	No.	Δ	–OH	=O	–CO₂R	glycoside with	AO–I	AO–II	Interval Region	BO–I	BO–II	BO–III	Ref.
Bredemolic acid	IV	12	3,x		28		1385s	1364m	1347w	1320w	1301m	(1274) 1263s	380)
Primulagenin A	IV	12	3β,16α, 28				1388s	1376m	1362m				372)
7β–Hydroxy– A₁ –barrigenol	IV	12	3β,7β,15α 16β,27,28				1388s	1375m	1361m				381)
Sojasapogenol C	IV	12, 15	3β,24				(1390)	1381s	1361s				372)
Sojasapogenol B	IV	12	3, 16, 23 or24				(1389)	1381s	1364m				372)
allo–Eetulin	IV		3β		19, 28– oxide		1388s	1377s	1362m			1243s in CHCl₃	382)

Table 36 IR Spectra of Ursane Series (V)

(in pyridine)   (Tschesche *et al.*)

Sapogenin	No.	Δ	–OH	=O	–CO₂R	AU–I	AU–II	AU–III	Interval Region	BU–I	BU–II	BU–III	Ref.
Ursolic acid	V	12	3β		28	1392s	1383s	1364m	(1350)	1312m	(1285) 1273s	1250s	372)
Ursonic acid	V	12		3	28	1386s	1379m	1363w	(1349) 1338w	1308m	(1281) 1270s	1245s	383)
Asiatic acid	V	12	2α,3β, 23		28	1392s	1380s	1362s		1309m	(1282) 1272s	1246s	372)
Chinovaic acid	V	12	3β		27, 28	1388s	1370m	1360w	1336w	1312m	1276s	1247s	372)
β–Boswellic acid	V	12	3α		23 or 24		1380s	1360m	1344m (1333)	1305m	1276m		372)
Commic acid	V	12	1β, 2β, 2β		23	(1392)	1382s (1371)	1361w		1324m	1255s	18α– H	384)
Pyrochinovaic acid Me-ester	V	13	3β		28	1384s	1372m	1364w	1349w (1336)	1318m 1303m	1243s	27– nor	385)
α–Amyrenone	V	12		3		(1388) 1382s	1371m	1363m					
ψ–Taraxasterol	V	20	3β			1385s		1360m		1327w	1250s		372)

380) Tschesche, R. *et al*: *Ber.*, **93**, 1303 (1960).
381) Knight, J.O. *et al*: *Tetr. Letters*, **1961**, No.3, 100.   382) Davy, G.S. *et al*: *Soc.*, **1951**, 2702.
383) Simonsen, J. *et al*: The Terpenes V. p.118 (Cambridge Univ. Press, 1957).
384) Thomas, A.F. *et al*: *Tetrahedron*, **16**, 264 (1961).
385) =(383) p.76.   386) Ruzicka, L *et al*: *Helv. Chim. Acta*, **22**, 758 (1939).

Table 37 IR Spectra of Tetracyclic Triterpenoids (VI, VII, XV and XVI)
(in pyridine) (Tschesche)

Sapogenin	No.	Δ	–OH	=O	–CO₂R	=CH₂	A–Region	Interval Region	B–Region	Out of B-Region	Ref.
Eburicoic acid	VI	8	3β		21	24	1375m 1360w		1278s (1255)		372)
Tumulosic acid	VI	8	3β, 16α		21	24	1376s 1360m	1335w	1282w 1265m		372)
Masticadieno-nic acid	VII	7, 24		3	26		1386s 1370s	1328w	1312w	1244s	372)
Isomasticadie-nonic acid	VII	8, 24		3	26		1377s 1370s	(1315)	(1287) 1245s		372)
Polyporenic acid	VI	8	3α,12α		26	24	(1383) 1374s		1310m 1285m	1230s	372)
Polyporenic acid–Me-ester	VI	7, 9:11	16α	3	21	24	(1380) 1374s	(1351) 1336w	1311w 1285w	1252m	372)
Lanosterol	VI	8, 24	3β				(1384) 1376s				387)
Dipterocarpol	XV	24	20	3			1382s				391)
Cycloartenone	XVI	24		3			1384s (1370)				394)

Table 38 IR Spectra of Lupane Series and the isomeric Triterpenoids (VIII, IX, X and XVII)
(in pyridine) (Tschesche)

Sapogenin	No.	Δ	–OH	=O	–CO₂R	Other Group	A–Region	Interval Region	B–Region	Out of B-Region	Ref.
3-Isopropyliden-Δ^{12}-4, 23, 24-trisnoroleanen-28-acid Me-ester	VIII	3, 12			28		1384s 1362s		1318w 1299m (1271) 1257s		388)
3-Isopropyl-$\Delta^{3:5,\,12}$-4, 23, 24-trisnor-oleadien-28-acid Me-ester	VIII	3:5, 12			28		1378s 1359s		1320w 1300w 1256s		388)
Betulinic acid	IX	20:29	3β		28		1392m 1380m 1360m		1319m 1294w 1273m		372)
Melaleucic acid	IX	20:29	3β		25, 28		1386m 1362w	1351m	1321m (1305) 1295w 1274m	(1243)	372)
Betulin	IX	20:29	3β,28				1389m 1374s 1360w				372)

387) Khastgir, H.N. *et al*: *Chem. & Ind.*, **1961**, 945.   388) unpublished.
389) Huneck, S: *Ber.*, **94**, 1151 (1961)   390) Asahina, Y *et al*: *ibid.*, **71**, 980 (1938).
391) Ourisson, G. *et al*: *Tetrahedron*, **3**, 279 (1958).   392) Mechoulam, R: *Chem. & Ind.*, **1961**, 1835.
393) Khastgir, H.N. *et al*: *ibid.*, **1961**, 1077.   394) Barton, D.H.R.: *Soc.*, **1951**, 1444.

Sapogenin	No.	Δ	–OH	=O	–CO₂R	Other Group	A–Region	Interval Region	B–Region	Out of B-Region	Ref.
Zeorinone	XVII		22	6			1393m (1382) 1363m				389)
Leucotylin	XVII		6α, 22, x				1391s 1366m (1357)				390)
Emmolic acid	X	20:29	3α		2, 28		1390w 1375m	1349s	1333w 1315w		392)

IR Spectra of Friedelin and isomeric Triterpenoids (XI, XII, XIII and XIV)

Sapogenin	No.	Δ	–OH	=O	A–Region	Ref.
Friedelin (CHCl₃)	XI			3	1394s, (1387), 1365w	373)
Alnusenone (CHCl₃)	XII	5		3	1387s, 1365m	373)
Multiflorenol	XIII	7	3β		1387s, (1376), 1365s	393)
Bauerenol	XIV	7	3β		1388s, 1376s	387)

(3) Characteristic Hydroxyl Frequencies in Triterpenoids[395] (See Table 39)

Table 39 (in CCl₄ soln.)　　　　(Cole et al.)

Primary–OH	3640~3641 cm⁻¹	
Lup–20(29)–en–28–ol	3641	ε 66

Primary + Secondary (*equatorial*)		
Lup–20(29)–ene–3β: 28–diol (Betulin)	3637	ε 110
Urs–12–ene–3β: 28–diol (Uvaol)	3625	107

Secondary (*equatorial*) C(3) 3628~3630 C(11) 3623~3625		
28–Norolean–17–en–3β–ol (Oreanol)	3629	ε 53
Methyl 3β–hydroxyolean–12–en–28–oate (Oleanolate)	3629	61
Methyl 3β–hydroxyurs–12–en–28–oate (Ursolate)	3630	54
13: 27–Cycloursan–3β–ol (Phyllanthol)	3629	59
9: 19 Cycloeburic–25–en–3β–ol (Cyclolaudenol)	3629	58
Friedelan–3α–ol (Friedelinol)	3630	—

Secondary (*axial*) 3635~3638		
Friedelan–3β–ol (*epi*–Friedelinol)	3635	—

395) Cole, A.R.H. *et al*: *Soc.*, **1959**, 1218.　　396) Cole, A.R.H. *et al*: *ibid.*, **1959**, 1224.

Secondary (*equatorial+axial*)

Methyl 3β: 19α–dihydroxyolean–12–en–28–oate (Siaresinolate)	3632	ε 95

Tertiary (*equatorial*)   3613

3α–Methylcholestan–3β–ol	3613	ε 73

Tertiary (*axial*)   3617∼3619

3β–Methylcholestan–3α–ol	3617	ε 70

(4) Intramolecular H-Bonding Diols and the Stereochemistry of Triterpenoid and related Compounds[396] (Table 40)

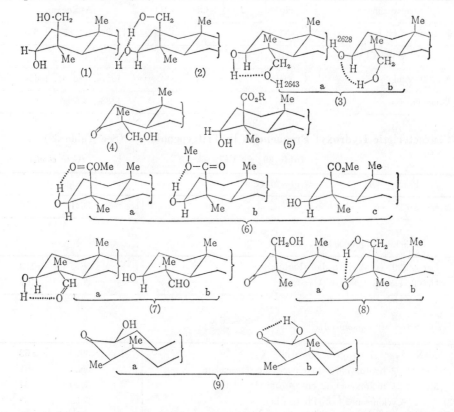

Table 40 (in CCl₄ soln.)   (Cole *et al.*)[396]

Compound	No.	Structure	Free OH	Bonded OH	Free C–O	Bonded C–O
Urs–12–ene–3α : 24–diol	(1)	(V) $\varDelta^{12}$, 3α : 24 diol	3641	—		
Urs–12–ene–3β : 24–diol	(2)	(V) $\varDelta^{12}$, 3β : 24 diol	3629	3550		

Compound	No.	Structure	Free OH	Bonded OH	Free C–O	Bonded C–O
Methylhederagenin	(3)	(IV) Δ^{12}, 3β, 23 diol, 28–CO$_2$Me	3643 (prim) 3628 (sec)	3532	1729	
Methyl hederagonate	(4)	(IV) Δ^{12}, 3–one, 23 ol 28–CO$_2$Me	3630 3602	3540	1728	1697
β–Boswellic acid	(5) R=H	(V) Δ^{12}, 3α–ol, 24– CO$_2$H	3638 3524	—	1737	1694
Methyl 3α–hydroxyurs–12–en–24–oate	(5) R=Me	(V) Δ^{12}, 3α–ol, 24– CO$_2$Me	3638	—	1723	
Methyl 3β–hydroxyurs–12–en–24–oate	(6)	(V) Δ^{12}, 3β–ol, 24– CO$_2$Me	3627	3548	1736	1709
Gypsogenin methylester	(7)	(IV) Δ^{12}, 3β–ol, 23–al, 28–CO$_2$Me	3623	3549	1729	1710
Ictreogenin methylester	(8)	(IV) Δ^{12}, 22β, 24–diol, 3–one, 28–CO$_2$Me	3635	3518	1746 1718	1694
Cerin	(9)	(XI) 2–ol, 3 one	3615	3491 3436	1715	1709

(5) Tetracyclic Triterpenoids[397)]

(XVIII)

(a)

Table 41 Average characteristic Frequencies (cm^{-1}) of Tetracyclictriterpenoid Ketones
(in CCl$_4$)   (Cole *et al.*)[397)]

Compound (XVIII)	C=C	C=O	CH$_2$–CO	CH$_2$–C=C	Cyclic-Ketone
3–Ketone		1708	1426~1427		Hexa-
Δ^8–3 : 7–Diketone		1714			"
11–Ketone		1704	1432~1433		"
7 : 11–Diketone		1709	1431~1433		"
A–Nor–3–ketone		1741	1407		Penta-
17–Ketone		1747			"
Δ^8–7–Ketone		1667	1417~1420	1425~1427	Conj. hexa-
Δ^8–11–Ketone		1657	1418	1428	"
Δ^8–7 : 11–Diketone	—	1677, 1686	1426~1428		"
$\Delta^{5:8}$–7 : 11–Diketone	1623	1673, 1654	1426		"
$\Delta^{5:8}$–7 : 11 : 12–Triketone	1629	1738, 1685, 1655			"
Δ^{24}–26–Aldehyde	1650	1692		1420	Conjtd. side-chain aldehyde

397) Cole, A.R.H. *et al* : *Soc.*, **1959**, 1212.

(b)   Ethylenic Doublebonds

Table 42   Characteristic Frequencies (cm⁻¹) of Ethylenic Doublebonds (in CCl₄)

(Cole *et al.*)[397]

$\varDelta$	Compounds (XVIII)	ν_{C-H}	$\nu_{C=C}$	$CH_2-C=C$	δ_{C-H} (CS₂)	Note
			Non–conjugated C=C			
$\varDelta^7$	Lanost–7–en–3β–yl acetate	3030	1662sh	1435	822	
$\varDelta^8$	Lanost–8–en–3β–ol	—	—	1435	—	
$\varDelta^8$	Lanosta–8 : 24–diene	—	—	1435	—	
$\varDelta^{9(11)}$	Lanost–9(11)–en–3β–ol	3053	—	1435	815	
$\varDelta^{24}$	Lanosta–8 : 24–diene	—	1673	—	—	
$\varDelta^{24}$	9 : 19–Cyclanost–24–en–3–one	—	1673	—	836	Tricyclene
$\varDelta^{25}$	9 : 19–Cycloeburic–25–en–3β–yl acetate	3040 3071	—	—	887	Tricycl., Vinylidene
			Conjugated C=C			
$\varDelta^{7\,:\,9(11)}$	Lanosta–7 : 9(11) diene	3030	1629	1432	814, 800	
"	*"* –3β–ol	3032	1629	1432	814, 800	

(6)   Characteristic ν_{C-H} Frequencies of Cyclopropane Ring[398]

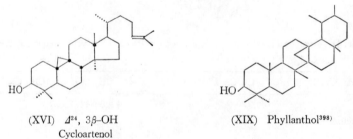

(XVI) $\varDelta^{24}$, 3β–OH   (XIX)   Phyllanthol[398]
Cycloartenol

(a)   3040∼3058 cm⁻¹ Region
Cycloartenol (3042 cm⁻¹), phyllanthol (3055 cm⁻¹)

398)   Cole, A.R.H. *et al*: *Soc.*, **1954**, 3807.

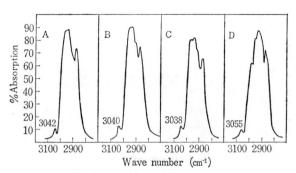

A: Cycloartenol (3042) C: Cycloartenone (3038)
B: Cycloartane (3040) D: Phyllanthane (3055)

Fig. 81 Characteristic ν_{C-H} of Cyclopropyl Methylene (in CCl$_4$ soln.)
(Cole *et al.*)[398]

(**b**) 750～1500 cm^{-1} Region

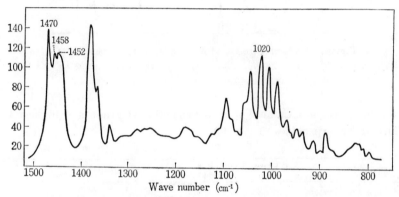

Fig. 82 IR Spectrum of Cycloartenol (in CS$_2$ soln.) (Cole *et al.*)[398]

(7) IR Spectra of oxidized Products of Triterpenoids and Steroids by RuO$_4$[399]

Triterpenoids and steroids, having tetrasubstituted double bond in the nucleous are oxidized to diketones by RuO$_4$ oxidation as following:

399) Snatzke, G. *et al*: *Ann.*, **663**, 123 (1963).

Characteristic IR spectra of those diketones are useful to confirm the structure of original triterpenoids and steroids.

[C] NMR Spectra[400]

Determined at 40 Mc in tetrachloroethylene, CHCl₃ as an internal standard, ppm of TMS=0, CHCl₃=7.25

(1) Typical Spectrum (Ursolic acid methylester acetate) (Fig. 83)

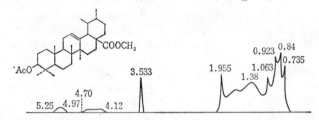

Fig. 83 NMR of Ursolic acid methylester acetate (Shamma *et al.*)
5. 11 br (11 c/s) (C₁₂–H), 4. 41 br (23 c/s) (C₃–H), 1. 955 s (–COOCH₃), 3. 533 (–OCOCH₃), 1. 25~2. 00 br (–CH₂–), 0. 735, 0. 84, 0. 923, 1. 065 (overlapping of seven C–CH₃)

(2) Relationship between Existance of 28-Carbomethoxyl group and highest C-Me chemical Shift of Triterpenoides (Table 43).

Table 43   (Shamma *et al.*)

Compound	C–Me Shift	28–COOMe	Compound	C–Me Shift	28–COOMe
Arjunolic acid Me–ester diacetate	0.683	+	7β–Hydroxy–A₁–barrigenol hexaacetate [Note]	.820	—
Melaleucic acid Me–ester	.695	+	Uvaol diacetate	.823	—
Echinocystic acid Me–ester diacetate	.713	+	Longispinogenin triacetate	.825	—
Oleanolic acid Me–ester acetate	.715	+	Chinchipegenin tetraacetate	.845	—
Oleanolic acid Me–ester	.730	+	α–Amyrin benzoate	.865	—
Ursolic acid Me–ester acetate	.735	+	Erythrodiol diacetate	.875	—
Cochalic acid Me–ester	.735	+	β–Amyrin benzoate	.905	—
Asiatic acid Me–ester triacetate	.763	+	11–Keto–α–boswellic acid Me–ester acetate	.848	—
Morolic acid Me–ester acetate	.765	+	Lupanol	.803	—
A₁–Barrigenol pentaacetate	.783	—	Betulin diacetate	.840	—
α–Boswellic acid Me–ester acetate	.788	—	Friedelin	.703	—
Glycyrrhetic acid Me–ester acetate	.808	—	Thurberogenin acetate	.810	—
Soyasapogenol B triacetate	.808	—	Dumortierigenin diacetate	.860	—

Examples of NMR spectra of some triterpenoids are shown in Fig. 84

---

400) Shamma, M. *et al*: *J. Org. Chem.*, **27**, 4512 (1962).
[Note] Revised structure.   a) S. Itō, T. Ogino, H. Sugiyama, and M. Kodama: *Tetrahedron Letters*, 2289 (1967).
   b) S.G. Errington, D.E. White, and M.W. Fuller: *ibid.*, 1289 (1967).

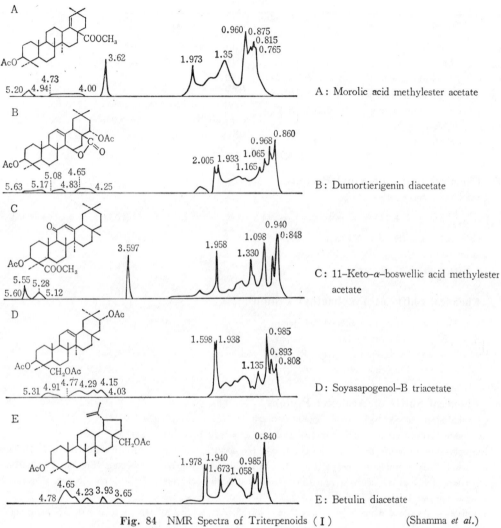

Fig. 84 NMR Spectra of Triterpenoids（Ⅰ）   (Shamma *et al.*)

A : Morolic acid methylester acetate

B : Dumortierigenin diacetate

C : 11-Keto-α-boswellic acid methylester acetate

D : Soyasapogenol-B triacetate

E : Betulin diacetate

(3) Chemical Shifts of Methylesters

In oleanane or ursane series :

-COOMe at 28 position	<3. 595 ppm
〃 24 or 30 position	3. 595～3. 650 ppm

See Table 44

Table 44 (Shamma *et al.*)

Compound	-OMe shift	Position of grp.	Compound	-OMe shift	Position of grp.
Ursolic acid Me-ester	3.578	28	Echinocystic acid Me-ester	3.595	28
Asiatic acid Me-ester triacetate	3.538	28	11-Keto-α-boswellic acid Me-ester acetate	3.597	24
Ursolic acid Me-ester acetate	3.535	28	Melaleucic acid Me-ester	3.608	28
Arjunolic acid Me-ester triacetate	3.538	28	Morolic acid Me-ester acetate	3.618	28
Oleanolic acid Me-ester acetate	3.555	28	α-Boswellic acid Me-ester acetate	3.640	24
Cochalic acid Me-ester	3.570	28	Glycyrrhetic acid Me-ester	3.645	30
Oleanolic acid Me-ester	3.580	28			

(4) Chemical Shifts of Vinyl Protons

Ursane or oleanane series:

C_{12}-H $\left(^H_{} {\scriptstyle >}={\scriptstyle <} \right)$ at 4.93~5.50 ppm region is not distinct, when conjugate to 11-keto, it appears as a sharp peak at 5.55 ppm (See Fig. 84 (c)).

Lupane series:

C_{29}-H at 4.30~5.87 (2) ppm.

(5) Chemical Shifts of Vinylmethyl Protons

Usual Me protons shift to 0.625~1.500 ppm

$CH_3C=C$ Me protons shift to 1.63~1.80 ppm

Example——Betulin diacetate (Fig. 84, E): 1.67, melaleucic acid, methylester: 1.64,
thurberogenin acetate: 1.80

(6) Chemical Shifts of Acetoxyl Protons

Chemical shifts of acetoxyl protons appear distinctly at 1.82~2.07 (1.92~1.97).

Difference between primary and secondary hydroxyl groups is not distinguishable.

Example——α-Boswellic acid methylester: 1.993, echinocystic acid methylester diacetate: 2.073, 11-keto-α-boswellic acid methylester acetate (Fig. 84, C): 1.958, sojasapogenol-B triacetate (Fig. 84, D): 1.958.

Diacetates of 1,2-glycols show a signal of doubletts at 1.85~1.92 ppm ($J=2$~3 c/s), higher than that of monoacetates at 1.92~1.97 ppm.

Example——Asiatic acid methylester triacetate (Fig. 85, A): 1.938, 1.903, arjunolic and, A_1-barrigenol

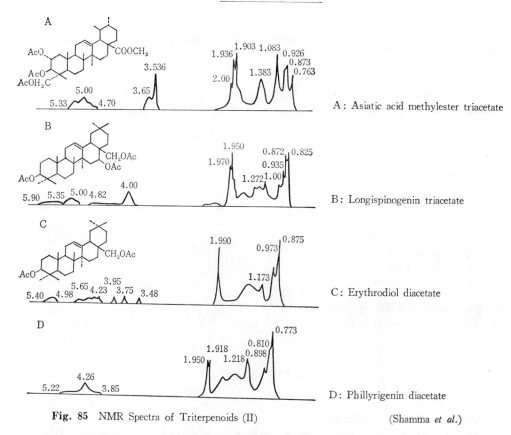

A : Asiatic acid methylester triacetate

B : Longispinogenin triacetate

C : Erythrodiol diacetate

D : Phillyrigenin diacetate

Fig. 85 NMR Spectra of Triterpenoids (II)   (Shamma *et al.*)

(7) Chemical Shifts of α-Proton to secondary Acetoxyl Group

$$\left[\quad \rangle C \langle \begin{matrix} \mathbf{H} \\ OCOCH_3 \end{matrix} \quad \right]$$

axial C_3–H   4.00~4.75 ppm
equatorial C_3–H   5.00~5.48 ppm

When strong steric hindrance exists between *axial*–H and angular–Me, the signal of *axial*–H is shifted to lower region.

Example——Chichipegenin tetraacetate : 5.81 br (25 c/s), longispinogenin triacetate (Fig. 85, B) : 5.62 br (22 c/s), dumortierigenin diacetate (Fig. 84, B) : 5.44 br (*equatorial*–H at 15), 4.45 br (2) (two *axial*–H at 3 and 22), soyasapogenol–B triacetate (Fig. 84, D)>4.75 cps

(8) Chemical shifts of α-Protons to Acetoxyl Groups in acetylated 1, 2-Glycoles (See Table 45)

Table 45   (Shamma *et al.*)

Triterpenoid	Positions of AcO–	Fig. 85	Widths of Absorption	Peak
Arjunolic acid–Me–ester triacetate	2, 3	—	4.70~5.18	4.98
Asiatic acid–Me–ester triacetate	2, 3	A	4.70~5.18	5.00
A₁–Barrigenol pentaacetate	15, 16	—	5.23~5.69	5.54

(9) Chemical Shifts of Methylene Protons of Acetoxymethyl Groups (See Table 46)

Table 46 Shifts of Methylene Protons of Acetoxymethyl Groups (Shamma *et al.*)

Triterpenoid	Position	Fig.	Shift
Arjunolic acid Me–ester triacetate	23	—	3.65 s
Asiatic acid Me–ester triacetate	23	85 A	3.65 s
A_1–Barrigenol pentaacetate	28	—	3.82 q
Uvaol diacetate	28	—	3.88 q
Erythrodiol diacetate	28	85 C	3.85 q
Phillysapogenol hexaacetate	28	—	3.89 q
Longispinogenin triacetate	28	85 B	4.00 s
Betulin diacetate	28	84 E	4.05 q
Sojasapogenol–B triacetate	24	84 D	4.22 d
Phillyrigenin diacetate	(24)	85 D	4.26 s
A_1–Barrigenol pentaacetate	28 [Note]	—	5.08 d
11–Keto–A_1–barrigenol pentaacetate	28	—	5.20 d

(10) Examples of NMR Spectra

See the item [E]

[D] Mass Spectra[401)402)]

(1) Oleanane and Ursane Series

Δ^{12}–Oleananes and Δ^{12}–ursanes produce mass fragments **a** and **b** as the result of reverse Diels–Alder fragmentation as following. The fragment **a** is characteristic for identification.

Fig. 86 shows the mass spectrum of methyl ursolate (XX: $R_1=H_2$, $R_2C=H_3$, $R_3=CO_2$–CH_3 $R_4=H$, $R_5=CH_3$): $a^+=262$, $c^+=(a-R_3)^+=203$

Other fragments are assigned as following.

---

401) Djerassi, C. *et al*: *Tetr. Letters*, **1962**, No.7, 263.
402) Djerassi, C. *et al*: *J.A.C.S.*, **85**, 3688 (1963).
[Note] a) S. Itō, T. Ogino, H. Sugiyama, and M. Kodama: *Tetrahedron Letters* 2289 (1967).
　　　 b) S.G. Errington, D.E. White, and M.W. Fuller: *ibid.*, 1289 (1967).

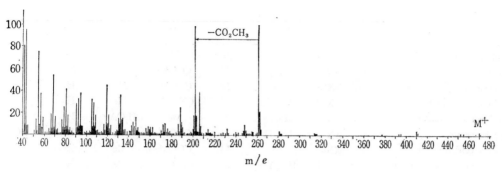

Fig. 86 Mass Spectrum of Methyl Oleanolate (Djerassi *et al.*)[401]

c m/e 203   f m/e 133   d

e   g

h m/e 216   i m/e 291

j m/e 317 R=COOCH₃

Example:

		R_1	R_2	R_3	R_4	R_5	
Me–β–boswellonate	XX–a	O	COOMe	Me	H	Me	
Me–ursolate	XX–b	H_2	Me	COOMe	H	Me	
Me–oleanolate	XX–c	O	Me	COOMe	Me	H	
Erythrodiol diacetate	XX–d	(H)OAc	Me	CH$_2$OAc	Me	H	
3–O–Hydroxy–β–amyrin diacetate	XX–e	(H)OAc	Me	Me	CH$_2$OAc	H	
Me–machaerate	XX–f	(H)OH	Me	COOMe	Me	H	C_{21}=O
Me–dioxoechinocystate	XX–g	O	Me	COOMe	Me	H	C_{16}=O
15–Oxoerythrodiol diacetate	XX–h	H(OAc)	Me	CH$_2$OAc	Me	H	C_{15}=O
Me–glycyrrhetate	XX–i	H(OH)	Me	Me	COOMe	H	C_{11}=O

XX–a   M$^+$ 468, M–Me (453), a (**218**), e,g (205), c (203), d (189), f (133)

XX–b   See Fig. 86

XX–c   M$^+$ 468, M–Me (453), M–COOMe (409), M–COOCH$_2$–H (408), a (**262**), e (249), g (205), c (**203**), d (**189**), f (133)

XX–d   M$^+$ 526, M–HOAc (**466**), M–CH$_2$OAc (453), a (**276**), e (263), g (249), a–HOAc (**216**), c (**203**), d,g–HOAc (**189**)

XX–e   M$^+$ 526, M–HOAc (466), M–CH$_2$OAc (453), a (**276**), g (249), a–HOAc (216), a–CH$_2$OAc (203), g–HOAc (189)

XX–f   M$^+$ 484, M–COOCH$_2$–H (424), a (**276**), a–CH$_2$OH (**244**), c (**217**), a–COOCH$_2$–H (**216**), g (**207**), d (**203**)

XX–g   M$^+$ 482, M–COOCH$_2$–H (**422**), a (**276**), a–COOCH$_3$ (**217**), M–COOCH$_2$–H (**216**), d (203)

XX–h   M$^+$ 540, M–HOAc (480), i (**291**), i–HOAc (**231**)

XX–i   M$^+$ 484, M–COOCH$_3$ (425), j (**317**), a (**276**)

(2) Bauerene Series[402]

(XXI)
Bauerenone

(XXII)
Multiflorenone

(XXIII)
Isomultiflorenone

(XXIV)
Arborenone

Fragments:

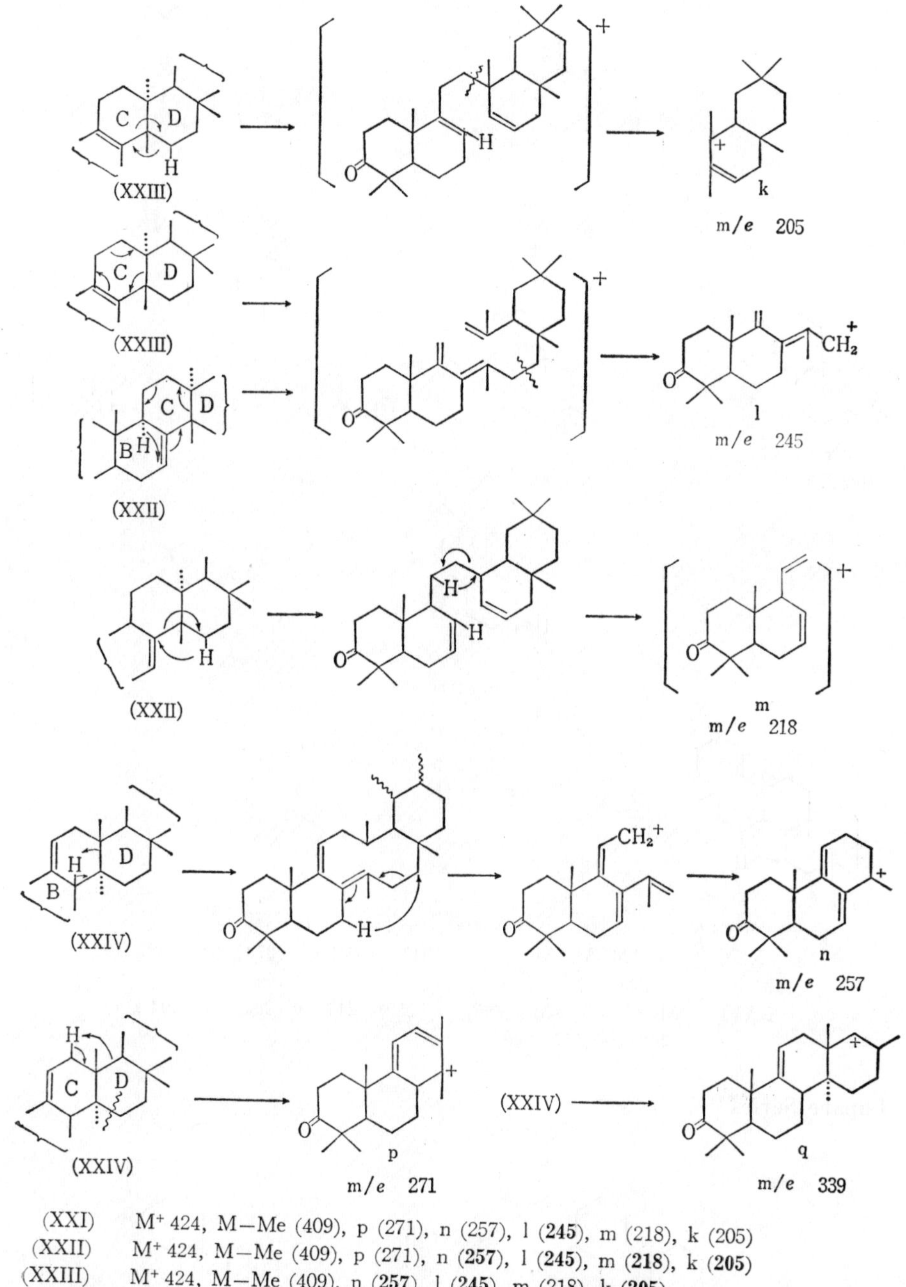

(XXI)   M+ 424, M—Me (409), p (271), n (257), l (**245**), m (218), k (205)
(XXII)   M+ 424, M—Me (409), p (271), n (**257**), l (**245**), m (**218**), k (**205**)
(XXIII)   M+ 424, M—Me (409), n (**257**), l (**245**), m (218), k (**205**)
(XXIV)   M+ 424, M—Me (**409**), q (**339**), p (**271**), n (**257**), l (245)

(3)　Friedelane Series[402)403)]

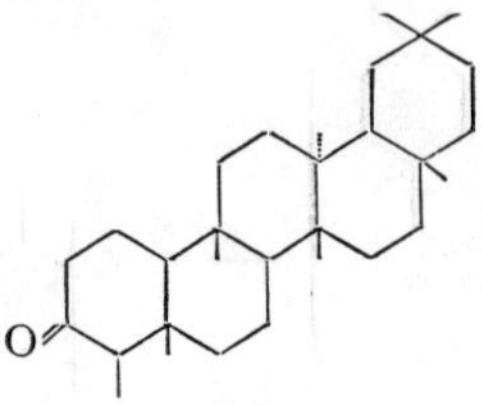

(XXV) 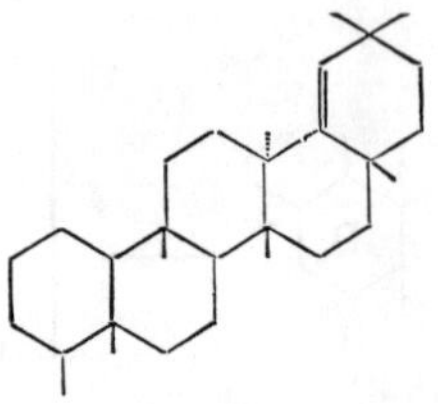　　　　　　　　　　　　(XXVI)

Fragments :

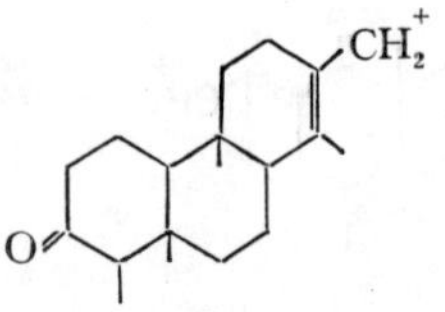

r　　　　　　　　s　　　　　　　t　　　　　　　u
m/e 273　　　　m/e 302　　　　m/e 341　　　　m/e 205

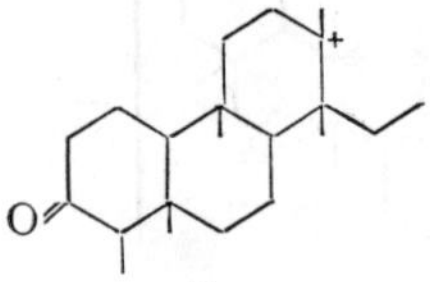

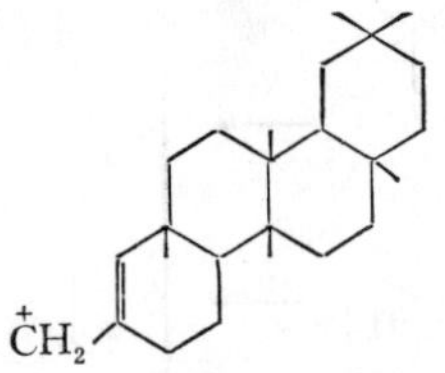

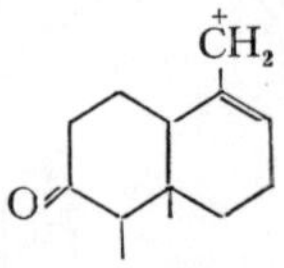

(XXVI) ⟶

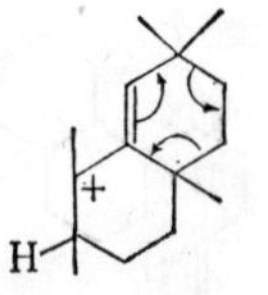

v　　　　　　　w
m/e 205　　　m/e 177

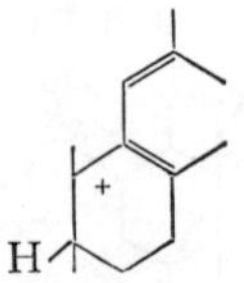

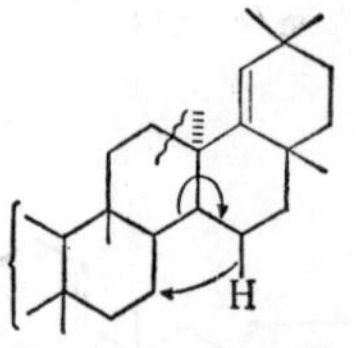

v″　　　　　　v′　　　　　　x
m/e 203　　　m/e 204　　　m/e 189

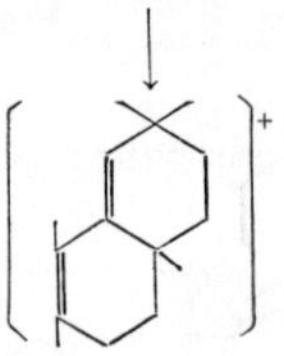

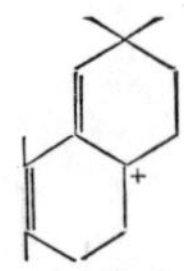

XXV　　M+ **426**, M−Me (411), t (341), s (302), r (**273**), 246, 232, 218, y (**205**) etc.

XXVI　　M+ 410, M−Me (**395**), r (259), **217**, v (**205**), v′ (**204**), v″ (**203**), w (**177**), etc.

(4)　Lupane Series[402)]

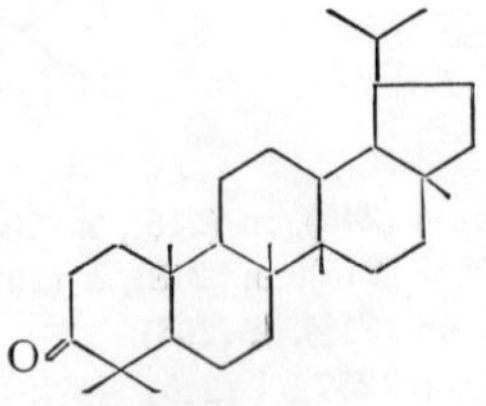

(XXVII)　　　　　　　　　　(XXVIII)
Lupane-3-one

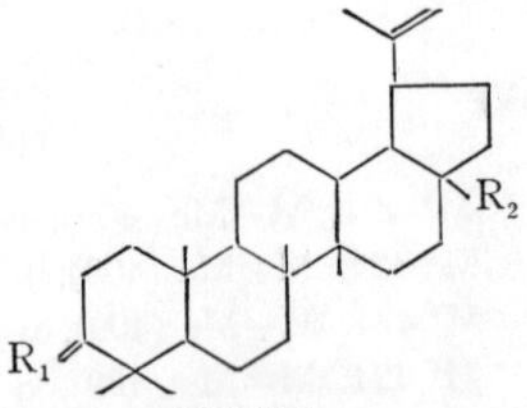

a : R_1=O, R_2=Me Lupene-3-one
b : R_1=(H)OH, R_2=CO_2Me Me-betulinate

403)　Courtney, J. L. *et al* : *Tetr. Letters*, **1963**, No. 1, 13.

Fragments :

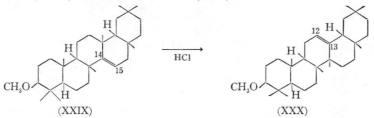

y m/e $R_1=O$ 205
      $R_1=(H)OH$ 207

z m/e 191

aa m/e 218

bb m/e $R_1=O$. 218
       $R_1=H(OH)$ 220

(XXVII) M^+ 426, M–Me (**411**), M–isoPr. (**383**), y (**205**), z (**191**)

(XXVIII–a) M^+ 424, M–Me (**409**), M–43 (381), aa, bb (**218**), y (**205**), 189. etc.

(XXVIII–b) M^+ 470, M–Me (**455**), M–H_2O (**452**), M–Me–H_2O (**437**), M–$COOCH_3$
         (411), aa (**262**), z (233), bb (220), y (**207**), y–H_2O (**189**). etc.

〔E〕 Oleanane-series Triterpenoids

(1) Sawamilletin (XXIX) and Isosawamilletin (XXX)[402]

$\xrightarrow{\text{HCl}}$

(XXIX)   (XXX)

《Occurrence》 The seed oil of *Echinochloa crusgallis* L. (ヒエ),   mp. 278°, $[\alpha]_D$ +8.2°, $C_{31}H_{52}O$

Taraxerol $\xrightarrow{\text{methylation}}$ Sawamilletin

IR See Fig. 87

(2) Miliacin (XXXI)[405]

$\xrightarrow{\text{HCl}}$ $\xleftarrow{\text{HCl}}$ Isosawamilletin

(XXX)

(XXXI)   3β–Methoxyolean–13(18)–ene

《Occurrence》 The seeds of *Panicum miliaceum* L. (キビ),   mp. 282°, $[\alpha]_D$ +8.0°, $C_{31}H_{52}O$

IR See Fig. 87.

404) Obara, T. *et al* : *Bull. Agr. Chem. Soc. Japan*, **21**, 388 (1957) ; *J. Chem. Soc. Japan*, **80**, 677, 1487, 1491 (1959).

405) Abe, S. *et al* : *Bull. Chem. Soc. Japan*, **33**, 271 (1960) ; *J. Chem. Soc. Japan*, **82**, 1051, 1057 (1961).

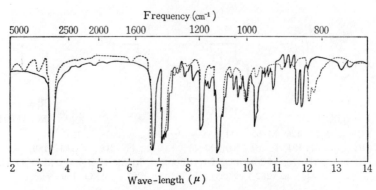

Fig. 87 IR Spectra of Miliacin (——) and Sawamilletin (······)
(in nujol)　　　　　　　　　(Abe *et al.*)

Relationship between the position of double bond in the molecule of oleanenes and the optical rotation is
indicated in Table 47.

Table 47　　　　　　　　　　　　　　　　(Abe)

Triterpenoid	Δ	$[\alpha]_D$	$[M]_D$	$\Delta[M]_D$
Germanicol	18(19)	+7.0	+29.9	−357
β-Amyradienol	11(12) ; 13(18)	−77.0	−327.0	
Milliacin	18(19)	+8.0	+34.1	−302
Dehydromilliacin	11(12) ; 13(18)	−63.0	−267.5	
Methylmorate acetate	18(19)	+37.0	+189.7	−843
Methyldehydrooleanolate acetate	11(12) ; 13(18)	−128.0	−653.7	
Methylmorate acetate	18(19)	+37.0	+189.7	+878
Methylisodehydrooleanolate acetate	12(13) ; 18(19)	+209.0	+1067	

(3) Oleanolic aldehyde and Erythrodiol[406]

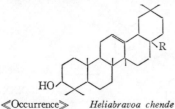

R=CHO　　(XXXII) Oleanolic aldehyde
R=CH₂OH　(XXXIII) Erythrodiol

≪Occurrence≫　　*Heliabravoa chende*

(a) Oleanolic aldehyde (XXXII)
mp. 230~231°, $[\alpha]_D$ +75°, $C_{30}H_{50}O_2$
Acetate oxime mp. 189~200°

　IR　λ_{max}^{KBr} μ: 2.93 (OH), 5.74 (ester), 8.07 (CH_3COO), 6.1 (C=C), 12.11, 12.24, 12.46 ($\Delta_{12\sim13}$)

(b) Erythrodiol　(XXXIII) mp. 208~217°
NMR See the item [C], (2), Table 43, (9) Table 46.
Mass See the item [D], (1) XX–d.

406) Shamma, M. *et al*: *J. Org. Chem.*, **24**, 726 (1959).

(4) "Triterpene A" (XXXIV)[408]

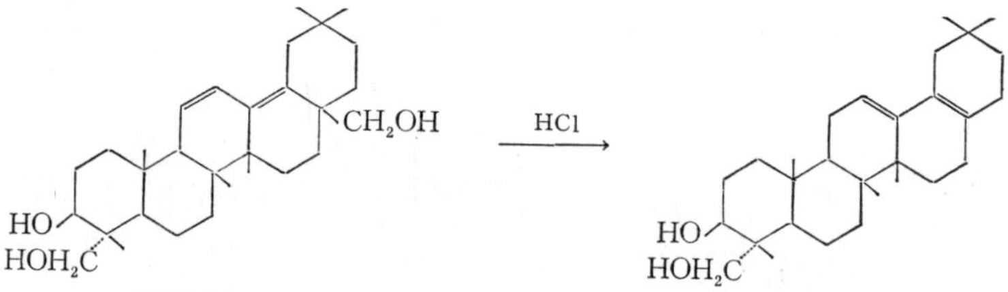

(XXXIV)
"Triterpene A"

(XXXV)
"Triterpene B"

≪Occurrence≫ *Scrophularia smithii* WYDLER

mp. 295~299°, $[\alpha]_D$ +65.4 (pyridine), $C_{30}H_{48}O_3$

Triacetate mp. 180~181°, $[\alpha]_D$ −85°, UV λ_{max}^{EtOH} mμ (log ε): 244 (4.47), 252 (4.5), 262 (4.33)

(5) Sophoradiol (XXXVI)[408]

≪Occurrence≫ *Sophora japonica* L. (エンジュ)

Sophoraside $C_{42}H_{68}O_{13}\cdot H_2O$ mp. 246~248° (decomp.)

⟶ Sophoradiol $C_{30}H_{50}O_2$ mp. 224~226°, $[\alpha]_D$ +88.5°

IR λ_{max}^{nujol} μ: 2.80 (OH), 6.06 (⊿), 12.34 (⟩C=C⟨$_H$)

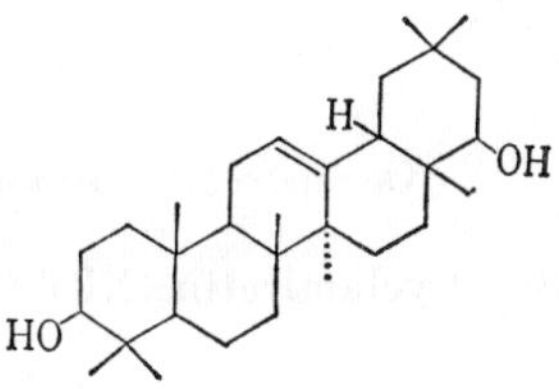

(6) Aegiceradiol (XXXVII)[409]

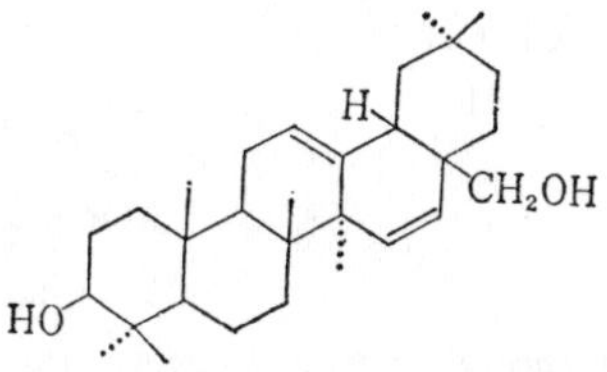

(XXXVI)

(XXXVII)

≪Occurrence≫ *Aegiceras majus* GAERTN., mp. 236~238°, $[\alpha]_D$ +40.3°, $C_{30}H_{48}O_2$

(7) 7-β-Hydroxy-A₁-barrigenol and A₁-Barrigenol[381]

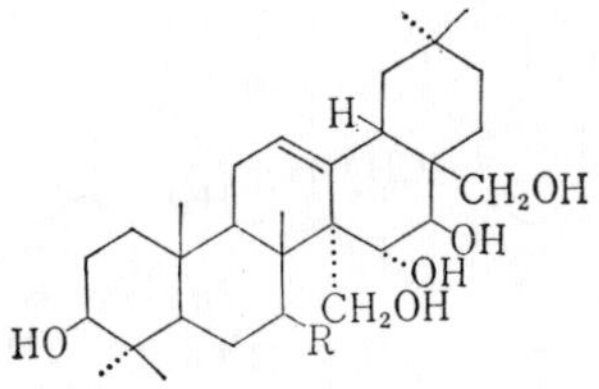

(XXXVIII)

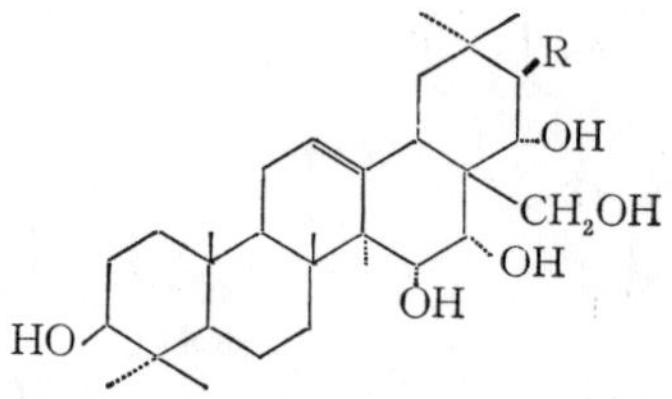

Revised structure

R=OH: 7β-Hydroxy-A₁-barrigenol
R=H: A₁-Barrigenol [Ref. p 146. Table 46]

≪Occurrence≫ *Pittosporum undulatum* VENT.

(XXXVIII) R=OH IR See the item [B], (2), Table 35, NMR See the item [C], (2) Table 43, (8) Table 45, (9) Table 46.

407) King, F. E. *et al*: *Soc.*, **1960**, 4738.
408) Ishimasa *et al*: *Y. Z.*, **78**, 1090 (1958), **80**, 304, 698 (1960).

(8) **Aralosides A** and **B** (XXXIX)[410)411)]

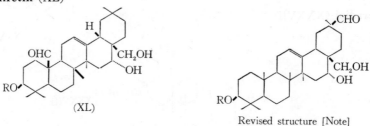

(XXXIX)

R= ;R′=H Araloside A

R= R′= Araloside B

《Occurrence》 *Aralia manschurica*, Araloside A mp.195~196°, [α]$_D$ −26.7° (MeOH), $C_{47}H_{74}O_{18}$

(9) **Cyclamiretin** (XL)[412)]

(XL)

Revised structure [Note]

《Occurrence》 *Cyclamen europaea*, mp. 239~240°, [α]$_D$ +48°, $C_{30}H_{48}O_4$

IR ν$_{max}$ cm^{-1}: 3460, 3378 (OH), 1712 (CHO)

(10) **Triterpenoids of *Terminalia* spp.**[413)]

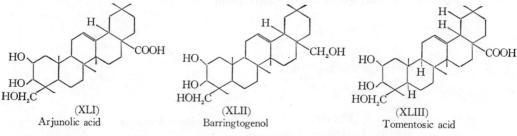

(XLI)
Arjunolic acid

(XLII)
Barringtogenol

(XLIII)
Tomentosic acid

409) Bose, P.K. *et al*: *Tetrahedron*, **18**, 461 (1962).
410) Kochetkov, N.K. *et al*: *Tetr. Letters*, **1962**, (16) 713.
411) Kochetkov, N.K. *et a*: *Izvest. Akad. Nauk. U.S.S.R. Khim.*, **1963**, (8), 1398; *Index. Chem.*, **11**, 33706, 33707 (1963).
412) Barton, D.H.R. *et al*: *Soc.*, **1962**, 5176.
413) Ramachandra Row, L. *et al*: *Tetrahedron*, **18**, 827 (1962); *Tetr. Letters* **1962**, No.4, 129.
[Note] R. Tschesche, H. Striegler and H. Fehlhaber: *Ann.*, **691**, 65 (1966).

≪Occurrence≫　　*Terminalia tomentosa*

(a) **Arjunolic acid** (XLI) ((IV) Δ^{12}, 2α, 3β, 23-tri OH, 28–COOH) mp. 332~336°, $[\alpha]_D$ +63° (EtOH), $C_{30}H_{48}O_5$

　　IR　See the item [B], (2), Table 35.

　　NMR　See the item [C], (2), Table 43, (3), Table 44, (8), Table 45, (9), Table 46

(b) **Barringtogenol** (XLII) ((IV) Δ^{12}, 2α, 3β, 23-tri OH, 28–CH$_2$OH) mp. 290~291°, $[\alpha]_D$ +20° (EtOH), $C_{30}H_{50}O_4$

(c) **Tomentosic acid** (XLIII) ((IV) Δ^{12}, 2α, 3β, 19, 23-*tetra.* OH, 28–COOH) mp. 328~330°, $[\alpha]_D$ +64° (EtOH), $C_{30}H_{48}O_6$

　　IR　See the item [B], (2), Table 35.

(11)　Bassic acid (Bassiaic acid) (XLIV)[414]

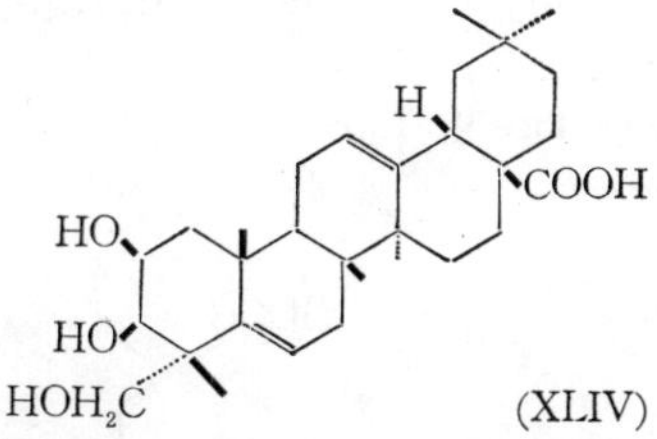

(XLIV)

≪Occurrence≫　　*Sapotaceae* spp.,　　$C_{30}H_{46}O_5$ Methylester mp. 180~190°, $[\alpha]_D$ +58°, $C_{31}H_{48}O_5 \cdot 1/2 H_2O$

IR　See the item [B], (2), Table 35.

(12)　Triterpenoids of *Albizzia* spp.

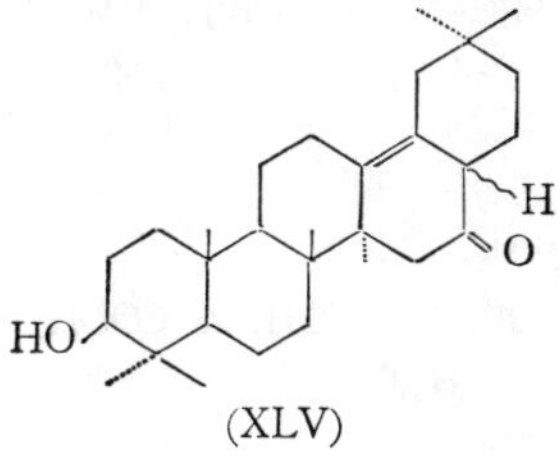

(XLV)
Albigenin

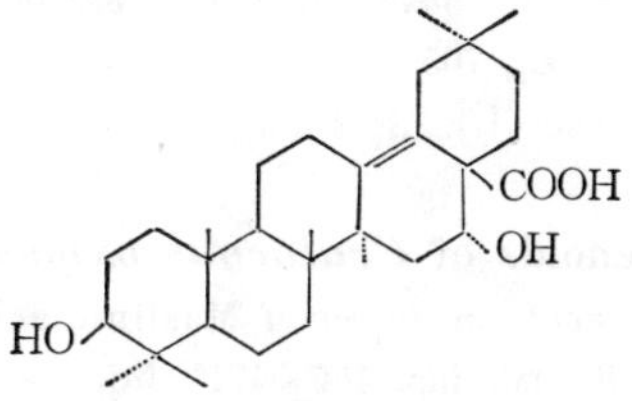

(XLVI)
Albigenic acid

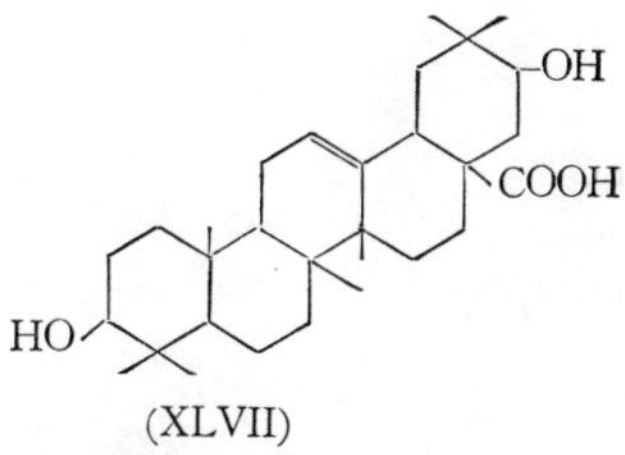

(XLVII)
Machaerinic acid

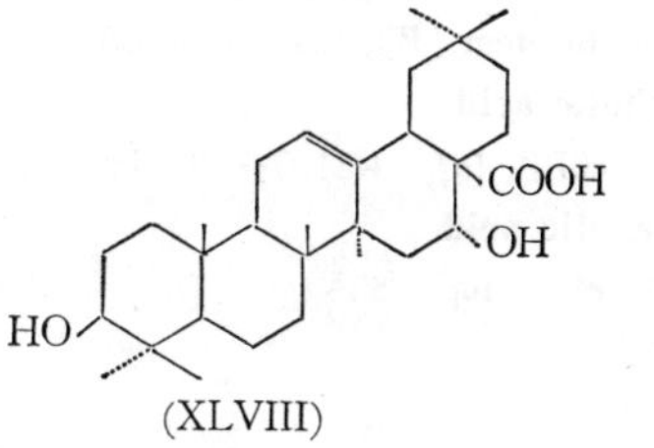

(XLVIII)
Echinocystic acid

(a) **Albigenin** (XLV)[415]

　　≪Occurrence≫　　*A. lebbeck*,　　mp. 226~228°, $[\alpha]_D$ −114°, $C_{29}H_{46}O_2$

　　IR　ν_{max} cm^{-1}: 3655 (OH), 1695 (C=O), 1365, 1385 (gem di Me)

---

414) King, T. J. *et al*: *Proc. Chem. Soc.*, **1959**, 393; *Soc*, **1961**, 4308.

415) Barua, A. K. *et al*: *Tetrahedron*, **18**, 155 (1962).

(b) Albigenic acid (XLVI)[415) 416)]

《Occurrence》   *A. lebbeck*, mp. 246~248°, $[\alpha]_D$ −13° (EtOH), $C_{30}H_{48}O_4$

(c) Machaerinic acid (XLVII)[417)]

《Occurrence》   *A. procea*[418)],   mp. 266~270°, $C_{30}H_{48}O_4$

IR ν_{max}^{nujol} cm^{-1}: 3400, 3240, 1741, 1689

Acetate mp. 260~265°,

IR ν_{mxa}^{nujol} cm^{-1}: 1723, 1689, 1257

Mass See the item [D], (1)

(d) Echinocystic acid (XLVIII)[417)]

《Occurrence》   *A. lebbeck*[419)], *A. anthelmintica*, *Elvira biflora*[419a)],
                mp. 319~324°, $[\alpha]_D$ +37.7° (EtOH), $C_{30}H_{48}O_4$

IR See the item [B], (2), Table 35
NMR See the item [C], (2), Table 43, (3), Table 44.

(13) Katonic acid (XLIX)[407)]

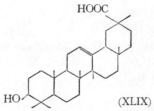

HOOC

HO

(XLIX)

《Occurrence》   *Sandoricum indicum* (*Meliaceae*),   mp. 285~287°, $[\alpha]_D$ +47°, $C_{30}H_{48}O_3 \cdot CH_3OH$
UV ε_{210} 3740, ε_{220} 705
IR See the item [B], (2) Table 35.

(14) Triterpenoids of *Crataegus oxyacanthus* L.[419b)]

(a) Crataegolic acid, an isomer of **Maslinic acid** (L) ((IV) Δ^{12}, 2, 3–diOH, 28–COOH)

Methylester diacetate mp. 166~171°, $[\alpha]_D$ +40.4° (CHCl$_3$), $C_{35}H_{40}O_6$

Dehydro——methylester diacetate mp. 184~187°

UV See the item [A], (3) Table 34.
IR See the item [B], (2), Table 35.

(b) Acantholic acid

mp. 274~279°, $[\alpha]_D$ +24.6±4° (MeOH), $C_{30}H_{48}O_5 \cdot 1/2$ H$_2$O

(c) Neotaegolic acid

Methylester   mp. 285~288°, $[\alpha]_D$ +63.8±3° (CHCl$_3$), $C_{31}H_{50}O_4$

416) Barua, A.K. *et al*: *Tetrahedron*, **7**, 19 (1959).   417) Varshney, I.P. *et al*: *J.P.S.*, **50**, 923 (1961).
418) Varshney, I.P. *et al*: *Arch. Pharm.*, **292**, 57 (1959).
419) Varshney, I.P. *et al*: *Bull. Soc. Chim. France*, **1957**, 1440.
419a) de Oliveira, M.M. *et al*: *J.P.S.*, **50**, 780 (1961).

(15) Maslinic acid (L)[419b]

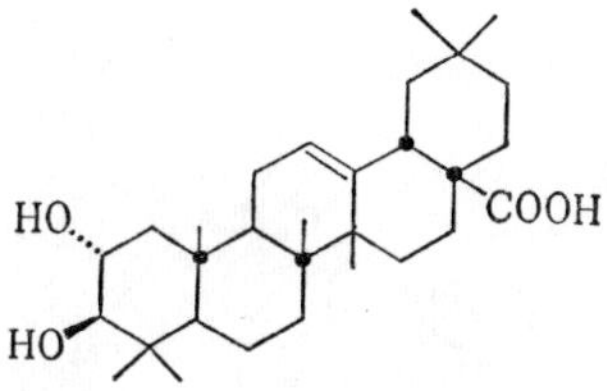

《Occurrence》 *Olea europaea*
Acetate   mp. 235~237°, $[\alpha]_D$ +31° (CHCl$_3$), $C_{34}H_{52}O_6$
IR ν_{max} cm^{-1}: 1735, 1690, 1656

(16) Treleasegenic acid (LI)[420]

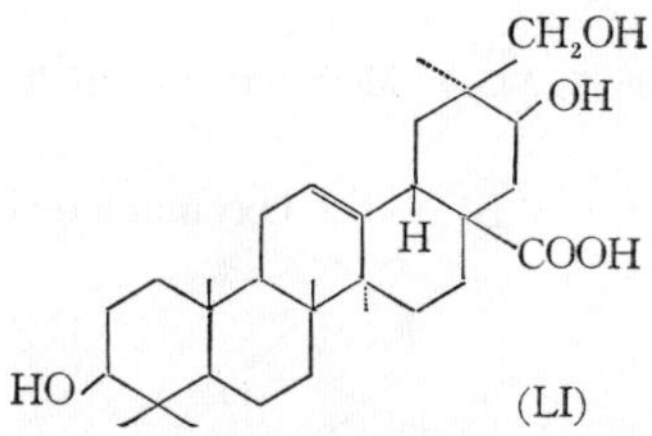

(LI)

《Occurrence》   *Cactus* spp.   mp. 270~280°, $C_{30}H_{48}O_5$
IR λ_{max}^{nujol} μ: 2.85, 3.06, 5.83

(17) Commic acid C (LII-a) and D (LII-b)[421]

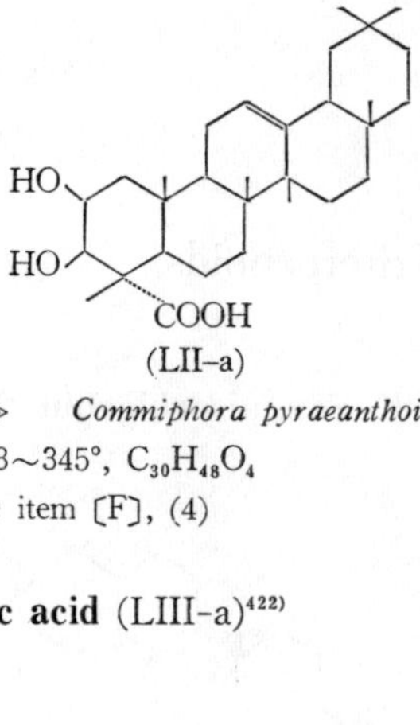

(LII-a) 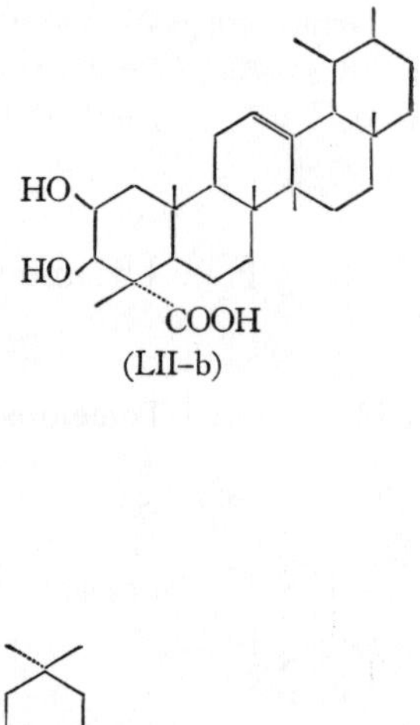   (LII-b)

《Occurrence》   *Commiphora pyraeanthoides* ENGL.
(LII-a) mp. 343~345°, $C_{30}H_{48}O_4$
(LII-b) See the item [F], (4)

(18) Polygalacic acid (LIII-a)[422]

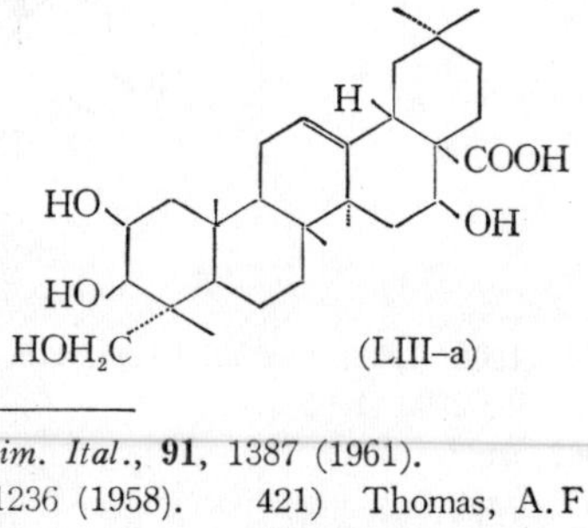

(LIII-a)

419b) Caglioti, L. *et al*: *Gazz. Chim. Ital.*, **91**, 1387 (1961).
420) Djerassi, C: *J.A.C.S.*, **80**, 1236 (1958).   421) Thomas, A.F: *Tetrahedron*, **15**, 212 (1961).
422) Rondest, J. *et al*: *Bull. Soc. chim. France*, **1963**, 1253; *Index Chem.* **10**, 32423 (1963).

≪Occurrence≫   *Polygala paenea* L..,   mp. 305°, $C_{30}H_{48}O_6$

(19) Senegin (LIII-b)[423]

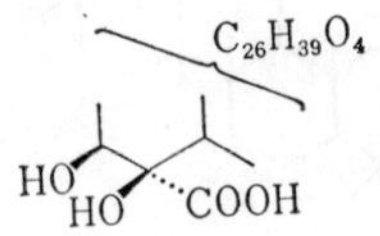

(LIII-b)

≪Occurrence≫   *Polygala senega* $C_{30}H_{44}O_8$

UV λ_{max}^{EtOH} mμ(log ε): 210 (3.7)

IR λ_{max} μ: 5.6 (5 membered lactone)

Dimethylester diacetate $C_{36}H_{52}O_{10}$

NMR cps: 210, 214 (CH_3CO), 146.5, 148.7 (Me–ester), 64~91 (vinylic unsatur)

(20) Machaerinic lactone (Substance B) and 2-Oxymachaerinic lactone (Substance X)[424]

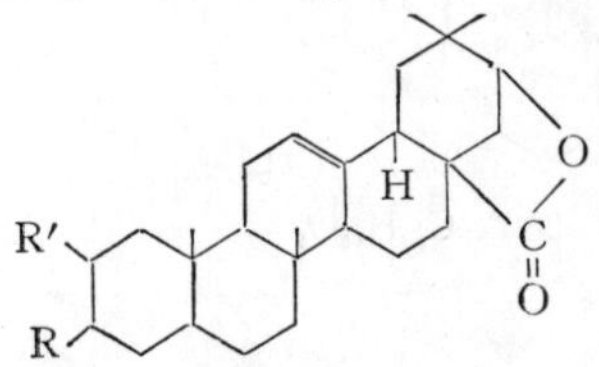

(LIV)

R=OH, R′=H   Substance B
R=R′=OH   Substance X

≪Occurrence≫   *Stryphnodendron coriaceum*

Substance B   mp. 240~243°, $[\alpha]_D$ −16°, $C_{30}H_{46}O_3$
Substance X   mp. 265~267°, $[\alpha]_D$ +5°, $C_{30}H_{46}O_4$

[F] Ursane-series Triterpenoids

(1) Vanguerolic acid (LV) and Tomentosolic acid[425] (Sanguisorbigenin)[425a, b] (LVI)

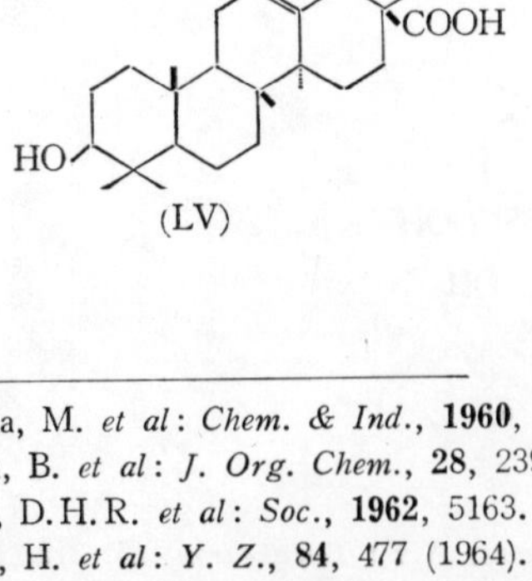

(LV)

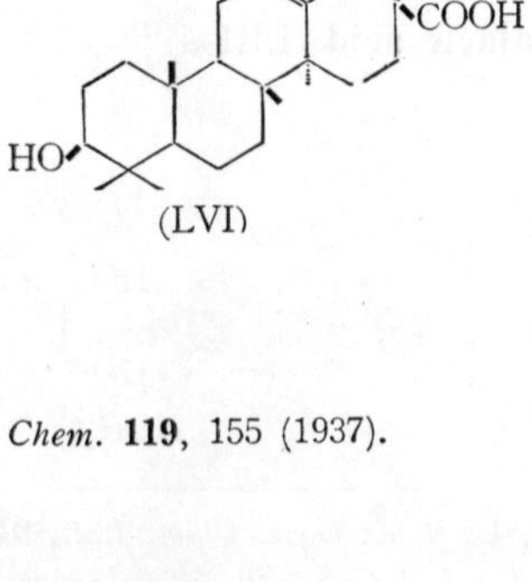

(LVI)

423) Shamma, M. *et al*: *Chem. & Ind.*, **1960**, 1272, *J. Biol. Chem.* **119**, 155 (1937).
424) Tursch, B. *et al*: *J. Org. Chem.*, **28**, 2390 (1963).
425) Barton, D.H.R. *et al*: *Soc.*, **1962**, 5163.
425a) Wada, H. *et al*: *Y. Z.*, **84**, 477 (1964).
425b) Takemoto, T. *et al*: *ibid.*, **84**, 367 (1964).

≪Occurrence≫ *Vangueria tomentosa*

(a) **Vanguerolic acid** (LV)

mp. 273°, $[\alpha]_D$ +306° (CHCl₃), $C_{30}H_{46}O_3$

UV λ_{max}^{EtOH} mμ(ε): 227~228 (7600)

IR ν_{max}^{nujol} cm⁻¹: 3375 (OH), 1690 (COOH)

Methyl vanguerolate mp. 161~163°, $C_{31}H_{48}O_3$

UV λ_{max}^{EtOH} mμ (ε): 225 (7900)

IR ν_{max}^{nujol} cm⁻¹: 3510 (OH), 1700 (Me–ester), 1660 (C=C)

Methyl vanguerolate acetate mp. 188~191°

UV λ_{max}^{EtOH} mμ (ε): 225 (7800)

IR ν_{max}^{nujol} cm⁻¹: 1720 (OAc and Me–ester) 1650 (C=C)

NMR TMS, τ: (in CDCl₃) 4.60 (one >C=C<^H) 7.13 (one diallylic H), (in CCl₄) 8.52, 8.36 (one >C=C<^{Me})

(b) **Tomentosolic acid** (LVI) (Sanguisorb genin)[425a) b)]

≪Occurrence≫ *Vangueria tomentosa, Sanguisorba officinalis* L. (ワレモコウ)[425a) b)]

mp. 284.5~286°, $[\alpha]_D$ +18° (CHCl₃), $C_{30}H_{46}O_3$

UV λ_{max}^{EtOH} mμ (ε): 210 (8700)

IR ν_{max}^{nujol} cm⁻¹: 3415 (OH), 1695 (COOH)

Methyltomentosolate acetate mp. 234~236° or 246~248°, $[\alpha]_D$ +9° (CHCl₃)

UV λ_{max}^{EtOH} mμ (ε): 215 (8000)

IR ν_{max}^{CCl4} cm⁻¹: 1725 (OAc and Me ester), 1655 (C=C)

NMR TMS, τ: ppm (CDCl₃) 0.82, 0.85, 0.93, 0.95 (C–CH₃), 1.55, 1.63 (C=C–CH₃)

(2) **Neoilexonol** (LVII)[426)]

≪Occurrence≫ *Ilex goshiensis* HAYATA, *I. buergei* MIQUEL.

mp. 205~206°, $[\alpha]_D$ +99.26° (CHCl₃), $C_{30}H_{48}O_2$

IR See Fig. 88.

---

426) Yagishita, K. *et al*: *Agr. Biol. Chem.*, (*Tokyo*) **25**, 844 (1961).

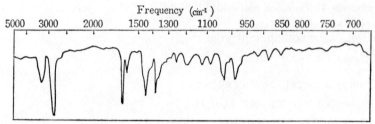

Frequency (cm⁻¹)

Fig. 88 IR Spectrum of Neoilexonol (in nujol)   (Yagishita *et al.*)

(3) **Urs-20-ene-3-ol-30-al-acetate** (LVIII)[427]

(LVIII)

≪Occurrence≫   *Stemmadenia donell-smithii* WOODSON

(4) **Commic acid D** (LII-b) and **E** (LIX)[428]

(LII–b)   (LIX)

≪Occurrence≫   *Commiphora pyracanthoides*

Comic acid E (LIX)   mp. 328~330° (decomp.), [α]$_D$ +104° (pyridine)
Triacetate mp. 329~331°, (decomp.)

IR $λ^{CH_2Cl_2}_{max}$ μ: 2.87, 5.70, 5.85

(5) **Asiatic acid** (Dammarolic acid) (LX)[428)429]

(LX)

≪Occurrence≫   Dammar Resin [Note]

Methylester   mp. 223~225°, [α]$_D$ +50° (CHCl$_3$), $C_{31}H_{50}O_5$·1/2 MeOH

427) Estrada, H. *et al*: *Biol. Inst. Quim. Univ. Mexico* **14**, 19 (1962); *Index Chem.*, **9** 28339 (1963).
428) Arigoni, D. *et al*: *Soc.*, **1960**, 1900.   429) Halsall, T. G. *et al*: *ibid.*, **1961**, 646.
[Note] Dammarenolic acid and nyctanthic acid are described in the item (11) Diterpenoids.

UV λ_{max} mμ (ε): 222 (600), 215 (1730), 210 (3080), 205 (4050)

IR See the item [B], (2), Table 36.

NMR See the item [C], (2), Table 43, (3), Table 44, (6) Fig. 85, (8), Table 45, (9), Table 46.

(6) Ifflaionic acid (LX-a)[430]

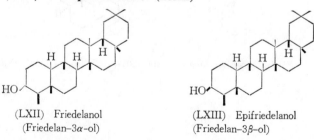

(LX-a)

≪Occurrence≫ *Flindersia* spp., $C_{30}H_{46}O_3$

[G] Friedelane, Bauerane-series Triterpenoids

(1) Friedelin (LXI)

(LXI)

≪Occurrence≫ *Ceratopetalum apetalum*[431]~[433], the leaves of *Euonymus japonica*(マサキ)[434], *Dendropanax trifidus*[435], *Aster tataricus* L.(シオン)[436]

mp. 257~265°, [α]$_D$ −22~27.8° (CHCl$_3$), $C_{30}H_{50}O$.

IR See the item [B], (2), Table 39.   NMR See the item [C], (2) Table 43.

Mass See the item [E], (3).

(2) Friedelanol (LXII) and Epifriedelanol (LXIII)

(LXII) Friedelanol
(Friedelan–3α–ol)

(LXIII) Epifriedelanol
(Friedelan–3β–ol)

430) Taylor, W.C. *et al*: *Australian J. Chem.*, **16**, 491 (1963); *Index Chem.*, **10**, 30805 (1963).

431) Jefferies, P.R. *et al*: *Soc.*, **1954**, 473.   432) Corey, E.J. *et al*: *J.A.C.S.*, **77**, 3667 (1955).

433) Ruzicka, L. *et al*: *Helv. Chim. Acta*, **38**, 1268 (1955).

434) Mazaki, T. *et al*: *Y.Z.*, **77**, 1357, 1354 (1957), **79**, 980 (1959).

435) Kimura, K *et al*: *4 th Symposium on the Chemistry of natural Products* p. 150. (Tokyo, 1960).

436) Takahashi, M. *et al*: *Y.Z.*, **79**, 1281 (1959).

(a) Friedelanol (LXII)

≪Occurrence≫ *Euonymus alata* Sieb. (ニシキギ)[434] *E. alata* Sieb. *f. Striata* Makino
(コマユミ)[434], *E. radicans* Sieb[435],

mp. 309~310°, $C_{30}H_{52}O$, acetate mp. 317~319°

IR See the item [B], (3), Table 39.

(b) Epifriedelanol (LXIII)

≪Occurrence≫ *Euonymus alata* Sieb.[434], *E. alata* Sieb., *f. Striata* Makino[434], *E. japonica* (マサ
キ)[435] *Aster tataricus* L. (シオン)[436], *Ceratopetalum apetalum*

mp. 282~284°, $[\alpha]_D$ +13.6° $(CHCl_3)$, $C_{30}H_{52}O$

IR See the item [B], (3), Table 39.

(3) Multiflorenol (LXIV) and **Bauerenol** (LXV)[437]

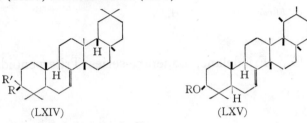

(LXIV)　　　　　　　　　　　(LXV)

≪Occurrence≫ *Gelonium multiflorum* A. Juss.

(a) Multiflorenol (LXIV)

mp. 188~190°, $[\alpha]_D$ −28° $(CHCl_3)$, $C_{30}H_{50}O$,

UV $\lambda_{max} m\mu$ (ε): 207 (4900)

IR See the item [B], (2), Table 39.

(b) Bauerenol (LXV)

mp. 206~207°, $[\alpha]_D$ −25° $(CHCl_3)$, $C_{30}H_{50}O$,

IR See the item [B], (2) Table 39.

(4) Epoxyglutinane (LXVI)[438]

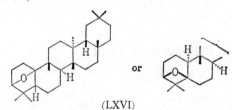

or

(LXVI)

≪Occurrence≫ *Rhododendron westlandii*,　　mp. 201~202°, $[\alpha]_D$ +74.8°, $C_{30}H_{50}O$

---

437) Sengupta, P. *et al*: *Tetrahedron*, **19**, 123 (1963).
438) Arthur., H.R. *et al*: *Soc.*, **1961**, 551.

(5) Alnusenone (LXVII)[439)440)]

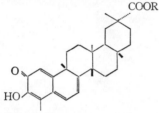

(LXVII)

《Occurrence》 *Alnus glutinosa*, mp. 247°, [α]$_D$ +31°, $C_{30}H_{48}O$

IR See the item 〔B〕, (2), Table 39.

(6) Pristimerin (LXVIII) and **Celastrol** (LXIX)

COOR

$R=CH_3$ Pristimerin (LXVIII)
$R=H$ Celastrol (LXIX)

(a) Pristimerin[441)] (LXVIII)

《Occurrence》 *Pristimeria indica, P. grahmi*[440)] *Maytenus dispermus*[440)]

mp. 219~220° (orange needles),

UV λ_{max}^{EtOH} mμ (log ε): 420~425 (4.10), λ_{min}^{EtOH} mμ (log ε); 291~293 (3.13), See Fig. 89

IR ν_{max}^{nujol} cm^{-1}: 3320, 1724s, 1654, 1586s, 1547, 1517, 1300s, 1243, 1218, 1204, 1185, 1152, 1140, 1082s, 1030, 867, 860, 848, 770

NMR[442)] 60 Mc, TMS, τ: 9.48, 8.90, 8.85, 8.75, 8.55 (5-$\overset{|}{\underset{|}{C}}$–Me), 7.85

(–$\overset{|}{C}$=$\overset{|}{C}$–Me), 6.53 (COOMe), 3.87d (J=6.8), 3.29d. (J=6.8), 3.70s

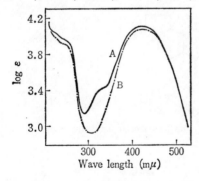

Fig. 89 UV Spectra of
―― Pristimerin
―·― Celastrol
(in EtOH) (Johnson *et al.*)

---

439) Kimura, K. *et al*: *4 th Symposium on the Chemistry of Natural Products.* p.150, (Tokyo, 1960); Spring, F.S. *et al*: *Soc.*, **1955**, 2616, *Tetrahedron*, **2**, 246 (1958); *Chem. & Ind.*, **1956**, 1054.
440) Bhatnager, S.S. *et al*: *J. Sci. Ind. Res.* (*India*) **10B**, 56 (1951).

(b) Celastol (Tripterine)[441]~[446]

≪Occurrence≫   *Celastrus scandens* L.,[443] *C. strigillosus* NAKAE (オニツルウメモドキ)[442], *Tripte-rigium Wilfordii* HOOKF.[445], *T. regellii* SPRAGUE et TAKEDA (クロツル)[442]

UV See Fig. 89.

(7) Arborinol (LXIX-a) and **Isoarborinol** (LXIX-b)[447]

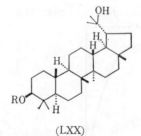

R=⟨$^{OH}_{H}$ Arborinol (LXIX–a)

R=⟨$^{OH}_{OH}$ Isoarborinol (LXIX–b)

≪Occurrence≫   *Glycosmis arborea*

(a) Arborinol (LXIX–a)

mp. 274~274.5°, [α]$_D$ +34.2°, $C_{30}H_{50}O$

Acetate $C_{32}H_{52}O_2$, mp 242~243°

Mass m/e 468, 453, 408, 393, 315, 301, 241

(b) Isoarborinol (LXIX–b)

Acetate $C_{32}H_{52}O_2$, mp. 287~288°

Arborenone (XXIV) (LXIX: R=O) Mass see the item [D], (2)

[H] Lupane-series Triterpenoids

(1) Lupane-3β-20-diol (LXX)

OH

H

H

H

RO

H

(LXX)

≪Occurrence≫   *Melodinus monogynus* ROXB.[448], *Maciura pomifera* (Osage Orange)[449]

mp 232°, [α]$_D$ +24°

441) Johnson, A. W. *et al*: *Soc.*, **1957**, 4079, 4669, **1960**, 549, **1963**, 2884.

442) Nakanishi, K. *et al*: *J. A. C. S.*, **77**, 3169, 6729; *Tetr. Letters*, **1962**, 603; *6 th Symposium on the Chemistry of Natural Products* p. 99 (Sapporo, 1962).

443) Gisvold, O: *J. A. P. A.*, **28**, 440 (1939).

444) Chou, T. Q: *Chinese J. Physiol.*, **10**, 529 (1936).

445) Kulkarni, A. B. *et al*: *Nature*, **173**, 1231 (1954).

446) Haller, H. L. *et al*: *J. A. C. S.*, **64**, 182 (1942).   447) Vorbrüggen, H. *et al*: *Ann.*, **668**, 57 (1963).

448) Chatterjee, A. *et al*: *J. Sci. Ind. Res. India*, **13B**, 546 (1954).

449) Lewis, K. G: *Soc.*, **1961**, 2330.

(2) 23-Hydroxybetulin (LXX-a)[450]

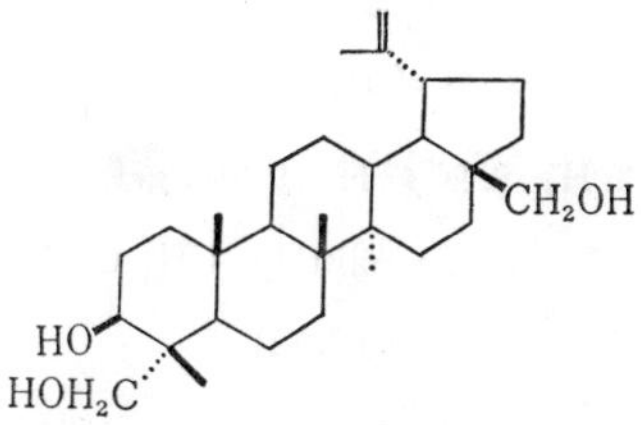

(LXX-a)

≪Occurrence≫   *Sorbus aucuparia* L.,   mp. 259~260°, $[\alpha]_D$ +24.6°, $C_{30}H_{50}O_3$

(3) Ceanothic acid (LXX-b)[451]

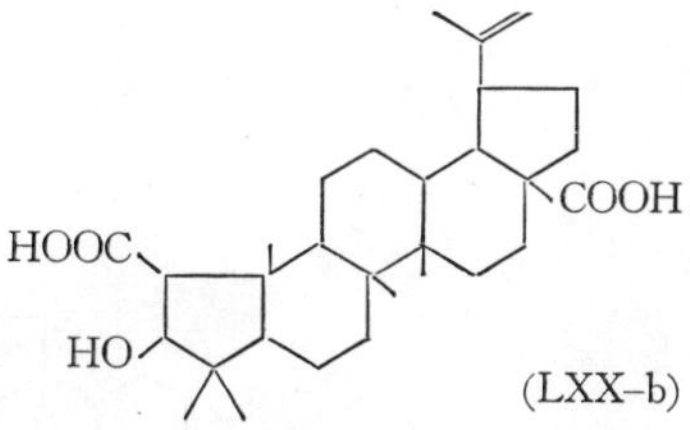

(LXX-b)

≪Occurrence≫   *Ceanothus americanus* (*Rhamnaceae*),   mp. 356~358°, $[\alpha]_D$ +37° (EtOH), $C_{30}H_{46}O_5$
IR ν_{max} cm^{-1}: 1642, 885 (Vinylidene)

〔I〕 Hopane-series Triterpenoids

(1) Hydrocarbons

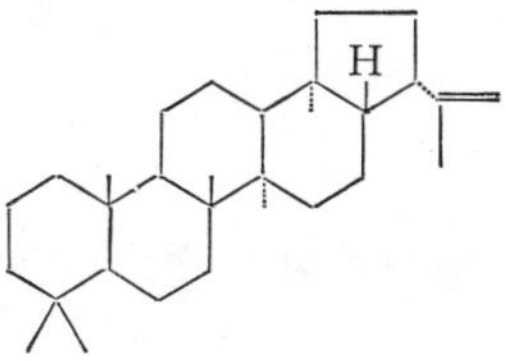

(LXXI) Hopene–b
(Diploptene)

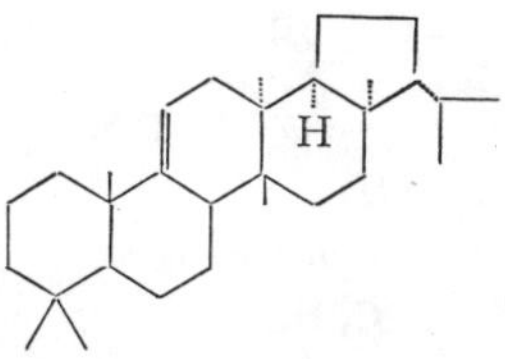

(LXXII) 9(11)–Fernene
(Fernene)

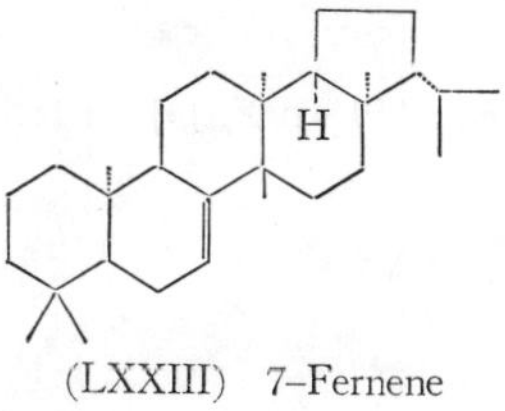

(LXXIII) 7–Fernene

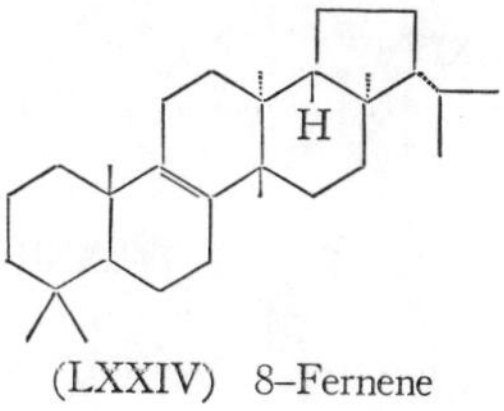

(LXXIV) 8–Fernene

450) Taylor, G.R. *et al*: *Soc.*, **1960**, 4303.
451) Mechoulam, R: *J. Org. Chem.*, **27**, 4070 (1962).
452) Ageta, H. *et al*: *Chem. Pharm. Bull.*, **11**, 408, 409 (1963).

(a) Hopene-b (Diploptene) (LXXI)[452]

≪Occurrence≫   *Dryopteris crassirhizoma* NAKAI (オシダ), *Pyrrosia lingua* FARWELL (*Polypodiaceae*)
mp. 207～208°, 211～212°, $[\alpha]_D$ +65.7°, 61.1° (CHCl₃), $C_{30}H_{50}$

IR ν_{max}^{KBr} cm⁻¹: 3067, 1770, 1639, 886 (>C=CH₂)

NMR in CCl₄, TMS, τ: 9.28 (3 H), 9.20 (3 H), 9.19 (3 H), 9.15 (3 H), 9.05 (6 H) singlets (>C–CH₃),

8.29 (3 H) w fine splitting (CH₃–C=C), 5.29 s (2 H of >C=CH₂)

Hydroxyhopanone (LXXVI)——→ (LXXI)

(b) Fernene (LXXII)[452) 453]

≪Occurrence≫   *Dryopteris crassirhizoma* NAKAI, mp. 170～171°, $[\alpha]_D$ −16.5° (CHCl₃), $C_{30}H_{50}$

IR ν_{max}^{KBr} cm⁻¹: 812, 716 (>=<_H)

NMR in CCl₄, δ: 5.28 (olefinic H)
Mass m/e 410, 243

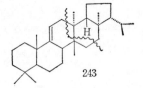

243

UV of Heteroannular diene

Fernene $\xrightarrow{SeO_2}$ Fernadiene
(LXXII)

Euphadiene
Baueradiene   Lanostadiene
Multifloradiene   Arboradiene
Fernadiene
(A)   (B)

(A) UV λ_{max}^{EtOH} mμ: 232, 239, 248

(B)   〃   : 236, 243, 252

(c) 7–Fernene (LXXIII)[453]

≪Occurrence≫   *Adiantum monochlamys* EATON (ハコネシダ), mp. 158.5～160.0°, $[\alpha]_D$ +26° (CHCl₃),
$C_{30}H_{50}$

IR ν_{max}^{KBr} cm⁻¹: 822, 796 (>C=C<_H)

Mass m/e 410

(b) 8–Fernene (LXXIV)[453]

≪Occurrence≫   *Adiantum monochlamys* EATON,   mp. 189～190°, $[\alpha]_D$ +18.0° (CHCl₃), $C_{30}H_{50}$
IR, NMR (>C=C<),

Fernene (LXXII) $\xrightarrow[\text{boil}]{\text{HOAc+HCl}}$ (LXXIV)

453) Ageta, H. *et al*: *7 th Symposium on the Chemistry of Natural Products* p. 193 (Fukuoka, 1963).

(2) Diplopterol (LXXV)[454)455)]

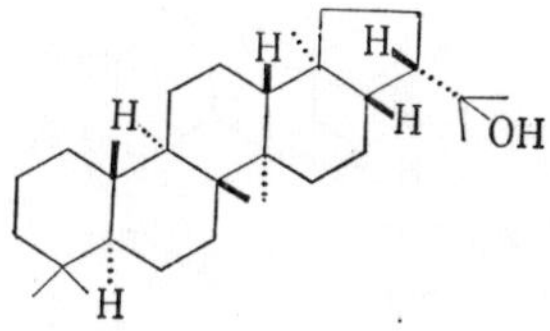

(LXXV)

《Occurrence》　*Diplopterygium glaucum* NAKAI,　mp. 254.0~256.0°, $[\alpha]_D$ +44.5° (CHCl₃), $C_{30}H_{50}O$

Hydroxyhopanone (LXXVI) $\xrightarrow{\text{Red.}}$ (LXXV)[456)], (LXXV) $\xrightarrow{\text{Red.}}$ Diploptene (LXXI)

(3) Hydroxyhopanone (LXXVI)[456)457)]

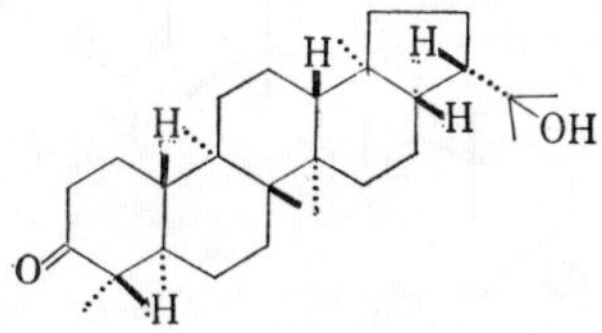

(LXXVI)

《Occurrence》　Dammar resin,　mp. 252~256°, $[\alpha]_D$ +64° (CHCl₃), $C_{30}H_{50}C_2$

IR　ν_{max}^{CS2} cm⁻¹: 3610, 1706

(4) Zeorin (LXXVII) and *epi*-Zeorin (LXXVIII)

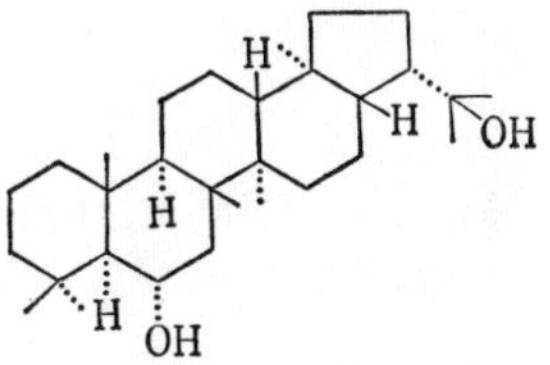

(LXXVII) 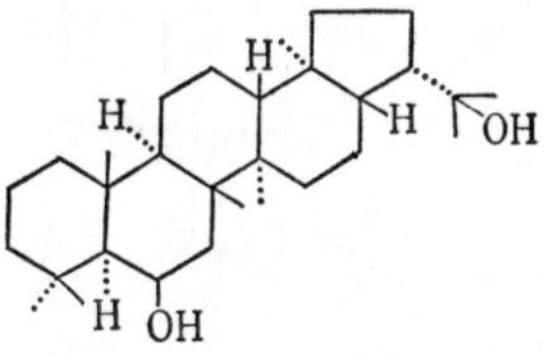　　　　　　　　(LXXVIII)

(a) **Zeorin** (LXXVII)[459)460)]

　　《Occurrence》　*Zeora sordida, Lecanora mauralis* (Lichen),　mp. 236~242°, $[\alpha]_D$ +63.3° (CHCl₃),

　　　　　　$C_{30}H_{52}O_2$

Zeorinone (6-ketone) mp. 251~255°

IR See the item 〔B〕, (2), Table 38

(b) ***epi*-Zeorin** (LXXVIII)[461)]

　　《Occurrence》　Synthetic

　　　　Zeorin $\xrightarrow{\text{CrO}_3}$ Zeorinone $\xrightarrow{\text{LiAlH}_4}$ *epi*-Zeorin　mp. 261~265°, $[\alpha]_D$ +21.5° (CHCl₃)

454)　Ageta, H. *et al*: *Chem. Pharm. Bull.*, **11**, 407 (1963).
455)　Jones, E.R.H. *et al*: *Soc.*, **1961**, 3891.　　456)　Mills, J.S. *et al*: *Soc.*, **1955**, 3132.
457)　Jeger, O. *et al*: *Helv. Chim. Acta*, **41**, 152 (1958).
458)　Jones, E.R.H. *et al*: *Soc.*, **1961**, 3891.　　459)　Huneck, S: *Ber.*, **94**, 614 (1961).
460)　Barton, D.H.R. *et al*: *Soc.*, **1958**, 2239.　　461)　Huneck, S: *Ber.*, **94**, 1151 (1961).

(5) **Leucotylin** (LXXIX)[461)462)]

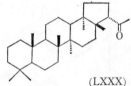

(LXXIX)[462)]

≪Occurrence≫　　*Parmelia leucotyliza* NYL.[463)],　　mp. 333°, $[\alpha]_D$ +48.4°, $C_{30}H_{52}O_3$

IR　$\nu_{max}^{CCl_4}$ cm^{-1}: 3676 (free OH), 3413, 3247 (bonded two–OH at C_{16} and C_{28}), see the item [B], (2), Table 38

(6) **Adiantone** (Nortriterpene) (LXXX)[464)]

(LXXX)

≪Occurrence≫　　*Adiantum capillus Veneris* L.,　　mp. 220~224°, $[\alpha]_D$ +81°, $C_{29}H_{46}O$

(7) **Davallic acid** (LXXX-a)[465)]

(LXXX–a)

≪Occurrence≫　　*Davallia divaricata* (タカサゴシノブ),　　mp. 283°, $C_{30}H_{48}O_2$
Methylate $C_{31}H_{50}O_2$, mp. 243°

UV　λ_{max}^{EtOH} mμ (ε): 207 (5100),　　IR　ν_{max}^{KBr} cm^{-1}: 1740

NMR in CHCl$_3$, δ: 5.32 (olefinic H)

Mass M$^+$ 454, M$-$15 (439), M$-$COOMe (395), M$-$(R+15+H=379), a (301), b (287), c (275)

462) Yoshioka, I. *et al*: *Chem. Pharm. Bull.*, **11**, 1468 (1963).
463) Asahina, Y. *et al*: *Ber.*, **71**, 980 (1938).
464) Ourisson, G. *et al*: *Tetr. Letters*, **1963**, 1283.
465) Nakanishi, K. *et al*: *7 th Symposium on the Chemistry of Natural Products*, p. 198 (Fukuoka, 1963);
Tetr. Letters, **1963**, 1451.

〔J〕 Tetracyclic Triterpenoids

(1) Euphol and Tirucallol (LXXXI)[460]

(LXXXI)

a : $R=Me-\overset{H}{\underset{|}{C}}-CH_2-CH_2-CH=C\overset{Me}{\underset{Me}{<}}$

Tirucallol

b : $R=H-\overset{Me}{\underset{|}{C}}-CH_2-CH_2-CH=C\overset{Me}{\underset{Me}{<}}$

Euphol

c : $R=Me-\overset{H}{\underset{|}{C}}-CH_2-CH_2-CH=C\overset{Me}{\underset{Me}{<}}$

Lanosterol

(a) Euphol[467]~[470]

《Occurrence》 Latex from *Euphorbia triangularis* and other *E.* spp.,
mp. 115~117°, $[\alpha]_D$ +32° (CHCl$_3$), C$_{30}$H$_{50}$O
IR See the item 〔B〕, (5), Table 41.

(b) Tirucallol[471][472]

《Occurrence》 Resin from *Euphorbia tirucalli*, *E. triangularis*,
mp. 133~134.5°, $[\alpha]_D$ +4.5° (C$_6$H$_6$), C$_{30}$H$_{50}$O
IR See the item 〔B〕, (5), Tab. 41

(2) Obliquol (12-Hydroxylanosterol) (LXXXII)[473]

(LXXXII)

《Occurrence》 *Poria obliqua* (Fungus)
mp. 188~190°, $[\alpha]_D$ +61° (CHCl$_3$), C$_{30}$H$_{50}$O$_2$
UV λ_{max} mμ (E$_{1cm}^{1\%}$): 244 (270), 252 (215) [Note]
IR ν_{max} cm^{-1}: 3350 (OH), 1030, 1048 (2—C–O), 1410
NMR 56.4 Mc, in CDCl$_3$, CHCl$_3$ as internal reference, τ: See Fig. 90

466) Warren, F.L. *et al*: *Soc.*, **1958**, 179.   467) Ruzica, L. *et al*: *Helv. Chim. Acta*, **37**, 2306 (1954).
468) Barton, D.H.R. *et al*: *Soc.*, **1951**, 2540.   469) Spring, F.S. *et al*: *ibid.*, **1944**, 249.
470) Jeger, O. *et al*: *ibid.*, **1951**, 2540.   471) Ruzicka, L. *et al*: *Helv. Chim. Acta*, **38**, 222 (1955).
472) Barbour, J.B. *et al*: *Soc.*, **1955**, 2194.
473) Kier, L.B. *et al*: *J.P.S.*, **50**, 471 (1961), **52**, 465 (1963).
[Note] probably due to a diene present as an impurity (author).

ジ エ ン

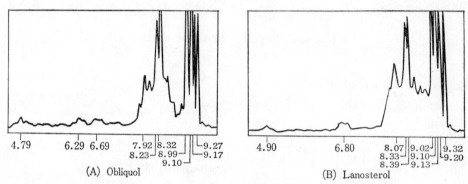

Fig. 90   NMR Spectra of (A) Obliquol and (B) Lanosterol   (Kier *et al.*)

(3)   **Betulafolientriol** and **Betulafolientetraol** (LXXXIII)[474]

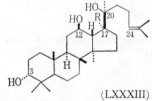

(LXXXIII)

R=H Betulafolientriol
R=OH Betulafolientetraol

《Occurrence》   The leaves of *Betula alba* (シラカンバ)

(a) **Betulafolientriol**

mp. 197～198°, $C_{30}H_{52}O_3$

UV λ_{max}^{EtOH} mμ (log ε): 212 (2.94)

IR   See Fig. 91

(b) **Betulafolientetraol**

mp 168～170°, $C_{30}H_{52}O_4$

λ_{max} mμ (ε): 238, 243, 249, 254, 261 (58～50) [Note]

IR   See Fig. 91

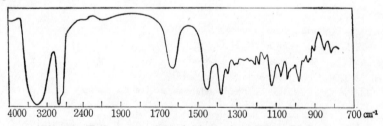

Fig. 91   IR Spectrum of Betulafolientriol (in KBr)   (Fischer)

---

474)   Fischer, F.G. *et al*: *Ann.*, **626**, 185 (1959), **644**, 146 (1961).
[Note]   probably due to an impurity (author).

(4) Dammarenediols I and II (LXXXIII-a)[474a]

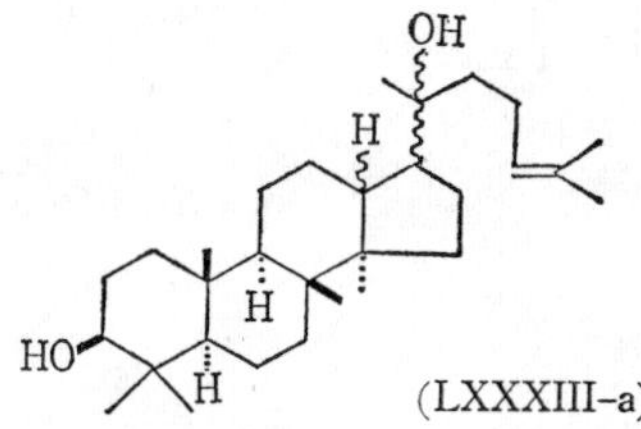

(LXXXIII-a)

《Occurrence》 Dammar resin

(a) Dammarenediol I
Monobenzoate mp. 166~168°, $[\alpha]_D$ +54° (CHCl$_3$)

(b) Dammarenediol II
Monobenzoate mp. 156~158°, $[\alpha]_D$ +59° (CHCl$_3$)

(c) Dammaradienyl acetate (LXXXIII-b)[474a][474b]

3-Acetates of dammarenediols $\xrightarrow{\text{POCl}_3}$ (XXXIII-b)

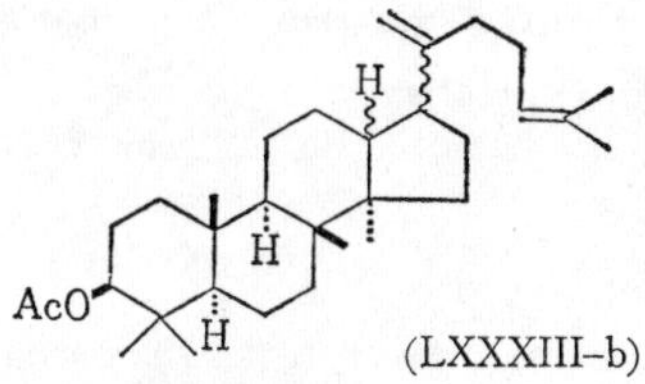

(LXXXIII-b)

《Occurrence》   *Inula helenium* L. (オオグルマ)[474b],   mp. 146~150°, $[\alpha]_D$ +48° (CHCl$_3$), C$_{32}$H$_{52}$O$_2$
IR ν_{max} cm^{-1}: 1721, 1244 (OAc), 1637, 884 ($>$C=CH$_2$)

(5) Cucurbitacins, Elatericins and Elaterin

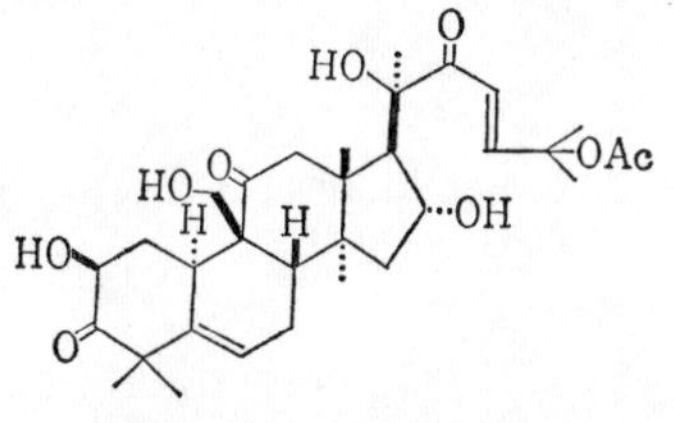

(LXXXIV) Cucurbitacin A[379]

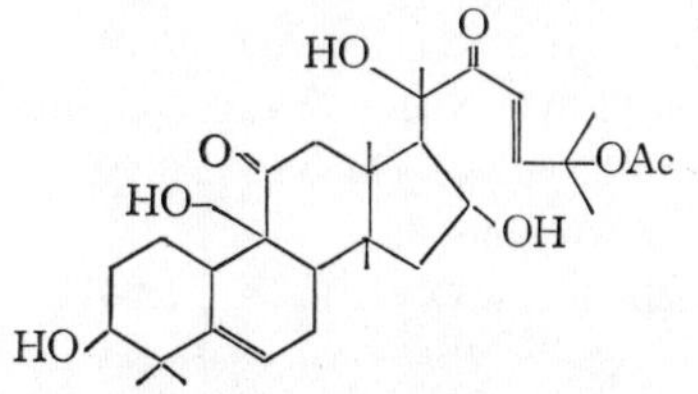

(LXXXV) Cucurbitacin C[475]

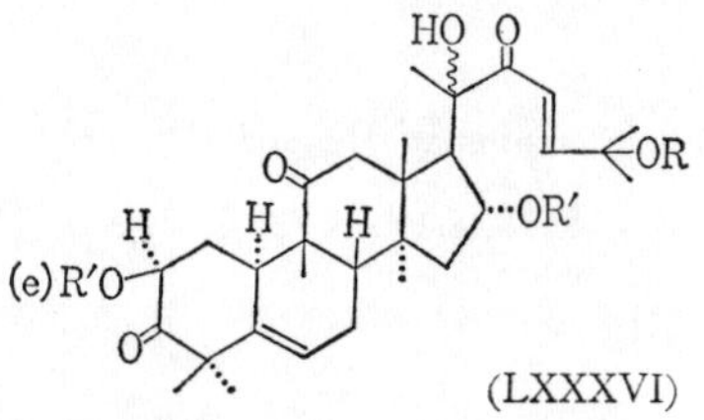

(LXXXVI)

R=Ac, R′=H Cucurbitacin B[475][476]
R=R′=H   Cucurbitacin D[475]
            (Elatericin A)[475]~[477]
R=Ac, R′=H, 2·OH *axial*
            epi-Cucurbitacin B[485]

474a) Mills, J.S.: *Soc.*, **1956**, 2196.   474b) Yoshioka, I *et al*: *Y.Z.*, **83**, 801 (1963).
475) Barton, D.H.R. *et al*: *Soc.*, **1963**, 3828.
476) Lavie, D. *et al*: *J. Org. Chem.*, **28**, 1790 (1963).   477) Lavie, D. *et al*: *ibid.*, **27**, 4546 (1962).

R=Ac, R′=H Cucurbitacin E[475]
(α–Elaterin=Elaterin[476])

R=R′=H Cucurbitacin I[475]
(Elatericin B[476])

R=Ac, R′=D–glucopyranosyl
α–Elaterin–D–glucoside[488]

(LXXXVII)

(LXXXVIII) Cucurbitacin F[479]   (LXXXIX)

(a) **Cucurbitacin A** (LXXXIV)

≪Occurrence≫   *Cucumis hookeri, C. leptodermis, C. myriocarpus*[478]

mp. 207~208°, $[\alpha]_D$ +97° (EtOH), $C_{32}H_{46}O_9$

UV λ_{max} mμ (ε): 229 (12200), 290 (206),   IR ν_{max} cm^{-1}: 1730 s, 1258 (OAc), 1715 (C=O), 1692, 1632 (CO–C=C)[478]

NMR in CDCl$_3$, 60 Mc, TMS, τ[475]:

2.98; 3.46 (olefinic protons at H_{23} and H_{24}, J=14.0), 4.20 br. olefin (H$_6$), 7.95 (OAc), 8.43 $\left(2 \right.$ ⟩C–Me$\left.\right)$,

8.56 $\left(2 \right.$ ⟩C–Me$\left.\right)$, 8.68 $\left(2 \right.$ ⟩C–Me$\left.\right)$, 8.94, $\left(\right.$⟩C–Me, C–13$\left.\right)$

(b) **Cucurbitacin C** (LXXXV)

≪Occurrence≫   *Cucumis sativus* var. *Hanzil* and other spp.[480]

mp. 207~207.5°, $[\alpha]_D$ +95° (EtOH), $C_{32}H_{48}O_8$

UV λ_{max} mμ (ε): 231 (11100), 298 (131)

IR ν_{max} cm^{-1}: 3448 (OH), 1731, 1256 (OAc), 1689, 1631[480]

NMR in CDCl$_3$, 60 Mc, TMS, τ[475]: 3.19, 3.26 (olefinic protons at H_{23} and H_{24}, neq. J=15.1), 3.77 m, (olefin H$_6$), 8.04 (OAc)

(c) **Cucurbitacin B** (LXXXVI, R=Ac, R′=H)

≪Occurrence≫   *Echinocystis fabacea*[481],   mp. 178~179°, $[\alpha]_D$ +87° (EtOH), $C_{30}H_{44}O_8$

UV λ_{max}^{EtOH} mμ (log t): 228 (4.02), 270~290 inf. (2.32)

IR λ_{max}^{KBr} μ: 2.92, 5.82, 5.92, 6.17[481]

NMR in CDCl$_3$, 60 Mc, TMS τ[475]: 2.95, 3.48 (olefinic protons at H_{23} and H_{24}, neq. J=16.2, 4.27 m, br. (olefin H$_6$), 8.01 (OAc), 8.45, 8.58 $\left(2 \right.$ ⟩C–Me$\left.\right)$, 8.64, 8.68, 8.72, 8.95, 9.04 $\left(5 \right.$ ⟩C–Me$\left.\right)$

478) Enslin, P.R. *et al*: *Soc.*, **1960**, 4779.   479) Van de Merwe *et al*: *ibid.*, **1963**, 4275.
480) Enslin, P.R. *et al*: *ibid.*, **1960**, 4787.
481) Noller, C.R. *et al*: *J. Org. Chem.*, **23**, 1984 (1958); **26**, 1211 (1961).

(d) Elatericin A (Cucurbitacin D) (LXXXVI, R=R'=H)

 ≪Occurrence≫ *Ecballium elaterium* L.,[476)483]   mp. 150~152°, $[\alpha]_D$ +46° (CHCl₃), $C_{28}H_{42}O_7$

UV $\lambda_{max}^{EtOH} m\mu$ (ε): 232 (9000)

IR ν_{max}^{CHCl3} cm⁻¹: 3425, 1689, 1626, 1377, 1088, 1058, 983

IR Spectra of C₂–H Epimers

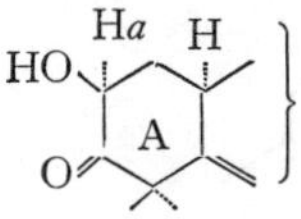

Elatericin A
Cucurbitacin B
their Dihydro–
compounds

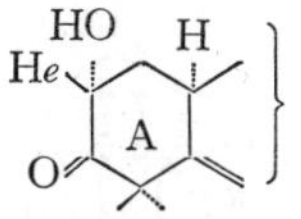

Tetrahydro–
elatericin B
Tetrahydro–
elaterin

C₂–OH (*equatorial*) 1125	>	(*axial*) 1100
C₃=O 1712	>	1705

NMR of 2, 16–Diacetate (See Fig. 92[476])

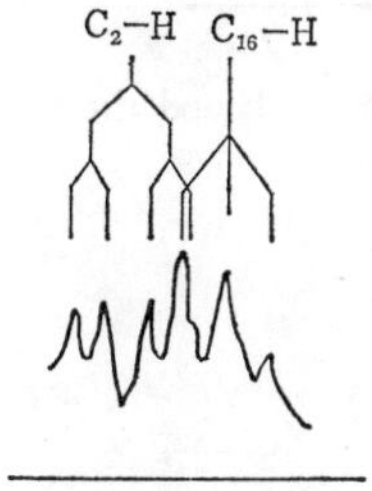

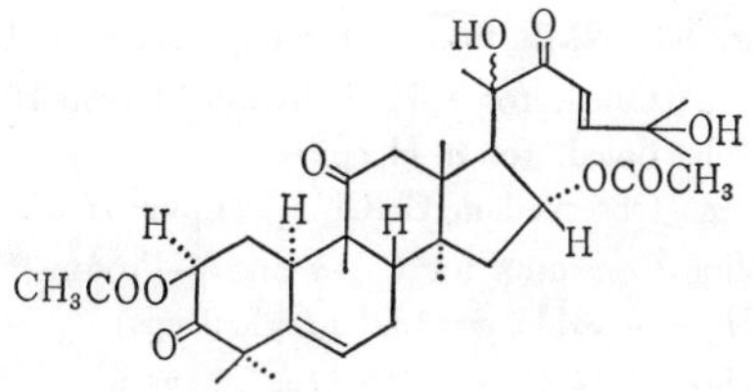

Fig. 92 NMR Spectrum for C₂–H and C₁₆–H Protons of Elatericin A diacetate (in CHCl₃) (Lavie *et al.*)

(e) Elaterin (α–**Elaterin**=**Cucurbitacin E**) (LXXXVII, R=Ac, R'=H)

 ≪Occurrence≫ *Ecballium elaterium* L.,[484] *Luffa echinata*[485]

mp. 232~233° (decomp.), $[\alpha]_D$ −59° (CHCl₃), $C_{32}H_{44}O_8$

UV $\lambda_{max}^{EtOH} m\mu$ (ε): 234 (11700), 267 sh (8350)

IR ν_{max}^{CHCl3} cm⁻¹: 3450, 1723, 1683, 1660, 1627, 1412, 1370, 1130, 1090, 990[486]

22–Dihydroelaterin   mp. 174~176°[476]
UV $\lambda_{max} m\mu$ (ε): 266 (6970).
IR ν_{max} cm⁻¹: 1724, 1720 (ester), 1700, 1694 (H₂₂ and H₁₁ C=O), 1664 (diosphenol)

2, 22–Tetrahydroelaterin   mp. 231~233°[476]
UV $\lambda_{max} m\mu$ (ε): 272 (857),
IR ν_{max} cm⁻¹: 1724 (ester), 1705, 1702 (H₃ and H₂₂ C=O), 1696 (C₁₁ C=O), 1100 (H₂ *axial* OH)
NMR of Diacetate (in deutero acetone), 60 Mc, TMS, τ: 2.85 (C₂₄–H)—3.18 (C₂₃–H) (neq. J=15.0),

  3.71 d (C₁–H) (J=3.0), 4.15 m (C₆–H), 7.85, 8.01, 8.13 (3 OAc), 8.43 s $\left(\rangle C\text{–Me}\right)$, 8.56

  $\left(2\ \rangle C\text{–Me}\right)$, 8.72 $\left(2\ \rangle C\text{–Me}\right)$, 8.95, 8.97 $\left(\rangle C\text{–Me}\right)$

(f) Elatericin B (Cucurbitacin I (LXXXVII R=R'=H))

 ≪Occurrence≫ *Ecballium elaterium* L.[483],   mp. 148~149°, $[\alpha]_D$ −52° (CHCl₃), $C_{28}H_{42}O_7$

UV $\lambda_{max}^{EtOH} m\mu$ (ε): 234 (11000), 266 (6850)

---

482) Lavie, D. *et al*: *J. Org. Chem.*, **28**, 1790 (1963).
483) Lavie, D. *et al*: *J.A.C.S.*, **80**, 710 (1958).   484) Lavie, D. *et al*: *ibid.*, **80, 707** (1958).
485) Lavie, D. *et al*: *Soc.*, **1962**, 3259.   486) Lavie, D. *et al*: *J.A.C.S.*, **81**, 3062 (1959).

IR ν_{max}^{CHCl3} cm^{-1}: 3410, 1685, 1660, 1629, 1606, 1413, 1090, 1005[484]

NMR of Dihydroelatericin B (mp. 158~160° decomp.) and it's C_{16}-epimer (in CHCl$_3$, 60 Mc, TMS, τ) (See Fig. 93)

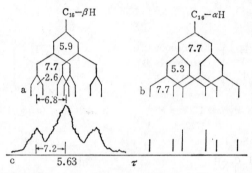

Fig. 93 NMR Spectra for C_{16}-Proton in Dihydroelatericin B and the Epimer.

a: Calcd. for β-H (Dihydroelatericin B)

b: Calcd. for α-H epimer

c: Observed in CHCl$_3$ (Lavie *et al.*)

Coupling Constants for C_{16}-α and β-Protons[487]

β-H——$C_{17}\alpha$H ϕ=137° (J=5.9 cps)

β-H——$C_{15}\beta$H   "   10° (J=7.7 ")

β-H——$C_{15}\alpha$H   "   115° (J=2.6 ")

(triplet 1:2:1)

α-H——$C_{17}\alpha$H ϕ= 10° (J=7.7 cps)

α-H——$C_{15}\beta$H   "   132° (J=5.3 ")

α-H——$C_{15}\alpha$H   "   10° (J=7.7 ")

(quartet 1:3:3:1)

(g) 2-*epi*-Cucurbitacin B (LXXXVI, R=Ac, R'=H, 2-OH *axial*)[485]

《Occurrence》   *Luffa echinata*,   mp. 229~231° (decomp.), [α]$_D$+41°, $C_{22}H_{46}O_8$

UV λ_{max} mμ (ε): 230 (11000)

IR ν_{max} cm^{-1}: 3350 (OH), 1730 (ester and 3-C=O), 1695 (12 and 22 C=O), 1630 (Δ^{23}), 1260 (OAc), 1102 (2-OH)

See IR Spectra of cucurbitacin B and Elatericin A

(h) Cucurbitacin F (LXXXVIII)[479]

《Occurrence》   *Cucumis dinteri* Cogn.,   mp. 244~245°, [α]$_D$ +38° (EtOH), $C_{30}H_{46}O_7$

UV λ_{max}^{EtOH} mμ (ε): 232 (11200) ($\alpha\beta$-unsatur. C=O), 300 (128) (isolated C=O)

IR ν_{max}^{KBr} cm^{-1}: 1690, 1629 ($\alpha\beta$-unsatur. C=O), 1700 (C=O)

NMR (in CDCl$_3$, 60 Mc, TMS, τ) of (LXXXIX)

7.97, 8.04 (two COCH$_3$), 7.73 (CO-CH$_3$), 8.87, 8.94, 8.96, 9.01, 9.08 $\left(\text{five } {>}\text{C-CH}_3\right)$, 6.99 ($C_{12}$-H$_2$), ABXY centr. at 5.19 ($C_2$-H, C_3-H), 4.28 m (C_6-H), 3.36 t (C_{16}-H) (J=2.7 cps). ABXY pattern in ring A: $\Delta\nu_{AB}$=20.9, J_{AB}=10.3; J_{AX}=10.3; J_{AY}=4.3, J_{BX}=J_{BY}=0 cps C_2-H, C_3-H are both. *axial* (∵ J_{ax-ax}=10.3, J_{ax-eq}=4.3 cps)

(i) α-Elaterin-D-glucoside (LXXXVII, R=Ac, R'=D-glucopyranosyl[488])

487) Karplus, M.: *J. Chem. Phys.* **30**, 11 (1959).

488) Khadem, H. El *et al*: *Tetr. Letters.*, **1962**, 1137.

≪Occurrence≫   *Citrullus colocynthis*,   mp. 158~160°, $[\alpha]_D$ +50°, $C_{38}H_{54}O_{13}\cdot2H_2O$

UV λ_{max}^{EtOH} mμ (log ε): 234~236 (4.11)

IR ν_{max}: 3570, 3540, 1725, 1685, 1625, 1425, 1415, 1390, 1370, 1250, 1220, 1120, 1070.

(6) Sapogenin of *Panax ginseng*

(XC)[489]
Protopanaxadiol

acid hydrolysis

(XCI)[490]
Panaxadiol

≪Occurrence≫   *Panax ginseng* (ニンジン)

Panaxadiol (XCI)   mp. 244~250°, $[\alpha]_D$ +1.0° (CHCl$_3$), $C_{30}H_{52}O_5$

IR ν_{max}^{KBr} cm^{-1}: 3480, 3288 (OH), 1117 (C–O–C)[491], ν_{max}^{CCl4} cm^{-1}: 3630, 3353 (OH)[492][493]

NMR[492] in CHCl$_3$, τ: 8.71 s $\left(1 \rangle C\text{–Me}\right)$, 8.76 s $\left(1 \rangle C\text{–Me}\right)$, 8.08 s $\left(1 \rangle C\text{–Me}\right)$, $\left(H_3C\text{–}\underset{|}{\overset{|}{C}}\text{–O–}\underset{|}{\overset{|}{C}}\langle{}^{CH_3}_{CH_3}\right)$

9.00 s $\left(2 \rangle C\text{–Me}\right)$, 9.10 s $\left(2 \rangle C\text{–Me}\right)$, 9.20 $\left(1 \rangle C\text{–Me}\right)$

Mass[492] m/e 460

$^{+}$OH

341

127

Non–hindered–3–OH and hindered–12–OH[492]

(XCI) $\xrightarrow{\text{Ac}_2\text{O, CrO}_3}$ 3–AcO, 12=O $\xrightarrow{\text{HOAc+H}_2\text{SO}_4}$ (XCII)
      Panaxanolone

(XCII)

(XCII) crystaline powder,   UV λ_{max}^{EtOH} mμ (log ε): 267.5 (3.85)

IR ν_{max}^{CS2} cm^{-1}: 1735 (OAc), 1617, 1608 (—C=C–C=O); ν_{max}^{KBr} cm^{-1}: 890 ($\rangle$C=CH$_2$)

489) Shibata, S *et al*: IUPAC Symposium p.**60** (Kyoto, 1964).
490) Shibata, S *et al*: *Tetr. Letters*, **1963**, 795.
491) Shibata, S *et al*: *Y.Z.*, **82**, 1634, 1638 (1962).
492) Shibata, S *et al*: *Chem. Pharm. Bull.*, **11**, 759, 762 (1963).
493) Shibata, S *et al*: *Tetr. Letters*, **1962**, 1239.

(7) Flindissol (XCIII)[494]

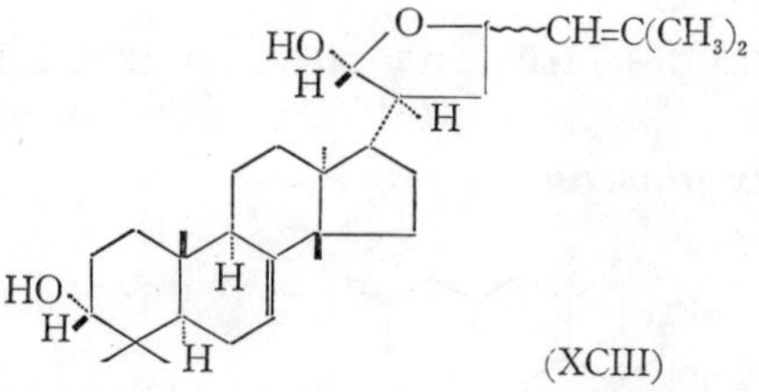

(XCIII)

≪Occurrence≫   *Flindersia dissosperma* DOMIN., *F. maculosa* LINDL

mp. 198° $[\alpha]_D$ −46° (CHCl$_3$), IR ν_{max}^{CS2} cm^{-1}: 3650, 2440, 1060, 1015, 982

(8) Shionone (XCIV)[495]

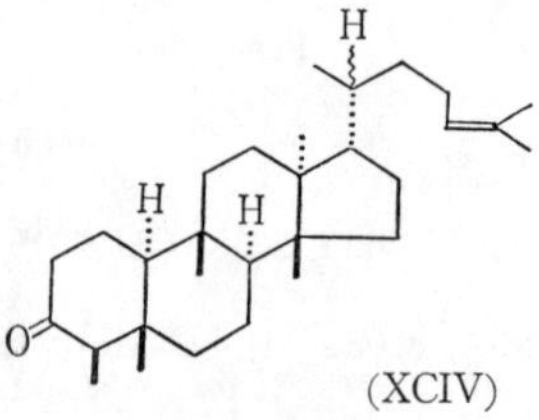

(XCIV)

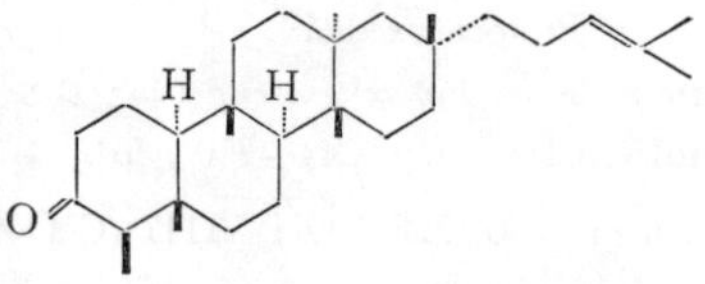

[Revised Structure Note]

≪Occurrence≫   *Aster tataricus* L. *fil.* (シオン),   mp. 161∼162°, $[\alpha]_D$ −56° (CHCl$_3$), C$_{30}$H$_{50}$O,

UV $\lambda_{max}^{Dioxane}$ mμ (ε): 290∼295 (50),   IR ν_{co} cm^{-1}: 1706

Conformation of A, B and C rings is similar to that of friedelin (O.R.D.)

(9) Alnus-Folienediolone (XCV)[496]

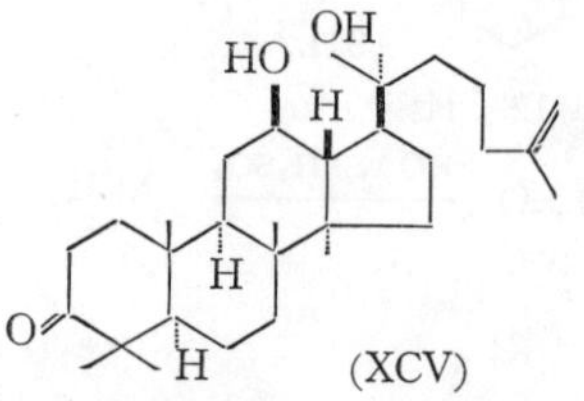

(XCV)

≪Occurrence≫   *Alnus glutinosa*,   mp. 199∼200°, $[\alpha]_D$ +52°, C$_{30}$H$_{50}$O$_3$

UV λ_{max}^{EtOH} mμ (log ε): 285 (1.61)   IR ν_{max}^{KBr} cm^{-1}: OH absorption, 1708 (C=O), 1640 885 (>C=CH$_2$),

(10) Dipterocarpol (XCV-a)[497]

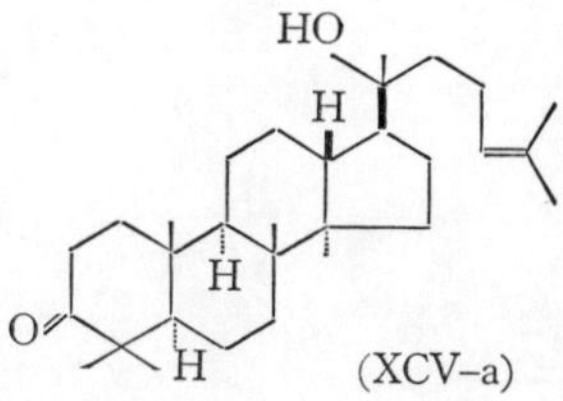

(XCV-a)

494) Birch, A.J. *et al*: *Soc.*, **1963**, 2762.
495) Ourisson, G. *et al*: *7 th Symposium on the Chemistry of Natural Products*, p.187 (Fukuoka, 1963).
496) Fischer, F.G. *et al*: *Ann.*, **644**, 162 (1961).   497) Crabbe, P *et al*: *Tetrahedron*, **3**, 279 (1958).
[Note] T. Takahashi, Y. Moriyama, Y. Tanahashi and G. Ourisson: *Tetr. Letters*, 2991 (1967).

《Occurrence》 Oleoresin of *Dipterocarpus* spp., mp. 135~136°, $[\alpha]_D$ +65° (CHCl₃), $C_{30}H_{52}O_2$

UV λ_{max}^{EtOH} mμ (ε): 290 (32), IR ν_{max} cm⁻¹: 3620 (OH), 1695 (C=O), 1420 (δ_{CH} of–CH₂–CO), 1370

$(\delta_{C-H}$ of $\rangle C\,Me_2)$, 815 $(\delta_{C-H}$ of $\rangle = \langle^H)$

Dammarenediol–II (LXXXIII–a) ⟶ Hydroxydammarenone–II=(XCV–a). See the item (4)

(11) Pachymic acid (XCVI)[498] and Polyporenic acid-B (XCVII)[499]

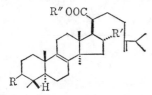

R=OAc, R′=OH, R″=H
Pachymic acid (XCVI)
R=R′=OH, R″=H
Polyprorenic acid–B (XCVII)

(a) Pachymic acid

《Occurrence》 *Poria cocos* (Schw.) Wolf. (ブクリョウ).
mp. 296~299°, $[\alpha]_D$ +17.7° (pyridine), $C_{33}H_{52}O_5$

UV λ_{max}^{EtOH} mμ (log ε): 235 (3.08), 242 (3.12), 251 (2.95) [Note]

IR ν_{max}^{nujol} cm⁻¹: 1738 (ester C=O), 1705 (COOH), 1645 (C=C), 1256 (acetyl C–O), 895 (δ $\rangle$C=CH₂)

(b) Polyporenic acid (XCVII)

《Occurrence》 *Polyporus betulinus* Fr.
3–β–Monoacetylmethylpolyporenate–B, $C_{34}H_{54}O_5$, mp. 180~184°, $[\alpha]_D$ +39° (CHCl₃)

UV λ_{max}^{EtOH} mμ (ε): 236 (2160), 243 (2420), 252 (1640) [Note]

Methylpolyporenate–B $C_{32}H_{50}O_4$, mp. 168~171°, $[\alpha]_D$ +28° (CHCl₃)

UV λ_{max}^{EtOH} mμ (ε): 236 (5250), 243 (6500), 252 (3550) [Note]

IR ν_{max}^{CCl4} cm⁻¹: 3645 (OH), 1737 (COOMe), 1640, 894 ($\rangle$C=CH₂)

IR See the item 〔B〕, (2), Table 37

(12) Helvolic acid (XCVIII)

(XCVIII) $C_{33}H_{44}O_8$
(Revised by Okuda, 1964)[500]

(XCIX) $C_{32}H_{42}O_8$
(Former. by Allinger. 1961)[501]

498) Shibata, S. *et al*: *Chem. Pharm. Bull.*, **6**, 608 (1958).
499) Jones, E.R.H. *et al*: *Soc.*, **1954**, 3234.
500) Okuda, S. *et al*: IUPAC Symposium p.62 (Kyoto, 1964).
501) Allinger, N.L. *et al*: *J. Org. Chem.*, **26**, 4522 (1961).
[Note] Author's comment: In view of the established structures, of these compounds, the UV maxima should originate the conjugate dicue(s) present in the specimens.

mp. 208~212° (decomp.),　　UV λ_{max} mμ ε: 299 (6200), 287 (8200), 275 (6900)[501]
NMR of Methyl helvolate

i) By Okuda[500] (τ): 2.70d, 4.17d (J=11.4) (C_1–H, C_2–H), 4.78 s (C_7–H), 7.90, 8.05 (two OCOCH$_3$), 8.72d

$(J=6.5)$ (C_4–CH$_3$), 8.30, 8.37 $\left(>=<^{CH_3}_{CH_3}\right)$, 8.55, 8.82, 9.08 $\left(\text{three} >C\text{–CH}_3\right)$

ii) By Allinger[501] (τ): 2.62d, 4.13d (J=5.2) $\left(-CO-\overset{2}{C}H=\overset{1}{C}H-C<\right)$, 4.74 s ($C_{12}$–H$_\beta$), 6.36 (COOCH$_3$),

7.92, 8.06 (two OCOCH$_3$), 8.36, 8.41 $\left(^H>=<^{CH_3(26)}_{CH_3(27)}\right)$, 8.58, 8.72, ($C_4$–gem di

Me), 8.87 (18–CH$_3$), 9.07 (19–CH$_3$) (See Fig. 94)

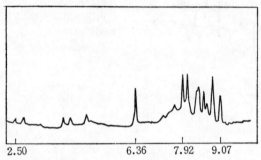

| 2.50 | | 6.36 | 7.92 | 9.07 |

Fig. 94　NMR of Methyl helvolate (Allinger *et al.*)

(13) Cephalosporin P$_1$ (C)[502]

(C)

≪Occurrence≫　　*Cephalosporium* (Mould),　　$C_{32}H_{48}O_8$

(14) Cycloartenol (Cycloart-23-ene-3β-ol (CI)[505]

≪Occurrence≫　　Linseed oil[503] (アマ二ン油), Rape oil[504],

Artocarpus integrifolia,[506] *Strychnos nuxvomica*[505]

mp. 112~114°, 116° $[\alpha]_D$ +49.2° (CHCl$_3$), $C_{30}H_{50}O$

IR ν_{max} cm^{-1}: 1005 (cyclopropane), 840, 820, 800 $\left(>=<_H\right)$

IR See the item [B], (6), Fig. 82.

Synthesis of cycloartane[507]

(CI)[505]

502)　Baird, B. M. *et al*: *Proc. Chem. Soc.*, **1961**, 257.
503)　Capella, P.: *Nature*, **190**, 167 (1961).　　504) Tamura *et al*: *J Chem. Soc. Japan.* **79**, 1053 (1958).
505)　Irvine, D.S. *et al*: *Soc.*, **1955**, 1316.　　506) Barton, D.T.R. *et al*: *ibid.*, **1951**, 1444.
507)　Barton, D.T.R.: *Proc. Chem. Soc.*, **1963**, 170.

(CI) 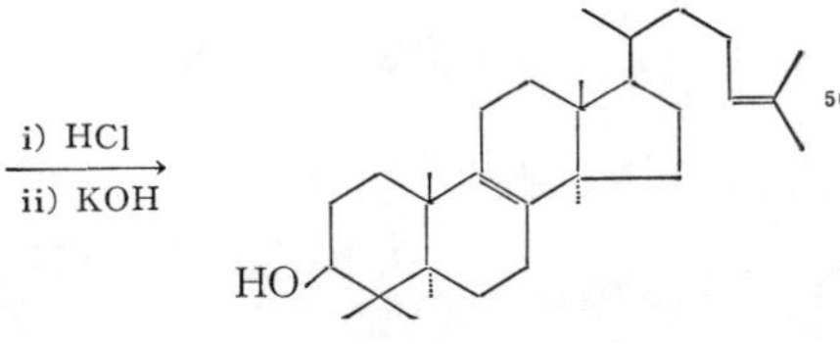 $\xrightarrow[\text{ii) KOH}]{\text{i) HCl}}$

504)

(15) Cyclolaudenol (CII)[508]

《Occurrence》 Opium marc (*Papaver somniferum*)

mp. 125°, $[\alpha]_D$ +46° (CHCl$_3$), C$_{31}$H$_{52}$O

UV λ_{max}^{EtOH} mμ (ε): 205 (1145).

[M]$_D$ of Lanostane series and Laudane series[508]

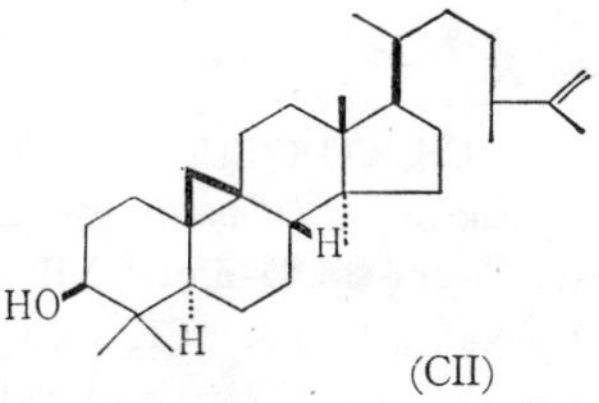

(CII)

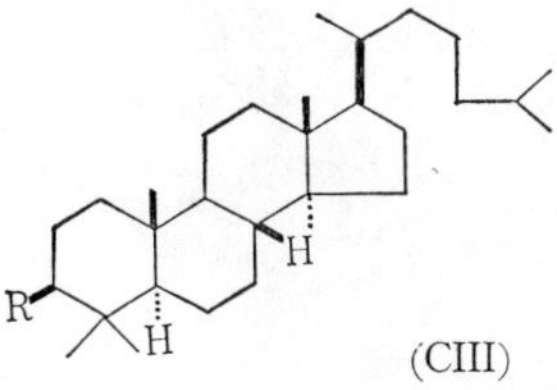

(CIII)

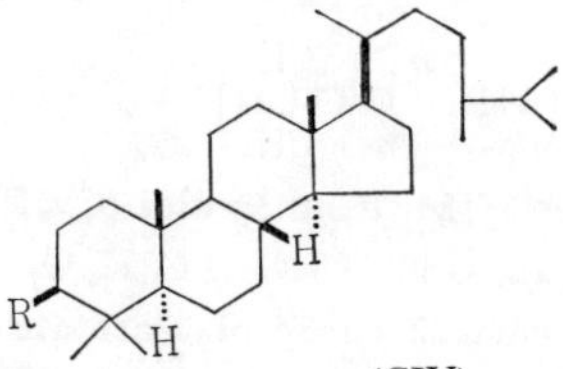

(CIV)

Lanostane	+145°	Laudane	+107°	−38°
Lanostanol	+150°	Laudanol	+93°	−57°
Lanostanylacetate	+193°	Laudanyl acetate	+155°	−38°
Lanostanone	+116°	Laudanone	+62°	−54°
Cycloartanol	+214°	Cyclolaudanol	+191°	−23°
Lanost–8–enol	+261°	Eburic–8–enol	+238°	−23°
Lanost–8–enylacetate	+275°	Eburic–8–enylacetate	+271°	−4°
Lanost–8–ene	+272°	Eburic–8–ene I	+234°	−38°

(16) Orysanol C (Alcohol C Ferulate) (CV)[509]

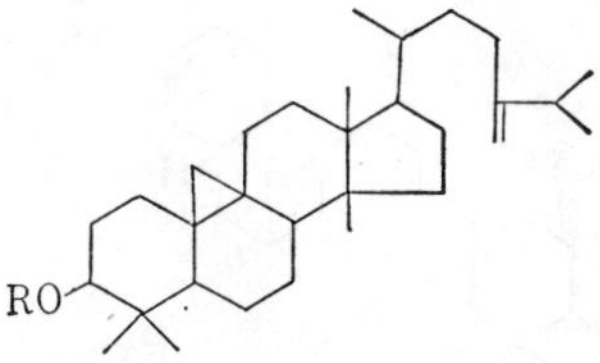

R= $\begin{array}{c}\text{H}_3\text{CO}\\\text{HO}\end{array}$〉-CH=CH-CO   Orysanol C (CV)

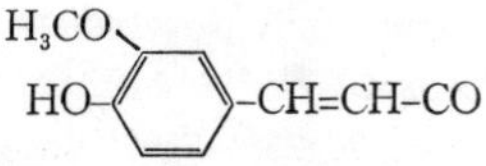

R=H   Alcohol C (CVI)

《Occurrence》 Rice bran oil (*Orysa sativa*)

(a) **Orysanol C (CV)**

mp. 162~164° / 193~194°, $[\alpha]_D$ +36° (CHCl$_3$), C$_{41}$H$_{60}$O$_4$

UV λ_{max}^{EtOH} mμ (log ϵ): 328 (4.29)

(b) **Alcohol C (CVI)**

mp. 121~122°, $[\alpha]_D$ +43° (CHCl$_3$), [M]$_D$ +189°, C$_{31}$H$_{52}$O

508) Spring, F.S. *et al*: *Soc.*, **1955**, 596, 1607.
509) Ohta, G: *Chem. Pharm. Bull.*, 8, 5, 9 (1960).

UV $\lambda_{\max}^{\text{EtOH}}$ mμ (log ε): 206 (1440),   IR $\nu_{\max}^{\text{CS2}}$ cm^{-1}: 3600, 3070, 3039, 1639, 1045, 1020, 1005, 988, 887

(17) Cycloartenols of Spanish Moss[510]

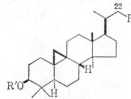

R′=H, R=CH=CH–C(CH$_3$)$_2$–OH **Cycloart–23–ene–3β, 25–diol** (CVII)

R′=H, R=CH$_2$–CH(OH)–C–(CH$_3$)=CH$_2$ **Cycloart–25–ene–3β, 24 (α or β) diol** (CVIII)

R′=Ac, R=CH=CH–C(CH$_3$)$_2$–OCH$_3$ **25–Methoxycycloart–23–en–3β–ol–3–acetate** (CIX)

R′=Ac, R=CH$_2$–CO–C(CH$_3$)=CH$_2$ **3β–Hydroxycycloart–25–en–24–one–3–acetate** (CX)

≪Occurrence≫   *Tillandsia usneoides* (Spanish moss)

(a) **Cycloart–23–ene–3β, 25–diol** (CVII)

mp. 200~204°, [α]$_D$ +38° (CHCl$_3$), C$_{30}$H$_{50}$O$_2$

UV: ε 200 mμ 3300, ε 210 mμ 330,   IR $\nu_{\max}^{\text{CHCl3}}$ cm^{-1}: 3600 (OH)

NMR τ: 8.69 s $\left(6\,\text{H of } \text{C}\underset{\diagdown\text{CH}_3}{\overset{\diagup\text{CH}_3}{-\text{C}-\text{OH}}}\right)$

(b) **Cycloart–25–ene–3β, 24 (α or β)–diol** (CVIII)

mp. 184~188°, [α]$_D$ +48° (CHCl$_3$), C$_{30}$H$_{50}$O$_2$

(c) **25–Methoxycycloart–23–en–3β–ol–3–acetate** (CIX)

mp. 152~154°, [α]$_D$ +48° (CHCl$_3$), C$_{33}$H$_{54}$O$_3$

IR $\nu_{\max}^{\text{CHCl3}}$ cm^{-1}: 1725

(d) **3β–Hydroxycycloart–25–en–24–one–3–acetate** (CX)

mp. 133~136°, [α]$_D$ +58° (CHCl$_3$), C$_{32}$H$_{50}$O$_3$

UV $\lambda_{\max}^{\text{EtOH}}$ mμ (ε): 218 (10000),   IR $\nu_{\max}^{\text{CHCl3}}$ cm^{-1}: 1710 (OAc), 1680 (—C=C–CO)

［K］ Onocerane-series

(1) α-**Onocerin** (α-onoceradienediol) (CXI)[511]

≪Occurrence≫   *Ononis spinosa*[512], *Lycopodium clavatum* L. (ヒカゲノカズラ)[513][514]

mp. 202~203°[511], 206.5~208°[513], 238~239°[514].

[α]$_D$+18° (pyridine)[511], +17° (CHCl$_3$)[513], C$_{30}$H$_{50}$O$_2$

IR $\nu_{\max}^{\text{KBr}}$ cm^{-1}: 3390, 1032 (OH), 3085, 1642, 885

(exocyclic methylene)

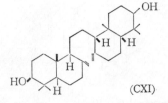

(CXI)

Total synthesis[514a]

Diacetate   mp. 222~224°[511], 223~225°[514], 227~229°[513]

[α]$_D$ +29° (CHCl$_3$)[511][513], +29.2° (CHCl$_3$), C$_{34}$H$_{54}$O$_4$

510) Djerassi, C. *et al*: *Soc.*, **1962**, 4034.   511) Barton, D.H.R. *et al*: *Soc.*, **1955**, 2639.

512) Zimmermann, J: *Helv. Chim. Acta*, **21**, 853 (1938), **23**, 1110 (1940).

513) Ageta, H. *et al*: *Chem. Pharm. Bull.*, **10**, 637 (1962).

514) Inubuse, Y. *et al*: *Y.Z.*, **82**, 1083, 1537 (1962).

514a) Stork G. *et al*: *J.A.C.S.*, **85**, 3419 (1963).

UV λ_{max}^{EtOH} mμ (ε): 205 (9500)[511]

IR ν_{max}^{KBr} cm^{-1}: 1730, 1371, 1268, 1026 (OAc), 3125, 1645, 879 (exocyclic methylene)[513]

(2) Terpenoids from *Lycopodium clavatum* L.[514]

(a) **Lycoclevanol**

mp 308~310°, $C_{30}H_{50}O_3$,

IR ν_{max} cm^{-1}: 3344 (OH)

Triacetate   mp. 197~198°, $[\alpha]_D$ −49.2° (CHCl$_3$)

(b) **Lycoclavanin**

mp 344~346°, $C_{30}H_{48}O_5$,

IR ν_{max} cm^{-1}: 3367 (OH), 1675 s (C=O)

Triacetate   mp. 236~237°, $[\alpha]_D$ −32.2° (CHCl$_3$)

UV λ_{max}^{EtOH} mμ (log ε): 247 (4.18),   IR ν_{max} cm^{-1}: 1733 s, 1658 s

[L] Miscellaneous

(1) Dammarenolic acid (CXII)[515)515a]

≪Occurrence≫   Dammar resin,   mp. 138~142°,

$[\alpha]_D$ +43° (CHCl$_3$), $C_{30}H_{50}O_3$

Methylate   mp. 89~92°, $[\alpha]_D$ +44° (CHCl),

ORD: +plain curve,

IR ν_{max}^{KBr} cm^{-1}: 3470, 1730, 1640, 894

(CXII)

(2) Nyctanthic acid (CXIII)[515a)515b)516]

≪Occurrence≫   *Nyctanthes arbortristis*,

mp. 222.5~223.5° $[\alpha]_D$ +86° (CHCl$_3$), $C_{29}H_{48}O_2$,

IR ν_{max}^{CCl4} cm^{-1}: 3050, 1635, 899 ($>$C=CH$_2$)

(CXIII)

(3) Sapogenin of Strawberry Clover[517]

≪Occurrence≫   *Trifolium fragiferum* (Strawberry Clover),   mp. 227~228°, $[\alpha]_D$ −35.5°, $C_{36}H_{54}O_7$,

Triacetate   UV λ_{max}^{EtOH} mμ (ε): 241.5 (23000), 249 (25900), 258 (20800) (conj. system);

IR λ_{max}^{KBr} μ: strong OH, 5.87, 5.76, 5.86, 8.1 (acetate)

515) Mills J.S. *et al*: *Soc.*, **1955**, 3132.   515a) Arigoni, D. *et al*: *ibid.*, **1960**, 1900.
515b) Turnbull J.H. *et al*: *ibid*, **1957**, 569.   516) Whitham G.H.: *Proc. Chem. Soc.*, **1959** 271.
517) Walter, E.D.: *J.P.S.*, **49**, 735 (1960).

(4) Serratenediol[518]

≪Occurrence≫   *Lycopodium Serratum* THUNB. var. *thunbergii* MAKINO (ホソバトウゲシバ)

mp. 300°, $C_{30}H_{50}O_2 \cdot 1/2 H_2O$

IR ν_{max}^{nujol} cm^{-1}: 3484 (OH)

Diacetate   IR ν_{max}^{nujol} cm^{-1}: 1724 (C=O), 1250 (C–O)

(5) Terebinthone[519]

≪Occurrence≫   *Schinus terebinthefolius* Radd. (*Anacardiaceae*) (Brazilian pepper tree)

mp. 178~179°, $[\alpha]_D$ −72.6° (CHCl$_3$), $C_{30}H_{46}O_3$

UV λ_{max} mμ (ε): 213 (7272)

IR ν_{max} cm^{-1}: 1708, 1690 (2C=O), 1640 (C=C), 1385, 1365 (gem Me$_2$ and angular-Me), 1095 (cyclic ether)

(6) Viminalol[520]

≪Occurrence≫   *Sarcostemma viminale* R. Br. (*Asclepiadaceae*),   mp. 176~178°, $C_{30}H_{50}O$

IR ν_{max} cm^{-1}: 3322 br. (assoc. OH), 1036, 1024, 903 (OH, *iso*-Pr. and ⟩C=C⟨H)

518) Inubuse, Y. *et al*: *Y. Z.*, **82**, 1339 (1962).
519) Kier, L. B. *et al*: *J. P. S.*, **51**, 245 (1962).
520) Hüttenrauch, R.: *Hoppe Seyler*, **326**, 166 (1961).

8 Steroids

Steroid sapogenins and cardenolide glycosides are described in the items 9 and 10, respectively.

[A] UV Spectra and Color Reaction

(1) UV Spectra

Cholesterol[521], ergosterol[521], cholesterylene[522], cholestadienol B-3-acetate, cholestatriene[523], neoergostatriene[524], neoergostatetraene, vitamine D_3[525], cholestenones[526] and oestrones[527].

(2) Color Reaction of conjugated Ketosteroids and its Absorption.

p-Aminodimethylaniline · $SnCl_2$ adduct, $C_8H_{12}N_2 \cdot H_2SnCl_4$ is added to the methanol solution of ketosteroids. A characteristic color appears depend on the structure. The colors and their absorption maxima are as follows:

(I)

Ring A						
Color	no	yellow	yellow	yellow	red	red
Developing	—	quick	quick	quick	slow	slow
λ_{max} mμ	—	415	420	420	455	455

521) Mohler: *Helv. Chim. Acta*, **20**, 811 (1937).   522) Skau: *J. Org. Chim.*, **3**, 166 (1938).
523) Müller *el al*: *Ber.*, **77**, 147 (1944).   524) Mayneord: *Proc. Roy. Soc.*, A **152**, 299 (1935).
525) Fuchs *el al*: *Pharm. Presse.*, **38**, 81 (1933).   526) Menschick *et al*: *Ann.*, **495**. 225 (1932).
527) Mayneord, W. V. *et al*: *Proc. Roy. Soc.*, A **158**, 634 (1937).

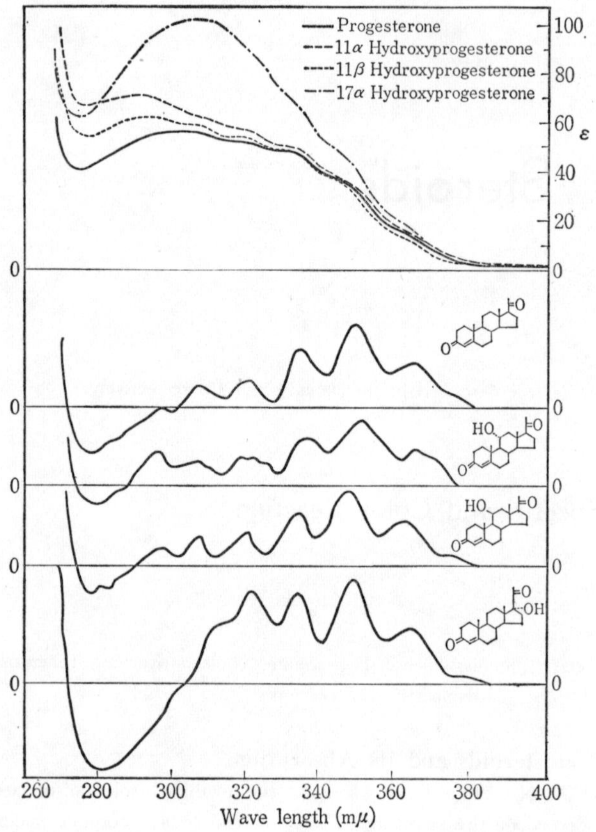

Fig. 95 Derivative UV Spectra of Progesterones (Olson *et al.*)

Fig. 96 Derivative UV Spectra of Progesterone, 6 α–Me, 6 β–Me epimers (Olson *et al.*)

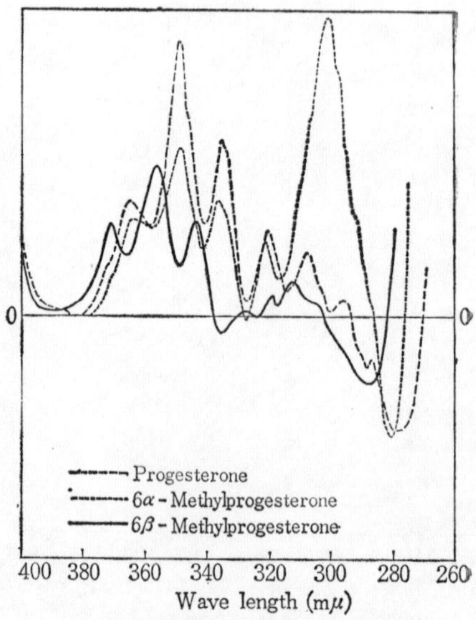

(3) Derivative UV Spectra[528]

Cary Model 11 Recording Spectrometer, attached with amplifier, synchronous rectifier and circuit derivative attachment, is used. The results using diethyleneglycole as a solvent are shown in Fig. 95 and 96. The spectrum, at the top in Fig. 95 shows usual UV spectrum and the others in the same figure are derivative spectra.

Fig. 96 shows the possibility of identification of α– from β–isomer.

(4) Absorption due to non-conjugated Dienes

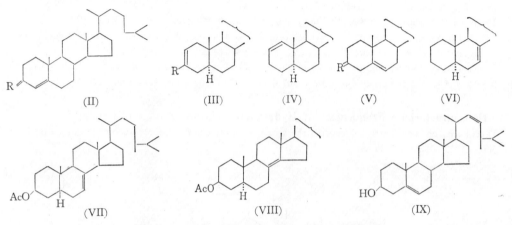

(II)   (III)   (IV)   (V)   (VI)

(VII)   (VIII)   (IX)

Table 48 UV Absorption due to non-conjugated Double bond of Steroids $\lambda_{max}^{cyclohexane}$ mμ (ε)

(Micheli *et al.*)

Steroids	No.	R	λ_{max}	ε	mp.
Δ^4–Cholestene	II	H⌒H	193	10,000	82.5～83
—— 3 β–ol	II	HO⌒H	199.5	8,700	129～130
—— 3 β–ol acetate	II	AcO⌒H	196	13,500	85～86
Δ^2–Cholestene	III	H	180	6,600	75.5～76.5
—— 3–ol acetate	III	AcO	182	8,740	90～90.5
Δ^1–Cholestene	IV	—	180 (EtOH) (190, 5)	10,300 (8,900)	69.5 90～91
Δ^5–Cholestene	V	H⌒H	189	8,900	
Cholesterol	V	HO⌒H	189.5	9,900	149
Cholesteryl acetate	V	AcO⌒H	189	10,000	115
Δ^7–Cholestene	VI	—	205	5,370	
		—	179.5	10,500	158～160
Δ^7–Ergostenyl acetate	VII	—	203.5	5,570	
$\Delta^{8(14)}$–Ergostenyl acetate	VIII	—	206	12,000	108
Stigmasterol	IX	—	188	21,700	169～170

528) Olson, E.C. *et al*: *Anal. Chem.*, **32**, 370 (1960).

[B] IR Spectra

(1) IR Spectra of

3-Hydroxysteroids (α or β), androstan-3 β-ol, coprostan-3 α-ol, colestan-3 α-ol, pregnan-3 β-ol[529], and ketosteroids (17, 3, 4, 6, 7, 11, 12, 20)[530][531]. Androstane, ergostane, androstanones (3 or 17), cholestanones (3 or 7), and ketocarboxylic acid esters[532].

Pregnanol, cholestanol, androstanol acetates (3 β or 17 β), Δ^4-19-norpregnene-3, 20-dione[532].

Etiocholan-3-one, androstan-3-one, Δ^1-androsten-17-β-ol-3-one, Δ^4-androsten-3-one, $\Delta^{1,4}$-cholestadien-3-one, $\Delta^{4,6}$-cholestadien-3-one, Δ^5-cholesten-3 β-ol-7-one acetate[533].

Pregnan-3 α-ol-20-one, allo-pregnal-3 β-ol-20-one, $\Delta^{1,4}$-androstadien-3, 17-dione, ——17 β-ol-3-one[533], 3-keto-$\Delta^{1,4}$-ethiocholandienic acid methylester. Cortisone[534], testosterone, progesterone[535].

(2) IR Characteristic Frequencies of Hydroxysteroids lower than 1350 cm^{-1} [536]

Useful data on the relationship between the structure of steroid and the absorption are reported by R. N. Jones (1952~1958) as follows.

(I) Androstan-3 α-ol   (II) Androstan-3 β-ol   (III) Etiocholan-3 α-ol

(IV) Etiocholan-3 β-ol   (V) Δ^5-Androsten-3 β-ol   (VI) Androstan-17 β-ol

(VII) Etiocholan-17 β-ol   (VIII) Allopregnan-20 α-ol   (IX) Allopregnan-20 β-ol

529) Cole, Jones, Dobriner: *J.A.C.S.*, **74**, 5571 (1952).   530) Jones, Herling: *J. Org. Chem.*, **19**, 1252 (1954).
531) Matui Y., Narisada M: Kagakuno Ryōiki Zokangō, **21**, 87 (1956) (Nankodo Pub. Inc. Tokyo).
532) Jones, Cole: *J. A. C. S.*, **74**, 5648 (1952).
533) Jones, Herling, Katzenellenbogen: *ibid.*, **77**, 651 (1955).
534) Dalinsky: *J. A. O. A. C.*, **34**, 758 (1951).   535) Carol, J: *J. A. O. A. C.*, **36**, 1001 (1953).
536) Jones, R. N *et al*: *J. A. C. S.*, **80**, 6121 (1958).

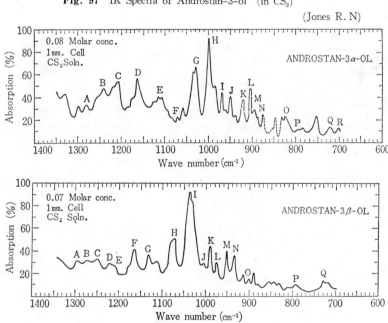

CH₃
H–C–OH

(X)
Pregnan–20 α–ol

CH₃
HO–C–H

(XI)
Pregnan–20 β–ol

Fig. 97 IR Spectra of Androstan–3–ol (in CS₂)

(Jones R. N)

Table 49 (I) Bands Useful for Characterizing
Hydroxysteroids (in CS₂)　　　　　(Jones *et al.*)

Steroids	No.	Stereo-str.	Main Band	Additional Bands
Androstan–3 α–ol	I	3 α–OH, 5α	1004∼999 (H)	1220∼1210 (C), 1167∼1162 (D), 1121∼1112 (E), 1033∼1027 (G), 976∼970 (I), 962∼952 (J), 935∼927 (K), 909∼904 (L)
Androstan–3 β–ol	II	3 β–OH, 5 α	1038∼1035 (I)	1170∼1163 (F), 1134∼1128 (G), 1078∼1072 (H), 993∼989 (K), 978∼976 (L), 957∼951 (M), 937∼934 (N)
Etiocholan–3 α–ol	III	3 α–OH, 5 β	1038∼1035 (I)	1175 (E), 1070∼1060 (H), 1014∼1007 (J), 952∼943 (K), 920∼908 (L), 900∼893 (M),

Steroids	No.	Stereo-str.	Main Band	Additional Bands
Etiocholan–3 β–ol	IV	3 β–OH, 5 β	1033～1030 (J)	1252～1249 (C), 1220～1215 (D), 1167～1162 (E), 986～980 (L), 962～959 (M), 916～908 (P)
Δ⁵–Androsten–3 β–ol	V	3 β, Δ⁵	1050 (J)	1217 (D), 1019 (K), 978 (M), 954 (N), 840 (P), 812 (Q), 796 (R)
Androstan–17 β–ol	VI	17 β–OH, 5 α	1046 (D)	1134 (B), 1085 (C), 1025 (E), 984 (G), 956 (H)
Etiocholan–17 β–ol	VII	17 β–OH, 5 β	1052 (H)	1135 (E), 1116 (F), 1067 (G), 1030 (I), 985 (J)
Allopregnan–20 α–ol	VIII	20 α–OH, 5 α	1012 (I) 957 (L)	1244 (B), 1174 (D), 1099 (G), 1067 (H)
Allopregnan–20 β–ol	IX	20 β–OH, 5 α	965 (K)	1100 (F),
Pregnan–20 α–ol	X	20 α–OH, 5 β	1008 (J) 952 (L)	1240 (B), 1172 (D), 1088 (H), 1064 (I)
Pregnan–20 β–ol	XI	20 β–OH, 5 β	1100 (D)	1032 (F), 998 (H), 970 (I)

(XII)
Cholestanol

(XIII)
Cholesterol

(XIV)
Ergostanol

(XV)
Δ²²–Stigmasten–3 β–ol

(XVI)
β–Sitosterol

Table 49 (II) Bands Useful for Characterizing Hydroxysteroids (in CS_2)

(Jones *et al.*)

Steroids	No.	Type No.	Stereo-str.	Main Band	Additional Bands
Cholestan–3 α–ol (α–Cholestanol)	XII	I	3 α–OH, 5 α	1002 (H)	1216 (C), 1164 (D), 1120 (E), 1033 (G), 974 (I), 960 (J), 934 (K), 909 (L)
Cholestan–3 β–ol (β–Cholestanol)	XII	II	3 β–OH, 5 α	1036 (I)	1168 (F), 1132 (G), 1072 (H), 990 (K), 975 (L), 954 (M), 396 (N)
Cholesterol	XIII	V	3 β–OH, Δ⁵	1049 (J)	1220 (D), 1023 (K), 986 (M), 954 (M), 954 (N), 840 (P), 804 (Q), 798 (R)

Steroids	No.	Type No.	Stereo-str.	Main Band	Additional Bands
Ergostanol	XIV	II	3 β–OH, 5 α	1037 (I)	1168 (F), 1130 (G), 1075 (H), 991 (K), 975 (L), 957 (M), 937 (M)
Δ^{22}–Stigmasten–3 β–ol	XV	II	3 β–OH, 5 α	1035 (I)	1168 (F), 1132 (G), 1074 (H), 991 (K), 957 (M), 935 (N)
Δ^5–Stigmasten–3 β–ol (β–Sitosterol)	XVI	V	3 β–OH, Δ^5	1051 (J)	1212 (D), 1022 (K), 986 (M), 953 (N), 838 (P), 805 (Q), 800 (R)

(3) 3-Hydroxy-5 α-pregnan-11, 20-diones and the Derivatives[537]

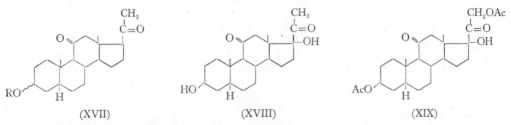

(XVII) (XVIII) (XIX)

(a) **3 α–Hydroxy–5 α–pregnan–11, 20–dione** (XVII)   R=H

　　IR λ^{CH2C12}_{max} μ: 2.76, 3.42, ca 3.48, 5.86 (11 and 20>C=O), 7.18, 7.35 (angular and C_{21} Me), 8.68
　　　　(20>C=O), 10.01 (HO *axial*, 3 α, 5 α), 10.82～10.91, ca 11.53 (3 α–OH, 5 α)

(b) **3β–Hydroxy–5α–pregnan–11, 20–dione** (XVII) R=H

　　IR λ^{CH2C12}_{max} μ: 2.75, 3.405, 3.48, 5.85 (11 and 20>C=O), 7.18, 7.34 (angular and C_{21} Me), 8.68, 9.27～
　　　　9.30, ca 9.62 (HO–*equatorial*), 9.68 (3β, 5α)

(c) **3 α–Acetoxy–5 α–pregnan–11, 20–dione** (XVII) R=Ac

　　IR λ^{CS2}_{max} μ: 3.40, 3.45～3.48, 5.735 (OAc), 5.83 (>C=O), 7.195, 7.35 (angular and C_{21} Me and AcO),
　　　　7.96, 8.06 (OAc *axial*), 8.61 (3α–OAc–5α–steroid), ca 8.7 (20>C=O), 9.83～9.85 (3α–
　　　　OAc–5α–steroid), 10.25～10.28 (3α–OAc–5α)

(d) **3α, 17 α–Dihydroxy–5 α–pregnan–11, 20–dione** (XVIII)

　　IR λ^{nujol}_{max} μ: 2.89, 2.99, 5.88 (11 and 20>C=O), 7.88～7.95, 8.16, 8.27, 8.67 (20>C=O), 10.03
　　　　(HO–*axial*, 3α, 5α) 10.84, 11.53 (3α–OH, 5α)

(e) **3 α, 21–Diacetoxy–17–hydroxy–5α–pregnan–11, 20–dione** (XIX)

　　IR λ^{CH2C12}_{max} μ: 2.78 (free OH), ca 2.80～2.90 (assoc. OH), 3.42, ca 3.48, ca 5.72 (3 and 21–OAc),
　　　　5.78 (21–OAc 20>C=O), 7.29, 7.35 (CH_3 of acetyl etc.), 8.15, ca 8.28, 8.42, 8.64
　　　　(3α–OAc–5α and 20>C=O), 9.56 (20>C=O), 9.84 (3α–OAc–5α), 10.31

(3) IR Spectra of Steroid Lactones[538]

　　Approximate values on $\nu_{C=O}$cm^{-1} for various steroid lactones, determined in CCl_4, $CHCl_3$, CH_2Cl_2 or CS_2 solutions, are as following :

(a) saturated–γ–lactones (1764～1789),　(b) saturated–δ–lactones (1723～1747),

(c) saturated–ε–lactones (1705～1727),　(d) unsaturated–γ–lactones (1723～1806)

537) Reichstein, T. *et al*: *Helv. Chim. Acta*, **42**, 1399 (1959).
538) Jones, N. R. *et al*: *J. A. C. S.*, **81**, 5242 (1959).

(a) Saturated-γ-Lactone

1778~1777 (CCl₄)

1786 (CCl₄)
1775 (CHCl₃)

1789 (CCl₄)
1776 (CHCl₃)

1782~1781 (CCl₄)
1777~1772 (CHCl₃)

1782 (CHCl₃)

1779 (CHCl₃)

1785 (CCl₄)

1773 (CS₂)
1764 (CH₂Cl₂)

1773 (CS₂)

(b) Saturated-δ-Lactone

1744~1742 (CS₂)
1747 (CCl₄)

1740~1739 (CCl₄)

1742 (CS₂)
1723~1718 (CHCl₃)
1744~1741 (CCl₄)

1744~1741 (CCl₄)

(c) Saturated-ε-Lactone

1727 (CS₂)
1705 (CHCl₃)

(d) Unsaturated-γ-Lactone

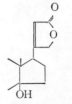

1783, 1756 (CS₂)

1784, 1758 (CS₂)
1787, 1752 (CHCl₃)

1783, 1755 (CS₂)
1790~1788, 1748 (CHCl₃)

1780, 1750 (CS₂)
1786, 1738 (CHCl₃)

(1740), 1718 (CHCl₃)   1751, 1736 (CS₂)   1740 (CS₂)   1757~1756 (CCl₄)
                   (1748), 1729 (CHCl₃)   1723 (CHCl₃)

[C] NMR Spectra

(1) Relationship between Chemical Structure and NMR Chemical Shifts[539].

Shoolery *et al* determined the NMR Spectra of 47 Steroids under the following conditions.
NMR in CDCl₃, 40 Mc, relative to C₆H₆ in external annulus, cps:
Three examples of them are indicated in Fig. 98.

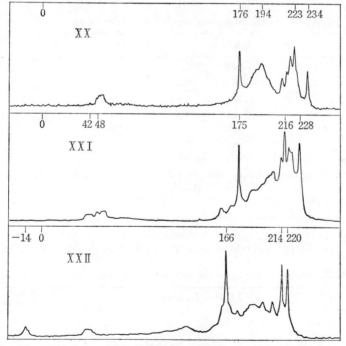

Fig. 98   NMR Spectra of 5, 6–Dihydroergosteryl acetate (XX), Stigmasteryl
acetate (XXI) and 16–Dehydropregnenolone (XXII)
(in CDCl₃, cps)   (Shoolery *et al.*)

---

539) Shoolery, J. N *et al*: *J. A. C. S.*, **80**, 5121 (1958).

(XX)

(XXI)

(XXII)

(a) Olefinic Protons

Signals appear at the low field, $-54 \sim +54$ cps.

Table 50 Shifts for Protons attached to Double Bonds (Shoolery)

Position	H to	Shift cps	Position	H to	Shift cps
Δ^1-3-keto	C_1	-54 d	Δ_5	C_6	$40 \sim 43$
	C_2	19 d	$\Delta^{17(20)}$	C_{20}	44 t
Δ^{16}-20-keto	C_{16}	-13	Δ^7	C_7	$47 \sim 48$
Δ^4-3-keto	C_4	$23 \sim 30$	Δ^{22}	C_{22}, C_{23}	$48 \sim 52$
$\Delta^{5,7}$	C_6, C_7	38			
$\Delta^{9(11)}$	C_{11}	39			

(b) Chemical Shifts of Hydroxyl Group and the Proton on substituted Carbon Atom

$$\left(> C < {OH \atop H} \right)$$

Chemical shift of hydroxyl proton is widely displaced depend on using solvent, concentration or temperature. Chemical shifts of the proton on the same carbon atom, connecting with hydroxyl group are indicated in Table 51. In general, the shifted value (cps) of *axial*–H is higher than that of *equatorial*–H.

Table 51 Shifts for Proton on substituted Carbon Atom in Steroids (Shoolery *et al.*)

Steroids	Proton Location	—OH	Proton Conformation	Shift cps
Androsterone	C_3	3α	*equatorial*	91
Epiandrosterone	C_3	3β	*axial*	113
5-Isoandrosterone	C_3	3α	*axial*	113
3-α-Hydroxypregnane-11, 20-dione	C_3	3α	*axial*	116
11-β-Hydroxyprogesterone	C_{11}	11β	*equatorial*	78
Corticosterone	C_{11}	11β	*equatorial*	82
11-α-Hydroxyprogesterone	C_{11}	11α	*axial*	95
11-α-Hydroxypregnane-3, 20-dione	C_{11}	11α	*axial*	95

(c) **Methoxyl Group** (–O–CH$_3$)   122 cps (Pregnenolone–3–methylether)

(d) **20–Keto–21–Methyl** $\left(\underset{17\ \ 20\ \ 21}{>\text{C}-\text{CO}-\text{CH}_3}\right)$ 171±1 c/s

(e) **21–Acetoxy–20–one** (–CO–CH$_2$–O–CO–CH$_3$)   168~169 cps (9/9) (–CH$_3$), 71 cps m (–CH$_2$–)

                                  20 21

(f) **3β–and 11α–Acetoxyl** (–O–CO–CH$_3$)   175±2 cps

(g) **The Shifts of 19–Angular Methyl**

    219~223 cps A, B, C rings without C=C, C=O Androstane–3β–ol–17–one, ——3α–ol–17–one, 5, 6–dihydroergosteryl acetate, etiocholane–3α–ol–17–one, ergosterol (218)

    214~216 cps type of Δ^5, 3–one, 20–one Δ$^{5(6)}$–Pregnene–3β–ol–20–one, pregnane–3,20 dione, allopregnane–3, 20–dione, Δ$^{5(6),16}$–pregnadiene–3β–ol–20–one, Δ$^{5(6)}$–pregne–3β–21–diacetoxy–20–one, Δ$^{5(6)}$ androstene–3β–ol–17–one, stigmasteryl acetate

    210~211 cps 11–α–Hydroxypregnane–3,20–dione, ——allopregnane–3,20–dione

    207~209 cps type of Δ^4–3–one, Δ^5–3–one, Δ^4–11–one, Δ^5–11–one, 3,11,20–trione Δ^4–Androstene–17β–ol–3–one,——17β–ol–17α–methyl–3–one, Δ^4–pregnene–3,20–dione–21–acetate, *allo*–pregnane or pregnane–3,11, 20–trione, Δ^4–pregnene–17α–ol–3,20–dione, Δ^5–cholestene–3–one

    204~206 cps Δ^4–Pregnane–11α–ol–3,20–dione, ——acetate

    199~200 cps type of Δ^4, 3,11–dione Δ^4–Pregnene–21–ol–3,11,20–trione, Δ^4–pregnene–3,11,20–trione, Δ^4–androstene–3,11,17–trione

    197 cps 11–β–ol, is lower 7 cps than the shift of corresponding 11–α–ol (204 cps). Δ^4–Pregnene–11β–ol–3,20–dione, ——11–β, 21–diol–3, 20–dione, Δ$^{4,9(11)}$–pregnadiene–11β, 21–diol–3–one.

(h) **The Shifts of 18–Angular Methyl**

    228~238 cps Most of 20–keto–C$_{21}$–Steroids, without 11–β–OH, 17–β–OH, Δ$_{17}$or 17–keto in the molecule. Δ$^{5(6)}$–Cholestene–3β–acetate, Δ$^{5(6)}$pregnene–3β–ol–20–one, Δ^4–pregnene–30,20–dione, *allo*–pregnane–3,11,20–trione, pregnane–3,20–dione, ——3,11,20–trione, Δ$^{5(6)}$–pregnene–3β–methoxy–20–one, ——3β,21–diacetoxy–20–one, ——3β,21–diol–20–one–21–acetate, *allo*–pregnane–3, 20–dione–21–acetate, Δ^4–pregnene–3, 11, 20–trione, *allo* pregnane–11α–ol–3,20–dione, stigmasteryl acetate, Δ$^{5(6)}$–pregnene–3β–acetate–20–one, 5,6–ergosteryl acetate, pregnane–11α–ol–3,20–dione, pregnane–3α–ol–11,20–dione,——acetate, ergosterol, Δ$^{5(6),7}$–cholestadiene–3β–acetate

    223~224 cps 17β–OH or 17–keto–steroids Δ^4–Androstene–17β–ol–3–one, 19–nortestosterone, androstane–3α–ol–17–one, etiocholane–3α–ol–17–one

    220~222 cps 17–keto, 11β–OH steroids ——Δ^4–Androstene–17β–ol–17α–methyl–3–one, Δ^4–pregnene–11β–ol–3, 20–dione, [Note], Δ$^{5(6),16}$–pregna–diene–3β–ol–20–one, Δ$^{5(6)}$–androstene–3β–ol–17–one, ——acetate, estrone, androstane–3β–ol–17–one

    220 cps Δ$^{5(6),16}$ Pregnadiene–3β–acetate–20–one, Δ^4–pregnene–11β, 21–diol–3,20–dione

(2) Chemical Shifts of Side-Chain[540].

Fig. 99 Shows the NMR spectra of the followed steroids. (in CDCl$_3$, 60 Mc, TMS as 282 c/s)

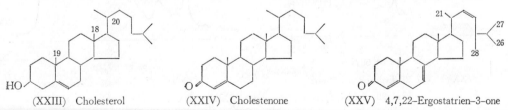

(XXIII) Cholesterol   (XXIV) Cholestenone   (XXV) 4,7,22–Ergostatrien–3–one

[Note] 11α–ol 228 cps

540) Slomp, G. *et al*: *J. A. C. S.*, **84**, 204 (1962).

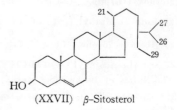

(XXVI)　5β–Stigmast–22–en–3–one

(XXVII)　β–Sitosterol

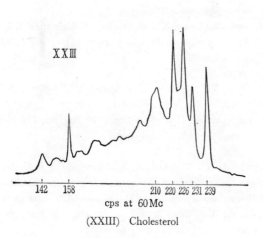

XXⅢ

cps at 60 Mc

(XXIII)　Cholesterol

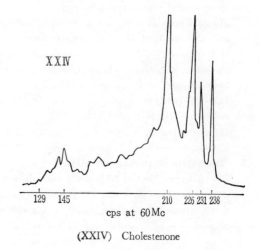

XXⅣ

cps at 60 Mc

(XXIV)　Cholestenone

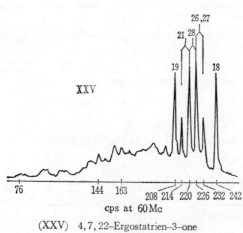

XXV

cps at 60 Mc

(XXV)　4,7,22-Ergostatrien-3-one

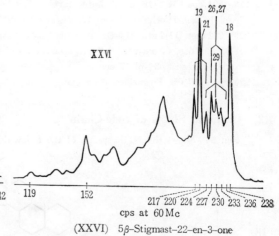

XXⅥ

cps at 60 Mc

(XXVI)　5β–Stigmast–22–en–3–one

Fig. 99　NMR Spectra of Steroides (XXIII)~(XXVII) (Slomp *et al.*)

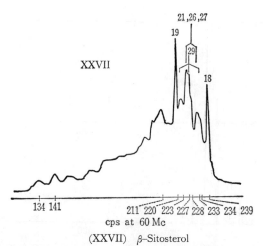

(XXVII) β-Sitosterol

Fig. 99 NMR Spectra of Steroids (XXIII)~(XXVII) (Slomp *et al.*)

(a) **Cholesterol** (XXIII) ca 210 (chain–CH$_2$–), 220 (19–Me), ca 226, 231 (21, 26, 27 3 Me), 239 (18–Me), 158 (OH), 129~145 (allylic H)

(b) **Cholestanone** (XXIV) ca 210 (chain–CH$_2$ and 19–Me), ca 226, 231 (21, 26, 27–3 Me), 238 (18–Me), 129~149 (allylic H)

(c) **4, 7, 22–Ergostatrien–3–one** (XXV) 208 (19–Me), 242 (18–Me), 226~232 d (iso Pr.) 214~220 d (21–Me), 220~232 d (28–Me)

(d) **5β–Stigmast–22–en–3–one** (XXVI) 220~238 (angularmethyl groups), 220 (19–Me), 238 (18–Me), 217 ~224 d (21–Me), 227~233 d (iso Pr.), 224~230~236 t (29–Me)

(e) **β–Sitosterol** (XXVII) 220 (19–Me), 239 (18–Me), 227~233 d (iso Pr.), 227~233 d (21–Me), 233~228 ~234 t (29–Me)

[D] Mass Spectra

(1) Mass Spectra of Steroid Ketones[541].

Fragmentation differs not only by the position of carbonyl radical but also by the conformation of A/B ring fusion.

Fig. 100 shows the mass spectra of 5α–androstan–11–one (XXX) and 5β–androstan–11–one (XXXI).

Fragmentation and the corresponding m/e of steroids (XXVIII)~(XXXIX) are described.

(a) **Cholestane** (XXVIII) M$^+$ 372, 315 (1), 272 (2), 262 (3), 259 (7), 232 (6), 217 (5), 149 (4), 109 (3′), 95 (2′), 55 (1′)

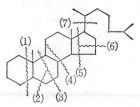

541) Djerassi, C. *et al*: *J. A. C. S.*, **84**, 1430 (1962).

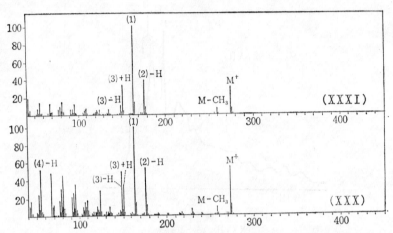

Fig. 100 Mass spectra of 5α–Androstan–11–one (XXX) and 5β–Androstan–11–one
(XXXI)   (Djerassi *et al.*)

(b) Androstane (XXIX) M⁺ 260, 245 s (M−CH₃), 232 w (6), 218 m ((5)−H), 203 s ((1)−H), 189 m, 175 w, 163 m ((2)−H), 145 s ((4)−H, (3)−H), 135 s ((8)−H), 121 s, 109 s ((4)−H, (3)−H), 95 s ((2)−H), 81 s ((7)−H), 67 s, 55 s ((1)−H)

(c) 5α–Androstan–11–one (XXX) M⁺ 274, 259 w (M−CH₃), 177 s ((2)−H), 164 s (1), 151 m ((3)+H), 149 s ((3)−H), 124 m, 109 w, 95 s, 80 s, 67 s, 55 s ((4)−H)

(d) 5β–Androstan–11–one (XXXI) M⁺ 274, 259 w (M−CH₃), 177 s ((2)−H), 164 s (1), 151 s ((3)+H), 149 w ((3)−H)

(e) **Cholestan-7-one** (XXXII) M^+ 386, 371 w (M—CH_3), 368 m (M—H_2O), 289 m (6), 255 s, 246 m **(5)**, 232 s (4), 191 m ((2)—H), 178 s (1), 165 w ((3)+1), 163 w ((3)—1), 150 m, 135 m, 109 w, 93 m, 81 s, **55 s**

(f) **3β-Acetoxycholestan-7-one** (XXXIII) M^+ 444, 426 w (M—H_2O), 384 w (M—CH_3COOH), 331 w **(6)**, 304 w (5), 290 s ((4)—H), 249 w ((2)—H), 236 s (1), 230 w ((3)+H), 93 m, 55 w

(g) **Cholestan-6-one** (XXXIV) M^+ 386, 371 s (M—CH_3) 330 s ((1)+1), 273 s (2), 246 m (4), **230 s** ((3)—H), 124 s, 119 s, 95 s, 81 s, 67 m, 55 s

(h) **Coprostan-6-one** (XXXV) M^+ 386, 371 s (M—CH_3) 330 s ((1)+1), 273 m (2), 246 (4), **230 s** ((3)—H), 124 s, 109 m, 95 m, 81 w, 67 w, 55 s

(i) **3β-Acetoxyergostan-12-one** (XXXVI) M^+ 458, 398 s (M—CH_3COOH), 331 w (2), 290 s ((1)+H), 271 s ((2)—CH_3COOH), 231 s ((1)—CH_3COOH+H), 81 s, 55 s

(j) Cholestan–4–one (XXXVII)   M+ 386, [Note 1] 371 m (M−CH₃), 368 w (M−H₂O), 246 m (2), 230 s ((3)−H), 110 m ((1)+H), 95 m, 81 m, 67 m, 55 s

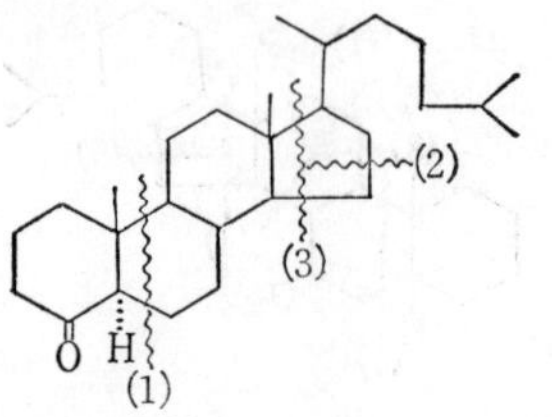

(k) Cholestan–1–one (XXXVIII)   M+ 386, 371 m (M−CH₃), 353 w, 300 w, 273 w (6), 246 s (5), 231 s (4), 137 m ((3)−H), 124 s (1), 107 m ((2)+H)

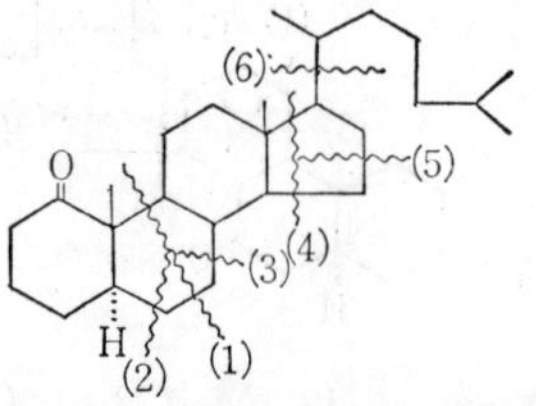

(l) Coprostan–3–one (XXXIX)   M+ 386, [Note 2] 371 w (M−CH₃), 316 s (3), 246 w (2), 231 s ((1)−H), 176 w, 161 m, 95 m, 81 w, 55 s

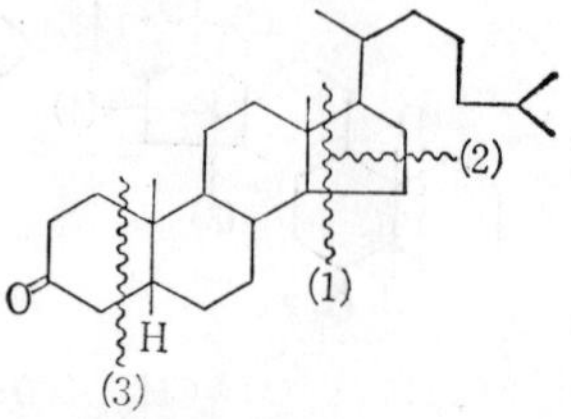

(2) Mass Spectra of Pregnanes and Pregnenes[542]

L. Peterson[542] determined the mass spectra of 30 sorts of pregnanes and pregnenes. The intensities of each fragments were compared to the strongest peak as 100% in the spectrum.

s: 100~20%   m: 20~10%   w: 10~1%

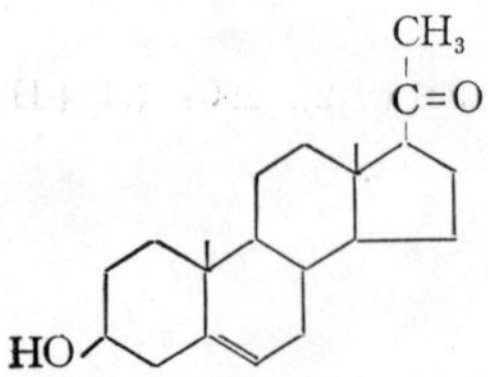

(XL)

Pregnenolone

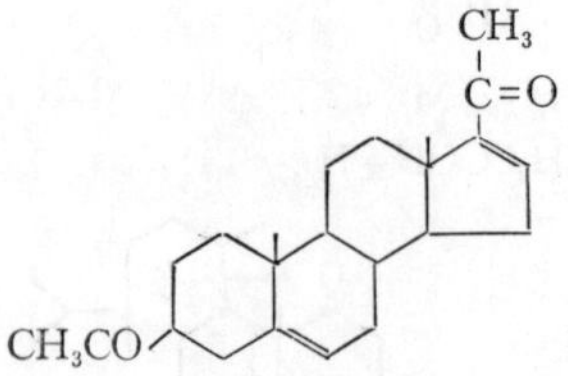

(XLI)

16–Dehydropregnenolone acetate

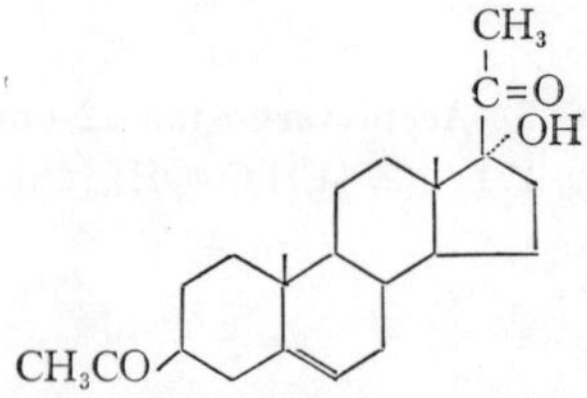

(XLII)

17–α–Hydroxypregnenolone acetate

[Note 1] Coprostan-4-one shows almost the same spectrum but m/e 110 is very strong.
[Note 2] Cholestan-3-one (5H–α) shows no m/e 316 but 231 is stronger.
542) Peterson L: *Anal. Chem.*, **34**, 1781 (1962).

(XLIII)

5α–Pregnane–3β, 17α–diol–20–one acetate

(XLIV)

Progesterone

(a): M^+
(b): $M-CH_3$
(c): $M-H_2O$
(d): $M-(H_2O+CH_3)$
(e): $M-2H_2O$
(f): $M-CH_3CO$
(g): $M-CH_3COOH$
(h): $M-(CH_3CO+H_2O)$
(i): $M-(CH_3COOH+CH_3)$
(j): $M-(CH_3COOH+H_2O)$

(k): $M-(CH_3CO+2H_2O)$
(l): $M-(CH_3COOH+CH_3+H_2O)$
(m): $M-(CH_3COOH+CH_3CO)$
(n): $M-(CH_3COOH+CH_3CO+H_2O)$
(o): $M-$ring D$+H$
(p): $M-$ring D$+H-CH_3COOH$
(q): $M-$ring D$+H-H_2O$
(r): Partial ring D cleavage
(s):   〃 $-CH_3CO_2H$
(t):   〃 $-H_2O$

(a) **Pregnenolone** (XL) M^+ 316 s, 301 s (b), 298 s (c), 283 s (d), 273 w (f), 255 s (h), 231 s (o), **213 s** (q), 205 s, 187 s, 161 s, 159 s, 147 s, 133 s, 119 s, 107 s, 105 s, 91 s

(b) **16–Dehydropregnenolone acetate** (XLI) M^+ 356(−), 297 s, 296 s (g), 281 s (i), 175 m, 109 m, **145s**, 133 m, 121 s, 105 s

(c) **17–α–Hydroxypregnenolone acetate** (XLII) M^+ 374 (−), 314 s (g), 313 w (h), 299 w (i), **298 m,** 297 s, 296 s (j), 281 s (l), 271 w (m), 268 w, 253 s (m), 227 w, 213 s (p), 173 w, 161 s, 145 s, **131s**, 121 s, 107 s, 105 s, 91 s

(d) **5α–Pregnane–3β, 17α–diol–20–one acetate** (XLIII) M^+ 402 (−), 387 w (b), 342 w (g), 327 w (i), 298 w, 213 w (p), 105 w, 87 s, 79 w

(e) **Progesterone** (XLIV) M^+ 315 s, 300m (b), 295 w (c), 281 w (d), 273 m, 272 s, 271 w (f), 257 w, 244 s, 230 m, 229 s (o), 227 w, 199 m, 150 s, 147 s, 133 s, 124 s, 121 s, 107 s, 105 s, 91 s, 79 s, **77 s,** 55 s

[E] ORD and Gas Chromatography

C. Djerassi *et al* reported on ORD of keto-sterols and cardenolides[543].

Analysis of C_{19}–, C_{21}–, C_{27}–steroids and cholic acid by high sensitive gas chromatography, using SE–30 or QF–1 and ionization detecter were reported[544]~[548]. K. Tsuda, *et al* reported on gas chromatography of C_{27}–, C_{28}– and C_{29}–phytosterols[549] (see the next item [F]).

543) Djerassi, C. *et al*: *Helv. Chim. Acta*, **41**, 250 (1958).
544) Lipsky, S. R. *et al*: *Anal. Chem.*, **33**, 318 (1961).
545) Bloomfield, D. K: *ibid.*, **34**, 737 (1962).
546) Brooks, C. J. W. *et al*: *Biochem. J.* **87**, 151 (1963).
547) Nicalaides, N: *J. Chromat.* **4**, 496 (1960).
548) Horning, E. C. *et al*: *J. A. C. S.*, **82**, 3481 (1960).
549) Tsuda, K. *et al*: *Chem. Pharm. Bull.*, **9**, 835 (1961).

(1) Examples of Gas Chromatography of Steroids[548]

<Condition> Argon ionization detector, 6 ft. × 4mm i. d. columns. Pressure, 20 p. s. i., 7/100 SE–30 on Chromosorb W, 80～100 mesh at 260°; Pressure 10 p. s. i., 2～3/100 SE–30 on Chromosorb W, 80～100 mesh at 222°.

Table 52 Gas Chromatography of Steroids (Relative Retention Time) (Horning *et al.*)

Compound	Relative Retention Time at 260°	222°
Androstane	0.17	0.11
Androstane–17–one	0.30	0.22
Androstan–3, 17–dione	0.56	0.47
4–Androsten–3, 17–dione	0.68	0.57
Pregnan–3, 20–dione	0.74	0.67
Allopregnan–3, 20–dione	0.82	0.74
Allopregnan–3, 20–diol		0.70
Allopregnan–3, 11, 20–trione	1.05	0.99
Coprostane		0.90
Cholestane	1.00	1.00
Cholestanyl methylether	1.58	1.78
Cholesteryl methylether	1.47	1.72
Cholestan–3–one	2.00	2.17
4–Cholesten–3–one	2.37	2.72
Cholestanol	1.70	1.99
Cholestenol	1.21(br.)	1.98
Cholestanyl acetate	1.15(v. br.)	2.84
Cholesteryl acetate	1.18(br.)	2.81
β–Sitosterol	1.82(v. br.)	3.26
β–Sitosteryl acetate		4.62
Stigmastane		1.65
Stigmasterol	1.62 2.29	2.84

〔F〕 Sterols

(1) Gas Chromatography of C_{27}, C_{28} and C_{29} Sterols[549]

C₂₇-Sterols   C₂₈-Sterols   C₂₉-Sterols

≪Condition≫　Barber–Colaman Model–10, Argon ionization Detector, 9 ft.×8 mm. i.d.　Column pressure, 35 lb/in², temp. 220°, 1% SE–30 on Chromosorb w, 60~80 mesh, Flash temp. 290°, cell temp. 170°, retention time of reference (XLV) 9.1 min.

Table 53　Relative Retention Time of Sterols　　　　(Tsuda *et al.*)

Sterols	R	Δ	No.	Time
C_{27}–Sterols				
Cholestane	H	—	XLV	1
Cholesterol	β–OH	5	XLVI (XIII)	1.69
$\Delta^{8(14)}$–Cholestenol	β–OH	8(14)	XLVII	1.73
Δ^{14}–Cholestenol	β–OH	14(15)	XLVIII	1.83
Δ^{7}–Cholesterol	β–OH	7	XLIX	1.93
20–Iso–22–dehydrocholesterol	β–OH	(20 α–)5, 22	L	1.35
22–Dehydrocholesterol	β–OH	5, 22	LI	1.57
$\Delta^{7,22}$–Cholestadien–3β–ol	β–OH	7, 22	LII	1.87
C_{28}–Sterols				
Δ^{22}–24β–Methylcholesterol (Brassicasterol)	β–OH	5, 22	LIII	1.91
24–Methylencholesterol	β–OH	5, 28	LIV	2.16
$\Delta^{8(14)}$–Ergosten–3β–ol (α–Ergosterol)	β–OH	8(14)	LV	2.17
5, 6–Dihydroergosterol	β–OH	8, 22	LVI	2.21
Δ^{24}–24–Methylcholesterol	β–OH	5, 24(25)	LVII	2.47
C_{29}–Sterols				
Stigmasterol	β–OH	5, 22	LVIII	2.38
Fucosterol	β–OH	5, 24(28)	LIX	2.76
Δ^{24}–24–Ethylcholesterol	β–OH	5, 24(25)	LX	2.98
Fucostadienone	=O	5, 24(28)	LXI	3.79
Δ^{24}–24–Ethylcholestenone	=O	5, 24(25)	LXII	4.06

Summary: (a) R.R.T. $_{20-Iso-22-dihydrocholesterol}$ < R.R.T. $_{22-dihydrocholesterol}$,　(b) Relative retention time increases 0.34~0.6 by the addition of –CH_2– radical (LI⇒LIII⇒LVIII), (c) $\Delta^5 < \Delta^{8(14)} < \Delta^{14} < \Delta^7$ (d) Δ^{22} decreases R.R.T. (XLVI⇒LI, XLIX⇒LII), (e) $\Delta^{22} < \Delta^{24(28)} < \Delta^{24}$

Murakami *et al*[550] reported on the separation of phytosterols by gas chromatography under the following condition.

≪Condition≫　Shimazu GC–1–B–Type (Dual Column, differential Flame), 150 cm×6 mm, 1.5% SE–30 on Chromosorb w (60~80 Mesh), Column temp. 240°, sample heater temp. 280°, detector carrier gas N_2, 150 *ml*/min.

　Relative retention times of phytosterols are indicated in parenthesis, when r.r.t of cholestane is taken as 1 (6 minutes)).　Campesterol (LXVI) (2.37), stigmasterol (LVIII) (2.60), β–sitosterol (LXVIII) (2.95), γ–sitosterol (LXIX) (2.38), β–sitostanol (3.00).

550)　Murakami, T. *et al*: *Y. Z.*, **83**, 427 (1963).

(2) Sterols See Table 54.

(a) C$_{27}$-Sterols

(XLVI)=(V)
Cholesterol

(LXIII)
Zymosterol

(LXIV)
Periocerol

(b) C$_{28}$ Sterols

(LXV)
Ergosterol

(LIII)
Brassicasterol

(LXVI)
Campesterol

(LIV) 24-Methylencholesterol

(LXVII) Ascosterol

(c) C$_{29}$ Sterols

(LVIII)=(IX) Stigmasterol

(LXVIII)=(XVI) β–Sitosterol (28α)
(LXIX) γ–Sitosterol (28β)

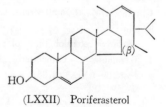

(LXX) α–Spinasterol

(LXXI) Fucosterol

(LXXII) Poriferasterol

Table 54 Naturally Occurring Sterols

Sterol	Str. No.	Formula	mp.	$[\alpha]_D$	Occurrence
Cholesterol	XLVI	$C_{27}H_{46}O$	149	$-$ 39	Gall
Zymosterol	LXIII	$C_{27}H_{44}O$	110	$+$ 49	Yeast
Peniocerol	LXIV	$C_{27}H_{46}O_2$	168~171	$+$ 59	*Peniocereus fosterianus*
Ergosterol	LXV	$C_{28}H_{44}O$	165	$-$130	Yeast, fungi
Brassicasterol	LIII	$C_{28}H_{46}O$	148	$-$ 64	*Brassica rapa* sojabeans, wheat,
Campesterol	LXVI	$C_{28}H_{48}O$	58	$-$ 33	*Brassica Campestris*, pollen, oyster
24–Methylencholesterol	LIV	$C_{28}H_{46}O$	142		
Ascosterol	LXVII	$C_{28}H_{46}O$	142	$+$ 45	Yeast
Stigmasterol	LVIII	$C_{29}H_{48}O$	170	$-$ 49	Soyabeans, calabar beans
β–Sitosterol	LXVIII	$C_{29}H_{50}O$	140	$-$ 36	see (7)
γ–Sitosterol	LXIX	$C_{29}H_{50}O$	148	$-$ 43	Soya oil
α–Spinasterol	LXX	$C_{29}H_{48}O$	172	$-$ 4	see (9)
Fucosterol	LXXI	$C_{29}H_{48}O$	124	$-$ 38	Brown algae
Poriferasterol	LXXII	$C_{29}H_{48}O$	156	$-$ 49	Sponge

(3) **Cholesterol** (XLVI)=(V)=(XIII)=(XXIII)

See Table 54, acetate 115°

UV See the item [A], (4) Tab. 48.

IR See the item [B], (2) Tab. 49 (II).

NMR See the item [C], (2), Fig. 99 (XXIII).

(4) **Peniocerol** (Cholest-8(9)-ene-3 β, 6 α-diol) (LXIV)[550)559)]

See Table 54.

UV: ε_{204} 4650, ε_{210} 3900, ε_{220} 1340 in EtOH

(5) **Ergosterol** (LXV)

See Table 54.

UV: (Δ^7-Ergostenylacetate (VII), $\Delta^{8)14)}$-ergostenyl acetate (VIII)), see the item [A], (4), Tab. 48.

IR: (Ergostanol (XIV)) See the item [B], (2), Table. 49 (II).

NMR: See the item [C], (1), (g), (h), (2) (XXV).

(6) **Stigmasterol** (LVIII)=(IX)

See Table 54.

UV: See the item [A], (4), Tab. 48.

IR: (Δ^{22}-Stigmasten-3β-ol (XV)) See the item [B], (2), Table 49 (II).

NMR: (Stigmasterol acetate) See the item [C], (1), (g), (h), (5β-Stigmast-22-en-3-one), [C], (2).

(7) β-**Sitosterol** (LXVIII)=(XVI)=(XXVII)

See table 54, acetate mp. 129~130°, 126~128°, $[\alpha]_D$-44° (CHCl$_3$), $C_{31}H_{52}O_2$.

≪Occurrence≫ *Pyrola japonica* (イチヤクソウ), *Panax gingseng*. C. A. MEYER (オタネニンジン)[550a)]

550a) Takahashi, M. *et al*: *Y. Z.*, **81**, 771 (1961).

Pinellia ternata (カラスビシャク), *Osmanthus fragrans* LOUR. var. *aurantiacus* MAKINO (キンモクセイ),
Ligustrum japonicum THUMB. (ネズミモチ), *Coix lacrymajobi* L. var. *ma-yuen* STAPF. (ハトムギ), *Aralia
elata* Seeman (タラノキ)[551], *Salvia officinalis*[552)553], *Ocimum bacilicum* (メボウキ), *Quassia amara* L[554].,
Aesculus hippocastanum[554a].

IR: See the item [B], (2), Tab. 49 (II).

NMR: See the item [C], (2) Fig. 99.

(8) β-Sitosterone (LXXIII)[554]

≪Occurrence≫   *Quassia amara* L.,   mp. 95~96.5°, $[\alpha]_D+81.3°$.

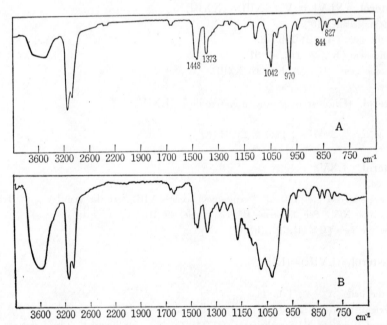

(LXXIII)

UV $\lambda_{max}\,m\mu$ (ε): 241 (16400).

IR ν_{max}^{KBr} cm^{-1}: 1680, 1622

A

1448 1373   1042 970   827
844

3600 3200 2600 2200 1900 1700 1500 1300 1150 1050 950 850 750 cm^{-1}

B

3600 3200 2600 2200 1900 1700 1500 1300 1150 1050 950 850 750 cm^{-1}

Fig. 101 IR Spectrum of (A) α-Spinasterol and (B) α-Spinasterol
glucoside in KBr   (Ito *et al.*)

551) Murakami, T. *et al*: *ibid.*, **83**, 427 (1963).
552) Nicholas, H. J: *J.P.S.*, **50**, 504 (1961).
553) Nicholas, H. J: *ibid.*, **50**, 645 (1961).   554) Lavie, D. *et al*: *Soc.*, **1963**, 5001,
554a) Fischer, F. G. *et al*: *Ann.*, **636**, 105 (1960).

(9) α-**Spinasterol** (LXX)

See Table 54.

≪Occurrence≫　*Bupleurum falcatum* L. (ミシマサイコ)[555], *Beta vulgaris* L. (サトウダイコン)[556], *Spinacia oleracea* L. (ホウレンソウ)[557], *Platycodon grandiflorum* DC. (キキョウ)[558]

mp. 166~168°, $[\alpha]_D$ −1.8°, $C_{29}H_{48}O \cdot 1/2H_2O$[559a], acetate mp. 183~185°, $[\alpha]_D$ −5.0.

Glucoside[556)558]　mp. 292°, $[\alpha]_D$ −34.1°, $C_{35}H_{58}O_6$

IR: See Fig. 101.

NMR: Acetate (5, 6–Dihydroergosteryl acetate) (XX)　See the item 〔C〕, (1), Fig. 98.

β–**Spinasterol**[560]　mp 148~150°, $[\alpha]_D$ +5.9°.

IR $\lambda_{max}\,\mu$: 9.51, 9.59, 12.01, 2.92 (OH), 6.26, 10.32, 12.01.

β–Spinasterol was a mixture of Δ^7–stigmasterol (LXXIII) with a small amounts of α–spinasterol (LXX)[560)561]

(10) Δ^7-**Stigmasterol** (**LXXIII**-a) and Δ^{22}-**Stigmasterol** (**LXXIV**)[562)563]

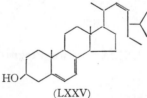

　　　　(LXXIII)　　　　　　　　　　　　　　　　(LXXIV)

≪Occurrence≫　Wheat, rye, *Bupleurum falcatum* L. (ミシマサイコ).

(a)　Δ^7-**Stigmasterol** (LXXIII-a)[561]　　mp. 145~147°, $[\alpha]_D$ +7.9~11° (CHCl₃)

Acetate mp. 157~158°, $[\alpha]_D$ +6.9° (CHCl₃).

(b)　Δ^{22}-Stigmasterol (LXXIV)=(XV)[562)564]　　mp. 158~159°, $[\alpha]_D$ +3.3° (CHCl₃).

Acetate, mp. 144~144.5°, $[\alpha]_D$ −7° (CHCl₃).

IR: See the item 〔B〕, (2), Table 49 (II).

NMR: (5β–Stigmast–22–en–3–one)　See the item 〔C〕, (2).

(11) **Dehydrostigmasterol** (LXXV)[564]

≪Occurrence≫　*Phellodendron amurense* RUPR. (オウバク)　　mp. 150°, $C_{29}H_{46}O$.

　　　　　　　　(LXXV)

---

555)　Takeda, K. *et al*: *Y. Z.*, **73**, 272 (1953).

556)　Matsumoto *et al*: *J. Chem. Soc. Japan.*, **75**, 346 (1954).

557)　Obata, Y. *et al*: *Bull. Agr. Chem. Soc. Japan.*, **19**, 189 (1955).

558)　Itō, M. *et al*: in press,

559)　Djerassi, C. *et al*: *Proc. Chem. Soc.*, **1961**, 450.

559a)　Takeda K. *et al*: *Y. Z.*, **73**, 272 (1953).

560)　Ball, C. D: *J. A. C. S.*, **64**, 2488 (1942).

561)　Takeda, K *et al*: *Chem. Pharm. Bull.*, **6**, 437 (1958).

562)　Takeda, K *et al*: *ibid.*, **6**, 536 (1958).

563)　Barton, D. H. R. *et al*: *Soc.*, **1948**, 783, 1354.

564)　Nishioka, *et al*: *Y. Z.*, **78**, 1433 (1958).

(12) 3α, 7α, 22α-**Trihydroxystigmastene** (5) (LXXVI) and 7α-**Hydroperoxystigmasten**-(5)-**diol**-(3α, 22α) (LXXVII)[565]

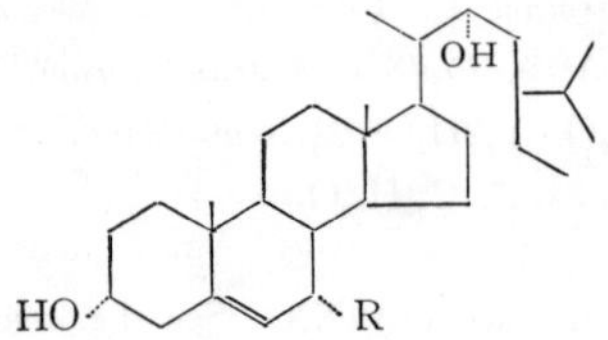

R=OH 3α, 7α, 22α–Trihydroxystigmastene (5) (LXXVI)

R=O–OH 7α–Hydroperoxystigmasten–(5)–diol–(3α, 22α) (LXXVII)

《Occurrence》 *Aesculus hippocastanum*

(a) 3α, 7α, 22α-**Trihydroxystigmastene** (5) (LXXVI) mp. 180~182°, $[\alpha]_D$ −61.5° (CHCl$_3$), C$_{29}$H$_{50}$O$_3$

UV λ_{max} mμ (log ε): 213 (2.41).

IR ν_{max} cm^{-1}: 3480, 1465, 1375.

(b) 7α–**Hydroperoxystigmastene**–(5)–**diol**–(3α, 22α) (LXXVII)

mp. 166~167°, $[\alpha]_D$ −78.0° (EtOH), C$_{29}$H$_{50}$O$_4$

UV λ_{max} mμ: <215

IR ν_{max} cm^{-1}: 1120, 1100, 965, 910

［G］ C$_{21}$ Steroids

C$_{21}$ Steroids, having skeleton (LXXVIII) occur in various plants as glycoside. They were reviewed by Teshesche[566] Amino steroids, belong to this type are listed in the item **20**, ［O］ Steroid Alkaloids.

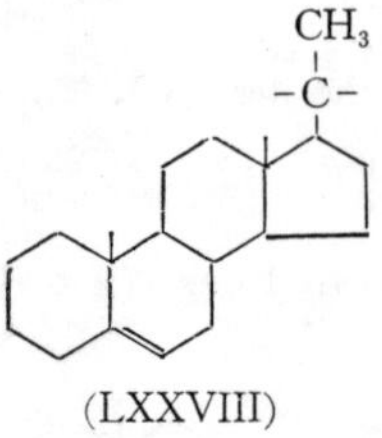

(LXXVIII)

(1) Δ^5-**Pregnenol**-(3 β)-**one** (20) (LXXIX) and 5 α-**Pregnanol** (3 β)-**one** (20) (LXXX)[566]

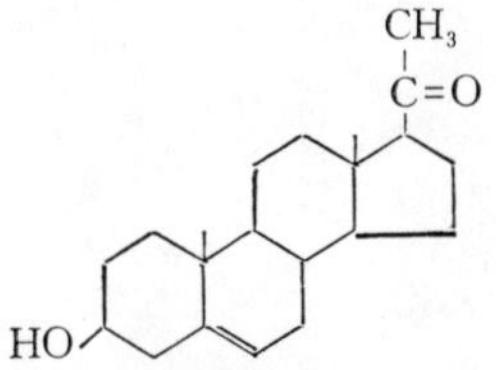

(LXXIX) 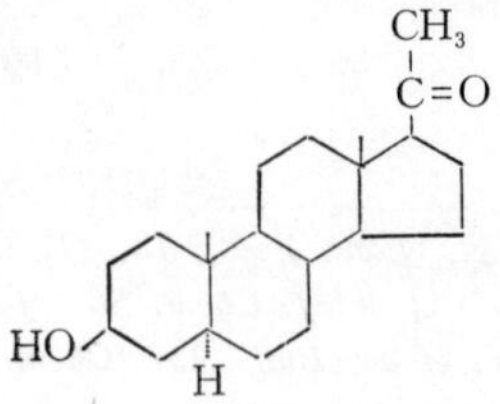   (LXXX)

565) Fischer, F. G. *et al*: *Ann.*, **636**, 88 (1960).

566) Tschesche, R: *Angew. Chem.*, **73**, 727 (1961).

≪Occurrence≫   *Schizoglossum Shireuse* BRWN. (Uzara), as a mixture " Uzarone " (1:2)–glucoside.
mp. 189~192°, $[\alpha]_D$ +66.5° ($CHCl_3$).

UV $\lambda_{max}^{MeOH}m\mu$ (E): 286 (66); IR $\lambda_{max}^{CHCl_3}\mu$: 9.31 (OH), 9.68, 9.93, 10.08, 10.24, 10.51.

(2) 14 α-**Digipronin** and **Digiprogenin**[567]~[569]

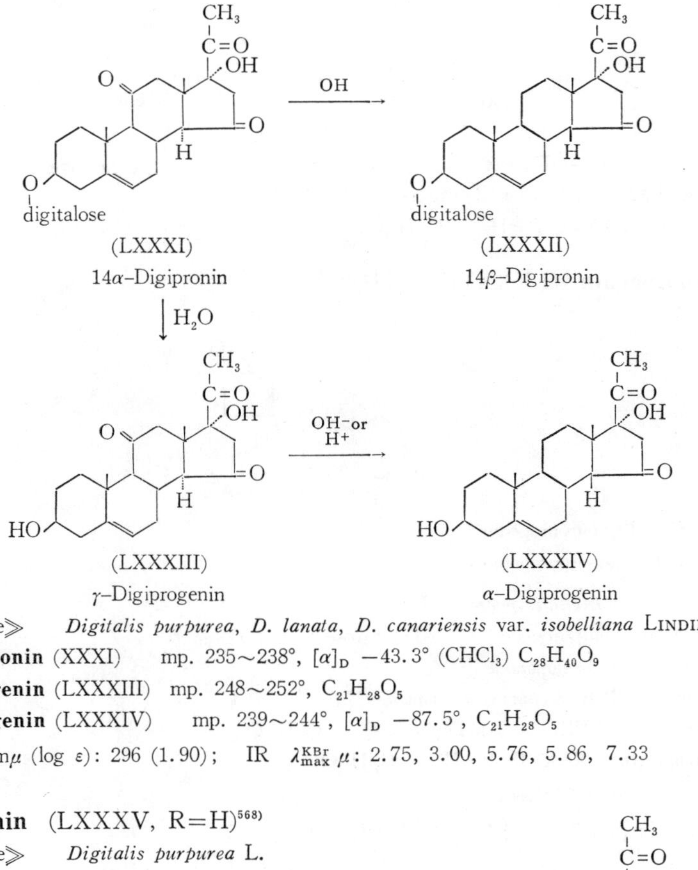

(LXXXI)
14α–Digipronin

(LXXXII)
14β–Digipronin

(LXXXIII)
γ–Digiprogenin

(LXXXIV)
α–Digiprogenin

≪Occurrence≫   *Digitalis purpurea, D. lanata, D. canariensis* var. *isobelliana* LINDINGER[569a]

(a) **14α–Digipronin** (XXXI)   mp. 235~238°, $[\alpha]_D$ −43.3° ($CHCl_3$) $C_{28}H_{40}O_9$

(b) **γ–Digiprogenin** (LXXXIII) mp. 248~252°, $C_{21}H_{28}O_5$

(c) **α–Digiprogenin** (LXXXIV)   mp. 239~244°, $[\alpha]_D$ −87.5°, $C_{21}H_{28}O_5$

UV $\lambda_{max}^{EtOH}m\mu$ (log ε): 296 (1.90);   IR $\lambda_{max}^{KBr}\mu$: 2.75, 3.00, 5.76, 5.86, 7.33

(3) **Purpnigenin** (LXXXV, R=H)[568]

≪Occurrence≫   *Digitalis purpurea* L.

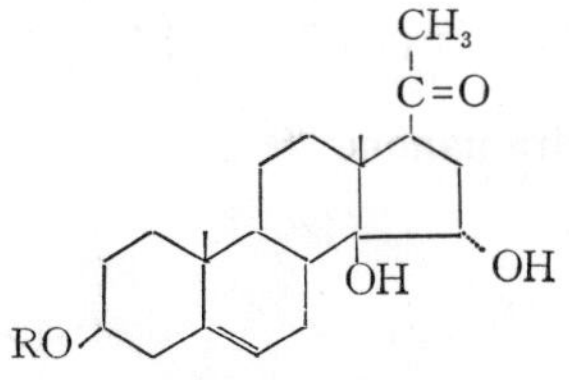

(LXXXV)

Purpnin⟶Purpnigenin+3–D–Digitoxose

Purpnin (LXXXV, R=3–D–Digitoxose)
mp. 282~287°, $C_{39}H_{62}O_{13}\cdot H_2O$

Purpnigenin (LXXXV, R=H)
mp. 239~243°, $C_{21}H_{32}O_4$

567) Tschesche, R: *Ann.*, **636**, 105 (1960).

568) Robertson, A. V. *et al*: *Soc.*, **1962**, 3610.

569) Satoh, D. *et al*: *Chem. Pharm. Bull.*, **10**, 37, 48 (1962).

569a) Pavanaram, S. K. *et al*: *Helv. Chim. Acta*, **46**, 1377 (1963).

(4) Digacetigenin (LXXXVI, R=H)[568]

《Occurrence》 *Digitalis purpurea* L.

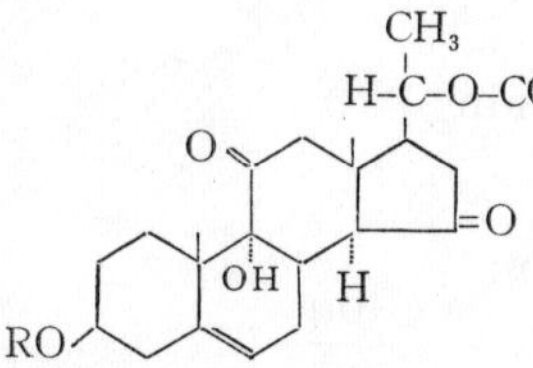

(LXXXVI)

Digacetinin——→Digacetigenin +3 D–Digitoxose——→Desacetyldigacetigenin +CH₃COOH

Digacetinin (LXXXVI, R=3 D–digitoxose) $C_{43}H_{64}H_{16}$

Digacetigenin (LXXXVI, R=H) $C_{21}H_{28}O_5$

(5) Digipurpurogenin (LXXXVII, R=H)[570)571)571a)b)]

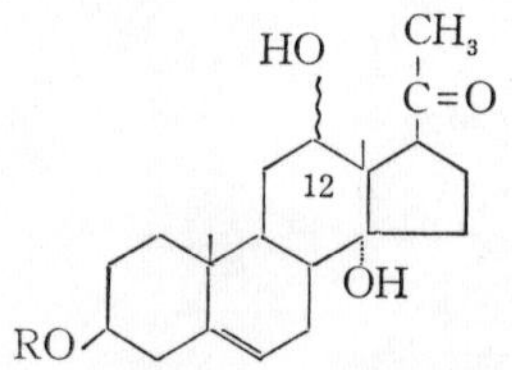

(LXXXVII)

R=H, 12–OHα Digipurpurogenin I
R=H, 12–OHβ Digipurpurogenin II

《Occurrence》 *Digitalis purpurea* L.

Digipurpurin——→Digipurpurogenin +3 D–Digitoxose
Digipurpurogenin II = Isoramanone[571c)]
Isodigipurpurogenin II (C₁₇–epimer) = Ramanone[571c)]

(a) Digipurpurin (LXXXVII, R=3 D–Digitoxose) $C_{39}H_{64}O_{14}$
(b) Digipurpurogenin (LXXXVII, R=H) $C_{21}H_{32}O_4$

IR ν_{max}^{KBr} cm⁻¹: 3400, 1675, 1367

Acetone adduct: mp. 167~175°, [α]_D +51° (CHCl₃), $C_{24}H_{38}O_5$

(6) Benzoylramanone[571c)]

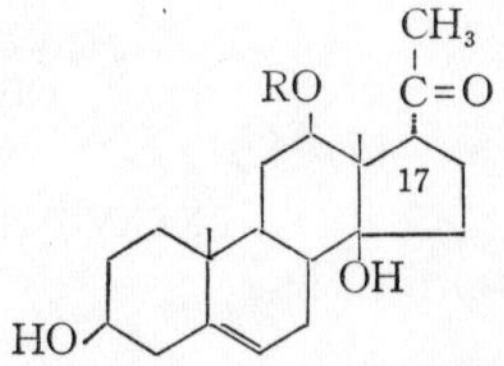

R=C₆H₅CO–: Benzoylramanone
R=H: Ramanone
R=H, 17–COMeβ: Isoramanone

(LXXXVII–a)

《Occurrence》 *Mitaplexis japonica* MAKINO

570) Tschesche, R. *et al*: *Ann.*, **648**, 185 (1961).
571) Shoppee, C. W. *et al*: *Proc. Chem. Soc.*, **1962**, 65.
571a) Tschesche, R. *et al*: *Ann.*, **603**, 59 (1957).
571b) Tschesche, R. *et al*: *Ber.*, **88**, 1569 (1955).
571c) Mitsuhashi, H. *et al*: *Chem. Pharm. Bull.*, **11**, 1333 (1963).

(a) **Benzoylramanone** $(R=C_6H_5CO-)$ mp. $222\sim226°$, $C_{28}H_{36}O_5$

UV λ_{max}^{EtOH} mμ (log ε): 233 (4.11), 276 (3.18)

IR ν_{max}^{nujol} cm^{-1}: 3600, 3500, 1720, 1690, 1580, 1270, 715.

(b) **Ramanone** $(R=H)$ mp. $184\sim196°$, $C_{21}H_{32}O_4$

IR ν_{max}^{nujol} cm^{-1}: 3500, 1680

(c) **Isoramanone** $(R=H, 17-COMe\beta)$ mp. $220\sim234°$, $C_{21}H_{32}O_4$

IR ν_{max}^{nujol} cm^{-1}: 3500, 1680, (21–Me)

(7) **Diginigenin** (LXXXVIII, R=H)[571)571a]

(LXXXVIII)　　　　(Former Structure)[572]

≪Occurrence≫　*Digitalis purpurea* L., *D. Lanata*, *D. Canariensis* L. var. *isobelliana*.

Diginin $\longrightarrow$ Diginigenin + D–Diginose

Digitalonin $\longrightarrow$ Diginigenin + D–Digitalose

(a) **Diginin** (LXXXVIII, R=D–Diginose) mp. $155\sim183°$, $[\alpha]_D$ $-176°$, $-224.3°$ (CHCl$_3$), $C_{28}H_{40}O_7$

UV λ_{max} mμ (log ε): 310 (2.00)

IR $\nu_{max}^{CHCl3(KBr)}$ cm^{-1}: 3585 (3500) (OH), 1735 (1740) (C=O), 1712 (1710) (C=O), 1655 (1635) (C=C), 1095, 1060, 1032 (1090, 1065, 1030) (C–O–C), 891, 872, 853 (880, 850, 810) (–CO–C–O –C)[568]

NMR τ: 9.00 s (18–Me), 8.67 d (J=6.5 c/s), 8.72 d (J=6.5 c/s) (diginose terminal Me and 21–Me), 6.60 s (diginose OCH$_3$), 8.45 s (19–Me)[571]

(b) **Diginigenin** (XXXVIII, R=H) mp. $115°$, $[\alpha]_D$ $-226°$, $C_{21}H_{28}O_4$

UV λ_{max} mμ (log ε): 310 (1.93)

IR ν_{max}^{CHCl3} cm^{-1}: 3585 (OH), 1735 (C=O), 1712 (C=O), 1655 (C=C), 1065, 1040 (C–O–C).

NMR of acetate[568] τ: 4.55 m (1 $C=\overset{6}{C}<_H$), 5.40 m (3–H and 20–H), 6.08 s (12–H), 8.99 s (18–Me), 8.73 d (J=6.5 c/s) (21–Me) 8.46 s (19–Me), 8.02 s (OAc)

Mass 344[571]

(8) **Digifologenin** (LXXXIX, R=H)[571a)573]

(LXXXIX)

---

572) Reichstein, T *et al*: *Helv. Chim. Acta*, **23**, 975 (1940).

573) Schoppee, C. W *et al*: *Soc.*, **1963**, 3281.

≪Occurrence≫   *Digitalis lanata, D. Canariensis* L. var. *isobelliana*

Digifolein ⟶ Digifologenin + D–Diginose

Lanafolein ⟶ Digifologenin + D–Oleanose

(a) **Digifolein** (LXXXIX, R=D–Diginose)   mp. 198~202°, $[\alpha]_D$ −214.5° (CHCl$_3$), $[\alpha]_{500}$−170°, $C_{28}H_{40}O_8$

UV λ_{max}^{MeOH} mμ (log ε): 310 (1.89)[569a]

IR ν_{max}^{KBr} cm^{-1}: 3500 (OH), 1743 (5 ring C=O), 1723 (6 ring C=O), 1655 (C=C), 1092, 1083,
1070 (C–O–C), 1032, 898, 870, 845 (CO–C–O–C)

NMR τ: 8.42 s (18–Me), 8.78 s (19–Me), 8.72 d, J=6.5 c/s (21–Me), 8.63 d, J=6.4 (5′–Me of diginose),
6.6 s (CH$_3$O of diginose)

(b) **Digifologenin** (LXXXIX, R=H)   mp. 110~115°, $[\alpha]_D$ −270° (acetone), $C_{21}H_{28}O_5$

UV λ_{max} mμ (log ε): 310 (1.9)

IR ν_{max} cm^{-1}: 3610, 3560, 1733, 1710, 1140, 1092, 1070, 1052, 890, 870

NMR τ: 8.42 s (13–Me), 8.78 s (19–Me), 8.72 d J=6.5 (21–Me), 4.25 m (=C^6<$_H$), 6.45 br (3–H), 5.4 br
(20–H), 5.86 br (2–H), 6.04 s (12–H), 7.3 br (2 and 3–(OH)$_2$))

(9) **Purpurogenin**[569]

≪Occurrence≫   *Digitalis purpurea* L,   mp. 249~252°, $C_{21}H_{30}O_5$

Purpuronin ⟶ Purpurogenin +3 D–Digitoxose

Purpuronin   mp. 278~281°, $C_{39}H_{60}O_{14}$

(10) **Sarcostin** (XC)[574)575)576]

(XC)[575]

≪Occurrence≫   *Sarcostemma australe* R. Br., *Metaplexis japonica* MAKINO, *Marsdenia tomentosa*
(キジョラン)

mp. 150°/260°[575], 154°/248°[576], $[\alpha]_D$ +67° (MeOH), $C_{21}H_{34}O_6$

NMR of 3, 12, 20–triacetate[575] (60 Mc, in CDCl$_3$, TMS, ppm): 0.987 s, 1.40 s (18, 19–Me resp.) 1.163
~1.27 d, J=6.4 (21–Me), 1.97 s, 2.01 s, 2.07 s (3×COCH$_3$), 1.89 (?), 2.48 s, 3.05 s (2×OH,
disappear with F$_3$C–COOH)

(XC–a)

If sarcostin has the former structure (XC–a), the signal due to the secondary methyl at 20 should be
doublet.

574) Cornforth, J. W: *Chem. & Ind.*, **1959**, 602.
575) Reichstein, T. *et al*: *Helv. Chim. Acta*, **46**, 694 (1963).
576) Mitsuhashi, H. *et al*: *Chem. Pharm. Bull.*, **10**, 811 (1962).

Mass[576a] (anion spectrum) M$-1=381$ M/e: 381, 362, 344, 326, 316, 310, 300, 292, 284, 274, 267, 257, 243, 239, 233

(11) Metaplexigenin (XCI)[571c]

(XCI)

≪Occurrence≫　Metaplexis japonica MAKINO,　　mp. 268~275°, C$_{23}$H$_{34}$O$_7$

IR　ν_{max}^{nujol} cm^{-1}: 1745, 1720, 1240

NMR　in pyridine, TMS, τ: 8.57 s (3H), 8.05 s (3H), 7.90 s (3H), 7.50 s (3H)

Metaplexigenin $\xrightarrow{\text{NaBH}_4}$ Sarcostin (XC)

(12) Cynanchogenin and Desacylcynanchogenin[571c)577]

R= $\genfrac{}{}{0pt}{}{CH_3}{CH_3}$ > CH–C=CH–CO– : Cynanchogenin (XCII)
$\qquad\qquad$ |
$\qquad\qquad$ CH$_3$

R= H, 17–Hβ:　　　　　Desacylcynanchogenin (XCIII)

(Revised Str.)[571c]

≪Occurrence≫　Cynancum Caudatum Max. (イケマ)

(a) Cynanchogenin (XCII)　　mp. 167°, [α]$_D$ -39.5°, C$_{28}$H$_{42}$O$_6$

UV　λ_{max} mμ (log ε): 218 (4.2) ($\alpha\beta$ unsatur. C=O)

IR　ν_{max} cm^{-1}: 3460 (O–H), 1700, 1640 ($\alpha\beta$ unsatur. C=O)

(b) Desacylcynanchogenin (XCIII)　　mp. 242°, C$_{21}$H$_{32}$O$_5$

IR　ν_{max} cm^{-1}: 1680 (–CO–CH$_3$)

(13) Penupogenin (XCIV)[577)570]

{Sarcostin (XC)} –CO–CH=CH–⟨ ⟩

(XCIV)

≪Occurrence≫　Cynancum caudatum Max. (イケマ),　　mp. 145~150°, C$_{30}$H$_{40}$O$_7$

UV　λ_{max}^{EtOH} mμ (ε): 279 (22000)

IR　ν_{max}^{KBr} cm^{-1}: 3400 (OH), 1690 (C=O), 1630 (C=C–C=C), 1600 (C=C–C=O), 1580 (arom.)

576a)　Reichstein, T. et al: Helv. Chim. Acta, **47**, 1032 (1964).

577)　Mitsuhashi, H. et al: Chem. Pharm. Bull., **10**, 433, 719 (1962).

9 Steroidal Sapogenins

[A] UV Spectra

(1) Sapogenins

Diosgenin, hecogenin, yucagenin, tigogenin, sarsasapogenin, smilagenin, gitogenin, markogenin, samogenin, chlorogenin, kryptogenin, rockogenin, managenin and kammogenin, treated with sulfuric acid[580].

[B] IR Spectra

(1) Sapogenins

Normal sapogenin, iso-sapogenin, sarsa-sapogenin, yuccagenin diacetate[581], hecogenin, dioscin[582], 3-desoxy-sarsasapogenin, 3-*epi*-sarsasapogenin, neotigogenin, yamogenin, tigogenin, yamogenin acetate, diosgenin acetate, diosgenin, samogenin diacetate, hecogenin acetate[583] and willagenin[584]. Chlorogenin acetate[585].

Diosgenin acetate, gitogenin acetate, hecogenin acetate, kammogenin acetate, kryptogenin acetate, manogenin acetate, samogenin acetate, sarsasapogenin acetate, smilagenin acetate, tigogenin acetate and yuccagenin acetate[585]. Acetates of chlorogenin, diosgenin, gitogenin, hecogenin, manogenin, samogenin, sarsasapogenin, smilagenin and tigogenin[586].

580) Walens, Turner, Wall: *Anal. Chem.*, **26**, 325 (1954).
581) Rothman, Wall, Eddy: *J. A. C. S.*, **74**, 4013 (1952).
582) Wall, Krider, Rothman, Eddy: *J. Biol. Chem.*, **198**, 533 (1952).
583) Jones, N: *J. A. C. S.*, **75**, 158 (1953).
584) Kenney, Wall: *J. Org. Chem.*, **22**, 468 (1957).
585) Eddy, C. R., Wall, M. E., Scott, M. K: *Anal. Chem.*, **25**, 266 (1953).
586) Wall, M. E., Eddy, C. R., Mc Clennan, M. L., Klumpp, M. E: *ibid.*, **24**, 1337 (1952).

(2) Characteristic Absorption at the 3100~2750 cm⁻¹ Region (ν_{C-H})[587]

(a) 20 α- and 20 β- Sapogenin

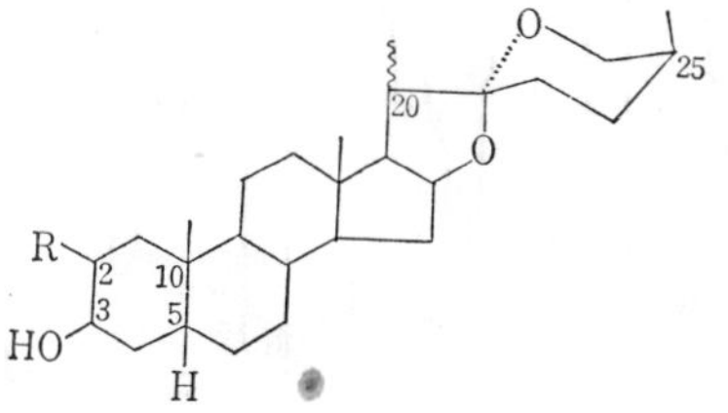

20 α or β, A/B *cis*, 25 L, 3 β–OH

(I)　Sarsasapogenin　(R=H)
(II)　Markogenin　(R=β–OH)

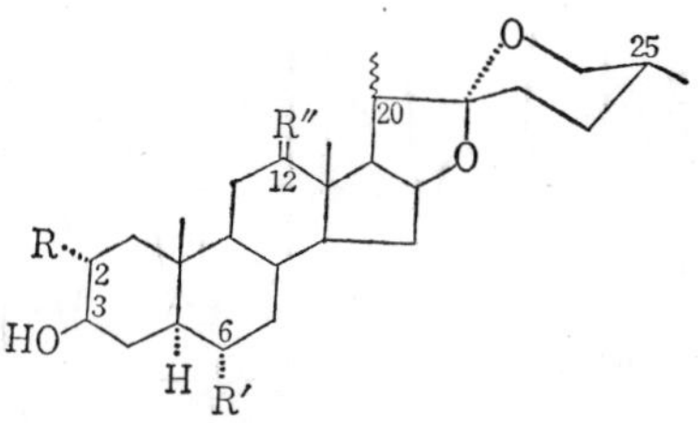

20 α or β, A/B *trans*, 25 D, 3 β–OH

(III)　Tigogenin　(R=R′=H, R″=H₂)
(IV)　Hecogenin　(R=R′=H, R″=O)
(V)　Gitogenin　(R=α–OH, R′=H, R″=H₂)
(VI)　Manogenin　(R=α–OH, R′=H, R″=O)
(VII)　Chlorogenin　(R=H, R′=α–OH, R″=H₂)

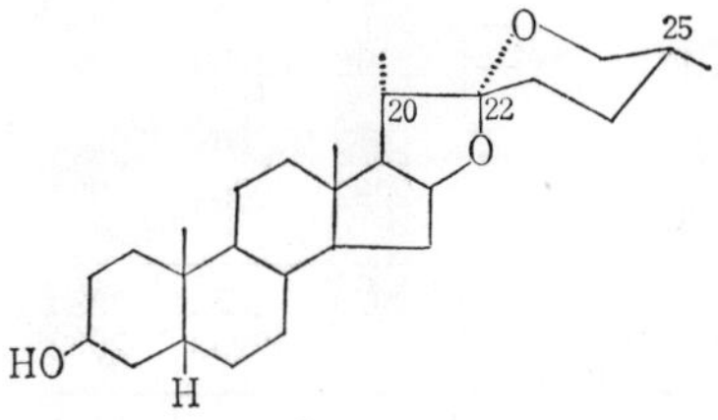

20 α, A/B *cis*, 25 D, 3 β–OH
(VIII)　Smilagenin

(b) Δ⁵-Sapogenin

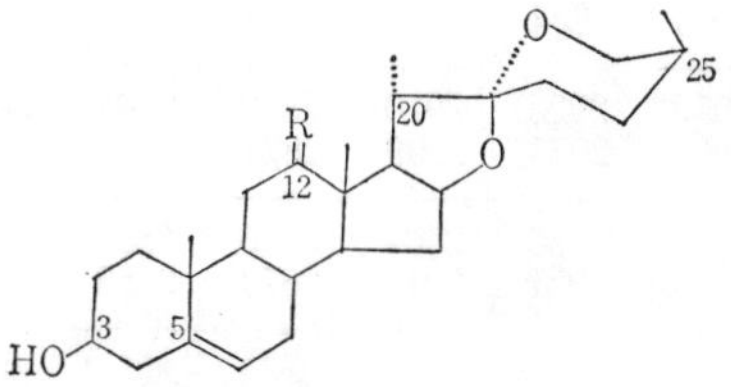

20 α, 25 L, 3 β–OH
(IX)　Yamogenin　(R=H₂)
(X)　Correllogenin　(R=O)

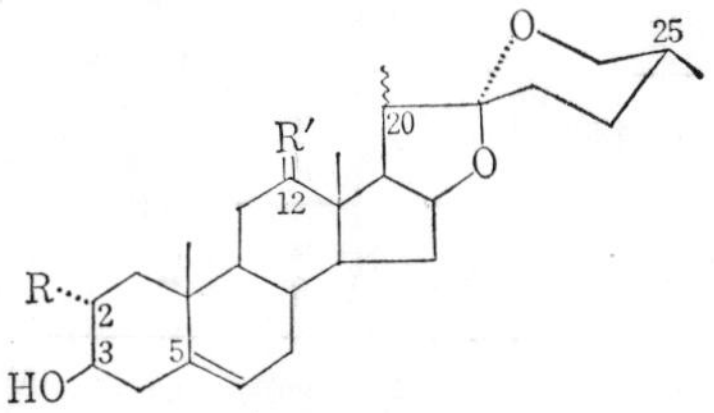

25 D, 3 β–OH
(XI)　Diosgenin　(20 α or β, R=H, R′=H₂)
(XII)　Gentrogenin　(20 α, R=H, R′=O)
(XIII)　Yuccagenin　(20 α, R=α–OH, R′=H₂)
(XIV)　Kammogenin　(20 α, R=α–OH, R′=O)

Fig. 102　and Table 55 show the IR absorption of sapogenin acetates at the region of 3100~2750 cm⁻¹

587)　Smith, A. M *et al*: *Anal. Chem.* **31**, 1539 (1959).

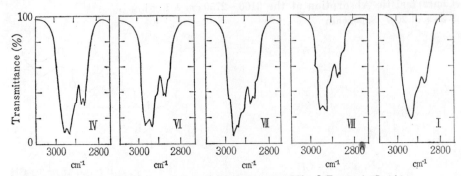

Fig. 102 IR Spectra of Steroid Sapogenins (in CCl₄, LiF prism) (Smith)
(IV) Hecogenin, (VI) Manogenin, (VII) Chlorogenin, (VIII) Smilagenin, (I) Sarsasapogenin

Table 55 Absorption Bands of Steroid Sapogenin Acetates at 3100~2750 cm⁻¹ (Smith)

Acetate of Sapogenin	C_{20}	No.	a	b	c	d	e	f	g	h	i
Tigogenin	α β	III	— —	2975 2992	2952 2953	2930 2932	2909 2920 2908	2873 2872	2861 2860	2846 2851	— —
Gitogenin	α β	V	— —	2972 2992	2952 2954	2929 2932	2909 2921 2907	2873 2872	2862 2863	2848 2848	— —
Hecogenin	α β	IV	— —	2979 —	2953 2953	2930 2932	2908 2907	2874 2872	2861 2861	— 2851	— —
Manogenin	α β	VI	— —	— —	2954 2953	2931 2932	2909 2907	2874 2872	2863 2863	2854 2852	— —
Chlorogenin	α β	VII	— —	2975 —	2954 2953	2930 2934	2909 2919	2874 2872	2861 2859	2851 —	— —
Smilagenin Neosmilagenin	α β	VIII	— —	2972 2993	2954 2953	2930 2934	2909 2909	2873 2873	2863 2864	2848 —	— —
Sarsasapogenin	α β	I	— —	— 3002	2953 2953	2935 2933	2910 2909	2873 2874	2864 2865	2848 —	— —
Markogenin	α β	II	— —	— 3005	2954 2954	2936 2931	2909 2911	2875 2874	2866 2865	2852 —	2839 —
Diosgenin	α β	XI	3034 3033	— 2992	2953 2953	2931 2933	2908 2907	2873 2873	2861 —	2851 2853	2829 2829
Yamogenin	α β	IX	3032 3032	2962 3008	2951 2954	2936 2936	2909 2908	2874 2873	— —	2852 2854	2831 2831
Yuccagenin	α	XII	3034	2972	2953	2929	2907	2873	2861	2848	2830
Gentrogenin	α	XII	3034	—	2955	2930	2908	2873	2860	—	2832
Kammogenin	α	XIV	3038	—	2957	2931	2907	2875	2863	2847	2833
Correllogenin	α	X	3033	—	2958	2937	2909	2873	2863	2853	2831

Characteristic bands (a～i) in Table 55 are assigned as following. Standard frequencies cm^{-1} are indicated in parenthesis.

a: =CH– (3035), b: 3000～2960, c: Assym. Me (2950), d: Assym. –CH$_2$– (2930), e: 2920～2910, f: Sym. Me (2870), g: Sym. –CH$_2$– (2860), h: 2850, i: 2830

Summary

(a) 3035 cm^{-1}　The weak band, characteristic of $\mathit{\Delta}^5$-sapogenin, shifts 10～20 cm^{-1} to high region from the corresponding band of unsaturated aliphatic hydrocarbon.

(b) 3000～2960 cm^{-1}　20 β, 25 L–Sapogenins (>3000), 20 β, 25 D– ″ (2992～2993 sh.), 20 α, 25 D– —— (ca 2975 sh.), 20 α, 25 L– —— (no)

(c) 2950 cm^{-1}　All compounds show absorption at 2952～2954 cm^{-1}

(d) 2930 cm^{-1}　All compounds show absorption at 2930～2936 cm^{-1}

(e) 2920～2910 cm^{-1}　All compounds show absorption at 2907～2911 cm^{-1}. Some of 20 β–sapogenins show further absorption at ca 2920 cm^{-1} region.

(f～i) 2880～2830 cm^{-1} $\mathit{\Delta}^5$–Sapogenins show absorption at ca 2830 cm^{-1} region.

〔C〕 NMR Spectra

(1) NMR and the Conformation at C$_{20}$, C$_{22}$ and C$_{25}$[587a)]

20 α, 22 β, 25 L
(I)　Sarsasapogenin

20 β, 22 α, 25 D
(XV)　Neosarsasapogenin

20 α, 22 β, 25 D
(VIII)　Smilagenin

20 β, 22 β, 25 D
(XVI)　Neosmilagenin

(VIII)　and (XI)　338 cps (signals due to 18–Me and 27–*equatorial*-methyl are overlapped), 324 (19–Me)

(XVI)　326, 328 cps (signals of 19–Me and 18–Me, low–shifted by steric hindrance of C$_{18}$/C$_{21}$), 337 d (27 –*equatorial*-Me)

587a)　Rosen, W. E. *et al*: *J. A. C. S.*, **81**, 1687 (1959).

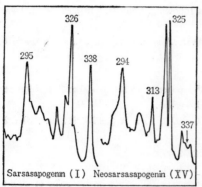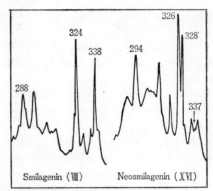

Fig. 103 NMR Spectra of Sapogenins (Rosen *et al.*)

NMR in CDCl$_3$, 60 Mc, C$_6$H$_6$ as an external annulus

(I) 326 cps (19–Me), 338 (18–Me), 321 d (21–Me), 326 d (27–*axial*–Me)

(XV) 325 cps (signal of 18–Me, lower–shifted by steric hinderance of C$_{18}$/C$_{21}$), 337 d (higher–shifted signal) of 27–*equatorial* methyl), 317 d

[D] 3–Hydroxysapogenins

(1) Timosaponin A–III (XVII)[588)589)]

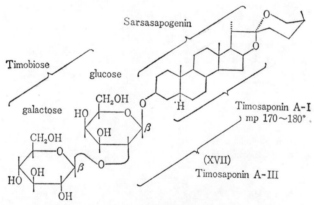

≪Occurrence≫   *Anemarrhena asphodeloides* BUNGE (知母, ハナスゲ)
mp. 317~322° (decomp.), [α]$_D$ −41.6° (dioxane), C$_{39}$H$_{64}$O$_{13}$

(2) Lumihecogenin acetate (XVIII)[590)]

[Source] Hecogenin (IV) acetate $\xrightarrow{\text{in dioxane, UV}}$ (XVIII)

UV λ_{max} mμ (ε): 204 (4800)

IR ν_{max}^{KCl} cm^{-1}: 2740 (CHO), 1739 (OAC), 1709 (CHO), 1240 (OAC)

588) Kawasaki, T. *et al*: *Chem. Pharm. Bull.*, **11**, 1221 (1963).
589) Kawasaki, T. *et al*: *Y. Z.*, **83**, 892 (1963).
590) Bladon, P *et al*: *Proc. Chem. Soc.*, **1962**, 225.

(3) **Iso-narthogenin** (XIX)[591] and **Narthogenin** (XX)

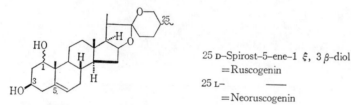

(XIX) 25 D
(XX) 25 L

≪Occurrence≫   *Metanarthecium luteo-viride* MAXIM.
(XIX) mp. 240~242°, $[\alpha]_D$ −110.2°, $C_{27}H_{42}O_4$

[E] 1,3-Dihydroxysapogenins

(1) **Ruscogenin** and **Neoruscogenin** (XXI)[592]

25 D–Spirost–5–ene–1 ξ, 3 β–diol
  =Ruscogenin
25 L– ————
  =Neoruscogenin

≪Occurrence≫   *Ruscus aculeatus* L.
Ruscogenin diacetate   mp. 190~192°, $[\alpha]_D$ −68°, $C_{31}H_{46}O_6$,   IR ν_{max} 899>921 cm⁻¹
Neoruscogenin diacetate   mp. 132~134°, $[\alpha]_D$ −64°, $C_{31}H_{46}O_6$,   IR ν_{max} 921>900 cm⁻¹

(2) **Saponins of *Reineckia Carnea* KUNTH.**[592a]

(XXII) Convallamarogenin

(XXIII) Iso–rhodeasapogenin

(XXIV) Isoreineckiagenin

(XXV) Reineckiagenin

591) Minato, H. *et al*: *Chem. Pharm. Bull.*, **11**, 876 (1963).   592) Burn, D. *et al*: *Soc.*, **1958**, 795.
592a) Takeda, K *et al*: *Tetrahedron*, **19**, 759 (1959); *6 th Symposium on the Chemistry of Natural Products*,
    p. 148 (Sapporo, 1962).

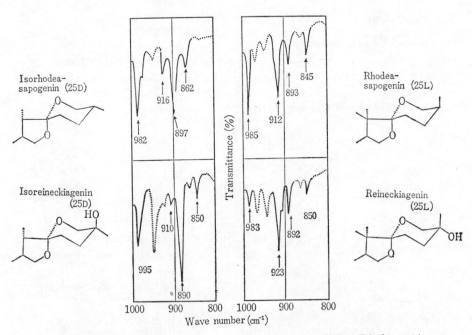

(XXVI)   Isocarneagenin

(a) **Convallamarogenin** (XXII)[592b]

mp. 259~261°, $[\alpha]_D$ −79.1°, $C_{27}H_{42}O_4$,   IR $\nu_{max}^{nujol} cm^{-1}$: 3350, 3295, 3095, 1655, 877

(b) **Isorhodeasapogenin** (XXIII)

mp. 239~240°, $[\alpha]_D$, −76.3° (CHCl₃) $C_{27}H_{44}O_4$,   IR $\nu_{max}^{CHCl_3} cm^{-1}$: 916<897 (see Fig. 104) [Note]

(c) **Isoreineckiagenin** (XXIV)

mp. 240~242°, $[\alpha]_D$ −65.7° (dioxane), $C_{27}H_{44}O_5$,   IR $\nu_{max}^{CHCl_3} cm^{-1}$: 910<890 (see Fig. 104)

NMR τ: 8.93 (27–CH₃)

(d) **Reineckiagenin** (XXV)

mp. 278~280°, $[\alpha]_D$ −70.6° (dioxane), $C_{27}H_{44}O_5$,   IR $\nu_{max}^{CHCl_3} cm^{-1}$: 923>892 (see Fig. 104)

NMR τ: 8.73 (27–CH₃)

(e) **Isocarneagenin** (XXVI)

mp. 242~244°, $[\alpha]_D$ −63.4° (MeOH. dioxane), $C_{27}H_{44}O_5$

Fig. 104 IR Spectra of 25 D–and 25 L–Sapogenins (in CHCl₃) (Takeda *et al.*)

[Note] Rhodeasapogenin (25 L). $\nu_{max} cm^{-1}$: 912>893.
592b) Tschesche, R *et al*: *Ber.*, **94**, 1699 (1961).

〔F〕 2, 3–Dihydroxysapogenins

(1) Yonogenin (XXVII) and Samogenin (XXVIII)[593]

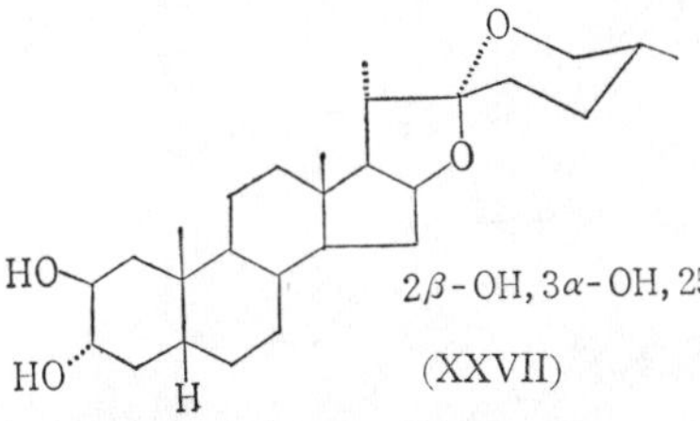

2β-OH, 3α-OH, 25D (XXVII)

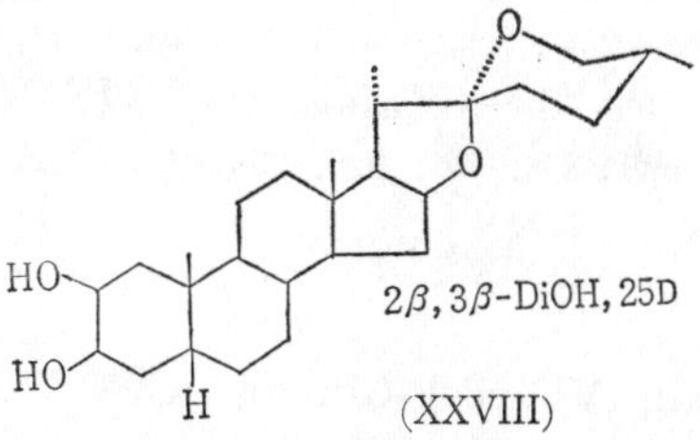

$2\beta,3\beta$-DiOH, 25D (XXVIII)

(a) Yonogenin (XXVII)[593]

≪Occurrence≫ *Dioscorea Tokoro* MAKINO (オ ニ ド コ ロ), mp. 240~243°, $[\alpha]_D$ −53° (CHCl₃), $C_{27}H_{44}O_4$
Diacetate mp. 212°, $[\alpha]_D$ −29° (CHCl₃)

(XXVII) ⟶ ⟶ (XXVIII)

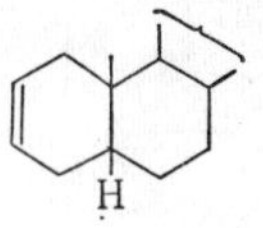

(b) Samogenin (XXVIII)[593][594]

≪Occurrence≫ *Yucca schottii, Agave funkiana,* mp. 207°, $[\alpha]_D$ −80° (CHCl₃), $C_{27}H_{44}O_4$

(c) Yononin (Yonogenin + α-L-Arabinose)[595]

mp. 238~240° (decomp.), $[\alpha]_D$ −14.5° (pyridine)

(2) Markogenin (II) and Neogitogenin (XXIX)[596]

Markogenin (II)
  5 β, 2 β–OH, 3 β–OH, –OH, 25 L
Gitogenin (V)
  5 α, 2 α–OH, 3 β–OH, 25 D

$5\alpha,2\alpha$-OH, 3β-OH, 25L

Neogitogenin (XXIX)

≪Occurrence≫ *Anemarrhena asphodeloides* BUNGE

593) Takeda, K. *et al*: *Chem. Pharm. Bull.*, **6**, 532 (1958).
594) Djerassi, C. *et al*: *J. A. C. S.*, **77**, 4291 (1955).
595) Kawasaki, T. *et al*: *Y. Z.*, **83**, 757 (1963).   596) Kawasaki, T. *et al*: *ibid.*, **83**, 892 (1963).

(a) Markogenin (II)[597]

≪Occurrence≫   *Yucca faxoniana, Y. schidigera,*   mp. 256~257°, $[\alpha]_D$ −70.3° (CHCl₃), $C_{27}H_{44}O_4$

IR $\nu_{max}^{CHCl_3}$ cm⁻¹: 1468, 1454, 1388, 1381, 1342, 1275, 1175, 1134, 1096, 1067, 1048, 1032, 1010, 1002, 986, 967, 950, 937, 915, 896, 872, 849 (915>896)

Diacetate   mp. 185~186°, $[\alpha]_D$ −84.2° (CHCl₃)

(b) Neogitogenin (XXIX)[596]

mp. 253~254°, $[\alpha]_D$ −66.4° (CHCl₃), $C_{27}H_{44}O_4 \cdot {}^1/_2$ H_2O

Diacetate   mp. 215~219°, $[\alpha]_D$ −112° (CHCl₃)

IR ν_{max}^{nujol} cm⁻¹: 854, 902, 928, 990 (928<902)

(XXIX) $\xrightarrow{HCl}$ Gitogenin (V) acetate (see the next item (3))

(3) Gitogenin (V) and **F-Gitonin** (XXX)[598]

(XXX) F–Gitonin

= Gitogenin β–D–glucopyranosyl (1→1 gluc.)→β–D–xylopyranosyl (1→3 gluc.)→β–D–glucopyranosyl (1→4) β–D–galactopyranoside

(a) Gitogenin (V)

≪Occurrence≫   *Digitalis purpurea* L., *Hosta lougipes* and other spp., *Agave* spp. *Yucca* spp., *Cestrum* spp., *Trigonella foenumgraecum*

mp. 270~272°, $[\alpha]_D$ −75° (CHCl₃), $C_{27}H_{44}O_4$

Diacetate   mp. 242~243°, $[\alpha]_D$ −97.2° (CHCl₃), IR ν_{max}^{nujol} cm⁻¹: 869, 901, 926, 979 (926<901)

(b) F-Gitonin (XXX)[598][599]

≪Occurrence≫   *Digitalis purpurea* L.,   mp. 252~255° (decomp.) $[\alpha]_D$ −58.5° (pyridine), $C_{50}H_{82}O_{23}$

IR ν_{max}^{nujol} cm⁻¹: 862, 898, 924, 982 (924>898)

〔G〕 3, 12–Dihydroxysapogenins

(1) Heloniogenin (XXXI)

≪Occurrence≫   *Heloniopsis orientalis* C. TANAKA (ショウジョウバカマ), mp. 212~213°, $[\alpha]_D$ −91°,

---

597) Wall, M.E. *et al*: *J.A.C.S.*, **75**, 4437 (1953).

598) Kawasaki, T. *et al*: *7 th Symposium on the Chemistry of Natural Products*, p. 82 (1963, Fukuoka).

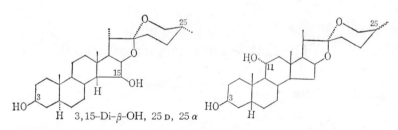

(XXXI)　Heloniogenin

$C_{27}H_{42}O_4$

UV　ν_{max} mμ (log ε): 205.6 (3.47),　IR　$\nu_{C=O}$ negative

〔H〕　3, 15–Dihydroxysapogenins

(1)　Digalogenin (XXXII)[599]

(XXXII)　Digalogenin　　　　(XXXIII)　25 D Nogiragenin
(XXXIV)　25 L Neonogiragenin

≪Occurrence≫　　*Digitalis purpurea* L.

Digalonin　(Digalogenin + 2 Glucose + 2 Galactose + Xylose)

(XXXII)　mp. 218～219°, $[\alpha]_D$ −75.5° (CHCl$_3$), C$_{27}$H$_{44}$O$_4$

Acetate　mp. 235～236°, $[\alpha]_D$ −82° (CHCl$_3$)

〔I〕　3, 11(2)–Di(Tri)-hydroxysapogenins

(1)　Nogiragenin (XXXIII) and Neonogiragenin (XXXIV)

≪Occurrence≫　　*Metanarthecium luteoviride* MAXIM. (ノギラン)

(a)　**Nogiragenin** (XXXIII)[600]

mp. 200～201°, $[\alpha]_D$ −70.6° (CHCl$_3$), C$_{27}$H$_{44}$O$_4$

IR　ν_{max}^{nujol} cm^{-1}: 978, 918, 890, 876 (890>918)

(b)　**Neonogiragenin** (XXXIV)[591]

mp. 128～131°, $[\alpha]_D$ −75.0° (CHCl$_3$), C$_{27}$H$_{44}$C$_4$

IR　ν_{max}^{nujol} cm^{-1}: 987, 917, 895, 848 (917>895)

599)　Tschesche, R. *et al*: *Ber.*, **94**, 2019 (1961).
600)　Takeda, K. *et al*: *Chem. Pharm. Bull.*, **9**, 388 (1961).

(2) **Metagenin** (XXXV)[591),601)~603)]

2 β, 3 β, 11 α–Tri OH, 5 β, 25 D
(XXXV) Metagenin

≪Occurrence≫   *Metanarthecium luteo-viride* MAXIM.

mp. 273~274°, [α]$_D$ −82° (CHCl$_3$, MeOH), C$_{27}$H$_{44}$O$_5$

IR λ$_{max}^{nujol}$μ: 11.46, 11.09, 10.88, 10.17

Triacetate mp. 250~252°, [α]$_D$ −77.5° (CHCl$_3$),   IR λ$_{max}^{cs2}$ μ:11.58, 11.14, 10.89, 10.19, 8.21, 8.08, 8.01

[J] 1, 2, 3–Trihydroxysapogenins

(1) **Tokorogenin** (XXXVI)[604)605)]

1–β, 2 β, 3 α–Tri–OH, 5 β, 25 D
(XXXVI) Tokorogenin

≪Occurrence≫   *Dioscorea tokoro* MAKINO (オニドコロ),   mp. 266~268°, [α]$_D$ −49.6° (EtOH),
C$_{27}$H$_{44}$O$_5$

Triacetate   mp. 255°, [α]$_D$ −8.2° (CHCl$_3$)   IR ν$_{max}^{nujol}$ cm^{-1}: 980, 917, 895, 862

Tokoronin (Tokorogenin + α–L–Arabinose)[595)]   mp. 270~274° (decomp.), [α]$_D$ −10.6° (pyridine),
C$_{32}$H$_{52}$O$_9$·$^1/_2$H$_2$O

601) Takeda, K. *et al*: *Y. Z.*, **77**, 175 (1957).
602) Takeda, K. *et al*: *Tetr. Letters*, **1960**, No. 3, 1; *Tetrahedron*, **15**, 183 (1961).
603) Hamamoto, K. *et al*: *Chem. Pharm. Bull.*, **8**, 1004, 1099 (1960), **9**, 32 (1961).
604) Morita, K. *et al*: *Bull. Chem. Soc. Japan*, **33**, 476, 791, 796, 800 (1959).
605) Nishikawa, M. *et al*: *Y. Z.*, **74**, 1165 (1954).

〔K〕 2, 3, 15–Trihydroxysapogenins

(1) Digitogenin (XXXVII)[606]

2 β, 3 α, 15 β–Tri–OH, 5 α, 25 D, 25 α
(XXXVII) Digitogenin

《Occurrence》 *Digitalis purpurea* L., mp. 288～291°, $[\alpha]_D$ −81° (CHCl₃), $C_{24}H_{44}O_5$

Desglucodigitonin (Digitoxigenin + Glucose + 2 Galactose + Xylose)

Digitonin (XXXVIII)[598][607]

(XXXVIII) Digitonin

《Occurrence》 *Digitalis purpurea* L., mp. 235°, $[\alpha]_D$ −54.3° (MeOH), $C_{56}H_{92}O_{29}$

〔L〕 1, 2, 3, 5 and 1, 3, 4, 5–Tetrahydroxysapogenins

(1) Kogagenin (XXXIX)[595][608]～[612]

《Occurreuce》 *Dioscorea tokoro* MAKINO, mp. 318～322°(decomp.), $[\alpha]_D$ −27° (pyridine), $C_{27}H_{44}O_6$

606) Djerassi, C. *et al*: *J.A.C.S.*, **77**, 3673 (1955), **78**, 3166 (1956).
607) Tschesche, R. *et al*: *Tetrahedron*, **19**, 621 (1963).
608) Takeda, K. *et al*: *ibid*., **7**, 62 (1959).
609) Kubota, T: *Chem. Pharm. Bull.*, **7**, 898 (1959).
610) Takeda, K. *et al*: *Y.Z.*, **77**, 822 (1957); *Chem Pharm. Bull.*, **6**, 532 (1958).
611) Nishikawa, M. *et al*: *Y.Z.*, **74**, 1165 (1954).
612) Morita, K: *Chem. Pharm. Bull.*, **5**, 494 (1957).

(XXXIX) 1β, 2β, 3α, 5β–Tetra-OH, 25-D

(2) Kitigenin (XL)[613)614)]

(XL) 1β, 3β, 4β, 5β–Tetra-OH, 25-D

≪Occurrence≫   *Reineckia carnea* KUNTH.,   mp. 298°, [α]$_D$ −35° (pyridine), $C_{27}H_{44}O_6$

3, 4–Diacetate   mp. 217~219°, [α]$_D$ −45.6 ± 2° ($CHCl_3$)

IR ν_{max}^{nujol} cm^{-1}: 3630, 3520~3300 (OH), 1745, 1730 (OAc), 1662 (H_2O)

〔M〕 Pentahydroxysapogenins

(1) Pentologenin (XLI)[615)]

(XLI) 1, 2, 3, 4, 5 Penta-OH, 25-D

≪Occurrence≫   *Reineckia carnea* KUNTH.,   mp. 320°(decomp), [α]$_D$ −54.5° ($CHCl_3$–MeOH),$C_{27}H_{44}O_7$

IR ν_{max}^{nujol} cm^{-1}: 3380~3500, 980, 915, 898, 860 (915<898)

Tetraacetate   mp. 165~168°, $C_{35}H_{52}O_{11}$

IR ν_{max}^{nujol} cm^{-1}: 3600 (OH), 1758, 1746 (OAc)

---

613) Takeda, K. *et al*: *Y. Z.*, **75**, 560 (1955).
614) Sasaki, K: *Chem. Pharm. Bull.*, **9**, 684, 693 (1961).
615) Takeda, K *et al*: *ibid.*, **9**, 631 (1961).

[N] Aromatic Sapogenins

(1) Luvigenin (XLII) and **Meteogenin** (XLIII)[616)~619)]

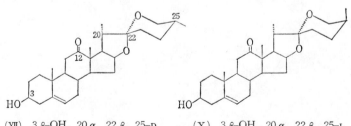

(XLII)   (XLIII)

≪Occurrence≫   *Metanarthecium luteoviride* MAXIM. (ノ ギ ラ ン)

(a) Luvigenin (XLII)

mp. 183~184°, $[\alpha]_D$ −34.9°, $C_{27}H_{38}O_2$

UV λ_{max} mμ (ε): 205 (16000), 215 (10200), 265 (260), 271 (200)

IR ν_{max}^{nujol} cm^{-1}: 3080, 1585 (arom.), 981, 923<900, 860 (ring F), 785, 740 (arom.)

(b) Meteogenin (XLIII)

mp. 157~158°, $[\alpha]_D$ −174.2° (CHCl$_3$), $C_{27}H_{38}O_3$

UV λ_{max}^{EtOH} mμ (log ε): 204 (4.24), 217 (sh), 263 (2.42), 270 (2.38)

IR ν_{max}^{CS2} cm^{-1}: 3613, 3540~3300 (OH), 3060 (C=C–H), 980, 920, 895, 860 (ring F, 895>920), 774, 749 (arom.)

[O] 12-Ketosapogenins

(1) Gentrogenin (XII) and **Correllogenin** (X)[620)]

(XII) 3 β-OH, 20 α, 22 β, 25-D   (X) 3 β-OH, 20 α, 22 β, 25-L

616) Igarashi, K *et al*: *5 th Symposium on the Chemistry of Natural Products*, p.5-1 (Sendai, 1961).

617) Takeda, K: *Tetrahedron*, **15**, 183 (1961).

618) Igarashi, K: *Chem. Pharm. Bull.*, **9**, 722 (1961).   619) Minato, H. *et al*: *ibid.*, **11**, 876 (1963).

620) Wall, M.E. *et al*: *J.A.P.A.* **46**, 155 (1957); *J. Org. Chem.*, **22**, 182 (1957), **23**, 1741 (1958).

≪Occurrence≫　　*Dioscorea spiculiflora*

(XII) $\xrightarrow{\text{Huang Ming Long reduction}}$ Diosgenin (XI)

(X) $\xrightarrow{\text{〃}}$ Yamogenin (IX)

(a)　Gentrogenin (XII)

　　mp. 215~216°, $[\alpha]_D$ −57°,　Acetate mp. 227°, $[\alpha]_D$ −56°, IR See Fig. 105

(b)　Correllogenin (X)

　　mp. 209~210°, $[\alpha]_D$ −69°, Acetate mp. 213~214°, $[\alpha]_D$ −60°, IR See Fig. 105

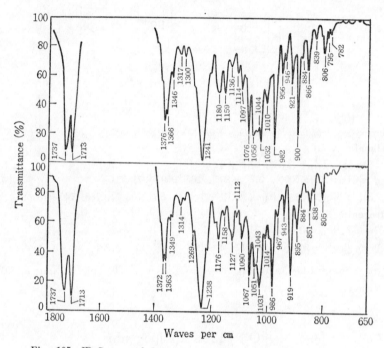

Fig. 105　IR Spectra of Gentrogenin acetate (top) and Correllogenin

　　acetate (bottom)　　　　　　　　　　　　　　(Wall *et al.*)

10 Cardenolides

[A] UV Spectra

(1) UV Spectra of Cardenolides and their Genins are summarized[621].

Glycosides of *Adenum multiforum* and glycosides of *Cryptostegia grandiflora*[622][623], strophanthidin-β-glucoside, *O*-acetylglucoside-B[624], 16-desacetyl-14,16-dianhydrobufotalin acetate, 14-monoanhydrobufotalin acetate[625], dianhydrogitoxigenin, gitoxigenin, gitoside[626], coloration of digitoxigenin by Baljet reaction[627], by Keller-Kiliani reaction[627], coloring of genins by sulfuric acid[628], (coloring of digitoxigenin, diginatigenin, digoxigenin, gitoxigenin and digitoxigenin by $FeCl_3$-sulfuric acid)[628].

(2) UV Absorption in Sulfuric Acid[629]

Thirty eight cardenolides and genins were measured their UV absorption at $210{\sim}600\,m\mu$ region in sulfuric acid solution. Spectra of the compounds, having cyclobutenolide moiety show λ_{max} at $230{\sim}240$ (centre 235). It is possible to distinguish followed the following three types of cardenolide by comparison of their spectra. See Table 56.

(A)	(B)	(C)
(I) Digitoxigenin	(II) Gitoxigenin	(III) Digoxigenin

$\lambda_{max}^{H_2SO_4}$ at : (A) 235, 350, 415 mμ, (B) 235, 320, 415, 490, 530 mμ (C) 235, 320, 390 mμ

621) Zeichmeister: Forschritt der Chemie der Organische Naturstoffe, 13 (1956).
622) Reichstein *et al*: *Helv. Chim. Acta*, **33**, 1993 (1950).   623) Reichstein *et al*: *ibid.*, **33**, 1013 (1950).
624) Reichstein *et al*: *Helv. Chim. Acta*, **40**, 284 (1957).
625) Meyers: *Helv. Chim. Acta*, **32**, 1993 (1949).
626) Murphy: *J. Am. Pharm. Assoc. Sci. Ed.*, **46**, 170 (1957).
627) Demoeu, Janssen: *ibid.*, **42**, 635 (1953).   628) Murphy: *ibid.*, **44**, 719 (1955).
629) Brown, B. T *et al*: *ibid.*, **49**, 777 (1960).

Table 56 Absorption Spectra of Cardenolides in Sulfuric acid Solution

$15\sim30$ mcg/ml $98\ \text{V}/\text{V}\ \%$ H_2SO_4, $\lambda_{max}^{H2SO4}m\mu$ $\left(E\ ^{1\ \%}_{1\ cm}\right)$ (Brown *et al.*)

Series	Compounds	$\lambda_{max}m\mu(E)$	$\lambda_{min}m\mu(E)$
A	Digitoxigenin	235(320), 350(290), 410(270)	270(115), 385(200)
A	20, 22 Dihydro ——	355(170), 420(280)	260(85), 385(130)
A	—— 3–one	230(180), 350(530), 420(50)	265(25), 390(15)
A	—— 3–acetate	235(370), 345(290), 415(270)	260(95), 385(170)
A	Lanatoside A	235(150), 325(175), 415(110), 480(95)	285(90), 375(90)
A	Digitoxin	235(270), 340(240), 420(240), 485(230)	265(175), 375(215), 440(165)
B	Gitoxigenin	225(230), 315(420), 415(120), 510(400) 540(440)	260(140), 365(90), 430(105)
B	Desacetyllanatoside B	315(180), 420(140), 500(310), 530(360)	255(115), 360(115), 445(120)
B	Gitoxin	235(210), 315(275), 415(185), 495(430) 530(505)	260(150), 375(140), 430(160), 510(405)
B	Digitalnium verum	235(85), 320(190), 495(210), 530(280)	260(150), 400(30)
B	Ditoside	235(140), 310(285), 500(385), 530(475)	265(105), 390(100)
C	Digoxigenin	230(410), 320(310), 390(350), 485(110)	260(95), 340(100), 430(90)
C	Lanatoside C	230(235), 320(130), 390(295), 480(160)	255(105), 430(105)
C	Digoxin	230(260), 320(225), 390(305), 490(210)	265(110), 340(130), 430(130)
C	Acetyl digoxin	230(195), 320(105), 390(280), 490(165)	260(95), 360(105), 440(105)
C	Desacetyllanatoside	235(215), 325(120), 390(260), 480(180)	265(95), 340(110), 440(105)
(IV)	Diginatigenin	230(160), 310(130), 390(210), 425(190) 490(85)	250(45), 345(85), 470(75)
	Diginatin	230(185), 310(175), 390(315), 480(230) 560(120)	220(170), 255(95), 335(160), 460(180) 540(105)
(V)	Sarmentogenin	230(460), 415(430)	215(360), 270(75)
	11–*epi*–Sarmentogenin	230(230), 415(410)	275(60)
(VI)	Ouabagenin	240(510), 295(190), 420(125), 510(210)	375(75)
	Ouabain	240(450), 325(215), 515(80)	215(270), 295(170), 365(45)
(VII)	Afroside acetate	235(520), 340(170), 360(205)	270(60)
	Gomphoside acetate	235(295), 355(250)	270(40)
	Strophanthin K	240(195), 325(220), 420(130)	270(90), 370(60)
(VIII)	Scillaren A	255(160), 327(425) 420(85), 535(95)	230(125), 270(145), 390(40)
	Scilloside	295(260), 505(315)	230(110), 330(70)

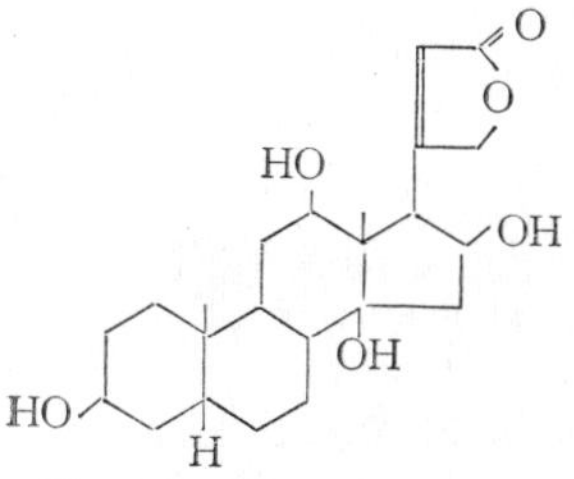

(IV) Diginatigenin 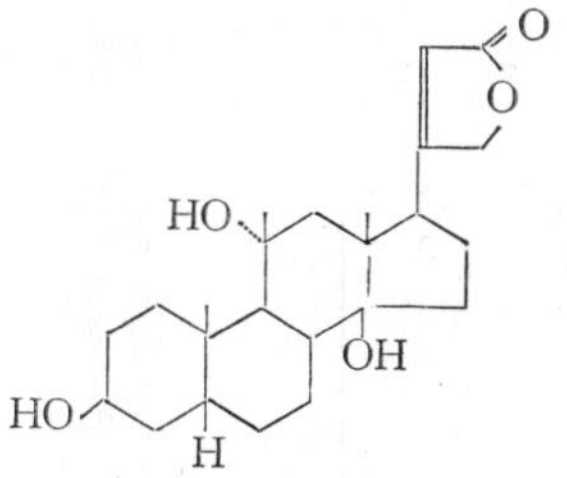   (V) Sarmentogenin 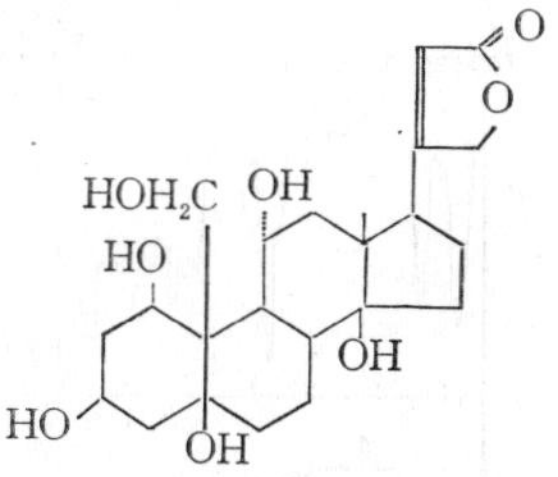   (VI) Ouabagenin

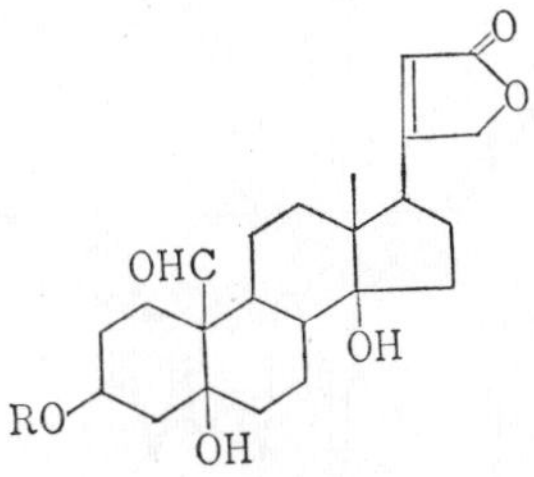

(VII)
R=H: Strophanthidin
R=Cymarose-β-glucose:
Strophanthin K

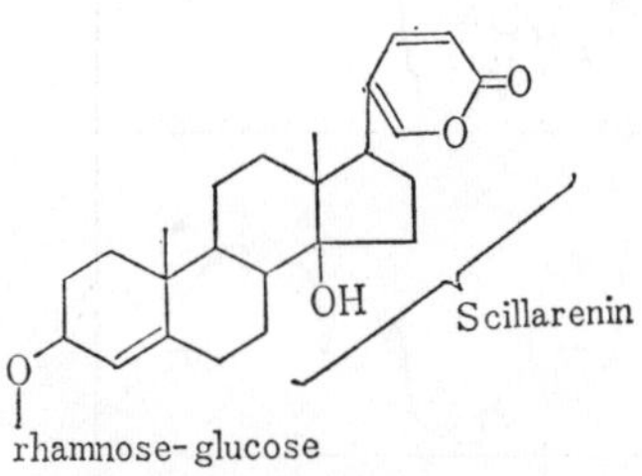

(VIII) Scillaren A

rhamnose-glucose

Scillarenin

[B] IR Spectra

(1) Strophantidin, hellebrigenin[630], ouabain[631]

(2) IR Spectra of Digitalis Glycosides and their Genins[632] (See Fig. 106)

630) Katz, A.: *Helv. Chim. Acta*, **40**, 831 (1957).   631) Sadtler **Card No.** 7549.
632) Bell, F. K: *J.A.P.A.*, **49**, 277 (1940).

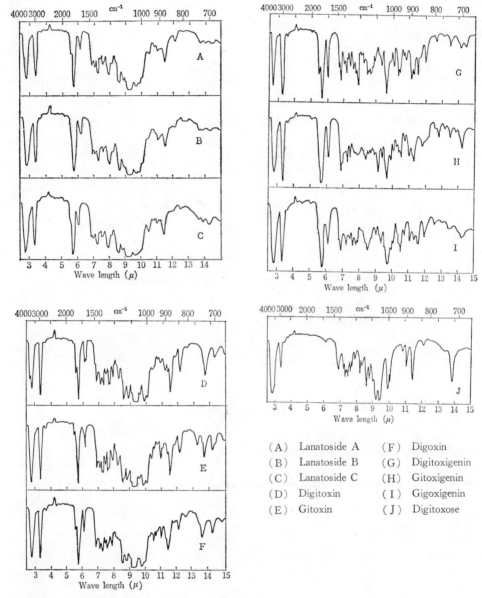

Fig. 106 IR Spectra of Digitalis Glycosides and their Genins
in KBr   (Bell)

(A) Lanatoside A   (F) Digoxin
(B) Lanatoside B   (G) Digitoxigenin
(C) Lanatoside C   (H) Gitoxigenin
(D) Digitoxin   (I) Gigoxigenin
(E) Gitoxin   (J) Digitoxose

〔C〕 Mass Spectra[632a)]

(1) Digitoxigenin (I)

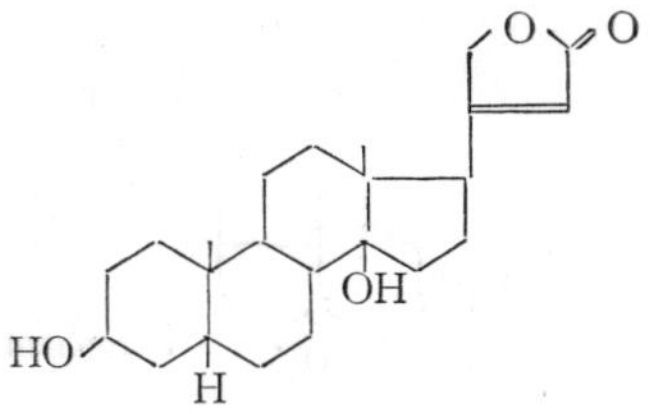

(I) Digitoxigenin

(Cation Spectrum) See Fig. 106a

m/e 374, $M - H_2O = 356$, $M - 2H_2O - CH_3 = 323$, $356 - H_2O = 338$, $338 - CH_3 = 323$,

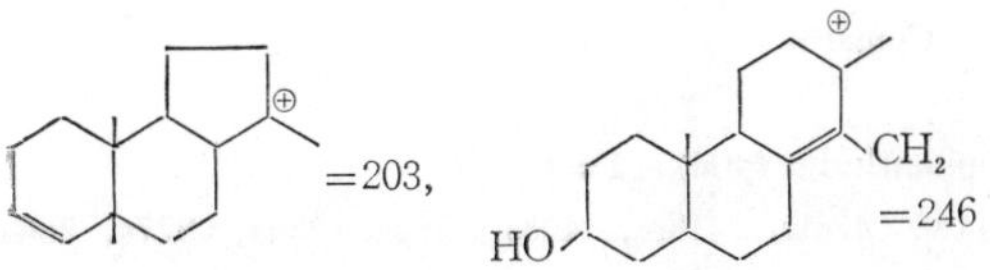

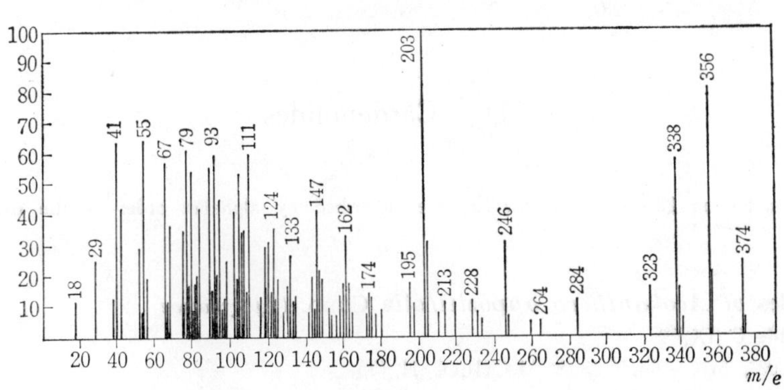

$= 203$,

$= 246$

Fig. 106a Mass Spectrum of Digitoxigenin (cation spectrum) (Reichstein *et al.*)

(2) Somalin (VIII-a)

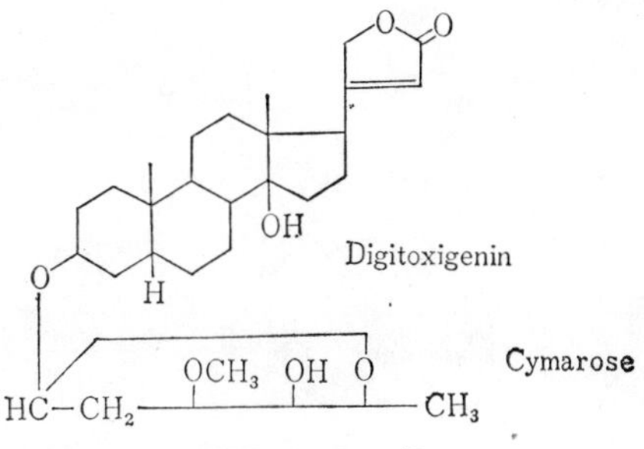

Digitoxigenin

Cymarose

$HC-CH_2$ —— OCH_3 OH O —— CH_3

(VIII a) Somalin

632a) Reichstein, T *et al*: *Helv. Chim. Acta*, **47**, 1032 (1964).

(Anion Spectrum)

M−1=517, M−1−H$_2$O=499, M−CH$_3$OH=486, M − CH$_3$OH−H$_2$O=468, Digitoxigenin − 1=373,

Digitoxigenin −1−H$_2$O=355, Digitoxigenin −1−2H$_2$O=337

Mass peaks: do

(3) Cymarin (LXXXVII)

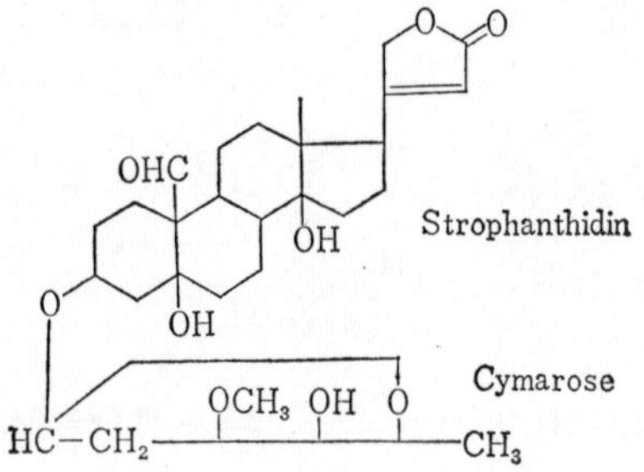

(LXXXVII) Cymarin

(Anion Spectrum)

M=548, M − H$_2$O=530, M − CH$_3$OH=516, strophanthidin (404) −1=403

Mass peaks: 548m, 530w, 514w, 501w, 470w, 444w, 428w, 418m, 403s, 392s, 384s, 377s, 366s,
354s, 347s, 328s, 313s, 303m, 278s, 261s, 256s

[D] Cardenolides

Cardenolides, having 22 (20)–butenolide ring are described by alphabet order of the name of original plants.

(1) Glycosides of Acokanthera oppositifolia CODD. (*Apocynaceae*)

(a) **Acovenoside C** (IX)[633]

　　mp. 201∼203°, [α]$_D$ −64.9° (80% MeOH), C$_{42}$H$_{66}$O$_{19}$

(b) **Acobioside A**

　　mp. 248∼258°, [α]$_D$ −73.5° (80% MeOH), C$_{36}$H$_{56}$O$_{14}$·2H$_2$O

(C) **Acovenoside A**

　　mp. 222°, [α]$_D$ −64.8° (dioxane)

---

633) Reichstein, T. *et al*: *Helv. Chim. Acta*, **45**, 2116 (1962).

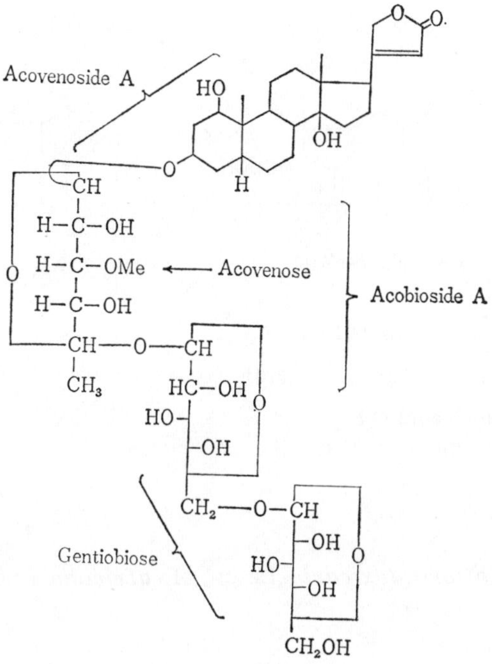

(IX) Acovenoside C

(2) Glycosides of *Acokanthera schimperi* BENT et HOOK (Ouabaio Baum)[634)]

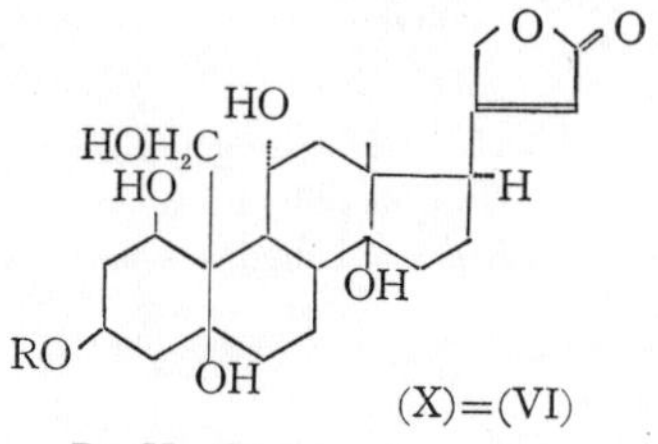

(X)=(VI)

R=H : Ouabagenin
R=L-rhamnosyl : Ouabain

(a) **Ouabagenin** (X)[635)]

mp. 255~256°, $[\alpha]_D$ +11° (H_2O), $C_{23}H_{34}O_8$ UV See the item [A] Table 56

(b) **Ouabain** (G-Strophanthin)[635)]

mp. 187~188°/241°, $[\alpha]_D$ −44° (MeOH), $C_{29}H_{44}O_{12}$ UV See the item [A] (2), Table 56

≪Other occurrence≫   *Acokanthera ouabaio, Strophanthus gratus, S. glaber*

(c) **Acoschimperosides**[634)]

Followed 15 glycosides are separated in crystals : acoschimperoside–Q, N, P, G, H, R1+R2, S, T, U, Y1, Y2, digluco–P, digluco–N, Z

---

634) Reichstein, T. *et al* : *Helv. Chim. Acta*, **42**, 2 (1959).
635) Taww, Ch *et al* : *ibid.*, **40**, 1469, 1860 (1957).

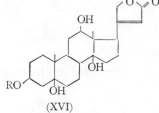

(XI)   Acoschimperoside P   (XII)   (XIII)

Acoschimperoside P (XI)

mp. 275~279°, $[\alpha]_D$ −35.6° (MeOH), $C_{32}H_{48}O_{10}$

UV λ_{max}^{EtOH} mμ (log ε): 217 (4.08), 270 (3.04)

14,16–Dianhydrogitoxigenin (XII)

mp. 210~213°, $[\alpha]_D$ +576.9° (MeOH)

L–Acofriose (XIII)

mp. 116~118°, $[\alpha]_D$ +38.8° (H$_2$O)

(3) **Glycosides of *Antiaris toxicaria* LESCH, *A. africana* ENGL. (Upas)[636)637)** (*Moraceae*)

(XIV)

R=H: Antiarigenin
R=Gu.: α–Antiarin
R=Rh.: β–Antiarin

(XV)

R=H: al–Dihydroantiarigenin
R=Gu.: al–Dihydro–α–antiarin
R=Rh.: al–Dihydro–β–antiarin

(XVI)

R=H: Antiogenin
R=Gu.: α–Antioside
R=Rh.: Antioside

(XVII)

R=H: Strophanthidin
R=Gu.: Desglucocheirotoxin
R=Rh.: Convallatoxin

(XVIII)

R=H: Strophanthidol
R=Gu.: Desglucocheirotoxol
R=Rh.: Convallatoxol

(XIX)

R=H: Digoxigenin

R=Rh.: Substance E

636) Reichstein, T *et al*: *Helv. Chim. Acta*, **45**, 1183, 2285 (1962), **46**, 117 (1963).
637) Martin, R.P *et al*: *ibid.*, **42**, 696 (1959).

(XX)

R=H: Cannogenin
R=Rh.: Malayoside (Substance C)

(XXI)

R=H: Cannogenol
R=Rh.: al–Dihydromalayoside

(XXII)

R=H: Digitoxigenin (I)
R=Rh.: Evomonoside

(XXIII) R=Gu.
D–Gulomethylose

(XXIV) R=Rh.
L–Rhamnose

Table 57 Antiaris Glycosides　　　　　　　　　　　(Reichstein)

Subst.	Name	mp.	$[\alpha]_D$ in MeOH	Formula	No.	λ_{max}^{EtOH} mμ (log ε)	Genin	Sugar
A″	Evomonoside	232~234	−30.9	$C_{29}H_{44}O_8$	XXII		Digitoxigenin	Rh.
C	Malayoside	220~230	−44.3	$C_{29}H_{42}O_9$ ·2H$_2$O	XX	216(4.16), 283 sh(1.85), 300sh(1.80)	Cannogenin	(Rh.)
E	Substance γ of Juslen	272 (decomp)	−23.9	$C_{29}H_{44}O_9$	XIX	217(4.09)	Digitoxigenin	(Rh.)
F	Desgluco–cheirotoxin	180~186	−3.5	$C_{29}H_{42}O_{10}$ ·4H$_2$O	XVII	217(4.12) 300(ca 1.5)	Strophanthidin	Gu.
H	Convallatoxin	232~238	+1.1	$C_{29}H_{42}O_{10}$ ·3H$_2$O	XVII	217(4.20) 299(1.75)	Strophanthidin	Rh.
M	——	220~234	−42.6	$C_{29}H_{42}O_{10}$ ·H$_2$O		217(4.16) 299(1.48)	?	(Rh.)
N	——	212~222	−37.7	$C_{29}H_{42}O_{10}$ ·2H$_2$O		217(4.14) 276sh(2.12)	?	(Rh.)
P	Convallatoxol	174~180	−10.0	$C_{29}H_{44}O_{10}$ ·1.5H$_2$O	XVIII	217(4.23)	Strophanthidiol	Rh.
Q	α–Antioside	192~202	−15.3	$C_{29}H_{44}O_{10}$ ·4H$_2$O	XVI	218(4.12)	Antiogenin	Gu.
R	Antioside	192~202 247~257	−9.8	$C_{29}H_{44}O_{10}$ ·3H$_2$O	XVI	218(4.19)	Antiogenin	Rh.
T	α–Antiarin	242~247	−6.5	$C_{29}H_{42}O_{11}$ ·H$_2$O	XIV		Antiarigenin	Gu.
V	β–Antiarin	244~249	+2.9	$C_{29}H_{42}O_{11}$ ·H$_2$O	XIV	218(4.25) 301(1.46)	Antiarigenin	Rh.

(4) Glycosides of *Apocynum Cannabinum* L.[638] (*Apocynaceae*)

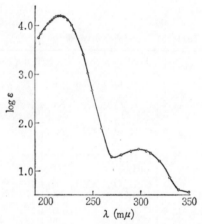

(XXV)

R=H : Cannogenin (XX)
R=D–Cy. : Apocannoside
R=Ol. : Cynocannoside

(XXVI) R=D–Cy.
D–Cymarose

(XXVII) R=Ol.
L–Oleandrose

(a) Cannogenin (XXV)

mp. 145°/185°/200~210°, $[\alpha]_D$ −15.0° (CHCl₃), $C_{23}H_{32}O_5$

UV See Fig. 107, λ_{max}^{EtOH} mμ (log ε): 216 (4.19), 298 (1.45)

IR See Fig. 108

Fig. 107 UV Spectrum of Cannogenin (in EtOH)
(Reichstein)

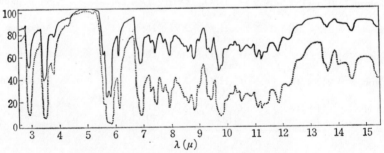

Fig. 108 IR Spectrum of Cannogenin in KBr
(Reichstein)

638) Reichstein, T. *et al*: *Helv. Chim. Acta*, **42**, 2418 (1959).

(b) Apocannoside (XXV)

mp. 122~132°/190~205°, $[\alpha]_D$ −8.1° ($CHCl_3$)

Cynocannoside (XXV)

mp. 166~176°/186°, $[\alpha]_D$ −44.3° (MeOH)

UV λ_{max}^{EtOH} mμ (log ε): 216 (4.14), 275~295 sh (1.68)

D-**Cymarose** (XXVI)

mp. 73~80°, $[\alpha]_D$ +54.3° (H_2O),

L-**Oleandrose** (XXVII)

Amorphous, $[\alpha]_D$ +12° (H_2O)

(5) Glycosides of *Asclepias fruticosa*[639] (*Asclepiadaceae*)

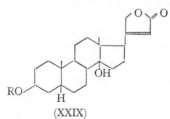

(XXVIII) Gomphoside

Gomphoside (XXVIII)[640]

《Other occurrence》 *Gomphocarpus fruticosus*[640], mp. 234~242°, $[\alpha]_D$ +16.3° (MeOH), $C_{29}H_{44}O_8$

NMR τ: 8.61 d, $J=7(CH_3-\overset{|}{\underset{H}{C}}-O)$[639]

Not hydrolysed by usual condition.

(6) Glycosides of *Beaumontia grandiflora* WALLICH[641]

(XXIX)

R=H :   Digitoxigenin (I)

R=L-Cy. : Wallichoside

R=L-Ol. : Beaumontoside

(XXX)

RH= :   Oleandrigenin

R=L-Cy. : Beauwalloside

R=L-Ol. : Oleandrin

---

639) Coombe R. G. *et al*: *Proc. Chem. Soc.*, **1962**, 214.

640) Watson, T. R. *et al*: *Australian J. Chem.*, **10**, 79 (1957); *C. A.*, **51**, 10550 a (1957).

641) Reichstein, T. *et al*: *Helv. Chim. Acta*, **46**, 1691 (1963).

(a) **Wallichoside**   (XXIX, R=L-Cy)
   mp. 193~196°, $[\alpha]_D$ −65.5° (CHCl₃)

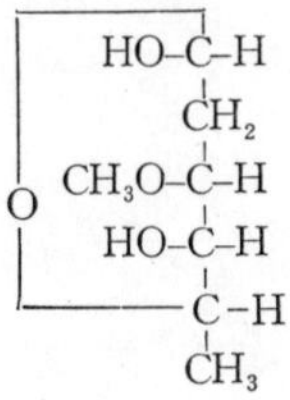

$$
\begin{array}{c}
\text{HO-C-H} \\
\text{CH}_2 \\
\text{CH}_3\text{O-C-H} \\
\text{HO-C-H} \\
\text{C-H} \\
\text{CH}_3
\end{array}
\qquad
\begin{array}{c}
\text{HO-C-H} \\
\text{CH}_2 \\
\text{H-C-OCH}_3 \\
\text{HO-C-H} \\
\text{C-H} \\
\text{CH}_3
\end{array}
$$

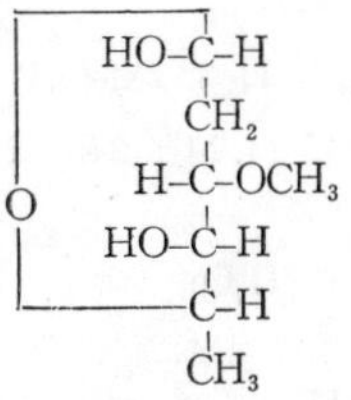

(XXXI) R=L-Cy.   (XXVII) R=L-Ol.
L-Cymarose   L-Oleandrose

(b) **Beaumontoside**   (XXIX, R=L-Ol.)
   mp. 202~203°, $[\alpha]_D$ −32.5° (CHCl₃)

(c) **Beauwalloside**   (XXX, R=L-Cy.)
   mp. 223~226°, $[\alpha]_D$ −80.7° (CHCl₃)

(d) **Oleandrin**   (XXX, R=L-Ol.)
   ≪Other occurrence≫   *Nelium oleander* L., *N. odorum*,   mp. 250°, $[\alpha]_D$ −52° (CHCl₃), $C_{32}H_{48}O_9$.

(e) **Oleandrigenin**   (XXX, R=H)
   mp. 223~225°, $[\alpha_D]$ −8.5° (MeOH), $C_{25}H_{36}O_6$

(f) **L-Cymarose**   (XXXI)
   mp. 87~91°, $[\alpha]_D$ −53.6° (H_2O)

(7) **Glycosides of *Calotropis procera*** (*Asclepiadaceae*)

Former Structures

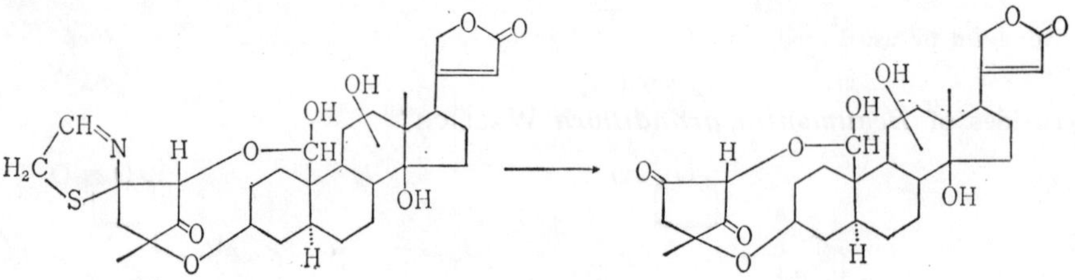

(XXXII)   Uscharin[642]   (XXXIII)   Uscharidin[642]

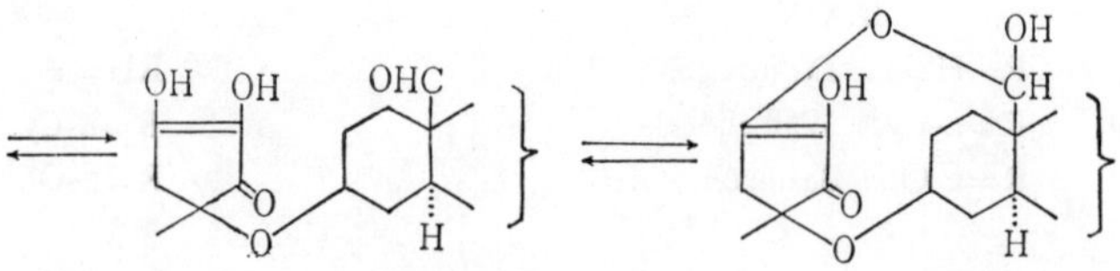

[(XXXIII)   Uscharidin

642) Hesse, G. *et al*: *Ann.*, **623**, 142, 146 (1959).

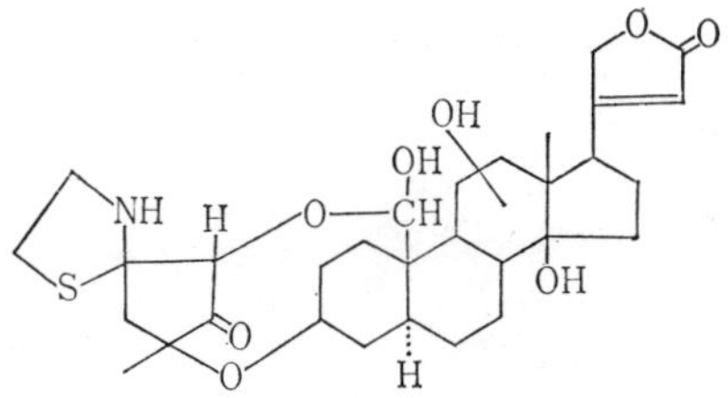

(XXXIV) Voruscharin[643]

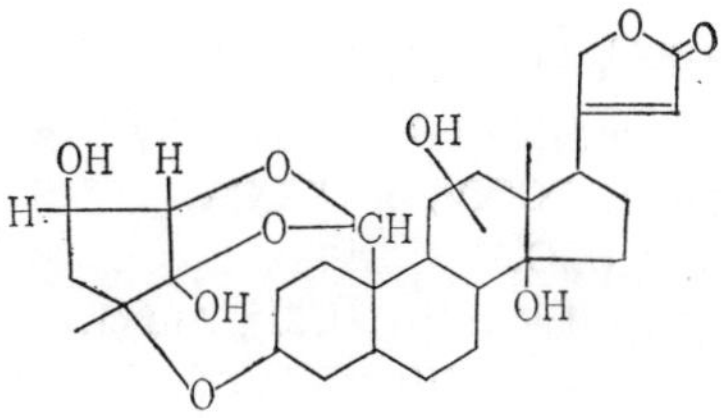

(XXXV) Calotropin[644]

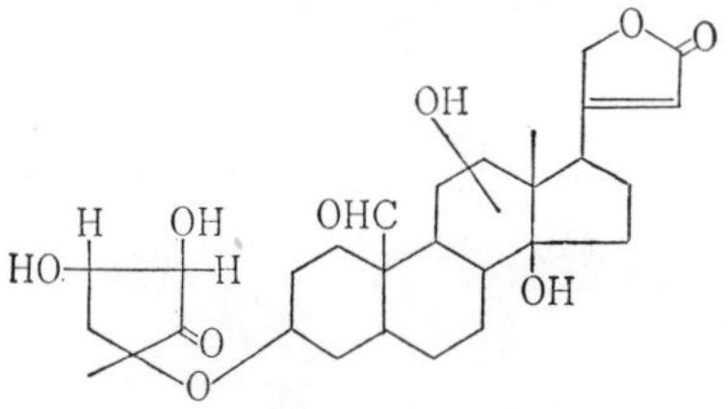

(XXXVI) Calactin[644]

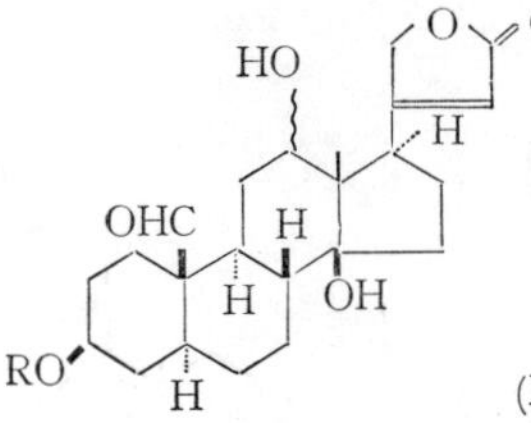

$$R = \begin{matrix} HO \\ H_3C \end{matrix} CH-CH_2-\overset{COOH}{\underset{O}{C}}\!-\!-\!CH\!- :$$

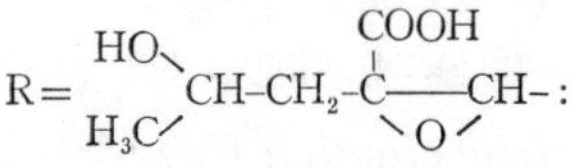

Calactinic acid[645]

R=H: Calotropagenin[646]

(XXXVII)

Revised Structures of T. L. Jones *et al* (1963)[647]

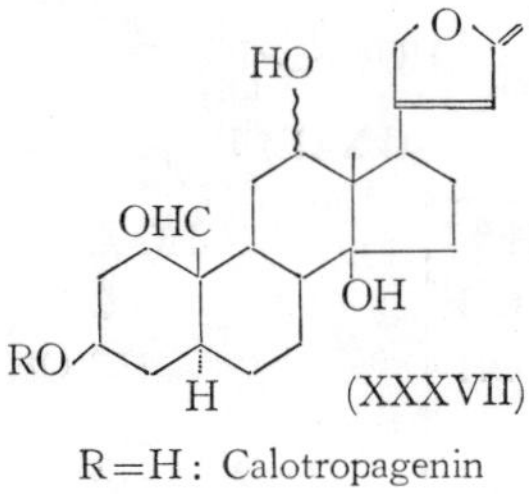

(XXXVII)

R=H: Calotropagenin

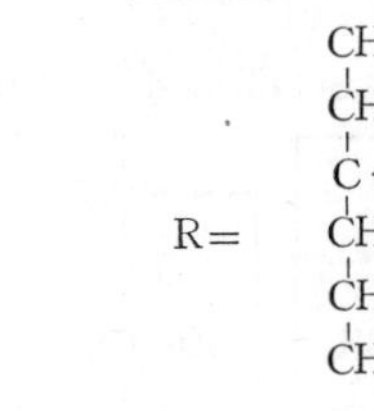

$$R = \begin{matrix} CH= \\ O=C \\ CH-OH \\ CH-OH \\ CH- \\ CH_3 \end{matrix} O$$

(XXXVIII)
Calotoxin

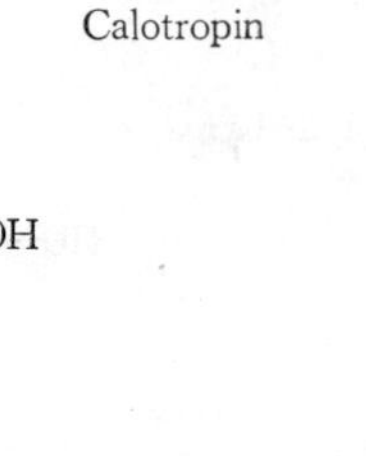

$$R = \begin{matrix} CH= \\ O=C \\ CH-OH \\ CH_2 \\ CH- \\ CH_3 \end{matrix} O$$

(XXXIX)
Calotropin

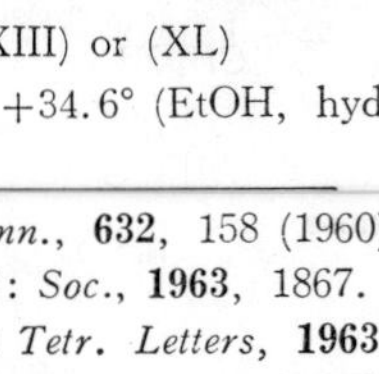

$$R = \begin{matrix} CH= \\ O=C \\ O=C \\ CH_2 \\ CH- \\ CH_3 \end{matrix} O$$

(XL)
Uscharidin

$$R = \begin{matrix} CH- \\ CH \\ C<COOH \\ CH_2 \\ CH-OH \\ CH_3 \end{matrix} O$$

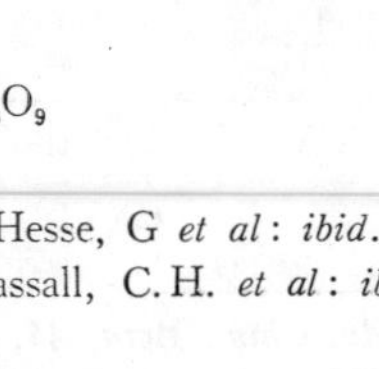

(XLI)
Calactinic acid

(a) Uscharidin (XXXIII) or (XL)

mp. 303~306°, $[\alpha]_D$ +34.6° (EtOH, hydrate), $C_{29}H_{38}O_9$

643) Hesse, G. *et al* : *Ann.*, **632**, 158 (1960). 644) Hesse, G *et al* : *ibid.*, **625**, 157 (1959).
645) Hassall, C.H. *et al* : *Soc.*, **1963**, 1867. 646) Hassall, C.H. *et al* : *ibid.*, **1959**, 85.
647) Jones, T.L. *et al* : *Tetr. Letters*, **1963**, No 2, 63.

(b) Voruscharin   (XXXIV)[643)]

  mp. 165～166°, $[\alpha]_D$ −60.8°, $C_{31}H_{42}O_8NS$

  UV λ_{max}^{EtOH} mμ (log ε): 217 (4.24), 303 (2.62)

(c) Calotropin   (XXXV) or (XXXIX)

  mp. 223°, $C_{29}H_{40}O_9$

  UV λ_{max} mμ (log ε): 217 (4.21), 310 (1.49)

(d) Calotoxin   (XXXVIII)

  mp. 265～271°, $[\alpha]_D$ +66° (CHCl$_3$–MeOH), $C_{29}H_{40}O_{10}$

  UV λ_{max}^{EtOH} mμ (log ε): 217 (4.22), 309 (1.49)[646)]

(e) Calactin   (XXXVI)

  mp. 215～218°, $[\alpha]_D$ +45° (MeOH), $C_{29}H_{40}O_9$

  UV λ_{max}^{HtOH} mμ (log ε): 219 (4.19), 301 (1.65)[646)]

(f) Calactinic acid   (XXXVII)=(XLI)

  mp. 170～172°, $[\alpha]_D$ −37.3° (pyridine), $C_{29}H_{40}O_{10}\cdot H_2O$

  UV λ_{max} mμ (log ε): 218, 307～309 (4.19)[645)]

(g) Calotropagenin   (XXXVII)[646)]

  mp. 238～250°, $[\alpha]_D$ +43° (CHCl$_3$), $C_{23}H_{32}O_6$

  UV λ_{max}^{EtOH} mμ (log ε): 218 (4.24), 309 (1.38)

(8) Glycosides of *Castilla elastica* CERV.[647a)] (*Moraceae*)

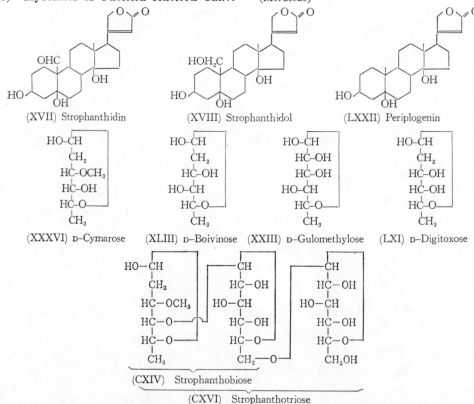

(XVII) Strophanthidin   (XVIII) Strophanthidol   (LXXII) Periplogenin

(XXXVI) D–Cymarose   (XLIII) D–Boivinose   (XXIII) D–Gulomethylose   (LXI) D–Digitoxose

(CXIV) Strophanthobiose

(CXVI) Strophanthotriose

647a) Reichstein T. *et al*: *Helv. Chim. Acta*, **44**, 904 (1961).

Table 57a Cardenolides of *Castilla elastica* (Reichstein *et al.*)

	Cardenolide	Str. No.	Composition	mp.	$[\alpha]_D$	See the item, Formula
A	Periplocymarin	LXXXVI	(LXXII)–(XXVI)	203~207	+29.0 (Me)	(21), (27)
B	Cymarin	LXXXVII	(XVII)–(XXVI)	146~148	+35.9 (Me)	(21), (27), (28)
C	Cymarol	CXI	(XVIII)–(XXVI)	218~235	+25.1 (Me)	(21), (27), (28)
D	Helveticoside	LXX	(XVII)–(LXI)	155~158	+29.0 (Me)	(15), (28)
E	Corchoroside	XLII	(XVII)–(XLIII)	188~190	+11 (Me)	(10)
F	Desglucocheirotoxin	XVII	(XVII)–(XXIII)	186~191	− 3 (Me)	(3), (9)
F′	Helveticosol		(XVIII)–(LXI)	147~152		(28)
F″	Subst. F″			165~171		
F‴	Subst. F‴			170~179		
G	K-Strophanthin-β		(XVII)–(CXIV)	207~212	+24.3 (Me)	(28)
H	Cheirotoxin		(XVII)–(XXIII)–gluc.	208~211	−17.2 (Me)	$C_{35}H_{52}O_{15}$ *
K	K-Strophanthoside		(XVII)–(CXVI)	199~200	+13.8 (Me)	(28)
C′	Periplogenin		(LXXII)	235~237	+27.6 (Me)	(17), (21)
C″	Strophanthidin		(XVII)	171~175/ 230~232	+44 (Me)	

* *Cheiranthus cheiri* L. [647b]

(9) Glycoside sof *Convallaria keisukei* MIG. (スズラン)[648] (*Liliaceae*)

(a) **Desglucocheirotoxin** (Strophanthidin–D–glucomethyloside)

(XVII, R=Gu.) See the item (3), Table 57, mp. 188~191°, $[\alpha]_D$ −7.8° (MeOH)

UV λ_{max} mμ: 217 (butenolide), 280~290° (CHO)

IR $\nu_{max}^{CHCl_3}$ cm^{-1}: 3360~3540 (OH), 1780, 1740 (C=C–C=O), 1710 (CHO), 1620 (C=C),

ν_{max}^{KBr} cm^{-1}: 1780 disappear

(b) **Convallatoxin** (Strophanthidine–L–rhamnoside) (XVII, R=Rh.) See the item (3) Tab. 57

(10) Glycoside of *Corchorus capsularis* L. [649] (*Tiliaceae*)

(a) **Corchoroside A** (Strophanthidin β–D–boivinoside) (XLII)

(XLII)=Strophanthidin (XVII, R=H)+D–Boivinose (XLIII)

$$
\begin{array}{l}
HO\text{-}C\text{-}H \\
CH_2 \\
H\text{-}C\text{-}OH \quad O \\
HO\text{-}C\text{-}H \\
H\text{-}C \\
CH_3
\end{array}
$$

(XLIII) D–Boivinose

(XLII) mp. 188~190°, $[\alpha]_D$ +11° (CH$_3$OH), $C_{29}H_{42}O_9$

(XLIII) mp. 100~103°, $[\alpha]_D$ −15° (acetone)

647b) Reichstein T. *et al*: *Helv. Chim. Acta*, **37**, 755 (1954).
648) Kimura M. *et al*: *Y.Z.*, **82**, 1320 (1962).
649) Reichstein, T. *et al*: *Helv. Chim. Acta*, **40**, 593, (1957).

(11) Glycosides of *Digitalis canariensis* L.[650] (*Scrophulariaceae*)

R=H: Canarigenin (XLIV)
R=Boivinose (XLIII): Canari–boivinoside (XLIV–a)
R=Digitoxose: Canari–digitoxoside (XLIV–b)

(a) **Canarigenin** (XLIV)　　mp. 192~196°, $[\alpha]_D$ +46° (CHCl₃), $C_{23}H_{32}O_4$

(12) Glycosides of *Digitalis grandiflora* MILL[651] (syn. *D. ambigua* MURR.)
(*Scrophulariaceae*)

Digitoxigenin (I), Gitoxigenin (II)

Diginigenin, anhydropurpurogenin and other C_{21}–steroids are also isolated from *Digitalis purpurea*. Those are described in the item **9**, 〔G〕 C_{21}–Steroids, (2)~(9).

(13) Glycosides of *Digitalis lanata* EHRH. (*Scrophulariaceae*)
(a) **Lanatosides A, B, C**[652]

R₁=H	R₂=H:	Acetyldigitoxin α (XLVIII)
R₁=H	R₂=OH:	Acetylgitoxin α (XLIX)
R₁=OH	R₂=H:	Acetyldigoxin α (L)
R₁=H	R₂=H:	Lanatoside A (XLV)
R₁=H	R₂=OH:	Lanatoside B (XLVI)
R₁=OH	R₂=H:	Lanatoside C (XLVII)

650) Tschesche, R. *et al*: *Ann.*, **663**, 157 (1963).
651) Tamm, Ch *et al*: *Helv. Chim. Acta*, **40**, 639 (1957).　　652) Kuhn, M. *et al*: *ibid.*, **45**, 881 (1962).

Acetyldigitoxin α (XLVIII) $\rightleftarrows$ R_1=H R_2=H : Acetyldigitoxin β (LI)
Acetylgitoxin α (XLIX) $\rightleftarrows$ R_1=H R_2=OH : Acetylgitoxin β (LII)
Acetyldigoxin α (L) $\rightleftarrows$ R_1=OH R_2=H : Acetyldigoxin β (LIII)

Lanatoside A (XLV)

mp. 245~248° (decomp.), $[\alpha]_D$ +31.4° (EtOH), $C_{49}H_{76}O_{19}$

UV: See the item [A], (2), Table 56, IR: See the item [B], (2), Fig. 106 (A)

Lanatoside B (XLVI)

mp. 245~248° (decomp.), $[\alpha]_D$ +36.7° (EtOH), $C_{49}H_{76}O_{20}$

UV: (Desacetyllanatoside B) See the item [A], (2), Table 56, IR: See the item [B], (2), Fig. 106 (B).

Lanatoside C (XLVII)

mp. 245~249° (decomp.), $[\alpha]_D$ +33.5° (EtOH), $C_{49}H_{76}O_{20}$

UV: See the item [A], (2), Table 56 IR: See the item [B], (2), Fig. 106 (C)

(b) Diginatigenin (IV)[653]

mp. 157°, $C_{22}H_{24}O_6$, UV: [A], (2), Table 56.

(c) Lanatoside E (LV)[654]

mp. 209~215°, $[\alpha]_D$ +26.8° (MeOH), $C_{50}H_{76}O_{21}$

R=H : 16-Formylgitoxigenin (Gitaloxigenin) (LIV)
R=2 Digitoxose+1 Acetyldigitoxose+1 Glucose : Lanatoside E (LV)

(14) Glycosides of *Digitalis purpurea* L. (*Scrophulariaceae*)

(a) Genins and Purpureaglycosides[655]

Purpurea glycoside A (LVI) (Desacetyllanatoside A (XLV)+CH₃COOH)

Purpurea glycoside B (LVII) (Desacetyllanatoside B (XLVI)+CH₃COOH)

Digitoxin (LVIII) (Desacetyllanatoside A (XLV)+CH₃COOH+Glucose)

Gitoxin (LIX) (Desacetyllanatoside B (XLVI)+CH₃COOH+Glucose)

Digitalinum verum (LX) (Gitoxigenin (II)+D-Digilanose+Glucose)

653) Ishidate, M. *et al*: *Chem. Pharm. Bull.*, **8**, 535 (1960).
654) Lenz, J. *et al*: *Helv. Chim. Acta*, **41**, 479 (1958).
655) Kaiser, F. *et al*: *Ann.*, **603**, 75 (1957).

Table 58 Physical Constants of Purpurea Glycosides and Genins

Compounds	No.	Formula	mp.	$[\alpha]_D$	UV	IR
Digitoxigenin	I	$C_{23}H_{34}O_4$	236~256°	+18~19°(MeOH)	λ_{max}^{H2SO4} Table. 56	Fig. 106(G)
Gitoxigenin	II	$C_{23}H_{34}O_5$	224~225° (decomp.)	+34~35°(MeOH)	″	Fig. 106(H)
Purpurea glycoside A	LVI	$C_{47}H_{74}O_{18}$	270~280° (decomp.)	+12°(MeOH)	(acetate) λ_{max}^{EtOH} 217(4.18)	
Purpurea glycoside B	LVII	$C_{47}H_{74}O_{19}$	240° (decomp.)	+15.5°(EtOH)	(acetate) ″ 216(4.16)	
Digitoxin	LVIII	$C_{41}H_{64}O_{13}$	233~236°	+17.5°(CHCl$_3$+ EtOH)	(acetate) ″ 217(4.21)	Fig. 106(D)
Gitoxin	LIX	$C_{41}H_{64}O_{14}$	282~285°	+22° (CHCl$_3$+EtOH)	(acetate) ″ 215(4.26)	Fig. 106(E)
Digitalinum verum	LX	$C_{36}H_{56}O_{14}$	241~246°	+1~2°(MeOH)	218	

(b) Digitoxigenin digitoxosides[655]

Digitoxigenin monodigitoxoside

mp. 181~184°, $C_{29}H_{44}O_7$

UV λ_{max} mμ (log ε): 218 (4.20)

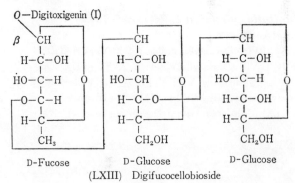

(LXI) D–Digitoxose

Digitoxigenin bisdigitoxoside

mp. 187~190°, $C_{35}H_{54}O_{10}$, UV λ_{max} mμ (log ε): 218 (4.19)

(c) Gitoroside (LXII)[656]

D–Boivinose (XLIII)—Gitoxigenin (II)

mp. 213~216°, $C_{29}H_{44}O_8$, Acetate $C_{35}H_{50}O_{11} \cdot H_2O$, mp. 128~131°

(d) Digifucocellobioside (LXIII)[657]

(LXIII) Digifucocellobioside

mp. 238~242°, $[\alpha]_D$ −1.4°, $C_{41}H_{64}O_{18} \cdot 2H_2O$, UV λ_{max}^{EtOH} mμ (log ε): 218 (4.22)

656) Sato, D. *et al*: *Chem. Pharm. Bull.*, **5**, 253 (1957).
657) Okano, A *et al*: *ibid.* **7**, 222 (1959).

(e) Gitorocellobioside (LXIV)[658]

mp. 250~254°, $[\alpha]_D$ +13.5° (MeOH),　$C_{41}H_{64}O_{18}$　　UV λ_{max}^{EtOH} mμ (log ε): 218 (4.26)

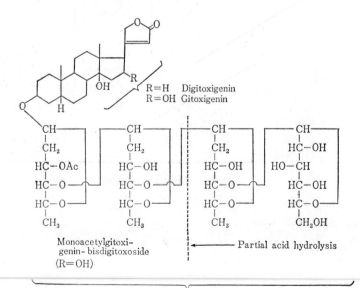

(LXIV)　Gitorocellobioside

(f) Purlanosides A (LXV) **and B** (LXVI)[658]

R=H　Digitoxigenin
R=OH　Gitoxigenin

Monoacetylgitoxi-
genin- bisdigitoxoside
(R=OH)

← Partial acid hydrolysis

(LXV)　R=H　　Purlanoside A
(LXVI)　R=OH　Purlanoside B

Purlanoside A (LXV)

Colorless syrupy substance, $[\alpha]_D$ +22.0° (EtOH), $C_{49}H_{76}O_{19}$

UV λ_{max}^{EtOH} mμ (log ε): 217 (4.23)

Purlanoside B (LXVI)

Colorless syrupy substance, $[\alpha_D]$ +27.8° (EtOH), $C_{49}H_{76}O_{20}$

UV λ_{max}^{EtOH} mμ (log ε): 218 (4.18)

(g) Gitoxin cellobioside (LXVII)[659]

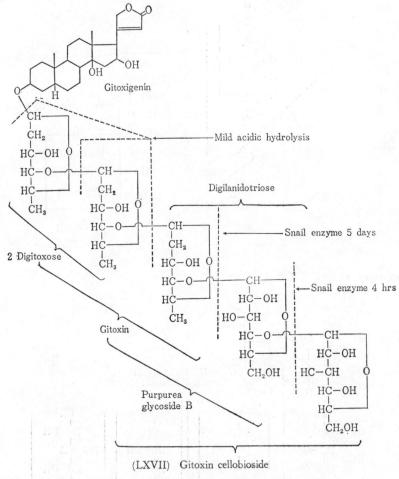

(LXVII) Gitoxin cellobioside

mp. 250~254°, $[\alpha]_D$ +13.5° (MeOH), $C_{41}H_{64}O_{18}$

UV λ_{max}^{EtOH} mμ (log ϵ): 218 (4.26)

658) Hoji. K: *Chem. Pharm. Bull.*, 9, 291, 296, 566 (1961).
659) Okano, A. *et al*: *Chem. Pharm. Bull.*, **7**, 226 (1959).

(h)　Odorobioside–G　(Digitoxigenin glucoside–digitaloside)　(LXVIII)[660]

mp. 240~242°, $[\alpha]_D$ −8.1° (MeOH),　　$C_{36}H_{56}O_{13}$

UV　λ_{max}^{EtOH} $m\mu$ (log ε): 218 (4.19)

《Other occurrence》　　*Nerium odorum*[661]

(i)　*allo*-Digitalinum verum　(LXIX)[662] and **(17β)-Digitalinum verum**　(LX)

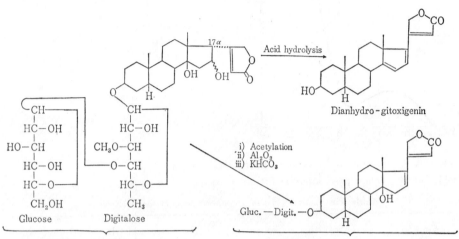

Glucose　　　　Digitalose

(LXIX)　*allo*-Digitalinum verum　　　　16–Anhydrodigitalinum verum monoacetate

allo–Digitalinum verum hexaacetate

mp. 237~239°, $[\alpha]_D$ +14.5° (CHCl₃),　　$C_{48}H_{68}O_{20}$

UV　λ_{max}^{EtOH} $m\mu$ (log ε): 217 (4.22)

(15)　Glycosides of *Erysimum crepidifolium*[663] (*Cruciferae*)

The following 8 glycosides (A~F) are isolated (see Table 58a)

Substance B (mp. 228~256°, $[\alpha]_D$ +11.8° (MeOH)), **C** (232~240°, −22.1°), **D** (218~229°, −19.0°), **E** (307~314°), **G** (230~237°), **H** (224~232°, +32.7°), **J** (180~187°, +48.9°) and **Substance I (Helveticoside)**　　(LXX)

mp. 153~157°, $[\alpha]_D$ +30.7° (MeOH),　　$C_{29}H_{42}O_9$

660)　Hoji, K: *Chem. Pharm. Bull.*, **9**, 289 (1961).
661)　Reichstein, T. *et al*: *Helv. Chim. Acta*, **35**, 687 (1952).
662)　Miyatake, K. *et al*: *Chem. Pharm. Bull.*, **9**, 375 (1961).
663)　Nagata, W. *et al*: *Helv. Chim. Acta*, **40**, 41 (1957).

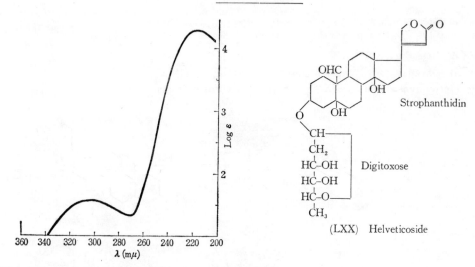

Fig. 109 UV Spectrum of Helveticoside (in EtOH) (Nagata *et al.*)

Glycosides of *Erysimum perofskianum* FISCH. et MEY.[663a] (*Cruciferae*)

Table 58a Cardenolides of *Erysimum perofskianum* (Reichstein)

	Cardenolide	Str. No.	Formula	mp.	$[\alpha]_D$(MeOH)	λ_{max}^{EtOH} (log ε)
A	Helveticoside	(LXX)	$C_{29}H_{42}O_9$	153~156	+29.6	216(4.18), 300(1.34)
B	Corchoroside A		$C_{29}H_{42}O_9$	165~167	+ 9.9	216(4.18), 300(1.34)
C	Kabuloside		$C_{29}H_{42}O_{10}$	amorph	+25.5	216(4.17), 308sh(2.59)
D	Perofskoside		$C_{29}H_{42}O_{10}$	amorph	+25.5	216(4.20)
E	Erysimoside		$C_{35}H_{52}O_{14}$	170~173	+16.8	216(4.18), 290sh(1.63)
E'	Eryperoside		$C_{35}H_{52}O_{14}$	178~180	+43.9	216(4.21)
F	Erycorchoside		$C_{35}H_{52}O_{14}$	238~240	+30.3	216(4.20)

(16) Glycosides of *Gomphocarpus fruticosus* (*Asclepiadaceae*)

(a) Gomphoside (XXVIII) See (5) *Asclepias fruticosa*

(b) Uzarigenin (LXX-a), R=H

《Occurrence》 *Gomphocarpus* spp. (Uzara), *Pergularia extensa*, *Roupellina bowinii*, *Xysmalobium undulatum*

mp. 240~256°, $[\alpha]_D$ +14°, (EtOH), $C_{23}H_{34}O_4$

R=H: Uzarigenin
R=Glucose: Uzarin

663a) Reichstein, T *et al*: *Helv. Chim. Acta*, **43**, 957 (1960).

IR See Fig. 109a[663b)]

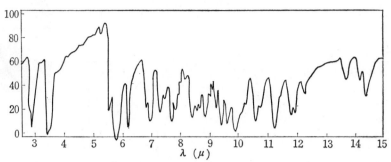

Fig. 109a IR Spectrum of Uzarigenin
(in KBr) (Reichstein *et al.*)

(c) **Uzarin** (LXX–a), R=2 Glucose

≪Occurrence≫ Same to uzarigenin, mp. 268～270°, $[\alpha]_D$ −27° (pyridine), $C_{35}H_{54}O_{14}$

(17) Glycosides of *Gongronema gazense* BULLOCK[664)] (*Asclepiadaceae*)

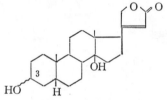

3β–OH: Sarmentogenin (V)
3α–OH: 3–*epi*–Sarmentogenin (LXX–b)

3β–OH: Digitoxigenin (I)
3α–OH: 3–*epi*–Digitoxigenin (LXXI)

(LXXII) Periplogenin

D–Digitoxose (LXI)

D–Cymarose (XXVI)

(a) **Sarmentogenin** (Genin J) (V)
mp. 265～275°, $[\alpha]_D$ +21.1° (MeOH), $C_{23}H_{34}O_5$
UV: See the item [A], (2), Tale 56.

(b) 3–*epi*–**Sarmentogenin** (Genin K) (LXX–b)
mp. 277～295° $[\alpha]_D$ +29.9° (MeOH), $C_{23}H_{34}O_5$
IR λ_{max}^{KBr} μ: 3.47, 3.50, 5.58, 5.75, 6.14, 6.90, 7.26, 7.34

(c) 3–*epi*–**Digitoxigenin** (Genin H) (LXXI)
mp. 286～290°, $[\alpha]_D$ +23.8° (MeOH), $C_{23}H_{34}O_4$

(d) **Periplogenin** (LXXII)
mp. 240～242° (decomp.), $[\alpha]_D$ +29.6° (MeOH), $C_{23}H_{34}O_5$
≪Other occurrence≫ *Strophanthus preussii* and other spp.

663b) Reichstein, T. *et al*: *Helv. Chim. Acta*, **42**, 1502 (1959).
664) Reichstein, T. *et al*: *ibid*., **46**, 505, 517 (1963).

Table 59 Gongronema Glycosides (Reichstein *et al.*)

Sm: Sarmentogenin, Ppl: Periplogenin, Cym: D–Cymarose, Dgx: D–Digitoxose, Ol: Oleandrose

Glycoside	Composition	mp.	$[\alpha]_D$(MeOH)	Formula	λ^{EtOH}_{max} mμ(log ε)
C	Sm–Cym–Cym–Cym	179~181°	+29.7°	$C_{40}H_{70}O_{14}$	217(4.16)
E	Sm–Cym–Cym	236~238°	+21.9°	$C_{37}H_{58}O_{11}$	217(4.19)
G	Sm–Dgx–Cym–Cym	217~220°	+21.8°	$C_{43}H_{68}O_{14}$	218(4.21)
H	3–*epi*–Digitoxigenin	286~290°	+23.8°	$C_{23}H_{34}O_4$	—
J	Sarmentogenin	262~265°	+20.1°	$C_{23}H_{34}O_5$	218(4.22)
K	3–*epi*–Sarmentogenin	276~284°	+29.9°	$C_{23}H_{34}O_5$	218(4.22)
L	Ppl–Dgx–Cym	150~153°	+31.5°	$C_{36}H_{56}O_{11}+3H_2O$	217(4.21)
M	Sm–Dgx–Cym	245~247°	+16.6°	$C_{36}H_{56}O_{11}$	218(4.21)
P	?	298~302°	+27.2°	$C_{23}H_{34}O_5+CH_3OH$	217(4.24)
R	Sm–Dgx–Dgx–Ol	263~266°	− 2.9°	$C_{42}H_{66}O_{14}+2H_2O$	218(4.21)

(18) Cardenolide of *Melia indica*[665] (*Meliaceae*)

Nimbin (LXXIII) See the item **11** Diterpenoids [I], (8)

(Former Str.) $C_{28}H_{38}O_8$[666] (Revised Str.)[668] $C_{30}H_{36}O_9$ (LXXIII)[667]

(19) Glycosides of *Nerium oleander* L. (*Apocynaceae*)[668]

The following 18 glycosides are separated in crystals by means of chromatography and counter current distribution from the ether-chloroform extracts of the seeds.

Table 60 Cardenolides from *Nerium oleander* seeds (Reichstein, *et al.*)

	Substance	Formula	Composition	mp.	$[\alpha]_D$	λ^{EtOH}_{max} mμ(log ε)	Ref.
B	Adigoside	$C_{30}H_{46}O_8$	(LXXIV)–(LXXVIII)	138~142	−16.8 (Me)	215(4.09)	new
γ	Digistroside	$C_{30}H_{46}O_7$	(I)–(LXXIX)	172/ 205~208	−16.6 (Me)	217(4.20)	
C	Oleandrin	$C_{32}H_{48}O_9$	(XXX)–(XXVII)	242~246	−48.2 (Me)		
δ	Cryptograndoside A	$C_{32}H_{48}O_9$	(XXX)–(LXXIX)	115~120	−31.2 (Me)		
D	16–Anhydrodesacetyl-cryptograndoside A	$C_{30}H_{44}O_7$	(LXXV)–(LXXIX)	230~232	+48.0 (Me)	215sh, 230sh, 270(4.20)	668)
E	Odoroside A	$C_{30}H_{46}O_7$	(I)–(LXXVIII)	180~185/ 200~206	− 4.3(Me)		669)
F	Nerigoside	$C_{32}H_{48}O_9$ $+C_4H_8O_2$	(XXX)–(LXXVIII)	155~163	−17.0 (Me)	216(4.22)	new

665) Narasimhan, N.S.: *Ber.*, **92**, 769 (1959).   666) Mitra: *J. Sci. Ind. Res., India*, **16B**, 477(1957).
667) Henderson, R. *et al*: *Proc. Chem. Soc.*, **1963**, 269.
668) Reichstein, T. *et al*: *Helv. Chim. Acta*, **33**, 1013 (1950).
669) Reichstein, T. *et al*: *Helv. Chim. Acta*, **32**, 939 (1949).

	Substance	Formula	Composition	mp.	$[\alpha]_D$	λ_{max}^{EtOH} mμ(log ε)	Ref.
G	16–Anhydrodesacetyl–nerigoside	$C_{30}H_{46}O_8$	(LXXV)–(LXXVIII)	182~186	+55.6(Me)	270(4.30)	new
H	Digitoxigenin	$C_{23}H_{34}O_4$	(I)	246~253	+13.8(Me)		
I	Oleandrigenin	$C_{25}H_{36}O_6$	(XXX)	225~230	− 6.9(Me)		
K	Desacetyl–oleandrin	$C_{30}H_{46}O_8$	(II)–(XXVII)	235~238	−22.2(Me)		
L	Desacetyl cryptograndoside A	$C_{30}H_{46}O_8$	(II)–(LXXIX)	203~206	− 4.6(Me)		668)
N	Desacetyl nerigoside	$C_{30}H_{46}O_8$	(II)–(LXXVIII)	211~216	+ 9.6(Me)		new
O	Odoroside H	$C_{30}H_{46}O_8$	(I)–(LXXX)	228~232	+ 5.9(Me)		670)
P	Neritaloside	$C_{32}H_{48}O_{10}$ $+C_4H_8O_2$	(XXX)–(LXXX)	135~140	−11.4(Me)	217(4.23)	new
π	16–Anhydro–strospeside	$C_{30}H_{44}O_8$	(LXXV)–(LXXX)	230~240	+62.2(Me)	270(4.19)	671)
ρ	Gitoxigenin	$C_{23}H_{34}O_5$	(II)	224~230	+28.0(Me)		
Q	Strospeside	$C_{30}H_{46}O_9$	(II)–(LXXX)	246~250	+17.0(Me)		671) 672)

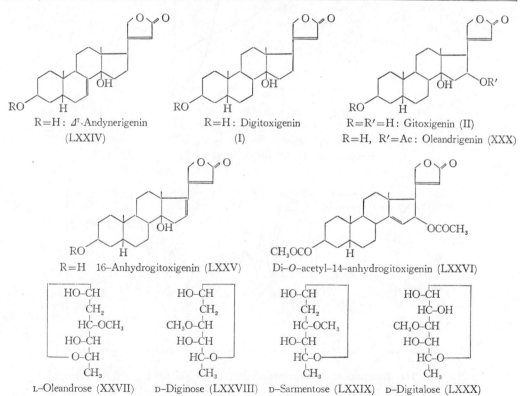

R=H: Δ^7-Andynerigenin (LXXIV)

R=H: Digitoxigenin (I)

R=R′=H: Gitoxigenin (II)
R=H, R′=Ac: Oleandrigenin (XXX)

R=H 16–Anhydrogitoxigenin (LXXV)

Di–O–acetyl–14–anhydrogitoxigenin (LXXVI)

L–Oleandrose (XXVII)   D–Diginose (LXXVIII)   D–Sarmentose (LXXIX)   D–Digitalose (LXXX)

670) Reichstein, T. *et al*: *Helv. Chim. Acta*, **35**, 687 (1952).
671) Reichstein, T. *et al*: *ibid.*, **33**, 1993 (1950).
672) Reichstein, T. *et al*: *ibid.*, **35**, 434 (1952).

≪Other occurrence of glycosides≫

Cryptograndoside A (Subst. δ): *Cryptostegia grandiflora*[668]

16–Anhydrodesacetyl cryptograndoside A (Subst. D): *Cryptostegia grandiflora*[668]

Odoroside A (Subst. E): *Nelium odorum* SOL.[669]

Oleandrin (Subst. C): *Nelium odorum* SOL.

Desacetylcryptograndoside A (Subst. L.)[668]: *Cryptostegia grandiflora*

Odoroside H (Subst. O): *Nerium odorum* SOL.

16–Anhydrostrospeside (Subst. π): *Adenium multiflorum*[671]

Strospeside (Subst. Q): *Nelium odorum* SOL., *Adenium multiflorum.*

UV See Fig. 110, IR See Fig. 111~114.

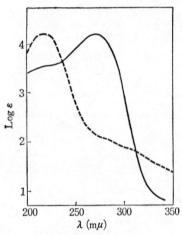

Fig. 110 UV Spectra of
------Subst. γ (Digistroside)
——Subst. D (16–Anhydrodesacetylcryptograndoside A)
(in EtOH) (Reichstein *et al.*)

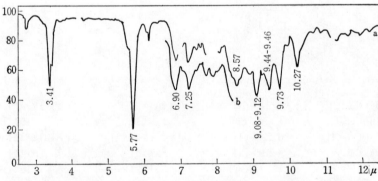

Fig. 111 IR Spectrum of Adigoside (Subst. B) (a) in CH_2Cl_2, (b) in KBr
(Reichstein *et al.*)

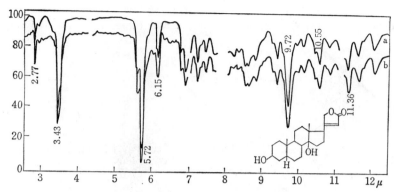

Fig. 112 IR Spectra of (a) Digitoxigenin and (b) Genin from Digistroside
(Subst. γ) (in CH₂Cl₂)   (Reichstein *et al.*)

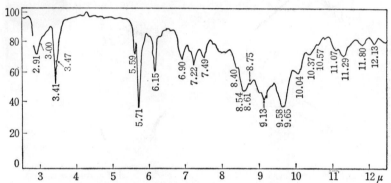

Fig. 113 IR Spectrum of 16–Anhydrodesacetyl–cryptograndoside A (Subst. D)
(in KBr) (Reichstein *et al.*)

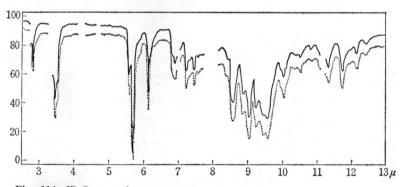

Fig. 114 IR Spectra of

——— 16–Anhydro–strospeside
——— Subst. π (in CH₂Cl₂)

(Reichstein *et al.*)

673) Reichstein, T. *et al*: *Helv. Chim. Acta*, **42**, 72 (1959) **43**, 102, 2035 (1960).
674) Reichstein, T. *et al*: *ibid.*, **42**, 1437 (1959).

(20) **Cardenolides of *Pachycarpus distinctus.* and *P. shinzianus*[673]** (*Asclepiadaceae*)

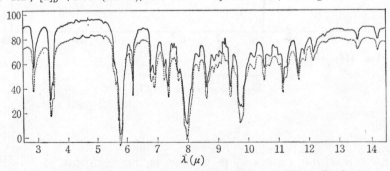

R=H: Digitoxigenin (I)=(LXXXI)

R=H: Xysmalogenin (LXXXII)

R=H: 3-*epi*-Digitoxigenin (LXXI)

R=H: Cannogenin (XXV)

R=H: Pachygenin (LXXXIII)

R=H: Sarmentogenin (V)

R=H: Pachygenol (LXXXIV)

D-Digitoxose—Cannogenin

(LXI)   (XXV)

Cannodigitoxoside (LXXXV)

(a) **Xysmalogenin (LXXXII)**

mp. 248~258°, [α]$_D$ −1.1° (pyridine), 3-*O*-acetate mp. 242~255°.

(b) **3-*epi*-Digitoxigenin (LXXI)**

mp. 272~282°, [α]$_D$ +25.3° (MeOH), 3-*O*-acetate mp. 205~210°, IR: Fig. 115

Fig. 115 IR Spectra of 3-*O*-Acetyl-3-*epi*-digitoxigenin (in KBr) (Reichstein *et al.*)

(c) **Cannogenin** (XXV)

　mp. 145°/185°/200~210°, [α]_D −13.5° (CHCl₃)

(d) **Pachygenin** (LXXXIII)

　mp. 216~226°, [α]_D −103.0° (CHCl₃), 3–O–acetate　mp. 272~282°, IR: Fig. 116

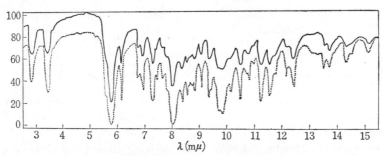

Fig. 116 IR Spectra of 3–O–Acetylpachygenin
(in KBr)　　　　　(Reichstein)

(e) **Cannodigitoxoside** (LXXXV)

　mp. 208~216°, [α]_D −7.1 (CHCl₃), C₂₉H₄₂O₈

　UV　λ_max^EtOH mμ (log ε): 216 (4.10)

(21) **Glycosides of _Pentopetia androsaemifolia_[675]** (_Asclepiadaceae_)

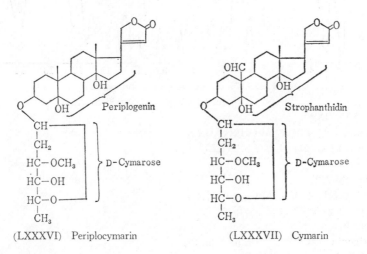

(LXXXVI) Periplocymarin　　　　　(LXXXVII) Cymarin

(a) **Periplocymarin** (LXXXVI)

　mp. 138~145°, [α]_D +24.9° (MeOH): mp. 143~145°, [α]_D +27.6° (MeOH)[676], C₃₀H₄₆O₈

(b) **Cymarin** (K–Strophanthin)　(LXXXVII)

　mp. 138~144°, 148°, 185°, [α]_D +39.2° (MeOH), C₃₀H₄₄O₉

(c) **Digitoxigenin** (I)　mp. 249~255°, [α]_D +14.6° (MeOH)

675) Schindler, O _et al_: _Helv. Chim. Acta_, **43**, 664 (1960).

676) Stoll, A _et al_: _ibid._, **22**, 1193 (1939).

(d) **Periplogenin** (LXXII) mp. 235~237°, $[\alpha]_D$ +27.6° (MeOH)

(22) **Glycosides of *Pergularia extensa*[677]** (*Asclepiadaceae*)

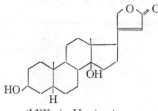

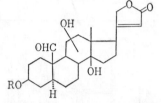

(LXX–a) Uzarigenin

R=H Calotropagenin (LXXXVIII)

R=$C_6H_9O_3$ Calactin (LXXXIX)

(a) **Uzarigenin** (LXX–a)

≪Other occurrence≫   See the item (16),   mp. 224~258°, $[\alpha]_D$ +12.2° (EtOH), $C_{23}H_{34}O_4$

UV $\lambda_{max}\,m\mu$ (ε): 217 (14800)[677a]

(b) **Calotropagenin** (LXXXVIII)

mp. 248~255°, $[\alpha]_D$ +45.6° (MeOH), $C_{23}H_{32}O_6$

UV $\lambda_{max}\,m\mu$ (log ε): 216 (4.22) (butenolide), 310 (1.55) (carbonyl),   IR: See Fig. 117

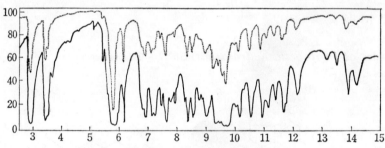

Fig. 117 IR Spectrum of Calotropagenin

(in KBr)   (Reichstein *et al.*)

(c) **Calactin** (LXXXIX)[678]

mp. 270~272°, $[\alpha]_D$ +70.4° (MeOH), $C_{29}H_{40}O_9$

UV $\lambda_{max}\,m\mu$ (log ε): 217 (4.23), 310 (1.55)

IR: See the reference[677]

(d) **Calotropin** (XC)[678]

mp. 234~240°, $[\alpha]_D$ +64.4° (MeOH), $C_{29}H_{40}O_9$

UV $\lambda_{max}\,m\mu$ (log ε): 217 (4.25), 305 (1.55)

IR: See ref.[677]

(e) **Coroglaucigenin** (XCI)[679][682] See the item (24)

mp. 248~256°, $[\alpha]_D$ +25.8° (MeOH)

677) Reichstein, T. *et al*: *Helv. Chim. Acta*, **45**, 907, 924 (1962).

677a) Mitsuhashi, H. *et al*: *Chem. Pharm. Bull.*, **11**, 1452 (1963).

678) Hesse, G. *et al*: *Ann.*, **566**, 130 (1950), **625**, 157, 174 (1959).

679) Reichstein, T. *et al*: *Helv. Chim. Acta*, **35**, 1073 (1952).

(23) Glycosides of *Rhodea japonica* ROTH. (オモト)[680)681)] (*Liliaceae*)

R=H: Oleandrigenin (XXX)
R=L-rhamnose: Rhodexin B (XCII)
R=L-rhamnose–D-glucose: Rhodexin C (XCIII)

R=H: Sarmentogenin (V)
R=L-Rhamnose: Rhodexin A (XCIV)

Anhydrorhodexigenin A (XCV)

- **(a) Rhodexin B (XCII)**
  mp. 262° (decomp.), $[\alpha]_D$ −39.5° (EtOH), $C_{29}H_{44}O_9$
  UV λ_{max} mμ (log ε): 215 (4.17), see Fig. 117
- **(b) Rhodexin C (XCIII)**
  mp. 275° (decomp.), $[\alpha]_D$ −17.7° (dil. EtOH), $C_{35}H_{54}O_{14}$
- **(c) Rhodexin A (XCIV)**
  mp. 265° (decomp.), $[\alpha]_D$ −20° (EtOH), $C_{29}H_{44}O_9 \cdot 2 H_2O$.
  UV λ_{max}^{EtOH} mμ (log ε): 220 (4.20)
- **(d) β-Monoanhydrorhodexigenin A (XCV)**
  mp. 207°, $[\alpha]_D$ −25.9° (CHCl₃), $C_{23}H_{32}O_4$ $\xrightarrow{HCl}$ α-Compound
- **(e) α-Monoanhydrorhodexigenin A (XCV)**
  mp. 240~243° (decomp.), $[\alpha]_D$ +36.1° (CHCl₃), $C_{23}H_{32}O_4$

(24) Glycosides of *Roupellina boivinii* PICHON (*Strophanthus boivinii* BAILL) (*Apocynaceae*)
Genins and Sugars

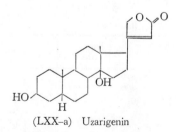

(LXX–a) Uzarigenin

(XCVI) 17 β–H–Uzarigenin

680) Nawa, H *et al*: *Chem. Pharm. Bull.*, **6**, 508 (1958).
681) Nawa, H: *Y. Z.*, **72**, 404, 407, 410, 414 (1952).

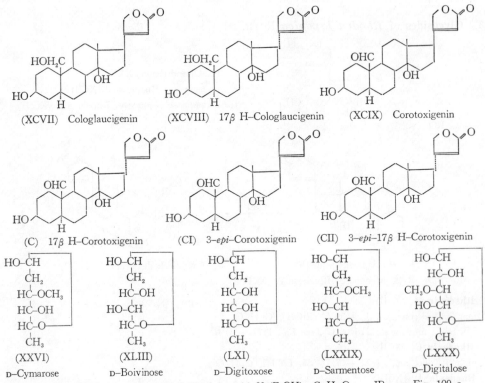

(XCVII) Cologlaucigenin   (XCVIII) 17β H–Cologlaucigenin   (XCIX) Corotoxigenin

(C) 17β H–Corotoxigenin   (CI) 3–epi–Corotoxigenin   (CII) 3–epi–17β H–Corotoxigenin

HO–CH	HO–CH	HO–CH	HO–CH	HO–CH
CH₂	CH₂	CH₂	CH₂	HC–OH
HC–OCH₃	HC–OH	HC–OH	HC–OCH₃	CH₃O–CH
HC–OH	HO–CH	HC–OH	HO–CH	HO–CH
HC–O	HC–O	HC–O	HC–O	HC–O
CH₃	CH₃	CH₃	CH₃	CH₃
(XXVI)	(XLIII)	(LXI)	(LXXIX)	(LXXX)
D–Cymarose	D–Boivinose	D–Digitoxose	D–Sarmentose	D–Digitalose

(a) **Uzarigenin** (LXX–a)   mp. 230∼246°, [α]$_D$ +14.0° (EtOH) C₂₃H₃₄O₄   IR: see Fig. 109–a

(b) **17–β H–Uzarigenin** (XCVI)   mp. 227∼229°, 231∼232°, [α]$_D$ +28.6°; +14.0° (CHCl₃), C₂₃H₃₄O₄

(c) **Cologlaucigenin** (XCVII)[682]   mp. 244°, [α]$_D$ +26.0° (MeOH), C₂₃H₃₄O₅

(d) **17–β H–Cologlaucigenin** (XCVIII)   mp. 238∼241°, [α]$_D$ +14.3° (MeOH), C₂₃H₃₄O₅

(e) **Colotoxigenin** (XCIX)   mp. 221°, [α]$_D$ +42.0° (MeOH), C₂₃H₃₂O₅

　　UV λ$_{max}^{EtOH}$ mμ (log ε): 216 (4.21), 310 (1.49)

(f) **17–β H–Corotoxigenin** (C)   mp. 218∼221°, [α]$_D$ +16.4° (MeOH), C₂₃H₃₂O₅

(g) **3–epi–Corotoxigenin** (CI)   mp. 187∼189°, C₂₃H₃₂O₅

(h) **3–epi–17–β H–Corotoxigenin** (CII)   mp. 200∼202°, C₂₃H₃₂O₅

(i) **Glycosides**

　　Cardenolides from the leaves

Table 61 Cardenolides from the leaves of *Roupellina boivinii*   (Reichstein)

	Cardenolides	Formula	Composition	mp.	[α]$_D$ (MeOH)	λ$_{max}^{EtOH}$ mμ(log ε)	UV
A	Madagascoside	C₃₀H₄₆O₇	(LXX–a)–(LXXIX)	195∼198/ 219∼222	−23.3	217(4.22) 275sh(2.12)	Fig.118
B	17β H–Madagascoside	C₃₀H₄₆O₇	(XCVI)–(LXXIX)	184∼190/ 210∼212	−6.6	218.5(4.20) 275sh(2.05)	"
γ	Tanagenin	—		259∼268	—	216(4.23) 274sh(2.15)	
E	Paulioside	C₃₀H₄₄O₈	(XCIX)–(LXXIX)	209∼212	−7.1	275sh(1.90) 305sh(1.60)	

682) Reichstein, T. *et al*: *Helv. Chim. Acta*, **44**, 1293, 1315 (1961), **35**, 673, 730, 442, (1952), **36**, 370 (1953).

	Cardenolides	Formula	Composition	mp.	$[\alpha]_D$ (MeOH)	$\lambda_{max}^{EtOH} m\mu(\log \varepsilon)$	UV
E'	17β H–Paulioside	$C_{30}H_{44}O_8$	(C)–(LXXIX)	190~192	—	—	
F	Roupelloside		Roupellogenin–(LXXX)	233~239	+4.8	217.5(4.18) 270sh(2.00)	
φ	Glycoside φ	$C_{30}H_{46}O_8$	(XCVII)–(LXXIX)	amorph.mix.	—	—	
φ'	Glycoside φ'	—	(XCVIII)–(LXXIX)	amorph.mix.	—	—	
G	Zettoside	$C_{29}H_{44}O_7$	(LXX–a)–(XLIII)	253~256	−27.9	216(4.21) 275sh(2.0)	Fig.118
I	17β H–Zettoside	$C_{29}H_{44}O_7$	(XCVI)–(XLIII)	135~138	−12.2	218(4.18) 275sh(2.0)	"
I'	Boistroside	$C_{29}H_{42}O_8$	(XCIX)–(LXI)	206~217	+5.4	216(4.20) 302sh(1.60)	
K	17β H–Boistroside	$C_{29}H_{42}O_8$	(C)–(LXI)	182~196	+1.9	218(4.16) 275sh, 305sh	
K″	Sadleroside	$C_{29}H_{42}O_8$	(CI)–(XLIII)	170~171	−9.1	275sh(1.68) 304(1.64)	
K‴	17β H–Sadleroside	$C_{29}H_{42}O_8$	(CII)–(XLIII)	244~248	+10.3	—	
L	Strospeside	$C_{30}H_{46}O_9$	Gitoxigenin(II)–(LXXX)	253~256	+18.0	218(4.19)	

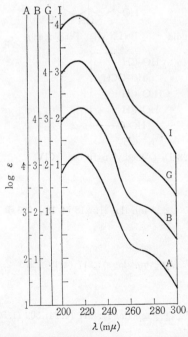

Fig. 118 UV Spectra of
A: Madagascoside
B: 17β H–Madagascoside
G: Zettoside   I: 17β H–Zettoside
· listed in Table 61 (Reichstein)

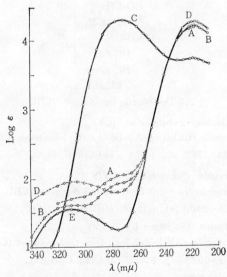

Fig. 119 UV Spectra of
A: Milloside   B: Pauloside
C: 16–Anhydrostrospeside
D: Stroboside   E: Boistroside
listed in Table 62 (Reichstein)

Cardenolides from the seeds.

Table 62 Cardenolides from the seeds of *Roupellina boivinii* (Reichstein)

	Cardenolides	Formula	Composition	mp.	$[\alpha]_D$ (MeOH)	λ_{max}^{EtOH} mμ(log ε)	UV
E	Boistroside	$C_{29}H_{42}O_8$	(XCIX)–(LXI)	213~219	+5.1	217(4.23) 309(1.54)	Fig.119
D	Stroboside	$C_{29}H_{42}O_8$	(XCIX)–(XLIII)	204~206	−13.5	217(4.23) 308(1.94)	"
B	Paulioside	$C_{30}H_{44}O_8$	(XCIX)–(LXXIX)	203~205	−10.1	217(4.21)	"
A	Milloside	$C_{30}H_{44}O_8$	(XCIX)–(XXVI)	142~146	−1.4	217(4.19)	"
F	Strospeside	$C_{30}H_{46}O_9$	Gitoxigenin (II)–(LXXX)	257~259	+15.9	—	
	Chrystioside	$C_{30}H_{44}O_9$	(XCIX)–(LXXX)	213	+13.8	217(4.22) 300sh(1.42)	

(25) Glycosides of *Streblus asper* LOUR[683] (*Moraceae*)

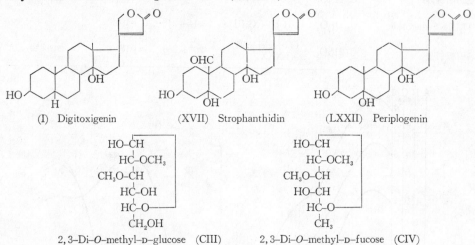

(I) Digitoxigenin   (XVII) Strophanthidin   (LXXII) Periplogenin

2,3–Di–*O*–methyl–D–glucose (CIII)   2,3–Di–*O*–methyl–D–fucose (CIV)

(a) Kamalaside (Substance α) (CV)

Periplogenin (LXXII)–β–2,3–Di–*O*–methyl–D–fucose (CIV)

mp. 174~178°, $[\alpha]_D$ +10.6° (MeOH), $C_{31}H_{48}O_9$,   UV λ_{max}^{EtOH} mμ (log ε): 218 (4.18), 295 sh (1.73)

(b) Asperoside (Substance A) (CVI)

Digitoxigenin (I)–β–2,3–Di–*O*–methyl–D–glucose (CIII)

mp. 198~205°, $[\alpha]_D$ −19.5° (MeOH), $C_{31}H_{48}O_9$,   UV λ_{max}^{EtOH} mμ (log ε): 217 (4.19), 270 (2.42)

(c) Strebloside (Substance B) (CVII)

Strophanthidin (XVII)–β–2,3–Di–*O*–methyl–D–fucose (CIV)

mp. 153~158°, $[\alpha]_D$ +25.3° (MeOH), $C_{31}H_{46}O_{10}$,   UV λ_{max}^{EtOH} mμ (log ε): 217 (4.24) 302 (1.50)

(d) Miscellaneous

Substance–β from ethereal extracts, substance C (indoroside), D (lucknoside), E, E′, E″, E‴, F, G, G′, and H from chloroform extracts, substance U, J, K, L, M and N from chloroform–methanol (2:1) extracts are also isolated by chromatography.

683) Reichstein, T. *et al*: *Helv. Chim. Acta*, **45**, 1515, 1534 (1962).

(26) Glycosides of *Strophanthus divaricatus* HOOK et ARN.[684] (Apocynaceae)

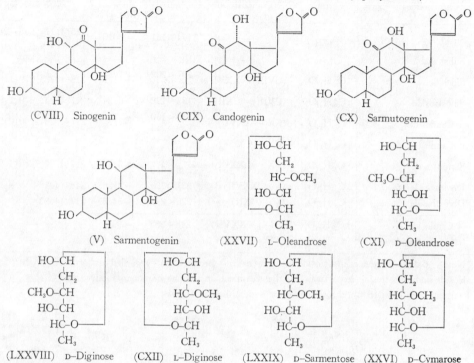

(CVIII) Sinogenin   (CIX) Candogenin   (CX) Sarmutogenin

(V) Sarmentogenin   (XXVII) L–Oleandrose   (CXI) D–Oleandrose

(LXXVIII) D–Diginose   (CXII) L–Diginose   (LXXIX) D–Sarmentose   (XXVI) D–Cymarose

(a) **Sinogenin** (CVIII)   mp. 222~231°, $[\alpha]_D$ +85.5° (MeOH), $[M]_D$ +342° (MeOH), $C_{23}H_{32}O_6$
UV λ_{max}^{EtOH} mμ (log ε): 217 (4.20), 270~280sh (1.83)

(b) **Caudogenin** (CIX)
mp. 220°, $[\alpha]_D$ −82.0° (MeOH) $[M]_D$ −332° (MeOH), $C_{23}H_{32}O_6$

(c) **Sarmutogenin** (CX)
mp. 258~262°, $[\alpha]_D$ +52° (MeOH), $[M]_D$ +210° (MeOH), $C_{23}H_{32}O_6$
IR: see Fig. 120

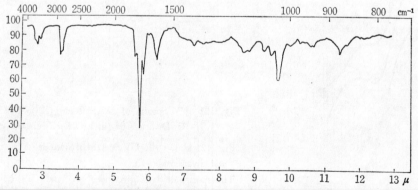

Fig. 120 IR Spectrum of Sarmutogenin (in CH_2Cl_2) (Reichstein *et al.*)

684) Reichstein, T. *et al*: *Helv. Chim. Acta*, 42, 160, 182 (1959).

Table 62 Cardenolides of *Strophanthus divaricatus* (Reichstein)

	Cardenolides	Formula	Composition	mp.	$[\alpha]_D$ (MeOH)	λ_{max}^{EtOH} mμ (log ε)
A_X	Subst A_X	—	—	174～181	−50.0	215(4.11), 288～289 (3.94)
A_Y	Subst A_Y	—	—	—	—	—
A	Sinoside	$C_{30}H_{44}O_9$	(CVIII)–(XXVII)	197～202/ 233～244	+11.9	216(4.22), 279(1.83)
A′	Sinostroside	$C_{30}H_{44}O_9$	(CVIII)–(CXII)	183～193	−6.3	217(4.24), 279(1.82)
B	φ–Caudoside	$C_{30}H_{44}O_9$	(CX)–(XXVII)	150～160/ 229～234	−7.7	217(4.22), 290(2.03)
X	φ–Caudostroside	$C_{30}H_{44}O_9$	(CX)–(CXII)	—	—	—
X	Sarmutoside	$C_{30}H_{44}O_9$	(CX)–(LXXIX)	150～152/ 233～245	+5.7	217(4.21), 288(1.99)
C	Divaricoside	$C_{30}H_{46}O_8$	(V)–(XXVII)	221～226	−32.6	218 (4.18)
D	Divostroside	$C_{30}H_{46}O_8$	(V)–(CXII)	225～231	−54.5	217 (4.22)
[Note]	Caudoside	$C_{30}H_{44}O_9$	(CIX)–(XXVII)	248～252	−99.6	—
	Caudostroside	$C_{30}H_{44}O_9$	(CIX)–(CXII)	amorph.	−74.9	—

[Note] Caudoside and caudostroside are not natural product but regarded as artifact, produced from sinoside and sinostroside respectively by the treatment of alumina chromatography.

Rearrangement, caused by chromatography, is described as follows:

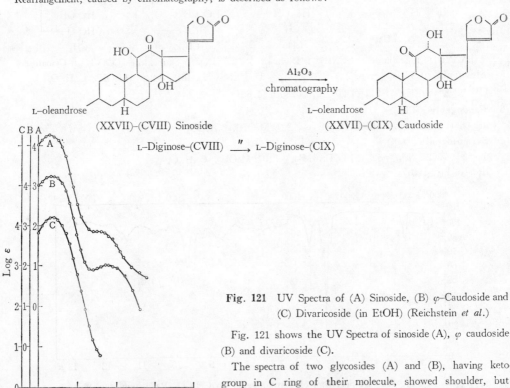

(XXVII)–(CVIII) Sinoside   (XXVII)–(CIX) Caudoside

L–Diginose–(CVIII) $\xrightarrow{\ ''\ }$ L–Diginose–(CIX)

Fig. 121 UV Spectra of (A) Sinoside, (B) φ–Caudoside and (C) Divaricoside (in EtOH) (Reichstein *et al.*)

Fig. 121 shows the UV Spectra of sinoside (A), φ caudoside (B) and divaricoside (C).

The spectra of two glycosides (A) and (B), having keto group in C ring of their molecule, showed shoulder, but glycoside (C), having no keto group in its molecule, showed no shoulder in its spectrum.

(27) Glycosides of *Strophanthus hispidus* P. D. C.[685] (*Apocynaceae*)

Table 63 Cardenolides of *Strophanthus hispidus* (Tamm *et al.*)

	Cardenolides	Formula	Str. No.	Composition	mp.	$[\alpha]_D$	see the item
A	Periplocymarin	$C_{30}H_{46}O_8$	(LXXXVI)	Periplogenin–D–cymarose	138~145	+27.6 (MeOH)	(21)
B	Cymarin (K–Strophanthin)	$C_{30}H_{44}O_9$	(LXXXVII)	Strophanthidin–D–cymarose	140~142/ 230~235 (lnhydr.)	+42.9 (MeOH)	(21)
C	Cymarol	$C_{30}H_{46}O_9$	(XII–a)	Strophanthidol (XVIII) R=D–cymarose	238~240	+19.1 (MeOH)	(21)
D	Cymarylic acid	$C_{30}H_{44}O_{10}$	(CXII–b)	Strophanthidinic acid (CXIII)–D–cymarose	154~157	+21.8 (CHCl₃)	λ_{max}^{EtOH} : 217(4.22)
E		M. W. (520)					217(4.24)
F		M. W. (520)					217(4.20)
G		$C_{29}H_{42\sim44}O_9$			140~143	+31.4	217(4.24)
H		M. W. (534)					217(4.22)
I		M. W. (727)			248~251	+7.4	217(4.23)

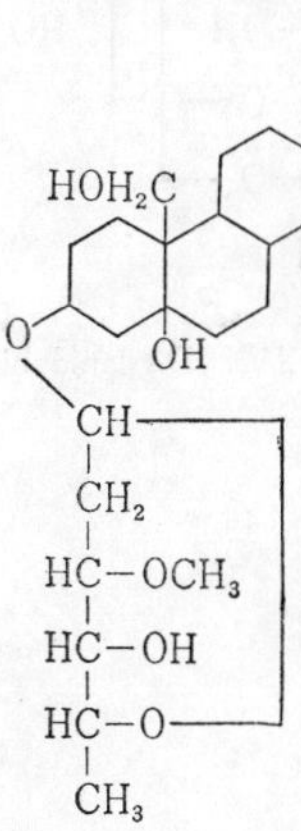

(CXII-a) Cymarol

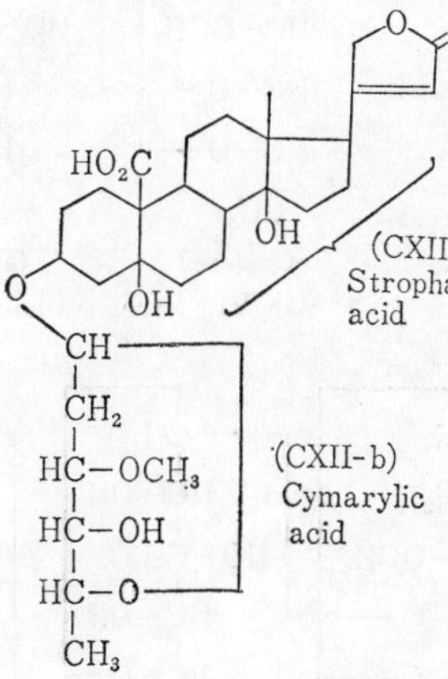

(CXIII) Strophanthidinic acid

(CXII-b) Cymarylic acid

Strophanthidinic acid (CXIII)

mp. 174~177°, $[\alpha]_D$ +54.4° (MeOH), $C_{23}H_{32}O_7 \cdot 1/2H_2O$

UV λ_{max}^{EtOH} mμ (log ε): 218 (4.22), anhydride mp. 185~190°

685) Karrer, W.: Konstitution und Vorkommen der organischen Pflanzenstoffe (Birkhäuser Verlag, 1958).

(28) Glycosides of *Strophanthus kombe* (*Apocynaceae*)

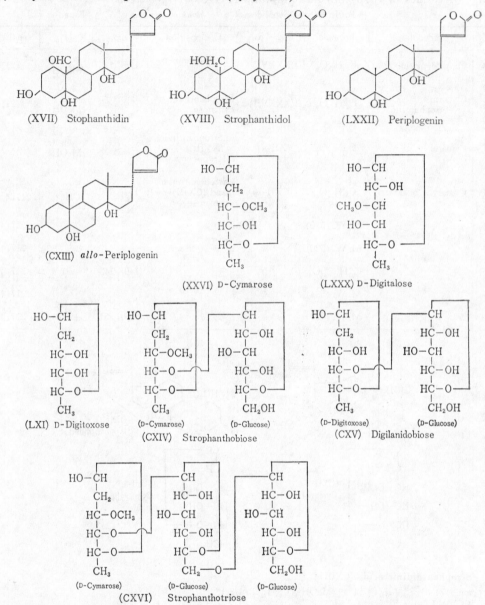

(XVII) Stophanthidin   (XVIII) Strophanthidol   (LXXII) Periplogenin

(CXIII) *allo*-Periplogenin

(XXVI) D-Cymarose   (LXXX) D-Digitalose

(LXI) D-Digitoxose   (D-Cymarose)   (D-Glucose)   (D-Digitoxose)   (D-Glucose)
(CXIV) Strophanthobiose   (CXV) Digilanidobiose

(D-Cymarose)   (D-Glucose)   (D-Glucose)
(CXVI) Strophanthotriose

(a) Table 64 shows the glycosides, summarized from the monographs of W. Karrer[685] and Kaiser[686]

686) von Kaiser, F. *et al*: *Ann.*, **643**, 192 (1961).

Table 64 Cardenolides of *Strophanthus kombé*   (Karrer and Kaiser)

Cardenolides	Composition	Formula	mp.	$[\alpha]_D$	See the item or ref.
Emicymarin (e–Strophanthin)	(LXXII)–(LXXX)	$C_{30}H_{46}O_9$	160~163	+13.5 (MeOH)	687)
allo–Periplocymarin	(CXIII)–(XXVI)	$C_{30}H_{46}O_8$	128~131	+48.3 (MeOH)	689)
allo–Emicymarin	(CXIII)–(LXXX)	$C_{30}H_{46}O_9$	160~162/ 260~265	+28 (MeOH)	688)
Cymarol	(XVIII)–(XXVI)	$C_{30}H_{46}O_9$	236~238	+28 (MeOH)	(CXI)
Cymarin (K–Strophanthin)	(XVII)–(XXVI)	$C_{30}H_{44}O_9$	148, 185, (anhydr.)205	+38 (EtOH)	(21)(27) (LXXVII)
K–Strophanthin β	(XVII)–(CXIV)	$C_{36}H_{54}O_{14}$	195	+32.6(H_2O)	689)
K–Strophanthoside	(XVII)–(CXVI)	$C_{42}H_{64}O_{19}$	199~200	+13.9 (MeOH)	690)
Helveticoside	(XVII)–(LXI)	$C_{29}H_{44}O_9$	153~157	+30.7 (MeOH)	(15)(LXX)
Erysimoside	(XVII)–(CXV)				686)
Glucoerysimoside	(XVII)–(CXVI)				686)
Helveticosol	(XVIII)–(LXI)				686)
Erysimosol	(XVIII)–(CXV)				686)
Glucocymarol	(XVIII)–(CXIV)				686)
K–Strophanthol–γ	(XVIII)–(CXVI)				686)

(b) β–D–**Glucosides of Strophanthidin** (XVII) and **Strophanthidol** (XVIII)[691]

≪Occurrence≫   Not naturally occurring compounds.

Strophanthidin–β–D–glucoside

mp. 234~238°, $[\alpha]_D$ +20.6° (H_2O)

UV λ_{max}^{EtOH} mμ (log ε): 217 (4.19) (butenoid ring), 300~305 (1.50) (aldehyde);

IR λ_{max}^{Nujol} μ: 2.85, 2.92~3.02 (OH), 5.75 (C=O butenoid ring), 5.84 (C=O, C_{19}–CHO), 6.17 (C=C butenoid ring)

Strophanthidol–β–D–glucoside

mp. 187~190°/206~210°, $[\alpha]_D$ +5.8° (MeOH)

UV λ_{max}^{EtOH} mμ (log ε): 217 (4.26):   IR λ_{max}^{nujol} μ: 2.88, 3.00~3.07 (OH), 5.75 (C=O, butenoid ring), 6.18 (C=C, butenoid ring)

(29) Glycosides of *Strophanthus sarmentosus* P. D C.[685)691]

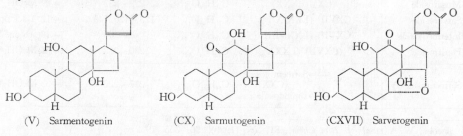

(V) Sarmentogenin   (CX) Sarmutogenin   (CXVII) Sarverogenin

687) Katz, A. *et al*: *Helv. Chim. Acta*, **28**, 476 (1945).   688) Katz, A. *et al*: *Chem. Zentr.*, **1945**, I, 1375.
689) Jacobs, W. A. *et al*: *J. Biol. Chem.*, **67**, 609, **69**, 153 (1926).
690) Stoll, A. *et al*: *Helv. Chim. Acta*, **20**, 1484 (1937).
691) Reichstein, T. *et al*: *ibid.*, **42**, 1448 (1959).

(CXVIII) Sarmentologenin

(CXIX) Bipindogenin

(CXX) Sarmentoside A

(CXXI) Tholloside

HO–CH	HO–CH	HO–CH	HO–CH	HO–CH
CH_2	CH_2	HC–OH	HC–OH	HC–OH
HC–OCH_3	CH_3O–CH	CH_3O–CH	HC–OH	HC–OH
HO–CH	HO–CH	HO–CH	HO–CH	HC–OH
HC–O	HC–O	HC–O	O–CH	O–CH
CH_3	CH_3	CH_3	CH_3	CH_3
(LXXIX)	(LXXVIII)	(LXXX)	(CXXII)	(CXXIII)
D–Sarmentose	D–Diginose	D–Digitalose	L–Rhamnose	L–Talomethylose

Table 65 Cardenolides of *Strophanthus sarmentosus* (Karrer and Reichstein)

Cardenolide	Composition	Formula	mp.	$[\alpha]_D$, Ref.
Sarmentocymarin	(V)–(LXXIX)	$C_{30}H_{46}O_8$	130~133, 205~208	−13(MeOH)[692]
Sarnovide	(V)–(LXXX)	$C_{30}H_{46}O_9$	223~225	+ 9(MeOH)[693]
Sargenoside	(V)–(LXXX)–glucose	$C_{36}H_{56}O_{14}$	amorph.	[694]
Sarmutoside	(CX)–(LXXIX)	$C_{30}H_{44}O_9$	250~252	+11.4(MeOH)[695] (Table. 62)
Musaroside	(CX)–(LXXX)	$C_{30}H_{44}O_{10}$	229~232	+29(MeOH)[695]
Sarveroside	(CXVII)–(LXXIX)	$C_{30}H_{44}O_{10}$	123~126/145	+12.1(acetone)[696]
Intermedioside	(CXVII)–(LXXVIII)	$C_{30}H_{44}O_{10}$	ca 200	+19(acetone)[697]
Panstroside	(CXVII)–(LXXX)	$C_{30}H_{44}O_{11}$	230~233	+30(MeOH)[698]
Sarmentoside A	(a)→Sarmentoloside (CXX) (b)→Bipindoside	$C_{29}H_{42}O_{11}$	217~222	−33(MeOH)[691]

692) Jacobs, W. A. *et al*: *J. Biol Chem.*, **81**, 765 (1929), **96**, 355 (1932).
693) Reichstein, T. *et al*: *Helv. Chim. Acta*, **34**, 1477 (1951).
694) Reichstein T. *et al*: *ibid.*, **35**, 1560 (1952).
695) Reichstein, T. *et al*: *ibid.*, **36**, 1073 (1953), **37**, 76 (1954).
696) Reichstein, T. *et al*: *ibid.*, **33**, 465 (1950).
697) Rosselet, J. P. *et al*: *ibid.*, **34**, 1036 (1951).
698) Reichstein, T. *et al*: *ibid.*, **33**, 2153 (1950), **34**, 2143 (1951).

Cardenolide	Composition	Formula	mp.	$[\alpha]_D$, Ref.
Tholloside	$\xrightarrow{(a)}$ Sarhamnoside (CXXI) $\xrightarrow{(b)}$ Lokundjoside	$C_{29}H_{42}O_{11}$	250~255	-16.1(MeOH)[691]
Sarmentoloside	(CXVIII)–(CXXIII)	$C_{29}H_{44}O_{11}$	176~179	-26.4(MeOH)[691]
Bipindoside	(CXIX)–(CXXIII)	$C_{29}H_{44}O_{10}$	163~166	-22.6(MeOH)[691]
Sarhamnoside	(CXVIII)–(CXXII)	$C_{29}H_{44}O_{11}$	238~241	-12.7(MeOH)[691]
Lokundjoside	(CXIX)–(CXXII)	$C_{29}H_{44}O_{10}$	252~258	-8.4 (MeOH)[691]
Bipindaloside	(CXIX)–(LXXX)	$C_{30}H_{46}O_{10}$	170~171	$+15.2$(MeOH)[691]

Genins	Str. No.	Formula	mp.	$[\alpha]_D$	λ_{max}^{EtOH} mμ(log ε), Ref.
Sarmentologenin	(CXVIII)	$C_{23}H_{34}O_7$	248~251	$+30.2$(MeOH)	218(4.21)[691]
Bipindogenin	(CXIX)	$C_{23}H_{34}O_6$	238~243	$+30.2$(MeOH)	218(4.14)[691]
Sarverogenin	(CXVII)	$C_{23}H_{30}O_7$	223~227	$+45$ (MeOH)	[699]

(a) Red. with NaBH$_4$, (b) CH$_2$(CH$_2$SH)$_2$, red. with Ni

(30) Glycosides of *Strophanthus Thollonii* FRANCH[700] (*Apocynaceae*)

Eighteen glycosides composed by the combination of 11 genins and 2 sugars are separated by Reichstein *et al*.

Among these the following glycosides were determined their structure.

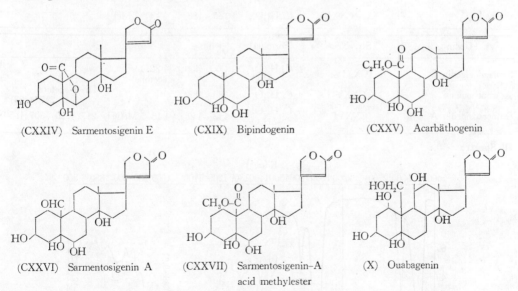

(CXXIV) Sarmentosigenin E   (CXIX) Bipindogenin   (CXXV) Acarbäthogenin

(CXXVI) Sarmentosigenin A   (CXXVII) Sarmentosigenin–A acid methylester   (X) Ouabagenin

699) Schindler, O: *Helv. Chim. Acta*, **39**, 375 (1956).
700) Reichstein, T. *et al*: *ibid*., **40**, 980 (1957), **41**, 736 (1958).

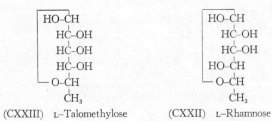

(CXXIII) L–Talomethylose   (CXXII) L–Rhamnose

Table 66 Cardenolides of *Strophanthus Thollonii*   (Reichstein *et al.*)

Cardenolide	Composition	Formula	mp.	$[\alpha]_D$	λ_{max}^{EtOH} mμ (log ε)
Sarmentoside E	(XXIV)–(CXXIII)	$C_{29}H_{40}O_{11}$	265~268	−37.3(MeOH)	*O*–Acetate 216(4.21)
Bipindoside	(CXIX)–(CXXIII)	$C_{29}H_{44}O_{10}$	*O*–Acetate 158~162	″ −36.4(CHCl₃)	*O*–Acetate 217(4.27)
Acarbäthoside	(CXXV)–(CXXIII)	$C_{31}H_{46}O_{12}$	*O*–Acetate 245~248	″ −24.3(CHCl₃)	Tetraacetate 217(4.22)
Sarmentoside A	(CXXVI)–(CXXIII)	$C_{29}H_{42}O_{11}$	236~240	−33.0(MeOH)	217(4.20), 303(1.42)
Sarmentoside A acidmethylester	(CXXVII)–(CXXIII)	$C_{30}H_{44}O_{12}$	*O*–Acetate 285~289	″ −26.8(CHCl₃)	*O*–Acetate 217(4.23)
Thollodiolidoside	(CXXIV)–(CXXII)	$C_{29}H_{40}O_{11}$	*O*–Acetate 321~321	″ −16.8(CHCl₃)	*O*–Acetate 217(4.20), 270~280sh
Lokundjoside	(CXIX)–(CXXII)	$C_{29}H_{44}O_{10}$	268~274	−11(Me)	217(4.24), 275~290sh
Tholläthoside	(CXXV)–(CXXII)	$C_{31}H_{46}O_{12}$	166~170	+6.1(Me)	217(4.29), 280~300sh
Tholloside	(CXXVI)–(CXXII) $C_{29}H_{42}O_{11}$ ·CH₃OH	259~265	−16.1(Me)	217(4.26), 305(1.41)
Ouabain	(X)–(CXXII)	$C_{29}H_{44}O_{12}$	185~188	−44.3(Me)	

Genin	Str. No.	Formula	mp.	$[\alpha]_D$	
Sarmentosigenin E	(CXXIV)	$C_{23}H_{30}O_7$ ·H₂O	169~175	+32.1 (acetone)	
Acarbäthogenin	(CXXV)	$C_{25}H_{36}O_8$	180~183		Di–*O*–acetate 217(4.23)
Sarmentosigenin A	(CXXVI)	$C_{23}H_{32}O_7$ ·3H₂O	143~149	+16.5(MeOH)	217(4.20), 307(1.50)

IR Spectra

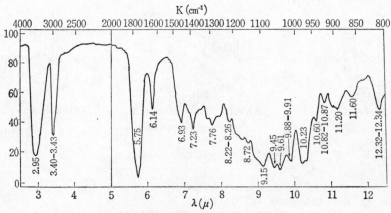

Fig. 122 Sarmentoside E in KBr

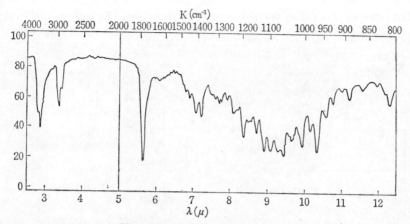

Fig. 123　Sarmentosigenin E in KBr

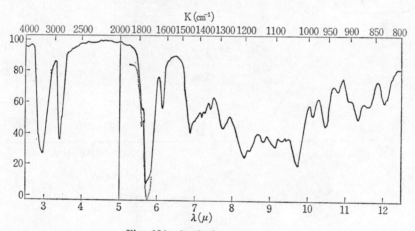

Fig. 124　Acarbäthogenin in KBr

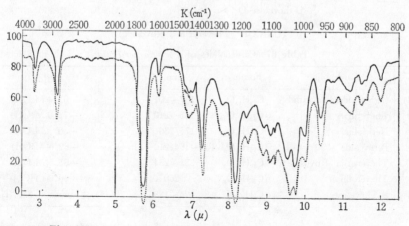

Fig. 125　(1)　Tetra-*O*-acetylsarmentoside A-acid methylester ━━━

(2)　Tetra-*O*-acetylacarbäthoside　　　·········

(in CH$_2$Cl$_2$)　　　　　　(Reichstein)

(31)　**Glycosides of _Thevetia nerifolia_** Juss. (_Th. peruviana_ K. Schum.)[701)702)] (_Apocynaceae_)

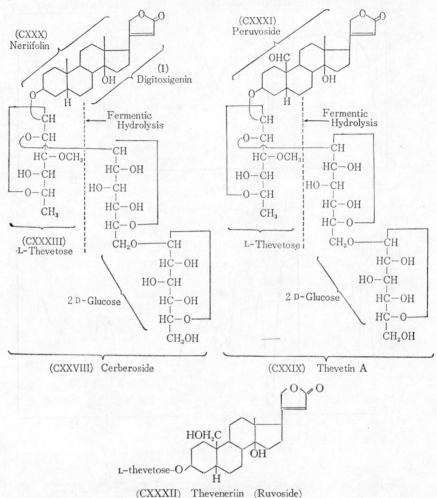

(CXXXII)　Theveneriin　(Ruvoside)

Table 67　Cardenolides of _Thevetia neriifolia_　　　　　(Reichstein _et al._)

Str. No.	Cardenolide	Formula	mp.	$[\alpha]_D$	ref.
(CXXVIII)	Cerberoside(Thevetin B)	$C_{42}H_{66}O_{18}$	197~201	−61.4(MeOH)	
(CXXIX)	Thevetin A	$C_{42}H_{64}O_{19}$	208~210	−72.0(MeOH)	
(CXXX)	Neriifolin	$C_{30}H_{46}O_8$	175/230	−50　(MeOH)	
(CXXXI)	Peruvoside	$C_{30}H_{44}O_9$	160~164/210~216	−69.6(MeOH)	
(CXXXII)	Theveneriin (Ruvoside)	$C_{30}H_{46}O_9$	232~234	−61.5(MeOH)	
	Thevefolin	$C_{30}H_{46}O_8$	260	−66.3($CHCl_3$)	
(CXXXIII)	L-Thevetose	$C_7H_{14}O_5$	124~129	−71.2(5 min)→−32.7(H_2O)	705)

701)　Schindler, O. _et al_: _Helv. Chim. Acta_, **43**, 652 (1960).
702)　Reichstein, T. _et al_: _ibid._, **45**, 938 (1962).

(32) Cardenolides of *Xysmalobium undulatum* R. Br[703] (*Asclepiadaceae*)

(LXXXIV) Pachygenol

(LXXXII) Xysmalogenin

(LXX–a) Uzarigenin

(XCVI) 17α–Uzarigenin (17βH–Uzarigenin)

(XCI) Coroglaucigenin

Teble 68 Cardenolides of *Xysmalobium undulatum*[703][704]　　(Reichstein *et al.*)

Str. No.	Cardenolide	Formula	mp.	$[\alpha]_D$	see the item λ_{max} mμ (log ε)
(LXXXIV)	Pachygenol	$C_{23}H_{32}O_5$	182~187/ 226~238	+10.2(MeOH)	(20) 217(4.22)(EtOH)
(LXXXII)	Xysmalogenin	$C_{23}H_{32}O_4$	232~249	+16.0(CHCl$_3$)	(20) 217(4.22)(EtOH)
(LXX–a)	Uzarigenin	$C_{23}H_{34}O_4$	230~240 227~248	+13.7(EtOH) +16.6(MeOH)	(16), (22), (24)
(XCVI)	17α–Uzarigenin	$C_{23}H_{34}O_4$	221~228	+25.6(MeOH)	(24)
(XCI)	Coroglaucigenin	$C_{23}H_{34}O_5$	241~248	+21.5(MeOH)	(22), (24)

703) Reichstein, T. *et al*: *Helv. Chim. Acta*, **43**, 102 (1960), **46**, 9 (1963).
704) Reichstein, T. *et al*: *ibid.*, **42**, 72 (1959).

[E] Bufadienolides

(1) Bufadienolides of *Bowiea volubilis* HARVEY (*Lilliaceae*)[705]

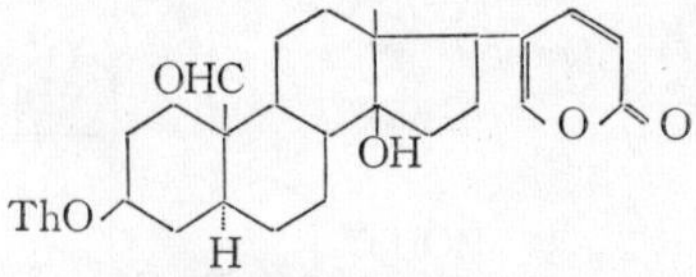

(CXXXIV) Bovoside A

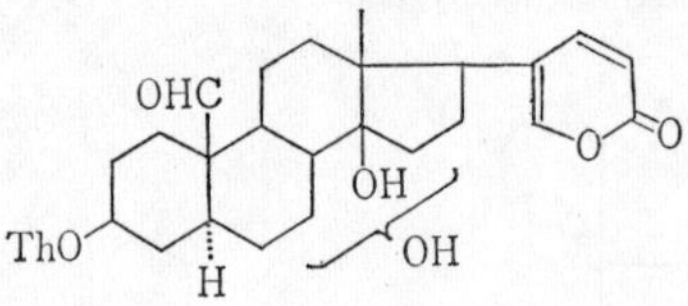

(CXXXV) Bovoside D

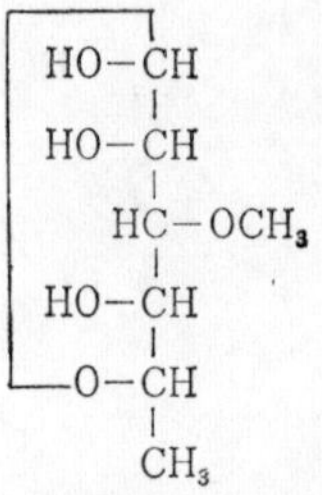

(CXXXIII) L–Thevetose
Th=Thevetosyl

Table 69 Glycosides of *Bowiea volubilis*   (Katz)[706]

Glycosides	Str. No.	Formula	mp.	$[\alpha]_D$ (MeOH)	λ_{max} mμ (log ε)
Bovoside A	(CXXXIV)	$C_{31}H_{44}O_9$	205~235	+0.68	300(8.73)
Bovoside D	(CXXXV)	$C_{31}H_{44}O_{10}$	284~298	−69.9	300(3.74)
L–Thevetose	(XCXXIII)	$C_7H_{14}O_5$	124~129	−71.2(5min) →−32.7(H₂O)	

(2) Bufadienolides of Chán Su (蟾蜍: センソ)

≪Occurrence≫   Secretion of *Bufobufo gargorizans*

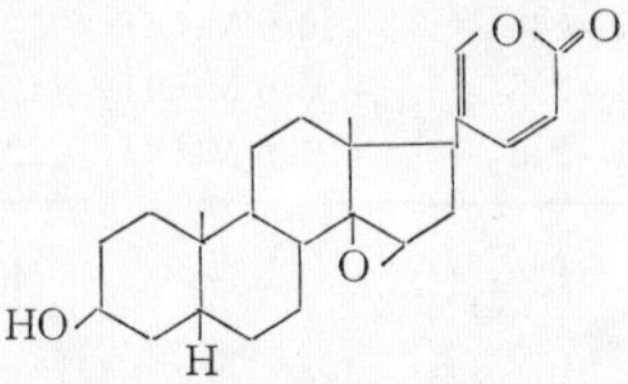

(CXXXVI) Resibufogenin

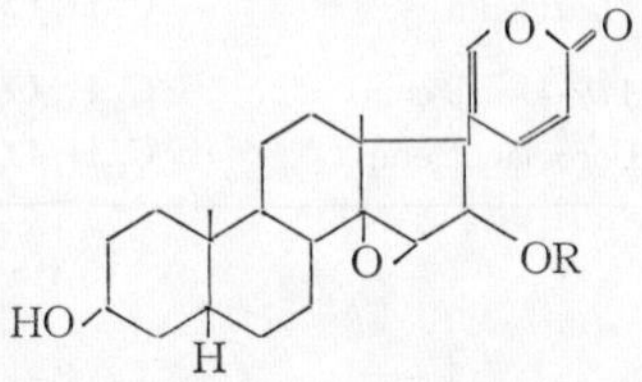

(CXXXVII) R=Ac: Cinobufagin[707]
(CXXXVIII) R=H: Desacetylcinobufagin
(Subst. A)

---

705) Katz, A. *et al*: *Helv. Chim. Acta*, **36**, 1344, 1417 (1953), **37**, 451, 833, (1954), **41**, 1399 (1958).
706) Katz, A. *et al*: *ibid.*, **33**, 1420 (1950).
707) Meyer, K. *et al*: *ibid.*, **43**, 1955 (1960).

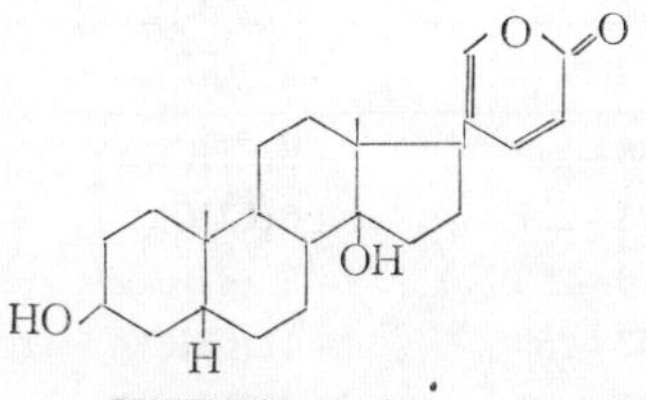

(CXXXIX) Bufalin

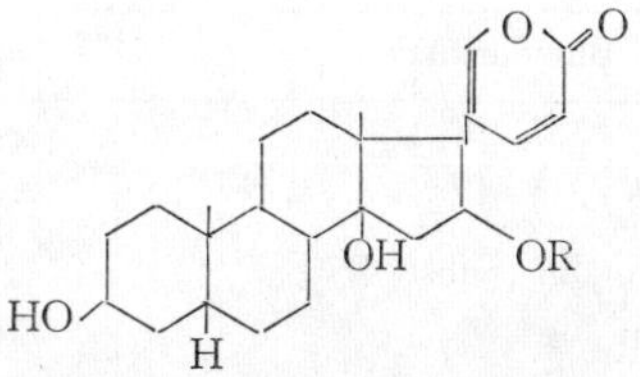

(CXL) R=Ac: Bufotalin
(CXLI) R=H: Desacetylbufotalin (Subst. B)

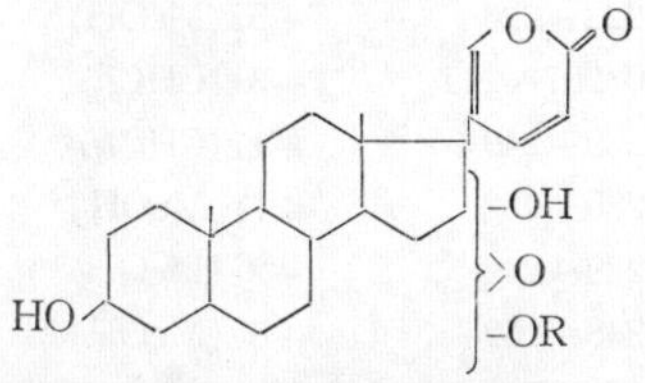

(CXLII) R=Ac: Cinobufotalin
(CXLIII) R=H: Desacetylcinobufotalin (Subst. D)

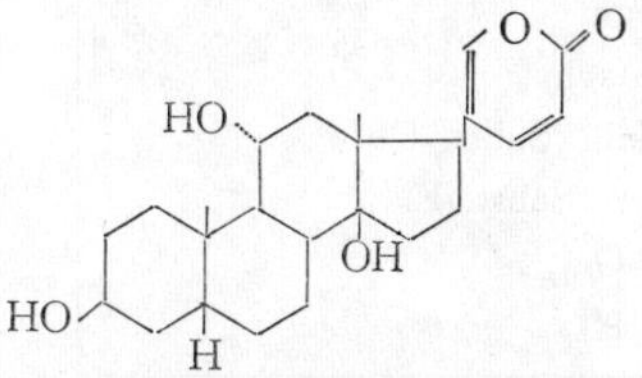

(CXLIV) Gamabufotalin

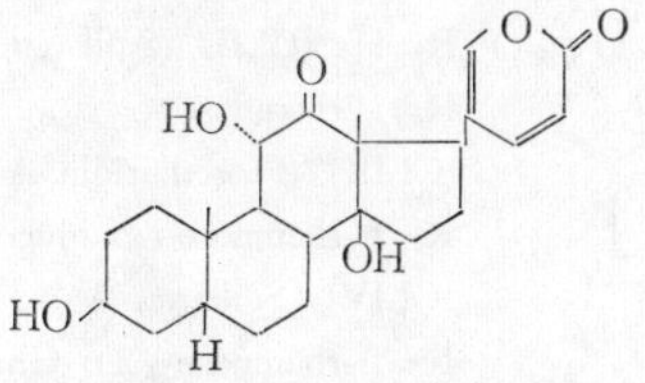

(CXLV) Arenobufagin[708]

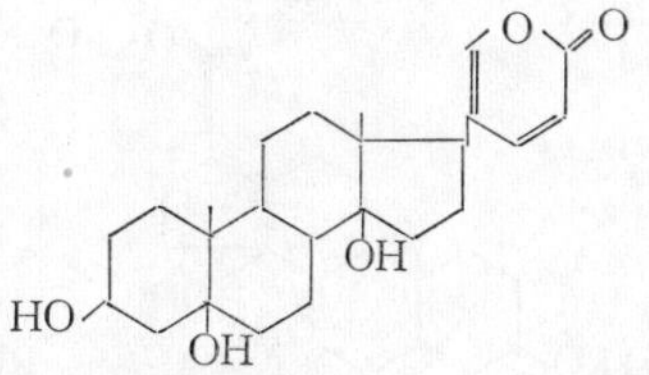

(CXLVI) Telocinobufagin

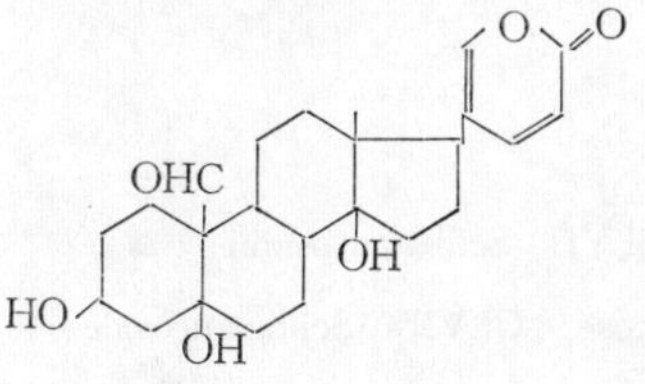

(CXLVII) Hellebrigenin (Subst. C)

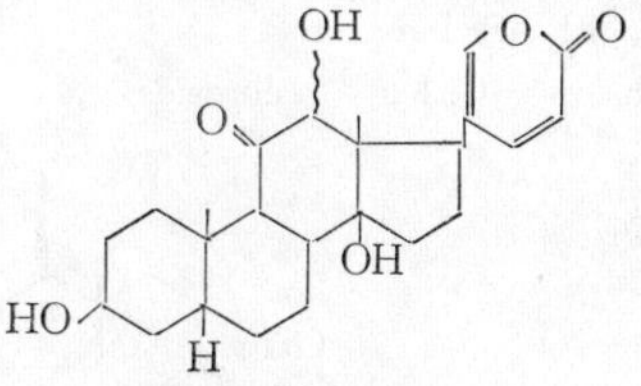

(CXLVIII) Bufarenogin (Subst. F)

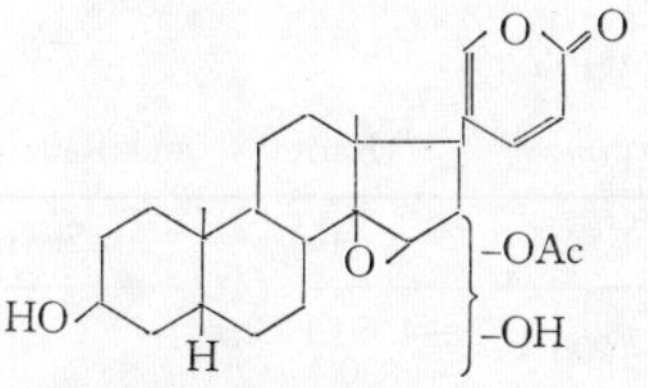

(CXLIX) Subst. G

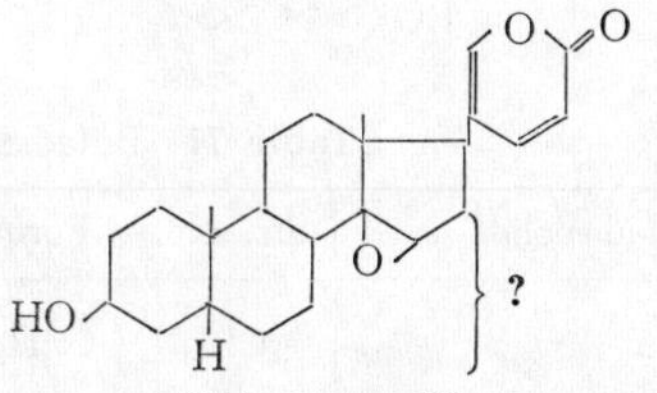

(CL) Subst. H

Table 70 Bufadienolides of Chán Su (Meyer et al.)

Bufadienolide	Str. No.	mp.	$[\alpha]_D$
Resibufogenin	(CXXXVI)	113~140/155~168	−5 (CHCl₃)
Cinobufagin	(CXXXVII)	213~215	−4 (CHCl₃)
Desacetyleinobufagin (Subst. A)	(CXXXVIII)	180~182	+22 (CHCl₃)

708) Meyer, K. et al: Helv. Chim. Acta, **43**, 1950 (1960).
709) Meyer, K. et al: ibid., **40**, 1270 (1957), **43**, 1955 (1960).

Bufadienolide	Str. No.	mp.	$[\alpha]_D$
Bufalin	(CXXXIX)	238~242	−9(CHCl₃)
Bufotalin	(CXL)	223~227	+5(CHCl₃)
Desacetylbufotalin (Subst. B)	(CXLI)	210~223	+30(dioxane)
Cinobufotalin	(CXLII)	257~259	+11(CHCl₃)
Desacetylcinobufotalin (Subst. D)	(CXLIII)	251~261	+34(MeOH)
Gamabufotalin	(CXLIV)	258~266	+1 (MeOH)
Arenobufagin	(CXLV)	222~228	+55(MeOH)
Telicinobufagin	(CXLVI)	160/207~211	+5 (CHCl₃)
Hellebrigenin (Subst. C)	(CXLVII)	239~249	+17(CHCl₃)
Bufarenogin (Subst. F)	(CXLVIII)	230~233	+11(MeOH)
Substance G	(CXLIX)	239~242	−2 (MeOH)
Substance H	(CL)	248~262	+6 (MeOH)

(3) Bufadienolides of *Scilla maritima* L.[710] (*Urginea maritima* BAKER)

R=H: (CLI) Scillirosidin
R=D-glucose: (CLI-a) Scilliroside

R=H: (CLII) Scillarenin A
R=L-rhamnose:
(CLIII) Proscillaridin A
R=L-rhamnose+D-glucose:
(CLIV) Scillaren A
R=L-rhamnose+2 D-Glucose:
(CLV) Glucoscillaren A

R=H: (CLVI) Scilliglaucosidin
R=D-glucose: (CLVII) Scilliglaucoside

Table 71 Bufadienolides of *Scilla maritima* (Karrer[685] and Renz *et al.*[710])

Bufadienolide	Str. No.	Formula	mp.	$[\alpha]_D$	See ref. and item
Scillirosidin	(CLI)	$C_{26}H_{34}O_7$	177~178/200~205	−4.6(CHCl₃) −23.0(MeOH)	
Scilliroside	(CLI-a)	$C_{32}H_{44}O_{12}$	168~170	−59 (MeOH)	
Scillarenin (Scillarenin A)	(CLII)	$C_{24}H_{32}O_{14}$	218~236	+17 (MeOH)	
Proscillaridin A	(CLIII)	$C_{30}H_{42}O_8$	219~222	−92 (MeOH)	
Scillaren A	(CLIV)	$C_{36}H_{52}O_{13}$	—	−74 (MeOH)	
Glucoscillaren A	(CLV)	$C_{42}H_{62}O_{18}$	228~232	−66 (MeOH)	
Scilliglaucosidin	(CLVI)	$C_{24}H_{30}O_5$	245~248	−49.5(MeOH)	[711] (4)
Scilliglaucoside	(CLVII)	$C_{30}H_{40}O_{10}$	164	−106 (MeOH)	[711] (4)

710) Renz, J. *et al*: *Helv. Chim. Acta*, **42.**, 1620 (1959).
711) Lichti, H. *et al*: *Helv. Chim. Acta*, **43**, 1666 (1960).

UV   Scillirosidin (Fig. 126),   IR   Scillirosidin (Fig. 127)

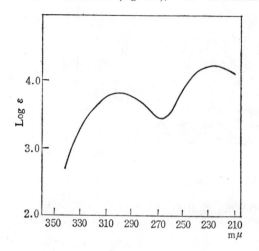

Fig. 126
UV Spectrum of Scillirosidin in MeOH
(Renz *et al.*)

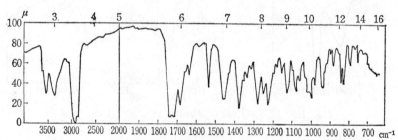

Fig. 127   IR Spectrum of Scillirosidin in nujol   (Renz *et al.*)

IR ν_{max}^{nujol} cm^{-1}: 3520, 3350 (OH), 1720, 1638, 1538, 1128, 840, 830 (α-pyrone), 1740, 1230, 1680 (acetate)

(4)   **Bufadienolides of *Urginea altissima*** Baker[711]

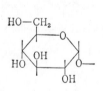

R=α-D-Glucoside:
(CLVII) Scilliglaucoside

R=-OH: (CLVI) Scilliglaucosidin

R=β-D-Glucoside:
(CLVIII) Altoside

(a)   **Scilliglaucosidin**   (CLVI)

mp. 244~248°, $[\alpha]_D$ +49.5° (MeOH), $[M]_D$ +197° (MeOH), $C_{24}H_{30}O_5$,

UV λ_{max}^{MeOH} mμ (log ε): 300 (3.72)

IR: Fig. 128

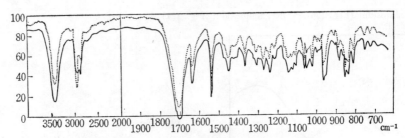

Fig. 128 IR Spectra of Scilliglaucosidin in KBr
(upper: from Altside, under: from Scilliglaucoside) (Lichti *et al.*)

(b) **Scilliglaucoside** (CLVII)

mp. 164°, $[\alpha]_D$ +106° (MeOH), $[M]_D$ +598° (MeOH), $C_{30}H_{40}O_{10}$

(c) **Altoside** (CLVIII)

mp. 222~228°, $[\alpha]_D$ +26.3° (MeOH), $[M]_D$ +151° (MeOH), $C_{30}H_{40}O_{10} \cdot 1/2 \ H_2O$

UV $\lambda_{max}^{MeOH} m\mu$ (log ε): 300 (3.74)

IR: Fig. 129

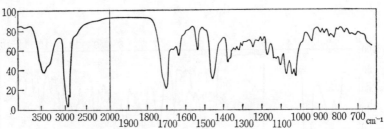

Fig. 129 IR Spectrum of Altoside in nujol (Lichti *et al.*)

(5) **Glycosides of *Urginea depressa*** BAKER[712]

(CLIX) Hellebrigenin-β-D-glucoside   (CLX) Hellebrigenol-β-D-glucoside

(a) **Hellebrigenin-β-D-glucoside** (CLIX)

mp. 240~244°/260~263°, $[\alpha]_D$ +0.05° (MeOH), $C_{30}H_{42}O_{11}$

Tetra-*O*-acetate   mp. 204~206°, $[\alpha]_D$ +1.8° (CHCl₃), $C_{38}H_{50}O_{15}$

UV: Fig. 130, IR: Fig. 131

IR $\lambda_{max}^{CH2Cl2} \mu$: 2.86 (O–H), 3.46 (C–H), ca 5.7 (CO–CH₃), ca 5.8, 6.11, 6.50 (α-pyrone),
    7.34 (CH₃), 8.23 (CH₃CO), 10.05, 10.23, 10.56, 12.00

---

712) Reichstein, T. *et al*: *Helv. Chim. Acta*, **42**, 1052 (1959).

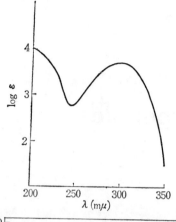

Fig. 130 UV Spectrum in EtOH

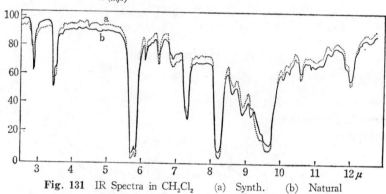

Fig. 131 IR Spectra in CH$_2$Cl$_2$ (a) Synth.   (b) Natural
of Tetra–*O*–acetylhellebrigenin–β–D–glucoside

(Reichstein *et al.*)

(b) Hellebrigenol–β–D–glucoside (CLX)

mp. 215~218°, [α]$_D$ −20.6° (MeOH), C$_{30}$H$_{44}$O$_{11}$

UV λ$_{max}^{EtOH}$ mμ (log ε): 300 (3.76),   IR λ$_{max}^{nujol}$ μ: 3.43 (C–H), 5.85, 6.12, 6.50 (α–pyrone), 6.90, 8.04, 9.34, 10.54, 11.15

11 Diterpenoids

[A] Hydrocarbons

(i) *Macromonocyclic Hydrocarbon*

   (1) **Thunbergene** (I–a or I–b)[713]

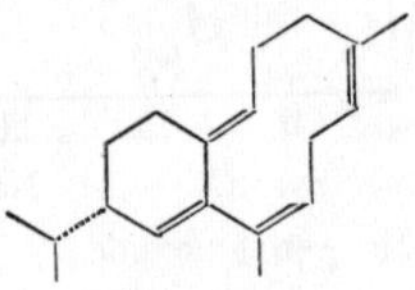

I–a   or   I–b

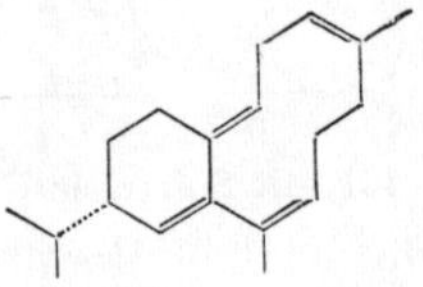

    ≪Occurrence≫   *Pinus densiflora*, *P. thunbergii*, *P. koraiensis* and other *Pinus* spp.

mp. 61°, $[\alpha]_D$ +260° (C_6H_6), $C_{20}H_{32}$

UV $_{max}^{hexane}$mμ (ε): 240 (14300), 246 (14300) (diene)

IR $\nu_{max}^{CCl_4}$cm^{-1}: 962 (*trans* $_H$>=<^{H}), 1365, 1385 (iso Pr.)

(ii) *Bicyclic Hydrocarbon*

   (2) **Biformene** (II)[714]

    ≪Occurrence≫   *Dacrydium biforme*

              Manool (XVI) $\xrightarrow{-2H_2O}$ (II)

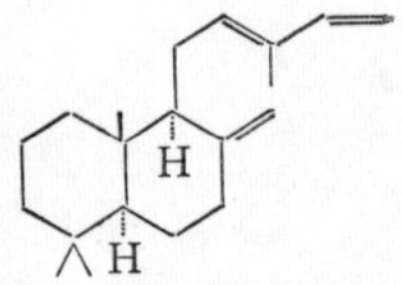

(II)

bp. 140~145°/0.5mm, n_D^{20} 1.5257, d $_4^{20}$ 0.952, $[\alpha]_D$ +12.2° (isooctane), $C_{20}H_{32}$

UV $\lambda_{max}^{isooctane}m\mu$ (ε): 228 (15300) (C=C–C=C)

IR ν_{max}^{film}cm^{-1}: 1596 (C=C–C=C), 3079, 1786, 1644, 888 (exocyclic=CH_2), 3034 inf. 990, 905 inf.
            (C=CH_2), 1388, 1366 (gem di-Me), 1413

---

713) Kobayashi, H. *et al*: *6th Symposium on the Chemistry of Natural Products*, p. 120 (Sapporo, 1962).
714) Carman, R.M. *et al*: *Soc.*, **1961**, 2187.

(iii)　*Tricyclic Hydrocarbons*

(3)　**Dolabradiene** (III)[715]

(III)

《Occurrence》　The leaves of *Thujopsis dolabrata* SIEB. et ZUCC. (ヒバ)

bp. 167∼170/7mm, d_4^{25} 0.9636∼0.9640, n_D^{25} 1.5235∼1.5244, $[\alpha]_D$ −67∼−71°, $C_{20}H_{32}$

UV　no λ_{max} from 200 mμ

IR　ν_{max}^{film} cm^{-1}: 908, 988, 1820 (vinyl), 890, 1783 (vinylidene), no gem Me_2

NMR　in CCl_4, 60 Mc, TMS, τ : 9.23 s, 9.08 s, 8.97 s (3〉C–CH$_3$), 5.60 s (1〉C=CH$_2$), 4.06∼5.31 m (–CH=CH$_2$), 7.89 (2〉C–H)

(4)　**Rimuene** (IV)[715][716]

(IV)

《Occurrence》　*Decrydium cupressimum, Podocarpus totara*

mp. 55°, $[\alpha]_D$ +44.7° ($CHCl_3$), $C_{20}H_{32}$

NMR　in $CDCl_3$, 60Mc, TMS, τ : Stereochemical structure of rimuene (IV) (9β-Me,13α-Me) was confirmed by the comparation of NMR of the three resinc acids (V), (VI) and (VII). The results are indicated in Table 72.

(V) Methyl isopimarate　(VI) Methyl sandaraco-　(VII) Methyl pimarate
　　　　　　　　　　　　　　　　pimarate

Table 72　NMR Shifts and Spin-coupling Constants for Resinic acids and Rimuene

Chem. Shifts : τ, J : cps　　　　　　　　　　　　　　　　　　　　　　　　　　　　(Wenkert)

Diterpenoids	Str. No.	H_9	H_{14}	H_A	H_B	H_C	J_{AB}	J_{AC}	J_{BC}
Rimuene	IV		4.64	4.28	5.10	5.21	9.0	15.5	1.4
Methyl isopimarate	V		4.74	4.24	5.22	5.18	9.5	17.0	1.4
Methyl sandaracopimarate	VI	7.80	4.83	4.28	5.32	5.16	10.5	17.4	1.7
Methyl pimarate	VII	7.78	4.89	4.31	5.10	5.15	10.1	17.1	1.9

715)　Kitahara, Y. *et al*: *5 th Symposium on the Chemistry of Natural Products* p. 7-1 (Sendai, 1961), *7 th Symposium* p. 232 (Fukuoka, 1963); *Tetr. Letters*, 1755 (1964).

716)　Wenkert, E: *J. A. C. S.*, **83**, 998 (1961).

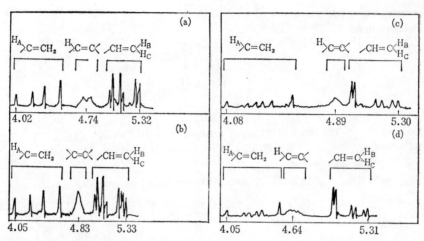

Fig. 132 NMR Spectra of (a) (V), (b) (VI), (c) (VII) and, (d) Rimuene (IV) (Wenkert)

(5) Sandaracopimaradiene (VII-a)[717]

(VII-b)
(VII-c) } See the item [B], (iii), (6),

(VII-a)

≪Occurrence≫ *Xylia dolabriformis*
mp. 39∼39.5°, $[\alpha]_D$ −12.4°, $C_{20}H_{34}$

(iv) *Tetracyclic Hydrocarbons*

(6) Mirene (VIII)[718]

(VIII)

≪Occurrence≫ *Podocarpus ferruginea*
mp. 59∼60°, $[\alpha]_D$ +43.8 (CHCl₃), $C_{20}H_{32}$

(7) Hibaene (IX)[719]

(IX)

717) Laidlaw, R. A. *et al*: *Soc.*, **1963**, 644.
718) Wenkert, E: *Chem. & Ind.*, **1955**, 282.
719) Kitahara, Y. *et al*: *6 th Symposium on the Chemistry of Natural Products*, p. 113 (Sapporo, 1962); *Tetr. Letters*, 1771 (1964).

≪Occurrence≫   *Thujopsis dolabrata* SIEB. et ZUCC. (ヒバ).

mp.   30°, $[\alpha]_D$ −49.9° (CHCl₃) $C_{20}H_{32}$, m/e 272

IR ν_{max} cm⁻¹: 750 (*cis* CH=CH), 1385, 1363 (gem Me₂)

NMR τ : 4.50q, J=6.0 cps (AB type, $\overset{\underset{C}{|}}{C}$-$\overset{\underset{C}{|}}{C}$-CH=CH-$\overset{\underset{C}{|}}{C}$-C), 9.01, 9.14, 9.17, 9.26 (4 $\overset{\underset{C}{|}}{C}$-$\overset{\underset{C}{|}}{C}$-CH₃)

(8) 4, 4, 10-Trimethyl-15-methylene-3, 13-cyclopentanoperhydrophenanthrene (X)[720]

(X)

≪Occurrence≫   *Erythroxylum monogynum*

bp. 149~150°/1.5mm, $[\alpha]_D$ +18°, $C_{20}H_{32}$

(9) (−) **Kaurene** (XI) and (+) **Phyllocladene** (XII)

(XI)[721]~[723]   (XII)[721][724]

(a) (−) **Kaurene** (XI)

≪Occurrece≫   *Agathis australis* SALISB. (New Zealand kauri)

mp. 51°, $[\alpha]_D$ −72° (CHCl₃), $C_{20}H_{32}$

NMR τ: 5.27 (C=CH₂), 8.97, 9.13, 9.18 (satur. Me)

(b) (+) **Kaurene**

≪Occurrence≫   *Podocarpus ferrugineus*

mp. 49~50°, $[\alpha]_D$ +74°, $C_{20}H_{32}$

Total synthesis of *dl*-Kaurene[725]

(c) (+) **Phyllocladene** (XII)

≪Occurrence≫   *Phyllocladus rhomboidalis, Dacrydium biforme, Sciadopitys verticillata, Podocarpus spicata, P. ferruginea* spp.

mp. 95~98°, $[\alpha]_D$ +15.8° (CHCl₃), $C_{20}H_{32}$

IR[726] ν_{max}: cm⁻¹ 3069 (=CH₂), 1657 (C=C), 872 (=CH₂ bending), see Fig. 133

720) Gupta, R. C. *et al*: *C. A.*, **48**, 7852 (1954), **49**, 6186 (1955).

721) Cross, B. E. *et al*: *Proc. Chem. Soc.*, **1963**, 17 ; *Tetr. Letters*, **1962**, 145.

722) Briggs, L. H. *et al*: *Soc.*, **1963**, 1345.

723) Djerassi, C. *et al*: *J. A. C. S.*, **83**, 3720 (1961).   724) Briggs, L. H. *et al*: *Soc.*, **1962**, 1840.

725) Bell, R. A. *et al*: *J. Org. Chem.*, **27**, 3742 (1962).

726) Bottomley, W. *et al*: *Soc.*, **1955**, 2624.

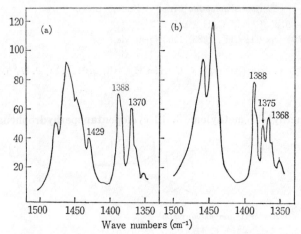

Fig. 133 IR Spectra of (a) Phyllocladene and (b) Isophyllocladene
in 1500~1350 cm⁻¹ region   (Bottomley *et al.*)

(10) Isokaurene (XIII) and **Isophyllocladene** (XIV)

(XIII)[722]   (XIV)[721) 724) 727)]

(a) Isokaurene (XIII)

≪Occurrence≫   *Agathis kauri*,   mp. 61°, $C_{20}H_{32}$

NMR τ: 4.93 (>C=C<^{H}), 8.29d (17–Me), 8.97, 9.13, 9.18 (satur. $C\!-\!\overset{C}{\underset{C}{C}}\!-\!CH_3$)

(b) Isophyllocladene (XIV)

≪Occurrence≫   *Cupressus macrocarpa*

mp. 111~112°, $[\alpha]_D$ +23.7° (CHCl₃), $C_{20}H_{32}$, from *Sciadopitys verticillata*,   mp. 111°, $[\alpha]_D$ −24.5°,
$C_{20}H_{32}$

Rimuene (XI) $\xrightarrow{\text{H}^+}$ Isophyllocladene (XIV)[728]

〔B〕 Alcohols

(i) *Marcromonocyclic Alcohol*

(1) α-and β-4, 8, 13-Duvatriene-1, 3 -diol (XV)[729]

(XV)

---

727) Grant, P. K. *et al* : *Tetrahedron*, **8**, 261 (1960).
728) Briggs, L. H. *et al* : *Tetr. Letters*, **1959** No. 8, 17.
729) Roberts, D. L. *et al* : *J. Org. Chem.*, **27**, 3989 (1962).

≪Occurrence≫ *Nicotiana tabacum*

α-form : mp. 65~66°, $[\alpha]_D$ +281.6° (CHCl₃) C₂₀H₃₄O₂

UV λ_{max}^{EtOH}mμ : <220

IR ν_{max}^{nujol}cm⁻¹ : 3300 s, 1665 w, 1345, 1190, 1160, 1118, 1024, 995, 974 s, 954, 818

NMR   τ : 4.81 (3), 5.07 (1), 5.63 (1) (broad), 8.36 (3) (d, *J*=1.2 c/s), 8.50 (3) (half peak width=3.2
　　　 c/s), 8.69 (3), 9.17 (6) (doublet with sec. splitting)

β-form : mp. 127~127.5°, $[\alpha]_D$ +162° (CHCl₃), C₂₀H₃₄O₂, (diastereoisomer of α-compound)

UV λ_{max}^{EtOH}mμ : <220

IR ν_{max}^{nujol}cm⁻¹ : 3280 s, 1667 w, 1166, 1142, 1117, 1095, 1083, 1035, 975 s, 946, 923, 889, 875

NMR   τ : 4.82 (2), 4.96 (2), 5.36 (1) (broad), 8.36 (3) (d, *J*=1.2 c/s), 8.53 (3) (half peak width=3.1
　　　 c/s), 8.65 (3), 9.18 (6) (doublet with. sec. splitting)

(ii) *Bicyclic Alcohols*

(2) (+) **Manool** (XVI)[730] and (+) *epi*-**Manool** (XVI-a)[730a]

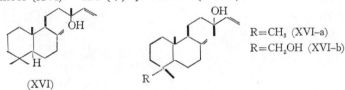

R=CH₃ (XVI–a)

R=CH₂OH (XVI–b)

(XVI)

(a) **Manool** (XVI)

　　≪Occurrence≫ *Cupressus sempervirens* L., *Dacrydium biforme*

mp. 53° bp. 142~144°/0.4mm, d₄²⁰ 0.9653, n_D 1.5184, $[\alpha]_D$ +37.4°, C₂₀H₃₄O

IR : See Fig. 133a

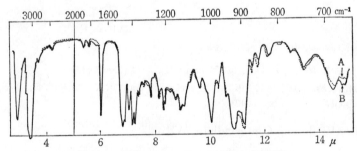

Fig. 133a IR Spectra of (+) Manool in KBr

(A) Derived from Di–*O*–acetylsclareol

(B) Extracted from *Dacrydium biforme* (Ohloff)

Acetate bp 155~156°/0.8mm, $[\alpha]_D$ +30° (CHCl₃), C₂₂H₃₆O₂

IR ν_{max}^{KBr}cm⁻¹ : Semicyclic double bonds—1650 m (C=C), 1415 m (CH₂), 890 s (CH₂), 1785 w (CH₂), 3095
　　　 m (ν_{C-H}), vinylic double bonds—1410 m (CH₂), 990 s (C–H), 915 s (CH₂), 1830 w (CH₂),
　　　 3095 m (ν_{C-H}), acetyl group—1740 ($\nu_{C=O}$), 1230 (ν_{C-C}).

(b) (+) *epi*-**Manool** (XVI–a)

　　≪Occurrence≫ *Pinus contorta* DOUGL.

---

730) Jeger, O. *et al* : *Helv. Chim. Acta*, **30**, 1853 (1947), **45**, 400 (1962).

730a) Rowe, J. W. *et al* : *J. Org. Chem.*, **29**, 1554 (1964).

mp. 36.5~38.5°, $[\alpha]_D$ +51° (CHCl$_3$), C$_{20}$H$_{34}$O

UV λ_{max}^{EtOH}mμ (ε) : 210 (3110)

IR $\nu_{max}^{CCl_4}$cm^{-1} : 1643, 1406, 992, 916, 886 (–CH=CH$_2$, >C=CH$_2$), 1385, 1364 (>C<$_{Me}^{Me}$), 3610 (OH),
3084 (C–H)

(c) (+)-18-Hydroxy-13-*epi*-manool (XVI-b)[730a]

 ≪Occurrence≫ *Pinus contorta* DOUGL.

mp. 113~114.5°, $[\alpha]_D$ +43° (CHCl$_3$), C$_{20}$H$_{34}$O$_2$

UV λ_{max}^{EtOH}mμ (ε) : 210 (3265)

IR ν_{max} cm^{-1} : 3050, 1640, 1410, 999, 919, 890 (–CH=CH$_2$, >C=CH$_2$), 3302, 1017 (OH) ;
$\nu_{max}^{CCl_4}$cm^{-1} : 3085 (=CH$_2$), 3643 (C$_{18}$–OH), 3617 (C$_{13}$–OH)

NMR in CDCl$_3$, 60 Mc, τ : 9.32, 9.03 (17–Me, 19–Me), 8.67 (16–allylic Me), 5.19, 5.48 (20 exocyclic
=CH$_2$), 3.8—5.1 AB$_2$ system (–$\overset{14}{C}$H=$\overset{15}{C}$H$_2$), 6.25 d, 6.36 d, J=11 c/s AB system (18–CH$_2$–OH)

(3) (−) Sclareol (XVII)[731)~733)]

 ≪Occurrence≫ *Salvia sclarea*

mp. 103.5~105°, $[\alpha]_D$ −4.25° (CHCl$_3$), C$_{20}$H$_{36}$O$_2$

(XVII)

(4) Labdane-8α, 15-diol (XVII-a)[733a)]

 ≪Occurrence≫ *Aeonium lindleyi* (*Crassulaceae*)

mp. 83~84°, $[\alpha]_D$ −10°, C$_{20}$H$_{38}$O$_2$

(XVII–a)

UV ν_{max}^{KCl} cm^{-1} : 3460, 3380, 1388, 1375, 1080

IR $\nu_{max}^{CCl_4}$ cm^{-1} : 3637, 3605 (free OH), 3500 (intra OH), 3370 (inter OH)

NMR τ : 5CH$_3$ (8.84, 9.03, 9.12, 9.2), 16 skeletal protons (8.27, 8.38, 8.45, 8.65), 2 OH (7.94),
–CH$_2$–OH (6.21, 6.32, 6.42)

(5) Contortolal (XVII-b) **and Contortadiol** (XVII-c)[730a)]

R=CHO (XVII–b)

R=CH$_2$OH (XVII–c)

(a) Contortolal (XVII–b)

 ≪Occurrence≫ *Pinus contorta* DOUGL.

2,4–DNP mp. 157~159°, $[\alpha]_D$ +22° (CHCl$_3$), C$_{21}$H$_{35}$O$_2$N$_3$

IR ν_{max} cm^{-1} : 3280~3510 (OH, NH), 1681 (C=O), 1580 (C=C, C=N), 995 (—CH$_2$OH),
885 (>C=CH$_2$) 763 (CH=N)

731) Ohloff, G : *Helv. Chim. Acta*, **41**, 845 (1958). 732) Jeger, O *et al* : *ibid.*, **30**, 1853 (1947).
733) Soucek, M. *et al* : *Chem. & Ind.*, **1962**, 1946. 733a) Baker, A. J. *et al* : *Soc.*, **1962**, 4705.

NMR　in $CDCl_3$, 60 Mc, τ : 9.45, 8.97 (17–Me, 19–Me), 8.33 (⟋⟍⟍⟍), 5.82, 5.94 (15–CH_2OH),

5.16, 5.48 (20=CH_2), 4.63 t 14–vinylic H), 3.98 s (18 CH=N), 1.2 (NH),
4.5 (NH_2)

(b)　Contortadiol (XVII–c)

《Occurrence》　*Pinus contorta* DOUGL.

mp.　106~107.5°, $[\alpha]_D$ +31° ($CHCl_3$), $C_{20}H_{34}O_2$

UV　$\lambda_{end}^{EtOH}m\mu(\varepsilon)$: 210 (5600)

IR　ν_{max} cm^{-1} : 3270, 1027 (CH_2OH), 3050, 1641, 898 (>C=CH_2)

(iii)　*Tricyclic Alcohols*

(6)　Sandaracopimaradien-3β-ol (VII–b) **and ——3β, 18-diol** (VII–c)[716a]

《Occurrence》　*Xylia dolabriformis* (VII–c)

mp. 152~153°, $[\alpha]_D$ −18.5°, $C_{20}H_{32}O_2$

See the item [A], (iii), (5), (Vll–a)

R=CH_3 : (VII–b)
R=CH_2OH : (VII–c)

(iv)　*Tetracyclic Alcohols*

(7)　Phyllocladenol (XVIII)[734)748]

(XVIII)

《Occurrence》　*Cryptomeria japonica* D. DON (スギ)

mp. 182~183°, $[\alpha]_D$ +14.52°, $C_{20}H_{34}O$

IR　ν_{max}^{KBr} cm^{-1} : 3436, 1126 (*tert.* OH), 1397, 1383, 1172, 1152, 920 (gem Me$_2$)

(8)　Grayanotoxin I (XX), **II** (XIX) **and III** (XXI)[735)~740]

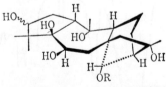

(503)

(XIX) Grayanotoxin II
= Grayanol[735)~737]

R=Ac : (XX) Grayanotoxin I[735)738)740]
　　　Andromedotoxin=Rhodotoxin
　　　=Acetylandromedol[738)~739]
R=H : (XXI) Grayanotoxin III[740]
　　　=Andromedol[739]

734)　Kondo, T *et al* : *Agr. Biol. Chem.*, **24**, 65 (1960).
735)　Kakizawa, H *et al* : *5 th Symposium on the Cemistry of Natural Products* p. 10–1 (Sendai, 1961).
736)　Takemoto, T *et al* : *ibid.*, p. 8–1.
737)　Nakajima, M *et al* : *ibid.*, p. 9–1 ; *Chem. & Ind.*, **1961**, 511 ; *Agr. Biol. Chem.*, **25**, 782, 793(1961).
738)　Tallent, W. H : *J. Org. Chem.*, **27**, 2968 (1962).
739)　Kakizawa, H *et al* : *4 th Symposium on the Chemistry of Natural Products*, p. 124 (1960).
740)　Kakizawa, H : *J. Chem. Soc. Japan*, **82**, 1096, 1216 (1961).

《Occurrence》 *Leucothoe grayana* MAM. (ハナノヒリキ), *Pieris japonica* D. DON. (アセビ), *Rhododendron matterrichii, Rh. maximum.* (シャクナゲの類)

(a) Grayanotoxin II (XIX)

mp. 198°, $[\alpha]_D$ −42°, $C_{20}H_{32}O_5$

IR ν_{max}^{KBr} cm⁻¹ [736] : ca 1190 d (gem Me₂), 889.7, 1621, ca 1730 ($\rangle$C=CH₂) ; 3390 (OH)

Tetraacetate mp. 172~173°

NMR[736] ref. C_6H_6, cps : 91.1, 198.2, 212.9, 215.4, 219.9, 237.4, 242.7

(b) Grayanotoxin I (XX)

mp. 238.5°, 260°, 272°, $[\alpha]_D$ −8.8 (EtOH), $C_{22}H_{36}O_7$

IR ν_{max}^{KBr} cm⁻¹ [740] : 3600~3200, 1730 (Fig. 134)

(c) Grayanotoxin III (XXI)

mp. 205°, $C_{20}H_{34}O_6$[739]

UV λ_{max} (ε) : 205 (500)

IR ν_{max}^{KBr} cm⁻¹ [740] : 3600~3200 (Fig. 135)

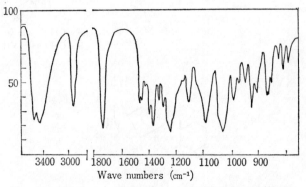

Fig. 134 IR Spectrum of Grayanotoxin I in KBr (Kakizawa)

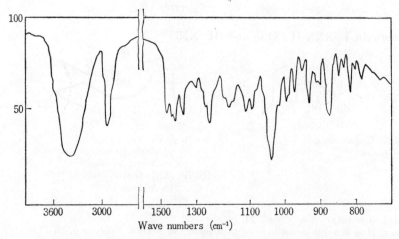

Fig. 135 IR Spectrum of Grayanotoxin III in KBr (Kakizawa)

IR NMR Spectra of derivatives and degradation products were discussed to confirm the structures of grayanotoxin–I and III[740]

(v) *Pentacyclic Alcohols*

(9) Cafestrol (XXII)[741)742)]

    ≪Occurrence≫   Coffee oil (*Coffea arabica*)

mp. 158～159°, $[\alpha]_D$ −101°, $C_{20}H_{28}O_3$

Tetrahydrocafestrol   mp. 157～159°

(XXII)

(10) Kahweol (XXIII)[743)744)]

    ≪Occurrence≫   Coffee oil (*Coffea arabica*), $C_{20}H_{26}O_3$

UV λ_{max} mμ : 290

IR ν_{max} cm^{-1} : 1735 w

(XXIII)

NMR in CDCl$_3$, TMS, δ : 5.79, 5.97, 6.19, 6.34 q (differs from cafestol (XXII), $\text{H}\!\!>\!\!\text{C=C}\!\!<\!\!^{\text{H}}$)

[C] Phenols

(1) Ferruginol (XXIV)[745)～748a)]

    ≪Occurrence≫   *Podocarpus ferruginea*[745)], *P. dacrydioides,*

                     P. totara, Cryptomeria japonica D. Don (スギ),

Total synthesis of *dl*–ferruginol[749)750)]

mp. 57～59°, $[\alpha]_D$ +40.6° (MeOH), $C_{20}H_{30}O$

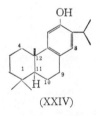

(XXIV)

Acetate   mp. 81～82°

NMR[516)] in CHCl$_3$, 60Mc, TMS, τ : 9.04 (6 H of gem Me (*eq*) and angular Me), 8.83—8.70 d (iso Pr.),

        8.78 (3H of gem Me (*ax*), 7.67 s, 3.12—3.02 d,

Δ^9–Dehydroferruginol[747)]

741) Finnegan, R. A : *J. Org. Chem.*, **26**, 3057 (1961).

742) Djerassi, C. *et al* : *J. A. C. S.*, **81**, 2386 (1959), **82**, 4352 (1960), **80**, 247 (1958) ; *Chem. & Ind.*, **1955**, 1481.

743) Kaufmann, H. P *et al* : *Ber.*, **96**, 2489 (1963).

744) Kaufmann, H. P *et al* : *Fette Seifen einschl. Anstrichmittel*, **64**, 206 (1962), **65**, 529 (1963).

745) Brandt, C. W. *et al* : *Soc.*, **1939**, 1031.   746) Todd, D *et al* : *J. A. C. S.*, **64**, 928 (1942).

747) Briggs, L. H *et al* : *Tetrahedron*, **7**, 270 (1959).

748) Kondo, Y *et al* : *Y. Z.*, **79**, 1298 (1959) ; *Chem. Pharm. Buli.*, **11**, 678 (1963).

748a) Kondo, Y *et al* : *Agr. Biol. Chem.*, **23**, 233 (1959).

749) Narasimha Rao, P *et al* : *Tetrahedron*, **4**, 294 (1958).

750) King, F. Er *et al* : *C. A.*, **49**, 2379 (1955).

(2) **Sugiol** (XXV)[747)748)]

(XXV)

《Occurrence》 *Podocarpus dacrydioides*[747)], *Cryptomeria japonica* D. Don (スギ)[748)]
mp. 298~299° (decomp.), $[\alpha]_D$ +30.6° (pyridine), $C_{20}H_{28}O_2$

UV λ_{max}^{MeOH} mμ (log ε)[748a)] : 232 (4.08), 283 (3.99), 304 sh (3.93)

IR ν_{max}^{nujol} cm^{-1} : 3145 m, 3096 m, 1634 s, 1587 s, 1572 s, 1506 m, 1460 m, 1385 m, 1353 m, 1318 s, 1274 s, 1247 m, 1182 m, 1088 w, 869 w, 771 w.

Acetate mp. 162~163°

NMR[748)] in CHCl$_3$, 60 Mc, TMS, τ : 9.01 (gem Me (*eq*)), 9.04 (angular Me), 8.83–8.75 (iso Pr, gem Me (*ax*)) 7.68 s (acetyl), 3.04, 2.04 (C$_8$ ring proton shifts to)

(3) **Xanthoperol** (XXVI)[747)748)]

(XXVI)

《Occurrence》 *Podocarpus dacrydioides*[747)], *Cryptomeria japonica* D. Don[748)]
Yellow prism mp. 246° (decomp.), 273~275°, $[\alpha]_D$ +142.6°, $C_{20}H_{26}O_3$

UV λ_{max}^{MeOH} mμ (log ε) : 252 (3.70), 357~358 (br) (4.03)

IR ν_{max}^{KBr} cm^{-1}[748)] : 3413 s, 1595 s, 1572 s, 1508 m, 1475 s, 1439 w, 1403 w, 1383 w, 1342 s, 1297 s, 1271 s, 1208 m, 1186 m, 1166 m, 1151 m, 1109 m, 1087 w, 1049 m, 917 m, 885 m, 869 m

Acetate mp. 156~157°

NMR in CHCl$_3$, 60 Mc, TMS, τ : 9.56 (one of gem Me shifts to), 9.10, 8.81 (another gem Me and angular Me), 8.69, 8.71 (iso Pr), 7.63 (acetyl), 2.91, 1.97 (arom. proton)

(4) **Totarol** (XXVII) and **16-Hydroxytotarol** (XXVIII)

R=H : (XXVII) Totarol
R=OH : (XXVIII) 16–Hydroxytotarol

≪Occurrence≫   *Podocarpus totara*[751)752], *Cryptomeria japonica* (スギ)[753], *Dacrydium cupressinum*

Total synthesis[754]

(a) Totarol (XXVII)

mp. 132°, $[\alpha]_D$ +41.4° (EtOH), $C_{20}H_{30}O$.

UV $\lambda_{max}m\mu$ (log ε) : 280~285 (3.29)[753] ; $\lambda_{max}m\mu$ (ε) : 278 (1970), 285 sh (1910)[754]

NMR δ : 6.70 q (AB type, J=8.9) ($\overset{H}{\underset{11}{}}\rangle\!\!\!=\!\!\!=\!\!\!\langle\underset{12}{}^H$), 4.35 br (13–OH), 3.07 quintet (J=7.2) (–CH$\langle{}^{CH_3}_{CH_3}$),

2.78 m (–CH$_2$–), 1.29 d (J=7.2) (2Me of iso Pr), 1.12 (10–Me), 0.94, 0.92 (4–gem Me$_2$) (See Table 73)

(b) 16–Hydroxytotarol (XXVIII)

mp. 230~231°, $[\alpha]_D$ +29° (EtOH)[752], +41.9° (EtOH)[751]

UV $\lambda_{max}^{EtOH}m\mu$ (ε) : 279 (1990)[752] ; $\lambda_{max}m\mu$ (log ε) : 256 (3.21), 283 (3.39)[751] ; $\lambda_{max}^{CHCl_3\ or\ nujol}\mu$: 2.99 m ;

9.80 s, 10.87 m (ident. with totarol)[752]

IR ν_{max} cm^{-1} : 3390 (OH), 1189 (phenol), 1014 (alc. OH)[751]

NMR[752] δ : 1.04 (Me), 3.66 q (AB type, J=10.9) (–CH$_2$–OH), 1.32 d (2Me of iso Pr), 1.16(10–Me)

(5) Cryptojaponol (XXIX)[753]

(XXIX) (Suggested Str.)

≪Occurrence≫   *Cryptomeria japonica*

mp. 204~205°, $[\alpha]_D$ 25.3°, $C_{21}H_{30}O_3$

UV $\lambda_{max}m\mu$ (log ε) in neutral soln. : 318 (3.47), 274 (4.16), 230 (4.14) —in alkaline soln. : ~380
   (3.69), ~300(3.84), 258 (4.29), 230 (4.14)

IR ν_{max}^{KBr}cm^{-1} : 3420, 2960~2860, 1680, 1640, 1600, 1475, 1425, 1392, 1378, 1324, 1255, 1215, 1185,
   1172, 1143, 950, 884, 877

(6) Hinokiol (XXX) and Hinokione (XXXI)[755]

R=OH, R'=H (XXX) Hinokiol

R, R'=O (XXXI) Hinokione

(Revised Str.)[755]

≪Occurrence≫   *Tetraclinis articulata* MASTERS, *Chamaecyparis obtusa* SIEB. et ZUCC. (ヒノキ)

(a) Hinokiol (XXX)

mp. 234~235°, $[\alpha]_D$ +74.4°, $C_{20}H_{30}O_2$

751) Cambie, R. C. *et al* : *Tetrahedron*, **18**, 465 (1962).

752) Wenkert, E. *et al* : *Tetr. Letters*, **1961**, (11), 358.

753) Kondo, Y. *et al* : *Y. Z.*, **82**, 1252 (1962).

754) Barltrop, J. A. *et al* : *Soc.*, **1958**, 2566.

755) Erdtman, H. *et al* : *Proc. Chem. Soc.*, **1960**, 174.

(b) Hinokione (XXXI)

mp. 188~189°, $[\alpha]_D$ +103.38°, $C_{20}H_{28}O_2$

(7) Nimbiol (XXXII)[756)757)]

(XXXII)[756)]

≪Occurrence≫ *Melia Azadirachta* L. (インドセンダン).

mp. 244°, $[\alpha]_D$ +32.3°, $C_{18}H_{24}O_2$

UV λ_{max}^{EtOH} mμ (log ε): 231 (4.07), 286 (4.06); IR $\lambda_{max}^{CHCl3}\mu$: 3.05 (assoc. OH), 6.03 (conj. C=O)

Methyl ether mp. 141~142°, $C_{19}H_{26}O_2$

Synthetic *dl*–methyl ether mp. 118~119°[756)]

(8) Podototarin (XXXIII)[758)759)]

(XXXIII)

≪Occurrence≫ *Podocarpus totara* G. Benn.

Totarol (XXVII) $\xrightarrow{\text{enzymic coupling by } \textit{Polyporus vericolor}}$ (XXXIII)[760)]

Totarol (XXVII) $\xrightarrow{\text{alkaline } K_3Fe(CN)_6}$ (XXXIII)[761)]

mp. 225~226°, $[\alpha]_D$ +76.1°, $C_{40}H_{58}O_2$

UV λ_{max} mμ (log ε): 216 (4.61), 254 (4.10), 290 (3.79)

IR ν_{max}^{CS2}cm^{-1}: 3472 (non–bonded OH), 1350, 1181 (phenolic OH)

NMR (Table 73)

Table 73 NMR Shifts and Spin-coupling Constants for Totarol and Podototarin (Cambie *et al.*)

Totarol (XXVII)		Podototarin (XXXIII)		Protons of
τ	J c/s	τ	J c/s	
2.97	8.4			}Aromatic
3.47	8.4	3.00		
5.57		4.91		OH

756) Dutta, P.C. *et al*: *Soc.*, **1960**, 4766.
757) Sen Gupta *et al*: *Chem. & Ind.*, **1958**, 861; *C. A.*, **53**, 1400 (1959).
758) Cambie, R.C. *et al*: *Chem. & Ind.*, **1962**, 1757.
759) Cambie, R.C. *et al*: *Tetrahedron*, **19**, 209 (1963).
760) Cambie, R.C. *et al*: *Proc. Chem. Soc.*, **1963**, 143.
761) Falshaw, C.P. *et al*: *Chem. & Ind.*, **1963**, 451.

Totarol (XXVII)		Podototarin (XXXIII)		Protons of
τ	J c/s	τ	J c/s	
6.74	7.2	6.66	7.8	$-CH{<}^{CH_3}_{CH_3}$
7.19		7.09		$C_7, C_{7'}$ $-CH_2-$
8.67	7.2	8.63	7.8	$\left.\begin{array}{c} \\ \\ \end{array}\right\}-CH{<}^{CH_3}_{CH_3}$
		8.65	7.8	
8.85		8.80		Angular Me
9.08		9.05		$\left.\begin{array}{c} \\ \end{array}\right\}$ gem Me_2
9.10		9.08		

(9) Macrophyllic Acid (XXXIV)[762]

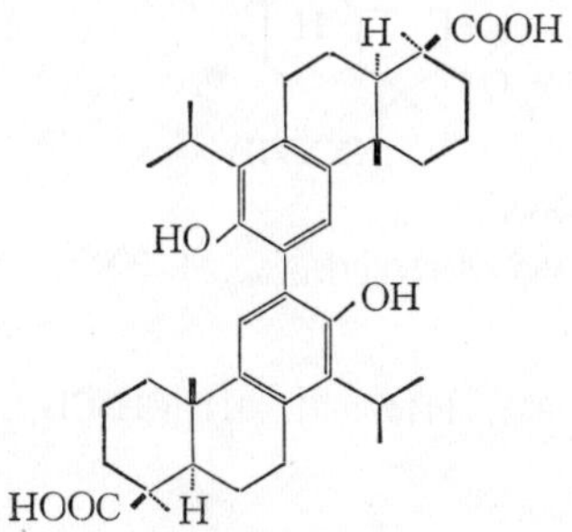

(XXXIV)

≪Occurrence≫　　*Podocarpus macrophyllus* D. DON.,　　mp. 237~238°, $[\alpha]_D$ +79°, $C_{40}H_{54}O_6$

UV　$\lambda_{max}^{neutral\,soln}$ mμ (log ε) : 290 (3.85), 254 (4.18), 220 (4.68) ; $\lambda_{max}^{0.01N-alc\,KOH}m\mu$ (log ε) : 318 (3.95),

　　260 sh (4.07), 226 (4.32)

IR　$\nu_{max}^{CS_2}$ cm^{-1} : 3521 (OH), 2700~2550 br (OH of COOH), 1704 (–COOH), 1182 (phenolic OH) ;

　　ν_{max}^{nujol} cm^{-1} : 3510 (OH), 2700~2550, 1696 (COOH). 1180

[D]　Oxides

(1) Manoyloxide(XXXV)[717][763]

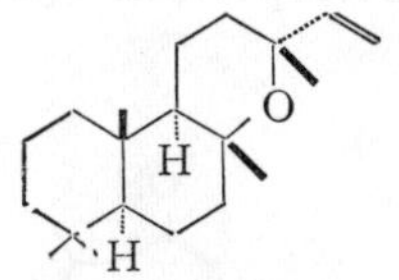

(XXXV)

≪Occurrence≫　　*Xylia dolabriformis, Dacrydium colensoi*

mp. 27°, 29°, $[\alpha]_D$ +21°, $C_{20}H_{34}O$

762)　Bocks, S.M. *et al* : *Tetrahedron*, **19**, 1109 (1963).

763)　Hosking, J.R : *Ber.*, **69**, 780 (1936).

(2) 3-Oxomanoyloxide (XXXVI)[717a]

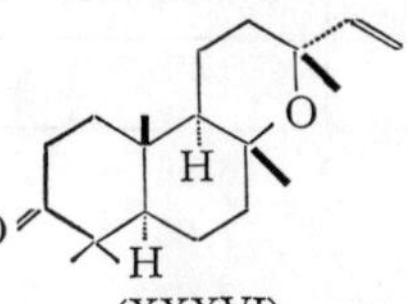

(XXXVI)

≪Occurrence≫ *Xylia dolabriformis*

mp. 99.5°, $[\alpha]_D$ +54°, $C_{20}H_{32}O_2$

(3) Colensenone (XXXVII)[764]

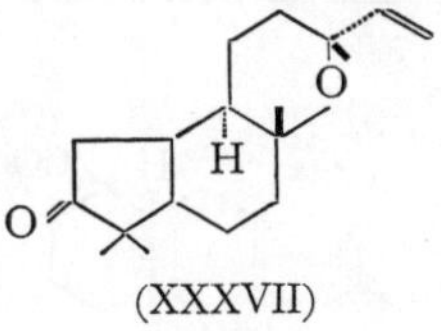

(XXXVII)

≪Occurrence≫ *Dacrydium colensoi*

mp. 99~100°, $C_{19}H_{30}O_2$, Pos. Cotton effect, $[\alpha]_{317.5m\mu}$ +3900°

UV $\lambda_{max}^{EtOH}m\mu$ (ε) : 299 (31), ε_{210}140

IR $\nu_{max}^{nujol}cm^{-1}$: 1733 (CO), 3070, 1643, 1414, 991, 911 (CH=CH$_2$), 1118, 1088 (C–O)

(4) Grindelic acid (XXXVIII)[765]

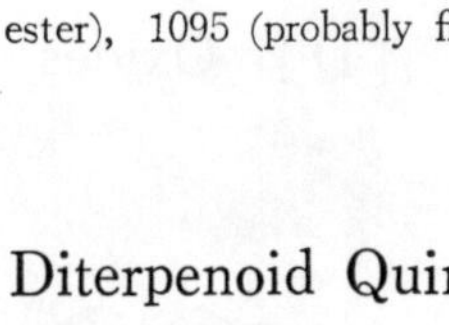

(XXXVIII)

≪Occurrence≫ *Grindelia robusta*, mp. 70.5°, $[\alpha]_D$ −134.1°, $C_{21}H_{34}O_3$

IR $\nu_{max}^{nujol}cm^{-1}$: no OH, 1740 (unconj. ester), 1095 (probably five or six membered ring ether),

835 ($>=<^H$)

[E] Diterpenoid Quinones

(1) Xanthoperol (XXVI)

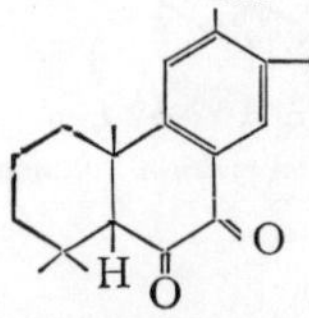

(XXVI)

See the item [C], (3) (p. 288 Ref.)

764) Grant, P. K. *et al* : *Soc.*, **1962**, 3740.
765) Panizzi, L *et al* : *Tetr. Letters*, **1961**, No. 11, 376.

(2) Biflorin (XXXIX)[766]

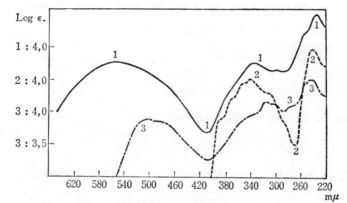

(XXXIX)

≪Occurrence≫   *Caparia biflora* L.,
Dark brownred cryst., mp. 159~160° $C_{20}H_{20}O_3$
UV   See Fig. 136, λ_{max} mμ (log ε) : 555 (3.76)
IR   See Fig. 137, λ_{max} μ : 5.94 w, 6.12 m, 6.22 s, 6.28 s

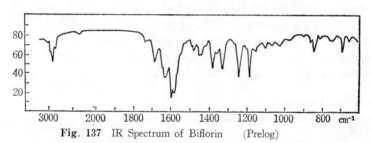

Fig. 136   UV Spectra of (1) Biflorin, (2) Di–O–acetyldihydrobiflorin,
(3) Chinoxaline derivative $C_{26}H_{24}ON_2$ (Prelog)

Fig. 137   IR Spectrum of Biflorin   (Prelog)

NMR δ : 2.70 s (CH$_3$), 1.94 s (CH$_3$), 1.55 d (3H, J=1.5), 1.71 d (3H. J<1) (C=C$<^{Me}_{Me}$), 2.4 m

(4H of C=C–CH$_2$–CH$_2$–C=C), 5.15 (1 H of $>$C=C$<^H$), 7.05 s (1 H of $>$C=C$<^H$),

7.2~7.5 q (AB system) (arom. 2 H)

766) Prelog, V *et al* : *Helv. Chim. Acta*, **41**, 1386 (1958), **46**, 409, **413**, 415 (1963).

(3) **Tanshinone I** (XL), **Tanshinone II** (XLI) and **Cryptotanshinone** (XLII)[767)~770)]

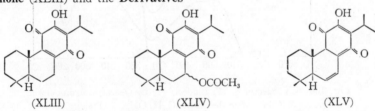

    (XL)   (XLI)   (XLII)

≪Occurrence≫   *Salvia miltiorrhiza*

(a) **Tanshinone I** (XL)   mp. 231~234°, $C_{18}H_{12}O_3$

(b) **Tanshinone II** (XLI)   mp. 211°, $C_{19}H_{18}O_3$

UV $\lambda_{max}^{EtOH}m\mu$ (log ε)[766)] : 223 (4.34), 250 (4.30), 266 (4.42), 347 (3.27), 464 (3.48), 510 sh (3.32),

IR $\nu_{max}^{KBr}cm^{-1}$ [769)] : 3157, 1701, 1650, 1584, 1539, 1505

$$(XLI) \xrightarrow{Pd, H_2} Cryptotanshinone (XLII)$$

(c) **Cryptotanshinone** (XLII)   mp. 191°, $C_{19}H_{20}O_3$

Chinoxaline derivative   mp. 150~151°, $C_{25}H_{24}ON_2$

(4) **Royleanone** (XLIII) and the **Derivatives**[771)]

    (XLIII)   (XLIV)   (XLV)

≪Occurrence≫   *Inula royleana* D.C. (*Compositae*)

$$Ferruginol (XXIV) \xrightarrow{Synth} (XLIII)$$

(a) **Royleanone** (XLIII)   mp. 181.5~183°, $C_{20}H_{28}O_3$

(b) **Acetoxyroyleanone** (XLIV)   mp. 212~214.5°, $C_{22}H_{30}O_5$

(c) **Dehydroroyleanone** (XLV)   mp. 168~171°, $C_{20}H_{26}O_3$

(5) **Coleone A** and **Coleone B**[772)]

   ≪Occurrence≫   *Coleus igniarius* (*Labiatae*)

(a) **Coleone A** (XLVI)

mp. 136~136.5°, $[\alpha]_D$ ca+100° (EtOH), $C_{20}H_{22}O_6$

UV $\lambda_{max}^{EtOH}m\mu$ (log ε) : 212 (4.32), 230 (4.28), 252.5 (4.23)~270 (4.08), 315 (4.04), 435 (3.78)

IR $\nu_{max}^{CCl4}cm^{-1}$: 3600, 3534, 3322, 3251, 2967, 1669, 1626

NMR ppm : **1.22** d, 1.35 d (iso Pr), 1.53 s ($\ni$C–CH$_3$), 2.65 s(CH$_3$–$\overset{|}{C}$=C)

767) Takiura, K. *et al* : *Y. Z.*, **63**, 40 (1943), **61**, 9, 475, 483 (1941); *Chem. Pharm. Bull.*, **10**, 112 (1962).
768) Wessely, F. *et al* : *Ber.*, **73**, 19 (1940).
769) Okumura, Y. *et al* : *Bull. Chem. Soc. Japan*, **34**, 895 (1961).
770) King, T. J : *Soc.*, **1961**, 5090.
771) Edwards, O. E. *et al* : *Canad. J. Chem.*, **40**, 1540 (1962).
772) Eugster, C. H. *et al* : *Helv. Chim. Acta*, **46**, 530 (1963).

(b) Coleone B (XLVII)

mp. 258~259°, $[\alpha]_D$ ca+130° (EtOH), $C_{19}H_{20}O_6$

UV $\lambda_{max}^{EtOH}m\mu$ (log ε): ~250 (3.96), 308 (4.22), 423 (3.75)

IR $\nu_{max}^{CHCl^3}cm^{-1}$: 1661, 1618, 1600

NMR ppm : 1.57~1.68 d (iso Pr), 1.95 s ($>$C–CH$_3$), 2.62 s (CH$_3$–$\overset{|}{C}$=C)

〔F〕 Aldehydes and Ketones

The following ketones have already been mentioned.

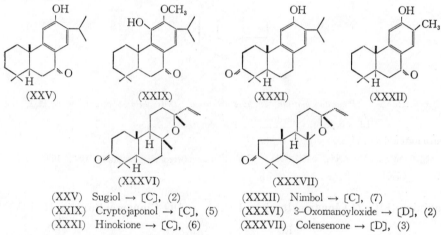

(XXV)   (XXIX)   (XXXI)   (XXXII)

(XXXVI)   (XXXVII)

(XXV) Sugiol → 〔C〕, (2)   (XXXII) Nimbol → 〔C〕, (7)
(XXIX) Cryptojaponol → 〔C〕, (5)   (XXXVI) 3–Oxomanoyloxide → 〔D〕, (2)
(XXXI) Hinokione → 〔C〕, (6)   (XXXVII) Colensenone → 〔D〕, (3)

(i) *Tricyclic Compounds*

(1) Cryptopinone (XLVIII) and **Isodextropimarinal** (XLIX)[773]

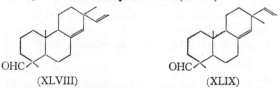

(XLVIII)   (XLIX)

≪Occurrence≫ *Pinus palustris, P. elliotti* and other spp.

(a) Cryptopinone (Dextropimarinal) (XLVIII)

2,4–DNP mp. 195~196°(decomp.), $[\alpha]_D$ −26°[774], $C_{20}H_{30}O$

(b) Isodextropimarinal (XLIX)

mp. 50~52°, $C_{20}H_{30}O$

---

773) Karrer, W : Konstitution und Vorkommen der organischen Pflanzenstoffe, p. 787 (Birkhäuser, 1958).
774) Barton, D.H.R. *et al* : *Acta Chem. Scand.*, **5**, 1356 (1951); *C.A.*, **47**, 118 (1953).

(2) Sandaracopimaradiene-3-one (L)[717]

(L)

≪Occurrence≫   *Xylia dolabriformis*
mp. 59~60°, $[\alpha]_D$ −56°, $C_{20}H_{30}O$

(3) Pleuromutilin (LI)[775]

OCOCH₂OH

(LI)

≪Occurrence≫   *Pleurotus mutilus, P. passeckerianus, Drosophila subatrata,* $C_{22}H_{34}O_5$

(LI) (antibiotics) $\xrightarrow{\text{hydrolysis}}$ Pleuromutenol

Pleuromutenol $C_{20}H_{32}O_3$
IR ν_{max}^{nujol} cm⁻¹ : 3552 (OH), 3460 (OH), 1730 (five-membered ring ketone), 3080, 1642, 912 (vinyl)
NMR in CDCl₃, 60 Mc, τ : 3.8 q, 4.7 q

(4) Taxinine (LII), *O*-Cinnamoyltaxicin-I (LIII) and *O*-Cinnamoyltaxicin-II (LIV)[776]~[782]

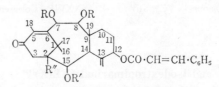

$OCO \cdot CH=CH \cdot C_6H_5$

R=R′=Ac, R″=H : (LII) Taxinine[777][779]
R=R′=H, R″=OH : (LIII) *O*-Cinnamoyltaxicin–I
R=R′=R″=H : (LIV) *O*-Cinnamoyltaxicin–II[776]~[778]

Taxine=Taxinine (LII)+HN(CH₃)₂

Crude taxine $\xrightarrow{\text{Hofmann degradation}-HN(CH_3)_2}$ "Desdimethylaminotaxine" $\longrightarrow$ Triacetates of *O*–Cinnamoyl-

taxicin–I and II $\xrightarrow[\text{of former}]{\text{desacetyl}}$ *O*–Cinnamoyltaxicin–I (LIII)[782]

Taxine–I $C_{33}H_{45}O_8N \rightarrow$ *O*–Cinnamoyltaxicin I $C_{29}H_{36}O_7$[781]

≪Occurrence≫   *Taxus baccata* L. (European),[779]~[781] *T. baccata*[776]~[778] subsp. *cuspidata* (イチイ)[776]~[778]

775) Birch, A. J. *et al* : *Chem. & Ind.*, **1963**, 374.
776) Uyeo, S. *et al* : *Y. Z.*, **82**, 1081 (1962).
777) Uyeo, S. *et al* : *7th Symposium on the Chemistry of Natural Products*, p. 226 (Fukuoka, 1963).
778) Nakanishi, K. *et al* : *ibid.*, p. 219 (1963).   779) Eyre, D. H *et al* : *Proc. Chem. Soc.*, **1963**, 271.
780) Langley, B. W. *et al* : *Soc.*, **1962**, 2972.
781) Baxter, J. N. *et al*: *Proc. Chem. Soc.* **1958**, 9.   782) Baxter, J. N. *et al* : *Soc.*, **1962**, 2964, 2972.

(a) Taxinine (*O*–Cinnamoyltaxicin II triacetate) (LII)

mp. 264~265°, M=611±10), $C_{55}H_{42}O_9$

UV $\lambda_{max}^{MeOH}m\mu$ (log ε)[778] : 218 (4.28), 223 (4.22), 280 (4.45)

IR ν_{max}^{KBr} cm^{-1} [778] : 1745 (acetate), 1720, 1644 (cinnamate), 1674 (conj. ketone), 911 ($>$C=CH$_2$)

NMR[778] δ, J (c/s)

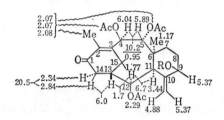

X–ray analysis

(b) *O*–Cinnamoyltaxicin–I (LIII)[782]

mp 233~234°, $[\alpha]_D$ +285°, $C_{29}H_{36}O_7$

UV $\lambda_{max}^{EtOH}m\mu$ (ε) : 282 (28200)

IR $\nu_{max}^{CHCl_3}$ cm^{-1} : 1703 (cinnamate C=O), 1671 (conj. C=O), 1645 (cinnamate C=C)

Triacetate mp. 237~239°, $[\alpha]_D$ +218°, $C_{35}H_{42}O_{10}$

UV $\lambda_{max}^{EtOH}m\mu$ (ε) : 281 (26500)

IR $\nu_{max}^{CHCl_3}$ cm^{-1} : 3663 (OH), 1742 (acetate), 1709 (cinnamate C=O), 1675 (conj. C=O),
                    1647 (cinnamate C=C)

NMR[779]

τ : 4.0 q (J=9) (2H of 7 and 8), 4.4-6.5 q (J=6.5) (2H of 15 and 14), 4.7 s (2 H of 20), 5.3 m (1 H of 12), 7.2 q (J=20) (2H of 3), 7.7 s (3 H of 18), 8.25 m (4 H of 10 and 11), 8.3 s (3 H of 16 or 17), 8.75 s (3 H of 16 or 17), 9.05 s (3 H of 19)

(c) *O*–Cinnamoyltaxicin–II (LIV)[782]

——Triacetate=(LII)[782]

mp. 265~267°, $[\alpha]_D$ +137°, $C_{35}H_{42}O_9$

UV $\lambda_{max}^{EtOH}m\mu$ (ε) : 279 (28500)

IR $\nu_{max}^{CHCl_3}$ cm^{-1} : 1735, 1704, 1675, 1645

(ii) *Tetracyclic Ketones*

(5) Stachenones (LV~LVII)[783]

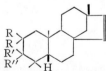

R=H, R′,R″=O : (LV) Stach–15–en–3–one (LVII) 2-Hydroxystacha–1,15–diene–3–one
RR=O, R′=OH, R″=Me : (LVI) 3-Hydroxystach–15–en–2–one

≪Occurrence≫ *Spirostachys africana* Sond. (Tamboti wood)

---

783) Johnson, R.F. *et al* : *Soc.*, **1962**, 4046.

(a) **Stach–15–en–3–one** (LV)

mp. 35~36.5°, $[\alpha]_D$ +22° (CHCl$_3$), C$_{20}$H$_{30}$O

UV λ_{max}^{EtOH} mμ (ε) : 285 (49)

IR ν_{max}^{CS2} cm^{-1} : 1706, 1381, 1363, 749

NMR in CCl$_4$, TMS, (δ=0.00) : 5.43 d, 5.67 d, J=6 c/s (*cis* CH=CH), peaks at 120~150 c/s (complex spin–spin coupling, CO–CH$_2$–CH$_2$), 1.00 (3 satur. $>$C–CH$_3$), 0.90 (1 satur. $>$C–CH$_3$)

(b) **3–Hydroxystach–15–en–2–one** (LVI)

mp. 129°, $[\alpha]_D$ +30°, C$_{20}$H$_{32}$O$_2$

UV λ_{max}^{EtOH} mμ (ε) : 285 (41), $\varepsilon_{203m\mu}$=7550

IR ν_{max}^{CS2} cm^{-1} : 3466, 1712, 739

NMR in CCl$_4$, TMS (δ=0.00) : 5.50 d, 5.58 d, J=6 c/s (*cis* CH=CH), 3.72 (CH·OH), 2.39 d, 1.98 d, J=13 c/s (–CO–CH$_2$–), 2.21 (OH), 0.69 (2 satur. $>$C–CH$_3$), 0.64 (2 satur. $>$C–CH$_3$)

(c) **2–Hydroxystacha–1, 15–diene–3–one** (LVII)

mp. 32°, $[\alpha]_D$ +49°, C$_{20}$H$_{28}$O$_2$

UV λ_{max}^{EtOH} mμ (ε) : 270 (10000)

IR ν_{max}^{CS2} cm^{-1} : 3420, 1670, 1648, 1405, 1379, 1362, 750

NMR in CCl$_4$, TMS (δ=0.00) : 6.15 s, 5.76 ($>$C–C=C–C$<$ HO H), 5.44 d, 5.65 d, J=6 c/s (*cis*CH=CH), 1.17 (satur. $>$C–CH$_3$), 1.07 (satur.$>$C–CH$_3$), 1.02 (2 satur. $>$C–CH$_3$)

(iii) *Polycyclic Ketones*

(6) **Aritasone** (LVIII)[784]

(LVIII)

≪Occurrence≫ *Chenopodium ambrosioides* (アメリカアリタソウ)

mp. 105~106°, $[\alpha]_D$ –118.6°

IR $\lambda_{max}^{nujol}\mu$: 5.80 (C=O), 5.93 (C=C), ca 8.7 (C–O–C)

784) Takemoto, T. *et al* : *Y. Z.*, **75**, 1036 (1955), **77**, 1157 (1957).

〔G〕 Resinic acids

(i) *Bicyclic acids*

(1) **Labdanolic acid** (LIX)[785]

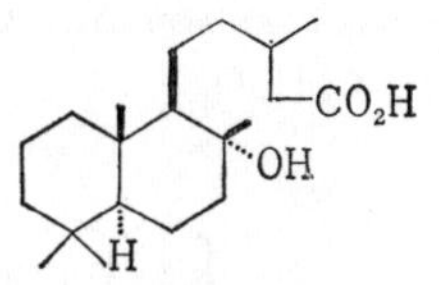

(LIX)

≪Occurrence≫   *Cistus labdaniferus* (Gum Labdanum)
Amorphous, $[\alpha]_D$ $-7°$ (CHCl$_3$), C$_{20}$H$_{36}$O$_3$
Methyllabdanolate acetate   mp. 84~84.5°, $[\alpha]_D$ $-29°$, C$_{23}$H$_{40}$O$_4$

(2) **Eperuic acid** (LX)[786][787]

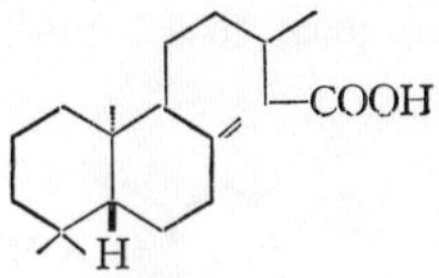

(LX)

≪Occurrence≫   *Eperua falcata* and other spp.
Oleyl ester   bp.   235~240°/0.002 mm, n_D 1.4945, $[\alpha]_D$ -7.1, C$_{38}$H$_{68}$O$_2$
Methyl epuruate   bp.   164°/0.4 mm, $[\alpha]_D$ $-28.2°$ (CHCl$_3$) n_D 1.4982, d 0.979, C$_{21}$H$_{36}$O$_2$

(3) **Isoeperuic acid** (LXI)[787]

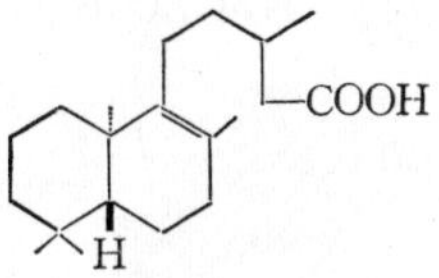

(LXI)

≪Occurrence≫   *Eperua falcata*
Methyl isoeperuate was separated by gas chromatography

(4) **Cativic acid** (LXII)[788]

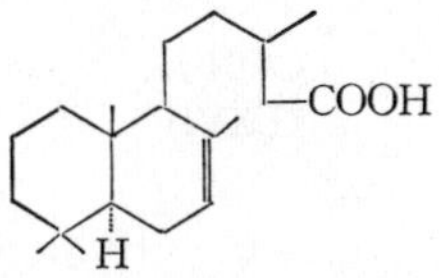

(LXII)

≪Occurrence≫   *Prioria copaifera* (Cativo Gum)
mp. 80~82°, $[\alpha]_D$ $-6.54°$ (EtOH), C$_{20}$H$_{34}$O$_2$

785) Cocker, J.D. *et al* : *Soc.*, **1956**, 4262.   786) King F.G., *et al* : *Soc.*, **1955**, 658.
787) Jones, G. *et al* : *ibid.*, **1963**, 430.   788) Zeiss H.H. *et al* : *J.A.C.S.*, **79**, 1201 (1957).

(5) **Agathendicarboxylic acid** (Agathic acid) (LXIII)[789]

(LXIII)

≪Occurrence≫　*Dammara australis* (*Agathis australis, A. dammara*)
mp. 203~204°, $[\alpha]_D$ +55~58° (EtOH), $C_{20}H_{30}O_4$

(6) **Alepterolic acid** (LXIV)[790]

(LXIV)

≪Occurrence≫　*Aleuritopteris argentea* FEE (ヒメウラジロ)
mp. 162.5~163.0, $[\alpha]_D$ ~45.5° (CHCl₃), $C_{20}H_{32}O_3$
IR ν_{max}^{KBr} cm⁻¹ : 3425, 1035 (OH), 3090, 1694, 1638, 1219 ($>$C=CH–COOH)

(7) **Copalic acid** (LXV)[791]

(LXV)

≪Occurrence≫　*Hymenaea courbaril* L. (*Leguminosae*) (Brazil copal)
bp. 160°/0.005 mm, $[\alpha]_D$ −6.9° (CHCl₃), $C_{20}H_{32}O_2$
Methyl copalate　bp. 160°/0.02 mm, $[\alpha]_D$ −11.4°, $C_{21}H_{34}O_2$
UV λ_{max}^{EtOH} mμ (log ε) : 225 (4.06)
IR λ_{max}^{CHCl3} μ : 5.80, 6.05, 11.20, 11.56
NMR shows a mixture with Δ^{7-8} isomer (ca 30%)

(8) **Communic acid** (LXVI)[792]

(LXVI)

≪Occurrence≫　*Juniperus communis* L.
Methyl communate　mp. 105~106°, $[\alpha]_D$ +48° (CHCl₃), $C_{21}H_{32}O_2$
IR ν_{max}^{CCl4} cm⁻¹ : 3110, 1615, 900 (–CH=CH₂), 1780, 886 ($>$C=CH₂), 1720 (ester)

789) Ruzicka, L. *et al* : *Helv. Chim. Acta*, **24**, 931, 1167 (1941), **26**, 2136 (1943), **31**, 2143 (1948).
790) Ageta, H. *et al* : *6th Symposium on the Chemistry of Natural Products*, p. 136 (Sapporo, 1962).
791) Djerassi, C. *et al* : *J. Org. Chem.*, **26**, 167 (1961).
792) Erdtman, H. *et al* : *Tetrahedron*, **16**, 255 (1962).

(9) Methyl sciadopate (LXVII)[793)794)]

《Occurrence》 *Sciadopytis verticillata* SIEB. et ZUCC.

mp. 108°, 108.5°, $[\alpha]_D$ −0.7°[793)], +0.36° (CHCl₃)[794)]

(LXVII)

UV $\lambda_{max}^{15\%EtOH}$ mμ (log ε)[794)] : 202 (4.20)

IR ν_{max}^{nujol} cm⁻¹: 3344 (OH), 1733 (ester), 1641, 896 (vinylidene), 816 $\left(>=<^H\right)$[793)] ; ν_{max}^{KBr} cm⁻¹: 3330, 1022,

1000 (OH), 3060, 1639, 834 $\left(>=<^H_H\right)$, 847 $\left(>=<^H\right)$, 1735, 1222, 1155 (ester)[794)]

NMR in CCl₄, 60 Mc, TMS, τ[794)] : 9.52 (17–Me), 8.83 (18–Me), 6.42 (−COOMe), 5.50, 5.16 ($>$C=CH₂),

4.46 t, J=6.0 $\left(>C=C<^H_{CH_2-}\right)$, 5.97 d, J=6.0 $\left(=C<^H_{CH_2OH}\right)$, 6.04 s $\left(=C<^C_{CH_2OH}\right)$

(10) Grindelic acid (XXXVIII)

See the item [D], (4)

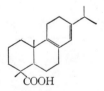

(XXXVIII)

(ii) *Abietic acid type*

(LXVIII) Abietic acid   (LXIX) Neoabietic acid   (LXX) Palustric acid

(LXXI) Levopimaric acid   (LXXII) Dehydroabietic acid   (LXXIII) Dihydroabietic acid

---

793) Sumimoto, M. *et al*: *7 th Symposium on the Chemistry of Natural Products*, p. 243 (Fukuoka, 1963).
794) Miyasaka, S: *ibid.*, p. 238 (1963); *Chem. Pharm. Bull.*, **12**, 744 (1964).

(LXXIV) Dihydropalustric acid   (LXXV) Tetrahydroabietic acid

Table 74 Mass Peaks[795] and Physical Constants of Resinic Acid Methylates (I)   (Genge, Karrer)

Resinic Acid	Str. No.	Formula	mp.	$[\alpha]_D$	pK$_{MCS}$*	Mass Peaks of Methylate		Occurrence	Ref.
						Parent	Base		
Abietic acid	(LXVIII)	$C_{20}H_{30}O_2$	170~174	−106~−122	7.93	316	256	*Pinus abies*, other *Pinus* sps.	[796], [797]
Neoabietic acid	(LXIX)	$C_{20}H_{30}O_2$	167~169	+159	7.94	316	316	*Pinus palstris* etc.	[798], [799]
Palustric acid	(LXX)	$C_{20}H_{30}O_2$	162~167	+71.6 (EtOH)		316	301	Pinegum resin *Pinus palstris*	[795], [808]
Levopimaric acid	(LXXI)	$C_{20}H_{30}O_2$	150~152	−276	7.90	316	146	*Pinus palstris* *P. maritima*	[800], [801]
Dehydroabietic acid	(LXXII)	$C_{20}H_{28}O_2$	173~173.5	+62 (EtOH)	7.92	314	299	*Pinus palstris*, *P. silvestris*	[802], [804]
Dihydroabietic acid	(LXXIII)	$C_{20}H_{32}O_2$	174~176	+108 (EtOH)		318	275	*Pinus palstris*	[803]
Dihydropalustric acid	(LXXIV)	$C_{20}H_{32}O_2$				318	243		[795]
Tetrahydroabietic acid	(LXXV)	$C_{20}H_{34}O_2$				320	277		[795]

* pK$_{MSC}$[805] : pK$_{methylcellosolve}$

ORD[806]

Enone ester (LXXVI) derived from abietic acid (LXVIII) is an antipode of that (LXXVII) from agathic acid (LXII).

(LXVIII)   (LXXVI)   (LXXVII)   (LXII)
Abietic acid   Agathic acid

← Antipode →

795) Genge, C. A: *Anal. Chem.*, **31**, 1750 (1959).
796) Ruzicka, L. *et al*: *Helv. Chim. Acta*, **24**, 504 (1941).
797) Burgstakler, A. W. *et al*: *J.A.C.S.*, **83**, 2587 (1961).
798) Harris, G. C. *et al*: *ibid.*, **70** 334, 339 (1948).
799) Gopinath, K. W. *et al*: *Helv. Chim. Acta*, **44**, 1040 (1961).
800) Ruzicka, L. *et al*: *ibid.*, **23**, 1346 (1940).   801) Klyne, W: *Soc.*, **1953**, 3072.
802) Stork, G. *et al*: *J.A.C.S.*, **78**, 250 (1956), **84**, 284 (1962).
803) Ruzicka, L. *et al*: *Helv. Chim. Acta*, **24**, 1389 (1941).   804) Bruun, H. H: *C. A.*, **49**, 12385 (1955).
805) Sommer, P. F. *et al*: *Helv. Chim. Acta*, **46**, 1734 (1963).
806) Tahara, A. *et al*: *7 th Symposium on the Chemistry of Natural Products*, p. 201 (Fukuoka, 1963);
Chem. Pharm. Bull., **11**, 1328 (1963).

(LXXVI) mp. 114~116°, $[\alpha]_D$ −48.9° (EtOH)

IR ν_{max}^{KBr} cm^{-1}: 1718, 1680, 1618

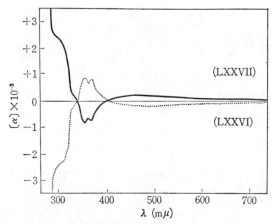

Fig. 138 ORD Curves of Enone ester (LXXVI) and (LXXVII) (Tahara *et al.*)

(10) Neoabietic acid (LXIX)[807)]

mp. 173~173.5°, $[\alpha]_D$ +161° (EtOH)

UV λ_{max}^{EtOH} mμ: 251; IR $\lambda_{max}^{CCl_4}$ μ: 5.82; λ_{max}^{nujol} μ: 3.10, 5.82, 5.93

(11) Palustric acid (LXX)[808)]

UV λ_{max} mμ: 265~266 (Fig. 139)

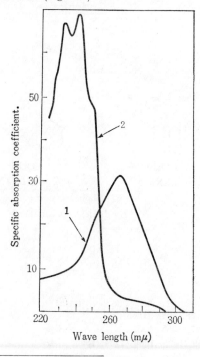

Fig. 139 UV Spectra of
(1) Palustric acid and
(2) Isomerized palustric acid (crude abietic acid) (Lawrence *et al.*)

807) Schuller, W. H. *et al.*: *J. A. C. S.*, **83**, 2563 (1961).
808) Lawrence, R. V. *et al.*: *J. A. C. S.*, **82**, 1734 (1960), **77**, 2823 (1955).

(LXX) $\xrightarrow{\text{HCl}}$ Abietic acid (LXVIII)

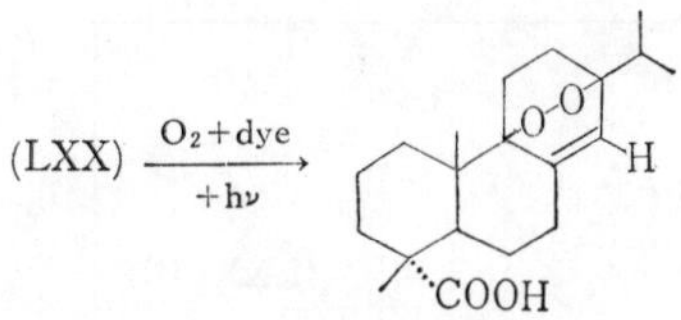

(LXX) $\xrightarrow[+h\nu]{O_2+dye}$

(LXXVIII)

7,13-Peroxido-$\Delta^{8(14)}$-dihydroabietic acid (LXXVIII)

mp. 111°/154°, $[\alpha]_D$ −73.9° (EtOH), $C_{20}H_{30}O_4 \cdot H_2O$

(12) **Levopimaric acid** (LXXI)[809]

UV See Fig. 139[808]

An example of reaction process to compare the stereochemistry of the derivatives by ORD is:

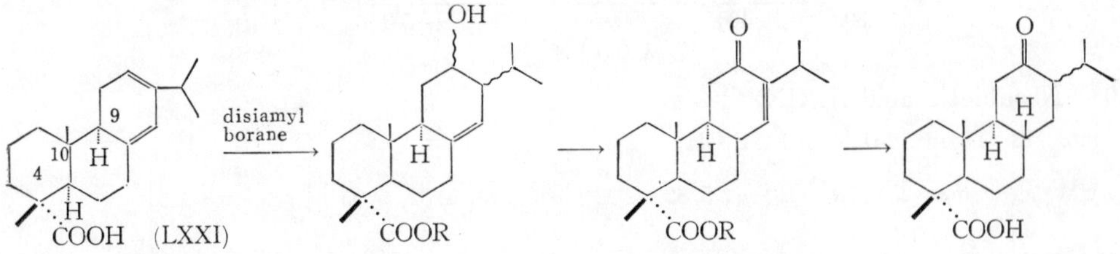

Diels–Alder's adduct[810]

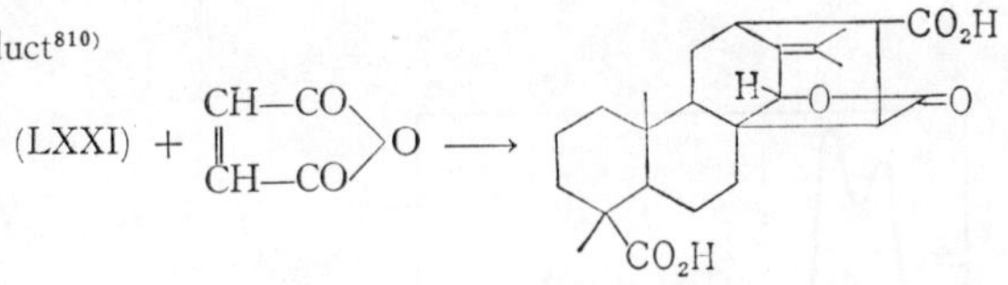

(LXXI) + $\underset{\text{CH—CO}}{\overset{\text{CH—CO}}{\|}}\!\!\!\!O \longrightarrow$

(iii) *Pimaric acid Type*

(LXXX) Pimaric acid
(Dextropimaric acid)

(LXXXI)

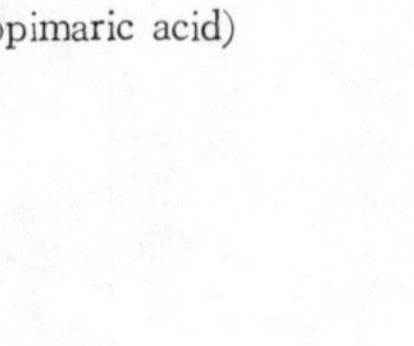

(LXXXII)

Isopimaric acid
(Isodextropimaric acid)

---

809) Dauben, W. G. *et al*: *J. Org. Chem.*, **28**, 1698 (1963).

810) Zalkow, L. H. *et al*: *J. Org. Chem.*, **27**, 3535 (1962).

811) Bose, A. K. *et al*: *Chem. & Ind.*, **1959**, 1628, **1960**, 1104, **1963**, 254.

812) Wenkert, E. *et al*: *J.A.C.S.*, **81**, 688 (1959).

813) Edwards, O. E. *et al*: *J. Org. Chem.*, **27**, 1930 (1962).

814) Ireland, R. E. *et al*: *ibid.*, **27**, 1931 (1962).

[G] Resinic acids

(LXXXIII) Dihydro-
pimaric acid

(LXXXIV) Tetrahydro-
pimaric acid

(LXXXV) (Desmethyl (VI))
Sandaracopimaric acid [715)805)815)816)]

(LXXXVI) 6 β–Hydroxy-
sandaracopimaric acid

(LXXXVII) Dihydrosanda–
racopimaric acid

Table. 75 Mass Peaks[795)] and Physicalic Constants
of Resinic acid methylesters (Ⅱ) (Genge, Karrer)

Resinic acid	Str. No.	Formula	mp.	$[\alpha]_D$	pK [805)]	Mass Peaks of Methyl ester		Occurrence	Ref.
						Parent	Base		
Pimaric acid	LXXX	$C_{20}H_{30}O_2$	217~219	+79 (EtOH)	7.90	316	180	*Pinus maritima P. palustris*	812) 818)
Isopimaric acid	LXXXII	$C_{20}H_{30}O_2$	162~164	0	7.98	316	287	*P. palustris, Dac –rydium biforme*	813) 814)
Dihydropimaric acid	LXXXIII	$C_{20}H_{32}O_2$	240~241	+17 (EtOH)		318	289		795)
Tetrahydropimaric acid	LXXXIV	$C_{20}H_{34}O_2$				320	291		795)
Sandaracopimaric acid	LXXXV	$C_{20}H_{30}O_2$	170~171	−19.7	7.94			*Cryptomeria japonica*	715) 805) 815) 816)
6 β–Hydroxysandara –copimaric acid	LXXXVI	$C_{20}H_{30}O_3$			7.85				805) 817)
Dihydrosandaraco –pimaric acid	LXXXVII	$C_{20}H_{32}O_2$			8.05				805) 817)

(13) Pimaric acid (Dextropimaric acid) (LXXX) and **Isopimaric acid** (Isodextropimaric acid)
(LXXXII)

(a) **Conformation of the derived Lactams**

815) Edwards, O. E. *et al*: *Canad. J. Chem.*, **38**, 663 (1960).
816) Galik, V. *et al*: *Chem. & Ind.*, **1960**, 722.
817) Edwards, O. E. *et al*: *Canad. J. Chem.*, **39**, 2543 (1961).
818) Edwards, O. E.: *Canad. J. Chem.*, **37**, 760 (1959); *Chem. & Ind.*, **1959**, 537.

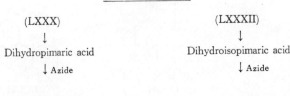

(LXXX)   (LXXXII)

↓   ↓

Dihydropimaric acid   Dihydroisopimaric acid

↓ Azide   ↓ Azide

Lactam (LXXXVIII)   Lactam (LXXXIX)

NMR 60 Mc, τ (LXXXVIII) 6.2 multiplet (six lines) (6–H), (LXXXIX) doublet (6–H)

(b) **Pimaric acid** (LXXX)

NMR See the item [A], (4), Table 72

(c) **Isopimaric acid** (LXXXI)[819]

≪Occurrence≫ *Cryptomeria japonica* D. DON (スギ), mp. 166~168°, $[\alpha]_D$ −20.0°, $C_{20}H_{32}O_2$

UV λ_{max} mμ (log ε): 212 (3.89)

IR ν_{max}^{nujol} cm^{-1}: 1818 w, 1681 s, 1634 w, 1524 w, 1412 m, 1348 w, 1319 w, 1280 s, 1250 sh, 1217 w, 1188 m, 1152 sh, 1145 w, 1022 w, 997 m, 990 m, 938 m, 907 m, 800 m, 733 w, 716 w

NMR See the item [A], (4), Table 72

[Note] Another structure (LXXXII) was presented for isopimaric acid[805)813)814]

(14) **Sandaracopimaric acid** (LXXXV)[715)805)816)820)821] (Cryptopimaric acid) (LXXXV)[821a]

≪Occurrence≫ *Cryptomeria japonica* D. DON.,   mp. 170~171°, $[\alpha]_D$ −19.7°, $C_{20}H_{30}O_2$

NMR See the item [A], (4), Table 72.

(iv) *Other Tricyclic Carboxylic acids*

(15) **Eriostemoic acid** (XC)[822]

≪Occurrence≫ *Eriostemon tomentellus, E. tryptomenoides, Galeznowia verrucosa*, $C_{20}H_{24}O_5$

CH_2–CH_2–COOH

OCH_3

(XC)

(16) **Podocarpic acid** (XCI)[823)~826]

≪Occurrence≫ *Podocarpus dacrydioides* A. RICH., *P. totara* G. BENN.,

mp. 187~193°, $[\alpha]_D$ +136° (EtOH), $C_{17}H_{22}O_3$, pK$_{MCS}$ 8.44, total synthesis[827]

819) Kondo, T. *et al*: *Agr. Biol. Chem.*, **23**, 233 (1959).

820) Galik, V. *et al*: *Tetrahedron*, **7**, 223 (1959).   821) Apsimon, J. W. *et al*: *Soc.*, **1961**, 752.

821a) Arya, V. P. *et al*: *Acta Chem. Scand.*, **15**, 682 (1961); *C. A.*, **56**, 1486 g (1962).

822) Duffield, A. M. *et al*: *Australian J. Chem.* **16**, 123 (1963); *Index Chem.*, **9**, 27064 (1963).

823) Todd, D. *et al*: *J. A. C. S.*, **64**, 928 (1942).   824). Haworth, R. D. *et al*: *Soc.*, **1946**, 633.

825) Briggs, L. H. *et al*: *Tetrahedron*, **7**, 270 (1959).   826) Cambie, R. C. *et al*: *ibid.*, **18**, 465 (1962).

827) Wenkert, E. *et al*: *J. A. C. S.*, **82**, 3229 (1960); **83**, 2320 (1961).

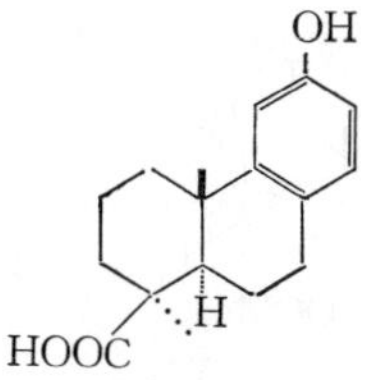

[G] Resinic acids

(XCI)

(17) Cassaic acid (XCII)[828]

≪Occurrence≫   *Erythrophleum quineense*

Cassaine (XCIII) + H_2O = Cassaic acid (XCII) + $C_4H_{11}ON$ (dimethylaminoethanol)

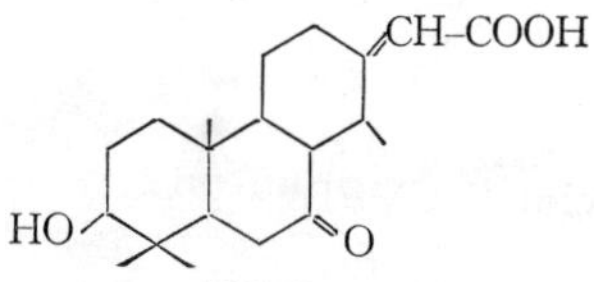

(XCII)

(a) **Cassaine** (XCIII)[829]

mp. 142.5°, $[\alpha]_D$ −113° (EtOH), $C_{24}H_{39}O_4N$

UV λ_{max}^{EtOH} mμ(log ε): 223 (4.26)[828]

Bisulfate[829]   mp. 305~306° (decomp.) $[\alpha]_D$ −86° (0.1 N H_2SO_4)

(b) **Cassaic acid** (XCII)[828]

mp. 205.5~207°, $[\alpha]_D$ −120° (EtOH), $C_{20}H_{30}O_4$

UV λ_{max}^{EtOH} mμ (log ε): 219 (4.1); IR λ_{max}^{nujol} μ: 2.85, 5.85, 5.97, 6.13

(18) Cassamic acid (XCIV)[830]

≪Occurrence≫   *Erythrophleum quineense*

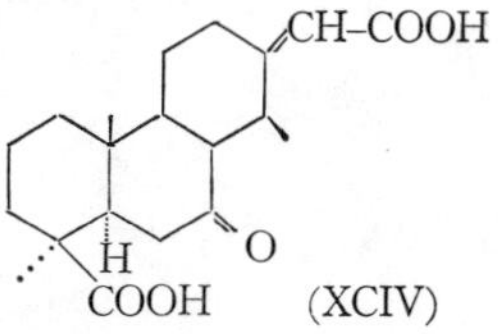

COOH   (XCIV)

Cassamine (XCV) + H_2O = Cassamic acid (XCIV) + $C_4H_{11}ON$ (dimethylamino ethanol)

(a) **Cassamine** (XCV)[829]

mp. 86~87°, $[\alpha]_D$ −56° (EtOH), $C_{25}H_{39}O_5N$

(b) **Cassamic acid** (XCIV)[830]

mp. 218~219°, $[\alpha]_D$ −70° (CHCl$_3$), $C_{21}H_{30}O_5$

(c) **7-Hydroxy-7-desoxocassamic acid** (XCVI)[830]

≪Occurrence≫   *Erythrophleum quineense*

(XCIV) $\xrightarrow{\text{NaBH}_4}$ (XCVI)

mp. 246~247°, $[\alpha]_D$ −50° (MeOH), $C_{21}H_{32}O_5$

828) Gensler, W. J. *et al*: *J. A. C. S.*, **81**, 5217 (1959).
829) Dalma, G. Manske, R. H. F: The Alkaloids Vol. IV, p. 266~270 (Academic Press, 1954).
830) Chapman, G. T. *et al*: *Soc.*, **1963**, 4010.

CH–COOH

14

7

H
COOMe₃

(XCVI)

UV λ_{max} mμ (log ε): 222 (4.18)

IR ν_{max}^{KBr} cm⁻¹: 3460 (OH), 1736 (ester CO), 1701 (acid CO), 1647 (conj. C=C), 1153 (C–O–C of ester)

Arya et al[831] reported that a reduction product of cassamine (XCV) shows almost same physicalic constants with those of (XCVI)[830], but it's configurations at 7 β–OH and 14–α–OH differ from (XCVI).

(v) *Tetracyclic diterpenoids*

(19) **Steviol** (XCVII) and **Isosteviol** (XCVIII)[832]~[838]

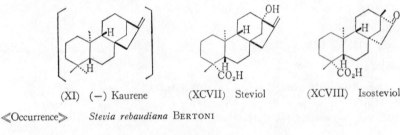

(XI) (−) Kaurene   (XCVII) Steviol   (XCVIII) Isosteviol

《Occurrence》 *Stevia rebaudiana* BERTONI

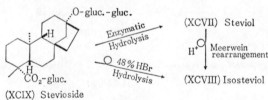

(XCIX) Stevioside

(a) **Stevioside** (XCIX)
mp. 196~198°, [α]$_D$ −39.3° (H₂O), C₃₈H₆₀O₈

(b) **Steviol** (XCVII)
mp. 212~213°, [α]$_D$ −93.6° (EtOH), C₂₀H₃₀O₃

(c) **Isosteviol** (XCVIII)
mp. 226~228°, C₂₀H₃₀O₃

831) Arya, V.P. *et al*: *Helv. Chim. Acta*, **44**, 1650 (1961).
832) Wood, H.B. *et al*: *J. Org. Chem.*, **20**, 875 (1955).   833) Mosettig, E. *et al*: *ibid.*, **20**, 884 (1955).
834) Mosettig, E. *et al*: *J.A.C.S.*, **85**, 2305 (1963).   835) Dolder, F. *et al*: *ibid.*, **82**, 246 (1960).
836) Cross, B.E. *et al*: *Proc. Chem. Soc.*, **1963**, 17.   837) Mosettig, E. *et al*: *J.A.C.S.*, **83**, 3163 (1961).
838) Djerassi, C. *et al*: *ibid.*, **83**, 3720 (1961).

〔H〕 Diterpenoid Lactones

Furanoterpenoid lactones are described in the item 〔I〕.

(i) *Tricyclic Lactones*

(1) **Andrographolide** (C)[839]~[842]

(Former Str.)[839]

(C) (Revised Str.)[840]

《Occurrence》 *Andrographis paniculata*

mp. 230~231°, $[\alpha]_D$ −127°, $C_{20}H_{30}O_5$

UV λ_{max}^{EtOH} mμ (ε): 223 (12300)

IR ν_{max}^{KBr} cm^{-1}: 3448, 3390~3279 (OH), 1828, 1647, 906 (C=CH$_2$), 1727 ($\alpha\beta$-unsatur. γ-lactone), 1672
(conj. C=C)

NMR of Triacetate in CDCl$_3$, TMS, δ: 0.75, 1.02 $\left(2 \right.$ $\left.\rangle C-CH_3\right)$, 2.03 (6), 2.10 (3) (3 COCH$_3$), 5.91

$\left(\rangle=\left\langle\begin{smallmatrix}H\\OAc\end{smallmatrix}\right.\right)$, 6.97 t, J=6.5 c/s (exocyclic β-proton, $-CH_2-CH=C\langle$)

NMR of 2-butenolide (CI), a model compound of the former structure: δ=7.63, J=1.7 c/s (endocyclic β-proton).

(CI)

axial-CH$_2$OH at C$_4$ position

NMR of ketoaldehydes (CII) and (CIII), derived from andrographolide show δ=9.64 and 9.73 respectively, which assigned for *axial*-CHO from comparison of the shifts of following compounds.

axial-voacapenal (9.77), isodihydroiresin (9.73), aldehyde (CIV)

equatorial-vinhital (9.23), (CV) derived from hederagenin (9.35).

839) Cava, M. P. *et al*: *Tetrahedron*, **18**, 397 (1962).
840) Cava, M. P. *et al*: *Chem. & Ind.*, **1963**, 167, 495.
841) Chan, W. R. *et al*: *Chem. & Ind.*, **1959**, 851, **1960**, 22.
842) Arya, V. P. *et al*: *J. Sci. Ind. Res.* **21 B**, 281 (1962); *C. A.*, **57**, 11248 i (1962).

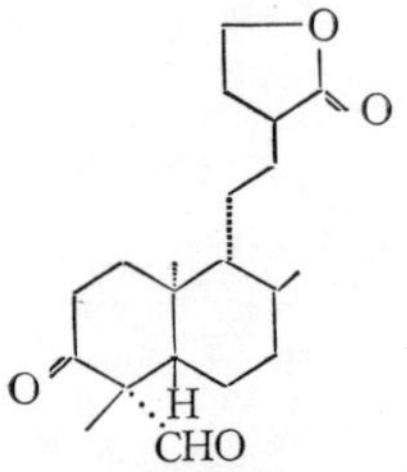

(CII) 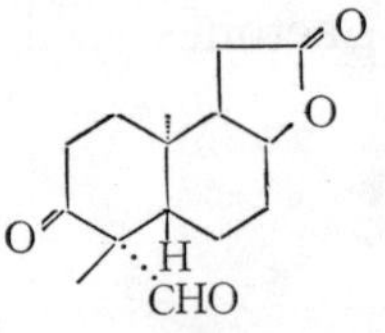 (CIII) 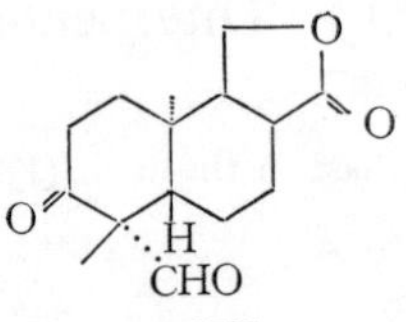 (CIV) 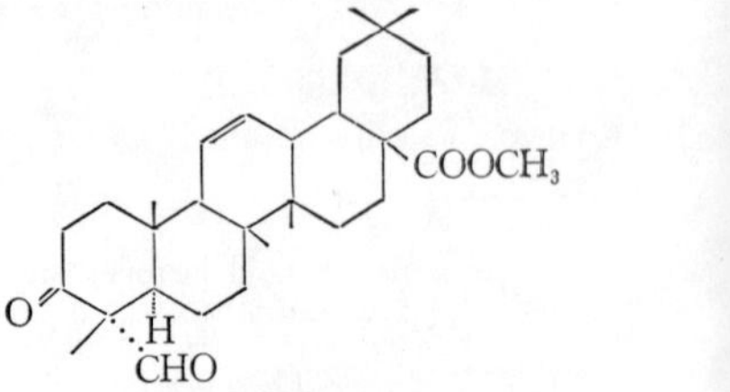 (CV)

axial–OH at C_3 Position

Andrographo–lide (C) $\longrightarrow$ (CVI) $\xrightarrow[+ZnCl_2]{C_6H_5CHO}$ (CVII)

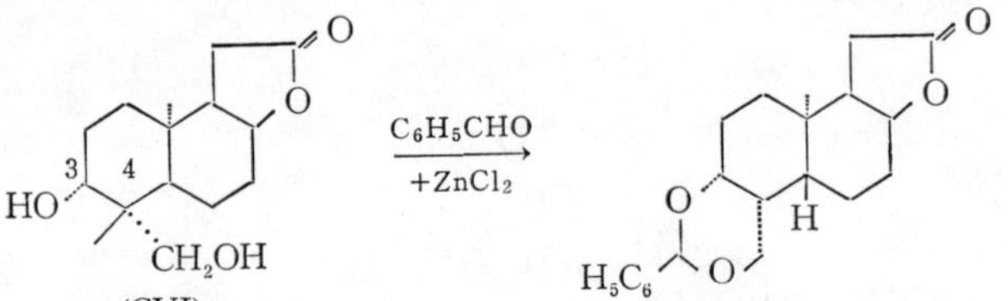

(ii) *Tetracyclic Lactones*

(2) Rosenonolactone (CVIII)[842) 844) 844a)]

《Occurrence》 A metabolite of *Trichothecium roseum* LINK.
mp. 208°, $[\alpha]_D$ −116° (CHCl₃), $C_{20}H_{28}O_3$
UV $\lambda_{max} m\mu$ (log ε:) 289 (1.6), 216 (1.98)

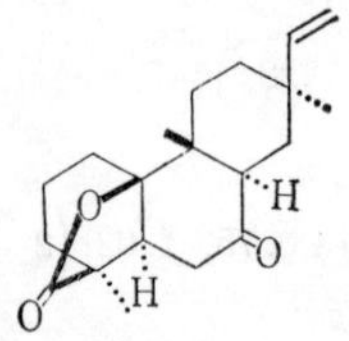

(CVIII)

(3) α-Levantenolide (CIX) and **β-Levantenolide** (CX)[845)]

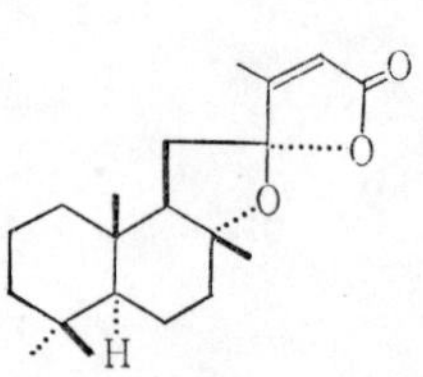

(CIX) 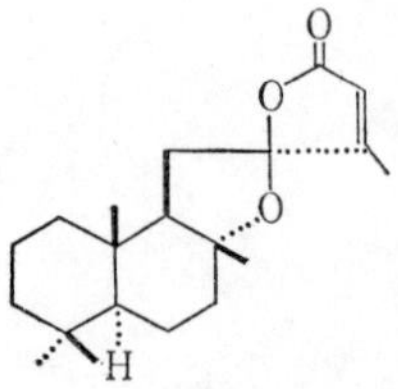 (CX)

《Occurrence》 *Nicotiana tabacum* (Turkish leaves)
(a) α–Levantenolide (CIX)
mp. 210°, $[\alpha]_D$ +60.4° (CHCl₃), $C_{20}H_{30}O_3$

843) Robertson, A. *at al*: *Soc.*, **1958**, 1799, 1807.
844) Whalley, W. B. *et al*: *J.A.C.S.*, **81**, 5520 (1959).
844a) Sim, G. A. *et al*: *Proc. Chem. Soc.*, **1964**, 19.
845) Giles, J. A. *et al*: *Tetrahedron*, **14**, 246 (1961).

IR λ_{max} μ: 12.1 $\left(\begin{smallmatrix}R\\R\end{smallmatrix}\!\!>\!\!C=CHR\right)$, 5.75 ($\alpha\beta$-unsatur. lactone)

NMR one $>\!\!=\!\!<_H$, five CH_3 $\left(\text{including one } \begin{smallmatrix}H_3C\end{smallmatrix}\!\!>\!\!C=CH-CO\right)$

(b) β-**Levantenolide** (CX)

mp. 208~209°, $[\alpha]_D$ −59.6° (CHCl$_3$), $C_{20}H_{30}O_3$

(4) Picrosalvin (CXI)[846]

《Occurrence》 *Salvia officinalis* L., mp. 221~226°, $C_{20}H_{26}O_4$

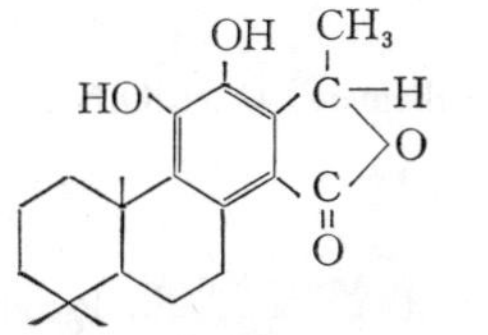

(CXI)

UV λ_{max} mμ (log ε): 285 (3.33)

IR ν_{max} cm^{-1}: 3521, 3289 (no assoc. 2-OH)

(CXI) $\xrightarrow{Br}$ Quinone

UV λ_{max} mμ (log ε): 425 (3.34), IR $\nu_{max}^{CCl_4}$ cm^{-1}; 1712, 1742 ($>\!\!C=O$)

(5) Quassin (CXII) and **related Compounds**

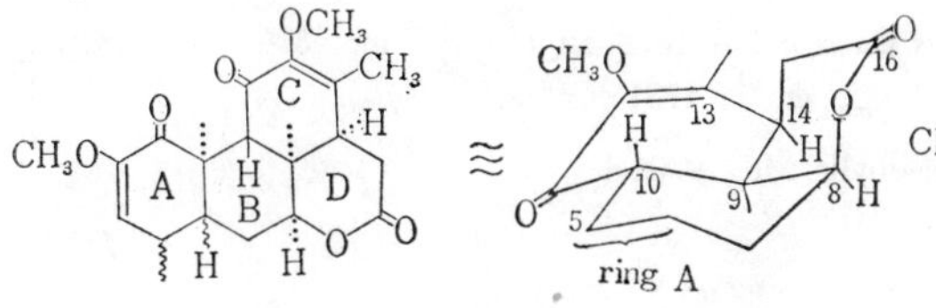

(CXII) Quassin[847]

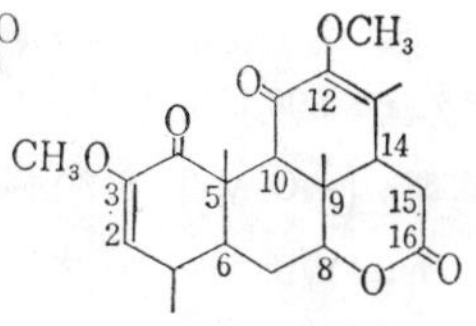

(CXIII) Alloquassin[847]

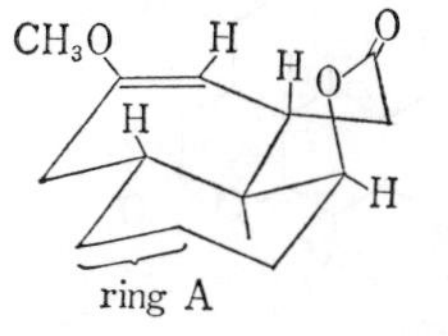

(CXIV) Isoquassin[847] 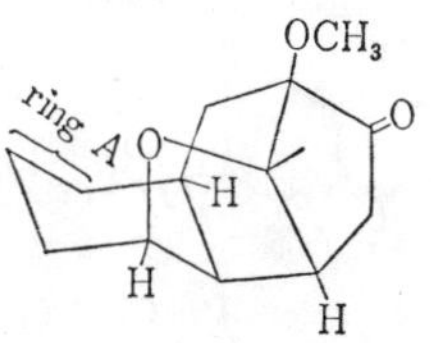   (CXV) Pseudoquassin[847] 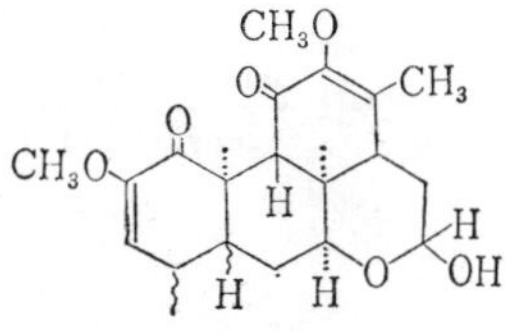   (CXVI) Neoquassin[848]

《Occurrence》 *Quassia amara* L., *Picrasma quassioides* BENN.

(a) Quassin (CXII)

mp. 222°, $[\alpha]_{355m\mu}$ −690°,[847] $C_{22}H_{28}O_6$

UV λ_{max}^{EtOH} mμ (ε)[847]: 255 (11650)

IR ν_{max}^{KBr} cm^{-1} [847]: 1745 (lactone), 1702, 1688 (ketone), 1636 (C=C)

846) Brieskorn, C. H. *et al*: *Ber.*, **95**, 3034 (1962).
847) Valenta, Z. *et al*: *Tetrahedron*, **18**, 1433 (1962).
848) Carman, R. M. *et al*: *Tetr. Letters*, **1961** 317.

NMR[849]   56.4 Mc, TMS, τ: 4.71 d $(J=2)$ $\left(>\text{C=C}<^\text{H}\right)$, 5.70 t (8-H), 6.38, 6.44 (CH$_3$O), 8.15

$\left(>\text{C=C}<^{\text{CH}_3}\right)$, 8.45, 8.79 $\left(>\text{C-CH}_3\right)$, 8.86 d $(J=5.6)$ $(>\text{CH-CH}_3)$

(b) Allo-quassin (CXIII)[850]

Quassin (CXII) $\longrightarrow$ Alloquassinolic acid, mp. 240°, C$_{22}$H$_{30}$O$_7$ $\xrightarrow[\text{Ac}_2\text{O+NaOAc}]{\text{heating or}}$ Allo-quassin

mp. 267°, [M]$_D$ +705° (CHCl$_3$),   C$_{22}$H$_{28}$O$_6$

UV $\lambda^{\text{EtOH}}_{\text{max}}$ mμ (ε): 262 (12600)

(c) Neo-quassin (CXIV)[850]

mp. 228°,   C$_{22}$H$_{30}$O$_6$, p–nitrophenylcarbamate C$_{29}$H$_{34}$O$_9$N$_2$, mp. 180°, α–O–methylate,

mp. 174°, β–O–methylate, mp. 213°

(d) Pseudoquassin (CXV)

Quassin (CXII) $\longrightarrow$ Alloquassinolic acid $\xrightarrow{\text{Ac}_2\text{O+NaOAc}}$ Pseudoquassin[847]

(6) Chaparrin (CXVII) and **Glaucarubol** (CXVIII)[851] [852]

R=CH$_2$OH, R'=H (CXVII) Chaparrin

R=CH$_2$OH, R'=OH (CXVIII) Glaucarubol

(a) Chaparrin (CXVII)

《Occurrence》   *Simaroubaceous Castela* NICHOLSONI HOOK, mp. 306~308°, C$_{20}$H$_{28}$O$_7$

UV λ_{max} mμ (ε): 308 (38),   IR ν_{max} cm^{-1}: 1731 (C=O)

NMR τ: 8.58 s $\left(>\text{C-Me}\right)$, 8.30 (vinylic Me), 9.01 d

Triacetate   mp. 191°

Tetraacetate   NMR: *sec.* Me (C$_{13}$), *tert.* Me (C$_{10}$), allylic Me (C$_4$)

R=CH$_2$OH, R'=H

(CXIX) Chaparrol   (CXX) Anhydro-chaparrin

NMR (CXIX) τ: 8.91 d, $J=7$ (3 H of sec. Me), 7.72 s, 7.80 s, (2 arom. Me), 3.10 s (vinyl 2 H)

(b) Glaucarubol (CXVIII)

《Occurrence》   *Simaroubaceous glauca*,   C$_{20}$H$_{28}$O$_8$

---

849) Valenta, Z. *et al*: *Tetrahedron*, **15**, 100 (1961).   850) Hanson, K. R. *et al*: *Soc.*, **1954**, 4238.
851) Geissman, T. A. *et al*: *Tetr. Letters*, **23**, 1083 (1962).
852) De Mayo, *P. et al*: *ibid.*, **23**, 1089 (1962).

(7) 7-Hydroxykaurenolide (CXXI)[853]

≪Occurrence≫   *Gibberella fujikuroi*

mp. 187~188°, $[\alpha]_D$ −25°, $C_{20}H_{28}O_3$

(CXXI)

IR $\nu_{max}^{CHCl_3}$ cm⁻¹: 3595, 3495 (OH), 1765 (γ–lactone), 1656, 888 (exocyclic=CH₂)

NMR in CHCl₃, 40 or 60 Mc, τ: 9.13 s, 8.72 s $\left(\!\!>\!\!C\text{–Me}\right)$.

(8) **Picrotoxin** (CXXII), **Picrotoxinin** (CXXIII) and **Picrotin** (CXXIV)[854~857)860)] [Note]

Picrotoxin (CXXII) (a mixture) ⟶ Picrotoxinin (CXXIII) + Picrotin (CXXIV)

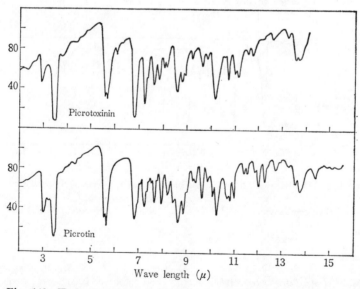

(CXXIII) Picrotoxinin[854)860)]   (CXXIV) Picrotin[855)]

≪Occurrence≫   *Anamirta paniculata* COLEB. (*A. cocculus* WIGHT et ARN)

(a) Picrotoxinin (CXXIII)

mp. 206~207°, $C_{15}H_{16}O_6$, IR[856)]: See Fig. 140

Fig. 140 IR Spectra of No. 1, Picrotoxinin and No. 2 Picrotin (Conroy)

853) Cross, B. E. *et al*: *Soc.*, **1963**, 2944.   854) Craven, B. M. *et al*: *Tetr. Letters*, **1960** 21.
855) Holker, J. S. E. *et al*: *Soc.*, **1958**, 2987.   856) Conroy, H: *J. A. C. S.*, **74**, 491 (1952).

[Note] The items (8) picrotoxin, picrotoxinin and picrotin should be replaced in the chapter-Sesquiterpenes.

(b) **Picrotin** (CXXIV)

    mp. 248~253°, $C_{15}H_{18}O_7$ IR[856] : See Fig. 140

$$(CXXIV) \longleftarrow \xrightarrow[855)]{POCl_3}$$

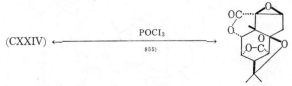

        (CXXV) Anhydropicrotin[855]

(c) **Anhydropicrotin** (CXXV)

    mp. 322~324° (decomp.) $[\alpha]_D$ −99° (AcOH), $C_{15}H_{16}O_6$

    IR ν_{max}^{nujol} cm^{-1}: 1783, 1767 sh.

(9) **Tutin** (CXXVI) and **Coriamyrtin** (CXXVII)[859)~864)] [Note]

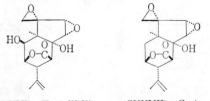

    (CXXVI) Tutin[859)860)]   (CXXVII) Coriamyrtin[861]

(a) **Tutin** (CXXVI)

    ≪Occurrence≫ *Coriaria* spp. of New Zealand, *C. japonica* mp. 212~213°, $C_{15}H_{18}O_6$

(b) **Coriamyrtin** (CXXVII)[861)~863)]

    ≪Occurrence≫ *Coriaria japonica* A. GRAY (ドクウツギ)

  mp. 229~230°, $C_{15}H_{18}O_5$, IR: See Fig. 141.

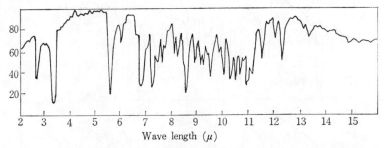

Wave length (μ)

Fig. 141 IR Spectrum of Coriamyrtin in nujol   (Kariyone *et al.*)

(CXXVII) $\longrightarrow$ Dihydrocoriamyrcin $C_{15}H_{20}O_3$ $\xrightarrow{Na_2CO_3}$ Isohydrocoriamyrcin (CXXVIII)

(CXXVII) $\xrightarrow{Na_2CO_3}$ Isocoriamyrcin (CXXIX)

---

857) Holker, J.S.E. *et al*: *Soc.*, **1957**, 3746.   858) Conroy, H: *J.A.C.S.*, **79**, 5550 (1957).

859) Mackay, M.F. *et al*: *Tetr. Letters*, **1963**, 1399.   860) Craven, B.M. *et al*: *Nature*, **1963**, 1193.

861) Okuda, T. *et al*: *Tetr. Letters*, **1964**, 439.

862) Kariyone, T. *et al*: **50**, 106, 659 (1930), **51**, 988 (1931), **54**, 203 (1934), **57**, 800 (1937), **63**, 510 (1942), **71**, 924 (1951), **73**, 925, 928 (1953).

863) Okuda, T: *Pharm. Bull.*, **3**, 185 (1954).   864) Okuda, T. *et al*: unpublished.

[Note] The item (9) tutin and coriamyrtin should be replaced in the chapter-Sesquiterpenes.

(c) Isohydrocoriamyrcin (CXXVIII)[861]

UV λ_{max}^{H2O} mμ (log ε): 232 (3.97)

IR ν_{max}^{KBr} cm^{-1}: 1688

(CXXVIII)[861]   (CXXIX)[862]

NMR in CHCl$_3$, τ: 0.40 s, 3.35 d, $J=3$ c/s (OHC–C=CH–CH–).

(10) Enmein (CXXX)[865)~871)]

≪Occurrence≫   *Isodon tricharpus* KUDO (クロバナヒキオコシ)

mp. 297~299° (decomp.) [α]$_D$ −156° (acetone)[868], −131° [867],   C$_{20}$H$_{26}$O$_6$

(CXXX) Enmein[870)871)]

UV λ_{max}^{EtOH} mμ (ε)[865]: 232.5 (7100), 256 sh (400), 300 (310); IR λ_{max}^{nujol} μ[865)869]: 2.89 (OH), 5.725

(γ or δ lactone), 5.88 (ketone ?), 6.07, 10.58, 10.66 (>C=CH$_2$)

IR ν_{max} cm^{-1} [867)868]: 3460 (OH), 1750, 1700

NMR[638] in CDCl$_3$, 60 Mc, TMS, τ:

Acetylenmein (Enmeindiacetate)

8.94 s (gem Me$_2$), 8.8 sh, 5.93 s $\left(-\overset{|}{\underset{|}{C}}-\overset{18}{C}H_2-O\right)$, 5.1 t (?), 5.4 m $\left(-CH_2-\overset{H}{\underset{1}{\overset{\diagup}{C}}}\overset{O-}{\underset{10}{\overset{|}{C}}}-\right)$,

3.85 s $\left(_H>\overset{6}{C}<\overset{O-}{OAc}\right)$, 3.92 s, 4.47 s (>$\overset{16}{C}$=CH$_2$)

Dihydroenmeindiacetate

8.94 s (gem Me$_2$), 8.86 d, $J=9$ c/s (>CH–Me), 5.94 s $\left(-\overset{|}{\underset{|}{C}}-\overset{18}{C}H_2-O\right)$, 5.11 t (?), 5.34 q (?)

$\left(-CH_2-\overset{H}{\overset{\diagup}{C}}\overset{O-}{\underset{|}{C}}-\right)$, 3.84 s$\left(_H>C<\overset{O-}{OAc}\right)$

865)   Ikeda, T. *et al*: *Y. Z.*, **78**, 1128 (1958).

866)   Takahashi, M. *et al*: *ibid.*, **80**, 594, 696 (1960).

867)   Kanatomo, S. *et al*: *ibid.*, **81**, 1049, 1807, 1437 (1961).

868)   Kubota, T. *et al*: *Bull. Chem. Soc. Japan*, **34**, 1737 (1961); *J. Chem. Soc. Japan*, **84**, 353 (1963).

869)   Kubota, T. *et al*: 5 th Symposium on the Chemistry of Natural Products, p. 13—5 (Sendai, 1961).

870)   Kubota, T. Ikeda, T. Uyeo, S. Fujita, T. Kosuge, T. Okamoto, T. *et al*: *7 th Symposium on the Chemistry of Natural Products*, p. 207 (Fukuoka, 1963).

871)   Natsume, M. *et al*: *ibid.*, p. 213 (1963).

Dihydroenmein   mp. 274~276° (decomp.) 282° (decomp.) negative Cotton Effect, $C_{20}H_{28}O_6$

UV λ_{max}^{EtOH} mμ (ε):$^{867)}$ 295 (57)

IR ν_{max} cm^{-1}: 1755

Dihydroenmeinmonoacetate$^{870)}$ mp. 238~240°, $C_{22}H_{30}O_7$

IR ν_{max} cm^{-1}: 1762 (15–C=O), 1720 (7–C=O), 1730 (3–OAc)

[I]   Furano-terpenoids and Tetranortriterpenoids

(1) Followed compound, accompanying furano ring have already been mentioned.

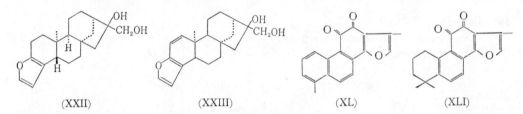

(XXII)   (XXIII)   (XL)   (XLI)

See the items:
(XXII)   Cafestrol   ⟶ [B], (9)
(XXIII)   Kahweol   ⟶ [B], (10)
(XL)   Tanshinone I   ⟶ [E], (3)
(XLI)   Tanshinone II   ⟶ [E], (3)

(2) Mass Spectra$^{872)}$

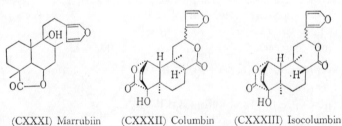

(CXXXI) Marrubiin   (CXXXII) Columbin   (CXXXIII) Isocolumbin

Mass peaks m/e and their relative intensities are described. Relative intensities are indicated as m (10~20%), s (>20%), when the intensity of dominant peak (Gothic letter) is 100%. Weak peaks, having less than 10% of intensity are negrected.

(a) Marrubiin (CXXXI)   28 s, 41 m, 43 m, 44 m, 55 m, 57 m, 67 m, 69 m, 71 m, 79 m, 81 s, 82 s. 83 s, 91 m, 93 s, 94 s, 95 s, 96 s, 97 s, 105 m, 107 s, 108 m, 110 m, 111 m, 119 m, 121 s, 123 s, 133 m, 134 s, 135 s, 136 s, 137 s, 139 m, 147 m, 149 s, 150 s, 151 s, 152 s, 153 m, 163 s, 164 s, 165 s, 173 m, 175 m, 177 s, 181 s, 189 s, 190 s, 191 s, 205 m, 209 s, 210 m, 219 m, 220 s, 237 m, 246 w

(b) Columbin (CXXXII)   28 m, 39 m, 41 s, 43 s, 44 s, 53 w, 55 w, 67 m, 69 m, 77 m, 79 s, 81 s, 82 s, 84 s, 91 s, 92 m, 93 s, **94**, 95 s, 97 s, 105 s, 106 m, 107 s, 108 s, 109 s, 121 s, 131 w, 133 w, 137 w, 149 s, 153 w, 161 m, 178 w, 204 w, 231 m, 259 w, 275 w, 313 m

---

872) Reed, R. I *et al*: *Soc.*, **1963**, 5933.

(c) **Isocolumbin** (XXXIII)   28 s, 39 m, 41 s, 43 s, 44 s, 53 m, 55 s, 67 m, 68 s, 69 s, 77 s, 79 s, 80 m, 81 s, 82 s, 84 m, 91 s, 92 m, 93 s, **94**, 95 s, 97 s, 105 m, 106 m, 108 s, 109 s, 110 s, 112 s, 117 m, 119 s, 120 m, 121 s, 122 s, 131 s, 133 s, 134 m, 135 m, 145 m, 147 s, 148 m, 153 m, 160 m, 161 m, 188 m, 205 m, 313 m.

(3) Cedrelone (CXXXIV) and **Anthothecol** (CXXXV)

R=R′=H   Cedrelone (CXXXIV)

R=OAc, R′=H   Anthothecol (CXXXV)

(a) **Cedrelone** (CXXXIV)[873)883)]

  ≪Occurrence≫   *Cedrela toona* Roxb.

mp. 209~214°, [α]_D −64.5° (CHCl_3), $C_{26}H_{30}O_5$

UV λ_{max}^{EtOH} mμ (ε): 217 (11800), 279 (9100), 327 (5530)

IR $\nu_{max}^{CHCl_3}$ cm⁻¹: 3425 (OH), 1678, 1685 sh(αβ–unsatur. C=O); ν_{max}^{nujol} cm⁻¹: 3130, 1505, 878 (furan ring)

(b) **Anthothecol** (CXXXV)[874)874a)]

  ≪Occurrence≫   *Khaya anthotheca* C. DC. (*Meliaceae*), $C_{28}H_{32}O_7$

UV λ_{max}^{MeOH} mμ (ε): 219 (13000), 281 (11000); $\lambda_{max}^{MeOH \cdot NaOH}$ mμ (ε): 224 (8320), 326 (4600)

IR ν_{max}^{nujol} cm⁻¹: 3400, 1740, 1670, 1655, 1620, 1504, 1235, 877

Acetate   mp. 234°.

(4) Limonin (Obaculactone) (CXXXVI), **Obacunone** (CXXXVIII) and **Nomilin** (CXXXVII)[874)875)]

(a) **Limonin** (CXXXVI) (Obaculactone, evodine, dictamnolactone)

  ≪Occurrence≫   Valencia orange (*Citrus* spp.), *Phellodendron amurense* Rupr. (オウバク), *Evodia rutae-carpa* Hook f. & Thoms., *Dictamnus albus* L.

mp. 293~294°, 298°, [α]_D −128°, (acetone), $C_{26}H_{30}O_8$

UV: See Fig. 142, λ_{max}^{EtOH} mμ (ε): 207 (6300)

IR ν_{max}^{nujol} cm⁻¹ [876)]: 1760, 1710, 1608, 1542, 1514; ν_{max}^{nujol} cm⁻¹ [875)]: 3124, 1603, 876 (furan), 1755~1760 (δ-lactones), 1706 (ketone)

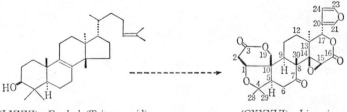

(LXXXI) Euphol (Triterpenoid)   (CXXXVI) Limonin (Obaculactone)

873) Hodges, R. *et al*: *Soc.*, **1963**, 2515.   874) Taylor, D. A. H. *et al*: *ibid.*, **1963**, 983.

874) Kubota, T *et al*: *4 th Symposium on the Chemistry of Natural Products*, p. 134 (1960); *Tetr. Letters*, **1960**, 1

874a) Taylor, D. A. H: *Soc.*, **1963**, 983.   875) Barton, D. H. R. *et al*: *Soc.*, **1961**, 255.

(LXXXI)   (LXXXVI)

(CXXXVII) Nomilin   (CXXXVIII) Obacunone
                                                        (Casimirolide)

Limonin (CXXXVI) ———HI———→ Deoxylimonin (CXXXIX)

Deoxylimonin (CXXXIX)[876]   mp. 331~336°, $[\alpha]_D$ −39° (CHCl₃)

UV λ_{max}^{EtOH} mμ (ε): 214 (17000)

IR ν_{max}^{nujol} cm⁻¹: 1747 (δ–lactone), 1714 (αβ–unsatur. δ–lactone), 1699 (ketone)

(CXXXIX)

(**b**) **Obacunone** (Casimirolide) (CXXXVIII)[877)~879)]

    ≪Occurrence≫   *Phellodendron amurense, Casimiroa edulis* (*Rutaceae*)

mp. 229~230°, $[\alpha]_D$ −50.2° (CHCl₃), $C_{26}H_{30}O_7$[877)]

UV   See Fig. 143 λ_{max} mμ (ε): 213 (14000)[874)]

IR λ_{max} μ: 3.19, 5.80, 5.90, 11.44[874)]; λ_{max}^{KBr} μ: 2.87, 3.31, 5.78, 5.88, 6.68, 6.85, 7.20, 7.45, 7.63,

    7.83, 8.17, 8.62, 8.95, 9.35, 9.40, 9.75, 10.16, 10.24, 10.92, 11.46, 12.17, 12.48 [881)]

(**c**) **Nomilin** (CXXXVII)

    ≪Occurrence≫   Orange and lemons (*Citrus* spp.)

mp. 278~279°, $[\alpha]_D$ −95.7° (acetone), $C_{28}H_{32}O_9$[880)]

UV   See Fig. 142

    Obacunone (CXXXVIII) ⎫ mild alkali   Obacunoic acid

    Nomilin (CXXXVII) ⎬————————→   (CXL)

(**d**) **Obacunoic acid** (CXL)[875)]

    mp. 205~208°, $[\alpha]_D$ −99° (acetone), $C_{26}H_{32}O_8$

876) Jeger, O. *et al*: *Helv. Chim. Acta*, **40**, 1420 (1957).
877) Emerson, O. H: *J.A.C.S.*, **73**, 2621 (1951).   878) Kubota, T. *et al*: *Chem. & Ind.*, **1957**, 1298.
879) Geissman, T. A. *et al*: *J. Org. Chem.*, **23**, 596 (1958).
880) Emerson, O. H: *J.A.C.S.*, **70**, 545 (1948).
881) Sondheimer, F. *et al*: *J. Org. Chem.*, **24**, 870 (1959).

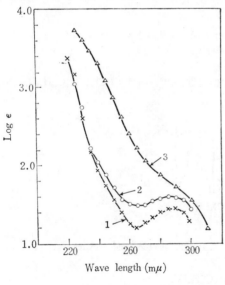

(CXL)

UV $\lambda_{\max} m\mu$ (ε): 204~205 (18000)

IR $\nu_{\max} cm^{-1}$: 3220 (OH), 2800~3400 (COOH), 1745 (δ–lactone), 1710 (cyclohexanone and C=C–COOH), 1621 (C=C).

NMR of methylate τ: 8.88, 8.97, 9.01, 9.04 $\left(\text{four} \gtrdot\text{C–Me}\right)$, 8.62 $\left(\overset{13}{\gtrdot}\text{C–Me}\right)$, 4.24, 4.49 ($\alpha$–H),

3.80, 4.08 (β–H) J=13.4 c/s (cis–β–subst. acrylic ester), 2.51, 2H(α–H), 3.66 (β–H) (furan ring), 4.$\overset{13}{63}$ ($\gtrdot C_{17}$H–O–)

Fig. 142 UV Spectra of (1) Limonin, (2) Nomilin and (3) Nomilic acid in EtOH (Emerson)

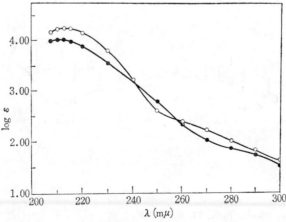

Fig. 143 UV Spectra of –o–o–o– Obacunone and –•–•–•– Obacunoic acid in MeOH   (Emerson)

(5) Evodol (CXLI)[882]

≪Occurrence≫   *Evodia rutaecarpa*, autoxydation of limonin (CXXXVI) → (CXLI)

mp. 281～282°, $[\alpha]_D$ −199° (acetone), $C_{26}H_{28}O_9$

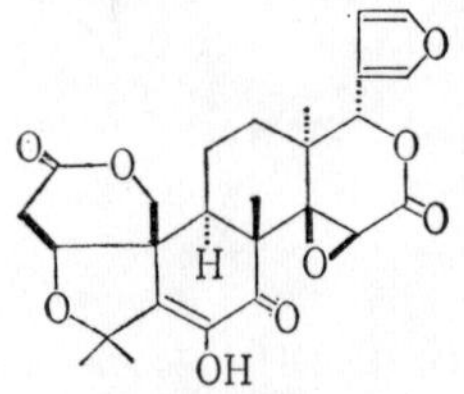

(CXLI)

UV λ_{max}^{EtOH} mμ (ε): 280 (9500); $\lambda_{max}^{0.1N-NaOH}$ mμ (ε): 340 (5500)

IR ν_{max}^{nujol} cm⁻¹: 3450, 3320, 3200, 1748, 1738, 1691, 1664

Acetate   mp. 299～331°,

UV λ_{max}^{EtOH} mμ (ε): 246 (12300); IR ν_{max}^{nujol} cm⁻¹: 1753, 1739, 1695, 1650

(6) Khivorin (CXLII)[884]

≪Occurrence≫   *Khaya ivorensis*,   mp. 256～263°, $[\alpha]_D$ −42° (CHCl₃), $C_{32}H_{42}O_{10}$

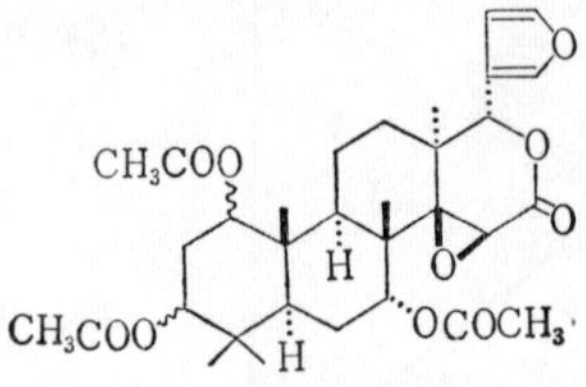

(CXLII)

UV λ_{max}^{EtOH} mμ (ε): 209 (6085)

IR $\nu_{max}^{CH_2Cl_2}$ cm⁻¹: 1500, 875 (furan), 1720～1750 and 1235 (lactone and acetate)

(7) Gedunin (CXLIII) and Dihydrogedunin (CXLIV)

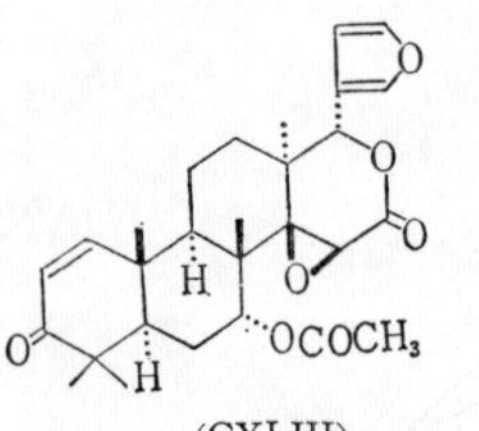

(CXLIII) 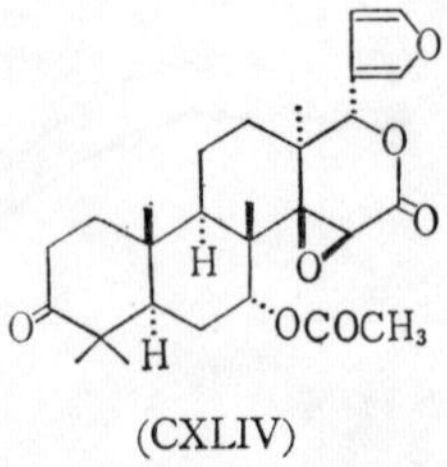   (CXLIV)

882) Hirose, Y.: *Chem. Pharm. Bull.*, **11**, 535 (1963).

883) Grant, I. G. *et al*: *Proc. Chem. Soc.*, **1961**, 444.

884) Taylor, D. A. H. *et al*: *Soc.*, **1962**, 768, **1963**, 980.

(a) **Gedunin** (CXLIII)[885)886)]

    ≪Occurrence≫   *Entandrophragma angolense* and other spp. (*Meliaceae*)

mp. 218°, $[\alpha]_D$ −44° (CHCl₃), $C_{28}H_{34}O_7 \cdot 1/2\ CH_3OH$

UV λ_{max}^{MeOH} mμ (log ε): 215 (4.12), 335 (1.8)   IR ν_{max}^{nujol} cm⁻¹: 3500, 1740, 1668, 1500, 875

(b) **Dihydrogedunin** (CXLIV)[887)]

    ≪Occurrence≫   *Guarea thompsonii* (Nigerian pearwood), *G. cedrata* (African cedar)

mp. 237~238°, $[\alpha]_D$ +3.7° (CHCl₃), $C_{28}H_{36}O_7$

UV λ_{max}^{EtOH} mμ (ε): 207 (6600), 289 (25)

(8) **Nimbin** (CXLV)[667)888)]

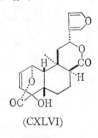

(CXLV) Nimbin (Partial structure)

    ≪Occurrence≫   *Melia indica* (nim tree),   mp. 205°, $C_{30}H_{36}O_9$

UV λ_{max}^{EtOH} mμ (ε): 335 (54)

NMR τ: 3.62, 4.10, J_{AB}=10 c/s (typical AB quartet) (−CO−CH=CH−C−)

(CXLV) $\xrightarrow{-CO_2}$ (CXLVI)

(CXLVI) τ=3.52~4.20 (complicated AB quartet, coupled with C₄-proton), 8.73 d (〉CH−Me)

Nimbin had been regarded as a cardenolide, having $C_{28}H_{38}O_8$ (±CH₂) by Mytra[889)] or $C_{30}H_{36}O_9$ by Nara-imhan[665)], which has been revised to a furanoterpenoid by Henderson *et al.*[888)]

See the item (10), [C], (18).

(9) **Columbin** (CXLVI)[890)891)]

    ≪Occurrence≫   *Jateorrhiza palmata* Miers (*Menispermaceae*)

mp. 201°, $[\alpha]_D$ +52.5° (pyridine), $C_{20}H_{22}O_6$

(CXLVI)

885) Robertson, J. M. *et al*: *Proc. Chem. Soc.*, **1962**, 222.

886) Taylor, D. A. H. *et al*: *Soc.*, **1960**, 3872, **1961**, 3705.   887) Housley, J. R. *et al*: *ibid.*, **1962**, 5095.

888) Henderson, R. *et al*: *Proc. Chem. Soc.*, **1963**, 269.

889) Mitra: *J. Sci. Ind. Res. India*, **16B**, 477 (1957).

890) Overton, K. H. *et al*: *Proc. Chem. Soc.*, **1961**, 211.

891) Barton, D. H. R. *et al*: *Soc.*, **1956**, 2085, 2090.

IR $\lambda_{max}\,\mu$: 3.19, 6.65, 11.42, 12.73, 13.30[892]

Mass   See the item (2) (CXXXII)

Dihydrocolumbin   UV $\lambda_{max}\,m\mu$ (ε): 210 (5700)

Columbin ──── alkali ────→ Isocolumbin
(CXLVI) C_{12}-epimerization   (CXLVII)

(CXLVII)

Mass   See the item (2) (CXXXIII)

(10) Palmarin (CXLVIII) and Isojateorin (CXLIX)[893]

(CXLVIII)   (CXLIX)

≪Occurrence≫   *Jateorrhiza palmata* Miers

(a) **Palmarin** (CXLVIII)

mp. 253~258°, $[\alpha]_D$ +17° (pyridine), $C_{20}H_{22}O_7$

Purified from a mixture " Chamanthin "

Methylether   mp. 261~263°

(b) **Isojateorin** (CXLIX)

mp. 165~167°, $[\alpha]_D$ +30° (pyridine), $C_{20}H_{22}O_7$

IR ν_{max}^{nujol} cm^{-1}: 3460 (OH), 1765~1715 (two lactones); $\nu_{max}^{CHCl_3}$ cm^{-1}: 1774, 1747 (two lactones)

(11) Cascarillin (CC)[894]

≪Occurrence≫   *Croton eleuteria* (*Euphorbiaceae*), $C_{22}H_{32}O_7$

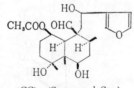

(CC) (Suggested Str.)

IR, NMR, Degradation

Subst. furan ring, 1 *axial*–OAc, 2 *sec.*–OH, 1 *tert.*–OH, 1 *tert.*–CHO (forms a hemiketal link with OH)

–CH$_2$–CH·OH–,   2 $>$C–CH$_3$,   1 $>$CH–CH$_3$, $>$C$<^{OH}_{CH_3}$

---

892) Kubota, T: *Tetrahedron*, **4**, 68 (1958).   893) Barton, D.H.R. *et al*: *Soc.*, **1962**, 4809, 4816.

894) Birtwistle, J.S. *et al*: *Proc. Chem. Soc.*, **1962**, 329.

(12) Clerodin (CCI)[895) 896)]

≪Occurrence≫　*Clerodendron infortunatum* (*Verbenaceae*)

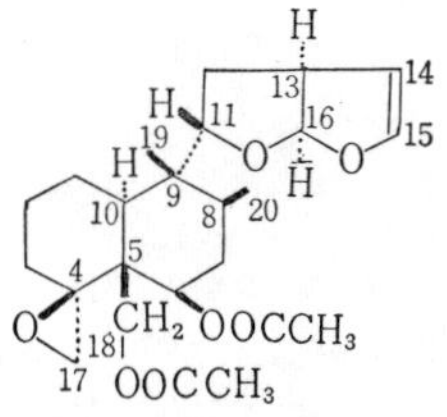

(CCI)

mp. 164~165°, $[\alpha]_D$ −47°,　$C_{24}H_{34}O_7$

UV　$\lambda_{max} m\mu$ (ε): 208 (4550)

IR　$^{nujol}_{max}$ cm^{-1}: 1727, 1252 (acetates), 1615, 738 (vinylether); ν^{CCl4}_{max} cm^{-1}:3025 $\left(>C=C<_H\right)$, 3045 (epoxide)

NMR　τ: 9.05 (19–Me), 9.05 (20–Me), 8.14, 7.98 (O·COCH$_3$), 6–H (overlap), ca 6.1 (11–H), 14–H (overlap), 3.61, J=ca 2.5 c/s (15–H), 4.09, J=6.0 c/s (16–H), 7.14 (17–CH$_2$–), 5.80, 5.18, J=12.8 c/s (18–CH$_2$–)

(13) Marrubiin (CCII)[897) 898)]

≪Occurrence≫　*Marrubium vulgare* (マルバハッカ) (*Labiatae*)

mp. 159.5~160.5°, $[\alpha]_D$ +44.80°,　$C_{20}H_{28}O_4$

UV　$\lambda_{max} m\mu$ (ε): 212 (5620)[899)]

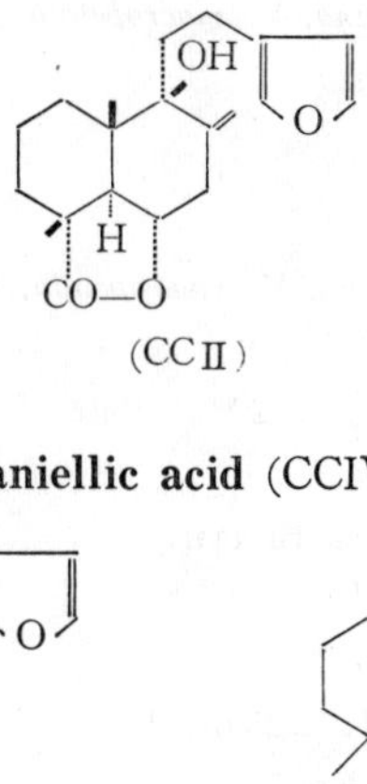

(CC II)

(14) Polyalthic acid (CCIII) and Daniellic acid (CCIV)

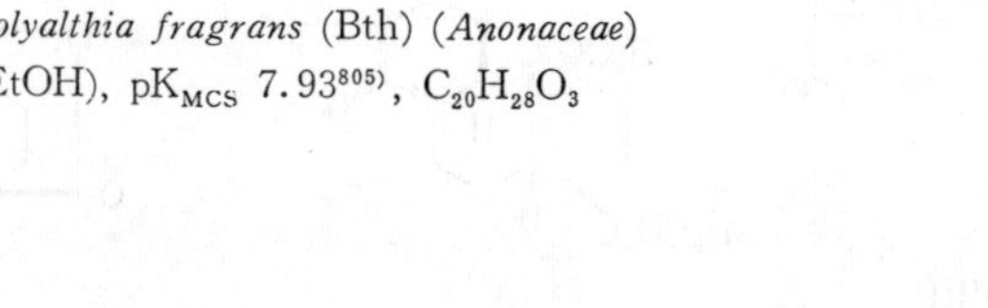

(a) Polyalthic acid (CCIII)

≪Occurrence≫　*Polyalthia fragrans* (Bth) (*Anonaceae*)

mp. 102°, $[\alpha]_D$ −26° (EtOH), pK$_{MCS}$ 7.93[805)], $C_{20}H_{28}O_3$

895)　Barton, D. H. R. *et al*: *Soc.*, **1961**, 5061.
896)　Robertson, J. M. *et al*: *ibid.*, **1963**, 4133.　　897)　Cocker, W. *et al*: *Chem. & Ind.*, **1955**, 772.
898)　Ghigi, E. *et al*: *Gazz. Chim. Ital.*, **84**, 428 (1954); *C.A.*, **49**, 5406 i (1955).
899)　Gopinath, K. W. *et al*: *Helv. Chim. Acta*, **44**, 1040 (1961).

UV λ_{max} mμ (ε): 213 (6200); IR λ_{max}^{nujol} μ: 6.7, 11.4 (furan ring), 3.25, 6.1, 11.2 (exocyclic methylene), 5.9 (carboxyl)

NMR in CDCl$_3$, 60 Mc, TMS, τ: 2.69 (1), 2.83 (1) (α-protons on furan ring), 3.79 (1) (β-proton on furan ring), 5.17 (1), 5.47 (1) ($>$C=CH$_2$), 8.37 (14) (cyclic methylene), 8.9 (3), 9.3 (3) $\left(\rightarrow\!\!\!\!\!\diagup\,\text{C-CH}_3\right)$

(b) Daniellic acid (CCIV)[900]

≪Occurrence≫　*Daniellia oliveri* (African copal tree)

mp. 129~130.5°, $[\alpha]_{589\,m\mu}$ $-58°$ (EtOH), pK$_{MCS}$ 8.60,　C$_{20}$H$_{28}$O$_3$

UV λ_{max}^{C7H16} mμ (ε): 225 (1650), IR ν_{max} cm^{-1}: 1600, 1558, 1495, 1241, 1170, 1065, 1022, 873

(15) Vouacapenyl acetate (CCV), Methyl vouacapenate (CCVI) and Methyl vinhaticoicate (CCVII)

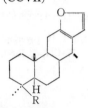

R=CH$_2$OAc (CCV) Vouacapenyl acetate　　R=COOCH$_3$ (CCVII) Methyl vinhaticoicate
R=COOCH$_3$ (CCVI) Methyl vouacapenate

(a) Vouacapenyl acetate (CCV)[901]

≪Occurrence≫　*Vouacapoua americana, V. macropetala, V. reticulata*

mp. 115°, $[\alpha]_D$ $+63°$ (CCl$_4$),　C$_{22}$H$_{32}$O$_3$

UV λ_{max}^{EtOH} mμ (log ε): 220 (4.33)

(b) Methyl vouacapenate (CCVI)[901]

≪Occurrence≫　*Vouacapoua americana, V. macropetala, V. reticulata*

mp. 103~104°, $[\alpha]_D$ $+101°$ (CCl$_4$),　C$_{21}$H$_{30}$O$_3$

UV λ_{max}^{EtOH} mμ (log ε): 222 (3.72), pK$_{MCS}$ 8.52[805]. Rate of hydrolysis in 0.5 N EtOH–KOH: 4.5 hr.

(c) Methyl vinhaticoicate (CCVII)[902]

≪Occurrence≫　*Plathymenia reticulata* BENTH.

mp. 108°, $[\alpha]_D$ $+67°$ (CHCl$_3$),　C$_{21}$H$_{30}$O$_3$

λ_{max}^{EtOH} mμ (log ε): 220 (3.88), pK$_{MCS}$ 8.01[805]

Rate of hydrolysis in 0.5 N–EtOH–KOH: 2.25 hr.[901]

(16) Sciadin (CCVIII), Dimethyl sciadinonate (CCIX) and Sciadinone (CCX)

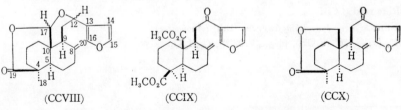

(CCVIII)　　　(CCIX)　　　(CCX)

≪Occurrence≫　*Sciadopitys verticillata* SIEB. et ZUCC. (コウヤマキ)

900) Haeuser, J. *et al*: *Bull. Soc. Chim. France*, **1959**, 1447; *C. A.*, **55**, 24815 e (1961).
901) King, F. E. *et al*: *Soc.*, **1955**, 1117.　902) King, F. E. *et al*: *ibid.*, **1953**, 1055, 4158.

(a) **Sciadin** (CCVIII)[903) 904)]

　mp. 160°, $[\alpha]_D$ +10.3°, +13.77° (CHCl$_3$),　　$C_{20}H_{24}O_4$

　UV　λ_{max}^{EtOH} mμ (ε): 206 (9100)[904)] ; λ_{max}^{EtOH} mμ (log ε): 205 (4.02)[903)]

　IR　ν_{max}^{KBr} cm^{-1}: 1742, 1636, 1607, 1507, 895, 875[903)] ; ν_{max}^{nujol} cm^{-1}: 1736 (lactone), 1636, 1606, 1505, 870
(furan ring), 1636, 894 (vinylidene)[904)]

　NMR　$\tau^{904)}$: 8.77 s (18–Me), 2.60, 3.53 (α,β–H of furan ring), 4.56 (C$_{16}$–H), 5.17~4.99 ($\rangle$=CH$_2$)

(b) **Dimethyl sciadinonate** (CCIX)[903)]

　mp. 122°, $[\alpha]_D$ −45.0° (CHCl$_3$),　　$C_{22}H_{28}O_6$

　UV　λ_{max}^{EtOH} mμ (log ε): 203 (4.32), 255 (3.60)

　IR　ν_{max}^{KBr} cm^{-1}: 1730 (ester), 1689 (C=C–C=O), 1647 (C=C), 1592, 1568, 1508 (furan), 879 (exocyclic
$\rangle$C=CH$_2$), 879 (furan), 822, 766

　NMR　τ: 1.93 (H$_{\alpha\prime}$), 2.56 (H$_{\alpha}$), 3.22 (H$_{\beta}$)

(c) **Sciadinone** (CCX)[903)]

　mp. 207°, $[\alpha]_D$ −59.9° (CHCl$_3$),　　$C_{20}H_{24}O_4$

　UV　λ_{max}^{EtOH} mμ (log ϵ): 201 (4.28), 254 (3.62)

　IR　ν_{max}^{KBr} cm^{-1}: 1730 (lactone), 1674 (C=C–C=O), 1642 (C=C), 1606, 1563, 1517 (furan), 893 (exocyclic
　　　　　$\rangle$C=CH$_2$), 872 (furan), 841, 749

　NMR　τ: 1.86 (H$_{\alpha\prime}$), 2.58 (H$_{\alpha}$), 3.29 (H$_{\beta}$)

903) Ishikawa, M *et al*: *Chem. Pharm. Bull.*, **11**, 271, 1346 (1963).
904) Sumimoto, M *et al*: *Tetrahedron*, **19**, 643 (1963); *6 th Symposium on the Chemistry of Natural Products*,
　　p. 125 (Sapporo, 1962).

〔J〕 Pyrano-terpenoids

(1) Miroestrol (CCXI)[904a) 904b)]

≪Occurrence≫ Oestrogenic substance from *Pueraria mirifica* (*Leguminosae*)

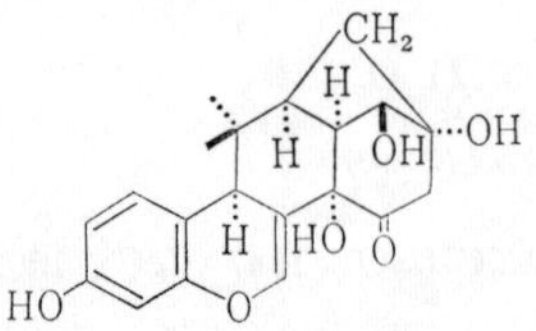

(CCXI)

mp. 268~270° (decomp.), $[\alpha]_D$ +301° (EtOH), $C_{20}H_{22}O_6$

UV $\lambda_{max}^{EtOH} m\mu$ (ε): 217 (21650), 285 (4575)

IR ν_{max}^{KBr} cm^{-1}: 3533, 3450 (O–H), 2950 (C–H), 1706 (nonconj. six membered ketone), 1661$\left(-O-CH=C\diagdown\right)$, 1623, 1597, 1511 (arom. ring), 1464, 1397, 1389, 1364, 1337, 1285 sh, 1178 sh, 1160, 1132, 1090 sh, 1075, 1064, 1045, 1030, 1015, 995, 971, 955, 918 906, 890, 871, 861, 828, 813, 780, 758, 735, 725, 686.

(2) Monascoflavin (CCXII)[904c)]

≪Occurrence≫ A yellow pigment, produced by *Monascus purpureus* WENTII.

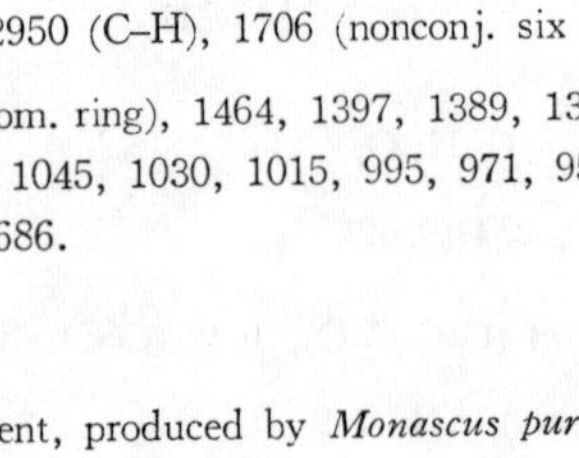

(CCXII)

mp. 143~145°, $C_{21}H_{26}O_5$

UV $\lambda_{max}^{McOH} m\mu$ (log ε): 228 (3.41), 385 (4.21)

IR ν_{max}^{KBr} cm^{-1}: 1788, (—CO-O—), 1720 ($\diagup$C=O), 1670 (–C=C–CO), 1600, 1522

NMR τ J=cps: 8.54 s (c), 8.12 d, J_{do}=6.8 (d), 7.4 m (e), 6.7 q, J_{fh}=5.0, J_{gh}=8.3 (h), 6.25 d, J_{if}=13.0 (i), 5.2 (j), 4.94 q, J_{jk}=12.6 (k), 4.04 d, J_{no}=15.6 (n), 3.42 d, J_{do}=6.8 (o)

904a) Cain, J. C: *Nature*, **198**, 774 (1960). 904b) Bounds, D. G. *et al*: *Soc.*, **1960**, 3696.
904c) Nakanishi, K. *et al*: *Tetrahedron*, **18**, 1195 (1963).

12 Sesquiterpenes

[A] UV Spectra

α–Cyperone[905] azulene[906], and the derivatives[907)908]. Artemisin, l–β–santonin, l–α–santonin, α–hydroxysantonin,[909], santonin, lumisantonin, umbellulone and dihydroumbellulone[910].

[B] IR Spectra

Santalol[911], azulene, 5–and 6–methylazulenes, 1, 4–and 4, 8–dimethylazulenes, 4, 8–dimethyl–6–isopropylazulene, S–guaiazulene[912] and carpesia lactone[913]. α–Hydroxysantonin, artemisin, l–α–santonin, l–β–santonin, l–α–desmotroposantonin[909], photosantonin, photosantonic acid, O–acetylisophotosantonic lactone, and lumisantonin[914].

[C] Hydrocarbons and Azulenes

(i) *Monocyclic Hydrocarbons*
(1) **Humulene**[915] (I)

905) Bradfield *et al*: *Soc.*, **1936**, 667, 676.
906) Gillam, Stern: An Introduction to Electronic Absorption Spectroscopy in Organic Chemistry p. 129 (Edward Arnold, 1954), II Ed, p. 147 (Edward Arnold 1958).
907) Plattner: *Helv. Chim. Acta.*, **31**, 804 (1948).   908) Gordon, M: *Chem. Review.*, **50**, 185~194 (1952).
909) Shibata, S., Mitsuhashi, H: *Pharm. Bull.*, **1**, 75 (1953).
910) Arigoni, D., Bosshard, H., Bruderer, H. Büchi, G., Jeger, O., Krebaum, L. J: *Helv. Chim. Acta,* **40**, 1732 (1957).
911) Sadtler Card No. 3878.   912) Günthard, Plattner: *Helv. Chim. Acta*, **32**, 284 (1949).
913) Kariyone, T., Naito, S: *Y. Z.*, **75**, 39 (1955).
914) Arigoni, D., Bosshard, H., Bruderer, H: *Helv. Chim. Acta*, **40**, 1732 (1957).
915) Sukh Dev.: *Tetrahedron*, **9**, 1 (1960).

(I)

≪Occurrence≫　　*Zingiber zerumbet*,　　bp. 114〜115°/5 mm, n_D 1.5015, $[\alpha]_D$ −0.31, $C_{15}H_{24}$
Humulene mono(di) oxide　See the item [E], (4)

(2)　α-Elemene (II)[916]

≪Occurrence≫　　Elemol (XLIa) $\xrightarrow[\text{dehydration}]{\text{HClO}_4-\text{HOAc}}$ (II)　$C_{15}H_{24}$

[Note]

(II)

UV　$\lambda_{\max}$ mμ (ε): 250 (17780)
IR　$\nu_{\max}^{\text{liq. film}}$ cm^{-1}: 1818, 1634, 1002, 912 (–CH=CH$_2$), 1604 (conj C=C), 877 $\left(\!\!\diagup\!\!\diagdown\text{C}=\text{C}\diagdown\!\!\diagup\right)$,

809 $\left(\diagup\!\!\diagdown\text{C}=\text{C}\diagdown_{\text{H}}\right)$

NMR　τ: 3.65 $\left(\diagdown_{\text{H}}\right)$, 3.98〜4.43 $\left(=\text{C}\diagdown_{\text{H}}\right)$, 4.92, 5.16 $\left(_{\text{H}}\!\diagdown\text{C}=\text{CH}_2\right)$, 8.20, 8.27

$\left(\diagup\!\!\diagdown\text{C}=\text{C(Me)}_2\right)$ 8.97d, J=6.6(–CH(Me)$_2$) 8.84 $\left(\diagup\!\!\rightarrow\text{C–Me}\right)$

(3)　δ-Elemene (III)[917]

(III)

≪Occurrence≫　　*Dysoxylon frazeranum*,　　bp. 107°/10 mm, n_D 1.4828, d_4^{25} 0.8590, $[\alpha]_D$ ±0°, $C_{15}H_{24}$
IR　$\nu_{\max}$ cm^{-1}: 907, 1001 (isolated vinyl), 894 (vinyliden), 818 $\left(\diagup\!\!\diagdown\text{C}=\text{C}\diagdown_{\text{H}}\right)$

(4)　Bisabolene (IV) and Isobisabolene (V)[918]

(IV)　　　　　　　　　　(V)

(a)　Bisabolene (IV)

≪Occurrence≫　　Bisabol–Myrrh oil (*Commiphrara erythraea*), bergamot oil, lemon oil, camphor oil etc.
　　　　　　bp. 262〜263°, $[\alpha]_D$ −41.3°, $C_{15}H_{24}$
IR　$\nu_{\max}$ cm^{-1}: 1308, 1252, 1208, 1176, 1125, 1095, 1057, 1017, 970, 947, 933, 917, 885, 875, 828,
　　　　　763, 725[918]

---

916) Bhattacharyya, S. C. *et al*: *Tetrahedron.*, **18**, 1509 (1962).　917)　　Gough, J. *et al*: *Tetr. Letters*,
1961, 763.　918) Bhattacharyya, S. C. *et al*: *Tetrahedron*, **18**, 1165 (1962).
[Note] See the item [D] Alcohols (2′) Elemol

(b) Isobisabolene (V)

≪Occurrence≫   Vetiver oil (*Vetiveria zizanoides* L.)
bp. 90~102°/8 mm, n_D 1.4966, d_4^{26} 0.8859, $[\alpha]_D$ −47°, $C_{15}H_{24}$

IR ν_{max} cm^{-1}: 1640, 892 ($>C=CH_2$), 833, 813, 791 ($>C=C<^H$) (See Fig. 144)

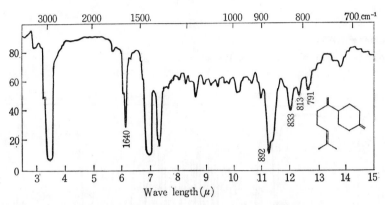

Fig. 144   IR Spectrum of Isobisabolene in CHCl₃
(Bhattacharyya)

(ii) *Bicyclic Hydrocarbons*

(5) Cadinenes[919]

(VI) β–Cadinene   (VII) γ–Cadinene   (VIII) γ₁–Cadinene

(IX) δ–Cadinene   (X) ε–Cadinene

Table 76   Physicalic constansts of Cadinenes   (Surveyed by Herout *et al.*)

Compound	Str. No.	d_4^{20}	n_D^{20}	$[\alpha]_D$	Freq. of C=C cm^{-1}
β–Cadinene	(VI)	0.9239	1.5059	−251	790, 812, 822
γ–Cadinene	(VII)	0.9125	1.5075	+148	794, 836, 890
γ₁–Cadinene	(VIII)	—	1.5155	−19	
δ–Cadinene	(IX)	0.9175	1.5086	+94	836
ε–Cadinene	(X)	0.9107	1.5038	+47	890

919)   Herout, V. *et al*: *Tetrahedron*, **4**, 246 (1958).

(a) (+) γ-**Cadinene** (VII) (Antipode of (XI))

   ≪Occurrence≫   Citronella oil, $[\alpha]_D + 148°$

(XI)

(b) (−) γ-**Cadinene** (XI)[920]

   ≪Occurrence≫   Khusol (XLV) ⟶ Tosylate $C_{22} H_{30}O_3S$ $\xrightarrow{\text{LiAlH}_4}$ (XI)

   bp. $100°/_{0.5\,mm}$, n_D^{31} 1.5060, d_4^{24} 0.9182, $[\alpha]_D$ −145°, $C_{15}H_{24}$

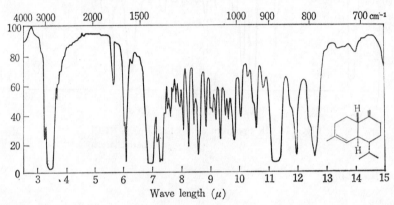

Fig. 145 IR Spectrum of (−) γ-Cadinene

(6) δ-**Selinene** (XII)[921]

   ≪Occurrence≫   β-Selinene ⟶ (XII)   bp. 119°/10 mm, n_D 1.5185, $[\alpha]_D$ +235.2°, $C_{15}H_{24}$

(XII)

   UV $\lambda_{max}^{EtOH} m\mu$ (ε): 241 (22626), 247 (24035), 255 (18436)

   IR ν_{max} cm⁻¹: 1613, 1449, 1366, 1325, 1282, 1258, 1205, 1163, 1149, 1111, 1053, 1026, 990, **950**,

   913, 873

(7) **Caryophyllene** (XIII) (β-Caryophyllene)[922]

   ≪Occurrence≫   Clove oil   bp. 129∼130°/14 mm, $[\alpha]_D$ −9°, $C_{15}H_{24}$

(XIII)

920) Bhattacharyya, S. C. *et al*: *Tetrahedron*, **19**, 1073 (1963).   921)= 920)
922) Clunie, J. S. *et al*: *Proc. Chem. Soc.*, **1960**, 82.

(8) β-**Bergamotene** (XIV)[923]

≪Occurrence≫　Bergamot oil,　$C_{15}H_{24}$

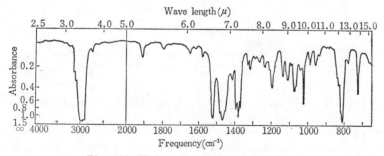

(XIV)

(9) **Cuparene** (XV)[924]

≪Occurrence≫　*Chamaecyparis thyoides*, *Biota orientalis*, *Widdringtonia* spp.

bp. 138°/19 mm, d_{21}^{21} 0.9374, n_D^{25} 1.5202, $[\alpha]_D$ +65.3° (pure oil), +65° (CHCl₃), $[R_L]_D$ 65.9°, $C_{15}H_{22}$

(XV)

UV　λ_{max}^{EtOH} mμ (ε): 259 (260), 265 (360), 273 (350), IR: See Fig. 146.

Total synth of (±)[925]

Fig. 146　IR Spectrum of Cuparene (Erdtman)

(10) **Azulenes**

(a) **UV and Visible Spectra of 1, 4–Dimethyl–7–alkyl (aryl) azulenes**[926]

R=Isoamyl　　(XVI)

R=Dodecyl　　(XVII)

R=Phenyl　　(XVIII)

UV & Visible　$\lambda_{max}^{cyclohexane}$ mμ (log ε):

(XVI)　245 (4.40), 287 (4.61), 350 (3.72), 368 (3.50) 604 (2.64), 664 (2.57), 738 (2.15),

(XVII)　245 (4.50), 286 (4.71), 350 (3.76), 367 (3.52) 608 (2.65), 663 (2.57), 738 (2.15),

(XVIII)　245 (4.27), 292 (4.50), 378 3.67) 608 (2.71), 635 (2.67), 666 (2.65), 739 (2.24),

923) Bhattacharyya, S. C. *et al*: *Tetr., Letters*, **1963**, 505.
924) Erdtman, H. *et al*: *Tetrahedron*, **4**, 361 (1958).　　925) Parker, W. *et al*: *Soc.*, **1962**, 1558.
926) Treibs, W. *et al*: *Ann.*, **634**, 111 (1960).

UV and IR spectra of alkyl, phenyl, benzyl, hydroxy, carboxy, aldehyde, aldoxim, nitro and nitril derivatives of azulene[927].

(b) NMR of Azulene[928]

Fig. 147 shows the NMR spectra of azulene in various solutions at 60 Mc, c/s.

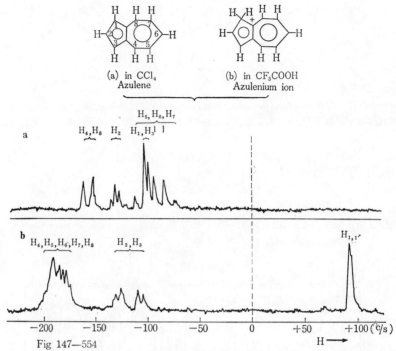

Fig 147—554

Fig. 147 NMR Spectra of Azulene: (a) in CCl_4 soln. (b) in CF_3COOH
soln. (Danyluk *et al.*)

Results are summarized as (i) the signal $+93$ cps due to $H_{1,1'}$ protons appears in (b) which not appears in (a), (ii) typical AB–type spin–spin coupling -117 cps, $J=1$ cps due to C_2, C_3 and C_1 protons in pentacyclic ring appears in (b), (iii) signals of C_4, C_5, C_6, C_7 and C_8 protons in (b) are remarkably lower-shifted.

(c) S–Guaiazulene (XX)[929]

≪Occurrence≫   *Eucalyptus globulus*, Gurjunbalsam oil etc.,   bp. 167～168/12 mm, mp. 31.5°, $C_{15}H_{18}$.

UV $\lambda_{max}^{isooctane}$ mμ (log ε)[929] : 245 (4.395), 284 (4.64), 289 (4.63), 305 (4.05), 336 (3.515),
349 (3.69), 367 (3.59).

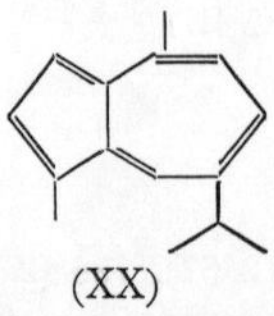

(XX) 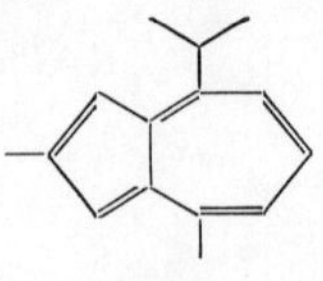   (XXI) 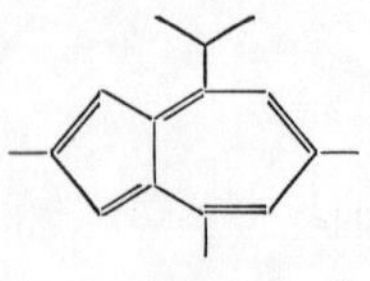   (XXII)

(d) Zierazulene (XXI) and **2,4,6–Trimethyl–8–isopropylazulene** (XXII)[930]

927) Reid, D. H. *et al* : *Soc.*, **1958**, 1100, 1110, 1118.
928) Danyluk, S. S. *et al* : *J.A.C.S.*, **82**, 997 (1960).
929) Naves, Y. R. *et al* : *Helv. Chim. Acta*, **42**, 1375 (1959).
930) Collins, D. J. : *Soc.*, **1959**, 531.

≪Occurrence≫　　Synthetic

$$\text{Zierone} \xrightarrow{\text{Red., Dehydrogenation}} \text{(XXI)}$$
$$\text{Hydroxymethylenezierone} \longrightarrow \text{(XXII)}$$

Zierazulene (XXI)

Visible　$\lambda_{max}^{cyclohexane}$ mμ:　545,　582,　632 (synth.)

IR　ν_{max}^{CHCl3} cm^{-1}:　1575 s, 1510 s, 1475 m, 1458m, 1393 w, 1384 w, 1371 w, 1321 s, 1192 w, 1125 w, 1032 w, (synth.)

S–Trinitrobenzene adduct mp. 122.5°, $C_{21}H_{21}O_6N_3$

2, 4, 6–Trimethyl–8–isopropylazulene (XXII)

UV　$\lambda_{max}^{cyclohexane}$ mμ (visible):　534,　570,　622

IR　ν_{max}^{CHCl3} cm^{-1}:　1586 s, 1510 s, 1470 m, 1458 m, 1394 m, 1384 m, 1374 m, 1334 m, 1316 w, 1204 w, 1155 w, 1087 w, 1034 w, 859 m

S–Trinitrobenzene adduct mp. 155~157°, $C_{22}H_{23}O_6N_3$

(11)　α-Himachalene (XXIII) and β-Himachalene (XXIV)[931]

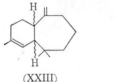

(XXIII)　　　　　　　　(XXIV)

≪Occurrence≫　*Cedrus deodara* LOUD.

(a)　α-Himachalene (XXIII)

bp. 93~94°/2 mm, n_D^{25} 1.5082, d_4^{25} 0.9206, $[\alpha]_D$ −187.1°

IR　ν_{max} cm^{-1}:　3060, 1770, 1625, 885 $\left(\!>\!C=CH_2\right)$, 1665, 865 $\left(\!>\!C=C\!<_H\right)$, 1388, 1377, 1362 (gem Me$_2$).

NMR　in CCl$_4$, 60 Mc, relative to H$_2$O cps:　−45s, $\left(\!>\!C=C\!<_H\right)$, +1 $\left(\!>\!C=CH_2\right)$

(b)　β-Himachalene (XXIV)

bp. 121~122°/4 mm, n_D 1.5130, d^{25} 0.9330, $[\alpha]_D$ +225.8°

IR　ν_{max} cm^{-1}:　1665, 857 $\left(\!>\!C=C\!<_H\right)$, 1360 (gem Me$_2$)

NMR　−37s (1H) $\left(\!>\!C=C\!<_H\right)$

(iii)　*Tricyclic Hydrocarbons*

(12)　Thujopsene (XXV)[932)~935) 935a) 936)]

≪Occurrence≫　*Thujopsis dolabrata* SIEB. et ZUCC. (ヒバ)

931)　Joseph, T. C. *et al*: *Tetr. Letters*, **1961**, 216.
932)　Erdtman, H. *et al*: *Chem. & Ind.*, **1960**, 622.
933)　Nozoe, T. *et al*: *Chem. Pharm. Bull.*, 8, 936 (1960).
934)　Kobayashi, H. *et al*: *Bull. Chem. Soc. Japan*, **34**, 1123 (1961).
935)　Shishido *et al*: *4 th Symposium on the Chemistry of Natural Products*, p. 118 (1960).
935a)　Norin, T.: *Acta. Chem. Scand.*, **1963**, 738.

$(XXV)^{935a)}$

bp. $125°/_{2\,mm}$, d_4^{26} 0.9315, n_D 1.50259, $[\alpha]_D$ −91.2° $(CHCl_3)$, $C_{15}H_{24}^{935a)}$

$(XXV) \longrightarrow$ Widdrol [D] $(16)^{935b)}$

$(XXV) \xrightarrow{O_3 \text{ or } KMnO_4} (XXVI)^{933)}$

(XXVI)

(XXVI) mp. 164°.

UV λ_{max}^{MeOH} mμ: 282.

IR ν_{max}^{KBr} cm^{-1}: 1691~1700

(13) Gurjunene $(XXVII)^{937)}$

≪Occurrence≫　*Dipterocarpus dyeri*,　bp. 76~77°/3 mm, n_D 1.5010, $[\alpha]_D$ −214°, $C_{15}H_{24}$

(XXVII)

UV λ_{max} mμ (ε): 204 (10000)
IR ν_{max} cm^{-1}: 1670 (C=C)

(14) Sesquiterpenes of *Bulnesia sarmienti*$^{938)}$

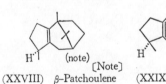

(XXVIII) β–Patchoulene　(XXIX) α–Guaiene　(XXX) α–Bulnesene　(XXXI) β–Bulnesene

　or

(XXXII) Bulnesol=(LXXVIII)　(XXXIII) Guaiol=(LXXV)　(XXXIV) Guaioxide

935b) S. Nagahama: *Bull. Chem. Soc. Japan*, **33**, 1467 (1960).
936) Kawamura, J: *Ringyô-Shikensho Hôkoku, Japan*, **30**, 59 (1930).
937) Ourisson, G. *et al*: *Tetr. Letters*, **1962**, 677; *Bull. Soc. Chim. France*, **1958**, 886.
938) Bates, R. B., *et al*: *Chem & Ind.*, **1962**, 1715.
　[Note] See the item [C], (24).

Gaschromatography of the essential oil was carried out on 20% carbowax 20 M on fire brick at 193°.

β-Patchoulene (0.2, 4), α-guaiene (0.8, 5.5), α-bulnesene (4.5, 9), guaioxide (2.9) 〔Note〕, β-bulnesene (2.5, 12), guaiol (33, 32), and bulnesol (45, 53), were separated.

Numbers in parenthesis indicate parcentage and r.r.t.

(15) Sesquiterpenes of *Pogostemon patchouli*[938]

As the same, β-patchoulene (2,4), α-guaiene (21, 5.5), α-bulnesene (21, 9), patchouli alcohol (35, 50) were separated.

(16) Copaene (XXXV)[939)940]

≪Occurrence≫ *Oxystigma mannii* HARMS (African copaibabalsam), *Sindora Wallichii*, *Cedrela toona* Roxb., *Chloranthus spicatus*(?), *Phyllocladus trichomanoides*.

bp. 114~114.5°/10 mm, d_4^{25} 0.9055, n_D^{25} 1.4880~1.4895, $[\alpha]_D$ −0.44°~+1.20° [940], −6.3° [939], $C_{15}H_{24}$

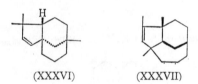

(XXXV)

(17) Clovene (XXXVI) and Isoclovene (XXXVII)[941]

≪Occurrence≫ Clove oil, $C_{15}H_{24}$

(XXXVI)   (XXXVII)

(18) Longifolene (XXXVIII)

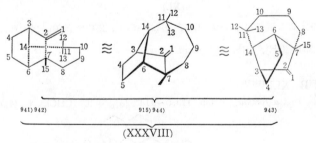

941)942)   915)944)   943)

(XXXVIII)

≪Occurrence≫ *Pinus longifolia* ROXB.

bp. 118~119°/10 mm, n_D^{26} 1.5015, d_4^{26} 0.9319, $[\alpha]_D$ +43.1°, $C_{15}H_{24}$

939) Büchi, G.B. *et al*: *Proc. Chem. Soc.*, **1963**, 214.
940) Briggs, L.H. *et al*: *Soc.*, **1947** 1338.
941) Dupont, G. *et al*: *Bull. Soc. Chim. France*, **1954**, 1075.
942) Sandermann, W. *et al*: *Ber.*, **95**, 1863 (1962).
943) Nayak, U.R. *et al*: *Tetrahedron*, 8, 42 (1960).   944) Corey, E.J: *J.A.C.S.*, **83**, 1251 (1961).

〔Note〕 α-Bulnesene and guaioxide could be separated on 20% Corning QF-1 at 190°.

Total synthesis of *dl*–longifolene[944], biogenesis[942].

NMR　in CCl_4, 40 Mc, relative to H_2O, cps[915]: +1.9(1), +11(1) $\left(\rangle C=CH_2\right)$, 84.2(1), 102.6(2), $\left(\rangle C-H\right)$, 121(10) (—CH_2—), 143(9) $\left(\rangle C-Me\right)$

(iv)　*Tetracyclic Hydrocarbon*

(19)　**Longicyclene** (XXXIX)

　　《Occurrence》　*Pinus longifolia* Roxb.,　　bp. 82°/2 mm, n_D^{30} 1.4888, d_4^{30} 0.9307, $[\alpha]_D$ +33.6°, $C_{15}H_{24}$

(XXXIX)

UV　λ_{max}^{EtOH} mμ (ruled out), ε_{210} (171), ε_{215} (113), ε_{220} (63), ε_{225} (38).

［D］ Alcohols

(i)　*Monocyclic Alcohols*

(1)　**Nuciferol** (XL)[945]

　　《Occurrence》　*Torreya nucifera*, S. et Z.,　　bp. 131~132°/0.05 mm, $[\alpha]_D$ +41.06°, $C_{15}H_{22}O$

(XL)

UV　λ_{max}^{EtOH} mμ (ε): 252.5 (374), 259 (478), 264.5 (570), 276.5 (560), 273 (622)

IR　$\nu_{max}^{liq.}$ cm^{-1}: 3340 (OH), 1518, 819 (arom.)

(2)　**Farnesiferol B** (XLI)[170) 946]

　　《Occurrence》　*Asa foetida*,　　mp. 113.5~114°, $[\alpha]_D$ +10° ($CHCl_3$), $C_{24}H_{30}O_4$

(XLI)

UV　λ_{max}^{EtOH} mμ (log ε): 242 (3.60), 252 (3.48), 298 (3.93), 326 (4.18).

IR　$\nu_{max}^{CHCl_3}$ cm^{-1}: 3590 (none assoc. OH), 1726, 1605, 1100 (coumarin).

Acetate   mp. 70.5~71.5°, $[\alpha]_D$ −18° (CHCl$_3$), C$_{26}$H$_{32}$O$_5$

IR $\nu_{max}^{CHCl_3}$ cm^{-1}: 1725, 1645, 1615, 900, 865, 840

(2′) Elemol (XLI-a)[944a]

≪Occurrence≫ *Canarium luzonicum* (Java-citronella oil),

mp. 48~49°, bp. 100~105°/0.1mm, $[\alpha]_D$ −3° (CHCl$_3$), C$_{15}$H$_{26}$O.

(XLI-a) (Revised Str.)

IR ν_{max}^{film} cm^{-1}: 3300, 3075, 1640, 910, 893

(XLIa) $\xrightarrow{-H_2O}$ α-Elemene (II)[916]

(ii) *Bicyclic Alcohols*

(3) Farnesiferol A (XLII)[170]

≪Occurrence≫ *Asa foetida*,   mp. 155~155.5°, $[\alpha]_D$ −55° (CHCl$_3$), C$_{24}$H$_{30}$O$_4$

(XLII)

UV ν_{max}^{EtOH} mμ (log ε): 250 (3.34), 2.98 sh (3.94), 324 (4.2)

IR $\nu_{max}^{CHCl_3}$ cm^{-1}: 3590, 1725, 1645, 1615, 1090, 890, 860, 840

(4) Drimenol (XLIII)[947) 946]

≪Occurrence≫ *Drimys winteri* FORST,   mp. 97~98°, $[\alpha]_D$ −18° (C$_6$H$_6$)

(XLIII)

UV λ_{max} mμ (ε): 210 (2140), 215 (950), 220 (250)

IR $\nu_{max}^{CS_2}$ cm^{-1}: 3570 (free OH), 3450 (bonded OH); ν_{max}^{nujol} cm^{-1}: 814 $\left({>} C{=}C {<}_H \right)$

(5) Khusinol (XLIV) and Khusol (XLV)[920) 921) 948]

≪Occurrence≫ *Vetiveria zizanoides* L.

944a) Halsall, T. G. *et al*: *Soc.*, **1964**, 1029.   945) Sakai, T. *et al*: *Tetr. Letters*, **1963**, 1171.

946) Arigoni, D. *et al*: *Helv. Chim. Acta*, **42**, 2557 (1959).

947) Appel, H. H. *et al*: *Soc.*, **1959**, 3322.

948) Bhattacharyya, S. C. *et al*: *Tetrahedron*, **19**, 223 (1963).

(a) Khusinol (XLIV) mp. 87°, $[\alpha]_D$ −174.4° (CHCl$_3$), C$_{15}$H$_{24}$O

IR ν_{max}^{nujol} cm^{-1} $^{948)}$: 3400, 1074 (OH), 1642, 909, 898, ($>$C=CH$_2$), 1667, 838, 797, ($>$C=C$<^H$),

1340, 1372 (iso Pr), 1709, 1372, 1282, 1241, 1215, 1166, 1116, 1045, 1029, 1010, 877, 851, 838.

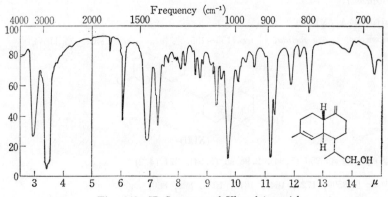

(XLIV)   (XLV)   (VII) (+) γ–Cadinene

(b) Khusol (XLV)   mp. 101~102°, $[\alpha]_D$ −137° (CHCl$_3$), C$_{15}$H$_{24}$O

IR: See Fig. 148, ν_{max}^{nujol} m^{-1}: 3340, 1028 (OH), 1642, 892 (=CH$_2$), 798, 840 ($>$C=C$<^H$)

NMR See Fig. 149.

Fig. 148 IR Spectrum of Khusol in nujol

(Bhattacharyya *et al.*)

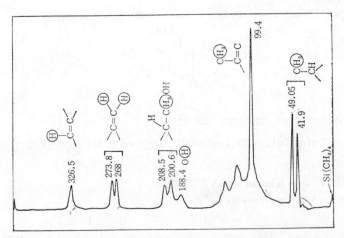

Fig. 149 NMR Spectrum of Khusol

(Bhattacharyya *et al.*)

(6) δ-**Cadinol** (XLVI)[949)950)]

$\ll$Occurrence$\gg$ *Citronella oil* (Java), *Chamaecyparis lawsoniana,*
                    Juniperus communis

mp. 139~140°, $[\alpha]_D$ −109°, $C_{15}H_{26}O$

NMR τ: 4.43 d, J=6.5 c/s $\left(>C=C<^{\mathbf{H}}_{CH_2} \right)$

(XLVI)
(Revised Str.)[950)]

(7) α-**Cadinol** (XLVII)[949)950)]

$\ll$Occurrence$\gg$ Same to (∂) δ–Cadinol.

mp. 74.8~75.4°, $[\alpha]_D$ −38.5°, −47°, $C_{15}H_{26}O$

NMR τ: 4.52 s $\left(>C=C<^H_{C\leftarrow}{}^H \right)$

(XLVII)[950)]

(8)   **Juneol** (XLIX) and **Laevojuneol** (L)[951)]

(XLIX)   (L)

(a)   **Juneol** (XLIX)

Santanolide C $\xrightarrow{LiAlH_4}$ $\xrightarrow[\text{reduction}]{\substack{\text{Huang}\\ \text{Minlon}}}$ Dihydrojuneol[951)] (LI)

$\ll$Occurrence$\gg$ *Juniperus communis* L.[952)].

         bp. 121~122°/$_{1\,mm}$, mp. 62.5~63°, $C_{15}H_{26}O$, phenylurethan mp. 186.5°[952)]

(b)   **Laevojuneol** (L)[951)]

$\ll$Occurrence$\gg$ *Vetiveria zizanoides* L.,   mp. 65°, bp. 90~102°/$_{0.5mm}$, $[\alpha]_D$ −57° (EtOH), $C_{15}H_{26}O$

IR See Fig. 150, ν_{max}^{nujol} cm^{-1}: 3400 (OH), 1780, 1639, 887 ($>C=CH_2$)

NMR τ: 5.18, 5.45 (vinylic protons)

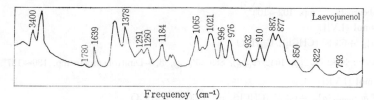

Frequency (cm^{-1})

Fig. 150   IR Spectrum of Laevojuneol (Bhattacharyya *et al.*)

949)   Herout, V. *et al*: *Tetrahedron*, **4**, 246 (1958).

950)   Dauben, W.G. *et al*: *ibid.*, **15**, 217 (1961).

951)   Bhattacharyya, S.C. *et al*: *Tetrahedron*, **18**, 969 (1962).

952)   Sorm, F. *et al*: *Chem. Listy.*, **50**, 1282 (1956); *C.A.*, **51**, 296h (1957).

(9) Juniper camphor (LII)[953)]

≪Occurrence≫   *Juniperus communis* L.[952)],   mp. 139~140°, $C_{15}H_{26}O$

Zingiberol (from. *Zingiber officinale*)

(A mixture)[953)] ⟶ Juniper camphor+β-Eudesmol (LVII)

HO $\langle$ H

(LII)

(10) Intermedeol (LIII)[954)]

≪Occurrence≫   *Bothriochloa intermedia*

mp. 47~48°, $[\alpha]_D$ +10.7° (EtOH), $C_{15}H_{26}O$

IR ν_{max}^{film} cm⁻¹: 3480, 1645, 910, 890

HO $\langle$ H

(LIII)

(11) α-Verbesinol (LIV) and **β-Verbesinol** (LV)[955)]

(LIV)   (LV)

As a mixture of (LIV)+(LV)

≪Occurrence≫   *Verbesina virginica* L.,   mp. 116~122°, $[\alpha]_D$ +49°, $C_{24}H_{32}O_3$

UV λ_{max}^{EtOH} mμ (ε): 230 (8300), 314 (18200); $\lambda_{max}^{0.1N-NaOH}$ mμ (ε): 313 (2500), 367 (26000)

(12) Eudesmols[956)]

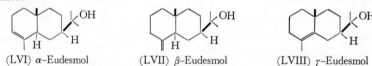

(LVI) α-Eudesmol   (LVII) β-Eudesmol   (LVIII) γ-Eudesmol

≪Occurrence≫   *Eucalyptus macarthuri, Machilus kusanoi, Melaleuca uncinata, Leptospermum flavescens,*

Balsamorrhiza sagittata

(a) α-Eudesmol (LVI)

mp. 75°, $[\alpha]_D$ +28.6° (CHCl₃), $C_{15}H_{26}O$,

IR ν_{max}^{film} cm⁻¹: 3016 (ν_{C-H}), 1645 ($\nu_{C=C}$), 802 (712) (δ'_{C-H})

3,5-Dinitrobenzoate   mp. 102~103°

(b) β-Eudesmol (LVII)

mp. 76°, $[\alpha]_D$ +63.8° (CHCl₃), $C_{15}H_{26}O$

IR ν_{max}^{film} cm⁻¹: 3072 (ν_{C-H}), 1642 ($\nu_{C=C}$), 889 (δ'_{C-H}), 3,5-Dinitrobenzoate mp. 136~137°

(c) γ-Eudesmol (LVIII)

bp. 83~86°/0.1mm, $[\alpha]_D$ +62.5° (CHCl₃), n_D^{20} 1.5087, $C_{15}H_{26}O$

IR ν_{max}^{film} cm⁻¹: (1640) ($\nu_{C=C}$), (818, 795, 763) (δ'_{C-H})

NMR τ[957)]: 9.00 s (angular Me), 8.86 (two Me of side chain), 8.40 $\left(\!\!>\!C=C\!<_{Me}\right)$, no vinyl absorption

953) Bhattacharyya, S.C. *et al*: *Tetrahedron*, **18**, 979 (1962).

954) Zalkow, L.H. *et al*: *Chem. & Ind.* **1963**, 38.

955) Gardner, P.D. *et al*: *J.A.C.S.*, **83**, 1511 (1961).   956) Mc Quillin, J. *et al*: *Soc.*, **1956**, 2973.

(13) **Cryptomeridiol** (LIX)[958]

≪Occurrence≫ *Cryptomeria japonica* D. Don (スギ),

mp. 134.5~135.5°, $[\alpha]_D$ −33.3° (CHCl₃), $C_{15}H_{28}O_2$

IR ν_{max} cm⁻¹: 3400, 1124 $\left(\text{>C-OH}\right)$

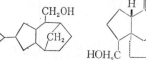

(LIX)

(14) **Occidentalol** (LX)[959][960]

≪Occurrence≫ *Thuja occidentalis* L.,

mp. 95°, $[\alpha]_D$ +361°, $C_{15}H_{24}O$

UV λ_{max} mμ (log ε): 266 (3.6) (endocyclic conj. system)
IR ν_{max} cm⁻¹: 1590, 845, 720 (conj. double bond)

(LX)

(15) **Bicyclovetivenol** (LXI) and **Tricyclovetivenol**[961]

CH₂OH

CH₃

(LXI)   [Revised Structure Note a)]

≪Occurrence≫ *Vetiveria zizanoides*

CH₂OH

CH₂

HOH₂C

(LXII)   [Revised Structure Note b)]

(16) **Widdrol** (LXIII)

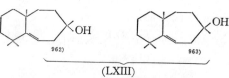

OH   OH

962)   963)

(LXIII)

≪Occurrence≫ *Widdringtonia cupressioides* and other spp.,

mp. 98°, $[\alpha]_D$ +105°[964][965]

Spectral data of the derivatives[963]

(a) (LXIV) mp. 154°

UV ε_{220}^{EtOH} 6.5, IR ν_{max} cm⁻¹: 3500 (OH, conctn.–independent) 3620 sh. (*cis* epoxyalcohol)

NMR 60Mc, in CCl₄, TMS, τ: 7.00 dd, $J=6$ and 7 (1H attached to epoxide ring)

(b) (LXV) mp. 119°

UV ε_{210}^{EtOH} 14, IR ν_{max} cm⁻¹: 3540, 3680 (free 2OH) (*trans* epoxyalcohol)

NMR τ: 7.20 dd, $J=3$ and 6 (1H attached to epoxide ring, configurational isomer of (LXIV))

957) Bates, R.B. *et al*: *Chem. & Ind.*, **1962**, 1759.
958) Sumimoto, M: *ibid.*, **1963**, 780, 1436.
959) Nakatsuka, T *et al*: *Bull. Agr. Chem. Soc.* **20**, 215 (1956), **23**, 140 (1959).
960) Erdtman, H. *et al*: *Tetrahedron*, **18**, 1315 (1962). 961) Decot, J. *et al*: *ibid.*, **4**, 1 (1958).
962) Enzell, C.: *Acta Chem. Scand.*, **16**, 1553 (1962); *Index Chem.*, **8**, 24656 (1963); *Tetr. Letters*, **1962**, 185.
963) Nozoe, T. *et al*: *Chem. Pharm. Bull.*, **11**, 132 (1963).
964) Erdtman, H. *et al*: *Acta Chem. Scand.*, **12**, 267 (1958); *C.A.*, **53**, 18977b (1959).
965) Nagahama, S: *Bull. Chem. Soc. Japan*, **33**, 1467 (1960).
[Note] a) I.C. Nigam, H. Komae, C.A. Neville, C. Radecka, and S.K. Paknikar: *Tetr. Letters*, 2497 (1968); F. Kido, H. Uda, and A. Yoshikoshi: *ibid.*, 2815 (1967).
   b) H. Takahashi: 12th Symposium on the Chemistry of perfumery, Terpene and Essential Oils p. 35 (1968).

(LXIII) $C_6H_5CO_3H$

$H_2SO_4 + HOAc$

(LXIV)   (LXVI)   (LXIX)

(LXVII)

(LXVIII)

(LXV)   (LXX)

(c) **(LXVI)** mp. 118~119°

UV ε_{210}^{EtOH} 1122,   IR ν_{max}^{nujol} cm^{-1}: 1623, 826

NMR τ: 9.13, 9.00 $\left(2 \; {>}C{-}Me\right)$, 8.75 $\left({>}C{<}^{OH}_{Me}\right)$, 8.33 $\left({>}C{=}\overset{|}{C}{-}Me\right)$, 6.12 dd, J=5.2 and 10.4

$({>}CHOH)$, 4.19 (unresolved m, $-CH_2{-}CH{=}C$)

(d) **(LXVII)** mp. 133°,

UV ε_{210}^{EtOH} 2076,   IR ν_{max}^{nujol} cm^{-1}: 1634, 905

NMR τ: 9.09, 8.96 $\left(2 \; {>}C{-}Me\right)$, 8.78 $\left(O{-}\overset{|}{C}{-}Me\right)$, 5.70 (unresolved m, ${>}CHOH$), 5.04 d, 4.84 d,

J=1.3 $({>}C{=}CH_2)$

(e) **(LXVIII)** mp. 141°

UV ε_{210}^{EtOH} 3073,   IR ν_{max}^{nujol} cm^{-1}: 1645, 816

NMR τ: 9.11, 9.11, 8.86 $\left(3 \; {>}C{-}Me\right)$, 8.79 $\left(O{-}\overset{|}{C}{-}Me\right)$, 5.9 $({>}CHOH)$, 4.64 t, J=4 $({-}CH{=}C{<})$

(f) **(LXIX)** mp. 138°

UV ε_{max}^{EtOH} mμ (ε): 225 (1137),   IR ν_{max}^{nujol} cm^{-1}: 1698

NMR τ: 9.92, 8.89 $\left(2 \; {>}C{-}Me\right)$, 8.67 $\left(O{-}\overset{|}{C}{-}Me\right)$ 8.38 d, J=1.5 $\left(Me{-}\overset{|}{C}{=}CH{-}\right)$, 4.53 (unresolved m,

$-CH_2CH{=}C{<})$

(g) **(LXX)** mp 100°

UV ε_{210}^{EtOH} 23.4,   NMR τ: 5.19 dd, J=4 and 11 $({>}CHOH)$

(17) Partheniol (LXXI)[966)967)]

<Occurrence>   *Parthenium argentatum*

mp. 127~128°, $[\alpha]_D$ +116.5° (CHCl$_3$), $C_{15}H_{24}O$

NMR τ: 5.02 br. $\left({>}C{=}C{<}^H\right)$, 5.45, 5.65 d

(LXXI)[966)]

966) Hendrickson, J.B. *et al*: *Chem. & Ind.*, **1962**, 1424.
967) Haagen-Smit, A.J. *et al*: *J.A.C.S.*, **70**, 2075 (1948).

(18) Carotol (LXXII) and **Daucol** (LXXIII)

(LXXII)[968) 969)]

(LXXIII)[969)]

≪Occurrence≫ *Daucus carota* L. (＝ンジン)

(a) Carotol (LXXII)[970)]

bp. 109°/1.5 mm, n_D^{26} 1.4944, $[\alpha]_D$ +28° (EtOH), $C_{15}H_{26}O$

NMR δ [Note a)] 9.050 $\left(\text{>CH-CH}\underset{\text{Me}}{\overset{\text{Me}}{<}}\right)$, flat peak $\left(\underset{-CH_2-\overset{H}{C}=C<}{}\right)$

(LXXII) $\xrightarrow{KMnO_4}$ Triol $\xrightarrow{CrO_3}$ acid $C_{15}H_{26}O_6$ (LXXIV)

(LXXIV) mp. 137~139°

NMR δ: 9.083 d $\left(-CH\underset{\text{Me}}{\overset{\text{Me}}{<}}\right)$

(LXXIV)

(b) Daucol (LXXIII)[969)]

Carotol (LXXII) $\xrightarrow{C_6H_5CO_3H}$ (LXXIII)

mp. 117~118°, $C_{15}H_{26}O_2$, Acetate mp. 80~81°, $C_{17}H_{28}O_3$, IR $\lambda_{max}^{CHCl_3} \mu$: 5.78, 8.00,

NMR δ: 8.932 s, 8.932 s $\left(2 \text{ >C-Me}\right)$, 9.184 d, 8.950 d $\left(-CH\underset{\text{Me}}{\overset{\text{Me}}{<}}\right)$, 6.265 dd $\left(-CH_2-CH\cdot OH-\overset{|}{\underset{|}{C}}-\right)$

(19) **Guaiol** (LXXV)[971)]

(LXXV)   (LXXVI)   (LXXVII)

≪Occurrence≫ Guaiac resin, champaca wood, *Eucalyptus maculate*,

mp. 91° $[\alpha]_D$ −30°, $C_{15}H_{26}O$, See the item [C] (14).

(20) **Bulnesol** (LXXVIII)[972)]

≪Occurrence≫ *Bulanesia sarmienti* LOR., $C_{15}H_{26}O$

See the item [C], (14).

(LXXVIII)

(21) **Hinesol** (LXXIX)[973)]

≪Occurrence≫ *Atractylodes lancea* DC. (茅蒼求)

mp. 59~60°, $[\alpha]_D$ −40.2° (CHCl_3), $C_{15}H_{26}O$

(LXXIX) [Former Str.] [Revised Str. Note b)]

968) Sorm, F. *et al*: *Tetr. Letters* No. 14, 24 (1959).

969) Zalkow, L. H. *et al*: *J. Org. Chem.*, **26**, 981 (1961).

970) Chiurdoglu, G. *et al*: *Tetrahedron*, **7**, 271 (1959).

971) Minato, H. *et al*: *Tetrahedron*, **13**, 308 (1961), **18**, 365 (1962); *Chem. Pharm. Bull.*, **9**, 625 (1961).

972) Dolejs, L. *et al*: *Tetr. Letters*, **1960**, 18.

973) Yoshioka, I. *et al*: *Chem. Pharm. Bull.*, **7**, 817 (1959), **9**, 84 (1961).

[Note] a) In the original report[969)], δ means $10.00-10^6 \times (\nu-\nu_{tms}/\nu_{tms})$. This value means τ value in usual.

b) J. A. Marshall, P. C. Johnson: *J.A.C.S.*, **89**, 2750 (1967).

J. A. Marshall, S. F. Brady: *Tetr. Letters*, 1387 (1969).

(22) α and β-**Betulenols** (LXXX) and (LXXXI)[974]

(LXXX) *trans*
α–Betulenol

(LXXXI) *cis*
β–Betulenol

≪Occurrence≫   *Betula lenta* L.

(**a**) α–**Betulenol** (LXXX)

d^{20} 0.978, n_D 1.5148, $[\alpha]_D$ −19.5, $C_{15}H_{24}O$

IR (α–or β–) ν_{max} cm^{-1}: 3350 (assoc. OH), 3068, 1635, 893 ($>$C=CH$_2$), 3025, 1670 $\left(>C=C<^H\right)$

(**b**) β–**Betulenol** (LXXXI)

bp. 157~158°/ 2.0 mm, d^{20} 0.975, n_D 1.5132, $[\alpha]_D$ −36°, $C_{15}H_{24}O$

(LXXX), (LXXXI)⟶Caryophyllene (XIII)

(iii) *Tricyclic alcohols*

(23) (+)-**Maaliol** (LXXXII)[975][976]

≪Occurrence≫   *Canarium samonense, Nardostachys jatamansi* DC.

mp. 103.5~105°, $[\alpha]_D$ +18.35° (EtOH), $C_{15}H_{26}O$

IR ν_{max}^{KBr} cm^{-1} [975]: 3333, 2857, 1449, 1379, 1163, 1136, 1099, 1031,

970, 935, 917, 900, 877, 869, 833, 791, 752   (LXXXII)

ν_{max} cm^{-1} [976]: 1456, 1381, 1337, 1326, 1303, 1270, 1240, 1229, 1218, 1198, 1175, 1158, 1143, 1123,
1105, 1069, 1054, 1037, 1018, 1005, 985, 976, 954~948 etc.

(24) **Spathulenol** (LXXXIII)[977]

≪Occurrence≫   *Eucalyptus spathulata* var. *grandiflora*,

$[\alpha]_D$ +56°, $C_{15}H_{24}O$

IR ν_{max} cm^{-1}: 3605, 3080, 1635, 890 (*tert.*–OH and $>$C=CH$_2$)

NMR in CDCl$_3$, 60Mc, τ: 5.32 (2) ($>$C=CH$_2$), 8.97 (6) $\left(\!\!>\!C\!-\!Me\right)$,

8.73 (3) $\left(>C<^{OH}_{Me}\right)$, 9.4 m $\left(\!\!>\!C\!-\!H \text{ in cyclopropane}\right)$

(LXXXIII)

(25) **Patchouli alcohol** (LXXXV)[978][878] [Note]

(LXXXIV)[978]
(Former Str.)

(LXXXV)[979]
(Revised Str.)

974) Treibs, W. *et al*: *Ann.*, **634**, 124 (1960).   975) Büchi, G. *et al*: *J. A. C. S.*, **81**, 1968 (1959).

976) Yves-René Naves: *Helv. Chim. Acta*, **46**, 2139 (1963).

977) Bowyer, R. C. *et al*: *Chem. & Ind.*, **1963**, 1245.

978) Büchi, G. *et al*: *J. A. C. S.*, **83**, 927 (1961). **84**, 3205 (1962), **78**, 1262 (1956).

979) Dobler, M. *et al*: *Proc. Chem. Soc.*, **1963**, 383.

[Note] (LXXXIV) had been proposed by Büchi *et al.*[978], which was revised to (LXXXV) as the results of X-ray analysis by Dobler *et al.*[979].

≪Occurrence≫   *Pogostemon patchouli* var. *suavis*,   mp. 55~56°, $[\alpha]_D$ −129° (CHCl₃), C₁₅H₂₆O

Acetate bp. 101~101.5°/0.2mm, mp. 24°, n_D 1.5010, d_4^{20} 1.043, $[\alpha]_D$ −71°

IR ν_{max}^{liq} cm⁻¹: 2941~2841, 1715, 1460, 1361, 1307, 1250, 1186, 1160, 1105, 1031~1000, 961, 934, 877,

855, 781, 746, 724, 690

(LXXXV) $\xrightarrow{-H_2O}$ α–Patchoulene + β–Patchoulene

NMR (τ) of α–Patchoulene[978] : 4.95 br $\left(-\overset{|}{C}=\overset{|}{C}-H\right)$, 8.37 $\left(CH_3-\overset{|}{C}=\overset{|}{C}-\right)$, 9.10 d, J=6c/s (−CHCH₃)

9.05, 9.11 (gem Me₂)

(26) Cedrol (LXXXVI)[980]

≪Occurrence≫   *Juniperus virginiana, J. procera, J. chinensis,*

J. occidentalis, Dacrydium elatum, Sciadopitys verticillata,

mp. 84~87°, $[\alpha]_D$ +9.3~10.5°, C₁₅H₂₆O

Total synthesis

(LXXXVI)

[E]   Ethers and Furano Compounds

(1) The following Compounds are described in other items

(XLI) Farnesiferol B   (XLII) Farnesiferol A   (LXXIII) Daucol

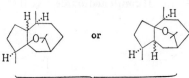

or

(XXXIV) Guaioxide

Farnesiferol B (XLI) ⟶ [D], (2)

Farnesiferol A (XLII) ⟶ [D], (3)

Daucol (LXXIII) ⟶ [D], (18)

Guaioxide (XXXIV) ⟶ [C], (14)

980) Stork, G. *et al*: *J.A.C.S.*, **83**, 3114 (1961), **75**, 3291, 3292 (1953).

(2)   **Dendrolasin** (LXXXVII) and **Torreyal** (LXXXVIII)[945)981)]

R=CH$_3$ (LXXXVII) Dendrolasin
R=CHO (LXXXVIII) Torreyal

(a)   **Dendrolasin** (LXXXVII)

≪Occurrence≫   *Lasius fulginosus* LATR. (ant)[982)], sweet potato fusel oil[983)] *Torreya nucifera*[981)]
bp. 148°/6 mm, d_{23} 0.9108, n_D^{20} 1.4849, $[\alpha]_D$ ±0, $C_{15}H_{22}O$

(b)   **Torreyal** (LXXXVIII)[981)]

≪Occurrence≫   *Torreya nucifera* S. et Z.,   bp. 124~126°/0.05 mm, $[\alpha]_D$ +1.90°, $C_{15}H_{20}O_2$

UV $\lambda_{max}^{EtOH} m\mu$ (ε): 224 (15940)

IR ν_{max}^{liq} cm^{-1}: 2720 (CHO), 1690 (conj. C=O), 1645 (C=C), 1570, 1504, 1164, 1028, 874, 779 (furan)

(LXXXVIII) $\xrightarrow{\text{Kishner Red.}}$ Dendrolasin

(3)   **Farnesiferol C** (LXXXIX)[170)946)]

≪Occurrence≫   *Asa foetida*,
mp. 84~85°, $[\alpha]_D$ −29° (CHCl$_3$), $C_{24}H_{30}O_4$
(LXXXIX)⟶(XC)
NMR   of (XC) in CDCl$_3$, 60Mc, TMS ppm: 6.23 d (gem Me$_2$), 5.95 s

$\left(\text{>C-Me}\right)$, 5.56 $\left(\text{Me-}\overset{|}{\text{C}}=\overset{|}{\text{C}}-\right)$, 2.15 $\left(\text{H-}\overset{|}{\text{C}}=\overset{|}{\text{C}}-\right)$, 3.60 d

$\left(\text{-CH}_2\text{-CH}<\overset{O^-}{\text{C}}\right)$

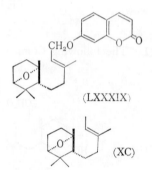

(LXXXIX)

(XC)

(4)   **Humulenemonoxide** (XCI) and **Humulenedioxide** (XCII)[983a)]

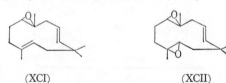

(XCI)   (XCII)

≪Occurrence≫   *Zingiber zerumbet* SMITH

Humulenemonoxide (XCI) n_D^{25} 1.4955, d 0.9575, $[\alpha]_D$ −43.3°, $C_{15}H_{24}O$

IR ν_{max} cm^{-1}: 1388, 1369 (gem Me$_2$), 968 $\left(_H\text{>C=C<}^H\text{-}trans\right)$, 820 $\left(\text{>C=C<}^H\right)$, See Fig. 151

Humulenedioxide (XCII)   mp. 105°, $C_{15}H_{24}O_2$,   IR ν_{max} cm^{-1}: 975 $\left(\text{disubst. }_H\text{>C=C<}^H\right)$

See Fig. 151

981)   Sakai, T. *et al*: *Tetr. Letters*, **1963**, 1171.
982)   Quilico, A: *Tetrahedron*, **1**, 177 (1957); *Ricerca Sci.* **26**, 177 (1956); *C.A.*, **51**, 361h (1957).
983)   Hirose, Y. *et al*: *J. Chem. Soc. Japan*, **82**, 725 (1961).
983a)   Bhattacharyya, S.C. *et al*: *Tetrahedron*, **18**, 575 (1962).

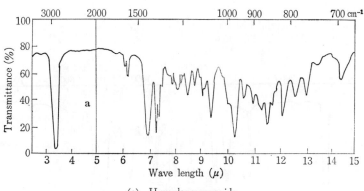

(a) Humulenemonoxide

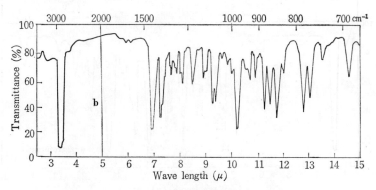

(b) Humulenedioxide

Fig. 151 IR Spectra of (a) Humulenemonoxide and (b) Humulenedioxide in CHCl₃ (Bhattacharyya *et al.*)

Humulene See the item 〔C〕, (1).

(5) α-and β-Agarofuran[984]

(XCIII) β-Agarofuran   (XCIV) α-Agarofuran   (Revised Str.) [Note]

≪Occurrence≫ Agar oil, obtained from decayed agarwood (*Aquillaria agallocha* ROXB.) by fungi.

(a) β-Agarofuran (XCIII)

bp. 130°/8 mm, n_D^{28} 1.4973, d 0.9646, $[\alpha]_D$ −127.1° (CHCl₃), $C_{15}H_{24}O$

IR See Fig. 152, NMR See Fig. 153.

984) Bhattacharyya, S.C. *et al*: *Tetrahedron*, **19**, 1079, 1519 (1963).
[Note] H.C. Barrett and G. Büchi: *J.A.C.S.*, **89**, 5665 (1967).

(b) α-**Agarofuran** (XCIV)

bp. 134°/4mm, n_D^{30} 1.5062, $[\alpha]_D$ +39.8°, $C_{15}H_{24}O$

IR See Fig. 152   NMR See Fig. 153

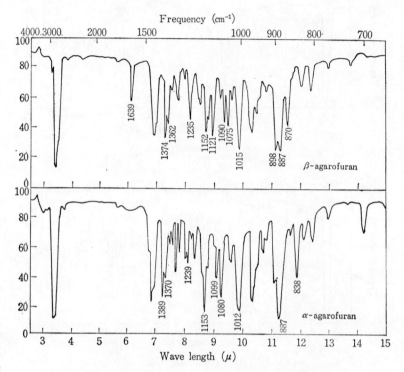

Fig. 152 IR Spectra of β- and α-Agarofurans with liquid film
(Bhattacharyya *et al.*)

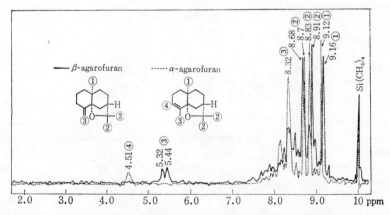

Fig. 153 NMR Spectra of β- and α-Agarofurans (in CCl₄, 60Mc,
TMS ppm) (Bhattacharyya *et al.*)

(6) Norketoagarofuran (XCV), **4-Hydroxydihydroagarofuran** (XCVI) and **3,4-Dihydroxy-dihydroagarofuran** (XCVII)[984]

(XCV)   (XCVI)   (XCVII)

≪Occurrence≫   Same to (5) α–and β–Agarofuran

(a) Norketoagarofuran (XCV)

mp. 56~57°, $[\alpha]_D$ −118.86° (CHCl$_3$), C$_{14}$H$_{22}$O$_2$

IR ν_{max}^{nujol} cm^{-1}: 1712, 1418, 1379, 1361, 1295, 1229, 1149, 1110, 1067, 1013, 888

NMR in CCl$_4$, 60Mc. TMS, ppm: 0.86 s (angular Me), 1.37 s, 1.11 s (gem Me$_2$) etc.

(b) 4-Hydroxydihydroagarofuran (XCVI)

mp. 130~131°, $[\alpha]_D$ −75.7° (CHCl$_3$), C$_{15}$H$_{26}$O$_2$

IR ν_{max}^{nujol} cm^{-1}: 3460, 1374, 1359, 1323, 1290, 1242, 1229, 1195, 1148, 1136, 1114, 1099, 1082, 1054,

1028, 1011, 1001, 985, 966, 954, 939, 929, 910, 885, 875, 852, 831, 810, 789, 768, 700.

NMR in CCl$_4$, 60Mc, TMS, ppm: 1.15 s (angular Me and one Me of gem Me$_2$), 1.24 (another Me of gem

Me$_2$), 1.32 $\left(>C<{}^{OH}_{Me}\right)$, 1.65 $\left(>C\text{–}OH\right)$

(c) 3,4-Dihydroxydihydroagarofuran (XCVII)

mp. 176°, $[\alpha]_D$ −40.98° (CHCl$_3$), C$_{15}$H$_{26}$O$_3$

IR ν_{max}^{nujol} cm^{-1}: 3520, 1431, 1397, 1389, 1316, 1295, 1235, 1140, 1106, 1089, 1075, 1038, 1015, 1003,

983, 962, 945, 933, 909, 881, 836, 828, 811.

(7) Zederone (XCVIII)[985]

(XCVIII)   (XCVIIIa)
(Suggested Str.)

The structure (XCVIII–a) seems to be better from the view of biogenesis[985].

≪Occurrence≫   *Curcuma zedoaria* ROSCOE.,   mp. 153.5~154°, $[\alpha]_D$ +265.8° (CHCl$_3$), C$_{15}$H$_{18}$O$_3$

IR ν_{max}^{KBr} cm^{-1}: 1660 ($\alpha\beta$–unsatur. C=O), 1524, 898 w, 882 w (β–furyl ketone), 1570~1560, 1520~1500

(two bands) (C=C), 1155~1150 s, 875~870 (sharp)

NMR in CDCl$_3$, 100Mc, TMS, ppm: 7.10 q, $J=1.3$ c/s (1)

$\left(\underset{H}{\overset{}{\underset{O}{\bigsqcup}}}\right)$, 2.12 d, $J=1.3$ (3) $\left(\begin{matrix}H_3C\\ \\H\end{matrix}\underset{O}{\bigsqcup}\right)$, 5.49 dd, ($J=3.7,\ 10.9$) (1) $\left(_{-CH_2}^{H}>C=C\right)$,

1.60 s (3) $\left(>C=C<^{CH_3}\right)$, 1.53 s (3) $\left(>C<^{O-}_{CH_3}\right)$, 3.72 s (2) $\left(>C\text{–}CH_2\text{–}C<\right)$, 3.81 s (1) $\left(>C\underset{O}{\diagdown\diagup}C<^{H}_{CO-}\right)$

984) Bhattacharyya, S.C. *et al*: *Tetrahedron*, **19**, 1079, 1519 (1963).
985) Takemoto, T. *et al*: *7 th Symposium on the Chemistry of Natural Products*, p. 174 (Fukuoka, 1963).

(8) Linderane (XCIX) and Linderene (C)[986]

(XCIX)

(C) (NMR Signals)

≪Occurrence≫   *Lindera strychnifolia* VIEL (天台烏薬) (*Lauraceae*)

(a) **Linderane** (XCIX)[987]

mp. 190~191°, $[\alpha]_D$ +180. 3°, $C_{15}H_{16}O_4$

UV $\lambda_{max} m\mu$ (ε): 214 (6770) (furan deriv.)

IR $\nu_{max} cm^{-1}$: 3130, 3060, 1616, 1556 (furan deriv.), 1779 (γ–lactone)

NMR τ: 2.87 , 8. 00

(b) **Linderene** (C)[988]

mp. 145°, bp. 164°/4.5 mm, $[\alpha]_D$ +15. 14° (CHCl₃), $C_{15}H_{18}O_2$,

UV $\lambda_{max} m\mu$ (ε): 206. 1 (10950)

NMR   Chemical shifts of protons of linderene are indicated as τ (J cps) on the structure (C) beside each protons.

(9) Verrucarol (CI)[989]

(Former Structure) (CI)   (Revised Structure)

Alkaline hydrolysis of Verrucarin A

Verrucarin A + 2H₂O = Verrucarol + Muconic acid + Verrucanic acid

($C_{27}H_{34}O_9$)   ($C_{15}H_{22}O_4$)   ($C_6H_6O_4$)   ($C_6H_{10}O_3$)

(a) **Verrucarin A**[990]

≪Occurrence≫   Metabolite of *Myrothecium verrucaria, M. roridum,*

mp. >360°, $[\alpha]_D$ +206° (CHCl₃), $C_{27}H_{34}O_9$

UV $\lambda_{max}^{EtOH} m\mu$ (log ε): 260 (4. 25);   IR $\lambda_{max}^{CH2Cl2} \mu$: 2.84, 5.83, 6.12, 6.29

---

986)   Takeda, K. *et al*: *7 th Symposium on the Chemistry of Natural Product*, p. 180 (1963).

987)   Kondo, H. *et al*: *Y. Z.*, **40**, 1047 (1925), **50**, 714 (1930), **59**, 504 (1939).

988)   Takeda, K *et al*: *Y. Z.*, **64**, 154 (1944); *Pharm. Bull.*, **1**, 164, 244 (1953).

989)   Tamm, Ch *et al*: *Helv. Chim. Acta*, **46**, 1786 (1963).

990)   Tamm, Ch. *et al.*: *ibid*, **45**, 839, 1726 (1962).

(b) Verrucarol (CI)[990]

mp. 155~156°, $[\alpha]_D$ −39° (CHCl₃), $C_{15}H_{22}O_4$

UV λ_{max}^{EtOH} mμ (log ε): 195 (3.90), (isolated. C=C); IR λ_{max}^{CH2Cl2} μ: 2.78, 2.89 (OH), 5.96 $\left(>C=C<^H_{}\right)$,

9.28 (cyclic −O−), 10.36 (?), 12.16 $\left(>C=C<^H_{}\right)$; (CaF₂ prism) 3.28 $\left(\text{C–H of }>C\diagup^{O}_{\diagdown}CH-\right)$

NMR[989] 60Mc, TMS, τ: 5.3 br. t (C₄–H), 6.9 d (J=4), 7.2 d (J=4) AB–system (–$\overset{13}{C}H_2$), ca 6.3 (overlap) (C₂–H, C₁₁–H) 4.55 d (J=5) (C₁₀–H), 9.08 s (Me₁₄), 8.28 s (Me₁₆), 6.35 (–$\overset{15}{C}H_2$–OH), ca 7.6 (OH)

(Revised Str.) [Note]

(10) Furanosesquiterpenes of *Petasites* spp.[1016]

(CI–a)
Furanoeremophilane

(CI–b)
Furanoeremophilone

R=H (CI–c): Petasalbin
R= —CO–C=CHCH₃ (*cis*) (CI–d):
　　　　　　　　CH₃ Albopetasin

R=H (CI–e): Albopetasol
R= —CO–C=CH–CH₃ (*cis*) (CI–f):
　　　　　　　　CH₃　Furanopetasin

≪Occurrence≫　*Petasites officinalis* MOENCH., *P. albus* GAERTN.

Table 77 Components of *Petasites officinalis* (Sorm *et al.*)

Compounds	Str. No	Formula	mp. (bp.)	$[\alpha]_D$	λ_{max} mμ (log ε)	ν_{max} cm⁻¹
Furanoeremophilane	CI–a	$C_{15}H_{22}O$	(148°/₁₆)	−8.4°	222 (3.78)	1576, 1660, 1776, 1810
Furanoeremophilone	CI–b	$C_{15}H_{20}O_2$	150°	±0°	280 (4.18)	1537, 1667, 3000
Petasalbin	CI–c	$C_{15}H_{22}O_2$			220 (3.84)	1567, 3485, 3615
Albopetasin	CI–d	$C_{20}H_{28}O_3$	106~107°	±0°	224 (4.08)	1572, 1653, 1718
Albopetasol	CI–e	$C_{15}H_{22}O_3$	178~180°	−24.1°	220 (3.82)	1568, 1652, 3320, 3370
Furanopetasin	CI–f	$C_{20}H_{28}O_4$	105~106°	±0°	222 (4.17)	1567, 1655, 1718, 3450, 3608

(11) Atractylon (CI–g)[990a]

≪Occurrence≫　*Atractylodes japonica* KOIDZ. (和蒼朮)

mp. 38°, $[\alpha]_D$ +40° (CHCl₃), $C_{15}H_{20}O$

UV λ_{max}^{EtOH} mμ (log ε): 220 (3.89)

(CI–g)

IR ν_{max}^{nujol} cm⁻¹: 1134 (–O–), 3077, 1639, 886 (vinylidene)

NMR in CCl₄, 40Mc, H₂O ref. as 5.30 τ: 3.15 (α–H), 8.15 d, J=1 c/s (β–Me), 5.26, 5.39 (>C=CH₂)

990a) Yoshioka, I. *et al*: *Chem. Pharm. Bull.*, **12**, 755 (1964).
[Note] Ref. J. Gutzwillen, R. Mauli, H. P. Sigg and C. Tamm: *Helv.* **47**, 2234 (1964).

(12) Animal Sesquiterpenes[991]

(CII) Aplysin   (CIII) Debromoaplysin   (CIV) Aplysinol

≪Occurrence≫   Unsaponificable matter from *Aphysia kurodai* (sea animal, ウミウシ, アメフラシ)

(a) Aplysin (CII)

mp. 85~86°, $[\alpha]_D$ −85.4° (CHCl$_3$), C$_{15}$H$_{19}$OBr

UV λ_{max}^{EtOH} mμ (log ε): 294 (3.66), 234 (3.95)

IR $\nu_{max}^{CCl_4}$ cm^{-1}: 1620, 1586, 1486, 1277, 1240, no OH

NMR   in CCl$_4$, 40Mc 8.91 d (3H), 8.75 s (3H), 8.07 s (3H), 7.68 s (3H), 3.48 s (1H), 2.95 s (1H)

Mass [Note] m/e 296 s, 294 s, 282 m, **281**, 280 w, 279 s, 240 m, 239 s, 238 m, 237 s, 202 m, 201 s, **185 w,** 160 s, 159 m, 128 w, 115 m, 109 w, 91 w, 77 w, 55 w, 44 w, 43 w, 41 m

(b) Debromoaplysin (CIII)

oil, mononitrodebromoaplysin mp. 107~108°

NMR   in CCl$_4$, 40Mc, τ: 9.00 d (3H), 8.73 s (3H), 8.66 s (3H), 7.67 s (3H), 3.42 s (1H), 3.10 q (2H)

Mass m/e 216 s, 202 m, **201**, 187 w, 173 m, 160 s, 159 s, 145 m, 91 w, 55 w, 43 m, 41 m

(c) Aplysinol (CIV)

mp. 158~160°, $[\alpha]_D$ −55.6° (CHCl$_3$), C$_{15}$H$_{19}$O$_2$Br

UV λ_{max}^{EtOH} mμ (log ε): 292 (3.60), 233 (3.89)

IR $\nu_{max}^{CCl_4}$ cm^{-1}: 1615, 1577, 1482, 1277, 1240

NMR   in CS$_2$, τ: 8.93 d (3H), 8.57 s (3H), 7.70 s (3H), 6.22 s (2H), 3.49 s (1H), 2.99 s (1H)

(13) Kessane (CV)[992]

≪Occurrence≫   *Valeriana officinalis* L. var. *latifolia* MIQ. and other Japanese *Valeriana* spp.,

bp. 110~112°/6 mm, d$_4^{25}$ 0.970, n$_D^{25}$ 1.491, $[\alpha]_D$ −7.2° (CHCl$_3$), C$_{15}$H$_{26}$O

IR ν_{max} cm^{-1}: 1095 (ether)

NMR   in CCl$_4$, 60Mc, TMS τ: 9.23 d, J=6.0 c/s (3H) (CH$_3$–CH<), 8.98 s (3H),

(CV)

$$8.82 \text{ s}(6H)\left(\text{>C-CH}_3 \text{ and } -\overset{CH_3}{\underset{CH_3}{C}}-O-\right)$$

See the item [C], (14), (XXXIV) Guaioxide

---

991) Yamamura, S. *et al* : *Tetrahedron*, **19**, 1485 (1963); *6 th Symposium on the Chemistry of Natural Products*, p. 76 (Sapporo, 1962).

992) Takemoto, T *et al* : *Chem. Pharm. Bull.*, **11**, 547 (1963).

[Note]   Relative intensity of each peak is indicated in parenthesis as s>20%, m>20~15%, w<15%, when the intensity of the peak m/e **281** is 100% in the mass spectrum of (CII).

(14) α-**Kessylalcohol** (CVI) and **Kessylglycol** (CVII)[993)~597)]

(CVI)[993) 994)]   (CVII)[993) 994)]

≪Occurrence≫   As acetates from the essential oil of *Valeriana officinalis* L. var. *latifolia*

α–**Kessylalcohol** (CVI)

mp. 85~86°, $[\alpha]_D$ −38.4°, $C_{15}H_{26}O_2$

Kessylglycol (CVII)

mp. 128°, (monohydrate) 58~59°, $[\alpha]_D$ −24.38°, $C_{15}H_{26}O_3$

Determination of stereochemical structure of α–kessylalcohol (CVI) by means of NMR, chemical reaction and ORD[994)].

(a) NMR [Note] (CVI): 0.8 d, J=6.0 (3H) $\left(CH_3-C\diagdown\right)$ 1.23 s, 1.26 s, 1.33 s (each 3H) (12–Me, 13–Me, 15–Me)

(b) Configuration of 2α–(R)–H; 2β–(S)–OH of (CVI) was confirmed by $[M]_D$ as following:

(CVI)   (CVIII) 2–*epi*–α–Kessyl-
α–Kessylalcohol   alcohol

$[M]_D^{(CVI)} \sim [M]_D^{(CVI-benzoate)}$ $[M]_D^{(CVIII)} \sim [M]_D^{(CVIII-benzoate)}$

$= \Delta [M]_D = -192°$ $= \Delta [M]_D' = +298°$
2–α–Series (R)   2–*epi*–Series (S)

(c) **Configuration of** C_1**–OH**

993) de Mayo, P: *Perfumery & Essential Oil Records*, **48**, 18 (1957).
994) Nozoe, T. *et al*: *7 th Symposium on the Chemistry of Natural Products* p 168 (Fukuoka, 1963).
995) Kanaoka, K. *et al*: *Y. Z.*, **61**, 123 (1941).
996) Ukita, Ch: *ibid.*, **64**, 285, (1944), **65**, 458 (1945).
997) Treibs, W: *Ann.*, **570**, 165 (1952).
[Note] in CCl_4, 60Mc, ppm (δ)

Lactone (CXII) $C_{14}H_{22}O_3$, mp. 65~66.5°, $[\alpha]_D$ +24.0°

IR ν_{max} cm^{-1}: 3597 (OH), 1770 (γ-lactone), NMR δ: 0.93 d, J=6.3 (CH$_3$CH$\langle$), 1.28 $\left(\text{CH}_3-\text{C}\langle^{\text{OH}}_{\text{C=O}}\right)$,

1.62, 1.73 br $\left(^{\text{CH}_3}_{\text{CH}_3}\rangle\text{C=C}\langle^{\text{H}}\right)$

The junction of all known lactones, fused to pentacyclic ring is *cis*, then 2β–OH/C$_1$–H must be *trans*

	(CVI) $\xrightarrow{\text{CrO}_3}$	(CXIII) α–Kessyl –ketone	$\xrightarrow{\text{OH}^-\text{ or Al}_2\text{O}_3}$	(CXIV)
		(CXIII)		(CXIV)
C$_1$–H		α		β
Cotton E.		[M]$_{317}$ +9060° [M]$_{278}$ −5700°		[M]$_{315}$ −9840° [M]$_{280}$ +9300°

(d) Configuration of C$_4$

Stereochemical relation between 14–Me/2–OH of (CVI) was confirmed by the comparison of the NMR shift (δ) of methyl protons with those of 2–*epi*–α–kessyl alcohol (CXV) and of 18–Me of steroid (CXIV–a)

(CXIV–a) 18–Me/16–OH *trans* 0.71 *cis* 0.95

(CVI)   14–Me/ 2–OH *trans* 0.80

(CXV)   "   *cis* 0.91

(CXIV–a)

$$(\text{CXIII}) \underset{\text{CrO}_3}{\overset{\text{LiAlH}_4}{\rightleftarrows}}$$

(CXV)

(e) Configuration of C$_5$

(CXIII) ―――――――――――――→ (CXIV)

Bayer-Billiger Oxidation
C$_6$H$_5$CO$_3$H

(CXVI)   (CXVII)

NMR in CHCl$_3$ δ:

Stereochemical correlation of C$_1$ and C$_5$ was elucidated from following data, calculated by Karplus's equation.

Calculated value: [998)~1000)]

$\phi°$	0	30	60	90	120	150	180
J_{AB} c/s	8.2	6.0	1.7	−0.28	2.2	6.9	9.2

Experimental results:

	Chemical shift δ	$J_{c/s}$	$\varphi°$	1–H/5–H
(CXVI)	4.12 d	10.2	180	*trans*
(CXVII)	4.31 d	4.0	~60	*cis*

(f) Configuration of C_7 and C_{10}

IR   Stereochemical correlation of oxide bridge at C_{10} and 2-OH was elucidated from the IR absorptions of (CVI), (CVIII) and (CXV) in various concentrations (0.2~0.001M) of carbontetrachloride solution.
Experimental results:

Compound	$\nu_{OH}^{CCl_4}$ in various concentrations		Configuration	
	indipentent	intensity is effected	C_2-OH	C_{10}-O-
(CVI)	3570		β	β
(CVII)		}3625, 3430	}α	β
(CXV)				

NMR   in CCl_4, 60Mc ppm: The influence of 2-OH to the chemical shifts of 12-Me, 13-Me and 15-Me is summarized as following: 1.20~1.23 (12-Me & 13-Me), 1.33, 1.15, 1.02 (15-Me of (CVI) (CVIII) and (CXVIII), resp.

Signals (1.33, 1.15 ppm) due to 15-Me protons of (CVI)

(CXIII) $\xrightarrow[\text{Red}]{\text{W.K.}}$

(CXVIII)

and (CVIII) are shifted to lower field from 1.02 ppm of (CXVIII) by the effect of hydroxyl proton. This shift disappears in their acetates should be interpreteted that 15-Me and 2-OH in (CVI) and (CVIII) are in parallel direction (*trans*), then oxide bridge at C_{15} of (CVI) must be β to C_2-OH.

(g) Configuration at C_8-OH of Kessylglycol (CVII)

(CVII)   (CIX)   (CXIX)
8-*epi*-Kessylglycol

$[M]_D^{(CVII)} \sim [M]_D^{(CVII-dibenzoate)}$

$= \Delta [M]_D = -184°$
Effect of 2β-OH $-192°$
$\Delta [M]^{(CVII)}$ $+8°$
8-α (R)

$[M]_D^{(CXIX)} \sim [M]_D^{(CXIX-dibenzoate)}$

$= \Delta [M]_D = -337°$
Effect of 2β-OH $-192°$ [Note]
$\Delta [M]^{(CXIX)}$ $-145°$
8-β (S)

(15) Kessanol (CXX)[1001]

《Occurrence》   As acetate, from *Valeriana officinalis* L. var.
latifolia (カノコソウ)

(a) Kessanol (CXX) mp. 115~116°, $[\alpha]_D$ +11.4° ($CHCl_3$), $C_{15}H_{26}O_2$

IR ν_{max}^{KBr} cm^{-1}: 3378 (OH), 1096 (-O-)

(CXX)

NMR   in CCl_4, 60Mc, TMS, τ: 9.18 d, $J=5.4$ (3H) (CH_3-CH<), 8.97 s, 8.85 s, 8.73 s (3H resp.)

$\left(CH_3-\overset{|}{\underset{|}{C}}-O-\right)$, 7.50 m (1H) $\left(H-\overset{|}{\underset{|}{C}}-OH\right)$

998)   Kawazoe. Y. *et al*: *Chem. Pharm. Bull.*, **10**, 338 (1962), **11**, 643 (1963).
999)   Karplus: *J. Chem. Phys.*, **30**, 11 (1959)
1000)   Jackman, L.M: Application of Nuclear Magnetic Resonance Spectroscopy in Organic Chemistry (Translated by Shimizu, H): p 126 (1962)
1001)   Takemoto, T. *et al*: *Chem. Pharm. Bull.*, **11**, 952 (1963).
   [Note]   On the effect of 2β-OH, see the item (b)

(b) Acetate d_4^{25} 1.051, n_D 1.488, $[\alpha]_D$ −15.6°, $C_{17}H_{28}O_3$

IR $\nu_{max}^{liq.}$ cm^{-1}: 1736, 1234 (OAc), 1081 (–O–)

NMR: 9.19 d, J=6.1 (3H) (CH$_3$–CH⟨), 8.97 s (3H), 8.76 s (6H) $\left(CH_3\text{–}\overset{|}{C}\text{–O–}\right)$, 7.98 (3H) (OAc), 4.88

m (1H) $\left(H\text{–}\overset{|}{C}\text{–OAc}\right)$

(16) Torilolone (CXXII)[1002]

Hydrolysis →

(CXIX,) (Suggested Str. of original compound)

(CXXII) (Suggested Str.)

≪Occurrence≫ Alkaline hydrolysate of the fruit extracts of *Torilis japonica*(Houtt.)DC. (ヤブジラミ)
mp. 136∼137°, $[\alpha]_D$ −145° (EtOH), $C_{15}H_{24}O_3$

UV λ_{max} mμ (log ε): 289 (1.49)
IR λ_{max} cm^{-1}: 3500 (OH), 1735 (five membered ring ketone)
NMR 60Mc, τ: 5.84 m (1H), 8.83 (6H), 9.00 d (3H), 9.12 d (3H)
ORD: antipodal to that of methyl allogeigerate (CXXIII)[1003]

(CXXIII)

(17) Bilobanone (CXXIV)[1004] **and α-Alantone** (CXXV)

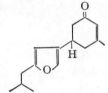

(CXXIV)   [Revised Str.][1004a]   (CXXV)

(a) Bilobanone (CXXIV)[1004]

≪Occurrence≫ Heart wood of *Ginkgo biloba* L. (イチョウ), bp. 123∼127°/1mm, $d^{22°}$ 0.978,
n^{20} 1.509, $[\alpha]_D$ +44.36°, $C_{15}H_{22}O_2$
IR λ_{max}^{nujol} μ: 5.85, 9.05, 10.65, 12.55   Semicarbazone   mp. 142∼143°

(b) α-Alantone (CXXV)   See the item [F], (2)

[F] Aldehydes and Ketones

(1) The following Compounds are described in other items

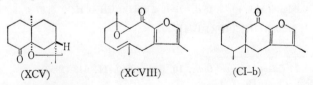

(XCV)   (XCVIII)   (CI–b)

1002) Nakazaki, M *et al*: *7 th Symposium on the Chemistry of Natural Products* p. 162 (Fukuoka, 1963).
1003) Barton, D.H.R. *et al*: *Soc.*, **1958**, 4518.
1004) Kimura, H. *et al*: *Y. Z.*, **78**, 1152 (1958), **82**, 214, 888 (1962).
1004a) Chem. Commun., 678 (1967).

Norketoagarofuran (XCV) ⟶ 〔E〕, (6)
Zederone (XCVIII) ⟶ 〔E〕, (7)
Furanoeremophilone (CI–b)⟶〔E〕, (10)

(i) *Monocyclic Ketones*

(2) *α*-**Alantone** (CXXV)[1005]

《Occurrence》 *Cedrus deodora* LOUD. (ヒマラヤスギ), *C. alantica* MANETTI
bp. 121~123° / 1 mm, d_{15}^{15} 0.9562, n_D^{20} 1.5181, $[\alpha]_D$ +2.5°, $C_{15}H_{22}O$

(CXXV)

(3) **Zerumbone** (CXXVI)[1006]

《Occurrence》 *Zingiber zerumbet*, mp. 66~67°, $C_{15}H_{22}O$
UV $\lambda_{max} m\mu$ (ε): 248 (8480), 325 (250), 233 (8000)

IR $\nu_{max}^{CCl_4} cm^{-1}$: 1662, 1650 sh, 1635 (C=C–CO); $\nu_{max}^{nujol} cm^{-1}$: 1658,

1650, 1635, 1390, 1370 (gem Me₂), 970 $\left(_H\!\!>=\!\!<^H\right)$, 830

$\left(>=\!\!<^H\right)$

(CXXVI)

NMR in CCl₄, 40Mc, standard H₂O 0 cps., CH₂Cl₂ 26 cps., CHCl₃ 103 cps : −37.9 (3H of C_3, C_{10}, C_{11}),
−23.7 (1H of C_7), −14.2 (1H), −6.3 (1H) (?), 104.3~112.2 (6H of–CH₂–4, 5, 8), 124.8 (3H),
132.7 (3H) $\left(_H\!\!>=\!\!<^{Me}\right)$, 147.7 (6H) (gem Me₂)

(ii) *Bicyclic Ketones and Aldehydes*

(4) **Valeranone** (Jatamansone) (CXXVII)

〔Former Str. (a)〕
Govindachari[1007]
Naves, Y. R.[1008]
Djerassi[1009]

〔Revised Str. (b)〕
Krepinsky[1010]
Takemoto[1011]

(CXXVII)

《Occurrence》 *Nardostachys jatamansi* DC., *Valeriana* spp. of European, Indian, Russian and Japanese
origin.
d_4^{25} 0.963[1011], n_D^{25} 1.4941[1011], 1.4936[1008], $[\alpha]_D$ −49.0°[1011], $C_{15}H_{26}O$, semicarbazone
mp. 206~207°[1008]

1005) Pfau, A. *et al*: *Helv. Chim. Acta* **15**, 1481 (1932), **17**, 129,384 (1934).
1006) Dev, S: *Tetrahedron*, **8**, 171 (1960).
1007) Govindachari, T. R. *et al*: *Chem & Ind.*, **1960**, 1059; *Tetrahedron*, **12**, 105 (1961).
1008) Naves, Y. R: *Helv. Chim. Acta*, **46**, 2139 (1963).
1009) Djerassi, C. *et al*: *Tetr. Letters* **1961**, 226. 1010) Krepinsky, J. *et al*: *ibid.*, **1962**, 169.
1011) Takemoto, T. *et al*: *Chem. Pharm. Bull.*, **11**, 1207 (1963).

UV $\lambda_{max}^{EtOH}\,m\mu$: 220, 350

IR $\nu_{max}\,cm^{-1}$: 1702, 1458~1442, 1385~1372, 1314, 1244, 1154, 1110, 1045, 935, 855~843, 828, 730[1008]

NMR in CCl_4, 60Mc, TMS, τ: 9.21 s (3H) $\left(CH_3-C\diagdown\right)$, 9.13 d, $J=7.1$ (6H) $\left(\underset{CH_3}{\overset{|}{CH_3-CH-CH_3}}\right)$, 9.02 s

(3H) $\left(CH_3-\overset{|}{\underset{|}{C}}-CO-\right)$

Configuration of A/B cis-Fusion in (CXXVII-b)[1011]

Difference between chemical shifts (δ ppm.) of C_{19}-Me[1012] of a steroid and its –1-one is indicated as $\varDelta$. Discrepancies are observed between $\varDelta$ values of steroides, depend on their A/B ring structure as following:

A/B trans: Cholestan-1-one ~Cholestane ($\varDelta=0.360\sim0.384$ ppm.)

A/B cis : Coprostan-1-one ~Coprostane ($\varDelta=0.212$ ppm)

A/B ring fusion in (CXXVII-b) has been confirmed as cis because $\varDelta$ from the signals of C_{14}-Me of valeranone and valerane was 0.13 ppm.

Configuration of 7-iso-Propyl-group in (CXXVII-b)[1011]

(CXXVII-b) →4 process→ (CXXVIII)

Enantiomer ↓

(LVII) β–Eudesmol →6 process→ (CXXIX)

(5) Kanokonol (CXXX)[1013]

≪Occurrence≫   Valeriana spp. (Japanese origin), as acetate

(a) **Kanokonyl acetate** (CXXX–Ac) d_4^{25} 1.050, n_D 1.490,

[α]$_D$ −54.2° (CHCl$_3$), C$_{17}$H$_{28}$O$_3$

IR $\nu_{max}^{liq}\,cm^{-1}$: 1740, 1227 (AcO–), 1702 (C=O)

(CXXX)

(b) **Kanokonol** (CXXX)

mp. 53~54°, [α]$_D$ −71° (CHCl$_3$), C$_{15}$H$_{26}$O$_2$

IR $\nu_{max}^{KBr}\,cm^{-1}$: 3515 (OH), 1692 (C=O), 1414 (–CH$_2$–CO)

NMR in CCl_4, 60Mc, TMS, τ: 9.16 d, $J=5.0$ (6H) $\left(-CH\diagdown_{CH_3}^{CH_3}\right)$, 8.99 s (3H) $\left(CH_3-\diagdown\right)$

(CXXX) →5 process→ (CXXVII-b) Valeranone

1012) Zürcher, R. F: *Helv. Chim. Acta*, **44**, 1380 (1961).
1013) Takemoto, T *et al*: *Chem. Pharm. Bull.*, **11**, 1210 (1963).

(6) Petasin (CXXXI)[1014)~1016)]

≪Occurrence≫   *Petasites hybridus*

mp. 96～98°, $[\alpha]_D$ +49° (CHCl₃), $C_{20}H_{28}O_3$

(CXXXI)

(7) Eremophilone (CXXXII)[1017)]

≪Occurrence≫   *Eremophila mitchelli*

mp. 41～42°, bp. 171/15 mm, d 0.9994,

n_D^{25} 1.5182, $[\alpha]_D$ −207°, $C_{15}H_{22}O$

(CXXXII)

(8) Carissone (CXXXIII)[1018)]

≪Occurrence≫   *Eucalyptus macarthuri, Carissia lanceolata*

mp. 78～79°, $[\alpha]_D$ +126°, 136.6° (CHCl₃), $C_{15}H_{24}O_2$

UV λ_{max} mμ (ε): 250 (14700), 311 (95),

2, 4–DNP, mp. 173～174°

(CXXXIII)

(9) Tadeonal (Polygodial) (CXXXIV)[1021)] and Isotadeonal (CXXXV)[1019~1023)]

(CXXXIV)[1019)]
(Revised Str.)

(CXXXV)

≪Occurrence≫   *Polygonum hydropiper* L. (ヤナギタデ)

(a) Tadeonal (CXXXIV)   bp. 138～140°/0.8 mm, $[\alpha]_D$ −210° (EtOH), n_D 1.5280, $C_{15}H_{22}O_2$, 2, 4–DNP

mp. 233～233.5°

UV λ_{max} mμ (ε): 229 (10100), 305 (106)

IR ν_{max} cm⁻¹: 2700 (—CHO), 1715 (satur. C=O), 1675

(αβ–unsatur. C=O), 1640 (conj. C=C), 825

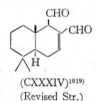

(CXXXVI)

1014) Aebi, A. *et al*: *Pharm. Weekblad*, **93**, 397 (1958); *C.A.*, **52**, 15834a (1958); *Pharm. Acta Helv.*, **30**, 277 (1955); *C.A.*, **50**, 2917d (1956).

1015) Aebi, A. *et al*: *Helv. Chim. Acta*, **42**, 1785 (1959).

1016) Šorm, F. *et al*: *Tetr. Letters*, **1961**, 697.

1017) Zalkow, L.H. *et al*: *J.A.C.S.*, **81**, 2914 (1959), **82**, 6354 (1960).

1018) McQuillin, J. *et al*: *Soc.*, **1956**, 2973.

1019) Kotake, M. *et al*: *7 th Symposium on the Chemistry of Natural Products*, p. 24. (Fukuoka, 1963).

1020) Kotake, M. *et al*: *6 th Symposium on the Chemistry of Natural Products*, p. 107 (Sapporo, 1962).

1021) Barnes, C.S. *et al*: *Australian J. Chem.*, **15**, 322, (1962).

1022) Kotake, M. *et al*: *J. Chem. Soc. Japan*, **83**, 757 (1962).

1023) Kawaguchi, *et al*: *Y.Z.*, **64**, 76 (1944), **57**, 767 (1937), **60**, 352 (1940).

Former structure of tadeonal (CXXXVI)[1020] has been revised to (CXXXIV) from NMR of the glycole (CXXXVII) which was obtained by LiAlH₄-reduction of tadeonal. The spectrum shows no signal due to olefinic methyl protons[1019].

$$(CXXXIV) \xrightarrow{\text{LiAlH}_4} (CXXXVII)$$

(CXXXVII)

(b) Isotadeonal (CXXXV)[1019]

bp. 145~153°/1 mm, $C_{15}H_{22}O_2$, 2,4-DNP, mp. 268° (decomp.)

IR ν_{max} cm^{-1}: 2710 (CHO), 1710 (satur. C=O), 1675 ($\alpha\beta$ unsatur. C=O), 1640 (conj. C=C), 1380, 1360 (gem Me₂), 820

(10) Zierone (Elleryone) (CXXXIX)[1024][1025]

≪Occurrence≫   *Zieria macrophylla,*

bp. 147~149°/18 mm, $[\alpha]_D$ −141.2°, $C_{15}H_{22}O$

UV λ_{max}^{EtOH} mμ (ε): 245 (8200)

IR ν_{max}^{CCl4} cm^{-1}: 1680, 1591 ($\alpha\beta$-unsatur. ketone)

(CXXXIX)

NMR: 1 $>$CH–CH₃, 3 CH₃–$\overset{|}{C}$=, no $>$C=$\overset{|}{C}$–H

Dihydrozierone

UV $_{max}^{EtOH}$ mμ (ε): 206 (5500), 227 sh (2200), 294 (330)

IR ν_{max} cm^{-1}: 1685

(11) Helminthosporal (CXL)[1026]

≪Occurrence≫   Toxic metabolite of *Helminthosporium sativum,*

$C_{15}H_{22}O_2$

(CXL)

(12) Acorone (CXLI) and **the Stereoisomers**[1027]~[1029]

(CXLI)   Acorone[1028]   (CXLII) Isoacorone   (CXLIV)
(CXLIII) Neoacorone   Cryptoacorone

≪Occurrence≫   *Acorus calamus* (ショウブ)

(a) Acorone (CXLI)[1027]

mp. 98.5~99°, $[\alpha]_D$ +143.9°, $C_{15}H_{24}O_2$

1024) Birch, A. J. *et al*: *Soc.*, **1962**, 792.   1025) Barton, D. H. R. *et al*: *Proc. Chem. Soc.*, **1961**, 308.
1026) de Mayo, P. *et al*: *J. A. C. S.*, 84, 494 (1962).
1027) Šorum, F. *et al*: *Chem. Listy.*, 51, 1704 (1957); *C. A.*, 52, 4560a (1958).
1028) Šorum, F. *et al*: *Chim. Listy.*, 52, 2102 (1958); *C. A.*, 53, 8190h (1959).
1029) Vrkoč, J. *et al*: *Coll. Czech. Chem. Comm.*, 27, 2709 (1962); *C. A.*, 58, 10243g (1963).

(b) Isoacorone (CXLII)[1027)]
mp. 97~98°, $[\alpha]_D$ −90.4°, $C_{15}H_{24}O_2$

(c) Neoacorone (CXLIII)[1027)]
mp. 83~84°, $[\alpha]_D$ +126.9°, $C_{15}H_{24}O_2$

(d) Cryptoacorone (CXLIV)[1029)]
mp. 107~108°, $[\alpha]_D$ +97.7° (EtOH), $C_{15}H_{24}O_2$

(iii) *Tricyclic Ketones*

(13) Germacrone (CXLV)[1030)~1032)]

(Former Str.)[1031)]   (Revised Str.)[1030)]
                                        (CXLV)

≪Occurrence≫   *Geranium macrorhizum* L.

UV λ_{max}^{EtOH} (ε): 312.5 (480), 242.5 (2800), 210 (12100)

IR ν_{max}^{KBr} cm^{-1}: 1670 (C=O), 1655 (C=C), 998, 857 (C=C)

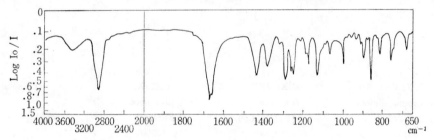

Fig. 154 IR Spectrum of Germacrone (Ohloff *et al.*)

(14) Aristolone (CXLVI)[1033) 1033a)]

≪Occurrence≫   *Aristolochia debilis* SIEB. et ZUCC.
(ウマノスズクサ)

mp. 100~101°, $C_{15}H_{22}O$, semicarbazone mp. 218°

UV λ_{max}^{EtOH} mμ (log ε): 235 (4.11), 310 (2.07) (typical
$\alpha\beta$ unsatur. C=O)   (CXLVI)[1033a)]

IR $\lambda_{max}^{CCl_4}$ μ: 6.03, 6.14 (See Fig. 155)

Dihydroaristolone   mp. 62.5°, $C_{15}H_{24}O$

1030) Ohloff, G. *et al*: *Ann.*, **625**, 206 (1959).
1031) Ognyanov, I. *et al*: *Chem. & Ind.*, **1957**, 820; *C. A.*, **51**, 16361d (1957).
1032) Naves, Y. R.: *C. A.*, **44**, 283 (1950).
1033) Furukawa, S. *et al*: *Y. Z.*, **81**, 559, 565, 570 (1961).
1033a) Büchi, G. G. *et al*: *Tetr. Letters*, **18**, 827 (1962).

UV λ_{max}^{EtOH} mμ (log ε): 213 (3.69), 278 (1.66) (conjugation of carbonyl to cyclopropane ring [Note])

IR λ_{max}^{nujol} μ: 5.92, 7.10 (conjugation remains)

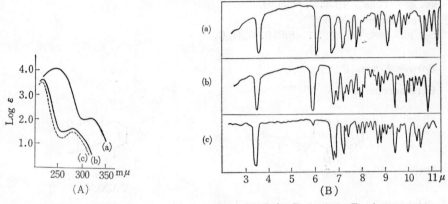

Fig. 155 UV (A) and IR (B) Spectra of Aristolone and the Derivatives (Furukawa *et al.*)
(A) in EtOH; (a) Aristolone, (b) Dihydroaristolone, (c) Dihydroumbellulone,
(B) (a) Aristolone (nujol), (b) Dihydroaristolone (nujol), (c) Deoxoaristolone (liq., film)

[Note] UV Absorptions of Cyclopropane Derivatives (λ_{max} mμ (ε))

(CXLVII)	(CXLVIII)	(CXLIX)	(CL)
Dihydroumbellulone	Carone	Umbellulone	
210(2470)	<220(>2680)	220(5000)	229 (log ε 4.06)
280(35)	288(34)	265(2900)	

(15) Illudin S (Lampterol, Lunamycin) (CLI) and Illudin M (CLII)

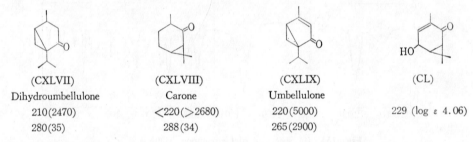

R=OH (CLI) Illudin S [Note]
R=H   (CLII) Illudin M

≪Occurrence≫ *Clitocybe illudens, Lampteromyces japonicus (Pleurotus japonicus)* (ツキヨタケ)

(a) **Illudin S**[1034] =Lampterol[1035] =Lunamycin[1036] (CLI)

mp. 124~126°, [ϕ]$_{589}$ −459° (MeOH), $C_{15}H_{20}O_4$

UV λ_{max}^{EtOH} mμ (ε): 233 (13200), 319 (3600) (crossconjugated dienone)[1034] ; λ_{max}^{MeOH} mμ (log ε): 235 (4.10), 320 (3.54)[1035], 235 (4.1), 325 (3.5)[1036]

1034) Mc Morries, T.C. *et al*: *J.A.C.S.*, **85**, 831 (1963).
1035) Nakanishi, K. *et al*: *Chem. Pharm. Bull.*, **12**, 853 (1964); *Y. Z.*, **82**, 377 (1963); *6 th Symposium on the Chemistry of Natural Products*, p. 105 (Sapporo, 1962).
1036) Shirahama H *et al*: *ibid.*, p 94 (Sapporo, 1962).
[Note] T. Matsumoto *et al*: *Tetrahedron*, **21**, 2671 (1965).

IR ν_{max} cm $^{-1}$: 1706, 1653, 1610$^{1034)}$; ν_{max}^{CHCl3} cm^{-1}: 3629, 3605 (free OH), 3500 (bonded OH of α-ketol), 1698, 1606 (*cisoid* α, β-unsatur. ketone), 1651 (C=C)$^{1035)}$

ν_{max} cm^{-1}: 3500, 3400, 3280, 1692, 1660, 1602, 1107, 1026, (3080, 1028, 868) (tricyclene)$^{1036)}$

NMR ppm$^{1036)}$: 0~0.8 br (2H) (ABX–system?)

$\left(\overset{}{\underset{H_2}{\diagdown}}H \right)$, 0.91, 0.95, 1.2 $\left(3\text{–Me} \right)$, 3.09s (2H), 4.5s (ca 2H), 5.0~5.5 splitted, ca 6.0q, no signal at ca 10 (no —CHO)

(b) Illudin M (CLII)$^{1034)}$ $C_{15}H_{20}O_3$

UV λ_{max}^{EtOH} mμ (ε): 228 (13900), 318 (3600)

IR ν_{max} cm^{-1}: 1695, 1661, 1595

(16) Cyclocolorenone (CLIII)$^{1039)\,1040)}$

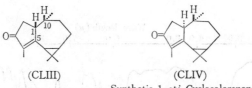

(CLIII)   (CLIV)
Synthetic 1–*epi*–Cyclocolorenone

《Occurrence》   *Pseudowintera colorata*

bp. 136~138°/5 mm, n_D 1.5270, d 1.0026, $[\alpha]_D$ −400° (EtOH), $C_{15}H_{22}O$

UV λ_{max}^{EtOH} mμ (ε): 264 (13260)

IR ν_{max}^{nujol} cm^{-1}: 1698 (—C=C–C=O)

NMR (τ) of C_{10}–Me: (CLIII) 9.25 d, J=6 c/s, (CLIV) 9.0 broad

(17) Patchoulenone (CLV)$^{1041)}$

《Occurrence》   *Cyperus rotundus* L., $C_{15}H_{22}O$

(CLV)

[G] Carboxylic Acids

(1) Valerenic acid (CLVI)$^{1042)\,1043)}$

《Occurrence》   *Valeriana officinalis* L.

mp. 140~142°, $[\alpha]_D$ −120° (EtOH), $C_{15}H_{22}O_2$

UV λ_{max}^{EtOH} mμ (ε): 217 (13000)

IR ν_{max}^{CHCl3} cm^{-1}: $\alpha\beta$-unsatur. acid, secur. doublebond.

(CLVI)

1037) Nakai: *Y. Z.*, **55**, 50 § (1959).
1038) Komatsu, *et al*: *Hakkô-Kyôkai-Shi* (*Tokyo*), **19**, 464 (1964).
1039) Corbett, R. E. *et al*: *Soc.*, **1958**, 3710.   1040) Büchi, G. *et al*: *Proc. Chem. Soc.*, **1962**, 280.
1041) Motl, O. *et al*: *Chem. & Ind.*, **1963**, 1284.   1042) Stoll, A. *et al*: *Ann.*, **603**, 158 (1957).
1043) Büchi, C. *et al*: *J. A. C. S.*, **82**, 2962 (1960).

Methyl valerenate

  bp. 98~100°/0.05 mm, n_D^{23} 1.5121, $[\alpha]_D$ −124° (EtOH)

UV λ_{max}^{EtOH} mμ (ε): 219 (13800)

IR ν_{max}^{CHCl3} cm^{-1}: 1710, 1640

NMR in CDCl$_3$, τ: 9.22 d, J=8 c/s (3H), 8.34 s (3H), 8.09 d, J=1 c/s (3H), 6.28 s (3H), 2.95 d, J=9 c/s with fine str.

(2) **Cuparenic acid** (CLVII)[1944]

  ≪Occurrence≫ *Widelringtonia* spp.

  mp. 158~160°, $[\alpha]_D$ +63° (CHCl$_3$) $C_{15}H_{20}O_2$

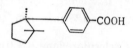

(CLVII)

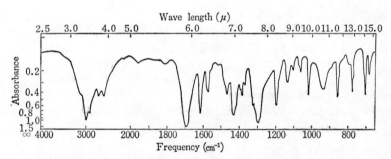

Fig. 156 IR Spectrum of Cuparenic acid (Erdtman *et al.*)

(3) **Hinokiic acid** (CLVIII)[1045) 1046)]

  ≪Occurrence≫ *Chamaecyparis obtusa* (ヒノキ)

(CLVIII)

(4) **Shellolic acid** (CLIX)[1047)~1049)]

  ≪Occurrence≫ Shellac,

  mp. 206~207° (decomp.) $[\alpha]_D$ +18° $C_{15}H_{20}O_6$

UV λ_{max} mμ (ε): 230 (6200)

IR ν_{max}^{nujol} cm^{-1}: 3450, 3200, 1725, 1680, 1625

Dimethyl shellolate

  mp. 149~150°

UV λ_{max} mμ (ε): 230 (6400)

IR ν_{max}^{CCl4} cm^{-1}: 3500, 1714, 1702, 1636

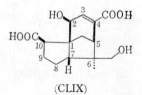

(CLIX)

1044) =924), 1045) Erdtman, H. *et al*: *Chem. & Ind.*, **1960**, 622.
1046) Shishido, *et al*: *4th Symposium on the Chemistry of Natural Products*, p 118 (1960).
1047) Yates, P: *J. A. C. S.*, **82**, 5764 (1960).   1048) Carruthers, W. *et al*: *Soc.*, **1961**, 5251.
1049) Cookson, R. C. *et al*: *Tetrahedron*, **18**, 547, 1321 (1962).

NMR in CCl_4+trace F_3CCO_2H, τ: 8.79 $\left(\!\!>\!\!C\text{--}CH_3\right)$ 6.50 (2×OH), 6.19, 6.22 (2×CO_2CH_3), 6.55 d, J=12 c/s, 6.78 d, J=12 c/s (—CH_2–O—), 5.44 d J=2.5 c/s $\left(\!^{HO}_{}\!\!>\!\!C\overset{H}{\underset{}{\text{—}}}\right)$ 3.33 d, J=2.5 c/s (–CH=)

〔H〕 Lactones

(1) Lactones of *Petasite* spp.[1016]

(CLX) Elemophilenolide[1050]

$R=\text{—}CO\text{--}C=CH\text{--}CH_3$ $\text{—}CO\text{--}CH=CH\text{--}SCH_3$
$\phantom{R=\text{—}CO\text{--}C}CH_3$

cis (CLXI) Petasitolide A   (CLXIII) S–Petasitolide A
trans (CLXII) Petasitolide B   (CLXIV) S–Petasitolide B

≪Occurrence≫   *Petasites officinalis* MOENCH, *P. hybrides* FL. WEST.

(a) Elemophilenolide (CLX)
mp. 125, $[\alpha]_D$ +16.6°, $C_{15}H_{22}O_2$
UV λ_{max} mμ (log ε): 220~224 (4.16)
IR ν_{max} cm^{-1}: 1760, 1750, 1699, 1038

(b) Petasitolide A (CLXI)
mp. 147°, $[\alpha]_D$ +48°, $C_{20}H_{28}O_4$
UV λ_{max} mμ (log ε): 218 (4.42)
IR ν_{max} cm^{-1}: 1762, 1712, 1650

(c) Petasitolide B (CLXII)
mp. 146°, $[\alpha]_D$ +31.8°, $C_{20}H_{28}O_4$
UV λ_{max} mμ (log ε): 219 (4.45)
IR ν_{max} cm^{-1}: 1761, 1713, 1650

(d) S–Petasitolide A (CLXIII)
mp. 201~202°, $[\alpha]_D$ −15.3°, $C_{19}H_{26}O_4S$
UV λ_{max} mμ (log ε): 219 (4.34), 228 (4.22)
IR ν_{max} cm^{-1}: 1750, 1695, 1572

(e) S–Petasitolide B (CLXIV)
mp. 199~200°, $[\alpha]_D$ −32.9°, $C_{19}H_{26}O_4S$
UV λ_{max} mμ (log ε): 219 (4.19), 289 (4.17)
IR ν_{max} cm^{-1}: 1750, 1695, 1571

(2) Ivalin (CLXV)[1051]

≪Occurrence≫   *Iva microcrocephala* NUTT., *I. imbricata* WALT.
mp. 130~132°, $[\alpha]_D$ +142°, $C_{15}H_{20}O_3$
UV λ_{max} mμ (ε): 208 (11000) ($\alpha\beta$–unsatur. lactone)
IR ν_{max} cm^{-1}: 3700, 3500, (OH), 1750 (γ–lactone),

(CLXV)

1050) Šorm, F. *et al*: *Tetrahedron.*, **19**, 1101 (1963).
1051) Herz, W. *et al*: *J. Org. Chem.*, **27**, 905 (1962).

1645, 1600 (2 C=C)

NMR in CDCl$_3$, 60 Mc, TMS, cps: 367.4 −336.7 d, J=1.6 c/s $\left(\text{CO–}\overset{|}{\text{C}}=\text{CH}_2\right)$, 291−271 d, J=1.5 $\left(\text{unconj. }>\text{C}=\text{CH}_2\right)$, 57 s (3H) $\left(>\text{C–CH}_3\right)$

(3) Asperilin (CLXV-a) and Ivasperin (CLXV-b)[1052a]

(CLXV-a)　　　　　　　　　　　(CLXV-b)

≪Occurrence≫　*Iva asperifolia* LESS.

(a) Asperilin (CLXV-a)

mp. 151~152°, $[\alpha]_D$ +149.6° (CHCl$_3$), C$_{15}$H$_{20}$O$_3$

UV λ_{max}^{EtOH} mμ (ε): 211 (8730)

IR ν_{max}^{CHCl3} cm^{-1}: 3700, 3500, 1755, 1655, 1645

Acetate mp. 176~178°

NMR in CDCl$_3$, 60 Mc, TMS., ppm: 6.12 d, 5.59 d, J=1 c/s (conj. =CH$_2$), 4.83 d, 4.55 d (unconj. =CH$_2$), superimposed on 4.7 m (1—H), 4.55 m (8—H), 2.06 (Ac), 0.89 (10—Me)

(b) Ivasperin (CLXV-b)

mp. 150~151°, $[\alpha]_D$ +140.5° (CHCl$_3$) C$_{15}$H$_{20}$O$_4$

UV λ_{max}^{EtOH} mμ (ε): 210 (7750)

IR ν_{max}^{CHCl3} cm^{-1}: 3650, 3450, 1760, 1660, 1650

(4) Pinnatifidin (CLXVI)[1052]

≪Occurrence≫　*Helenium pinnatifidum*
mp. 161~164°, C$_{15}$H$_{18}$O$_3$

(CLXVI)

UV λ_{max} mμ (ε): 237.5 (13550) $\left(>\text{C}=\text{CH–CO}—\right)$

IR ν_{max} cm^{-1}: 1675 (C=C–CO– is not in five membered ring), 1770 $\left(\gamma\text{–lactone conjtd w. }>\text{C}=\text{CH}_2\right)$, 1630

NMR in CDCl$_3$, 60 Mc, TMS, ppm: 6.09—5.59 d, J=1.5 c/s. $\left(—\text{O–CO–}\overset{|}{\text{C}}=\text{CH}_2\right)$, 5.83 s $\left(—\text{CO–}\overset{\text{H}}{\underset{|}{\text{C}}}=\text{C}<\right)$, 1.00 s $\left(>\text{C–Me}\right)$, 1.95 s $\left(=\text{C}<\overset{\text{C}}{\underset{\text{Me}}{}}\right)$

(5) Iresin (CLXVII)[1053]

≪Occurrence≫　*Iresine celosioides*,　　mp. 140~142°, $[\alpha]_D$ +21°, C$_{15}$H$_{22}$O$_4$
UV λ_{max}^{EtOH} mμ (log ε): 224 (4.16)

1052)　Herz, W. *et al*: *ibid.*, **27**, 4041 (1962).　　1052a)　Herz, W. *et al*: *ibid.*, **29**, 1022 (1964).
1053)　Djerassi, C. *et al*: *J.A.C.S.*, **76**, 2966 (1954), **79**, 3528 (1957), **80**, 2593 (1958); *Tetrahedron*, **7**, 37 (1959).

IR: See Fig. 157, $\lambda_{max}^{CHCl_3}$ μ: 2.95, 5.71, 5.92

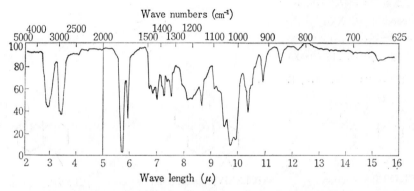

(CLXVII)

Wave numbers (cm⁻¹)

Fig. 157 IR Spectrum of Iresin in $CHCl_3$ (Djerassi *et al.*)

(6) Confertifolin (CLXVIII)[1019)1054)1055)]

≪Occurrence≫ *Polygonum hydropiper* L. (ヤナギタデ)[1019)],

Drmys confertifolia[1054)1055)]

mp. 153~154°, $C_{15}H_{22}O_2$

UV λ_{max} mμ (ε): 217 (11750)[1054)]

IR $\nu_{mam}^{CCl_4}$ cm⁻¹: 1769, 1677[1054)]; ν_{max} cm⁻¹: 1750, 1730

(C=O), 1765 (C=C), 1015, 1000 (C–O)
NMR τ: 5.55, 5.36[1054)]

(CLXVIII)

(7) Lactones of *Drimys* spp.[1054)1055)]

(CLXIX)
Valdiviolide

(CLXX)
Fuegin

(CLXXI)
Winterin

(CLXXII)
Futronolide

(CLXVIII) Confertifolin See the item (6).

(a) Valdiviolide (CLXIX)

mp. 177~178°, $[\alpha]_D$ +111° (CHCl₃), $C_{15}H_{22}O_3$
UV λ_{max} mμ (ε): 221 (10300)
IR $\nu_{max}^{CCl_3}$ cm⁻¹: 1769, 1680; ν_{max}^{KCl} cm⁻¹: 3360

(b) Fuegin (CLXX)

mp. 170~172°, $[\alpha]_D$ +76° (CHCl₃), $C_{15}H_{22}O_4$

1054) Appel, H. H. *et al*: *Tetrahedron*, **19**, 635 (1963). 1055) Appel, H. H. *et al*: *Soc*, **1960**, 4685.

UV λ_{max} mμ (ε): 217 (6500)

IR ν_{max}^{CHCl3} cm^{-1}: 3605, 3476; ν_{max}^{CCl4} cm^{-1}: 1767, 1687

(c) Winterin (CLXXI)

mp. 158°, $[\alpha]_D$ +109° (CHCl$_3$), C$_{15}$H$_{20}$O$_3$

UV λ_{max} mμ (ε): 257 (3820)

IR ν_{max}^{CCl4} cm^{-1}: 1847, 1776, 1668

(d) Futronolide (CLXXII)

mp. 215~217°, C$_{15}$H$_{22}$O$_3$

UV λ_{max}^{EtOH} mμ (ε): 218 (10000)

IR ν_{max}^{CHCl3} cm^{-1}: 3604, 3476, 1758, 1749, 1699

(CLXXIII) Drimenin　　　　(CLXXIV) Isodrimenin　　　(CLXVIII) Confertifolin

≪Occurrence≫　　*Drimys winteri* FORST.

(e) Drimenin (CLXXIII)[1055]

mp. 133°, $[\alpha]_D$ −42° (C$_6$H$_6$), C$_{15}$H$_{22}$O$_2$

(f) Isodrimenin (CLXXIV)

mp. 131~132°, $[\alpha]_D$ +87° (C$_6$H$_6$), C$_{15}$H$_{22}$O$_2$

(8) Balchanolides

R=H (CLXXV)　Balchanolide　　　(CLXXVII)　Balchanin　　　(CLXXVIII)　Millefolide
R=Ac (CLXXVI)　Acetylbalchanolide

(a) Ealchanolide (CLXXV)[1056]

≪Occurrence≫　*Artemisia balchanorum*'　　mp. 154°, $[\alpha]_D$ +105° (EtOH), C$_{15}$H$_{22}$O$_3$

(b) Acetyl balchanolide (CLXXVI)[1057]

≪Occurrence≫　*Achillea millefolium*,　　mp. 125°, $[\alpha]_D$ +128.1°, C$_{17}$H$_{24}$O$_4$

(c) Balchanin (CLXXVII)[1058]

≪Occurrence≫　*Artemisia balchanorum*,　　C$_{15}$H$_{22}$O$_3$

(9) Eupatoriopicrin (CXXIX)[1059]

≪Occurrence≫　*Eupatorium cannabium*　　mp. 157~161°, $[\alpha]_D$ +95° (CHCl$_3$), C$_{20}$H$_{26}$O$_6$

1056)　Krash, H. *et al*: *Coll. Czech. Chem. Comm.*, **26**, 2612 (1961); *C. A.*, **56**, 7368 c, f (1962).
1057)　Krash, H. *et al*: *ibid.*, **26**, 1826 (1961); *C. A.*, **55**, 27400a (1961).
1058)　Krash, H. *et al*: *ibid.*, **27**, 2925 (1962); *Index Chem.*, **8**, 26107 (1963).
1059)　Dolejs, L. *et al*: *ibid.*, **27**, 2654 (1962); *C. A.*, **58**, 10244b (1963).

(CLXXIX) $\xrightarrow{\text{Kydrolysis}}$ (CLXXX)

Eupatolide (CLXXX)
mp. 182~188°, $C_{15}H_{20}O_3$

R=−CO−C(CH$_2$OH)=CH−CH$_2$OH :
(CLXXIX) Eupatoriopicrin
R=H : (CLXXX) Eupatolide

(10) Scabiolide (CLXXXI)[1060]

≪Occurrence≫   *Centaurea scabiosa*
mp. 120°, $[\alpha]_D$ +101° (CHCl$_3$), $C_{19}H_{26}O_7$

(CLXXXI)

(11) Saussurea Lactone (CLXXXII)[1061]

≪Occurrence≫   *Saussurea lappa* CLARKE (木香)
mp. 148~149°, $[\alpha]_D$ +66°, $C_{15}H_{22}O_2$
UV : no peak in 210~340 mμ region.
Tetrahydrosaussurea lactone (CLXXXIII)
mp. 123~125°, total synthesis[1062]

IR ν_{max}^{nujol} cm^{-1}

(CLXXXII)

(CLXXXII)	1760,	1633,	1450,	1374,	1334,	1310,	1276		
(CLXXXIII)	1760		1444,	1374,	1333,	1315,	1285		
(CLXXXII)		1258,	1228,	1207,	1187,	1178,	1158,	1146	
(CLXXXIII)	1271,	1257,	1236,	1207,		1178,	1158,	1139	
(CLXXXII)	1127,		1080,	1060,	1041,	1007,	996,	974,	943
(CLXXXIII)	1124,	1102,	1073,	1057,	1037,	1005,	991,	968	
(CLXXXII)	909,	892,	856,	840,	826,	812,	774,	727	
(CLXXXIII)	914,	873,	853,		823,	811,	782,	742,	714

(12) Costunolide (CLXXXIV)[1063]

≪Occurrence≫   *Saussurea lappa* CLARKE
mp. 106~107°, $[\alpha]_D$ +128°
UV λ_{max} mμ (log ε): 213 (4.21) (γ–lactone)
IR ν_{max}^{nujol} cm^{-1}: 2833, 1764, 1666, 1443, 1376, 1323,
1287, 1244, 1202, 1174, 1133, 1054, 1019, 994,
967, 952, 942, 893, 874, 863, 850, 841, 815, 782

(CLXXXIV)

1060) Šorm, F. *et al*: *Coll. Czech. Chem. Comm.*, **27**, 1905 (1962); *C. A.*, **58**, 1497 (1963).
1061) Bhattacharyya, S.C. *et al*: *Tetrahedron*, **13**, 319 (1961).
1062) Bhattacharyya, S.C. *et al*: *ibid.*, **19**, 1061 (1963).
1063) Bhattacharyya, S.C. *et al*: *Tetrahedron*, **9**, 275 (1960).

(13) 12-Methoxydihydrocostunolide (CLXXXV)[1064]

≪Occurrence≫   *Saussurea lappa* CLARKE

(CLXXXIV) $\xrightarrow{\text{MeOH, KOH}}$ (CLXXXV)

mp. 127~128°, $[\alpha]_D$ +115°, $C_{16}H_{24}O_3$

UV λ_{max}^{EtOH} mμ (ε): 212 (8800), 220 (5900), 230 (2400)

IR ν_{max}^{nujol} cm^{-1}: 2850, 2350, 1760, 1660, 1461, 1444,

1378, 1346, 1310, 1263, 1226, 1207, 1188, 1178, 1142, 1128, 1106, 1082, 1065, 1042, 1017,

990, 979, 962, 933, 920, 892, 872, 854, 840

(CLXXXV)

(14) Gafrinin (CLXXXVI)[1065]

≪Occurrence≫   *Geigeria africana* GRIES

mp. 110°, $[\alpha]_D$ −16.1°, $C_{17}H_{24}O_5$

UV λ_{max} (log ε): 205 (4.05)

IR ν_{max} cm^{-1}: 3509 (OH), 1751 ($\alpha\beta$–unsat. γ–lactone)

1701, 1250 (OAc), 813 $\left(\!\!\begin{array}{c}\\ \end{array}\!\!C=C\!\!\begin{array}{c}H\\ \end{array}\!\!\right)$

(CLXXXVI)

(15) Aristololactone (CLXXXVII)[1066]

≪Occurrence≫   *Aristolochia reticulata, A. serpentaria*

mp. 110~111°, $[\alpha]_D$ −59°, $C_{15}H_{20}O_2$

(CLXXXVII)

(16) Pyrethrosin (CLXXXVIII)[1067]

≪Occurrence≫   *Chrysanthemum cinerariaefolium* (ジ
チュ ウギク)

Collapsed at198~200°, $[\alpha]_D$− 31° (CHCl$_3$), $C_{17}H_{22}O_5$

UV λ_{max}^{EtOH} mμ (ϵ): 204 (14600), mμ (ε): 210 (12200),

220 (6000), 230 (1700)

IR ν_{max}^{nujol} cm^{-1}: 1760 (γ–lactone), 1735, 1242 (O·Ac), 1670, 1650 (two C=C)

(CLXXXVIII)

(17) Vulgarin (CLXXXIX)[1068]

≪Occurrence≫   *Artemisia vulgaris* L.

mp. 174~175°, $[\alpha]_D$ +48.7°, $C_{15}H_{20}O_4$

UV λ_{max} mμ (ε): 215 (10400)

IR ν_{max} cm^{-1}: 3520 (OH), 1775 (γ–lactone), 1665
(C=C–CO)

NMR in CDCl$_3$, 60 Mc, TMS, τ: 3.42 d, 4.17 d, J=10.2 c/s

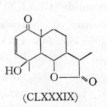

(CLXXXIX)

1064) Bhattacharyya, S.C. *et al*: *Tetrahedron.*, **12**, 178 (1961).
1065) Villiers, J.P: *Soc.*, **1961**, 2049.
1066) Stenlake, J.B. *et al*: *J. Ph. Ph.*, **6**, 1005 (1954); *Soc.*, **1959**, 3289.
1067) Barton, D.H.R. *et al*: *Soc.*, **1957**, 150, **1960**, 2263.
1068) Geissman, T.A. *et al*: *J. Org. Chem.*, **27**, 1856 (1962).

$-4.25\,\mathrm{d}$ $\left(\underset{\underset{|}{|}}{-\overset{H}{\underset{|}{C}}}\!\!>\!\!C=C\!\!<\!\!\overset{H}{CO-}\right)$, $8.46\,\mathrm{s}$, $8.81\,\mathrm{s}$ $\left(2\!\!>\!\!-Me\right)$, $8.78\,\mathrm{d}$, $J=6\,\mathrm{c/s}$ $\left(>\!CH\text{-}CH_3\right)$,

See Fig. 158

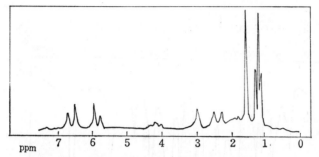

Fig. 158 NMR Spectrum of Vulgarin (Geissman *et al.*)

(18) Santonin, Isomers and **Derivatives**

(CXC) *l*–α–Santonin[1069~1070]   (CXCI) *l*–β–Santonin[1070]   (CXCII) Desoxy–φ–santonin[1073]

(CXCIII) Finitin[1073)1074]   (CXCIV) φ–Santonin[1074)1075)1088]   (CXCV) Artemisin[1075)~1080]

(a) *l*–α–**Santonin** (CXC)

≪Occurrence≫   *Artemisia cina, A. monogina, A. kurramensis* etc.

mp. 171~174°, $[\alpha]_D$ —173° (EtOH), $C_{15}H_{18}O_3$

UV λ_{max}^{EtOH} mμ (log ε): 240 (4.12), 260 (3.93) s

IR $\nu_{max}^{CHCl_3}$ cm^{-1}: 1780, 1665, 1635, 1615[1081], 1135 $\left(\overset{O}{\overset{\|}{C}}\text{-}O\right)$, 1030 (C–OCO), see Fig. 159[1083]

1069) Asher, J.D.M. *et al*: *Proc. Chem. Soc.*, **1962**, 111.
1070) Asher, J.D.M. *et al*: *ibid.*, **1962**, 335, 829
1071) Barton, D.H.R. *et al*: **1962**, 112.   1072) Nakazaki, M. *et al*: *ibid.*, **1962**, 151.
1073) Dauben, W.G. *et al*: *J.A.C.S.*, **82**, 2239 (1960).
1074) Kawatani, T. *et al*: *Y.Z.*, **74**, 793 (1954).
1075) Dauben, W.G. *et al*: *J.A.C.S.*, **82**, 2232 (1960).
1076) Sumi, M. *et al*: *ibid.*, **80**, 7504 (1958).
1077) Sumi, M: *J.A.C.S.*, **80**, 4869 (1958).   1078) Cocker, W. *et al*: *Soc.*, **1963**, 534.
1079) Barton, D.H.R. *et al*: *Proc. Chem. Soc.*, **1960**, 279.
1080) Cocker, W. *et al*: *Soc.*, **1963**, 5235.   1081) Hendrickson, J.B. *et al*: *Soc.*, **1962**, 1678.

NMR in CDCl$_3$, 40 Mc, TMS τ: 3.13, 3.38, 3.75, 4.01, 7.93, 8.67

(b) *l–β*-**Santonin** (CXCI)

≪Occurrence≫　*Artemisia salina, A. monogyna, A. finita*

mp. 216∼218°, $[\alpha]_D$ −137.2° (CHCl$_3$), C$_{15}$H$_{18}$O$_3$

UV　λ_{max}^{EtOE} mμ (log ε): [1074] : 240 (4.11)

IR　ν_{max} cm^{-1}[1083] : 1781 (C=O), 1158 $\left(\begin{smallmatrix} O \\ \| \\ C-O \end{smallmatrix}\right)$, 1025 (C–OCO); in CHCl$_3$, see Fig. 159

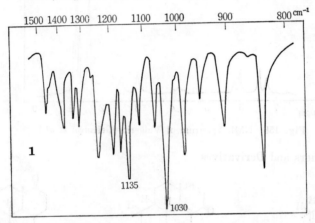

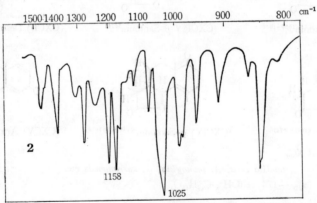

Fig. 159　IR Spectra of (1) α–Santonin and (2) β–Santonin in CHCl$_3$ (Kanzawa *et al.*)

(c)　Desoxy-φ-santonin (CXCII)[1072]

≪Occurrence≫　Mother liquor of santonin, extracted from *Artemisia* spp.

mp. 101.3∼101.9°, $[\alpha]_D$ −207° (EtOH), C$_{15}$H$_{20}$O$_3$

UV: $\varepsilon_{210\ m\mu}$ 10000, ε_{290} 40

IR　ν_{max} cm^{-1}: 1770 (γ–lactone), 1710 (satur. ketone)

$$\text{(CXCII)} \xrightarrow[]{\substack{K_2CO_3 \\ O \\ \| \\ C}} \text{(CXCIII) Finitin}$$

1082)　Barton, D. H. R : *Helv. Chim. Acta*, **42**, 2604 (1959).

1083)　Kanzawa, T. *et al* : *J. A. C. S.*, **80**, 3705 (1958).

(d) Finitin (CXCIII)[1073]

≪Occurrence≫ *Artemisia finita*, mp. 153~155°, $[\alpha]_D$ −167.7° (CHCl₃), $C_{15}H_{20}O_3$

UV λ_{max} mμ (log ε): 298 (1.95) (isolated ketone), 207 (3.99) $\bigg\rangle C{=}C\bigg\langle$

IR λ_{max}^{nujol} μ: 5.65 (none conj. five membered ring lactone), 5.85 (non conj. sixmembered ring ketone), 6.01 (isolated C=C)

(e) φ-Santonin (CXCIV)

≪Occurrence≫ *Artemisia* spp., mp. 183~184°, $[\alpha]_D$ −169° (CHCl₃), $C_{15}H_{20}O_4$

IR ν_{max}^{nujol} cm⁻¹: 3484, 909 (OH), 1706 (C=O), 1753 (1764 in CHCl₃) (butenolide), 1658 (C=C)[1084]

(f) Artemisin (CXCV)

≪Occurrence≫ *Artemisia maritima*, mp. 203°, $[\alpha]_D$ −84.3° (EtOH), $C_{15}H_{18}O_4$

IR ν_{max}^{nujol} cm⁻¹: 1780 (C=O)

Methanesulfonate[1080] mp. 225° d, $[\alpha]_D$ −49.4°, $C_{16}H_{20}O_6S$

UV λ_{max}^{EtOH} mμ (log ε): 238 (4.19)

IR ν_{max}^{nujol} cm⁻¹: 1792, 1670, 1637, 1620, 830

Acetate[1082] mp. 230~233°, $[\alpha]_D$ +120° (CHCl₃)

UV λ_{max}^{EtOH} mμ (ε): 239 (13900)

(CXCV) ⟶ [structure] =O ⟵ (CXCIV) φ-Santonin[1085]

(CXCVI)

(CXCV) ⟶ Iodide ⟶ *l*-α-Santonin (CXC)[1085]

(19) Alantolactone (CXCVII)[1078)1086)1087]

≪Occurrence≫ *Inula helenium* (オオグルマ), mp. 76°, $C_{15}H_{20}O_2$

Tetrahydroalantolactone (CXCVIII) mp. 142~143°, $[\alpha]_D$ +11.5° (CHCl₃)

(CXCVII) Alantolactone[1078]

(CXCVIII)[1078]

Artemisin (CXCV) ⟶ [structure] ⟵ (CXCVIII)

(CXCIX)[1078]

φ-Santonin (CXCIV)[1088] ⟶ (CXCIX) ≡ [structure]

1084) Cocker, W. *et al*: *Soc.*, **1955**, 588, **1956**, 1828.
1085) Sumi, M.: *Chem. Pharm. Bull.*, **5**, 187 (1957).
1086) Tsuda, K. *et al*: *J. A. C. S.*, **79**, 5721 (1957).   1087) Nakazawa, S: *ibid.*, **82**, 2229 (1960).
1088) Cocker, W. *et al*: *Soc.*, **1959**, 1998.

Helenin   A mixture of alantolactone, isoalantolactone, and dihydroisoalantolactone[1089]

(20) Photochemical Transformates of Santonin[1082]

(CC) Lumisantonin[1090]

(CCI) Isophotosantonic lactone[1069)1091]

R=H: (CCII) Photosantonic acid[1091)1092]
R=C$_2$H$_5$: (CCIII) Photosantonin[1093]

(CCIV) Photosantoninic acid[1094]

(a) Lumisantonin (CC)[1095]

≪Occurrence≫   UV Radiation of santonin in dioxane-ethanolic solution, *Artemisia kurramensis*
mp. 153.5~155°, [α]$_D$ −150.8° (CHCl$_3$), C$_{15}$H$_{18}$O$_3$

UV λ_{max}^{EtOH} (log ε): 237 (3.66), 340 (2.45)

IR ν_{max}^{KBr} cm^{-1}: 1774 (γ–lactone), 1711 (conj. ketone), 1753 (conj. double bond); ν_{max}^{nujol} cm^{-1}: 1777, 1715, 1575

(b) Isophotosantonic lactone (CCI)[1091]

Ref. UV irradiation of santonin in 45% acetic acid solution
mp. 165~167°, [α]$_D$ +129° (CHCl$_3$), C$_{15}$H$_{20}$O$_4$

UV λ_{max}^{EtOH} mμ (ε): 239 (13000)

IR ν_{max}^{nujol} cm^{-1}: 3740 (OH), 1776 (γ–lactone), 1693 (five membered ring ketone), 1660 (C=C–CO–
of five membered ring)

(c) Photosantonic acid (CCII)[1091]

Ref. Same to (b) Isophotosantonic acid,   mp. 154~155°, [α]$_D$ −129° (CHCl$_3$), C$_{15}$H$_{20}$O$_4$

UV λ_{max}^{EtOH} mμ (ε): 210 (6700)

(d) Photosantonin (CCIII)[1091]

≪Occurrence≫   UV irradiation of santonin in ethanolic solution
mp. 67~68.5°, [α]$_D$ −121° (EtOH), C$_{17}$H$_{24}$O$_4$

UV λ_{max}^{EtOH} mμ (ε): 204 (8500)

NMR[1092] of (CCII) or (CCIII) in CCl$_4$—CDCl$_3$, 60 Mc, C$_6$H$_6$ =0 c/s: 320 d (>CH–CH$_3$)

1089) Cocker, W. *et al*: *Soc.*, **1961**, 4721.   1090) Barton, D. H. R. *et al*: *ibid.*, **1960**, 4596.
1091) Barton, D. H. R. *et al*: *ibid.*, **1957**, 929, **1958**, 3314.
1092) Van Tamelen, *et al*: *J. A. C. S.*, **81**, 1666 (1959).
1093) Arigoni, D. *et al*: *Helv. Chim. Acta*, **40**, 1732 (1957).
1094) Satoda, I. *et al*: *Y. Z.*, **83**, 561, 566, 574 (1963), Yoshii, E: *ibid.*, **83**, 825, 833 (1963).
1095) Satoda, I. *et al*: *ibid.*, **79**, 267 (1959).

280, 290 $\left(\rangle C = C \underset{Me}{\overset{Me}{\diagdown}} \right.$, non equivalent $\Big)$, 200~280 (—CH_2— and lactone —CH–CH_3), 208

~216 d $\left(—CO–CH_2–CH=C\diagup \right)$, 148 d $\left(—O–\overset{H}{\underset{|}{C}}–\overset{H}{C}\diagup \right)$, 55 t $\left(\overset{—CH_2}{\underset{H}{\diagup}} C = C \diagup \right)$

(CCIII) $\longrightarrow$ ROOC image $\longleftrightarrow$ ROOC image $\longrightarrow$ image

(e) Photosantoninic acid (CCIV)[1093)1094)

≪Occurrence≫   UV radiation of santonin in aqueous KOH solution.

mp. 175~178° (decomp.), 278° (decomp.), $[\alpha]_D$ —18° (MeOH), $C_{30}H_{40}O_8 \cdot H_2O$

Oxime mp. 286°

Dimethyl ester   mp. 198~199°, $[\alpha]_D$ +13° (CHCl$_3$)

UV λ_{max}^{EtOH} mμ (ε): 215 (9550), 289 (120)

IR ν_{max}^{CHCl3} cm^{-1}: 3497 (OH), 1712 (COOCH$_3$, C=O); ν_{max}^{KBr} cm^{-1}: 3484, 1724, 1709

(21) Geigerin (CCV)[1096)1097)

≪Occurrence≫   *Geigeria aspera* HARV.

mp. 130~131°, $[\alpha]_D$ —102° (CHCl$_3$), $C_{15}H_{20}O_4$

UV λ_{max}^{EtOH} mμ (ε): 237 (16000)

IR ν_{max}^{CHCl3} cm^{-1}: 1776 (γ–lactone), 1751 (acetate), 1712

(cyclopentenone), 1658 (conj. ethylenic linkage)

(CCV)

Configuration correlation of santonin with artemisin[1076) C_{11}–Configurations of α–santonin (CXC) and artemisin (CXCV) are confirmed by the derivation of the two compounds to desoxygeigerin (CCVII), the configuration at C_{11} being already established[1097).

(CCVI)   (CCVII)

l–β–Santonin (CXCI) $\longrightarrow$ (CCVI), X=OAc, 13 β—Me

Artemisin (CXCV) $\longrightarrow$ (CCVI), X=OAc, 13 α—Me

(CCVI), X=OAc, 13 α—Me $\longrightarrow$ (CCVII), X=H, 13 β—Me, 1 α—H $\xrightarrow{\text{dil. H}_2\text{SO}_4}$ (CCVII), X=H,

13 α—Me, 1 α—H $\xrightarrow{\text{EtOH-KOH}}$ (CCVII), X=H, 13 α—Me, 1 β—H=Deoxygeigerin[1097)

1096) Barton, D. H. R. *et al*: *Soc.*, **1958**, 4518.   1097) Henbest, *et al*: *Soc.*, **1961**, 4472.

(22) Geigerinin (CCVIII)[1098) 1099)]

≪Occurrence≫   *Geigeria aspera* HARV.

mp. 202~203° $[\alpha]_D$ −10.7° (EtOH) $C_{15}H_{22}O_4$

UV λ_{max} mμ (log ε): 209 (4.0)

NMR of acetate: See Table 78

(Revised Str.)

(CCVIII)[1099)]

Table 78.   NMR of Geigerinin acetate (de Villers *et al.*)

Protons of	Position	Chemical sift τ	Multiplicity	Coupling Constant c/s
C=CH$_2$	13	3.83—4.56	AX	Δ_v=43.2, J_{AX}=3.25
CH–COCH$_3$	4	~4.7 ⎫	ABXY	⎧ Δ_v ~17, J_{AB}=6.7, J_{AX}=
CH–COCH$_3$	3	~4.9 ⎭		⎩ J_{AY}=0, J_{BX}≠J_{BY}≠0
>CH–O	8	5.75	octet	J_1=11.2, J_2=9.0, J_3=3.2
O–CO–CH$_3$	3 or 4	7.86	s	
O–CO–CH$_3$	4 or 3	7.94	s	
>C–CH$_3$	14	8.95	s	
>CH–CH$_3$	15	9.01	d	J=5.3

(23) Pulchellin (CCIX)[1100)]

≪Occurrence≫   *Gaillardia* spp. *(Compositae)*, *Cupressales* spp.[1101)]

mp. 165~168°, $[\alpha]_D$ −36.2°, $C_{15}H_{22}O_4$

UV λ_{max} mμ (ε): 209 (9400)

IR ν_{max} cm⁻¹: 3560 (OH), 1763 (γ–lactone), 1660

(C=C)

NMR ppm: 4.1 m (C$_2$–H), 3.65 d, J=4 (C$_4$–H), 4.2 m

(C$_8$–H), 5.47 d 6.15 d, J=3(C$_{13}$–H$_2$) 0.90 (C$_5$–Me), 1.37 d (C$_{10}$–Me)

(CCIX)

(24) Lactones of *Helenium* spp.

(Revised Str.)

(CCIX) Helenalin[1102) 1104)]

(Revised Str.)

(CCX) Neohelenalin[1103)] =Mexicanin D[1104)]

(CCXI) Mexicanin A[1104)]

1098) de Villers, J.P.: *Soc.*, **1959**, 2412.   1099) de Villers, J.P. *et al*: *ibid.*, **1963**, 4989.

1100) Herz, W. *et al*: *Tetrahedron*, **19**, 483 (1963).   1101) Herz, W. *et al*: *Angew. Chem.*, **74**, 947 (1962).

1102) Herz, W. *et al*: *J.A.C.S.*, **80**, 4876 (1958).   1103) Herz, H. *et al*: *ibid.*, **82**, 2276 (1960).

1104) Herz, W. *et al*: *J.A.C.S.*, **85**, 19 (1963).

(CCXII) Isohelenalin[1104]　　(CCXIII) Isomexicanin A[1104]　　(CCXIV) Mexicanin C[1105]

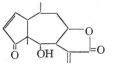

(CCXV) Mexicanin E[1106]　　　　　　(CCXVI) Mexicanin I[1107] [Note]

(a) **Helenalin** (CCIX)[1102) 1104]

≪Occurrence≫　　*Helenium flexuosum* RAF., *H. campestre* SMALL., *H. autumnale* L., *Balduina angustifolia* ROBINS.

mp. 169～172°, $[\alpha]_D$ −102° (EtOH), $C_{15}H_{18}O_4$,　　NMR: See Table 79

Methanesulfonate　　mp. 144°, $C_{16}H_{20}O_6S$

(b) **Neohelenalin** (Mexicanin D) (CCX)[1103]

≪Occurrence≫　　Same to (a) Helenalin,　　mp. 225～226°, $[\alpha]_D$ +143° (EtOH), $C_{15}H_{18}O_4$

UV　λ_{max}^{EtOH} mμ (ε): 235 (17800), 208 (13800)

IR　$\nu_{max}^{CHCl_3}$ cm^{-1}: 3400 (OH), 1760 (γ–lactone), 1695, 1640 (cyclopentenone), 1410

NMR: See Table 79

(c) **Mexicanin A** (CCXI)[1104]　　mp. 138～140°, $[\alpha]_D$ −27° (CHCl₃), $C_{15}H_{18}O_4$

UV　λ_{max}^{EtOH} mμ (ε): 212 (8400)

IR　$\nu_{max}^{CHCl_3}$ cm^{-1}: 3400 (OH), 1750 (lactone), 1652, 1630 (C=C)

NMR: See Table 79

(d) **Isohelenalin** (CCXII)[1104]　$C_{15}H_{17}O_4$

Dihydroisohelenalin　　mp. 238～240°, $[\alpha]_D$ −117°, $C_{15}H_{18}O_4$

UV　λ_{max}^{EtOH} mμ (ε): 228 (6500), 310～315 (74)

IR　$\nu_{max}^{CHCl_3}$ cm^{-1}: 3300 (OH), 1755 (γ–lactone), 1705, 1580 (cyclopentenone)

(e) **Mexicanin C** (CCXIV)[1105]　　mp. 251～252°, $[\alpha]_D$ −80°, $C_{15}H_{20}O_4$

UV　λ_{max}^{EtOH} mμ (ε): 226 (8600)

IR　ν_{max} cm^{-1}: 3400 (OH), 1760 (γ–lactone), 1705, 1785, 1580 (cyclopentenone)

NMR ppm: 4.1 br. $\left(\rangle\overset{6}{C}H\text{-}OH\right)$, 4.8 m ($C_8$–H)

Table 79. NMR of Helenalin and the Isomers　　　　(Herz *et al.*)[1104]

	Helenalin	Neohelenalin	Mexicanin A–acetate
Protons	(CCIX)	(CCX)	(CCXI)–Ac
H_2	7.72 dd (J=7.5, 2)		5.90 t (J=1.8)
H_3	5.90 dd (J=7, 2.5)		2.86 t (J=1.8)

[Note] One of the diastereomer of helenalin (CCIX).

1105) Herz, W. *et al*: *Tetrahedron*, **19**, 1359 (1963).
1106) Romo de Vivar, A. *et al*: *J.A.C.S.*, **83**, 2326 (1961).
1107) Dominguez, E. *et al*: *Tetrahedron*, **19**, 1415 (1963).

Protons	Helenalin (CCIX)	Neohelenalin (CCX)	Mexicanin A–acetate (CCXI)–Ac
H_6	4.46 br	4.24 br	5.16 d (J=5.5)
H_8	4.96 t (J=8)	5.01 q (J=7.5)	4.80 t (J=7)
H_{13}	6.24 d (J=3)	6.18 d (J=3)	6.35 d (J=1.9)
	5.78 d (J=3)	5.71 d (J=3)	5.91 d (J=1.9)
C_5–Me	0.98		1.00
C_{10}–Me	1.26 d (J=6)	1.23 d (J=7)	1.29 d (J=7)
Misc		1.68 d (J=2)	1.97

NMR in $CDCl_3$, 60 Mc, TMS, ppm., J. cps

(f) **Mexicanin D** (CCX): See (b) Neohelenalin

(g) **Mexicanin E** (CCXV)[1106)1108]

≪Occurrence≫ *Helenium mexicanum, H. ooclinum,* mp. 100.5~101.5°, $[\alpha]_D$ −47°, $C_{14}H_{16}O_3$

UV λ_{max}^{EtOH} mμ (ε): 218 (18200), 318 (77)

IR ν_{max}^{CHCl3} cm^{-1}: 1760, 1660 (conj.-γ-lactone), 1700, 1588 (cyclopentenone)

(h) **Mexicanin I** (CCXVI)[1107]

≪Occurrence≫ *Helenium mexicanum,* mp. 257~260°, $[\alpha]_D$ +42.5°, $C_{15}H_{18}O_4$

UV λ_{max} mμ (ε): 216 (14420)

IR ν_{max} cm^{-1}: 3560 (sec. OH), 1768, 1668 (γ-lactone)

NMR of acetate ppm: 6.23 d, 5.65 d, J=3.4 c/s ($>$C=CH$_2$), 7.63 dd, 6.08 dd (AB part of ABX system

$C_3H=C_2H–C_1H<$, J_{AB}=6, J_{AX}=3, J_{BX}=3), 5.92 d, J=4.8 (C_6–H), three pairs of doublets centred

at 4.8 (C_8—H), 2.07 s (COCH$_3$), 1.28 d (5—CH$_3$)

(25) **Balduilin** (CCXVII)[1104]

≪Occurrence≫ *Balduina unifolia* NUTT. $C_{17}H_{20}O_5$

ORD (in dioxane) $[\alpha]_{700}$ +29°, $[\alpha]_{589}$ +43°, $[\alpha]_{360}$ −612°,

$[\alpha]_{277.5}$ +2890°

NMR (See Tab. 79): 7.50 dd (J=6.5, 1.5) (H_2),

6.14 dd (6.5, 3.5) (H_3), 6.00 d, J=3 (H_6), 4.72,

(H_8), 6.33 d, J=2, 5.80 d, J=2 (H_{13}), 1.13

(C_5—Me), 1.27 d, J=7 (C_{10}—Me)

OCOCH$_3$

(CCXVII)

(26) **Linifolin A** (CCXVIII) and **Linifolin B** (CCXIX)[1108]

OCOCH$_3$
(CCXVIII)

OCOCH$_3$
(CCXIX)

≪Occurrence≫ *Helenium linifolium*

1108) Herz, W: *J. Org. Chem.,* **27**, 4043 (1962).

(a) Linifolin A (CCXVIII)   mp. 195~198°, $[\alpha]_D$ +30°, $C_{17}H_{20}O_5$

UV λ_{max} mμ (ε): 215 (12300), 320~325 (43)

IR ν_{max} cm^{-1}: 1710, 1595 (cyclopentenone), 1755, 1600 ($\alpha\beta$ unsatur. lactone and OAc)

NMR in CDCl$_3$, 60 Mc, TMS, ppm.: 7.51 dd, (H$_2$ $J_{2,2}$=2, $J_{2,3}$=6), 6.01 dd (H$_3$, $J_{1,3}$=3), 6.16 d, 5.60 d,

J=4 (C=CH$_2$), 1.21 s ($\rangle$C-Me), 2.02 s (—OCOCH$_3$), 1.26 d, J=6 (C$_{10}$—Me), 5.85 d, J=4.5

($\rangle$C$_6$$\langle$$^H_{OAc}$), 4.86 td, J=10, 10,3 ($\rangle$C$_8$$\langle$$^H_{O—}$)

(b) Linifolin B (CCXVIII) (Mexicanin A (CCXI) acetate)   mp. 149~151°, $C_{17}H_{20}O_5$

UV λ_{max} mμ: 205~210, 280

IR ν_{max} cm^{-1}: 1755, 1600

NMR ppm.: 6.05 t J=1.5 (H$_2$), 2.94 d—3.04 d, J=1.5 (—$\overset{3}{C}H_2$—), 4.61 t (H$_8$) 4.41 d (H$_6$), 1.02 s

($\rangle$C—Me), 1.31 d, J=7 ($\rangle$CH—Me), 2.12 s (OCOCH$_3$)

(27) Bigelovin (CCXX) and Isotenulin (CCXXI)[1109]

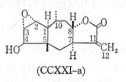

(CCXX)   (CCXXI)[1110]

《Occurrence》   *Helenium Bigelovii* GRAY

(a) Bigelovin (CCXX) mp. 190~191°, $[\alpha]_D$ +46.1° (EtOH), $C_{17}H_{20}O_5$

UV λ_{max}^{EtOH} mμ (ε): 215 (13000), 321 (48)

IR $\nu_{max}^{CHCl_3}$ 1755, 1730, 1705, 1660, 1585, 1240

NMR cps: 364—375 d (C$_3$—H), 463—472 d (C$_2$—H), 354—375 d, splitted ($\rangle$=CH$_2$), 72 s ($\rangle$C—Me),

73—82 d ($\rangle$CH—Me), 268, 271, 278, 281, 290, 293 (C$_8$—H) (almost identical to balduilin (CCXVII)),

334, 342 (C$_6$—H) (differs from balduilin (CCXVII) 358, 361)

(b) Isotenulin (CCXXI)[1109][1111][1112]

mp. 159~161°, $[\alpha]_D$ +10° (CHCl$_3$), $C_{17}H_{22}O_5$

UV λ_{max} mμ (ε)[1111]: 226 (7000),   IR ν_{max}^{nujol} cm^{-1} [1111]: 1778, 1748, 1705, 1588, 1238

NMR[1112] TMS, ppm: 7.57 dd, J=6,2 c/s (H$_2$), 6.12 dd, J=6, 2.5 (H$_3$), 5.60 d, J=4 (H$_6$), 4.70 t,

J=11 (H$_8$), 1.20 s (C$_5$—Me), 1.20 d, J=4 (C$_{10}$—Me), 1.20 d, J=4 (C$_{11}$—Me), 2.17 s (—OAc)

(28) Amaralin (CCXXI-a)[1112a]

《Occurrence》   *Helenium amarum* H. ROCK.

mp. 195~198°, $[\alpha]_D$ +5° (CHCl$_3$),

$C_{15}H_{20}O_4$, m/e 264

IR ν_{max}^{nujol} cm^{-1}: 3430 (OH), 1736 (γ–lactone), 1660,

940 (C=C)

(CCXXI-a)

1109) Geissman, T.A. *et al*: *J. Org. Chem.*, **27**, 4127 (1962).
1110) Rogers, D. *et al*: *Proc. Chem. Soc.*, **1963**, 92.   1111) Barton, D.H.R. *et al*: *Soc.*, **1956**, 142.
1112) Herz, W. *et al*: *J.A.C.S.*, **84**, 3857 (1962).
1112a) Lucas, R.A. *et al*: *J. Org. Chem.*, **29**, 1549 (1964).

NMR in $CDCl_3$, 60 Mc, TMS, ppm: 3.42 q (2 H—3 H), 3.81 d (4—H), 1.15 (5—CH_3), 4.33 overlap. q. (8—H), 1.26 d (10—Me), 6.17 d, 5.46 d (=CH_2).

(29) **Carbrone** (CCXXI-b)[1112b]

≪Occurrence≫   *Carpesium abrotanoids* L.

mp. 90~91°, $[\alpha]_D$ +116.9°

UV λ_{max} mμ (ε): 213 (8150)

IR ν_{max} cm^{-1}: 1712, 1758, 3100, 1665, 822

NMR τ: 9.56, 9.62, 9.77 (protons on cyclopropane ring)

(CCXXI-b)

(30) **Tenulin** (CCXXII)[1109)1112]

≪Occurrence≫   *Helenium bigelovii* GRAY, *H. amarum* (Raf.) and other spp.

mp. 195~197°, $[\alpha]_D$ −22.7°, $C_{17}H_{22}O_5$

UV λ_{max}^{EtOH} mμ (ε): 225 (7000)

IR $\nu_{max}^{CHCl_3}$ cm^{-1} 3520, 1765, 1700, 1580[1109]

(CCXXII)

NMR[1112] of acetate TMS, ppm: 7.54 dd, $J=5$, 1.5 c/s (H_2), 6.04 dd, $J=5$, 2.5 (H_3), 4.45 d, $J=5.5$ (H_6), 5.33 t, $J=11$ (H_8), 1.32 s (C_5—Me), 1.27 d, $J=5$ (C_{10}—Me), 1.31 (C_{11}—Me), 1.32 (O-C-Me)

(31) **Ambrosin** (CCXXIII) and **Parthenin** (CCXXIV)

(CCXXIII)[1114]

(Revised Str.)
(CCXXIV)[1114)1115]

(a) **Ambrosin** (CCXXIII)[1113)~1116]

≪Occurrence≫   *Ambrosia maritima*, *Parthenium incanum*

mp. 146°, $[\alpha]_D$ −154.5°, $C_{15}H_{18}O_3$

(CCXXIII) $\longrightarrow$ Tetrahydroambrosin
‖
Hexahydroanhydroparthenin $\longleftarrow$ Parthenin (CCXXIV)

NMR[1114] TMS, ppm: 6.14 dd, 7.50 dd, AB part of ABX, $J_{2,3}=6.2$, $J_{1,2}=3$, $J_{1,3}=1.9$ $\left(\begin{smallmatrix} H & H & H \\ -C_3=C_2-C_1 \end{smallmatrix}\right)$,

5.51 d, 6.29 d, $J=3$ (>C=CH_2), 4.69 d, $J=9.2$ (H_6), 1.17 s (>C-Me), 1.07 d (>CH-Me)

(b) **Parthenin** (CCXXIV)[1114]

≪Occurrence≫   *Parthenium hysterophorus* L.

mp. 163~166°, $[\alpha]_D$ +7.02° (CHCl$_3$), $C_{15}H_{18}O_4$

1112b) Minato, H. *et al*: *Proc. Chem. Soc.*, **1964**, 120.
1113) Herz, W. *et al*: *J.A.C.S.*, **81**, 6088 (1959).   1114) Herz, W. *et al*: *ibid.*, **84**, 2601 (1962).
1115) Herz, W. *et al*: *Tetr. Letters*, **1961** 82.   1116) Soine, T.O. *et al*: *J.A.P.A.*, **42**, 387 (1953).

UV λ_{max}^{EtOH} mμ (ε): 215 (15100), 340 (22)

IR $\nu_{max}^{CHCl_3}$ cm^{-1}: 3450, 1755, 1718, 1655, 1592

NMR[1114] TMS, ppm: 7.55 d, $J=6$ (H$_2$), 6.18 d, $J=6$ (H$_3$), 5.08 d, $J=7$ (H$_6$), 5.59 d, 6.29 d, $J=3$
$\left(\bigtriangleup C=CH_2\right)$, 1.28 s $\left(\bigtriangleup C-Me\right)$, 1.11 d, $J=8$ (C$_{10}$—Me)

(32) Coronopilin (CCXXV)[1117]

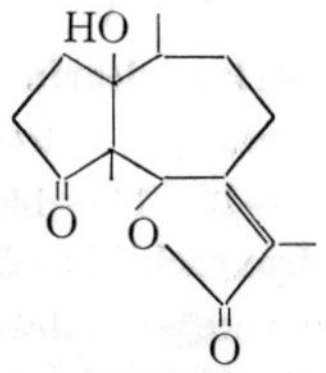

≪Occurrence≫   *Ambrosia psilostachya* DC. var.
coronopifolia (T. & G.) FARW.

mp. 177~178°, $[\alpha]_D$ −30.2° (EtOH), C$_{15}$H$_{20}$O$_4$

UV λ_{max}^{EtOH} mμ (ε): 213 (9800), 290 sh (29)

IR $\nu_{max}^{CHCl_3}$ cm^{-1}: 3600, 3400 (OH), 1750 (cyclopentanone
and γ–lactone), 1655 (C=C), 1408 (—CH$_2$–CO—,
—C=C-CO—)

(CCXXV)

(33) Achillin (CCXXVI)[1118][1119]

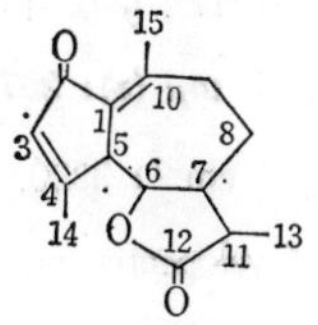

≪Occurrence≫   *Achillea lanulosa*
mp. 144~145°, $[\alpha]_D$ +160° (CHCl$_3$), C$_{15}$H$_{16}$O$_3$

UV λ_{max}^{EtOH} mμ (log ε): 255 (4.22) (cross conjtd. cyclo-
pentadienone):

IR $\lambda_{max}^{CCl_4}$ μ: 5.57 (γ–lactone), 5.92 (cross conjtd. cyclo-
pentadienone) 6.05, 6.16 (2C=C),

(CCXXVI)

NMR τ, c/s: 3.84 (C$_3$—H), 6.16 unsymm. t. $J=9$ (C$_6$—H), 6.60 d, $J=10$ (C$_5$—H), 7.57 s (15—Me),
7.70 s (14—Me), 8.87 d, $J=8$ (13—Me)

(34) Matricarin series[1118][1119]

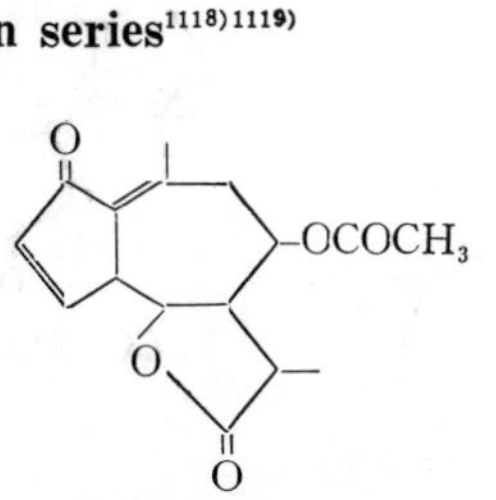

(CCXXVII) Matricarin   (CCXXIX) Desacetylmatricarin
(CCXXVIII) Artilesin A   (CCXXX) 8–Hydroxyachillin

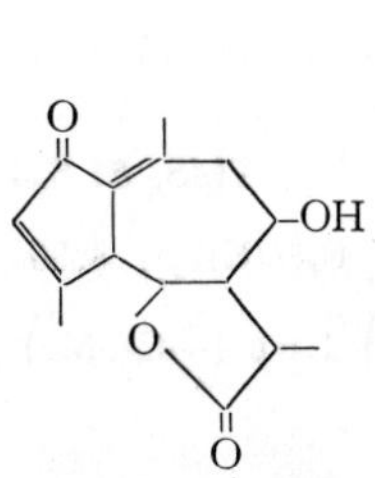

(a) Matricarin (CCXXVII)[1118][1119]

≪Occurrence≫   *Achillea lanulosa, Artemisia tilesii, Matricaria chamomilla*
mp. 193~195°, $[\alpha]_D$ +23.5° (CHCl$_3$), C$_{17}$H$_{20}$O$_5$

UV λ_{max}^{EtOH} mμ (ε): 255 (15100)

1117)   Herz, W. *et al*: *J. Org. Chem.*, **26**, 5011 (1961).
1118)   White, E. H. *et al*: *Tetr. Letters*, **1963**, 137.   1119)   Herz, W. *et al*: *J. A. C. S.*, **83**, 1139 (1961).

(b) Artilesin A (CCXXVIII)[1119]

<Occurrence> *Artemisia tilesii*

mp. 190~191°, $[\alpha]_D$ +23.5° (CHCl$_3$), C$_{17}$H$_{20}$O$_5$ (mixed w (CCXXVII) mp. 174~187°

UV λ_{max}^{EtOH} mμ (log ε): 255 (4.15)

IR $\nu_{max}^{CHCl_3}$ cm^{-1}: 1780 (γ–lactone), 1740 (OAc), 1690 (cyclopentenone), 1645, 1622, (2 C=C)

NMR in CHCl$_3$ or CDCl$_3$, 40 Mc, TMS, cps: 56.5 m, $J\doteqdot$1.5 (H$_3$, **H** coupled to non–adjacent–H), 204,

208~209 (split) $\left(2C=C\diagdown^{CH_3}\right)$, 215 (O–COCH$_3$), 240.5—247 (split) (11—Me)

(c) Desacetylmatricarin (CCXXIX)[1119]

<Occurrence> *Artemisia tilesii*

mp. 123~125°/143~146°, C$_{16}$H$_{18}$O$_4$ · H$_2$O

(d) 8–Hydroxyachillin (CCXXX)[1118]

<Occurrence> *Achillea lanulosa*

mp. 161~162°, $[\alpha]_D$ +110° (MeOH), C$_{15}$H$_{18}$O$_4$

(35) Lactucin (CCXXXI)[1120]

<Occurrence> *Lactuca virosa* L.

mp. 213~217°, $[\alpha]_D$ +49° (MeOH), C$_{15}$H$_{16}$O$_5$

UV λ_{max}^{EtOH} mμ (ε): 257 (14000)

IR ν_{max}^{nujol} cm^{-1}: 3320, 3240 (OH), 1755 (γ–lactone),

1662 (cyclopentenone ring), 1623, 1610 (C=C–)

(CCXXXI)

(36) Estafiatin (CCXXXII)[1121]

<Occurrence> *Artemisia mexicana* WILLD.

mp. 104~106°, $[\alpha]_D$ −9.9°, C$_{15}$H$_{18}$O$_3$

UV λ_{max} (ε): 214 (9850)

IR ν_{max} cm^{-1}: 1755, 1668 ($\alpha\beta$ unsatur. lactone), 1640,

903 (exocyclic methylene)

NMR in CHCl$_3$, 60 Mc, TMS, δ: 5.48 d, 6.18 d,

J=3.3 c/s $\left(\diagdown C_{13}=CH_2\right)$, 4.86 d, 4.96 d, J=2

$\left(\diagdown C_{15}=CH_2\right)$ 1.6 s $\left(\diagdown C—Me\right)$

(CCXXXII)

(37) Mokkolactone (CCXXXIII) and **Dehydrocostuslactone** (CCXXXIV)[1121a]

<Occurrence> *Jurinea aff. souliei* FRANCH., *Inula racemosa* HOOK. fil. (川木香).

1120) Barton, D. H. R. *et al*: *Soc.*, **1958**, 963.

1121) Sauchez-Viesca, F. *et al*: *Tetrahedron*, **19**, 1285 (1963).

1121a) Takemoto, T. *et al*: *Chem. Pharm. Bull.*, **12**, 632 (1964).

(CCXXXIII)　　　　　　　　　　(CCXXXIV)

(a) Mokkolactone (CCXXXIII)

mp. 35~37°, $[\alpha]_D$ +18.2° (CHCl$_3$), C$_{15}$H$_{20}$O$_2$

IR ν_{max}^{KBr} cm^{-1}: 1770 (γ–lactone), 3110, 1642, 889

NMR in CCl$_4$, 60 Mc, TMS, τ: 8.83 d, J=6.2 c/s $\left(\diagup\!\!\diagdown C=CH_2\right)$, 5.28—5.20, 5.02—4.84 (tow =CH$_2$)

(b) Dehydrocostuslactone (CCXXXIV)

mp. 61~61.5°, $[\alpha]_D$ −19.8° (CHCl$_3$), C$_{15}$H$_{18}$O$_2$

IR ν_{max}^{KBr} cm^{-1}: 1764 (γ–lactone), 3125, 1639, 893 (vinylidene)

NMR in CCl$_4$, 60 Mc, τ: 6.16 t $\left(\text{H}\!-\!\overset{|}{\underset{|}{C}}\!-\!O\!-\!CO\!-\!\right)$, 5.20—5.16, 5.03—4.78 $\left(CH_2\!=\!C\diagup\!\!\diagdown\right)$, 4.65, 3.94

$\left(CH_2\!=\!\overset{|}{C}\!-\!CO\!-\!O\right)$

13 Non-terpenic Essential Oils and Isoprenoides

[A] UV Spectra

(1) Non-terpenic Essential Oils (Aromatic Compounds)

Protocatechol, resorcinol, hydroquinone, guaiacol, thymol, pyrogallol, and phloroglucinol[1122]. Eugenol, safrol, isosafrol, and isoeugenol[1122]. Benzaldehyde[1123], salicylaldehyde[1124], acetophenone[1125], tropic acid[1125][1126], vanillin, veratric aldehyde, and syringaldehyde[1127]. Benzoic acid, ethylbenzoate[1128], phenylpropionic acid[1126], salicylic acid[1129], cis-cinnamic acid, trans-cinnamic acid, o-toluic acid methylester, m-toluic acid methylester, and p-toluic acid methylester[1130].

Methylprotocatechuate, syringaldehyde, syringic acid[1131], gallic acid trimethylester, sinapic acid, and sinapine iodide[1132]. Phenols[1140], eugenol, isoeugenol[1141], methylsalicylate[1142], 2, 4–DNP of aldehydes and ketones[1143], carbonyl compounds[1144][1145], benzaldehyde[1146], vanillin[1147], and cinnamic aldehyde[1148].

1122) Hillmer et al: Z. Physik. Chem., **167**, 407 (1934).　　1123) Heilbron et al: Soc., **1943**, 264, 268.

1124) Kiss: Z. Physik. Chem. **A 189**, 344 (1941).　　1125) Ley et al: Ber., **67**, 501 (1934).

1126) Castille et al: Bull. Soc. Chim. Biol., **10**, 623 (1928).

1127) Patterson et al: J.A.C.S., **65**, 1862 (1943).　　1128) Schauenstein et al: ibid., **59**, 1321, 2616.

1129) Landolt: Börnstein (I) Abb., 140 (1951).　　1130) Wolf: ibid., **B 21**, 383 (1933).

1131) Pearl: J.A.C.S., **72**, 1743 (1950).　　1132) Kung et al: ibid., **71**, 1836 (1949).

1133) Booker, Gillam: Soc., **1940**, 1453.

1134) Yves-René Naves et al: Helv. Chim. Acta, **31**, 1240 (1948).

1135) Buraway et al: Soc., **1941**, 20.　　1136) Yves-René Naves et al: Helv. Chim. Acta, **31**, 2057 (1948).

1137) Naves: ibid., **31**, 1427 (1948).　　1138) Doule et al: Z. Physik. Chem., **B 8**, 68 (1930).

1139) Mohler: Helv. Chim. Acta., **20**, 1183 (1937).　　1140) Pearson: Analyst., **80**, 656 (1955).

1141) Vespe, Balz: Anal. Chem., **24**, 664 (1952).

1142) Leighton, A. E.: " Standard Methods of Analysis of Dubbing " Australia Dept. of Defence, Munitions Supply Board. May (1926).　　1143) Gordon et al: Anal. Chem., **23**, 1754 (1951).

1144) Woodman, A. G. et al: J.A.C.S. **30**, 1607 (1908).

1145) Alyea, H. N. et al: ibid., **51**, 90 (1929).　　1146) Rees, H. L. et al: Anal. Chem., **21**, 989 (1949).

1147) Ensminger: J. A. O. A. C. **36**, 679 (1953).

1148) Wachsmuth, H. et al: J. Pharm. Belg (N. S). **1**, 65 (1946).

(2) Terpenic Essential Oils

Piperylene, myrcene, β–phellandrene, and zingiberene. Menthadiene, α–phellandrene, and α–terpinene[1133]. Citronellol, farnesol, geraniol[1134], citronellal, citral, farnesal[1134], citral, and β–cyclocitral[1135]. α–Ionone, β–ionone[1137], φ–ionone[1135)1136] α–ionone semicarbazone, β–irone[1136]. Menthone, fenchone, camphor[1138] thujone, and l–menthone[1139]. Piperitone, menthone, carvone, carvomenthone[1149], piperitenone, and pulegone[1150].

UV λ_{max} mμ (ε) of verbenene, piperitone, verbenone, umbellulone, β–phellandrene, myrtenal, phellandral, pinocarvone, pulegone, and nopadiene[1151]. Terpentine[1152], terpineol, terpinhydrate[1153)1154], menthol[1155] citral, geraniol[1156)1157], carvone[1158], camphor[1159], and ascaridole[1160].

(3) UV Spectra of the following compounds are listed in the item 12, [F], (14),. Structure numbers of the compounds are those in the item (12).

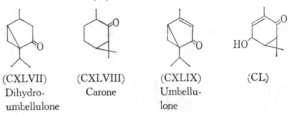

| (CXLVII) | (CXLVIII) | (CXLIX) | (CL) |
| Dihydro-umbellulone | Carone | Umbellu-lone | |

[B] IR Specta

(1) IR Spectra of

(a) Non-terpenic Essential Oil (Aromatic compounds)

Phenyl ethyl alcohol, benzaldehyde, acetophenone, vanillin, phenylethyl acetate, benzyl benzoate, thymol, eugenol, isoeugenol and diphenyloxide[1161)1162].

(b) Terpenic Essential Oil

Conjugated p–menthadienes[1163], dipentene, terpineol[1161],α–pinene, β–pinene, camphene, cymol, and limonene[1164]. Gem–dimethylhydronaphthalenes[1165], α, β, γ and δ–valenes, valeranone, valenol, valerianaphenol, and isovalerylester Ⅱ[1166]. Geraniol, linalool, geranylacetate and citral[1161]. Ascaridol[1167)1168], α–irone, β–irone, α–ionone, dihydro–α–ionone, β–ionone, dihydro–β–ionone and dihydro–γ–ioneone[1169].

1149) Cooke et al: Soc., **1938**, 1408.   1150) Yves-René Naves: Helv. Chim. Acta, **25**, 1023 (1942).
1151) Moore, Fisher: J.A.C.S., **78**, 4362 (1956).
1152) Bogatskii, et al: Z. Anal. Chem., **76**, 103 (1929).   1153) Perelmann: Pharm. Z., **77**, 1204 (1932).
1154) Platt, James: J. Am. pharm. Assoc. Sci. Ed., **44**, 666 (1955).
1155) Masamune, H: J. Biol. Chem (Japan), **18**, 277 (1933).
1156) Parker, C. E. et al: J. Biol. Chem., **130**, 149 (1939); Anal. Ed., **13**, 834 (1941).
1157) "Official and Tentative Methods of Analysis of the A. O. A. C" 7 th Ed. pp 311, (1950).
1158) Tattje: J. Pharm. Pharmacol., **9**, 629 (1957).
1159) Matérn et al: Svensk Farm Tids., **54**, 445 (1950).
1160) Cardoso do Vale: Noticias. farm. (Portugal) **14**, 391 (1948).
1161) Indo, G: Koryo, **36**, 1 (1955). (Nihon Koryo Kyokai).   1162) Ooi: Chem. Pharm. Bull., **5**, 149 (1957).
1163) Pines, H. et al: J.A.C.S., **77**, 6314 (1955).
1164) Takeshita, et al: Kogyo Kagaku Zasshi, **59**, 645, 648 (1956).
1165) Stoll, M. et al: Helv. Chim. Acta, **39**, 183 (1956).
1166) Stoll A. et al: Helv. Chim. Acta, **40**, 1205 (1957).
1167) Szmant, H., Halpern, A: J.A.C.S., **71**, 1133 (1949).
1168) Szmant, Halpern: J.A.C.S., **71**, 1133 (1949).
1169) Günthard, Ruzicka: Helv. Chim. Acta, **31**, 642 (1948).

Terpineol[1172], borneol[1173], 1, 4-cineol[1174], cineol[1175] camphor[1176] and α-ionone[1177].

(2) IR Spectra of Volatile Components of Crude Drugs[1170) 1171]

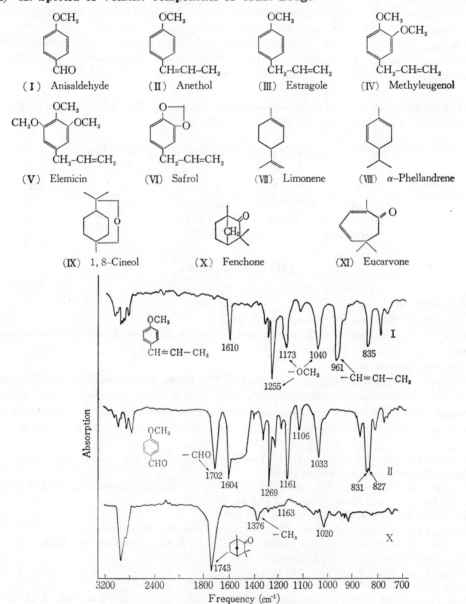

(I) Anisaldehyde   (II) Anethol   (III) Estragole   (IV) Methyleugenol

(V) Elemicin   (VI) Safrol   (VII) Limonene   (VIII) α-Phellandrene

(IX) 1, 8-Cineol   (X) Fenchone   (XI) Eucarvone

Fig. 160 IR Spectra of Anethole (II), Anisaldehyde (I)
and Fenchone (X) in CS_2 (Fujita *et al.*)

1170) Fujita, M. *et al* : *Y. Z.*, **80**, 589 (1960).
1171) Nagasawa, M. *et al* : *ibid.*, **81**, 129 (1961).   1172) Sadtler Card No. 1369.
1173) Sadtler Card No. 2143.   1174) Sadtler Card No. 8580.   1175) Sadtler Card No. 5848.
1176) Sadtler Card No. 244-B.   1177) Sadtler Card No. 5906.

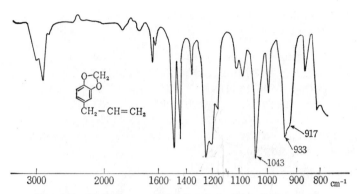

Fig. 161 IR Spectrum of Safrol (Ⅵ) in CS₂ (Fujita *et al.*)

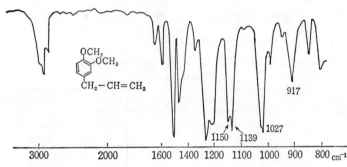

Fig. 162 IR Spectrum of Methyleugenol (Ⅳ) in CS₂ (Nagasawa)

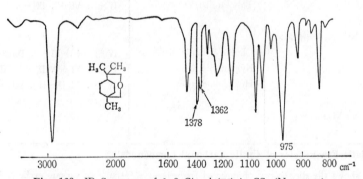

Fig. 163 IR Spectrum of 1, 8–Cineol (Ⅸ) in CS₂ (Nagasawa)

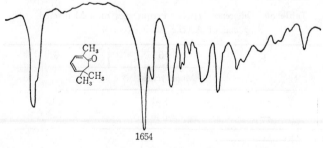

Fig. 164 IR Spectrum of Eucarvone (Ⅺ) in CS₂ (Nagasawa)

(a) Characteristic Absorption Bands of Components

IR $\nu_{max}^{CS_2}$ cm^{-1} :

Anisaldehyde (I) 1702, 1161 (–CHO), 1604,, 1268, 1183, 1033 (—O—, —O–CH), 831, 827

Anethol (II) 1610, 1255, 1173, 1040 (OCH$_3$), 961 (—CH=CH–CH$_3$), 835

Estragol (III) (—O—, —O–CH), 911 (—CH$_2$–CH=CH$_2$)

Methyleugenol (IV) 1150, 1139, 1027 (OCH$_3$), 917 (—CH$_2$–CH=CH$_2$)

Elemicin (V) 1125 (3, 4, 5-trimethoxyphenyl), 917 (—CH$_2$–CH=CH$_2$)

Safrol (VI) 1643, 1431, 1360, 1256, 1186, 1094, 1043 (—O–CH$_2$–O—), 991, 933 (—O–CH$_2$–O—), 917 (—CH$_2$–CH=CH$_2$), 854, 803, 770

Limonene (VII) 1375 (CH$_3$), 884 $\left(-C\diagdown\diagup\genfrac{}{}{0pt}{}{CH_2}{CH_3}\right)$

α–Phellandrene (VIII) 1375 (CH$_3$), 1371 $\left(-CH\diagdown\diagup\genfrac{}{}{0pt}{}{CH_3}{CH_3}\right)$

1, 8–Cineol (IX) 1378 (CH$_3$), 1362, 1224 (—O—, —O–CH), 975

Fenchone (X) 1743 $\left(\diagup\diagdown C=O\right)$, 1376 (–CH$_3$), 1163, 1020

(b) Qualitative Analysis of Essential Oils by IR Spectroscopy[1170)1171)] (See Table 79 a)

Table 79 a (Fujita, Nagasawa)

Essential Oil	Location in Japan	Components
Star anis oil		(II), (III), (VIII), (IX), terpineol
Anis oil		(III), etc. : (undetected)
Fennel oil		(I), (II), (X)
Bastard star oil		(VI), (IX), eugenol
Essential oil of *Asiasarum Sieboldii* F. MAEKAWA(ウスバサイシン)	Tōhoku-district	IV(卌), VI(+), IX(−), XI(卌)
	Chūbu & Sanin-districts	IV(卌～+,) VI(卌), IX(卌～+), XI(−)
	Kyūshū-district	IV(+), VI(卌), IX(+), XI(−)
Essential oil of *A. heteropoides* (Fr. SCHMIDT) MAEKAWA (オクエゾサイシン)	Hokkaido-district	IV(卌), VI(卄～卌), IX(−), XI(卌～+)

(3) Characteristic Frequencies of Cyclic Carbonyl Compounds[1178)].

Table 80 shows the discrepancy of carbonyl frequencies $\nu_{C=O}^{CCl_4}$ cm^{-1} of various cyclic carbonyl compounds from that of cyclohexanone, measured under the same condition.

Table 80 Difference between Frequencies of n–Membered
Ring and of Acyclic Analog (cm^{-1}) (Hall *et al.*)

n	4	5	6	7
		Monocyclics		
Lactams	—	+31	−13	−15
N-Me-Lactams	—	+46	—	0

1178) Hall, H. K *et al* : *J. A. C. S.*, **80**, 6429 (1958).

n	4	5	6	7
Monocyclics				
Lactones	+83	+40	+ 5	− 8
Carbonates	—	+72	+31	—
Ketones	+69	+30	0	−10
Ureas	—	+23	+23	− 6
Imides	—	+36	+32	—
N-Me-Imides	—	+ 5	− 1	—
N-Ac-Lactams	—	+37	− 7	− 7
Urethanes	—	+52	+12	—
Anhydrides	—	+37	−14	—
Average	+76±7	+37±11	+7±14	−8±3
Bicyclics				
Lactams	—	+37	− 1	−29
Lactones	—	+29	+ 4	—
Carbonates	—	—	+16	+ 7
Ketones	—	+35	—	—
Ureas	—	—	+17	—
Imides	—	—	+18	—
N-Me-Imides	—	—	− 4	—
N-Ac-Imides	—	+38	+11	—
Urethanes	—	—	− 6	−14
Anhydrides	—	—	−19	—
Average	—	+35±3	+4±10	−12±13

[C]   Mass Spectra of Terpenic Hydrocarbons[1178a)]

(1) Fragmentation

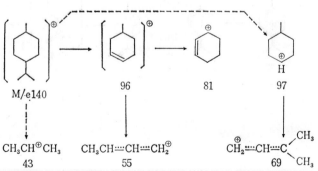

1178a)   Thomas, A. F *et al*: *Helv. Chim. Acta*, **47**, 475 (1964).

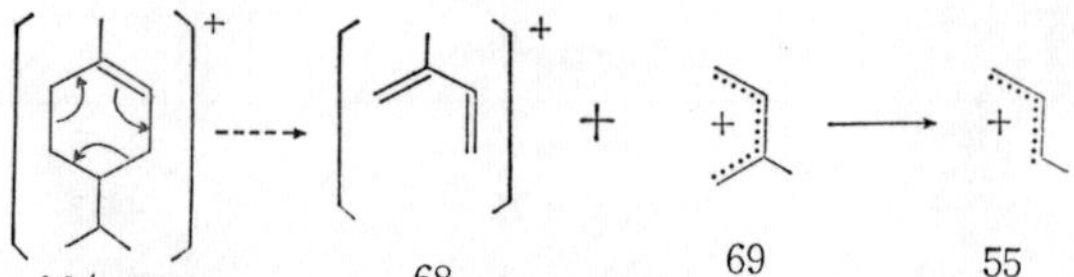

M/e 138   68   69   55

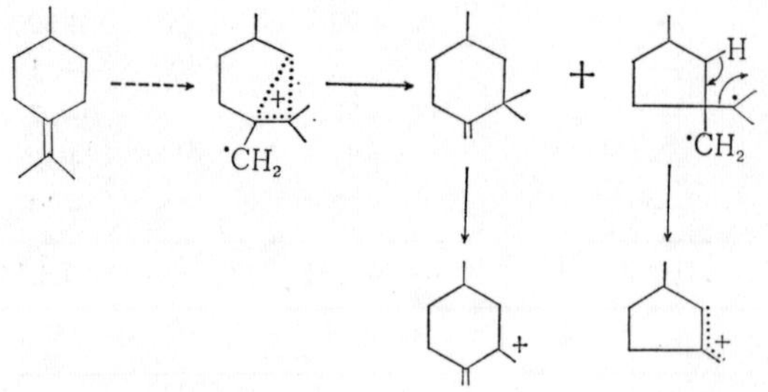

123   95

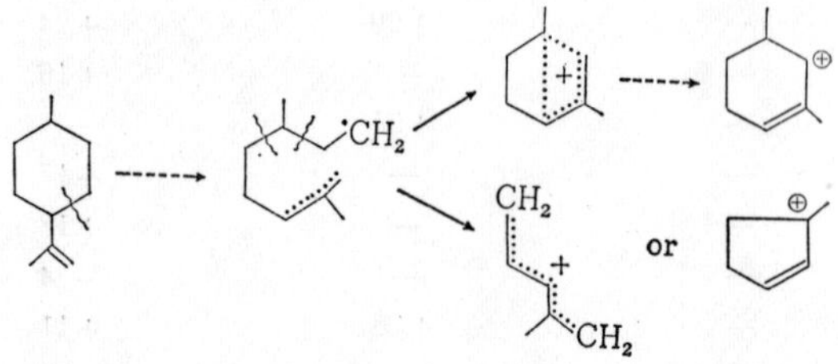

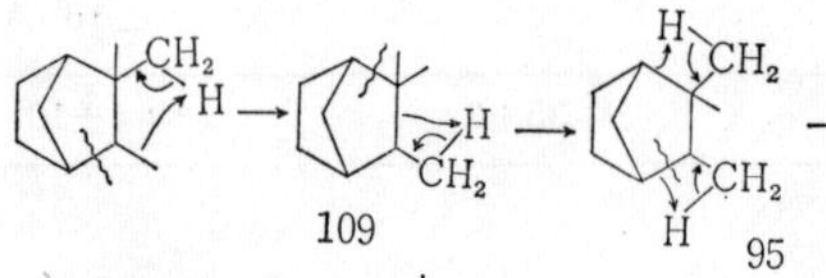

or

109   95

M/e 69
M/e 70

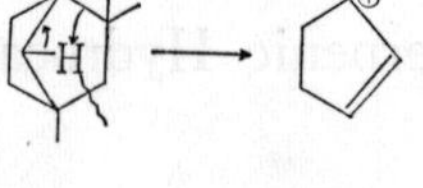

M/e 81
M/e 82

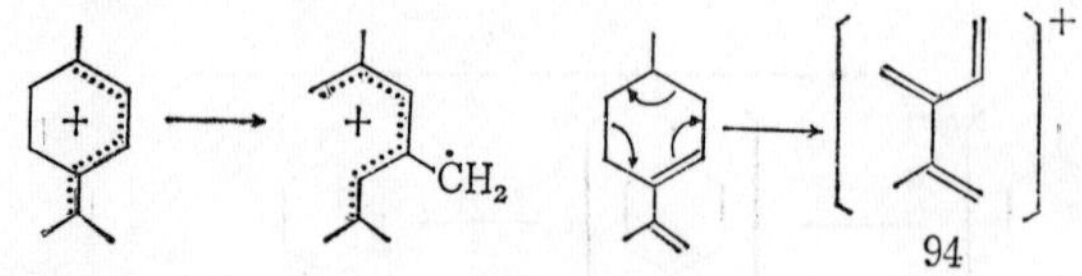

94

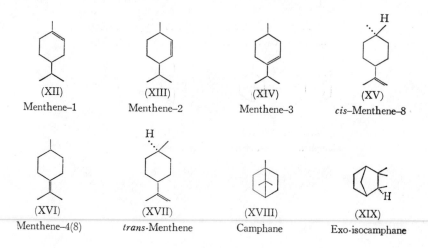

(2) Mass Peaks

Mass peaks of terpenic hydrocarbons (XII)~(XLIII) were compared by Thomas *et al.*[1178a] **as shown in** Table 80 a.

(XII)	(XIII)	(XIV)	(XV)
Menthene–1	Menthene–2	Menthene–3	*cis*–Menthene–8

(XVI)	(XVII)	(XVIII)	(XIX)
Menthene–4(8)	*trans*-Menthene	Camphane	Exo-isocamphane

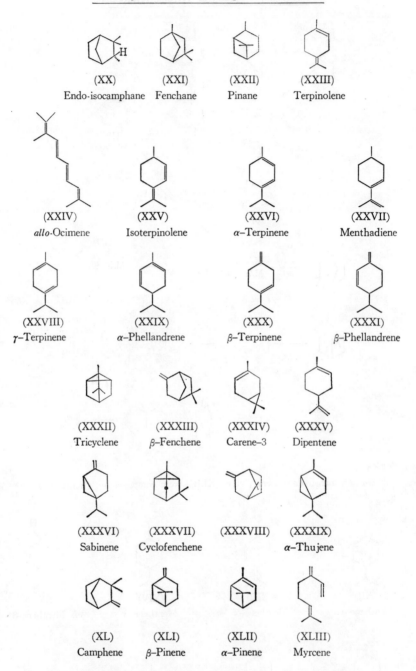

(XX)	(XXI)	(XXII)	(XXIII)
Endo-isocamphane	Fenchane	Pinane	Terpinolene

(XXIV)	(XXV)	(XXVI)	(XXVII)
allo-Ocimene	Isoterpinolene	α–Terpinene	Menthadiene

(XXVIII)	(XXIX)	(XXX)	(XXXI)
γ–Terpinene	α–Phellandrene	β–Terpinene	β–Phellandrene

(XXXII)	(XXXIII)	(XXXIV)	(XXXV)
Tricyclene	β–Fenchene	Carene–3	Dipentene

(XXXVI)	(XXXVII)	(XXXVIII)	(XXXIX)
Sabinene	Cyclofenchene		α–Thujene

(XL)	(XLI)	(XLII)	(XLIII)
Camphene	β–Pinene	α–Pinene	Myrcene

Table 80 a Mass Spectra of Terpenic Hydrocarbons

s: 100~20%, m: 20~10%, w: 10~1%   (Thomas *et al.*)

M/e \ Str.No.	138	123	110	109	96	95	94	93	91	83	82	81	80	79	77	70	69	68	67	55	53	43	41
XII	s	m	—	w	m	100	w	w	w	w	m	s	w	m	w	—	m	s	w	s	m	w	s
XIII	m	m	w	—	m	100	w	w	w	w	m	s	w	w	w	—	w	m	s	m	w	w	m
XIV	s	s	—	w	s	100	m	w	w	w	m	s	w	m	w	—	w	m	w	s	w	m	s
XV	s	s		m	s	100	w	w	w	m	s	100	w	m	w	w	s	s	s	s	m	w	w
XVI	s	s	w	w	m	100	w	w	w	w	s	s	w	m	w	w	s	m	s	s	m	w	s
XVII	s	s	w	s	s	s	w	w	w	w	s	100	w	m	w	w	s	s	s	s	m	w	w
XVIII	s	s	w	w	m	100	w	w	w	w	s	s	w	w	w	w	s	s	s	s	m	w	s
XIX	m	m	w	s	s	100	w	w	w	m	s	s	w	w	w	s	s	s	s	s	m	m	s
XX	m	m	w	s	m	100	w	w	w	m	s	s	w	w	w	s	s	s	s	s	m	m	s
XXI	w	s	w	s	m	s	w	w	w	s	s	100	s	m	w	m	m	m	s	s	m	s	s
XXII	w	s	w	w	s	s	m	w	w	s	s	s	s	m	w	w	s	s	s	100	m	m	s

M/e \ Str.No.	136	121	119	107	105	95	94	93	92	91	81	80	79	78	77	69	68	67	65	55	53	51	43	41
XXIII	s	100	w	m	s	—	w	s	w	s	w	w	s	w	s	—	w	m	w	w	m	w	s	s
XXIV	s	100	w	w	s	w	w	s	w	s	w	w	s	w	s	—	w	w	w	w	w	w	m	s
XXV	s	100	w	m	m	w	w	w	w	s	w	w	s	w	s	—	w	w	w	w	w	w	m	s
XXVI	s	100	m	m	m	—	w	s	m	s	w	w	s	w	s	—	—	w	w	w	w	w	m	m
XXVII	s	s	w	s	w	w	m	s	w	m	w	m	100	w	s	w	w	m	w	m	m	w	w	s
XXVIII	s	s	m	w	w	—	w	100	m	s	w	w	m	s	—	w	w	w	w	w	w	w	s	m
XXIX	m	w	w	—	w	—	w	100	s	s	—	w	w	w	s	w	—	—	w	w	w	w	w	w
XXX	s	m	w	w	m	w	w	100	w	s	w	w	s	w	s	w	w	w	w	w	w	w	w	m
XXXI	m	w	—	w	w	—	m	100	w	s	—	w	m	w	s	w	w	w	w	w	w	w	w	m
XXXII	m	s	w	w	w	w	m	100	m	m	w	w	m	w	m	—	w	w	w	w	w	w	w	m
XXXIII	s	s	w	s	w	m	s	s	w	m	s	s	100	m	s	m	w	m	w	m	m	w	w	s
XXXIV	m	m	w	w	w	w	m	100	s	s	w	m	m	w	s	—	w	w	w	w	w	w	m	m
XXXV	m	m	w	m	w	w	m	s	m	w	w	w	m	w	m	w	100	s	w	w	m	w	w	m
XXXVI	m	w	—	w	w	—	w	100	w	s	w	w	m	w	s	m	—	w	w	w	w	w	w	m
XXXVII	m	m	w	m	w	w	w	100	s	m	w	m	m	w	m	—	w	w	w	w	w	w	w	m
XXXVIII	s	s	—	s	w	m	s	100	m	s	s	s	s	w	s	m	w	m	w	m	m	w	w	s
XXXIX	w	w	—	—	w	—	w	100	s	s	—	w	s	—	—	w	w	w	w	w	w	w	w	m
XL	m	s	—	s	w	m	m	100	m	s	w	w	s	w	s	w	s	s	w	m	m	w	w	s
XLI	w	m	—	w	w	w	m	100	w	m	w	m	m	w	m	s	w	w	w	w	w	w	w	s
XLII	w	m	w	w	w	w	m	100	s	s	w	w	m	w	s	—	w	w	w	w	w	w	w	m
XLIII	w	w	—	w	w	w	m	100	w	m	w	w	m	w	m	s	w	m	w	w	m	w	w	s

[D] Hydrocarbons

(1) Osmane (XLIV)[1178b]

≪Occurrence≫ *Osmanthus fragrans* (キンモクセイ)

bp. 159.5°/763 mm d 0.7818, n_D 1.43676, $[\alpha]_D \pm 0°$, $C_{10}H_{20}$

(XLIV)

(2) 2,4-*p*-Menthadiene (XLV)[1178c]

≪Occurrence≫ Valenciaorange oil

bp. 56°/25 mm, n_D 1.4660, $C_{10}H_{16}$

UV λ_{max}^{EtOH} mμ: 260 (homoannular diene)

IR ν_{max} cm^{-1}: 795 $\left(\rangle=\left\langle^H\right.\right)$, 750 $\left(cis \;^H\rangle C=C\left\langle^H\right.\right)$

(XLV)

[E] Phenols

See the chapter **16** Phenolic Substances.

[F] Alcohols

(1) Artemisia alcohol (XLVI)[1179]

$$CH_3 \atop CH_3 \rangle C=CH-CH-\underset{\underset{CH=CH_2}{|}}{\overset{\overset{OH \; CH_3}{|}}{C}}-CH_3 \quad (XLVI)$$

≪Occurrence≫ *Artemisia annua* L. (クソニンジン)

bp. 71°/6 mm, d_4^{20} 0.865, n_D^{20} 1.4640, $[\alpha]_D$ −31.8, $C_{10}H_{18}O$

IR λ_{max} μ: 6.10, 10.12, 11.01 $\left(^H_H\rangle C=C\left\langle^H_R\right.\right)$, 5.97, 11.89 $\left(^R_R\rangle C=C\left\langle^H_R\right.\right)$

1178b) Ishiguro, T. *et al*: *Y. Z.*, **77**, 566 (1957).
1178c) Hunter, G. L. K: *J. Org. Chem.*, **29**, 498 (1964).
1179) Takemoto, T. *et al*: *Y. Z.*, **77**, 1307, 1310 (1957).

〔G〕 Ketones

(i) *Noncyclic Ketone*

(1) Artemisiaketone (XLVII)[1180]

　≪Occurrence≫　*Artemisia annua* L. (クソニンジン)

bp. 61°/7 mm, d_4^{25} 0.868, n_D 1.4631,　$C_{10}H_{16}O$

UV　λ_{max}^{EtOH} mμ (log ε): 241 (4.08), 330 (2.05)

　Semicarbazone　mp. 96~97°

IR: See Fig 165

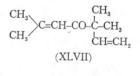

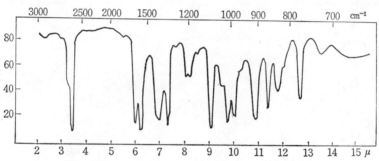

Fig. 165 IR Spectrum of Artemisiaketone (Takemoto *et al.*)

(ii) *Monocyclic Ketones*

(2) Linderone (XLVIII) and **Methyllinderone** (XLIX)[1180a]

　≪Occurrence≫　*Lindera pipericarpa* (*Lauraceae*)

(a) Linderone (XLVIII)

mp. 92~93° (orange),　$C_{16}H_{14}O_5$

UV　λ_{max}^{EtOH} mμ (log ε): 244 (4.29), 262 sh (4.15), 357 (4.47)

IR　ν_{max}^{KBr} cm^{-1}: 1715, 1630, 1595

(b) Methyllinderone (XLIX)

mp. 83~85° (yellow),　$C_{17}H_{16}O_5$

UV　λ_{max}^{EtOH} mμ (log ε): 244 (4.35), 368 (4.50)

IR　ν_{max}^{KBr} cm^{-1}: 1725, 1675, 1640, 1585

NMR　TMS, cps: 245, 251 (1:2) (1 CH$_3$O-, 2 CH$_3$O-)

1180) Nakajima, T. *et al*: *Y. Z.*, **77**, 1339 (1957), **82**, 1323 (1962).
1180a) Kiang, A. K. *et al*: *Proc. Chem. Soc.*, **1961**, 455.

(iii)   *Bicyclic Ketone*

(3)   **Rotundifolone** (L)[1181]   (Lippione=Mitglyoxal)[1182) 1183)]

   ≪Occurrence≫   *Mentha rotundifolia*[1181)], *Lippia turbinata*, *Mentha viridis*[1182)]

   mp. 27.5°, $[\alpha]_D$ +199.6°, n_D 1.5045, $C_{10}H_{14}O_2$

   UV λ_{max} mμ (log ε): 260 (3.95)

(L)

(iv)   *Furano Compounds*

(4)   **UV and IR Spectra of Furenidones**[1184)]

(LI)   (LII)   (LIII)   (LIV)

(LV)   (LVI)   (LVII)   (LVIII)

(LIX)   (LX)   (LXI)

Table 81   UV and IR Absorptions of Furenidones

(Eugster *et al.*)

Str. No.	λ_{max}^{EtOH} mμ (ε)		$\nu_{max}^{CCl_4}$ cm^{-1}		Str. No.	λ_{max}^{EtOH} mμ (ε)		$\nu_{max}^{CCl_4}$ cm^{-1}	
LI	260	(12200)	1712	1610	LVII	250	(4690)	1792, 1767, 1719, 1615	
LII	258	(12550)	1742 1707	1645 1608	LVIII	280	(16000)	1637, 1582	
LIII	—	—	1715	1648	LIX	226 270	(2400) (14000)	1681, 1570 (CHCl₃)	
LIV	220 241 303	(8900) (8170) (18300)	1701	1610	LX	210 277	(11650) (4090)	1681, 1567	
LV	212 260	(10160) (12200)	1751 1718 1709	1608	LXI	205 265	(13600) (12720)	1721, 1667, 1585 (KBr)	
LVI	231 263.5	(7280) (11470)	1754 1672	1595 (KBr)					

1181)   Shimizu: *J. Agr. Chem. Japan*, **33**, A 9 (1959).
1182)   Chakravarti *et al*: *Perfume et Essent. Oil Rec.* **45**, 217 (1954), **46**, 256 (1955).
1183)   Bhattacharyya *et al*: *ibid.*, **47**, 62 (1957).
1184)   Eugster, C. H *et al*: *Helv. Chim. Acta*, **46**, 1259 (1963).

(5) Egomaketone (LXII)[1185]

≪Occurrence≫ *Perilla frutescens* BRIT. (エ ゴマ)

bp. 122~126°/20 mm, 233°/764 mm, d 1.0097, 1.0312

n_D 1.4951, $C_{10}H_{12}O_2$

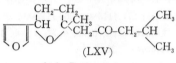

(LXII)

UV λ_{max}^{MeOH} mμ (log ε): 250 (4.18)

IR λ_{max} μ: 3.18, 6.40, 6.63, 11.50 (furan), 5.95 (conj. CO), 6.17, 11.85, 12.78 ($R_2C=CHR$)

(6) Naginataketone (LXIII)[1186]

≪Occurrence≫ *Perilla frutescens* BRIT. (エ ゴマ)

bp. 116~119°/20 mm, d_4^{30} 1.0130, n_D^{30} 1.5395, $C_{10}H_{12}O_2$

UV λ_{max}^{EtOH} mμ (ε): 295 (14400) (–C=C–CO–C=C–)

(LXIII)

IR λ_{max} μ: 6.02 (5.9~6.0 in usual–C=C–CO), 3.13, 6.31, 6.74, 11.28, 13.15 (furan ring), 6.17 (C=C), 11.88, 12.49 (δ_{C-H}).

(7) Volatile bitter Principles of Blackrotted Sweet Potato and the Related Compounds

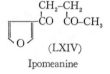

(LXIV)

Ipomeanine

(LXV)

(+) Ipomeamarone

(−) Ngainone

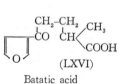

(LXVI)

Batatic acid

(LXVII)

Myoporone

(a) Ipomeanine (LXIV)

bp. 65~95°/0.003 mm, n_D^{22} 1.490~1.485, $[\alpha]_D$+10°, $C_9H_{10}O_3$

IR λ_{max} μ: 3.19 (C–H), 6.40, 6.62 (arom), 11.44, 13.30

(b) Ipomeamarone (LXV)

bp. 100~103°/0.003 mm, n_D^{22} 1.481~1.478, d_4^{25} 1.0290, $[\alpha]_D$ +22~28° (EtOH), $C_{15}H_{22}O_3$

IR λ_{max} μ: 3.19 (C–H), 6.40, 6.63 (arom.), 11.42, 13.62.

(c) Batatic acid (LXVI)

mp. 88.5~89.5°, $[\alpha]_D$ +17.5°, $C_{10}H_{12}O_4$

(d) Ngainone (LXV)

≪Occurrence≫ *Myoporum acuminatum*, M. *laetum*

bp. 159~160°/10 mm, n_D^{25} 1.4769, d_4^{25} 1.0231, $[\alpha]_D$ −25.1° (EtOH), $C_{15}H_{22}O_3$

(e) Myoporone (LXVII)

≪Occurrence≫ *Myoporum bontioides* (ハマジンチ ョウ)

bp. 117~119°/0.01 mm, n_D 1.4770, $[\alpha]_D$ ±0°, $C_{15}H_{22}O_3$

UV λ_{max} μ: 3.19 (C–H), 6.41, 6.64 (arom), 11.45, 13.00

1185) Ueda, T. *et al*: *J. Chem. Soc. Japan*, **81**, 1308, 1751 (1960); **84**, 425 (1963); *Chem. & Ind.*, **1962**, 1618.

1186) Kubota, T. *et al*: *Bull. Chem. Soc. Japan*, **31**, 491 (1958), *Tetrahedron*, **4**, 68 (1958).

(8) Toxol (LXVIII)[1187]

≪Occurrence≫ *Aplopappus heterophyllus*, pathogene of "milk sickness"

mp. 52~53°, $[\alpha]_D$ −25.1° (MeOH), $C_{13}H_{14}O_3$

UV λ_{max} mμ (log ε): 233 (4.02), 273 (4.13)

IR λ_{max}^{KBr} μ: 2.96, 5.95, 6.05, 6.21

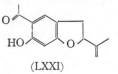

(LXVIII)

(9) Tremetone (LXIX), **Dehydrotremetone** (LXX) and **Hydroxytremetone** (LXXI)[1188]

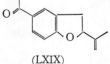

(LXIX)   (LXX)   (LXXI)

≪Occurrence≫ *Eupatorium urticaefolium*

(a) Tremetone (LXIX)

n_D 1.5658, d_4^{28} 1.080, $[\alpha]_D$ −59.6, $C_{13}H_{14}O_2$

UV λ_{max}^{EtOH} mμ (ε): 277 (11950), 280 (12600), 285 (12300)

IR λ_{max} μ: 5.98, 6.10 sh, 6.22, 6.72, 11.05, 12.20

(b) Dehydrotremetone (LXX)

mp. 87.5~88.5°, $[\alpha]_D$ ±0, $C_{13}H_{12}O_2$ UV λ_{max}^{EtOH} mμ (ε): 252 (39000), 280 (19000), 292 (15500)

(c) Hydroxytremetone (LXXI)

mp. 70~71°, $[\alpha]_D$ −50.7° (EtOH), $C_{13}H_{14}O_3$ UV λ_{max}^{EtOH} mμ (ε): 236 (38700), 280 (29100), **326**

(19300); $\lambda_{max}^{0.1N-NaOH}$ mμ (ε): 248 (38800), 280 (20300), 359 (20000)

(10) Trichothecolone (LXXIII)[1189],[1190]

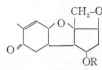   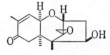

[Revised Str. Note]

R=−CO−CH≤CH·CH₃: (LXXII) Trichothecin
R=H: (LXXIII) Trichothecolone

≪Occurrence≫ Antifungal metabolite of *Trichothecium roseum* LINK.

(a) Trichothecin (LXXII)

mp. 118°, $[\alpha]_D$ +44°, $C_{19}H_{24}O_5$

UV λ_{max}^{Hexane} mμ (ε): 217 (18000), IR: See Fig. 166, 2,4-DPH. mp. 263~265°

(LXXII) $\xrightarrow{\text{alkaline hydrolysis}}$ (LXXIII)+Isocrotonic acid

(b) Trichothecolone (LXXIII)

mp. 183~184°, $[\alpha]_D$ +22.5° (CHCl₃), $C_{15}H_{20}O_4$

1187) Zalkow, L. H *et al*: *Chem. & Ind.*, **1963**, 292.
1188) De Graw, J. L. *et al*: *J. Org. Chem.*, **27**, 3917 (1962); *Tetrahedron*, **18**, 1295 (1962).
1189) Freeman, G. G. *et al*: *Soc.*, **1959**, 1105; *Nature*, **162**, 30 (1948); *Biochem. J.*, **44**, 1 (1949).
1190) Fischman, J. *et al*: *Soc.*, **1960**, 3948.
[Note] W. O. Godtfredsen and S. Vangedal: *Proc.* 188 (1964).

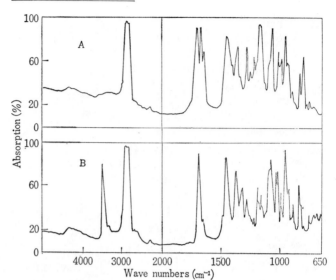

Fig. 166 IR Spectra of
(A) Trichothecin and
(B) Trichothecolone
(Solid, mulled) (Freeman *et al.*)

UV λ_{max}^{MeOH} mμ (ε).: 226 (8000), IR: See Fig. 166

Acetate mp. 148~149°, 2,4–DPH. mp. 261~262°

〔H〕 Pyran Derivatives

(1) Genipin (LXXIV)[1191]

≪Occurrence≫ *Genipa americana*

mp. 120~121°, $[\alpha]_D$ +135° (MeOH),

ORD +plain disp. curve, $[\alpha]_{275}$ +3400°, $C_{11}H_{14}O_5$

UV λ_{max}^{EtOH} mμ (log ε): 240 (4.12)

IR $\lambda_{max}^{CHCl_3}$ μ: 2.78, 2.99, 5.90s, 6.13 s

NMR in CDCl$_3$, 60Mc, TMS, δ: 7.52 s (3–H), 5.86 s (7–H), 4.81 d

(1–H), 4.29 s (—$\overset{10}{C}H_2$—), 3.73 s (—$\overset{11}{C}OOCH_3$)

See the item **6**, 〔C〕 Iridoids

(2) Constituents of *Acradenia franklinii*[1192]

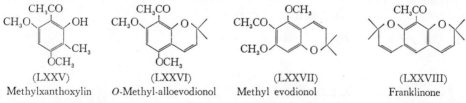

(LXXV)	(LXXVI)	(LXXVII)	(LXXVIII)
Methylxanthoxylin	*O*-Methyl-alloevodionol	Methyl evodionol	Franklinone

(a) Methylxanthoxylin (LXXV)

mp. 143°, $C_{11}H_{14}O$, UV λ_{max} mμ (log ε): 290 (4.3)

1191) Djerassi, C. *et al*: *J. Org. Chem.*, **25**, 2174 (1960), **26**, 1192 (1961).

1192) Baldwin, M. E. *et al*: *Tetrahedron*, **16**, 206 (1962).

(b) *O*-Methylalloevodionol (LXXVI)

mp. 108°, $[\alpha]_D \pm 0°$, $C_{15}H_{18}O_4$

UV λ_{max} mμ (log ε): 277 (4.2), 220 inf. (4.1)

IR ν_{max} cm^{-1}: 1706, 1638, 1609, 1587, 1383, 1367

(c) Methylevodionol (LXXVII)

mp. 78~79°, $C_{15}H_{18}O_4$

IR ν_{max}^{-1}: 1697, 1634, 1607, 1572, 1385, 1366

(d) Franklinone (LXXVIII)

mp. 128~129°, $[\alpha]_D \pm 0°$, $C_{17}H_{20}O_2$

UV λ_{max} mμ (log ε): 399 (3.5), 262 (4.7), 252 (4.7), 220 (4.0)

IR ν_{max} cm^{-1}: 1704, 1643, 1635, 1632, 1390

NMR τ: 6.3 (OCH$_3$), 7.6 (COCH$_3$), 8.7 (12 H) (2 gem (CH$_3$)$_2$, 3.6 d, 4.7 d $(2 -\overset{H}{\underset{|}{C}}=\overset{H}{\underset{|}{C}}-)$

[I] Lactones

(1) **Nepetalactone** (LXXIX)[1193][1194]

《Occurrence》 *Nepeta cataria* (イヌハッカ)

bp. 71~72°/0.05 mm, 129~130°/13 mm, $[\alpha]_D$ −13, $C_{10}H_{14}O_2$

IR λ_{max} μ: 5.67, 5.93

Dihydronepetalactone bp. 85°/0.35 mm, $C_{10}H_{18}O_2$

IR λ_{max} μ: 5.79

(LXXIX)

(2) **Matatabi-lactone** (LXXX) and **Actinidine** (LXXXI)[1195],[1196]

(LXXX): a mixture of

R$_1$=CH$_3$, R$_2$=H Isoiridomyrmecin[1196]

R$_1$=H, R$_2$=CH$_3$ Iridomyrmecin[1196]

(LXXX)

(LXXXI)

《Occurrence》 *Actinidia polygama* MIQ. (マタタビ)[1195], *Iridomyrmex* (Argentine ant)[1196]

(a) **Matatabilactone** (LXXX)[1195]

bp. 106~109°/2 mm, $[\alpha]_D$ +31.9° (CHCl$_3$), $C_{10}H_{16}O_2$

(b) **Actinidine** (LXXXI)[1195][1197][1198]

bp. 100~103°/9 mm, $[\alpha]_D$ −7.2 (CHCl$_3$), $C_{10}H_{13}N$

UV λ_{max}^{EtOH} mμ (ε): 262 (2400), IR λ_{max}^{liq} μ: 6.30 (C=N)

1193) Meinwald, J: *J.A.C.S.*, **76**, 4571 (1954).
1194) Mc Elvain, S. M. *et al*: *ibid.*, **77**, 1599 (1955), **80**, 3420 (1958).
1195) Sakan, T. *et al*: *Bull. Chem. Soc. Japan*, **32**, 315, 1154 (1959).
1196) Cavil, G.W.K. *et al*: *Australian J. Chem.*, **10**, 352 (1957).
1197) Sakan, T. *et al*: *Bull. Chem. Soc. Japan*, **33**, 712 (1960).
1198) Sakan, T. *et al*: *J. Chem. Soc. Japan*, **81**, 1445, 1447 (1960).

(3) Butylphthalides and the Derivatives[1199)~1203)]

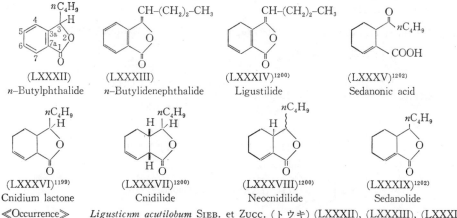

(LXXXII)
n–Butylphthalide

(LXXXIII)
n–Butylidenephthalide

(LXXXIV)[1200)]
Ligustilide

(LXXXV)[1202)]
Sedanonic acid

(LXXXVI)[1199)]
Cnidium lactone

(LXXXVII)[1200)]
Cnidilide

(LXXXVIII)[1200)]
Neocnidilide

(LXXXIX)[1202)]
Sedanolide

≪Occurrence≫   *Ligusticnm acutilobum* SIEB. et ZUCC. (トウキ) (LXXXII), (LXXXIII), (LXXXIV), *Cnidium officinale* MAKINO (センキュウ), (LXXXII), (LXXXIII), (LXXXIV), (LXXXVII), (LXXXVIII), *Apium graveolens* (セロリ) (LXXXII), (LXXXV), (LXXXIX)

(a) n-Butylphthalide (LXXXII)[1202)]

bp. 106~108°/0.1 mm, n_D^{22} 1.5228, $[\alpha]_D$ −57° (CHCl₃), $C_{12}H_{14}O_2$

UV λ_{max}^{EtOH} mμ (ε): 227 (9800), 274 (1740), 281 (1740)

IR $\nu_{max}^{CHCl_3}$ cm⁻¹: 1775 (γ-lactone), 1600, 1580, 1460 (arom.)

NMR in CCl₄, τ: 4.6 (C–CH–O–C=O), and $\left(\begin{smallmatrix}H\\\end{smallmatrix}\!\!>\!\!=\!\!<\!\!\begin{smallmatrix}H\\\end{smallmatrix}\right)$

(b) n-Butylidenephthalide (LXXXIII)[1201)]

bp. 182~184°/7 mm, $C_{12}H_{12}O_2$
4–Buthylphthalaz–1–one,   mp. 160~163°

(c) Ligustride (LXXXIV)

bp. 168~169°/6 mm, n_D^{25} 1.5649, $[\alpha]_D$ −0.59° ±1.77°, $C_{12}H_{14}O_2$
UV λ_{max} mμ (log ε): 320 (3.8),
IR: See Fig. 167

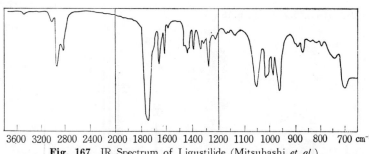

Fig. 167 IR Spectrum of Ligustilide (Mitsuhashi *et al.*)

1199) Mitsuhashi, H. *et al*: *5 th Symposium on the Chemistry of Natural Products*, p 16–1 (Sendai, 1961).
1200) Mitsuhashi, H. *et al*: *7 th Symposium on the Chemistry of Natural Products* p 18 (Fukuoka, 1963).
1201) Mitsuhashi, H. *et al*: *Chem. Pharm. Bull.*, **8**, 243 (1960).
1202) Barton, D.H.R. *et al*: *Soc.*, **1963**, 1916.
1203) Noguchi, K. *et al*: *Y.Z.*, **54**, 913 (1934), **57**, 769, 783 (1937).

(d) Sedanonic acid (LXXXV)[1202]

mp. 110~111°, $[\alpha]_D \pm 0°$, $C_{12}H_{18}O_3$

UV λ_{max}^{EtOH} mμ (ε): 214 (8900)

IR $\nu_{max}^{CHCl_3}$ cm^{-1}: 1710 (acyclic ketone), 1695 ($\alpha\beta$-unsatur. COOH), 1650 (conj. C=C)

NMR in CCl$_4$, τ: 2.71 (olefinic H)

(LXXXVII) $\xrightarrow[\text{ii) acid}]{\text{i) alkali}}$ (LXXXVIII) Isocnidilide $\xrightarrow{\text{NaBH}_4}$

$\updownarrow$ diastereomer   NaBH$_4$ $\Updownarrow$ CrO$_3$

(LXXXVIII) Neocnidilide $\longrightarrow$ (LXXXV)[1200)1202]

$\rightleftharpoons$ (LXXXIX) dl-Sedanolide[1202]

(e) Cnidiumlactone (LXXXVI)[1203]

bp. 178~180°/13 mm, $[\alpha]_D$ −72° (CHCl$_3$), $C_{12}H_{18}O_2$
IR ν_{max} cm^{-1}: 1775 (C=O), 1665 (C=C)[1199], a mixture?[1200]

(f) Sedanolide (LXXXIX) natural

bp. 185°/17 mm, $n_D^{24.5}$ 1.49234, $d_4^{24.5}$ 1.0383, $C_{12}H_{18}O_2$

(g) dl-Sedanolide (LXXXIX)[1202]   (see item (d))

bp. 116~118°/0.2 mm, n_D^{25} 1.4938, d_4^{27} 1.0335, $C_{12}H_{18}O_2$

UV λ_{max}^{EtOH} mμ (ε): 220 (9100)

IR $\nu_{max}^{CHCl_3}$ cm^{-1}: 1755 ($\alpha\beta$-unsatuu. lactone), 1680 (C=C)

(4) Digiprolactone (XC)[1203a]

《Occurrence》 *Digitalis purpurea* L.

mp. 149~151°, $[\alpha]_D$ −100.5° (CHCl$_3$), $C_{11}H_{16}O_3$

UV λ_{max}^{EtOH} mμ (log ε): 214 (4.15);

IR λ_{max} μ: 2.90 (OH), 5.78(C=O), 6.18 (C=C)

(XC)

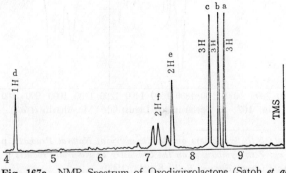

Fig. 167a NMR Spectrum of Oxodigiprolactone (Satoh *et al.*)

1203a) Satoh, D. *et al*: *Chem. Pharm. Bull.*, **12**, 752 (1964).

14   Pyrethroids

[A]   UV Spectra

(1) Pyrethrin I, pyrethrin II, tetrahydropyrethrin II, pyrethrolone, tetrahydropyrethrolone, dihydrojasmone, and their semicarbazones, (See Table 82). Tetrahydropyrethrolone, pyrethrolone-semicarbazones.[1204]

cis-and trans-Pyrethrolones and penta-2, 4-dienyl compounds[1205].

Pyrethrins[1205a)1205d)], allethrolone[1205b)], chrysanthemum monocarboxylic acid[1205c)], and α–dl–trans–allethrin[1205d)].

(2)   Absorption Maxima of Pyrethrins and Related Compounds[1206]

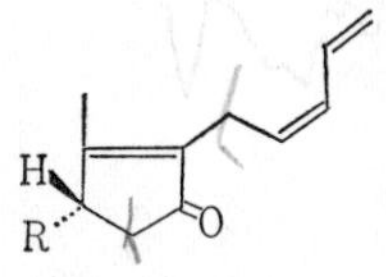

R=OH   Pyrethrolone (I)
R = (II) Pyrethrin I   (IV)
R = (III) Pyrethrin II   (V)
(Steric structure) [1207)1208)]

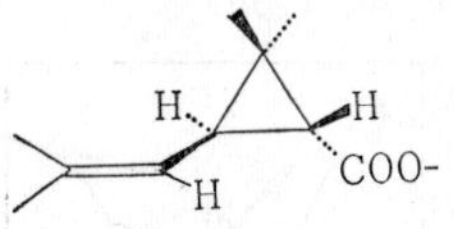

(II)
(+)–Chrysanthemum
monocarboxylic acid

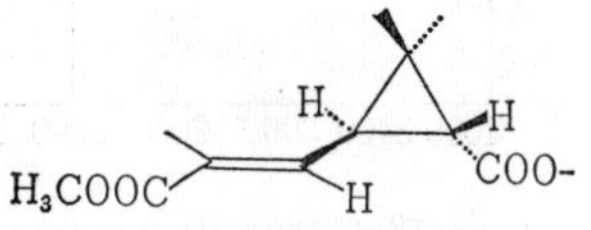

(III) (+)–Pyrethric acid

1204)   Gillam, West: *Soc.*, **1942**, 487, 671.
1205)   Crombie *et al*: *Soc.*, **1956**, 3963.
1205a)   Skukis, Cristi, Wachs: *Soap. Sanit. Chem.*, **27**, 124 (1951).
1205b)   Freeman: *Anal. Chem.*, **25**, 645 (1953).
1205c)   Schreiber, Mc Clellan: *Anal. Chem.*, **26**, 604 (1954).
1205d)   Oiwa, Shinohara, Takeshita, Ōno: *Bochu Kagaku*, (*Kyōto*) **18**, 142 (1953).
1206)   Gillam, A. E. *et al*: *Soc.*, **1942**, 487, 761. 1207) Godin, P. J. *et al*: *Proc. Chem. Soc.*, **1961**, 452.
1208)   Godin, P. J. *et al*: *Soc.*, **1963**, 5878.

Table 82 UV Absorptions of Pyrethroids

(Gillam *et al.*)

Compounds	K Band		R Band		Semicarbazone			
	λ_{max}	ε	λ_{max}	ε	λ_{max}	ε	λ_{max}	ε
Pyrethrin I	227	29,000	—	—	229	21,400	267	21,100
Pyrethrin II	231	35,400	—	—	236	31,200	266	24,000
Tetrahydropyrethrin II	236.5	24,600	—	—	238.5	20,600	262	26,200
Pyrethrolone	227	26,700	308	95	231	22,300	265	20,200
Tetrahydropyrethrolone	232	11,600	312	65	—	—	265	22,500
Dihydrojasmone	237	12,200	204	55	—	—	266.5	20,400

[B] IR Spectra

(**1**) Pyrethric acid (natural *dl-trans*–3(*trans*–2′–carboxypropenyl)–2, 2–dimethylcyclopropane–1–carboxilic acid) and its *cis*–3–isomer[1209]. Ethylchrysanthemum monocarboxylates[1210].

(2) **Pyrethric acid** (III)

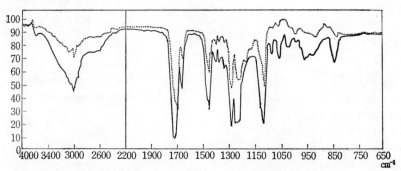

Fig. 168 IR Spectra of synthetic (±)–*trans*-Pyrethric acid (——) and natural (+)–*trans*-pyrethric acid (······) (Matsui *et al.*)[1211]

[C] Pyrethrins

≪Novel Occurrence≫   *Paeonia albiflora* PALL[1212].
Biosynthesis[1207) 1208) 1213)]

1209) Inoue, Takeshita, Ono: *Bochu Kagaku* (Kyōto) **20**Ⅲ, 102 (1955).
1210) Sadtler Card No. 5317.   1211) Matsui, M. *et al*: *Agr. Biol. Chem.* (*Tokyo*), **27**, 373 (1963).
1212) Chmielewska, I. *et al*: *Nature*, **196**, 776 (1962).
1213) Crowley, M.P. *et al*: *Nature*, **191**, 281 (1961).

15 Constituents of Male-fern and Hop-corn

[A] UV Spectra

Tri (dimethylallyl)-phloroacetophenone, acetylmethylfilicic acid, tetramethylphloacetophenone, butyrylfilicic acid[1214], hexahydrolupulone, hexahydrocolupulone, and colupulone[1215].

[B] IR Spectra

Flavaspidic acid[1216]

[C] Constituents of Male-Fern

≪Occurrence≫   *Dryopteris Filix-mas* SCHOTT., *D. crassirhizoma* NAKAI (オシダ), *Aspidium spinulosum*, *Dryopteris austriaca* (JACQ) WOYNAR., *D. erythrosora* O. KTZE, *D. nipponensis* KOIDZ, *D. fuscipes* C. CHR., *D. decipiesus* O. KTZE, *D. hondoensis* KOIDZ., *D. championi* C. CHR., *D. kinkiensis* KOIDZ., *D. bissetiana* C. CHR., *D. pacifera* TAGAWA., *D. sacrosancta* KOIDZ., *D. varia* O. KTZE[1218)1219)].

(1) **Filixic acid** (I)[1217)]

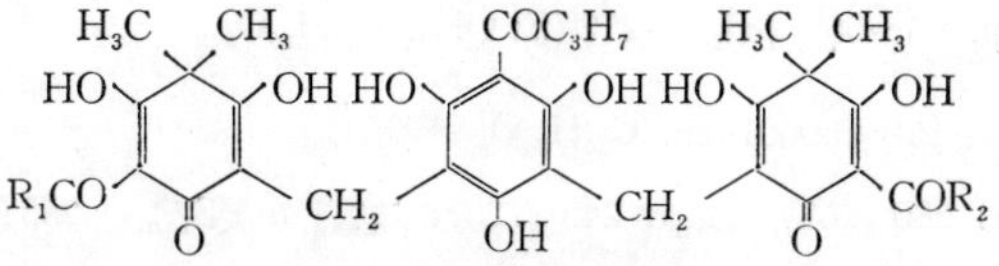

(I) Filixic acid

1214) Riedl, Risse: *Ann.*, **585**, 209 (1954).   1215) Riedl, Nickl: *Ber.*, **89**, 1863 (1956).

1216) Cänback: *J. Pharm. Pharmacol.*, **8**, 225 (1956).

1217) Penttilä, A. *et al*: *Acta Chem. Scand.*, **17**, 191 (1963).

1218) Hisada, S: *Y. Z.*, **81**, 301, 303 (1961).   1219) Hisada, S: *ibid.*, **81**, 1270 (1961).

≪Occurrence≫ *Dryopteris Filix-mas*

(a) **Natural Filixic acid** (I)[1217]   mp. 183~185° is a mixture of filixic acid BBB (II), filixic acid PBB (III) and filixic acid PBP (IV).

(b) **Filixic acid BBB** (II): $R_1 = R_2 = n$-C_3H_7[1217], mp. 172~174°, $C_{36}H_{44}O_{12}$

(c) **Filixic acid PBB** (III): $R_1 = C_2H_5$, $R_2 = n$-C_3H_7[1217], $C_{35}H_{42}O_{12}$

(d) **Filixic acid PBP** (IV): $R_1 = R_2 = C_2H_5$[1217], mp. 192~194°, $C_{34}H_{40}O_{12}$

(2) **Aspidin**[1218]

(V)

≪Occurrence≫ *Dryopteris erythrosora* and other *Dryopteris* spp.,   mp. 124°, $C_{25}H_{32}O_8$

UV $\lambda_{max}^{cyclohexane}$ mμ (log ε): 230 (4.42), 292 (4.33)

(3) **Known Constituents**

Filmalone[1220], filicin (filixic acid)[1220][1221], filicinic acid butanone[1222][1223], albaspidine[1221][1224], filicinic acid[1225], flavaspidic acid[1221], aspidin, dihydroflavaspidic acid, xanthone aspidinol, desaspidin[1225a].

[D]   Constituents of Hop-corn

≪Occurrence≫ *Humulus lupulus* L. (ホップ)

R=*iso*-Bu (VI) Lupulone   (VIII) Humulone   (IX) Hulupone
R=*iso*-Pr (VII) Colupulone

(1) **Lupulone** (VI)   mp. 93°, $C_{26}H_{38}O_4$

(2) **Colupulone** (VII)   mp. 91~93°, $C_{25}H_{36}O_4$

(3) **Humulone** (VIII)   mp. 63~65°, $[\alpha]_D$ −212°(Et OH), $C_{21}H_{30}O_5$

(4) **Hulupone** (IX)[1226]   bp. 110°/0.0001mm, $C_{20}H_{28}O_4$

UV $\lambda_{max}^{acid-EtOH}$ mμ($E_{1cm}^{1\%}$): 280 (268); λ_{min}: 245 (125); λ_{max}^{EtOH} mμ($E_{1cm}^{1\%}$): 255 (446), 325 (299); λ_{min}: 280

(110)

1220) Kraft: *Arch. Pharm.*, **242**, 489 (1904).   1221) Boehm: *Ann.*, **318**, 253 (1901).

1222) Boehm: *Ann.*, **318**, 230 (1901).   1223) Riedl, Risse: *Ber.*, **87**, 865 (1954).

1224) Böhm *et al*: *Ann.*, **318**, 301 (1901).   1225) Riedl, Risse: *Ann.*, **585**, 209 (1954).

1225a) Aebi *et al*: *Helv. Chim. Acta*, **40**, 266 (1957)   1226) Stevens R. *et al*: *Soc.*, **1963**, 1763.

16 Phenolic Substances

Phenols, occuring as glycosides in natural are described in the item **6**, [D] Phenolic Glycosides. In addition, the following compounds, having phenolic moiety are described in the item which numbers are indicated in parenthesis.

Stilbene glycosides (**6**, [E]), coumarin and furocoumarin glycosides (**6**, [F]), chalcone and flavanoid glycosides (**6**, [G]), anthocyan (**6**, [H]) and anthochlor pigments (**6** [H]), flavone and flavonoid glycosides (**6**, [I]), chromones, chromans, xanthones and their glycosides (**6**, [J]), rotenoids (**6**, [K]), coumaranochromone derivatives (**6**, [L]), pyrane derivatives (**6**, [M]), tannins (**6**, [N]), diterpenoids (**11**, some of nonterpenic essential oils (**13**), constituents of male-fern and hop-corn (**15**), naphtoquinones and anthraquinones (**17**, lignoids (**18**), and phenolic alkaloids (**20**).

UV and IR spectra of phenolic compounds (**18**, (XVIII)~(LIV)), the precursor or model compounds of lignoids are indicated in the item **18**, [E], (1), Table 87, Fig. 187~189 UV and (2), Fig. 189a (IR).

[A] Phenols and Naphthols

(1) ***p*- Hydroxybenzylmethylether** (I)[1227]

$$HO-\langle\ \rangle-CH_2-O-CH_3 \quad (I)$$

≪Occurrence≫ *Citrullus colocynthis* SCHRAD.

mp. 85~86° $C_8H_{10}O_2$

(2) **Musizin** (II)[1228]

≪Occurrence≫ *Maesopsis eminii*

(II)

≪Occurrence≫ *Maesopsis eminii*

UV $\lambda^{\text{light petroleum}}_{\text{max}}$ mμ (log ε): 219 (4.17), 266 (4.39), 402 (3.68)

1227) Watanabe, K. *et al*: *Agr. Biol. Chem.* (*Tokyo*), **25**, 269 (1961)
1228) Covell, C. J. *et al*: *Soc.*, **1961**, 702.

IR $\nu_{max}^{CHCl_3}$ cm^{-1}: 1630 (not influenced by concentration)

(3)   **Heyderiol** (III)[1299]

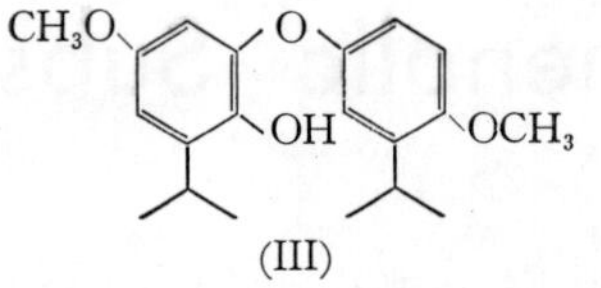

(III)

≪Occurrence≫   *Heyderia decurrence* TORREY. (Incense cedar)

mp. 62.2~63.2°, $C_{22}H_{30}O_4$

UV $\lambda_{max}^{isooctane}$ mμ (log ε): 287 (3.92)

IR $\nu_{max}^{Hexachlorobutadiene}$ cm^{-1}: 3520 (OH); ν_{max}^{KBr} cm^{-1}: 1595, 1620 (benzenoid)

(III) $\xrightarrow{FeCl_3}$ (IV)

Heyderioquinone (IV)

mp. 124~124.4°, $C_{21}H_{26}O_4$

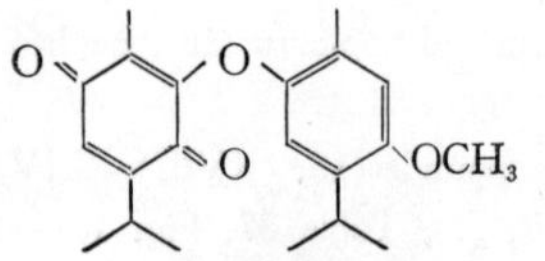

(IV)

UV $\lambda_{max}^{methylcyclohexane}$ mμ (log ε): 235 (4.15), 260 (4.27) (benzenoid A), 284~285 inf. (3.57) (benzenoid B), 363 (2.98), 455 (2.61), 475 inf. (2.59) (quinoid C)

IR ν_{max}^{KBr} cm^{-1}: 1660 (conj. C=O), 1615 (conj. C=C) 1673, 1640 (conj. C=C)

NMR   in CS$_2$, 40 Mc, EtOH band= 0 ppm: −1.45, −1.30, −1.15 (1:1:1.1) (3 arom. −H), +1.3
(3.0) (CH$_3$O−), +2.70, +3.00 (3:3.1) (arom. CH$_3$), +3.80, +3.95 (6:6) (iso-Pr. doublet)

(4)   **Constituents of Kousso**[1230]~[1230]

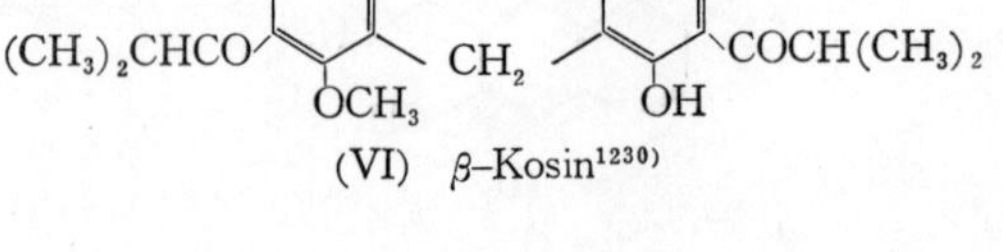

(V)   α–Kosin[1232]

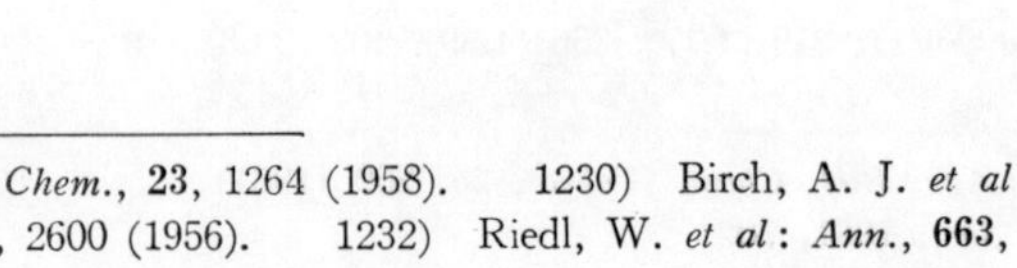

(VI)   β–Kosin[1230]

1229)   Zavarin E: *J. Org. Chem.*, **23**, 1264 (1958).   1230)   Birch, A. J. *et al*: *Soc.*, **1952**, 3102.

1231)   Riedl, W: *Ber.*, **89**, 2600 (1956).   1232)   Riedl, W. *et al*: *Ann.*, **663**, 83 (1963).

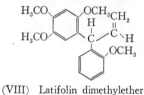

(VII) Protokosin

≪Occurrence≫ *Hagenia abyssinica* (Kousso)

(a) α–Kosin (V) $C_{25}H_{32}O_8$

UV $\lambda_{max}\,m\mu\,(\varepsilon)$: 230 (30000), 287 (23000); λ_{min}: 254; $\lambda_{max}^{acidic\ soln}\,m\mu\,(\varepsilon)$: 227 (30800), 290 (24400), 340 inf. (6370); λ_{min}: 253, 280, 325 inf.[1232]; $\lambda_{max}^{alkaline\ soln}\,m\mu\,(\varepsilon)$: 260 inf. (12650), 330 (16450)[1232]

IR $\lambda_{max}^{nujol}\,\mu$: 3.13 m, 6.24 s, 7.10 m, 7.40 s, 7.67 w, 7.87 s, 8.43 m, 8.63 w, 9.00 s, 9.35 w, 9.84 w, 10.10 m, 10.77 w, 12.3 w[1230]

(b) β–Kosin (VI)[1230] $C_{25}H_{32}O_8$

UV $\lambda_{max}m\mu\,(\varepsilon)$: 228 (30300), 292 (21600); λ_{min}: 254.5

IR $\lambda_{max}^{nujol}\,\mu$: 3.10 m, 6.24 s, 7.13 m, 7.34 m, 7.63 w, 7.89 m, 8.45 s, 9.02 s, 9.34 w, 9.84 m, 10.05 m, 10.80 w, 11.2 w, 12.1 w

(c) Protokosin (VII)[1230]

mp. 182°, $[\alpha]_D$ +8° (CHCl$_3$), $C_{25}H_{32}O_8$

UV $\lambda_{max}\,m\mu\,(\varepsilon)$: 223 (25860), 287 (19840); λ_{min}: 253,

IR $\lambda_{max}^{nujol}\,\mu$: 2.96 w, 3.23 w, 6.20 s, 6.45 s, 7.11 s, 7.38 s, 7.85 w, 8.15 w, 7.45 m, 8.67 m, 8.99 m, 9.12 m, 9.65 w. 10.01 w, 10.33 w, 10.63 m, 10.80 w, 11.02 m, 13.13 w, 13.77 w.

(5) **Latifolin and the related Compounds**

(VIII) Latifolin dimethylether   (IX) Dalbergione   (X) Dalbergin

≪Occurrence≫ *Dalbergia latifolia*[1233][1234]

(a) Latifolin

mp. 123~123.5°, $[\alpha]_D$ −26.2° (EtOH)[1234], +135.9°[1233], $C_{17}H_{18}O_4$

UV $\lambda_{max}^{EtOH}\,m\mu\,(\varepsilon)$: 213 (17250), 283 (5100), 292 (4700)[1233]

IR $\nu_{max}^{CHCl3}\,cm^{-1}$: 3558, 3424 (OH); $\nu_{max}^{KBr}\,cm^{-1}$: 2849 (OCH$_3$), 1640, 1001, 917 (CH=CH$_2$),763[1234];

$\nu_{max}^{KBr}\,cm^{-1}$: 3500, 3450 (OH), 1630 (C=C), 1600, 1510 (arom.), 1035 (O–CH$_3$), 920[1233]

(b) Latifolin dimethylether (VIII)[1233]

mp. 89°, $[\alpha]_D$ +21.9°, $C_{19}H_{22}O_4$

NMR τ: 3.46 s (3′–arom. H), 3.31 s (6′–arom. H)[1234]

(c) Dalbergione (IX) See the item **17**, [D], (7)

1233) Seshadri, T. R. *et al*: *Tetrahedron*, **18**, 1503 (1962).
1234) Dempsy, C, B. *et al*: *Chem. & Ind.*, **1963**. 491.

(6) (−)‑**Angolensin** (XI)[1235]

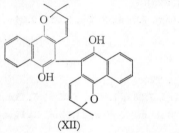

(XI)

≪Occurrence≫   *Afrormosia elata, Pterocarpus angolensis, P. indicus P. erinaceus*
mp. 120.5∼121°, $[\alpha]_D$ −115° (CHCl₃), $C_{16}H_{16}O_4$

(±)‑**Angolensin diacetate**   mp. 88.5∼89.5°
IR ν_{max}^{CCl4} cm⁻¹: 1680, 1760

(±)‑**Angolensin enolic triacetate**   mp. 131°
IR ν_{max}^{CHCl3} cm⁻¹: 1750

(7) **Techtol** (XII) **and Dehydrotechtol** (XII‑a)[1236]

(XII)   (XII‑a)

≪Occurrence≫   *Techtonia grandis* L. f. (Teak)
(a) **Techtol** (XII)
mp. 216∼218°, $C_{30}H_{26}O_4$
UV λ_{max} mμ (log ε): 275 (4.50), 347.5 (3.78), 362.5 (3.65)
(b) **Dehydrotechtol** (XII‑a)
mp. 195∼197°, $C_{30}H_{24}O_4$
UV λ_{max} mμ (log ε): 271 (4.53), 340 (3.09)
IR ν_{max} cm⁻¹: 1645 (C=O)

(8) **Gossypol** (XIII)
≪Occurrence≫   The seeds *of Gossypium arboreum* (ワタ)
mp. 184° (ether+petroleumether), 214° (ligroin),
$C_{30}H_{30}O_8$

(XIII)

1235) Foxall, C. D. *et al*: *Soc.*, **1963**, 5573.
1236) Sandermann, W. *et al*: *Tetr. Letters*, **1963**, 1269; *Ber.*, **97**, 588 (1964).

（ 9 ）　Leucodrin (XIII-a)[1236a)b)c)]

　　≪Occurrence≫　*Leucadendron concinnum, L. adscendens, L. stokoei*
　　mp. 212～212.5°, $[\alpha]_D$ −15.45°, $C_{15}H_{16}O_8$

　IR　ν_{max}^{KBr} cm⁻¹: 1800, 1770 (five membered lactones)

　NMR　of Tetraacetate $\tau: \tau_A$ 6.72, τ_B 7.13, τ_C 5.79, J_{AB}
　　=−17.2, J_{AC} 12.4, J_{BC} 7.8 (ABC system on ring
　　A), J_A 4.18, τ_B 6.23, τ=8.3 (AB system on ring B)

(XIII-a)

〔B〕 Marihuana Substances

　≪Occurrence≫　*Cannabis sativa* var. *indica*（インド大麻）

（ 1 ）　UV Spectra[1237)]

　　Cannabinol acetate, 3–acetoxy–2n–amyl–6, 6, 9–trimethyl–6–benzopyran, synthetic isomer of conjugated tetrahydrocannabinol, cannabidiol, cannabidiol dimethylether, and an isomer produced by isomerization of canna-bidiol.

(XIV)　Cannabinol

(XV)　Cannabidiol

(XVI)　Tetrahydrocannabinol

(XVI-a)　Cannabigerol

（ 2 ）　Cannabinol (XIV)[1237a)]

　　mp. 76～77°, $C_{21}H_{26}O_2$

　UV　$\lambda_{max}^{neutral}$ mμ: 283; λ_{min}: 248; $\lambda_{max}^{0.1N-NaOH-MeOH}$ mμ (log ε): 283 (4.04), 328; λ_{min}: 263 (3.91), 310
　IR　See Fig. 169

（ 3 ）　Cannabidiol (XV)[1238)]

　　mp. 66～67°, $[\alpha]_D$ −125°, −130° (EtOH), $C_{21}H_{30}O_2$
　UV　λ_{max} mμ (log ϵ): 212 (4.57), 273 (3.04), 280 (3.02)

1236a)　Rapson, W. S: *Soc.*, **1938**, 282.　　1236b)　Perold, G. W. *et al*: *Proc. Chem. Soc.*, **1964**, 62.
1236c)　Diamond, R. D. *et al*: *ibid.*, **1964**, 63.
1237)　De Ropp, R. S: *J. A. P. A.*, **49**, 756 (1960).
1237a)　Adams, R: *J.A.C.S.*, **62**, 732, 2201, (1940), **63**, 1971 (1941).
1238)　Korte, F. *et al*: *Ann.*, **630**, 71 (1960).

（4）　**Tetrahydrocannabinol** (XVI)[1237a]

$[\alpha]_D$ $-161°$, $C_{21}H_{30}O_2$

UV $\lambda_{max}^{neutral}$ mμ (log ε): 275 (3.26), 282 (3.28); λ_{min}: 251; $\lambda_{max}^{0.1N-NaOH-MeOH}$ mμ (log ε): 292 (3.53)

3.25; λ_{min}: 269

IR　See Fig. 169

（5）　**Cannabigerol** (XVI-a)[1238a]

mp. 51～53°, $C_{21}H_{32}O_2$

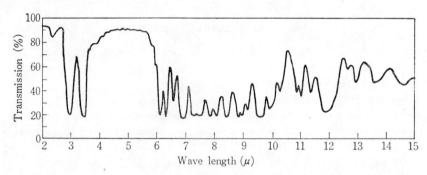

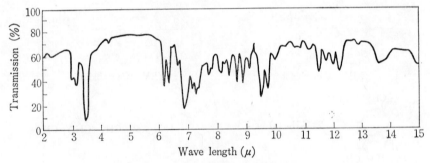

Fig. 169　IR Spectra of Cannabinol (upper) and
Tetrahydrocannabinol (under)　(De Ropp)

[C]　Tropolones

（1）　**UV Spectra**

(a)　Tropolone, tropolonemethylether and cupric tropolones[1239]. Purpurogallin, dimethylpurpurogallin, trime-thylpurpurogallin and tetramethylpurpurogallin[1240].

1238a)　Gaoni, Y. *et al*: *Proc. Chem. Soc.*, **1964**, 82.　　1239)　Cook *et al*: *Soc.*, **1951**, 508.

(b) On the UV Spectra of Tropolones[1241].

UV Spectra of tropolone, 2–substituted methyl–, ethyl–, phenyl–, and other homologues, synthetic tropolones, having α– or β– naphthyl group were investigated. Absorption is shifted to higher region by substitution of arylic groups.

(c) UV Absorption spectra of tropone, troponium ion, tropolone and 2,4,6–octatrienal[1242]. Measurements of polarized UV absorption of tropolones were studied.

(2) IR Spectra

(a) Cycloheptenone, 2, 3–benzocycloheptanone, 2, 6, 6–trimethylcyclopentadien–2, 4–one[1243], 3–hydroxytropone[1244], 4, 5–benzotropone[1243], tropone[1245], tropolone methylether[1249], tropolone[1248], 3, 7–di (p–methylbenzyl)–tropolone[1246], Cu–derivs. of tropolone[1248].

Characteristic absorption due to tropolone nucleous[1247], colchicine, colchiceine, trimethyl colchicinic acid, hexahydrocolchiceine, tetrahydrocolchiceine, γ–thujaplicin, and β–methyltropolone[1243].

(b) **Tropolone** (XVII)[1250]

IR ν_{max}^{solid} cm⁻¹: 3518 w, 3208 (O–H), 3049, 3003 vw (C–H), 1613 s (C=O), 1548 vs (C=C), 1480 vs (C–O–H), 1469 sh (C–C), 1425 vs, 1363 vw (2×675), 1310 m, 1266 vs, 1238 vs, 1208 vs, (C–N–H or C–H in plane deform.), 1049 m (357+679), 1009 vw 958m, 919m, 874 w (2×436), 860 m, 789 w (357+436), 754 s, 708 s (C—H out of plane deform., C–C strech., O-H out of plane deform.), 672 w (C–C–C ring deform.), 531 vw (″), 433 m (C–C–C ring deform.)

(XVII)

IR $\nu_{max}^{CCl_4}$ cm⁻¹: 3167, 3120 (O–H)

(c) Characteristic absorption of tropone, 2–chlorotropolone and tropolone at 1700~1530 cm⁻¹ region[1251]. Band I (1612~1622 cm⁻¹) and band II (1547~1573 cm⁻¹) are regarded as characteristic absorption of tropolone nucleous, due to $\nu_{C=O}$ and $\nu_{C=C}$.

(3) Dolabrinol (XVIII)[1252]

≪Occurrence≫ *Cupressus pygmaea* (LEMM.) SARG.

bp. 140~160°/1~2 mm, $C_{10}H_{10}O_3$

UV $\nu_{max}^{Isooctane}$ mμ (log ε): 248.5 (4.31), 325 (3.68), 356 (3.60) 365 (3.69), 374 (3.79)

(XVIII)

IR ν_{max}^{KBr} cm⁻¹: 3250 s, 2970 w, 2930 w, 1637 w, 1613 w, 1515 m, 1540 s, 1520 s, 1455 s, 1420 s, 1375 s, 1305 s, 1285 s, 1265 s, 1240 s, 1195 s, 1100 w, 1060 m, 1040 w, 1013 w, 962 m, 908 m, 895 m, 815 m, 795 m

NMR in CCl₄, TMS, ppm: +7.8 (3) (CH₃C=), +5.0, +4.8 (2) (C=CH₂), +3.0 (3) (arom. 3H), +1.15 (2) (2OH)

1240) Haworth *et al*: *ibid.*, **1948**, 1046.　　1241) Makai: *J. Chem. Soc. Japan.*, **79**, 1547 (1958).
1242) Hosoya, H. *et al*: *Tetrahedron*, **18**, 859 (1962).　　1243) Scott *et al*: *J. A. C. S.*, **72**, 240 (1950).
1244) Jones *et al*: *Chem. & Ind.*, **1954**, 192.　　1245) Doering *et al*: *J. A. C. S.*, **73**, 876 (1951).
1249) Doering *et al*: *J. A. C. S.*, **73**, 828 (1951).　　1248) Koch: *Soc.*, **1951** 512.
1248) Leonard *et al*: *ibid.*, **75**, 4989 (1953).　　1247) Pauson, P. L: *Chem. Rev.*, **55**, 21 (1955).
1250) Ikegami, Y: *Bull. Chem. Soc. Japan*, **34**, 94 (1961).
1251) Ikegami, Y: *ibid.*, **35**, 972 (1962).
1252) Zavarin, E: *J. Org. Chem.*, **27**, 3368 (1962).

(4) Isopygmaein (XIX)[1252]

    ≪Occurrence≫   *Papuacedrus lorricellenis* (SCHLECTER) L.

    mp. 110~112°, $C_{11}H_{14}O_3$

    UV $\lambda_{max}^{Isooctane}$ mμ (log ε): 373 (3.77), 358 (3.75), 326 (3.75), 313 inf.

      (3.63), 250 (4.47)

    IR ν_{max}^{KBr} cm^{-1}: 3220 s, 2980 m, 2890 m, 1590 m, 1555 s, 1495 m, 1475 s,

      1463 s, 1445 m, 1410 s, 1390 m, 1365 m, 1337 s, 1290 s, 1243 s, 1230 s, 1210 s, 1195 m, 1168 s, 1140 s,

      1095 s, 1065 m, 1035 m, 990 m, 920 m, 855 w, 800 m, 783 w, 765 m, 675m

(XIX)

(5) α-Thujaplicinol (XX)[1253]

    ≪Occurrenec≫   *Cupressus pygmaea* (LEMM). SARG.

    n_D 1.6323, d_4^{22} 1.184, $C_{10}H_{12}O_3$

    Benzylamine adduct mp. 110~111°, dicyclohexylamine adduct

    mp. 130~130.5°, Cu–chelate mp. 303.5~304.5°

(XX)

    UV $\lambda_{max}^{Isooctane}$ mμ (log ϵ): 247.5 (4.63), 324 (3.73), 356 (3.82), 364 (3.88), 372.5 (4.05)

    IR ν_{max}^{CCl4} cm^{-1}: 3230 m, 2950 m, 2860 w, 1615 w, 1585 w, 1538 s, 1520 s, 1455~1470 m, 1420 s, 1350 w,

      1292 s, 1262 s, 1220 s, 1200 s, 1180 s, 1140 w, 1110 w, 1075 w, 1060 w, 1027 w, 1010 w, 938 w, 962,

      950, 920, 910, 895, 866 (all w)

    NMR τ (J_{cps}): 8.76 d (J=7)$\left(CH<^{CH_3}_{CH_3}\right)$, 6.26 m$\left(-CH<^{CH_3}_{CH_3}\right)$

(6) Thujic acid (XXI, R=H)[1254]

    ≪Occurrence≫   *Thuja plicata* DON

    mp. 88~89°, $C_{10}H_{12}O_2$

(XXI) R=H

    NMR of methylate, τ: two peaks at ca. 2.2$\left(H\underset{COO}{\overset{\parallel}{\diagup}}H\right)$

    3.2~4.9 $\left(\overset{H\ H}{\diagdown\diagup}\right)$, 6.2 (ester Me), 8.9 (gem Me$_2$)

(7) Stipitatonic acid (XXII)[1255]

    ≪Occurrence≫   *Penicillium stipitatum* THOM

    mp. 237~237.5°, $C_9H_4O_6$

    UV λ_{max}^{H2O} mμ: 253, 333, 369, 432

(XXII)

1253) Zavarin, E. *et al*: *J. Org. Chem.*, **26**, 173 (1961).

1254) Davis, R. E. *et al*: *Tetr. Letters.*, **1962**, 839.   1255) Segal, W: *Soc.*, **1959**, 2847.

17 Quinones

[A] The Quinones, described in other Items.

(1) **Diterpenoid quinones**   See the item **11**, [E].

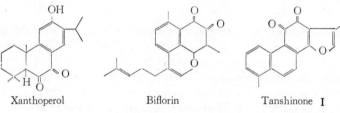

Xanthoperol

Biflorin

Tanshinone **I**

Tanshinone **II**

Cryptotanshinone

Royleanone

(2) **Phenolic substances**   See the item **16**, [A].

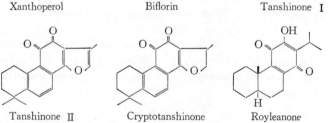

Heyderioquinone

[B] UV Spectra

1, 4–Benzoquinone, 1, 2–benzoquinone[1256)1257)], 1, 4–benzoquinone, 2–methylbenzoquinone, 2, 3–and 2, 5–dimethylbenzoquinones[1258]. Naphthoquinone, naphthazarin[1259]. Isonaphthazarin and diacetates[1260]. Juglone, —acetate[1260]. Hydroxyjuglone, —diacetate, droserone, —diacetate[1259]. Naphtha purpurin, —triacetate, phtiocol, —acetate, lawsone[1261]. Lomatiol, —diacetate, lapachol, —diacetate[1262]. Shikonin and alkanin[1263].

Anthraquinones. 1–hydroxyanthraquinone[1264], 2–hydroxyanthraquinone, 2–methoxyanthraquinone, 1–hydroxyanthraquinone, 1–methoxyanthraquinone anthraflavine, quinizarine, chrysazine, 1–hydroxy–8–methoxyanthraquinone, dimethoxyanthraquinone, arizarin, anthrone, 1, 2–dimethoxyanthraquinone and 1, 5–dihydroxyanthraquinone[1264]

Hydroxy-anthraquinones: 1, 2, 3–trihydroxyanthraquinone, 1, 4, 8–trihydroxyanthraquinone, and 1, 2, 8–trihydroxyanthraquinone[1265]. Chrysophanol, emodin, chrysophanol+$MgAc_2$, emodin+$MgAc_2$, chrysophanol anthrone+p–$NOC_6H_4NMe_2$, emodin anthrone+p–$NOC_6H_4NMe_2$ and chrysophanol anthrone[1266].

[C] IR Spectra

α, β–Unsaturated ketones, p–benzoquinone, 1, 4–naphthoquinone, and anthraquinone[1267]. 1, 4–Naphtoquinone, 2–methylnaphthoquinone, 2–hydroxynaphtoquinone, 2, 3–dihydroxynaphthoquinone, 3–hydroxy–2–methylnaphthoquinone and 2–methoxynaphthoquinone[1266a]. Anthraquinone, 1–hydroxyanthraquinone, 2–hydroxyanthraquinone, alizarin, xanthopurpurin, quinizarin, 1, 5–dihydroxyanthraquinone, chrysarin, purpurin, 1, 4, 5, 8–tetrahydroxyanthraquinone, emodin, triacetylemodin and chrysophanol[1267].

$\nu_{C=O}$ of anthraquinones[1266]. 1, 8–Dihydroxyanthraquinone, 1, 2–dihydroxyanthraquinone, 1, 3–dihydroxyanthraquinone[1268].

ν_{O-H} of Hydroxyanthraquinones. Anthrone, 1–hydroxyanthrone, 4–hydroxyanthrone, oxyanthrone, anthraquinol[1268]. 1, 7–Dihydroxy–5–methylanthraquinone[1269], emodin–3–glucoside[1270], emodin–8–glucoside[1271], 1, 7–dihydroxy–5–methylanthraquinone–7–glucoside[1272], 1, 8–dihydroxy–3–hydroxymethylanthrone–10–glucoside[1273] and chrysophanol–8–glucoside[1274].

1256) Schauenstein et al: Ber., 59, 2617 (1926). 1257) Goldschmidt et al: ibid., 61, 1858 (1928).
1258) Braude: Soc., 1945, 490. 1259) Moron: Soc., 1941, 159.
1260) Macbeth: ibid., 1935, 325. 1261) Lugg et al: Soc., 1937, 1597.
1262) Cooke et al: Soc., 1939, 878. 1263) Brockmann: Ann., 521, 1 (1936).
1264) Morton et al: Soc., 1941, 159.
1265) Ikeda, T., Yamamoto, Y., Tsukida, K., Kanatomo, S: Y. Z., 76, 217 (1956).
1266) Tsukida, K., Suzuki, N: Y. Z., 74, 1092 (1954).
1266a) Shibata, S., Tanaka, O: Kagaku no Ryoiki, Zokan (Nankodo Pub. Inc. Tokyo), 23, 157 (1956).
1267) Flett: Soc., 1948, 1441.
1268) Hoyer, H: Ber., 86, 1016 (1953).
1269) Sadtler Card No. 3593. 1270) Sadtler Card No. 3601. 1271) Sadtler Card No. 3602.
1272) Sadtler Card No. 3594. 1273) Sadtler Card No. 3555. 1274) Sadtler Card No. 3580.

(1) 1,4-Benzoquinones[1275) 1276)]

(2) Molecular compounds of 1,4-naphtoquinone and hydroquinone and a dimer of 2-methyl-1,4-naphthoquinone[1277)]

(3) Red-oxy. Potential and Carbonyl Frequency of Quinones[1277a)] See Table 83.

Table 83 $\nu_{C=O}$ cm^{-1} and Red-oxy. Potential of Quinones   (Kurosawa)

Benzoquinone	ν_{max}^{CS2} cm^{-1}	ν_{max}^{nujol} cm^{-1}	Red-oxy. potential (volt)
Chlorobenzoquinone	1678		0.736
Benzoquinone	1668	1663	0.711
Acetylbenzoquinone	1668	1645	0.564
Toluquinone	1663	1663	0.656
Acetyltoluquinone	1662	1645	0.539
o–Xyloquinone	1660	1660	0.588
Acetyl–o–xyloquinone	1654	1640	0.504
p–Xyloquinone	1660	1668	0.597
m–Xyloquinone	1657	1653	0.592
Thymoquinone	1660		0.589
Pseudocumoquinone	1657	1650	0.536
Duroquinone	1642	1639	0.471
Ubiquinone	1653		0.542

Naphthoquinone	ν_{max}^{CCl4} cm^{-1}	Red-oxy. potential (V)
1, 4–Naphthoquinone	1675	0.483
2–Methyl–1, 4–naphthoquinone	1670	0.422
2, 3–Dimethyl–1, 4–naphthoquinone	1660	0.340
2, 6–Dimethyl–1, 4–naphthoquinone	1668	0.405
2, 7–Dimethyl–1, 4–naphthoquinone	1673	0.407
2–Methoxy–1, 4–naphthoquinone	1660	0.369

〔D〕 Benzoquinones

(1) UV and IR Spectra of Alkylbenzoquinones[1278)]

1275) Edwards, R. L. et al: J. Appl. Chem., **10**, 246 (1960); Anal Abstr., **8**, 179 (1961).
1276) Flaig, W et al: Ann., **626**, 215 (1959).
1277) Kodera, K: Y. Z., **80**, 659 (1960).
1277a) Kurosawa, E: Bull. Chem. Soc. Japan, **34**, 300 (1961).
1278) Natori, S et al: Chem. Pharm. Bull., **12**, 236 (1964).

(a) 2-Methoxy-6-alkylbenzoquinones

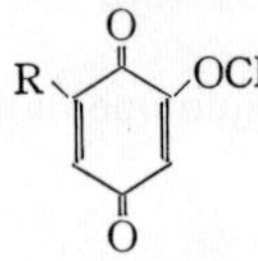

$R=CH_3$ (I)

$R=C_{11}H_{23}$ (II)

$R=C_{12}H_{25}$ (III)

(b) 2-Hydroxy-5-alkylbenzoquinone

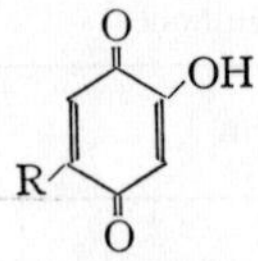

$R=C_{16}H_{33}$ (IV)

(c) 2-Hydroxy-3,6-dialkylbenzoquinones

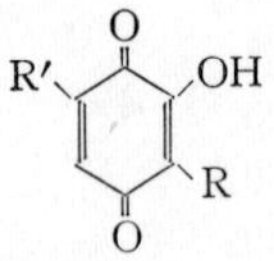

$R=CH_3, R'=C_8H_{17}$ (V)

$R=CH_3, R'=-CH(CH_2)_3-CH(CH_3)_2$ (VI)
$\qquad\qquad\quad CH_3$

$R=CH_3, R'=-CH(CH_3)_2$ (VII)

(d) 3-Alkyl-2,5-dihydroxybenzoquinones

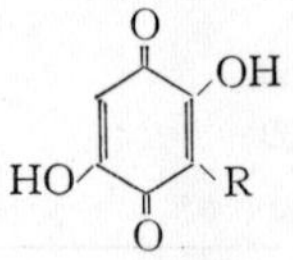

$R=C_{11}H_{23}$ Embelin (VIII)

$R=C_{13}H_{27}$ Rapanone (IX)

$R=C_{18}H_{37}$ (X)

(e) 2,5-Dihydroxy-3,6-dialkylbenzoquinones

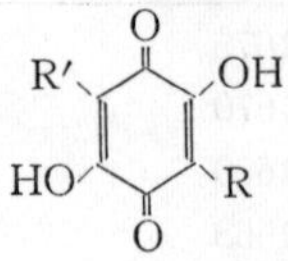

$R=CH_3, R'=-C=CH$
$\qquad\qquad H_3C \quad CH_3$ (XI)

$R=CH_3, \ R'=-CH(CH_3)_2$ (XII)

$R=CH_3, R'=-(CH_2)_{13}-CH=CH-(CH_2)_8-CH_3$ Maesaquinone (XIII)

$R=CH_3, R'=C_{19}H_{39}$ Dihydromaesaquinone (XIV)

$R=C_{10}H_{21}, R'=C_{10}H_{21}$ (XV)

(f) 2,6-Dihydroxy-3,5-dialkylbenzoquinone

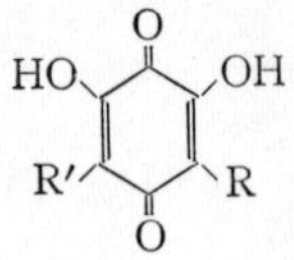

$R=CH_3, R'=CH_3$ (XVI)

(g) Helicobasidin (XVII)

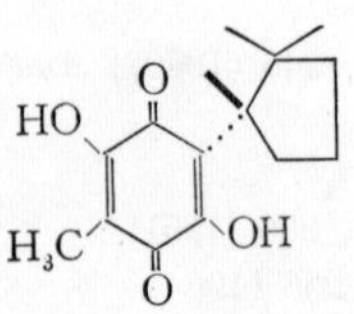

(XVII)

(h) Dimethoxytoluquinones

(XVIII)   (XIX)   (XX)

Table 84 UV and IR Spectra of Benzoquinones   (Natori *et al.*)

R	R'	Name	λ_{max}^{EtOH} mμ (log ε)	IR cm^{-1} ν_{O-H} solid[a] soln.[b]	$\nu_{C=O}$ solid[a]	($\nu_{C=C}$) soln.[b]
			(a) 2-Methoxy-6-alkylbenzoquinones			
CH$_3$			< 200 265 (3.94) 363 (2.82)	— —	1678 1649 1605	1680 1649 1606
C$_{11}$H$_{23}$			< 200 268 (4.09) 364 (2.90)	— —	1685 1651 1606	1680 1647 1605
C$_{12}$H$_{25}$			< 200 268 (4.03) 363 (2.91)	— —	1679 1646 1624 1599	1681 1650 1607
			(b) 2-Hydroxy-5-alkylbenzoquinone			
C$_{16}$H$_{23}$			209 (4.04) 268 (4.12) 383 (2.85)	3372 3452	1649 1633 1613	1656
			(c) 2-Hydroxy-3,6-dialkylbenzoquinones			
CH$_3$	C$_8$H$_{17}$		< 200 270 (4.10) 406 (2.99)	3272 3472	1666 1637 1613	1657 1636 1613
CH$_3$	—CH-(CH$_2$)$_3$-CH(CH$_3$)$_2$ CH$_3$		< 200 270 (4.08) 406 (2.99)	3267 3463	1663 1634 1608	1658 1640 1617
CH$_3$	—CH(CH$_3$)$_2$		— 267 (4.16) 404 (3.01)	3252 3431	1668 1643	1662 1645
			(d) 3-Alkyl-2,5-dihydroxybenzoquinones			
C$_{11}$H$_{23}$		Embelin	202 (4.16) 291 (4.25) 427 (2.45)	3318 3356	1637sh 1613	1636
C$_{13}$H$_{27}$		Rapanone	203 (4.18) 291 (4.26) 425 (2.46)	3324 3324	1635sh 1612	1639
C$_{18}$H$_{37}$			201 290 428	3330 3322	1614	1638 1613
			(e) 2,5-Dihydroxy-3,6-dialkylbenzoquinones			
CH$_3$	—C=CH H$_3$C CH$_3$	[1279]	— 287 (4.24) 436 (2.40)	3310[c] 3365	1617[c]	1642
CH$_3$	—CH(CH$_3$)$_2$	[1279]	— 293 (4.31) 435 (2.36)	3319[c] 3327	1616[c]	1640
CH$_3$	C$_{19}$H$_{37}$	Maesaquinone	209 (4.24) 294 (4.36) 440 (2.47)	3320 3360	1615	1635
CH$_3$	C$_{19}$H$_{39}$	Dihydromaesa -quinone	— 292 (4.37) —	3320 (315)[c] 3360	1615 (1609)[c]	1633
C$_{10}$H$_{21}$	C$_{10}$H$_{21}$		209 (4.14) 295 (4.29) 429 (2.30)	3325 3374	1614	1630

1279) Bycroft, B.W *et al*: *J. Org. Chem.*, 28, 1429 (1963).

R	R'	Name	λ_{max}^{EtOH} mμ (log ε)	IR cm^{-1}			
				ν_{O-H} solid[a]	soln.[b]	$\nu_{C=O}$ ($\nu_{C=C}$) solid[a]	soln.[b]

(f) 2,6-Dihydroxy-3,5-dialkylbenzoquinone

| CH$_3$ | CH$_3$ | 1279) | — | 297 (4.26) | 426 (2.26) | 3417[c] 3476 | (1660)[c] 1641 | 1653 1645 |

(g) Helicobasidin

| CH$_3$ | C$_8$H$_{15}$ | | 210 (4.13) | 297 (4.15) | 377 (2.61) | 430 (2.47) | 3319 (3300)[c] 3327 | 1613 (1609)[c] 1638 |

a) in nujol,   b) in CHCl$_3$,   c) in KBr

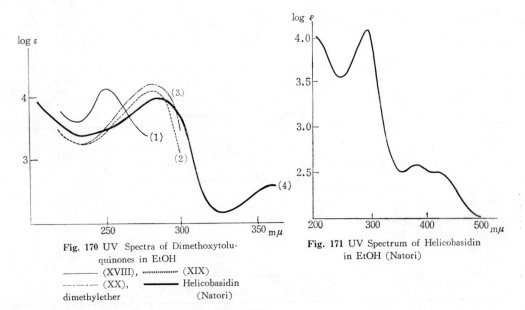

Fig. 170 UV Spectra of Dimethoxytoluquinones in EtOH
———— (XVIII), ·············· (XIX)
—·—·— (XX), ——— Helicobasidin
dimethylether   (Natori)

Fig. 171 UV Spectrum of Helicobasidin
in EtOH (Natori)

(2) **IR Absorption of 1,4-Benzoquinones**[1279a]

Table 85 IR Spectra of 1,4–Benzoquinone Derivatives   (Yates *et al.*)

Str. No.		XXI	XXII	XXIII	XXIV	XXV	XXVI	XXVII	XXVIII
	2	H	Me	Me	Me	Me	Me	Me	Me
	3	H	H	Me	H	H	Me	Me	OH
	5	H	H	H	Me	H	OMe	OH	Me
	6	H	H	H	H	Me	H	H	H

1279a) Yates, P *et al*: *J.A.C.S.*, **78**, 650 (1956).

Wave length of major I.R. bands

Str. No.	XXI	XXII	XXIII	XXIV	XXV	XXVI	XXVII	XXVIII
left: in CS₂	5.99	6.01	6.02	6.02	6.03	5.98		
right: in nujol	6.01	6.01	6.02	6.00	6.05		6.03	6.03
						6.09		
		6.28	6.27	6.24	6.09	6.20	6.14	6.12
	7.38	7.31	7.27	7.27	7.27	7.26		
	7.32	7.42	7.32	6.20			6.21	6.21
		7.48		7.46		7.36		
		7.42		7.40	7.60			6.40
	7.71	7.74	7.67		7.64	7.53		
	7.65	7.70	7.65				7.53	7.41
		7.85		8.05	7.80	7.89		
		7.80		7.98	7.76		7.76	7.68
								7.77
		8.85	8.80	8.71	8.51	8.41		
		8.78	8.80	8.69	8.30	8.48	8.14	8.36
					8.48		8.45	
							8.66	8.87
	9.42	9.19	9.43			9.39		
	9.22	9.12	9.42		9.65		9.65	9.47
	9.35							
	10.61	9.99		10.01				
	10.61	9.99		9.95	10.68			10.40
	11.38	11.08		11.01	10.71	11.05		
	11.16	10.81		10.80	10.96		11.38	11.21
		11.31	12.05		10.68			
			12.03		10.87			
		12.38	12.42	12.44	11.8	11.83		
		12.13	12.44	12.57	12.58		12.58	
						12.67		
		14.71		14.11			13.55	

(3) Helicobasidin (XVII)[1280]

≪Occurrence≫ *Helicobasidium mompa* TANAKA.

mp. 190~192°, $[\alpha]_D$ −123° (CHCl₃), $C_{15}H_{20}O_4$

UV: See Tab. 84, Fig. 171

IR $\nu_{max}^{CHCl_3}$ cm⁻¹: 3327, 1353 (OH), 2953, 2872, 1462, 1382 (aliph. side chain seems to be

not so large), 1684 w, 1638 br.s (conj. C=O and C=C of quinone)

NMR in CHCl₃, 60 Mc, TMS, τ: no olefinic proton, 8.07 s (3) (CH₃–C=C), 9.15 s (3),

$$8.90 \text{ s (3), } 8.65 \text{ s (3) (3 } CH_3\text{–}C\langle)$$

(4) Maesaquinone (XIII)

≪Occurrence≫ *Maesa japonica* MORITZI. (イズセンリョウ)

mp. 122°, $C_{26}H_{42}O_4$

UV, IR: See Tab. 84

NMR[1281] τ: no ring olefinic proton, 8.05 s

(XIII)

(3) (CH₃–C$\langle$), 4.68 t (2) (—CH=CH—)

1280) Natori, S. *et al*: *Chem. Pharm. Bull.*, **11**, 1343 (1963).

1281) Ogawa, H. *et al*: 19th Annual Meeting of the Pharmaceutical Society of Japan, p. 127 (Tokyo, 1964).

(5) Embelin (VIII) and Vilangin (XXIX)

(VIII) (XXIX)

≪Occurrence≫   *Embelia ribes* BURM. f.

(a) **Embelin** (VIII)

mp. 142~143°, $C_{17}H_{26}O_4$

UV, IR: See Tab. 84

(b) **Vilangin** (XXIX)[1282]

mp. 264~265°, $C_{35}H_{52}O_8$

2 Embelin (VIII) + CH_2O ⟶ (XXI)

(6) Rapanone (IX)

≪Occurrence≫ *Rapanea maximowiczii* KOIDZ.

(シマタイミンタチバナ)

mp. 139~140°, $C_{19}H_{30}O_4$

UV, IR: See Table 84

(IX)

(7) Dalbergione (Dalbergione I) (XXX)[1283]~[1285]

≪Occurrence≫   *Dalbergia nigra, D. latifolia* ROXB.

mp. 114~116°, $[\alpha]_D$ +13° (CHCl_3), $C_{16}H_{14}O_3$

NMR in $CDCl_3$, 60 Mc, TMS. τ: 4.09 $\left(\begin{array}{c}CH_3O \\ \diagdown = \diagup H\end{array}\right)$,

(XXX)

3.50 $\left(\begin{array}{c}O \\ \diagup\diagdown \\ H\end{array}\right)$,   4.87 $(\diagup C\text{–}H)$,   4.13–3.55

$\left(\begin{array}{c}H \\ H\diagup\diagup H\end{array}\right)$,   5.12–5.00,   4.80–4.62 $\left(\begin{array}{c}H \\ \diagup = \diagup H \\ H \quad H\end{array}\right)$,   6.17 (3) (CH_3O), 2.70 (5) (5H of ring B)

Related compounds are described in the item **16**, [A], (5).

(8) "Pflanzliches Chinon" (Plant quinone) (XXXI)[1286][1287]

≪Occurrence≫   The leaves of " Ross Kastanien ", " Blut buchen ",

" Ahorn, Walnuss ", " Pappeln " (German)

mp. 46~48° (yellow), $[\alpha]_D$ 0.00, $C_{58}H_{88}O_2$

UV $\lambda_{max}^{Petroleum\ ether}$ mμ (E$_{1cm}^{1\%}$): 254 (253),   (XXXI)

261 (232), 314 (11), See Fig. 172

1282) Bheemasankara Rao *et al*: *Indian J. Pharm.*, **24**, 262 (1962); *J. Org. Chem.*, **26**, 4529 (1961); *Tetrahedron*, **18**, 361 (1962).

1283) Eyton, W. B. *et al*: *Proc. Chem. Soc.*, **1962**, 301.

1284) Seshadri, T. R. *et al*: *Tetr. Letters*, **1963**, 211.

1285) Marini Bettolo *et al*: *Annali di Chimica*, **52**, 1190 (1962); *Index Chem.*, **8**, 26776 (1963).

1286) Planta, C. v.: *Helv. Chim. Acta*, **42**, 1278 (1959).

1287) Isler, O. *et al*: *ibid.*, **42**, 1283 (1959).

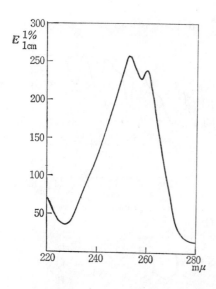

Fig. 172 UV Spectrum of " Pflanzliches Chinon " in Petroleum ether (Isler *et al.*)

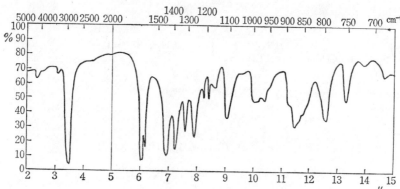

Fig. 173 IR Spectrum of " Pflanzliches Chinon " (Isler *et al.*)

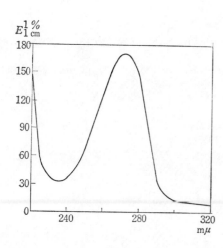

Fig. 174 UV Spectrum of Ubiquinone in cyclohexane (Morton *et al.*)

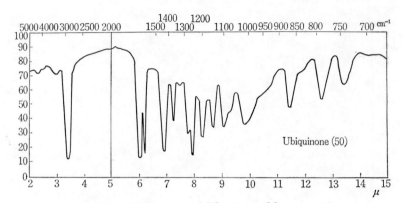

Fig. 174–a　IR Spectrum of Ubiquinone (Morton *et al.*)

IR　$\lambda_{max}\,\mu$: 6.07, 11.45, 12.57, 13.29, see Fig. 173

NMR　in $CDCl_3$, 60 Mc, TMS 0.00 Hz (cps): 388 (　), 309 ($-\overset{\text{H}}{\underset{\;}{C}}=C-CH_2-$), 187 d

(　$CH_2-CH=$), 121 (　, $-CH_2$), 97 ($=\overset{CH_3}{\underset{\;}{C}}-$)

(9)　**Ubiquinone** (Coenzyme Q_{10}) (XXXII)[1286) 1288) ~1296)]

$$CH_3O-\text{(quinone ring)}-(CH_2-CH=\overset{CH_3}{\underset{\;}{C}}-CH_2)_{10}-H\quad\text{(XXXII)}$$

≪Occurrence≫　Hearts of cow and pig,　mp. 49°, $C_{59}H_{90}O_4$

UV[1288)]　$\lambda_{max}^{cyclohexane}$ mμ (ε): 238 (2500), 272 (14500), 325 inf. (1120), 405~410 (750), see Fig. 174.

IR[1288)]　$\lambda_{max}\,\mu$: 6.05 (quinone), 7.94, 9.14 (ether), 11.47, 12.61, 13.40 (all *trans*-isoprene chain), see Fig. 175

NMR[1289)]　in CCl_4, 40 Mc, refer to H_2O, cps: +8 (10) (10-$\overset{\text{H}}{\underset{\;}{C}}$=), −34 (6) (2 CH_3O-), −64, −69

(2) ($=\overset{\;}{\underset{\;}{C}}-CH_2-CH=$), −113 (36+3) (9 $=\overset{\;}{\underset{\;}{C}}-CH_2-CH_2-CH=$ and $>C-CH_3$ of ring), −125 (33)

(10$=\overset{\;}{\underset{\;}{C}}-CH_3$ of chain)

1288)　Morton, R. A. *et al* : *Helv. Chim. Acta*, **41**, 2343 (1958).
1289)　Shunk, C. H. *et al* : *J. A. C. S.*, **80**, 4752, 4753 (1958).
1290)　Diplock, A. T. *et al* : *Nature*, **186**, 554 (1960).
1291)　Isler, O. *et al* : *Helv. Chim. Acta*, **42**, 2616 (1959).
1292)　Imada, I : *Chem. Pharm. Bull.*, **11**, 815 (1963).
1293)　Gale, P. H : *Biochem.*, **2**, 203 (1963); *Index. Chem.*, **8**, 26634 (1963).
1294)　Erickson, R. E. *et al* : *J.A.C.S.*, **81**, 4999 (1959).
1295)　Shunk, C. H. *et al* : *J. A. C. S.*, **81**, 5000 (1959).
1296)　Linn, B. O *et al* : *J.A.C.S.*, **82**, 1647 (1960).

NMR: See also[1286)1296)], synthesis[1291)1295)]

(10) Tauranin (XXXII-a)[1296a)]

≪Occurrence≫ *Oospora aurantia* S. et V.

mp. 150~160°(decomp,) $[\alpha]_D$ −148° (MeOH), $C_{22}H_{30}O_4$

UV λ_{max}^{MeOH} mμ (log ε): 266 (4.07), 415 (3.07)

IR $\nu_{max}^{CHCl_3}$ cm^{-1}: 3640, 3415 (OH), 1662 sh, 1644, 1622 (CO, C=C), 2940, 1476, 1469, 1390, 1367 (CH$_3$, CH$_2$), 899 (quinone, =CH$_2$)

NMR in CDCl$_3$, 60 Mc, TMS, ppm: 0.76 s, 0.81 s, 0.86 (3Me, a), 1.0~2.8 m (—CH$_2$–CH), 4.53 d, J=1.2 c/s (CH$_2$–O, b), 4.67 s (=CH$_2$, c), 6.65 t, J=1.2 c/s (ring H), 7.05 br, s (enolic OH, e)

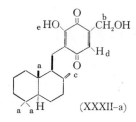

(XXXII-a)

[E] Naphthoquinones

(1) Spinochrome N (XXXIII) and **Spinochrome E** (XXXIV)[1297)1298~1303)]

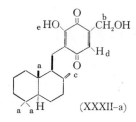

(XXXIII)   (XXXIV)

(a) Spinochrome N (XXXIII)[1298)]

≪Occurrence≫ *Anthocidaris crassispina* (AG.)[1299)], *Hemicentrothus pulcherrimus* (AG.).[1300)]

mp. >260°d, $C_{10}H_6O_6$

(b) Spinochrome E (XXXIV)[1301)]

≪Occurrence≫ *Paracentrotus lividus* (LAW.), *Psammechinus miliaris* (GMELIN).

mp. >300 d, $C_{10}H_6O_8$

IR ν_{max}^{KBr} cm^{-1}: 3520, 3150, 1652 sh, 1629, 1586, 1490, 1456, 1344, 1274, 1181, 1062 w, 1041 w, 992, 839, 803, 766, 716

Hexaacetate mp. 192°, $C_{22}H_{18}O_{14}$

UV: See Fig. 175

1296a) Nakanishi, K. *et al*: *Chem. Pharm. Bull*, **12**, 796 (1964).

1297) Smith, J. *et al*: *Soc.*, **1961**, 1008.

1298) Kuroda, Ch *et al*: *Proc. Japan Acad.*, **34**, 616 (1958); *Sci. Papers. Inst. Phys. Chem. Res. Tokyo*, **53**, 356 (1959).

1299) Kuroda, Ch *et al*: *Proc. Imp. Acad.* (*Tokyo*), **20**, 23 (1944).   1300) Kuroda, Ch *et al*: *ibid.*

1301) Smith, J. *et al*: *Tetr. Letters*, **1960**, 10.   1302) Lederer: *Biochim. Biophys. Acta*, **9**, 92 (1952).

1303) Yoshida: *J. Mar. Biol. Assoc. U. K.*, **38**, 455 (1959).

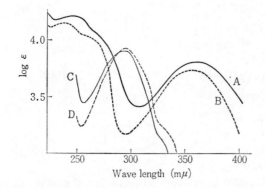

Fig. 175 UV Spectra of
A   Spinochrome E hexaacetate
B   Synth, 2,3,5,8–Tetraacetoxy–1,4–
    naphthoquinone in MeOH
C   Spinochrome E leucooctaacetate
D   Synth, 1,2,3,4,5,8–Hexaacetoxy–
    naphthalene in CHCl$_3$

(Smith *et al.*)

(2) **Mompain** (XXXV) or (XXXVI)[1304]

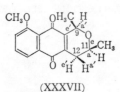

(XXXV)

(XXXVI)

≪Occurrence≫   *Helicobasidium mompa* TANAKA.,   mp. >300°, deep red purple, C$_{10}$H$_6$O$_6$

UV λ_{max}^{EtOH} mμ (log ε): 228 (4.53), 272 (4.06), 318 (3.93), 486 sh (3.78), 517 (3.84), 554 sh (3.65)

IR ν_{max}^{KBr} cm^{-1}: 3485, 3260, ca 2700 br (OH), 1657 sh, 1600 (C=O, C=C)

Tetraacetate   mp. 176~179°

UV λ_{max}^{EtOH} mμ (log ε): 248 (4.18), 266 sh (4.02), 352 (3.53)

Leucohexaacetate   mp. 229~232°

UV λ_{xam}^{EtOH} mμ (log ε): 295 (3.92)

Tetramethylether   mp. 169~171°

NMR τ: 6.18 s (3), 6.13 s (3), 6.04 s (6) (4CH$_3$O—), 4.04 s (1), 3.22 s (1)

(3) (+)-**Eleutherin** (XXXVII) and (−)-**Isoeleutherin** (XXXVIII)[1305]

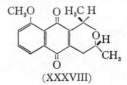

(XXXVII)

(XXXVIII)

≪Occurrence≫   *Eleutherine bulbosa* URB.

(a) (+)-**Eleutherin** (XXXVII)
  mp. 175°, [α]$_D$ +346° (CHCl$_3$), C$_{16}$H$_{16}$O$_4$
  IR: See Fig. 176

(b) (−)-**Isoeleutherin** (XXXVIII)
  mp. 176~177°, [α]$_D$ −46° (CHCl$_3$), C$_{16}$H$_{16}$O$_4$
  IR: See Fig. 177

1304) Natori, S. *et al*: *7th Symposium on the Chemistry of Natural Products.* p. 149 (Fukuoka, 1963).
1305) Schmid, H. *et al*: *Helv. Chim. Acta*, **41**, 2021 (1958).

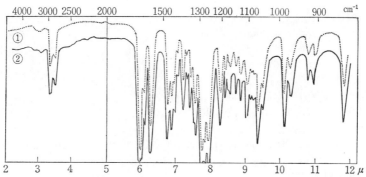

Fig. 176 IR Spectra, of (1) Synth. (±)–Eleutherin
and (2) natural (+)– Eleutherin in CCl₄ (Schmid *et al.*)

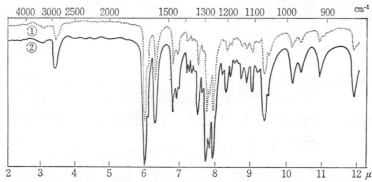

Fig. 177 IR Spectra of (1) Synth. (±)–Isoeleutherin
and (2) natural (−)–Isoeleutherin in CCl₄
(Schmid *et al.*)

NMR   in CHCl₃, 40 Mc, τ, J c/s :[1306a]

Eleutherin (XXXVII)——~8.64 d (J~6.7) (C₁₁–Me), ~8.47 d (J~6.7) (C₉–Me), 7.85 d, d, d (J: 17.8, 3.5, 9.2) 7.28 d, t (J: 17.8, 2×2.9) (12–CH₂–), ~6.38 complex (J~3), (C₁₁–H), ~5.15 complex (J~3.3) (C₉–H), 6.00 s (OMe), ~2.8 d (?), 2.43, 2.3 (arom. CH) ; coupling constants $J^{gem}_{12,12}$ 17.8, $J^{a'a}_{12,11}$ 9.2, $J^{e'a}_{12,11}$ 2.9, $J^{a'a'}_{9,12}$ 3.5, $J^{a'e'}_{9,12}$ 2.9.

Isoeleutherin (XXXVIII)——8.74 d (J~6) (C₁₁–Me), 8.55 d (J~7) (C₉–Me), 7.92 d, d, d (J 18.8, 8.8, 2.0), 7.41 d, d ; br (J 18.8, 4.5, <1) (12–CH₂–), ~6.1 (over lap.) (C₁₁–H), 5.13 q, br (J 7, ~2) (C₉–H), 6.08 s (OMe), 2.44, 2.37 (arom. CH) ; Coupling constants $J^{gem}_{12,12}$ 18.8, $J^{a'a}_{12,11}$ 8.8, $J^{e'a}_{12,11}$ 4.5, $J^{e'a'}_{9,12}$ 2.0, $J^{e'e'}_{9,12}$ <1

(4) 2-(γ,γ-**Dimethylpropenyl**)-1, 4-**naphthoquinone** (XXXIX)[1306]

≪Occurrence≫   *Tectonia grandis* (Teak)
mp. 56~58°, C₁₅H₁₄O₂

(XXXIX)

1306) Sandermann, W. *et al* : *Angew. Chem.*, **74**, 782 (1962).
1306a) Todd, L. *et al* : *Soc.*, **1964**, 98.

(5) Tingenone (XL)[1307]

≪Occurrence≫   *Euonymus tingens* WALL.

mp. 171~172°, orange red., $C_{25}H_{30}O_4$

UV λ_{max}^{EtOH} mμ (log ε): 250 (3.83), 263 (3.64), 285
(2.18), 425 (5.02)

IR $\nu_{max}^{CHCl_3}$ cm^{-1}: 1698 s, 1647 w, 1590 s, 1538 m,
1513 s, 1439 s, 1372 s, 1342 inf., 1316 m, 1307 inf.,
1285 s, 1080 s

NMR δ: 7.37 (phenolic OH), 3 arom.-H, 2.27 (COCH$_3$), 1.52, 1.35, 1.02 (9) (3 $>$C-CH$_3$), 2.97

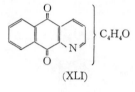

(6) Liriodenine (XLI)[1308]

≪Occurrence≫   *Liriodendron tulipifera* L. (*Magnoliaceae*)

mp. 282°, $C_{17}H_9O_3N$

UV λ_{max}^{EtOH} mμ (log ε): 247.4 (4.22), 268.2 (4.13), 309.2
(3.62), 413 (3.82); λ_{min}: 257.9 (4.08), 291.9 (3.51),
340 (3.16), 455~700 (0.00); $\lambda_{max}^{0.1N-HCl-EtOH}$ mμ (log ε):
256.7 (4.33), 277.3 (4.26), 329 (3.67), 392 (3.69), 455
(3.58); λ_{min}: 268.7 (4.20), 307 (3.53), 362 (3.55), 426 (3.52), 545~700 (0.00)

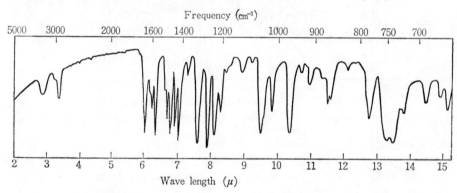

Fig. 178 IR Spectrum of Liriodenine in KBr (Buchanan *et al.*)

(7) Vitamine K₁ (XLII)

(XLII)

≪Occurrence≫   Alfalfa, tomato, sojabeans etc., yellowish viscous oil, $C_{31}H_{47}O_2$

NMR[1286] in CCl$_4$, 25 Mc, C$_6$H$_6$: 0.00Hz (cps): −12 m $\left(\begin{array}{c}H\\H\end{array}\right)$, 56 $\left(-\overset{CH_3}{\underset{H}{C}}=C-CH_2-\right)$, 101 d

---

1307) Seshadri, T. R. *et al*: *Tetr. Letters*, **23**, 1047 (1962).
1308) Buchanan, M. A. *et al*: *J. Org. Chem.*, **25**, 1389 (1960).

$\left(\begin{array}{c}\text{see structure}\end{array}\right)$, 129 s $\left(\begin{array}{c}\text{see structure}\end{array}\right)$, 138 s (=C—), 152 (—CH$_3$, —CH$_2$), 159, 162,

see Fig. 179.

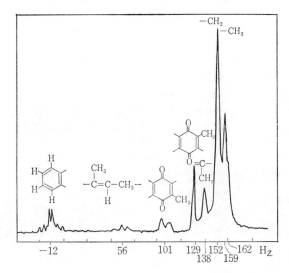

Fig. 179 NMR Spectrum of Vitamine K$_1$ (Planta *et al.*)[1286]

(8) **Vitamine K$_2$** (XLIII)[1286) 1309)]

(XLIII)

$n=4$: Vitamine K$_2$ (20), $n=7$: Vitamine K$_2$ (35)

(XLIII–a)

mp. 53.5~54.5°, C$_{41}$H$_{56}$O$_2$[1309)]

NMR[1286)] in CCl$_4$, 25 Mc, C$_6$H$_6$: 0.00 Hz (cps):

		H (benzene ring)	CH$_3$ —C=C-CH$_2$— H	CH$_2$-CH=	O ring (CH$_3$)	—CH$_2$	=C— CH$_3$
(XLIII)	$n=7$	—12 m	55 br	100 d	130 s	132	142 s
(XLIII)	$n=4$	—12 m	57 br	101 d	130 s	135	143 s

1309) Karrer, W: Konstitution und Vorkommen der organischen Pflanzenstoffe, 2636 (Birkhäuser Verlag, 1958).

(9) Chimaphilin (XLIV)[1309a)]

   ≪Occurrence≫   *Pyrola renifolia* Maxim.

mp. 112.5∼113.5°, $C_{12}H_{10}O_2$

Renifolin See the item 6, [D], (10)

H_3C CH_3

(XLIV)

[F] Anthraquinones

(1) UV Absorptions Shift by substituted Radicals.[1310)]

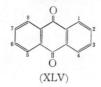

(XLV)

UV λ_{max} mμ : **Anthraquinone**

    (XLV) (327), 1–OH–anthraquinone (405), 1–OCH$_3$–anthraquinone (380), 2–OH–anthraquinone (365), 2–OCH$_3$–anthraquinone (363), 1,2–Di–OH–anthraquinone (416), 1,4–Di–OH–anthraquinone (476), 1,5–Di–OH–anthraquinone (428), 1,8–Di–OH–anthraquinone (430), 1,8–Di–CH$_3$O–anthraquinone (385)

(2) IR Spectra of Oxyanthraquinones[1311)]

See Table 86.

    Occurrence, melting point and formula of natural occurring anthraquinones, listed in Table 86 **are** described.

(a) 3–Hydroxy–2–methylanthraquinone (XLVIII)

    CH$_3$

    OH

(XLVIII)

   ≪Occurrence≫   *Coprosma lucida*

mp. 302° (sublim), $C_{15}H_{10}O_3$

(b) Alizarin (XLIX)

   O OH

       OH

(XLIX)

   ≪Occurrence≫   *Rubia tinctorum* L.

mp. 289∼290°, $C_{14}H_8O_4$

(c) Purpuroxanthin (L)

   O OH

       OH

(L)

   ≪Occurrence≫   *Rubia tinctorum* L.

mp. 268∼270°, $C_{14}H_8O_4$

(d) Anthragallol (LII)

   O OH

       OH

       OH

(LII)

   ≪Occurrence≫   *Coprosma lucida*

mp. 310∼314°, $C_{14}H_8O_5$

1309a) Inoue, H. *et al*: *Y. Z.*, **78**, 301 (1958); *Chem. Pharm. Bull.*, **12**, 255, 533 (1964).
1310) Labhart, H: *Helv. Chim. Acta*, **40**, 1410 (1957).   1311) Bloom, H *et al*: *Soc.*, **1959**, 178.

Table 86. IR Spectra of Oxyanthraquinones (Bloom *et al.*)

Values are IR absorption bands (cm⁻¹); s = strong, m = medium, w = weak. The original table is printed sideways; readings are listed per compound below.

Unsubstituted Anthraquinone

(XLV)	1675 s; 1572 m; 1333 s; 1305 w; 1284 s; 1209 w; 1163 w; 1098 w; 971 m; 936 s; 812 s; 693 w

Monosubstituted Anthraquinone

Compound	Bands
1-OH (XLVI)	1667 s; 1631 s; 1582 s; 1357 s; 1290 s; 1266 s; 1230 s; 1159 s; 1101 w; 1066 m; 1042 w; 1013 s; 927 w; 881 s; 838 s; 815 w; 775 s; 737 m; 708 s
2-OH (XLVII)	3344 s; 1667 s; 1587 s; 1339 s; 1304 s; 1282 s; 1239 s; 1224 s; 1175 m; 1145 w; 1088 s; 986 s; 932 s; 911 w; 883 s; 854 s; 811 w; 795 m; 749 m; 719 m; 711 s

Disubstituted Anthraquinone

Compound	Bands
2-OH, 3-Me (XLVIII)	3311 s; 1658 s; 1575 s; 1504 m; 1342 s; 1307 s; 1284 w; 1250 s; 1186 s; 1163 w; 1106 s; 1082 s; 1047 m; 1019 w; 963 s; 917 m; 904 m; 888 m; 797 s; 776 s; 736 s; 716 s
1,2-Di-OH (XLIX)	3367 m; 1658 s; 1634 s; 1587 s; 1351 m; 1333 w; 1289 s; 1195 s; 1178 s; 1048 m; 1034 w; 1011 s; 897 m; 854 s; 829 s; 770 s; 752 s; 714 s
1,3-Di-OH (L)	3413 s; 1675 s; 1637 s; 1590 s; 1340 s; 1319 s; 1266 m; 1195 w; 1163 s; 1098 w; 1063 w; 1030 w; 1006 m; 923 w; 879 w; 863 w; 801 w; 778 s; 728 m; 714 s
1,8-Di-OH (LI)	1678 s; 1621 s; 1600 w; 1348 m; 1300 w; 1269 s; 1205 m; 1159 m; 1080 m; 1058 m; 1032 m; 972 s; 922 w; 846 s; 816 s; 779 s; 741 m

Trisubstituted Anthraquinone

Compound	Bands
1,2,3-Tri-OH (LII)	3378 m; 1650 s; 1626 s; 1575 s; 1340 m; 1309 s; 1280 m; 1264 w; 1221 m; 1122 s; 1091 w; 1063 w; 1032 s; 989 s; 869 m; 834 m; 796 w; 752 w; 716 m
1,2,4-Tri-OH (LIII)	3226 m; 1621 s; 1580 s; 1337 s; 1309 s; 1264 s; 1183 s; 1064 m; 1029 m; 964 s; 904 w; 867 s; 831 w; 751 s; 732 s; 704 s
1,3-Di-OH-2-Me (LIV)	3413 s; 1658 s; 1623 s; 1582 s; 1339 s; 1312 s; 1277 w; 1233 m; 1196 s; 1124 s; 1098 m; 1076 w; 1044 m; 1015 s; 973 m; 943 m; 868 s; 831 s; 804 s; 772 w; 746 s; 734 s; 713 s
1,8-Di-OH-3-Me (LV)	1675 s; 1621 s; 1558 m; 1350 s; 1287 m; 1271 s; 1208 s; 1160 s; 1086 s; 1052 w; 1024 s; 996 w; 903 m; 870 w; 841 s; 818 w; 806 m; 752 s; 733 s

Tetrasubstituted Anthraquinone

Compound	Bands
1,2,5-Tri-OH-6-Me (LVI)	3460 m; 1623 s; 1605 m; 1355 m; 1311 w; 1293 s; 1261 w; 1225 s; 1174 m; 1156 m; 1078 s; 1016 s; 968 m; 870 w; 848 w; 800 s; 734 s

(e) Purpurin (LIII)

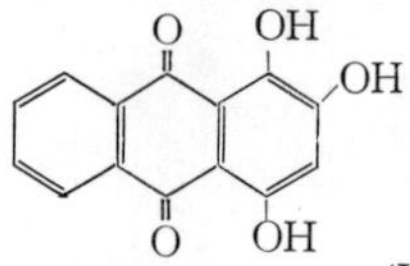

(LIII)

《Occurrence》　*Rubia tinctorum* L.

mp. 253~259°, $C_{14}H_8O_5$

(f) Rubiadin (LIV)

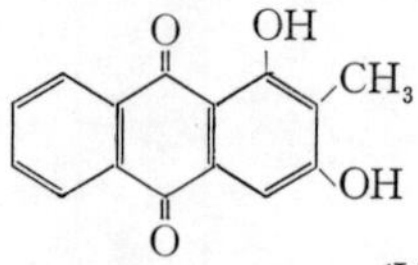

(LIV)

《Occurrence》　*Rubia tinctorum* L.

mp. 297°, $C_{15}H_{10}O_4$

(g) Chrysophanic acid (LV)

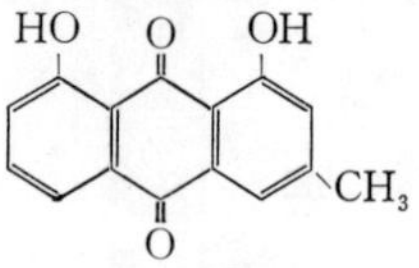

(LV)

《Occurrence》　*Rheum emodi, R. undulatum, Rumex patientia* and other *Rumex* spp., *Rhamnus cathartica, R. pursiana, Cassia tola, C. occidentalis, Polygonum multiflorum* and other *Polygonum* spp., *Penicillium islandicum.*

mp. 193~196°, $C_{15}H_{10}O_4$

(h) Morindon (LVI)

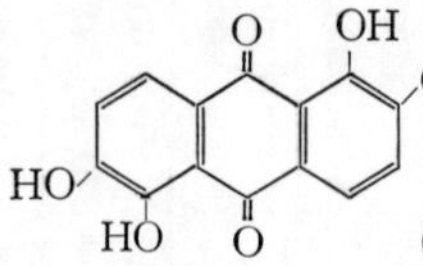

(LVI)

《Occurrence》　*Morinda umbellata, Coprosma grandiflora* L.

mp. 281~284°, 272~275°, $C_{15}H_{10}O_5$

(3) Rhein (LVII)[1312]

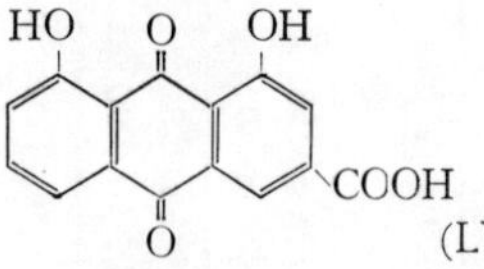

(LVII)

《Occurrence》　*Rheum coreanum* NAKAI and other *Rheum* spp.

mp. 318~319°, $C_{15}H_8O_6$

UV λ_{max}^{MeOH} mμ (ε): 229 (36800), 258 (20100), 435 (11100)

IR ν_{max}^{KBr} cm^{-1}: 1701 (—COO$^-$), 1637 (chelated $>$C=O)

(4) Anthraquinones of *Cassia obtusifolia* L.[1313]

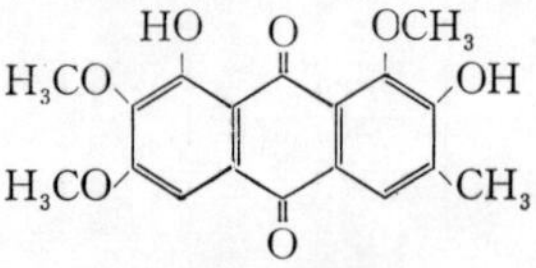

(LVIII) Obtusin 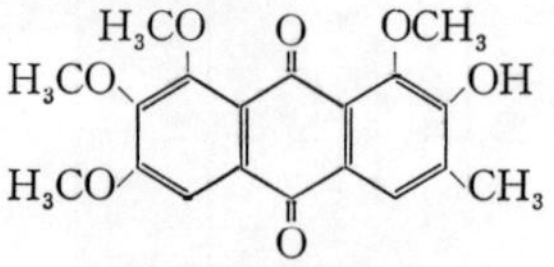　　　(LIX) Chryso-obtusin 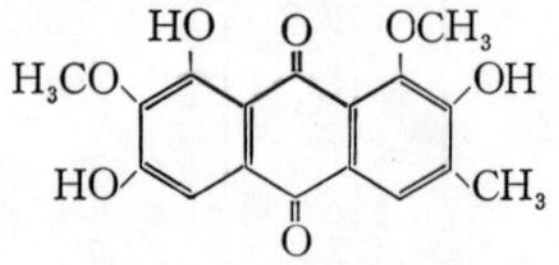　　　(LX) Auranti-obtusin

(a) Obtusin (LVIII)

mp. 242~243°, $C_{18}H_{16}O_7$

IR ν_{max}^{KBr} cm^{-1}: 3318 (OH), 1653 (non-chelated C=O), 1628 (chelated C=O)

(b) Chryso-obtusin (LIX)

mp. 214~215°, $C_{19}H_{18}O_7$

IR ν_{max}^{nujol} cm^{-1}: 3325 (OH), 1676 (non-chelated C=O), 1582 (arom, band)

1312) Nawa, H *et al*: *J. Org. Chem.*, **26**, 979 (1961).
1313) Takido, M: *Chem. Pharm. Bull.*, **8**, 246 (1960).

(c) Aurantio-obtusin (LX)

mp. 265~266°, $C_{17}H_{14}O_7$

IR ν_{max}^{nujol} cm^{-1}: 3325 (OH), 1663 (non-chelated C=O), 1629 (chelated C=O)

(5) Carminic acid (LXI)[1314]

(LXI)

≪Occurrence≫　*Dactylopius coccuscosta*,　mp. 120° (decomp.), $C_{22}H_{20}O_{13}$

IR ν_{max}^{nujol} cm^{-1}: 1708 s, 1693 s, 1677 m, 1648 m, 1632 m, 1606 s, 1566 s, 1509

(6) Anthraquinones of Natal Aloe[1315]

(LXII) Nataloe-emodin

(LXIII) Homonataloin

≪Occurrence≫　Natal Aloe—*Aloe candelabrum, A. distans, A. macracantha, A. plicaticatilis, A. ferox, A. vera, A. eru* and *A. perryi*

(a) Nataloe-emodin (LXII)

mp. 212~213°, (synth.) 216~217°, $C_{15}H_{10}O_5$

UV λ_{max}^{EtOH} mμ (log ε): 232 (4.30), 260 (4.30), 290 inf. (4.00), 432 (3.89)

IR ν_{max}^{KBr} cm^{-1}: 3550 m, 3200 m, 2930 m, 2320 m, 1660 m, 1626 s, 1555 w, 1537 w, 1455 s, 1435 m, 1423 m, 1335 s, 1212 w, 1200 w, 1135 w, 1075 w, 1033 m, 866 m, 822 m, 746 s, 695 w

(b) Homonataloin (LXIII)

mp. 202~204°, $[\alpha]_D$ −111.5° (EtOAc), $C_{22}H_{24}O_9 \cdot H_2O$

UV λ_{max}^{EtOH} mμ (log ε): 222 (4.38), 250 inf. (3.85), 273 inf. (3.85), 294 (4.12), 347 (3.85)

IR ν_{max}^{KBr} cm^{-1}: 3730 w, 3480 s, 3220 s, 2900 m, 2350 w, 1638 s, 1610 s, 1588 s, 1487 s, 1452 m, 1440 m, 1378 s, 1360 m, 1330 m, 1294 s, 1270 s, 1258 m, 1215 s, 1173 w, 1160 w, 1147 w, 1128 m, 1110 m, 1087 s, 1070 s, 1068 s, 1040 m, 1025 s, 984 w, 969 w, 935 m, 908 m, 882 w, 856 w, 842 w, 833 w, 776 m, 757 s, 727 w, 696 s

1314) Ali, M.A. *et al*: *Soc.*, **1959**, 1033.
1315) Haynes, L. J. *et al*: *Soc.*, **1960**, 4879.

(7) Cascarosides A and B[1316]

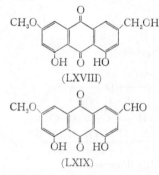

R₁=R₂=H : (+) Barbaloin (LXIV)
R₁,R₂=H or glucose: Cascaroside A (LXV)
R₁=R₂=H, 10–Epimar: (−) Barbaloin (LXVI)
R₁,R₂=H or glucose: Cascaroside B (LXVII)

《Occurrence》 *Rhamnus purshiana* DC.

(a) **Cascaroside A** (LXV)
mp. 168~169°, 180~181°, 189~190°, $[\alpha]_D$ −40° (EtOH)
UV λ^{N-HCl}_{max} mμ ($E^{1\%}_{1cm}$): 267 (108~112), 295 (153~163), 323.5 (131~141), see Fig. 180.

(b) **Cascaroside B** (LXVII)
mp. 165~167°, 181~182°, $[\alpha]_D$ −110° (EtOH)
UV λ^{N-HCl}_{max} mμ ($E^{1\%}_{1cm}$): 267 (106~113), 294 (146~158), 326 (129~141)

(8) Fallacinol (Fallacin B) (LXVIII) and Fallacinal (Fallacin A) (LXIX)[1317]

(LXVIII)

(LXIX)

《Occurrence》 *Xanthoria fallax* ARN.

(a) **Fallacinol** (LXVIII)
mp. 236~237°, $C_{16}H_{12}O_6$,
IR See Fig. 181

(b) **Fallacinal** (LXIX)

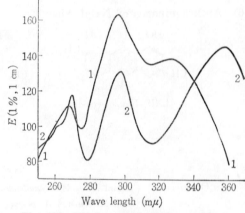

Fig. 180 UV Spectra of (1) Cascaroside A,
(2) after hydrolysis in N–HCl at 70°
(Barbaloin) (Fairbairn *et al.*)

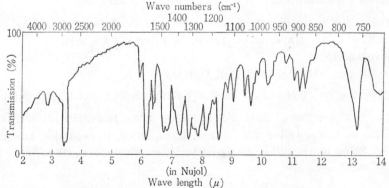

Fig. 181 IR Spectrum of Fallacinol (Murakami)

1316) Fairbairn, J. W *et al*: *J. Ph. Ph.*, **12**, 45 T (1960), **15**, 292 T (1963).
1317) Murakami, T: *Chem. Pharm. Bull.*, **4**, 298 (1956).

[F] Anthraquinones

IR ν_{max}^{nujol} cm^{-1}: 1713 (aryl-CHO), 1675 (non-chelated C=O), 1630 (chelated C=O), See Fig. 182.

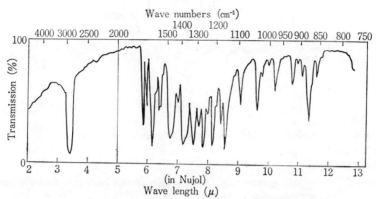

Fig. 182 IR Spectrum of Fallacinal (Murakami)

(9) Bisanthraquinones of *Penicillium* spp.

(LXX) Skyrin

(LXXI) Rugulosin (dimer of Flavoskyrin (LXXII))

(LXXIII) Rubroskyrin

conc.-H$_2$SO$_4$
$\xrightarrow{-2H_2O}$

(LXXIV) Iridoskyrin

(LXXV) Luteoskyrin

— 435 —

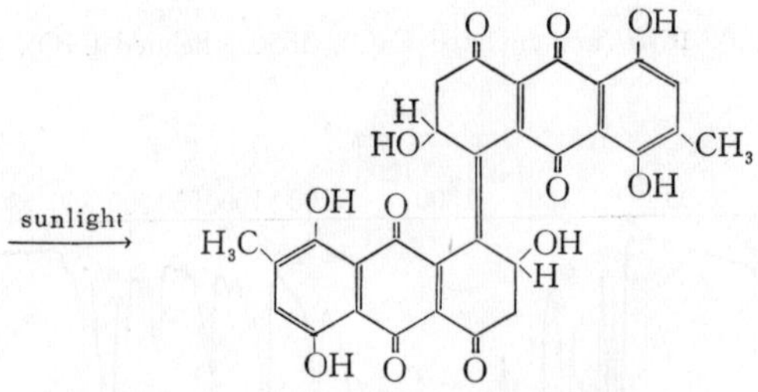

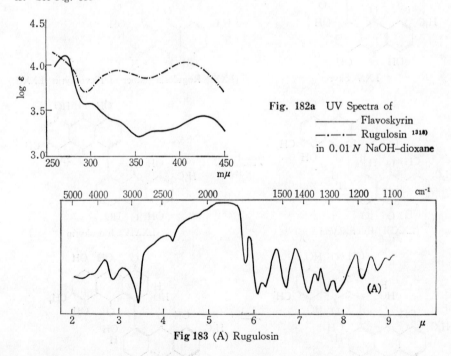

(LXXVI) Lumiluteoskyrin

(a) **Skyrin** (LXX)

《Occurrence》 *Penicillium islandicum*. Sopp., *P. rugulosum* THOM.

mp. $>380°$, $C_{30}H_{18}O_{10}$

(b) **Rugulosin** (Radicalisin) (LXXI)[1318)1320)1321)

《Occurrence》 *Penicillium rugulosum* THOM., *P. wortamanni* KLÖCKER, *Endothia parasitica* A. et A.

mp. 293° (decomp.) $C_{30}H_{22}O_{10}$

UV See Fig. 182a

IR See Fig. 183

(c) **Flavoskyrin** (LXXII)[1318)1321)

《Occurrence》 *Penicillium islandicum* Sopp.

mp. 208° (decomp.), $[\alpha]_D$ $-295°$ (dioxane), $C_{15}H_{12}O_5$

UV See Fig. 182a

IR See Fig. 183

Fig. 182a UV Spectra of

——— Flavoskyrin

—·—·— Rugulosin [1318)

in 0.01 N NaOH–dioxane

Fig 183 (A) Rugulosin

1318) Shibata, S *et al*: *Chem. Pharm. Bull.*, **8**, 889 (1960).

1319) Shibata, S *et al*: *ibid.*, **8**, 884 (1960).   1320) Shibata, S *et al*: *ibid.*, **4**, 111 (1956).

1321) Shibata, S *et al*: *ibid.*, **4**, 303 (1956).

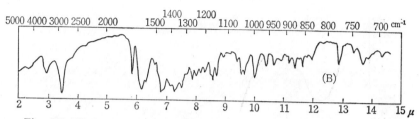

Fig. 183 IR Spectra of (A) Rugulosin and (B) Flavoskyrin [1321] (Shibata *et al.*)

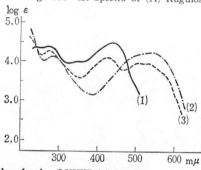

Fig. 184 UV Spectra of
(1) Luteoskyrin in EtOH
(2) Lumiluteoskyrin in dioxane
(3) Rubroskyrin in dioxane
(Shibata *et al.*)[1323]

(d) Rubroskyrin (LXXIII)[1319) 1322) 1323]

《Occurrence》 *Penicillium islandicum* Sopp.
mp. 281° (decomp.), $C_{30}H_{22}O_{12}$

UV See Fig. 184
IR See Fig. 185

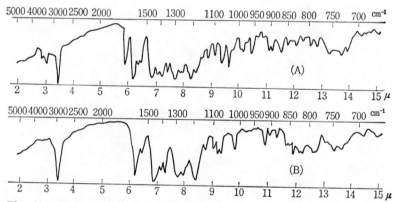

Fig. 185 IR Spectra of (A) Rubroskyrin and (B) Iridoskyrin[1322] (Shibata *et al.*)

(e) Iridoskyrin (LXXIV)[1320) 1322]

《Occurrence》 *Penicillium islandicum* Sopp.
Sublime at 320°, $C_{30}H_{18}O_{10}$
UV See Fig. 186
IR See Fig. 185 (B)

(f) Luteoskyrin (LXXV)[1319) 1322) 1323]

《Occurrence》 *Penicillium islandicum* Sopp.
mp. 281° (decomp.), $[\alpha]_D$ −880° (acetone), $C_{30}H_{20}O_{12}$

1322) Shibata, S. *et al*: *Chem. Pharm. Bull.*, **4**, 309 (1956). 1323) Shibata, S. *et al*: *ibid.*, **9**, 352 (1961).

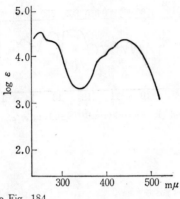

Fig. 186 UV Spectrum of Iridoskyrin in EtOH[1320] (Shibata *et al.*)

UV　See Fig. 184

IR　ν_{max}^{nujol} cm⁻¹: 1623 (C=O), 3378 (O–H)

(g) Lumiluteoskyrin (LXXVI)[1323]

　　≪Occurrence≫　　*Penicillium islandicum* Sopp.

mp. >360°, $C_{30}H_{20}O_{12}$

UV　See Fig. 184

IR　ν_{max}^{nujol} cm⁻¹: 3453 (non-bonded OH), 1689 (non-chelated ring C=O), 1614 (chelated ring C=O)

Diacetate　mp. >360°

IR　ν_{max}^{nujol} cm⁻¹: 1751 (alc. acetate C=O), 1713 (non-chelated ring C=O), 1617 (chelated ring C=O)

18   Lignoids

[A]   UV Spectra

Savinin[1324], hinokinin (cubebinolide)[1325], podophyllotoxin, α–, β–, and γ– apopicropodophyllins[1326], cubebinolide (hinokinin), podophyllotoxin, α–apopicropodophyllin, β–apopicropodophyllin and γ–apopicropodophyllin[1327]. α–and, β–peltatines[1328], podophyllotoxin–β–glucoside, β–peltatin–β–glucoside, 4–demethylpodophyllotoxin–β–glucoside and α–peltatin–β–glucoside.

[B]   IR   Spectra

IR ($\nu_{C=O}$) of podophyllotoxin, picropodophyllin, desoxypodophyllotoxin, desoxypicropodophyllin, isodesoxypodophyllotoxin and isodesoxypicropodophyllin[1329]. Savinin (hibalactone)[1324], podophyllotoxin–β–$\varDelta$–glucoside, β–peltatin–β–$\varDelta$–glucoside, 4–demethylpodophyllotoxin–β–$\varDelta$–glucoside and α–peltatin–β–$\varDelta$–glucoside[1331].

1324)   Kariyone, T., Isoi, K: *Y. Z.*, **74**, 1312 (1954).
1325)   Keimatsu, S., Isiguro, T: *ibid.*, **56**, 103 (1936).
1326)   Schrecker, Hartwell: *J.A.C.S.*, **74**, 5676 (1952).
1327)   Yamaguchi, K: Syokubutsu Seibun Bunsekiho, Vol. II p. 138. (1959, Nankodo Pub. Inc. Tokyo)
1328)   Press, Brun: *Helv. Chim. Acta*, **37**, 190 (1954).
1329)   Schrecker, Hartwell: *J.A.C.S.*, **75**, 5916 (1953).

[C] Classification by Freudenberg *et al.*[1330]

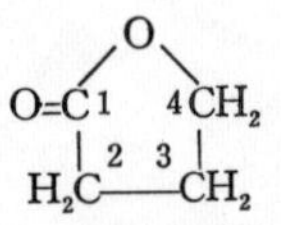

(I) Butanolide

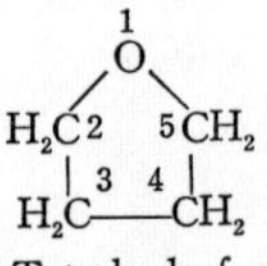

(II) Tetrahydrofuran

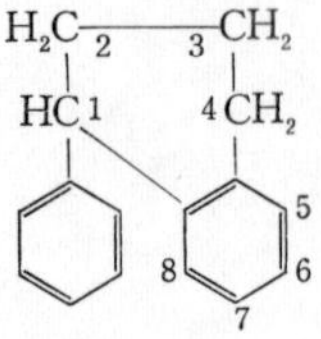

(III) 1–Phenyl–1, 2, 3, 4–
tetrahydronaphthalene

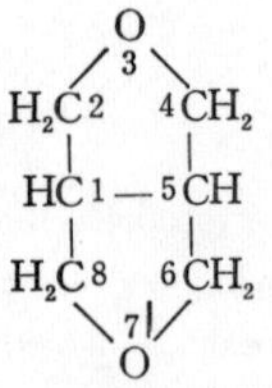

(IV) 3, 7–Dioxabicyclo–
(3, 3, 0)–octane

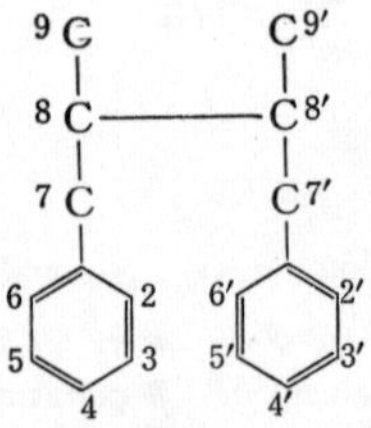

(V) Lignane

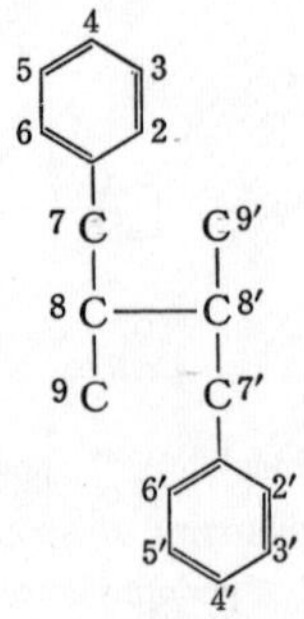

(VI) α–Lignane

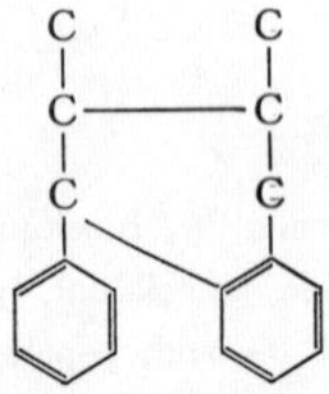

(VII) Cyclolignane

[D] Reviews

(1) Lignans[1332]

(2) The Chemistry of Lignification[1333]

Degradative techniques, oxydative degradation products, ethanolysis products and hydrogenesis products.
Conversion of CO_2 to lignin, polymerization reactions.
Products of the action of mushroom laccase on coniferyl alcohol.
The shikimic acid pathway of aromatization operating in *Escherichia Coli.*
Pathway from carbohydrates to lignin monomers.
A possible route for the formation of phenolic acids in the plant.
Other possible lignification pathways, factor involved in the control of lignification.
Cellwall carbohydrates and their relation to lignin.
Present status and future prospects.

1330) Freudenberg, K. *et al*: *Tetrahedron*, **15**, 115 (1961).
1331) Wartburg, Angliker, Renz: *Helv, Chim. Acta*, **40**, 1331 (1957).
1332) Hartwell, H *et al*: *Fort. Chem. Org. Naturstoffe*, **15**, 83 (1958).
1333) Brown, S. A: *Science*, **134**, 305 (1961).

〔E〕 Precursor or Model Compounds of Lignin

(1) UV Spectra[1334)

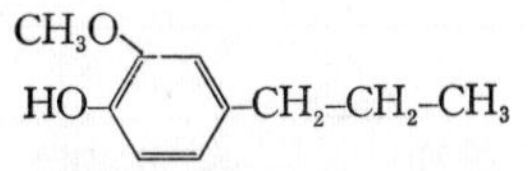

(VIII) 4–Propylguaiacol

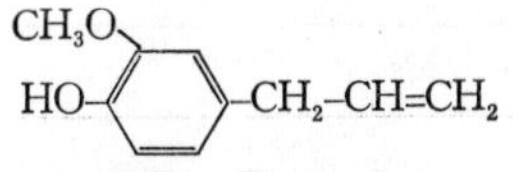

(IX) Eugenol

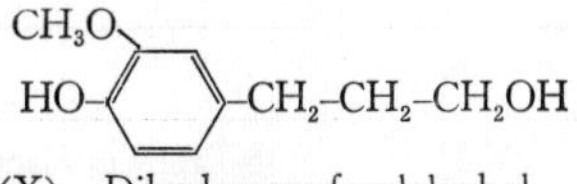

(X) Dihydroconyferylalcohol

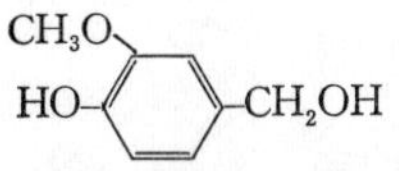

(XI) Vanillylalcohol

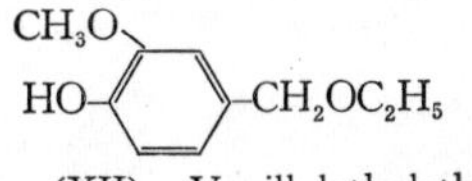

(XII) Vanillylethylether

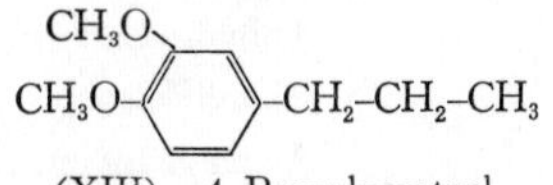

(XIII) 4–Propylveratrol

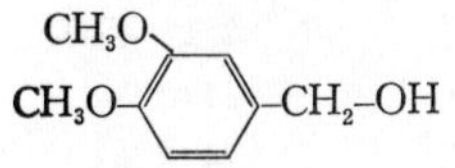

(XIV) Veratrylalcohol

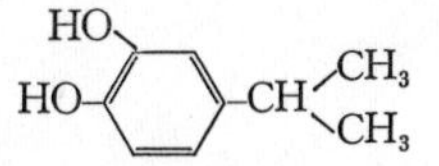

(XV) 4–Isopropylpyrocatechol

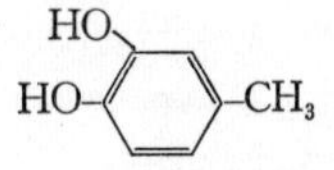

(XVI) Isocresol (Homocatechol)

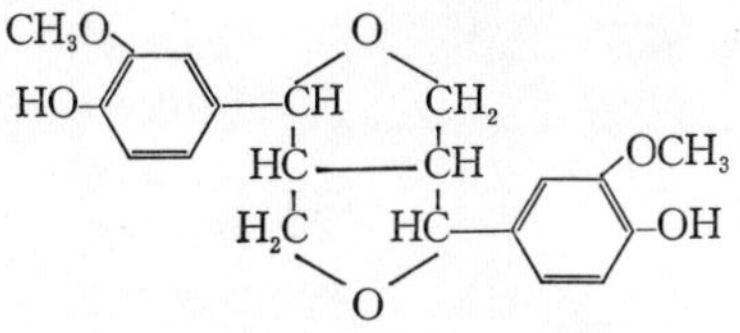

(XVII) Pinoresinol

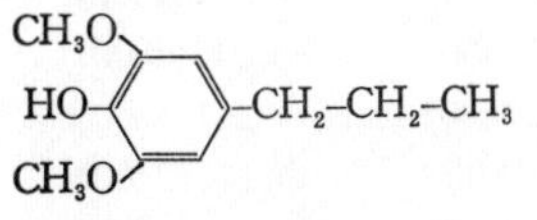

(XVIII) 3,5–Dimethoxy–
4–hydroxyphenyl–
propane

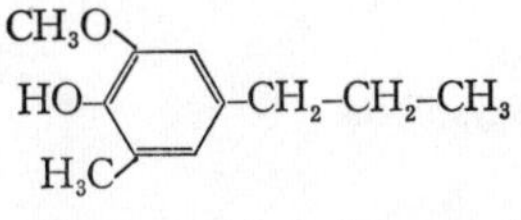

(XIX) 4–Methyl–6–propyl·
guaiacol

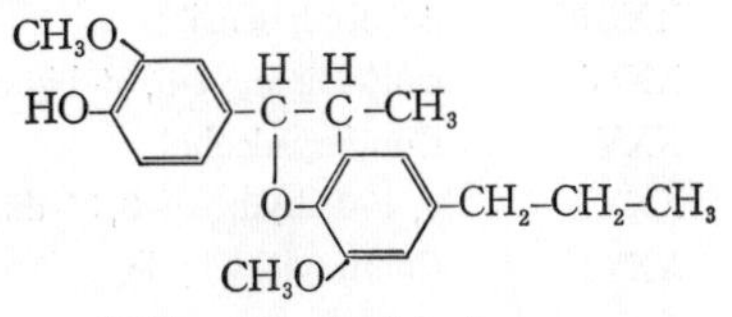

(XX) Dihydrodehydro–
diisoeugenol

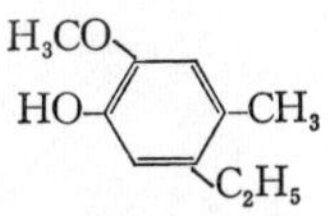

(XXI) 5–Ethylcreosol

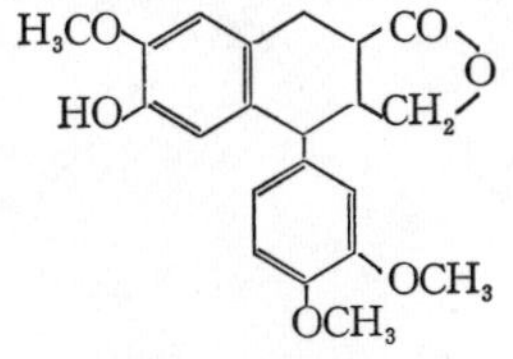

(XXII) Conidendrin

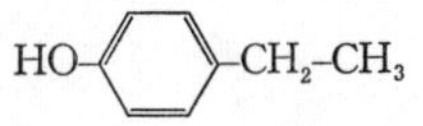

(XXIII) 4–Ethylphenol

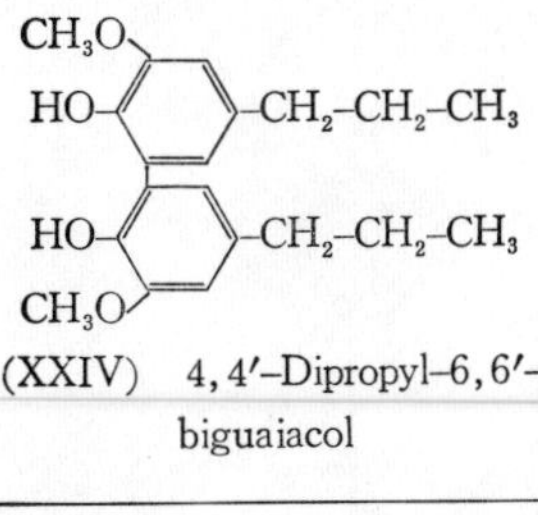

(XXIV) 4,4′–Dipropyl–6,6′–
biguaiacol

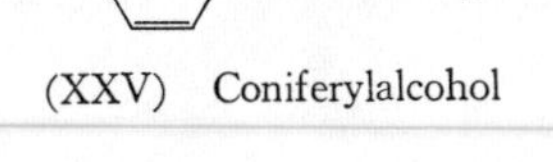

(XXV) Coniferylalcohol

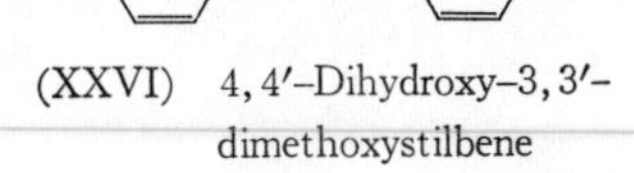

(XXVI) 4,4′–Dihydroxy–3,3′–
dimethoxystilbene

1334) Pew, J. C: *J. Org. Chem.*, **28**, 1048 (1963).

CH₃O, HO— —CH=CH–CHO

(XXVII) Coniferylaldehyde

CH₃O, HO— —C–CH–CH₃ (with O double bond, OC₂H₅)

(XXVIII) 2–Ethoxy–1–(4–hydroxy–3–
methoxyphenyl)–1–propanone

Table 87 UV Spectra of Lignin Model Compounds (in EtOH) (Pew J. C.)

Str. No.	Compounds	λ_{max} (ε)	λ_{min} (ε)
VIII	4–Propylguaiacol	280 (2960)	250 (240)
IX	Eugenol	281 (3120)	252 (400)
X	Dihydroconiferylalcohol	281 (3040)	250 (200)
XI	Vanillylalcohol	280 (2840)	251 (280)
XII	Vanillylethylether	280 (2920)	252 (280)
XIII	4–Propylveratrole	279 (2880)	251 (280)
XIV	Veratrylalcohol	279 (2720)	251 (400)
XV	4–Isopropylpyrocatechol	282 (2960)	249 (240)
XVI	Isocresol	281 (2920)	249 (200)
XVII	Pinoresinol	280 (3320)	252 (360)
XVIII	3, 5-Dimethoxy–4–hydroxyphenylpropane	273 (1280)	265 (560)
XIX	4–Methyl–6–propylguaiacol	280 (2280)	252 (320)
XX	Dihydrodehydrodiisoeugenol	281 (3000)	256 (560)
XXI	5–Ethylcreosol	284 (3320)	253 (240)
XXII	α–Conidendrin	283 (352)	255 (400)
XXIII	4–Ethylphenol	278 (1920)	245 (80)
XXIV	4-4′–Dipropyl–6, 6′–biguaiacol	290 (3000)	271 (1480)
XXV	Coniferylalcohol	265 (15400)	240 (5800)
XXVI	4, 4′–Dihydroxy–3, 3′–dimethoxystilbene	333 (23600)	261 (4000)
XXVII	Coniferylaldehyde	341 (23600) 240 (10400)	270 (2400)
XXVIII	2–Eth oxy–1–(4–hydroxy–3–methoxyphenyl)–1–propanone	280 (8600) 305 (7600)	250 (1800) 296 (7200)

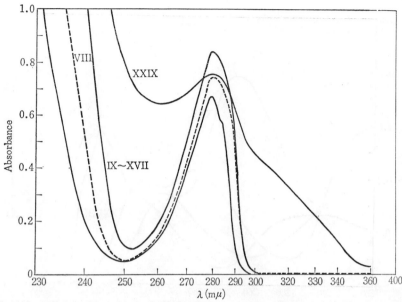

Fig. 187 UV Spectra of Spruce Lignin (XXIX) and Lignin Model Compounds (VIII~XVII) (Pew)

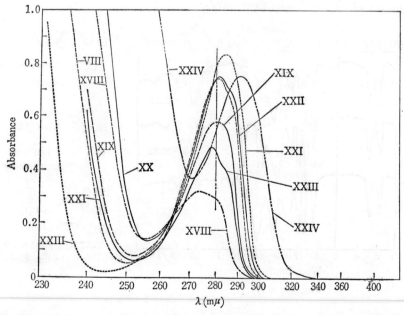

Fig. 188 UV Spectra of Lignin Model Compounds (VIII, XVIII~XXIV) in EtOH (Pew)

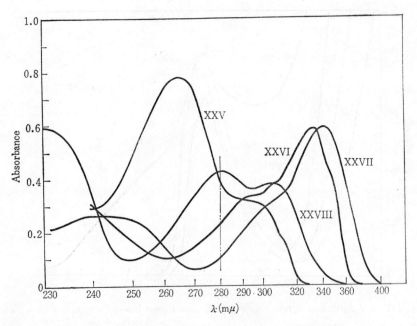

Fig. 189 UV Spectra of Lignin Model Compounds (XXV～XXVIII) (Pew)

(2) IR Spectra of Lignin Model Compounds[1334a] Fig. 189 a (1)～(9))

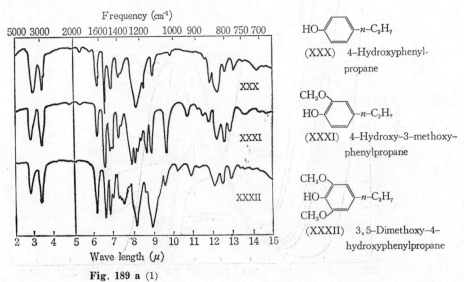

Fig. 189 a (1)

(XXX) 4-Hydroxyphenyl-
propane

(XXXI) 4-Hydroxy-3-methoxy-
phenylpropane

(XXXII) 3,5-Dimethoxy-4-
hydroxyphenylpropane

1334a) Pearl, I. A: *J. Org. Chem.*, **24**, 736 (1959).

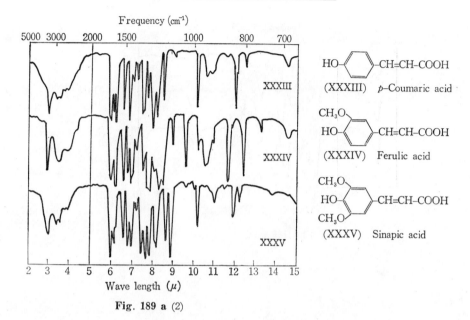

Fig. 189 a (2)

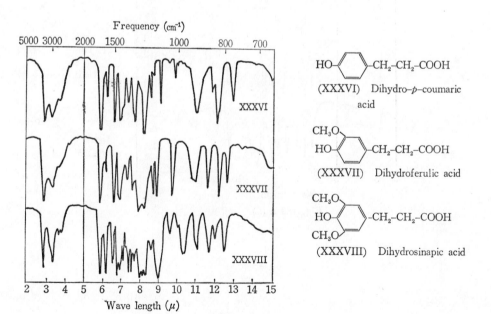

Fig. 189 a (3)

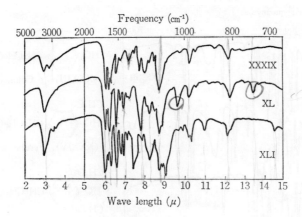

Frequency (cm⁻¹)

Wave length (μ)

Fig. 189 a (4)

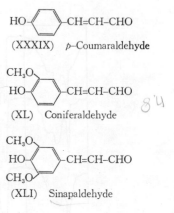

HO—⟨⟩—CH=CH–CHO

(XXXIX) *p*–Coumaraldehyde

CH₃O
HO—⟨⟩—CH=CH–CHO

(XL) Coniferaldehyde

CH₃O
HO—⟨⟩—CH=CH–CHO
CH₃O

(XLI) Sinapaldehyde

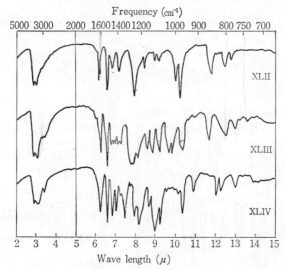

Frequency (cm⁻¹)

Wave length (μ)

Fig. 189 a (5)

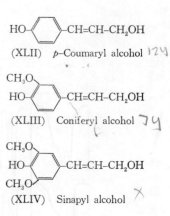

HO—⟨⟩—CH=CH–CH₂OH

(XLII) *p*–Coumaryl alcohol

CH₃O
HO—⟨⟩—CH=CH–CH₂OH

(XLIII) Coniferyl alcohol

CH₃O
HO—⟨⟩—CH=CH–CH₂OH
CH₃O

(XLIV) Sinapyl alcohol

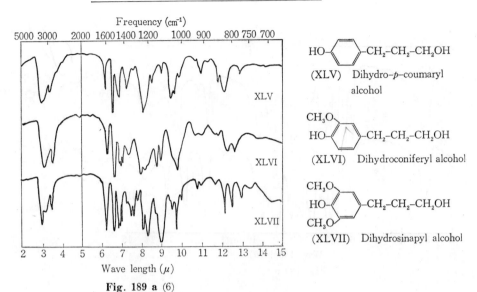

Fig. 189 a (6)

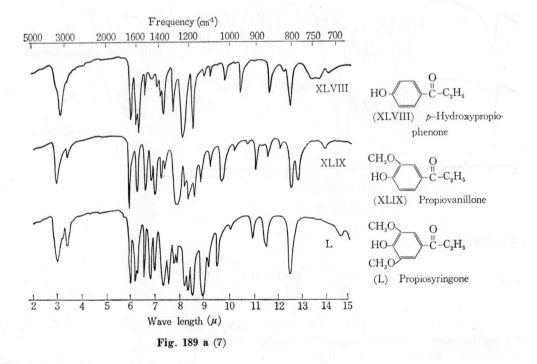

Fig. 189 a (7)

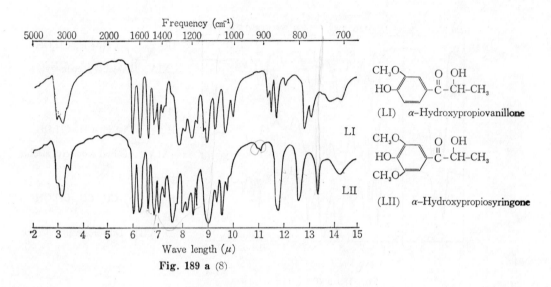

Fig. 189 a (8)

(LI)　α–Hydroxypropiovanillone

(LII)　α–Hydroxypropiosyringone

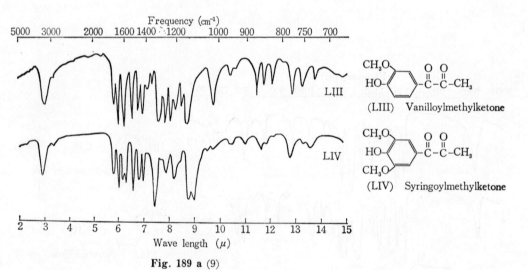

Fig. 189 a (9)

(LIII)　Vanilloylmethylketone

(LIV)　Syringoylmethylketone

（3）　**Hispidin**　(LV)

(LV)

≪Occurrence≫　　*Polyporus hispidus*[1335]，*Phaeolus schweinitzii*[1335a]．
mp. 256~258°, yellow needles, $C_{13}H_{10}O_5$

1335)　Bu'Lock, J. D. *et al*: *Soc.*, **1962**, 2085.
1335a)　Ueno, A *et al*: *Chem. Pharm. Bull.*, **12**, 376 (1964).

UV λ_{max}^{EtOH} mμ (log ε)[1335a] : 206 (4.48), 223 (4.50), 253 (4.14), 366 (4.24)

IR ν_{max}^{KBr}cm^{-1} [1335a] (hydrate): 3500, 1690, 1660, 1117, 812 (s); ν_{max}^{KBr}cm^{-1} [1335a] : 3300, 1656, 1125, 805~825 (m)

[F] Butanolides (I)

(1) **Trachelogenin** (LVI)[1337]

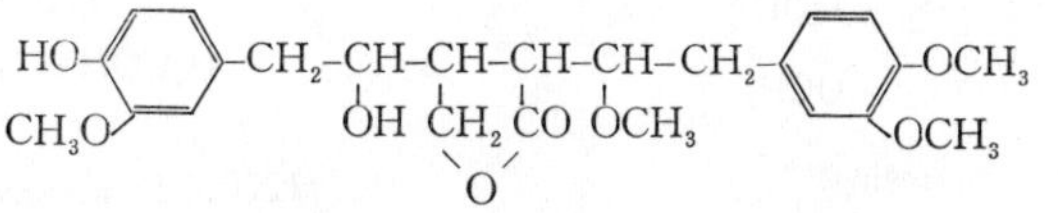

(XLI)

≪Occurrence≫ *Trachelospermum asiaticum* NAKAI var. *intermedium* NAKAI. (テイカカズラ)

Tracheloside (LVII)
- alkaline hydrolysis → Trochelonic acid (LVIII)
- acid hydrolysis → Trachelogenin (LVI)+Glucose
  - Alkali → Trochelogenic acid (LIX)

(**a**) **Tracheloside** (LVII)[1336]

mp. 176° (decomp.) $[\alpha]_D+78.0°$ (MeOH), $C_{36}H_{50}O_{18}$

UV λ_{max}^{MeOH} mμ (log ε): 280 (3.4).

IR ν_{max}^{KBr} cm^{-1}: 3350 br (glucose OH), 1760 (C=O), 1602, 1512 (arom. ring)

(**b**) **Trachelonic acid** (LVIII)[1336]

mp. 143°, pK'=4.70 (MeOH)

IR ν_{max}^{KBr} cm^{-1}: 3100, 3332 (OH); Na salt ν_{max}^{KBr} cm^{-1}: 3345, 3127 (OH), 1585 (-COO-)

(**c**) **Trachelogenin** (LVI)[1336)1337]

mp. 144°, $C_{24}H_{30}O_8$

UV λ_{max}^{MeOH} mμ (log ε): 283 (3.4); $\lambda_{max}^{NaOH-MeOH}$ mμ: 297

IR ν_{max}^{KBr} cm^{-1}: 3491, 3479 (OH), 1767 (γ-lactone), 1602, 1510 (arom. ring), 1458 (-CH_2-), 1373, 1277 (phenolic OH)

(**d**) **Trachelogenic acid** (LIX)

mp. 160°(decomp.) $C_{24}H_{32}O_9$, pK_1'=4.65 (MeOH)

1336) Takano, Ch *et al*: *Y. Z.*, **78**, 879, 882, 885 (1958).
1337) Takano, Ch *et al*: *ibid.*, **79**, 447, 1449 (1959).

〔G〕 3, 7–Dioxabicyclo–(3, 3, 0)–octanes (IV)

(1) Structural Correlation of Lignoids[1338]

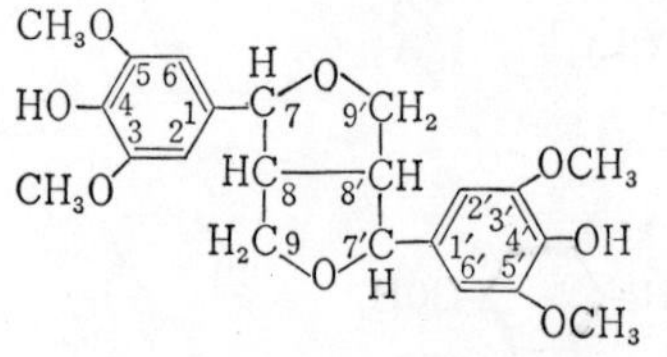

(LX) Syringaresinol

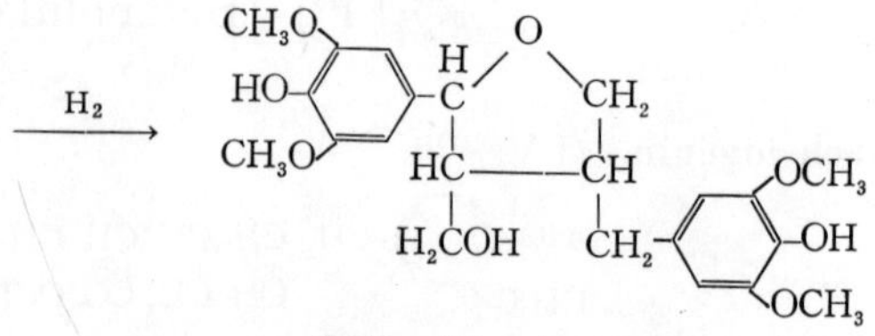

(LXI) Dimethoxylariciresinol

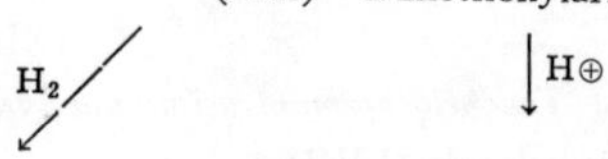

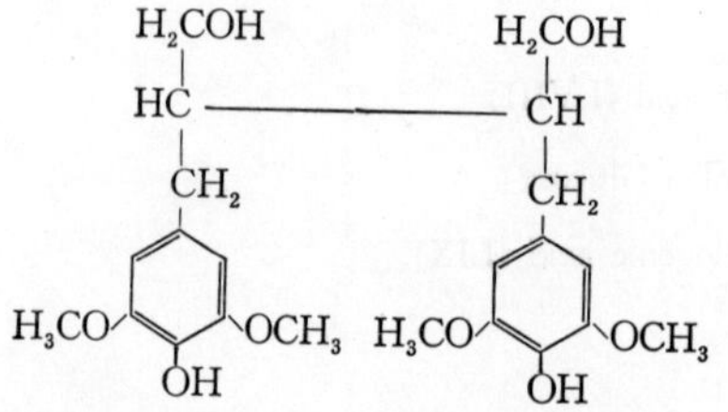

(LXII) Dimethoxy–secoiso–lariciresinol

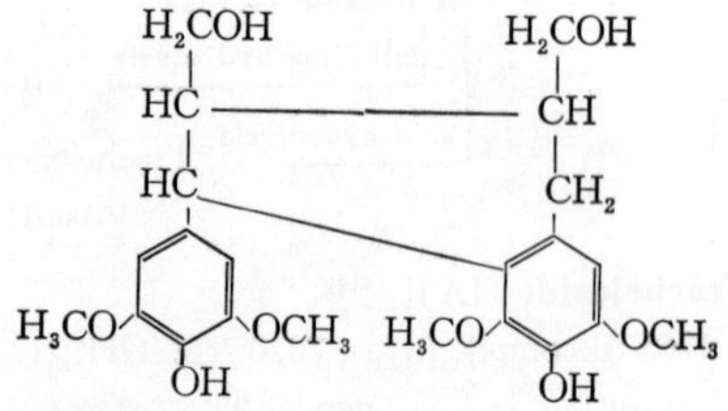

(LXIII) Dimethoxy–iso–lariciresinol

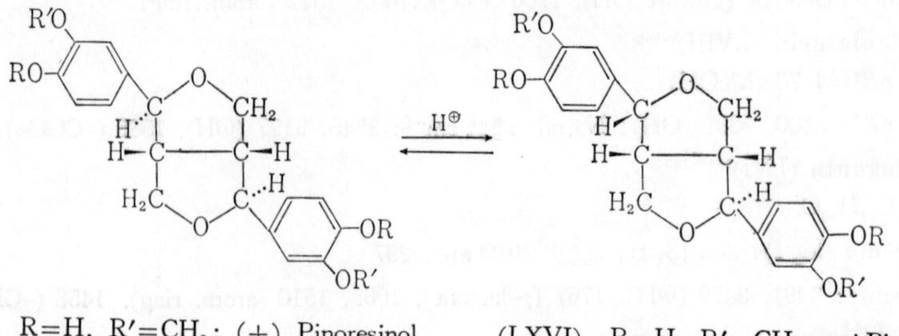

(LXIV) R=H, R'=CH₃: (+) Pinoresinol
(LXV) R, R'=−CH₂−: (+) Sesamin

(LXVI) R=H, R'=CH₃: epi-Pinoresinol
(LXVII) R, R'=−CH₂−: (+) Asarinin

1338) Weinges, K: Ber., 94, 2522 (1961).

(LXIV)
(LXV)

(LXVI)
(LXVII)

(LXVIII) α-Conidendrin

LiAlH₄

H₂ (left)

H₂ (right)

(LXIX) (+)-Isolarici-resinol

H⊕

H⊕

(LXX) R=H, R′=CH₃ : (+) Lariciresinol
      R, R′=—CH₂— : (LXXI)

(LXXII) R=H, R′=CH₃ : (+) Epi-larici-resinol
        R, R′=—CH₂— : (LXXIII)

H₂

H₂

(LXXIV) R=H, R′=CH₃ : (−)-Seco-iso-lariciresinol
(LXXV) R, R′=—CH₂ : (−)-Dihydrocubebin

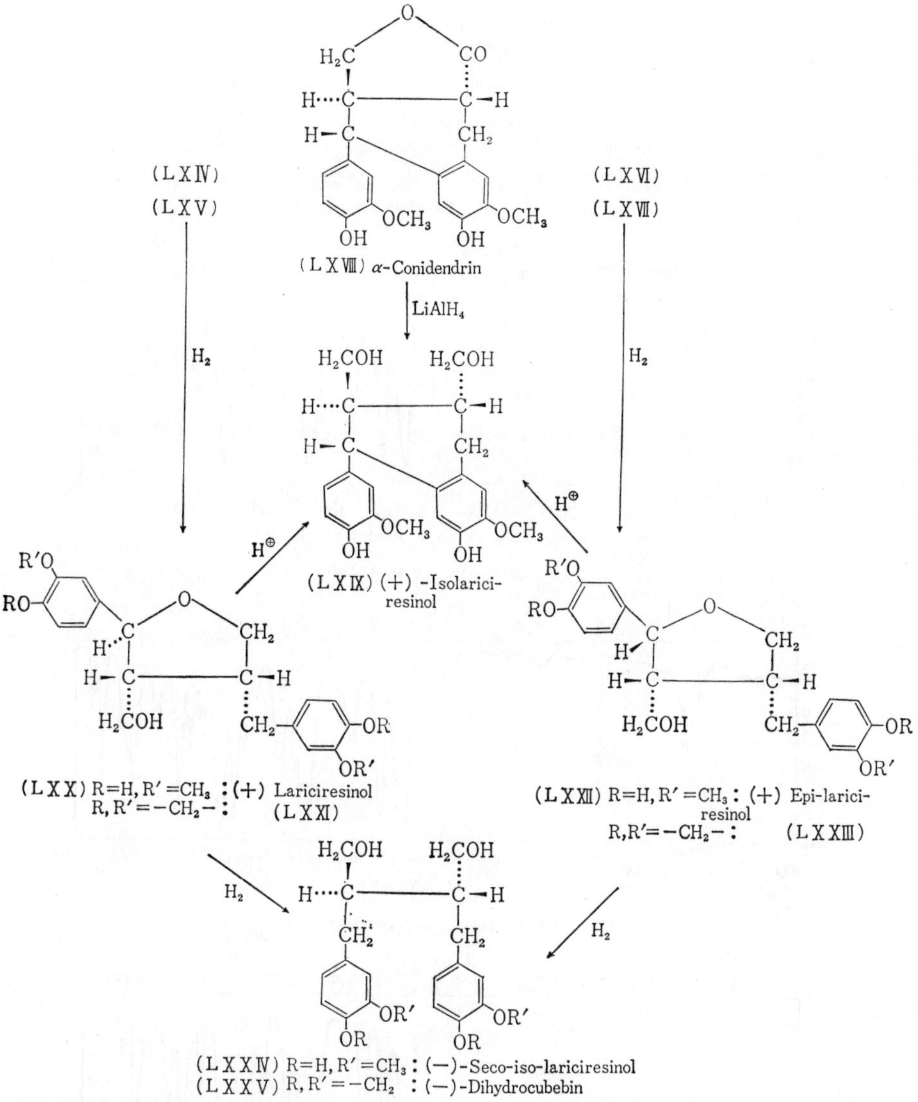

(+)-**Dimethoxy-iso-lariciresinol** (LXIII)

≪Occurrence≫ *Alnus glutinosa*

mp. 165~167°, $[\alpha]_D +52°$ (acetone), $C_{22}H_{28}O_8$

(2) **Lignoids of *Picea excelsa*[1338)**

(a) (+)-**Pinoresinol** (LXIV) mp. 120~121°, $[\alpha]_{589}^{25}+82.4°$, $C_{20}H_{22}O_6$

(b) (+)-*epi*-**Pinoresinol** (LXVI) mp. 137~138°, $[\alpha]_{589}^{25}+130.4°$, $C_{20}H_{22}O_6$

(c) (+)-**Isolariciresinol** (LXIX) mp. 115~117°, $[\alpha]_{578}^{25}+65.20°$

(d) (−)-**Seco-isolariciresinol** (LXXIV) mp. 112~113°, $[\alpha]_{578}^{25}-35.3°$, $C_{20}H_{26}O_6$, see the item 〔H〕

(e) (+)-**Lariciresinol** (LXX) mp. 99~101°, $[\alpha]_{578}^{25}+17.3°$, $C_{20}H_{24}O_6$

(3) (+)-Sesamin, (+)-Asarinin and (+)-epi-Asarinin[1339)~1342)]

(a) (+)-Sesamin (LXV)

《Occurrence》 *Sesamum indicum* (ゴマ), *Ocotea usambarensis*, *Fagara xanthoxyloides*, *Ginkgo biloba* (イチョウ), *Paulownia tomentosa* (桐)[1342)]

mp. 123°, $[\alpha]_D + 71°$ (CHCl₃), C₂₀H₁₈O₆

UV λ_{max}^{EtOH} m$\mu(\varepsilon)$: 237.5 (8108), 287.5 (7269)[1342)]   IR ν_{max}^{KBr} cm⁻¹: 2860, 1630 (br), 1505, 1450, 1370, 1255, 1185, 1100, 1060, 1040, 930, 855, 790, 750[1342)]   IR See Fig. 190 (A)

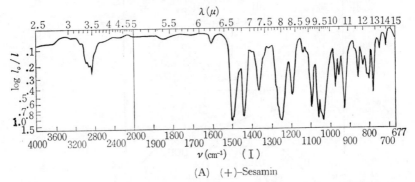

(A) (+)-Sesamin

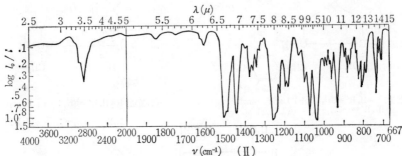

(B) (+)-Asarinin ((+)-epi-Sesamin)

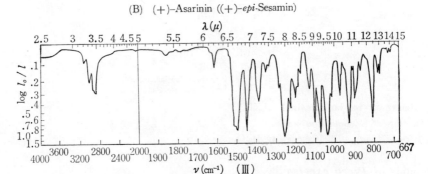

(C) (+)-epi-Asarinin ((+)-Diasesamin)

Fig. 190 IR Spectra of (A) (+)-Sesamin, (B) (+)-Asarinin and (C) (+)-epi-Asarinin
(Freudenberg *et al.*)

1339) Freudenberg, K *et al*: *Ber.*, **94**, 851 (1961).
1340) Jones, W. A. *et al*: *J. Org. Chem.*, **27**, 3232 (1962); Bocker, E. D: *Tetr. Letters*, **1962**, 157.
1341) Freudenberg, K *et al*: *Ann.*, **623.**, 129 (1960).
1342) Takahashi, K *et al*: *Y. Z.*, **83**, 1101 (1963).

NMR

(1) See Fig. 191 (A), in CDCl$_3$, internal standard, TMS=10 ppm[1338] ; 3.0~3.2 (6) 6 arom. protons at 2,3, 6,2′,3′,6′), 4.0 s (4) (2–O–CH$_2$–O–), 5.2 d (7,7′–H)

(2) See Fig. 192 (B), 60 MC, in CDCl$_3$, TMS, cps[1340] : 348,350 (2–O–CH$_2$–O–), 279 d (2 ϕ–CH $\overset{\displaystyle O}{\underset{\displaystyle |}{\diagdown}}$),

220~270 m (2–CH$_2$–), 179 m (2 H at 8,8′)

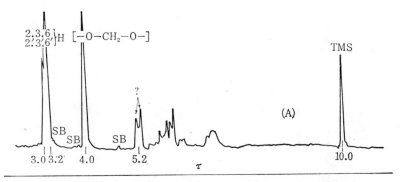

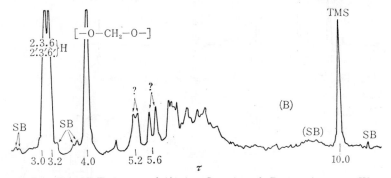

Fig. 191 NMR Spectra of (A) (+)–Sesamin and (B) (+)–Asarinin (Weinges)

(b) (+)–**Asarinin** (LXVII)

≪Occurrence≫ *Acronychia muelleri,* mp. 121.5°, [α]$_D$+124° (CHCl$_3$), C$_{20}$H$_{18}$O$_6$

IR: See Fig. 190 (B)

NMR: See Fig. 191 (B), in CDCl$_3$, internal standard, TMS=10 ppm[1338] : 3.0~3.2 (6) (6 arom. protons at 2,3,6,2′,3′,6′), 4.0 s (4) (2–O–CH$_2$–O–), 5.2 d 5.6 d (7–H, 7′–H)

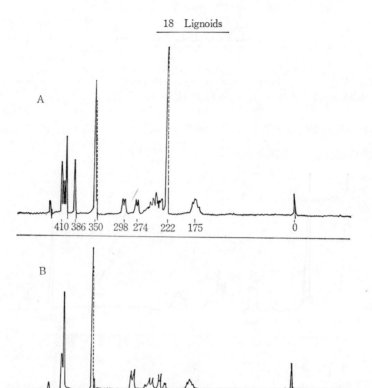

Fig. 192 NMR Spectra of (A) Sesangolin and (B) Sesamin (Jones *et al.*)

(c) (+)-*epi*-**Asarinin** ((+)-Diasesamin) (LXXVI)[1339]

    mp. 168~171°, $[\alpha]_D$+385° (CHCl$_3$), C$_{20}$H$_{18}$O$_6$

    IR See Fig. 190 (c)

(LXXVI).

(4) **Sesangolin** (LXXVII)[1340]

(LXXVII)

    《Occurrence》 *Sesamum angolense* WELW.

    mp. 87~88°, 101° (polymorphism), $[\alpha]_D$+48.5° (CHCl$_3$) C$_{21}$H$_{20}$O$_7$

— 454 —

UV $\lambda_{max}^{Isooctane}$ mμ (ε): 293 (7870), 236, λ_{min}: 257 (622), 222 (5830)

IR $\nu_{max}^{CS_2}$ cm^{-1}: 2850 m, 1245 s, 1190 s, 1155 m, 1040 s, 940 s, 860 m, 817 (m), 807 (m)

NMR: See Fig. 192 (A), 60 MC, in CDCl$_3$, TMS, cps[1340]: 348, 350 (2–O–CH$_2$–O–), 222 s (3) (–OCH$_3$), 402: 406 (2: 1) (3 H on ring B), 386 s: 410 s (1: 1) (2 H on ring

A must be *para*), 175 m (2 H at 8, 8′), 220~270 m (2–CH$_2$), 274 d, 298 d (2 ϕ–$\overset{\textstyle O}{\underset{|}{C}}$H   )

(5) (+) **Pinoresinol** (LXIV)

≪Occurrence≫ *Picea excelsa* (see the item (2)), *P. vulgaris*, *Pinus silvestris*
mp. 122°, $[\alpha]_D + 84.4$ (acetone), C$_{20}$H$_{22}$O$_6$
UV See Fig. 193
IR See Fig. 194

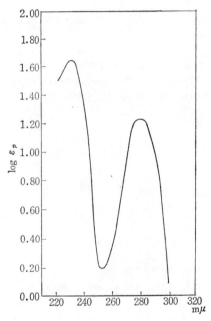

Fig. 193 UV Spectrum of Pinoresinol
(dioxane-water)
(Freudenberg *et al.*)

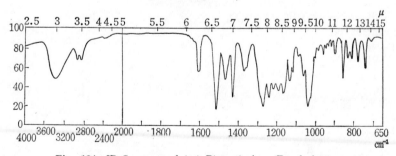

Fig. 194 IR Spectrum of (+)–Pinoresinol   (Freudenberg *et al.*)

(6) Paulownin (LXXVIII)[1342]

(LXXVIII)

≪Occurrence≫　*Paulownia tomentosa* STEUD. (桐).
(a) Methanolate (LXXVIII)·CH$_3$OH (in MeOH–extract)
　mp. 84°, $[\alpha]_D + 51.07°$, C$_{21}$H$_{22}$O$_8$
　UVλ_{max}^{EtOH} mμ (ε): 237 (10280), 287 (8700)
　IR　ν_{max}^{KBr} cm^{-1}: 2925, 2880, 1610, 1490, 1450, 1260, 1240, 940, 815, 750
(b) Paulownin (LXXVIII)
　mp. 104～105°, C$_{20}$H$_{18}$O$_7$
　NMR　in CHCl$_3$, 40 MC, internal ref. CHCl$_3$, τ of CHCl$_3$ 2.75 ppm, τ: 3.13 s, 3.23 s (6) (6 arom. pro-
　　　tons), 4.08 s (4) (2-O-CH$_2$-O-), 5.28 d, J=6.0 c/s (1) (C$_2$-H), 5.32 s(1) (C$_6$-H), 5.55～6.46
　　　complex multiplet (4) (2-CH$_2$), 7.03 m (C$_1$-H), 7.90 s (C$_5$-OH)

〔H〕 Lignanes (Ⅴ)

(1) (−)-Secoisolariciresinol (LXXIV)[1343]
　≪Occurrence≫　*Picea excelsa* (See the item 〔G〕, (2)),
　　　　　　　　Podocarpus spicatus.
　mp. 112.5～113.5°, $[\alpha]_D$ −35.6° (acetone), C$_{20}$H$_{26}$O$_6$
　UV　λ_{max}^{EtOH} mμ (log ε): 209 (4.5), 229 (4.3), 283 (3.9)
　IR　ν_{max}^{KBr} cm^{-1}: 3436, 3165, 2915, 1608, 1515, 1471, 1458,
　　1433, 1385, 1359 sh, 1339, 1312, 1272, 1241, 1199, 1189,
　　1155, 1129, 1120, 1098, 1064, 1050 sh, 1031, 1007, 977,
　　947, 937, 919, 910, 897, 844, 808, 798, 744

(LXXIV)

(2) Hibalactone (LXXIX)
　≪Occurrence≫　*Juniperus sabina, Chamaecyperis obtusa*
　mp. 146～148°, $[\alpha]_D$ −88° (CHCl$_3$) C$_{20}$H$_{16}$O$_6$
　UV　See Fig. 195
　IR　See Fig. 196

(LXXIX)

1343) Briggs, L. H *et al*: *Tetrahedron*, **7**, 262 (1959).

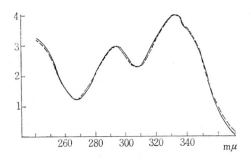

Fig. 195 UV Spectra of Hibalactone
.... Natural (−)-Hibalactone
——Synth. (±)-Hibalactone

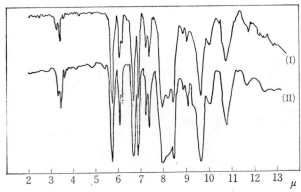

(I)

(II)

Fig. 196 IR Spectra of Hibalactone
(I) Natural (−)-Hibalactone
(II) Synth. (±)-Hibalactone

（3） **Matairesinol** (LXXX)[1344]

≪Occurrence≫ *Tsuga mestensiana, T. heterophylla, Abies amabilis, Podocarpus spicalus* (MATAI)

mp. 119°, $[\alpha]_D$ −48.6° (acetone)

(LXXX)

〔I〕 Cyclolignanes (VII)

（1） **Podophyllotoxin** (LXXXI)

≪Occurrence≫ *Podophyllum emodi* WALL, *P. peltatum* L., *P. pleidanthum* HANCE (八角連)[1345], *Callitrus drummondii*[1346], *Diphylleia grayi* (サンカヨウ)[1347]

mp. 112～117° (decomp.), 180～181° (after dried), $[\alpha]_D$ −139° (CHCl₃), $C_{22}H_{22}O_8$

IR ν_{max}^{nujol} cm⁻¹ [1347] : 3525 (OH), 1770 (unsatur. five membered lactone), 1600, 1507 (arom.), 1250, 1130, 1040, 933 (–O–CH₂–O–)

(LXXXI)

1344) Barton, G. M *et al* : *J. Org. Chem.*, **27**, 322 (1962).
1345) Shibata, S *et al* : *Y. Z.*, **82**, 777 (1962).
1346) Kier, L. B *et al* : *J. P. S.*, **52**, 502 (1963).
1347) Murakami, T. *et al* : *Y. Z.*, **81**, 1596 (1961).

Acetate mp. 202~203°[1347)

UV λ_{max}^{EtoH} mμ (log ε): 205 (4.76), 290 (3.63),

NMR: See Tab 88

(2) **Picropodophyllin** (LXXXII)

    ≪Occurrence≫ *Podophyllum emodi* var. *hexandrum, Diphylleia*
        grayi (サンカヨウ)[1347).

mp. 232°, [α]$_D$ +5.3° (CHCl$_3$), C$_{22}$H$_{22}$O$_8$

IR ν_{max}^{KBr} cm^{-1} [1347) : 3500 br, 1767, 1595, 1507, 1480, 1422, 1330,

  1250, 1130, 1040, 1005, 926

Acetate mp. 215~216°[1347)

UV λ_{max}^{EtOH} mμ (log ε): 205 (4.75), 290 (3.61)

(LXXXII)

(3) **Diphyllin** (LXXXIII)[1347)

    ≪Occurrence≫ *Diphylleia grayi* (サンカヨウ)

mp. 291°, C$_{21}$H$_{16}$O$_7$

IR ν_{max}^{nujol} cm^{-1}: 3320 br, 1709 (chelated $\alpha\beta$–unsatur. five membe

  red lactone), 1613, 1495, 1242, 1221, 1174, 1132, 1042, 1010,

  936

Acetate mp. 236~240°

UV λ_{max}^{EtOH} mμ (log ε): 260 (4.72), 295 (4.02), 302 sh (4.01), 312

  (4.04), 348 (3.71)

IR $_{max}^{KBr}$ cm^{-1}: 1768, 1621, 1605, 1498, 1485, 1346, 1228, 1169, 1034, 1014, 929

(LXXXIII)

(4) **Desoxypodophyllotoxin** (Anthricin) (LXXXIV)[1345)

    ≪Occurrence≫ *Anthriscus silvestris, Juniperus silicicola,*
        Podophyllum peltatum, P. pleidanthum HANCE
        (八角連)[1345).

mp. 167~168°, [α]$_D$ −119° (CHCl$_3$), C$_{22}$H$_{22}$O$_7$

(LXXXIV)

(5) **Desoxypodophyllinic acid-1β-D-glucopyranosylester** (LXXXV)[1348)

(LXXXV)

    ≪Occurrence≫ *Podophyllum peltatum* L., *P. emodi* WALL.

mp. 123~128°, [α]$_D$ −151.5° (MeOH), C$_{28}$H$_{34}$O$_{13}$

1348) Kuhn, M. *et al*: *Helv. Chim. Acta,* **46**, 2127 (1963).

UV $\lambda_{max} m\mu$ (log ε): 289 (3.68)

IR $\nu_{max} cm^{-1}$: 1490, 1590 (arom. ring), 1755 (ester)

(6) 4-Demethyl-7-dehydroxypodophyllotoxin (LXXXVI)[1349]

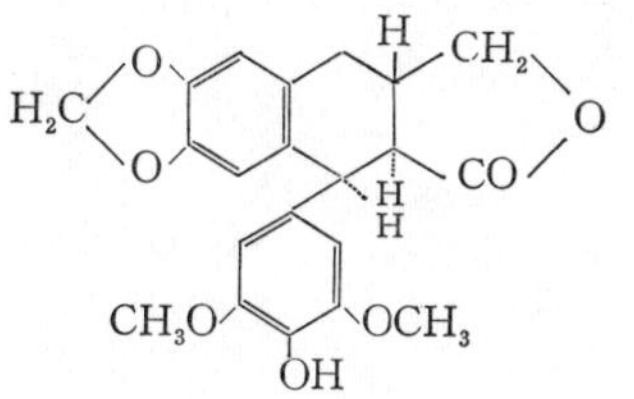

≪Occurrence≫ *Polygala paenea* L.

mp. 244～248°, $[\alpha]_D$ −128° (CHCl$_3$), $C_{21}H_{20}O_7$

(LXXXVI)

(7) NMR of the Stereo-isomers of Sikkimotoxin and Podophyllotoxin[1350]

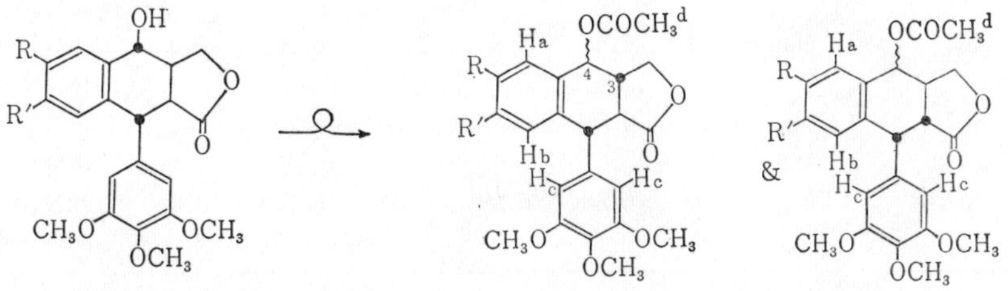

(A)　　　(B)

R=R′=OCH$_3$ (LXXXVII) Picrosikkimotoxin
R, R′=—O—CH$_2$—O— (LXXXII) Picropodophyllin

Table 88 NMR Spectra of the Stereoisomers of Podophyllotoxin acetate and Sikkimotoxin
acetate　　　　　　　　　　　　　　　　　(Schreier)

(A)

Type	R, R′	H_{C-3}/H_{C-4}	Compound	Str. No.
A	CH$_3$O	*trans*	O–Acetyl–DL–isosikkimotoxin	LXXXVIII
A	CH$_3$O	*cis*	O–Acetyl–DL–episikkimotoxin	LXXXIX
A	OCH$_2$O	*trans*	O–Acetyl–DL–isopodophyllotoxin	XC
A	OCH$_2$O	*cis*	O–Acetyl–DL–epipodophyllotoxin	XCI
B	OCH$_2$O	*trans*	O–Acetyl–podophyllotoxin	XCII
B	OCH$_2$O	*cis*	O–Acetyl–epipodophyllotoxin	XCIII

1349) Polonsky, J. *et al*: *Bull. Soc. Chim. France*, **1962**, 1722; *Index Chem.*, **8**, 24720 (1963).
1350) Schreier, E: *Helv. Chim. Acta*, **46**, 75 (1963).

NMR in 60 Mc, TMS, $\delta=0$ ppm, J=c/s

H_{C-3}/H_{C-4}	Str. No.	15% soln. of	H_a s (1H)	H_b s (1H)	H_c s (2H)	H_d s (3H)	H_{C-4} d (1H)	J	H_{CH_3O}	H_{O-CH_2-O}
trans	LXXXVIII	CDCl$_3$	6.74	6.35	6.41	2.24	6.17	9.0	s (3H) 3.87 s (3H) 3.85 s (6H) 3.81 s (3H) 3.63	—
cis	LXXXIX	CDCl$_3$	6.91	6.43	6.50	2.16	6.12	2.5	s (3H) 3.89 s (3H) 3.86 s (6H) 3.83 s (3H) 3.66	—
trans	XC	pyridine	7.05	6.66	6.89	2.19	6.42	9.5	s (3H) 3.89 s (6H) 3.73	d 5.98
cis	XCI	pyridine	7.11	6.68	6.85	2.15	6.26	3.0	s (3H) 3.88 s (6H) 3.74	d 5.97
trans	XCII	CDCl$_3$	6.81	6.56	6.43	2.20	~5.92	>7	s (3H) 3.84 s (6H) 3.80	s 6.01
cis	XCIII	CDCl$_3$	6.90	6.58	6.30	2.13	6.17	3.0	s (3H) 3.81 s (6H) 3.75	s 5.99

s : singlet,　　d : doublet

(8) **Lyoniside** (XCIV) and **Lyoniresinol** (XCV)[1351]

≪Occurrence≫ *Lyonia ovalifolia* SIEB et ZUCC. var. *elliptica* HAND-MAZZ.(ネジキ).

(a) Lyoniside (XCIV)

mp. 123.5°/165.1°, $[\alpha]_D+26.7°$ (EtOH), C$_{27}$H$_{36}$O$_{12}$·5H$_2$O

UV λ_{max}^{EtOH} mμ (log ε): 281 (3.60)

(XCIV) $\xrightarrow{\text{butanone, Me}_2\text{SO}_4,\ \text{K}_2\text{CO}_3}$ Lyoniresinol

Dimethyl ether

(b) Lyoniresinol (XCV)

mp. 170.7~171°, $[\alpha]_D$ +68.4° (acetone), C$_{22}$H$_{28}$O$_8$

NMR in pyridine, TMS, δ: 4.17 m (2 α–H, 3 α–H), 3.85, 3.72 (OCH$_3$), 3.17 m (1–H, 1′–H), 2.55 m (2–H, 3–H)

Dimethylether

mp. 158~160°, $[\alpha]_D$ +30.1°, (EtOH), C$_{24}$H$_{32}$O$_8$

UV λ_{max}^{EtOH} mμ (log ε): 273 (3.29),　　IR ν_{max}^{KBr} cm^{-1}: 3306, 3288 (OH)

NMR in pyridine, TMS, δ: 4.97 d, J=5.5 c/s (4–H), 4.11 m (2 α–H, 3 α–H), 3.88, 3.8, 3.67, 3.61 (OCH$_3$), 3.11 m (1–H, 1′–H), 2.5 m (2–H, 3–H)

R=xylose: Lyoniside (XCIV)

R=H: Lyoniresinol (XCV)

1351) Kato, Y: *Chem. Pharm. Bull.*, **11**, 823 (1963); Yasue., M., Kato, Y: *Y. Z.*, **81**, 526, 529 (1961), **80**, 1013 (1960).

（9） Otobain (XCVI)[1352]~[1355]

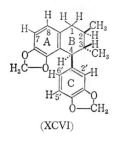

(XCVI)

≪Occurrence≫　*Myristica otoba*

mp. 137~138°, $[\alpha]_D$ −40.5°, −43° (CHCl₃), C₂₀H₂₀O₄

UV λ_{max}^{EtOH} mμ (ε)[1352] : 234 (9300), 287 (6700)

IR ν_{max}^{KBr} cm⁻¹ [1352] : 1850 m, 928 s (–O–CH₂–O–), 1362, 1242, 1130, 1045 (arom.)

NMR　(1)　See Fig. 197[1352],　　(2)　τ[1353] : 3.30 (H–7, H–8) 3.38 (H–2′, H–5′ H–6′), 4.12 (–O–CH₂–O– on ring C), 4.32, 4.34, 4.41, 4.43 (–O–CH₂–O– on ring A), 6.56 d, *J*=6 (H–4), 7.44, 7.54 (1–CH₂–), 8.5~9.0 (unresolved, H–2, H–3?), 9.16, 9.23 (2–CH₃, 3–CH₃ resp.)

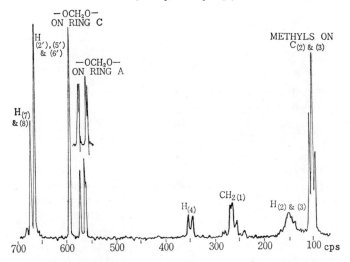

Fig. 197　IR Spectrum of Otobain　100 Mc, cps (Bhacca *et al.*)

（10） Hydroxyotobain (XCVII) and Isootobain (XCVIII)

(XCVII)

(XCVIII)

≪Occurrence≫　*Myristica otoba*

(a) Hydroxyotobain (XCVII)[1353]

mp. 116~117°, $[\alpha]_D$ −28° (CHCl₃), C₂₀H₂₀O₅

(b) Isootobain (XCVIII)[1354] [1355a]

mp. 106~108°, $[\alpha]_D$ +5.3°, C₂₁H₂₄O₄

1352) Bhacca, N. S *et al*: *J. Org. Chem.*, **28**, 1638 (1963).　　1353) Gilchrist, T. *et al*: *Soc.*, **1962**, 1780.
1354) Wallace, R. *et al*: *ibid.*, **1963**, 1445.　　1355) Stevenson, R: *Chem. & Ind.*, **1962**, 270.
1355a) Baughman, F *et al*: *Soc.* **1921**, 199.

NMR $\tau^{1354)}$: ~3.5, 3.83 (arom. H), 4.13 (–O–CH$_2$–O–), 6.20, 6.42 (OCH$_3$), 6.63 d, J=9 c/s (H–4), 7.35 d, J=6 c/s (2 H–1), 8.5 (2>CH–CH$_3$), 8.94 d, J=6~7 cps (>CH–CH$_3$), 9.12 d, J=5.4 c/s (>CH–CH$_3$)

(11) (–)-Galcatin (XCIX) and (–)-Galbulin(C)[1356)1357)]

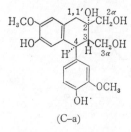

(XCIX)   (C)

≪Occurrence≫   *Himantandra* spp.

(a) (–)-Galcatin (XCIX)

mp. 117~118°, [α]$_D$ –8.8 (CHCl$_3$), C$_{21}$H$_{24}$O$_4$

NMR $\tau^{1354)}$: 3.85 (arom. H), 4.21 (–O–CH$_2$–O–), 6.15 (OCH$_3$), 6.63 d (H–4), 7.35 d (2 H–1), 8.4 (2>CH–CH$_3$), 8.95 d, 9.16 d (2>CH–CH$_3$)

(b) (–)-Galbulin (C)

mp. 135°, [α]$_D$ –8° (CHCl$_3$), C$_{22}$H$_{28}$O$_4$

(12) (+)-Isoolivil ((+)-Cycloolivil) (C-a)[1357a~c)]

≪Occurrence≫   *Olea cunninghamii*

mp. ca 98°, [α]$_D$ +61.8° (EtOH), C$_{20}$H$_{24}$O$_7$

Dimethylether   mp. 183.3~184°

NMR in pyridine, TMS, ppm[1357a)] : 4.75 d, J=12 c/s (*trans* two H at 3 and 4), 4.53~4.08 m (3 α, 2 α —CH$_2$—), 3.75, 3.73, 3.59, 3.51 (four CH$_3$O—), 3.11 d, J=16.8, 3.79 d, J=16.8 (1, 1′—CH$_2$—), 2.82~2.61 (3-H)

(C-a)

［J］ Miscellaneous

(1) Schizandrin (CI) and Desoxyschizandrin (CII)[1358)]

≪Occurrence≫   *Schizandra chinensis* BAILL. (チョウセンゴミシ).

1356) Birch, A. J. *et al* : *Soc.*, **1958**, 4471.   1357) Schrecker, A. W. *et al* : *J. A. C. S.*, **77**, 432 (1955).
1357a) Kato, Y : *Chem. Pharm. Bull.*, **12**, 512 (1964).   1357b) Smith, M : *Tetr. Letters.*, **1963**, 991.
1357c) Briggs, L. H. *et al* : *Soc.*, **1937**, 271.
1358) Kochetkov, N. K. *et al* : *Tetr. Letters*, **1961**, (20), 730 ; **1962**, (9), 361.

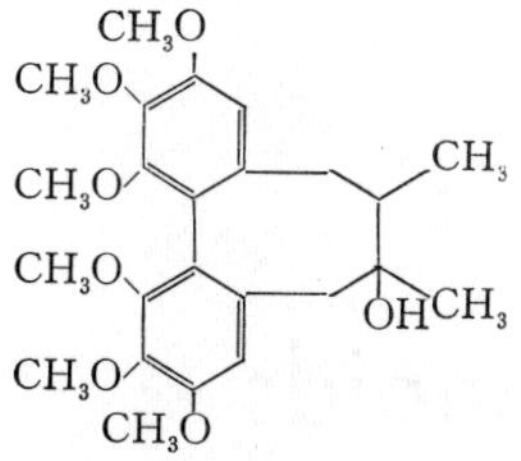

(CI)

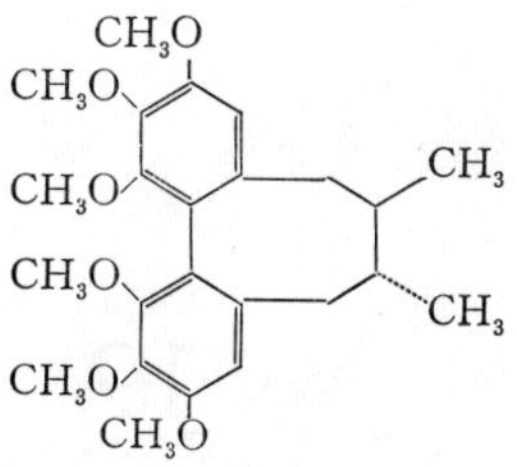

(CII)

(a) **Schizandrin** (CI)

C$_{24}$H$_{32}$O$_7$

IR ν_{max} cm^{-1}: 3495 (OH)

(b) **Desoxyschizandrin** (CII)

mp. 116~ 117°, [α]$_D$ +107° (CHCl$_3$), C$_{24}$H$_{32}$O$_6$

UV λ_{max} mμ (ε): 248 (16500), 273 inf. (3100), 283 (2400) (practical. ident. with schizandrin)

IR no peaks in the region 1800~1600 cm^{-1}

NMR cps: 18 (arom. H), 123, 135 (OCH$_3$), 177, 192 (benzylic –CH$_2$–), 239d, J=6, 250 d, J=6 (>CH–**CH$_3$**)

19 Carotenoids

[A] UV Spectra

(1) Steric isomerization and UV λ_{max}[1359)]

all-*trans*-β–Carotene and all-*trans*-lycopene, all-*trans*–, 6–mono-*cis*–, 1, 6–or 3, 6–di–*cis*–, 1, 3, 5, 7, 9, 11–hexa–*cis*–, 1, 3, 5, 6, 7, 9, 11– hepta-*cis*-lycopenes[1360)]. Lycopene, lycoxanthin, capsanthin, γ–carotene, bixin, β–carotene, kryptoxanthin, zeaxanthin, fucoxanthin, α–carotene, lutein, violaxanthin, azafrin, crocetin, flavoxanthin[1361)~1376)]. Synth.-all-*trans*-β–carotene, 15, 15′–*cis*–β–carotene[1377)], γ–carotene[1378)], lutein[1369)], synth-all-*trans*-lycopene, 15, 15′–*cis* lycopene[1379)] 15, 15′–*cis*–kryptoxanthin, all-*trans*-kryptoxanthin[1380)], flavoxanthin[1376)], methylbixin, and crocetin dimethylether[1381)].

(2) Number of conjugated Double-bonds and UV Absorption Maxima of 1st-Band[1359a)]

(I) all *trans*–α–Carotene

1359) Winterstein, A: *Angew. Chem.* **72**, 902 (1960).
1359a) Winterstein, A. *et al*: *Ber.*, **93**, 2951 (1960).
1360) Hata, K: *Kagaku no Ryoiki* (Nankodo Pub. Inc. Tokyo). **2** (4), 148 (1948).
1361) Gillam, Stern: An Introduction to Electronic Absorption Spectroscopy in Organic Chemistry p. 152~155, 213~218, 236~239 (E. Arnold, 1954).
1362) Karrer, Widmer: *Helv. Chim. Acta*, **11**, 751 (1928).   1363) Zechmeister: *Ber.*, **69**, 422 (1936).
1364) Zechmeister: *Ann.*, **454**, 54 (1927), **455**, 70 (1927), **465**, 288 (1928).
1365) Kuhn *et al*: *Ber.*, **66**, 407 (1933).   1366) Karrer *et al*: *Helv. Chim. Acta*, **12**, 741, 754 (1929).
1367) Kuhn: *Ber.*, **64**, 1349 (1931).   1368) Kuhn *et al*: *ibid.*, **66**, 1746 (1933).
1369) Kuhn *et al*: *Hoppe Seyler.* **197**, 161 (1931).   1370) Heilbron *et al*: *Biochem. J.*, **29**, 1369 (1935).
1371) Karrer *et al*: *Helv. Chim. Acta*, **16**, 641 (1933).   1372) Kuhn *et al*: *Hoppe Seyler*, **213**, 188 (1931).
1373) Kuhn *et al*: *Ber.*, **64**, 326 (1931), Karrer *et al*: *Helv. Chim. Acta*, **14**, 1044 (1931).
1374) Kuhn *et at*: *Ber.*, **66**, 883 (1933).   1375) Karrer *et al*: *Helv. Chim. Acta*, **16**, 643 (1933).
1376) Kuhn *et al*: *Hoppe Seyler*, **213**, 192 (1932).   1377) Isler *et al*: *Helv. Chim. Acta*, **39**, 249 (1956).
1378) Kuhn, Brockmann: *Ber.*, **66**, 407 (1933).   1379) Isler *et al*: *Helv. Chim. Acta*, **39**, 463 (1956).
1380) Isler *et al*: *ibid.*, **40**. 456 (1957).   1381) Isler, O. *et al*: *Helv. Chim. Acta*, **40**, 1242 (1957).

[A] UV Spectra

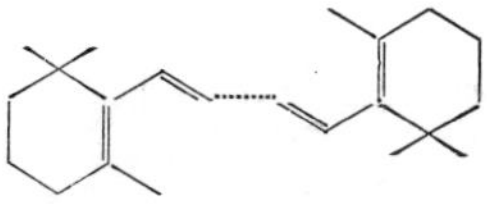

(II) β–carotene 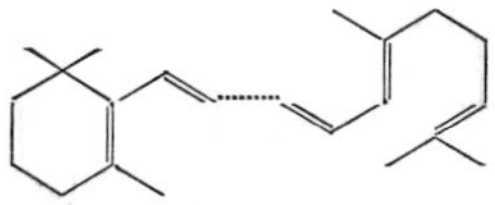   (III) γ–carotene 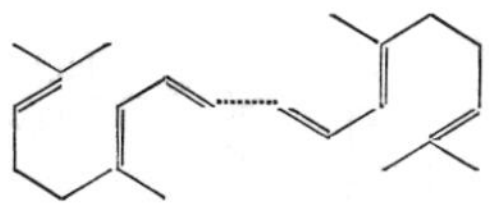   (IV) Lycopene

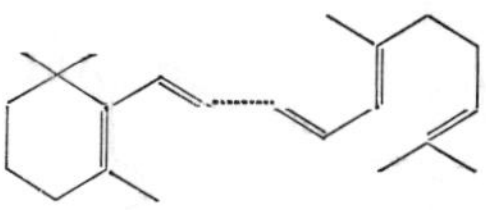

(V) Torulin 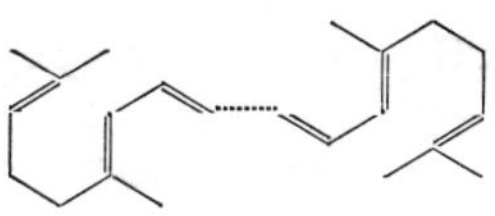   (VI) Monodehydro-lycopene 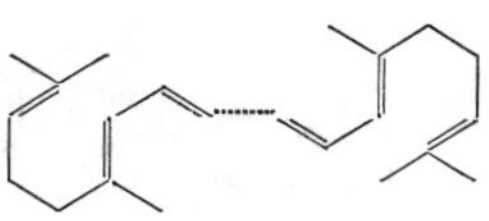   (VII) Bisdehydro-lycopene

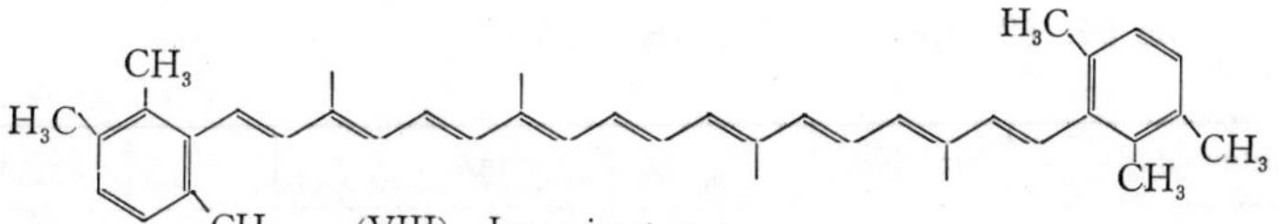

(VIII) Isorenieratene

Table 89 UV Absorption Maxima of natural Carotenoid Hydrocarbons (in CS_2)

(Winterstein *et al.*)

Carotenoid	Str.No	Number of conj. double bonds		$\lambda_{max}^{CS_2} m\mu$
		Aliphatic	Ring	(1 st Band)
α–Carotene	I	9 +	1	510
β–Carotene	II	9 +	2	520
γ–Carotene	III	10 +	1	533
Lycopene	IV	11		548
Torulin	V	12 +	1	563
Monodehydrolycopene	VI	13		574
Bisdehydrolycopene	VII	15		601[Note]
Isorenieratene	VIII	9 +	6	520

[Note] See the item [C], (12)

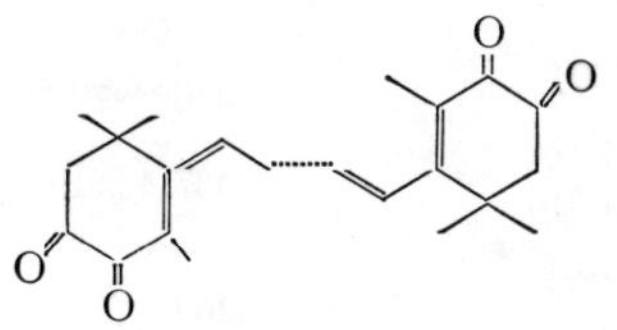

(IX) Astacin 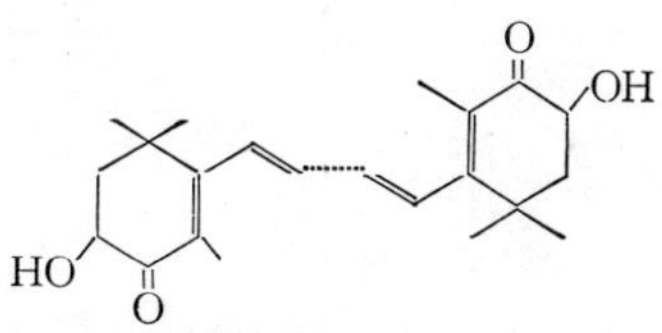   (X) Astaxanthin

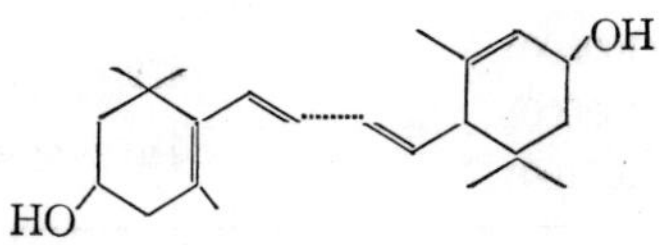

(XI) Luteine 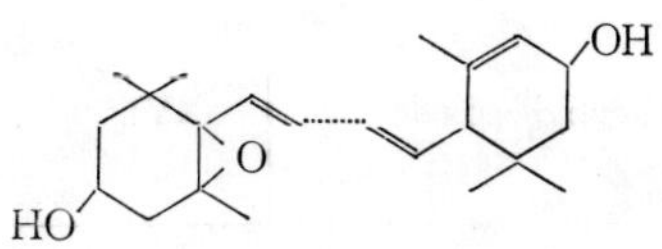   (XII) Lutein–5,6–epoxide

— 465 —

(XIII) Taraxanthin, stereoisomer of
(XIV) Trollixanthin

(XV) Violaxanthin

(XVI) Xanthophyllepoxide

(XVII) Trollichrom

(3) Carotenoids of *Scenedesmus* (Algae)[1382]

Table 90 UV Spectra of Carotenoid Pigments of *Scenedesmus*

(Iwata *et al.*)

Carotenoids	Str. No.	$\lambda_{max}^{hexane}m\mu$	$\lambda_{max}^{CS2}m\mu$
α–Carotene	I	~420, 446, 475	478, 506
β–Carotene	II	~423, 450, 478	~450, 483, 510
Poly-*cis*-lycopene	IV	~423, 445, 473	~450, 476, 507
Astacin	IX	462	498~501
Astaxanthin	X	462	498~503
Lutein	XI	423, 446, 476	445, 475, 506
Lutein–5,6–epoxide	XII	~424, 443, 473	~445, 471, 503
Taraxanthin	XIII	420, 443, 473	443, 470, 502
Violaxanthin	XV	420, 443, 473	443, 472, 502

(4) Epoxycarotenoids[1383]

Table 91 UV Spectra of Epoxycarotenoids (in benzene)

(Karrer *et al.*)

Carotenoids	Str. No.	$\lambda_{max}\,m\mu\ (\epsilon)$	$\lambda_{min}\,m\mu\ (\epsilon)$
Taraxanthin	XIII	485 (132300) 455 (138600) 428.5(91700)	472 (74600) 438 (82100)
Violaxanthin	XV	483 (128400) 453.5(134400) 428 (88500)	470 (77000) 437 (81900)
trans-Trollixanthin	XIV	482 (121800) 454 (127700) 427 (84400)	469 (61000) 437 (71700)
Xanthophyllepoxide	XVI	482.5(100800) 456 (133000) 430 (105100)	472 (73700) 440 (92800)

1382) Iwata, I. *et al*: *A.B.C.*, **27**, 259 (1963).
1383) Karrer, P. *et al*: *Helv. Chim. Acta*, **40**, 69 (1957).

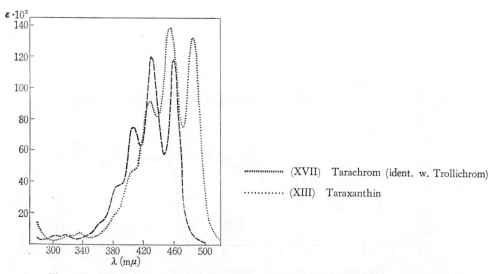

Fig. 198 UV Spectra of Tarachrom (identical with that of Trollichrom) and Taraxanthin (in benzene) (Karrer *et al.*)

.... (XVII) Tarachrom (ident. w. Trollichrom)

.... (XIII) Taraxanthin

〔B〕 IR Spectra

β–Carotene, 15,15′–*cis*–β–carotene[1384], neo–β–carotene U, neo–β–carotene B[1385], all-*trans*-zeaxanthin, neozeaxanthin A and neozeaxanthin B[1386].

Lycopene[1386a], all-*trans*-lycopene, neo-lycopene A and prolycopene[1385].

all-*trans*–α–Carotene, neo–α–carotene U, neo–α–carotene B[1385].

all-*trans*–γ–Carotene, neo–γ–carotene, mono *cis*–γ–carotene pro–γ–carotene poly–*cis*–γ–carotene all-*trans*-crocetin dimethylester, and mono-*cis*-crocetin dimethylester[1385].

all-*trans*-Dimethylbixin, natural dimethylbixin[1385], natural stable methylbixin and natural stable crocetin-dimethylester[1386b]

〔C〕 Natural and Synthetic Carotenoids

(1) Synth. all-*trans*-α-Carotene (I)[1387]

mp. 160~162°, $C_{40}H_{56}$

UV $\lambda_{max}^{petrolether}$ mμ (E$_{1cm}^{1\%}$: 422(1850), 446(2785), 474(2515)

(2) Synth. all-*trans*-γ-Carotene (III)[1387]

mp. 152~153.5°, $C_{40}H_{56}$

1384) Isler *et al*: *Helv. Chim. Acta*, **39**, 249 (1956).

1385) Lunde, Zechmeister: *J.A.C.S.*, **77**, 1647 (1955).

1386) Isler *et al*: *Helv. Chim. Acta*, 2041 (1956).

1386a) Isler *et al*: *ibid*, **39**, 463 (1956).   1386b) Isler *et al*: *Helv. Chim. Acta*, **40**, 1242 (1957).

1387) Isler, O. *et al*: *Helv. Chim.* Acta, **44**, 985 (1961).

UV $\lambda_{max}^{Petrolether}m\mu$ $(E_{1cm}^{1\%})$: 437(2055), 462(3100), 494(2720).

IR See Fig. 199.

NMR in CS_2, 56.4 Mc, TMS, τ: See Fig. 200

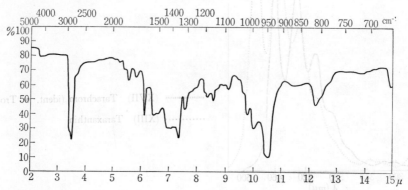

Fig. 199 IR Spectrum of γ–Carotene in $CHCl_3$ (Isler *et al.*)

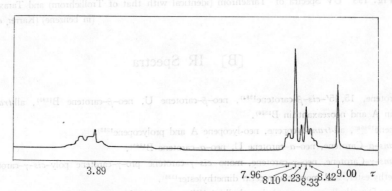

Fig. 200 NMR Spectrum of γ–Carotene (Isler *et al.*)

(3) Torulin (3′, 4′-Dehydro-γ-carotene) (XVIII)[1388]

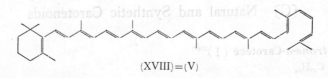

(XVIII)=(V)

《Occurrence》 *Torula rubra.*   mp. 183~184°, $C_{40}H_{54}$

UV $\lambda_{max}^{Petrolether}m\mu$ $(E_{1cm}^{1\%})$: 460(2315), 484(3240), 518(2680)

(4) Synth. (+)- and (−)-ε-Carotenes (XIX)[1389]

(a) (+)-ε-Carotene

mp. 196°, $[\alpha]_D$ +806° (C_6H_6), $C_{40}H_{56}$

1388) Isler, O. *et al*: *Helv. Chim. Acta*, **44**, 994 (1961).   1389) Karrer, P. *et al*: *ibid.*, **41**, 32 (1958).

(XIX)

UV $\lambda_{max}^{cyclohexane}$ mμ (logε): 268(4.55), 331(3.91), 420(5.01), 445(5.19), 475(5.18)

(b) (−)-ε-Carotene

mp. 197~198°, $[\alpha]_D$ −786° (C$_6$H$_6$), C$_{40}$H$_{56}$

UV $\lambda_{max}^{cyclohexane}$ mμ (logε): 268(4.53), 332(3.91), 420(5.00), 445(5.18), 475(5.17)

(5) **Kryptoxanthin** (XX)[1390]

≪Occurrence≫ *Carica papaya* L., *Physalis Franchetti*

mp. 158.5~159°, C$_{40}$H$_{56}$O

UV $\lambda_{max}^{petrolether}$ mμ (E$_{1cm}^{1\%}$): 452(2386), 480(2080)

IR See Fig. 201

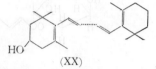

(XX)

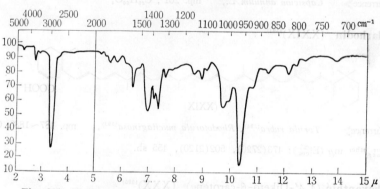

Fig. 201 IR Spectrum of natural Kryptoxanthin in CS$_2$ (Isler *et al.*)

(6) **Capsanthin** (XXI) **and Kryptocapsin** (XXII)

R=OH (XXI) Capsanthin[1391]
R=H (XXII) Kryptocapsin[1392]

(a) Capsanthin (XXI)

≪Occurrence≫ *Capsicum annuum* (トウガラシ). *Lilium tigrinum*, *Asparagus officinalis*.

1390) Isler, O. *et al* : *Helv Chim Acta*, **40**, 456 (1957).
1391) Karrer, P. *et al* : *ibid.*, **44**, 1257 (1961), **47**, 741 (1964).
1392) Cholnoky, L. *et al* : *Tetr. Letters*, **1963**, 1857.

mp. 175~176°, $[\alpha]_{cd}$ −64~−70° (CHCl$_3$), C$_{40}$H$_{56}$O$_3$

NMR: almost identical with that of kryptocapsin

(b) Kryptocapsin (XXII)[1392]

≪Occurrence≫   *Capsicum annuum*

mp. 160~161°, C$_{40}$H$_{56}$O$_2$

UV λ_{max} mμ (ε): 520 inf (87000), 486(112000)

IR ν_{max} cm^{-1}: 1664, 1582, 971

NMR in CDCl$_3$, τ: 9.16, 8.97, 8.80, 8.65, 8.29, 8.02 (relative intensities 1: 2: 1: 1: 1: 4)

(7) Capsorubin (XXIII)

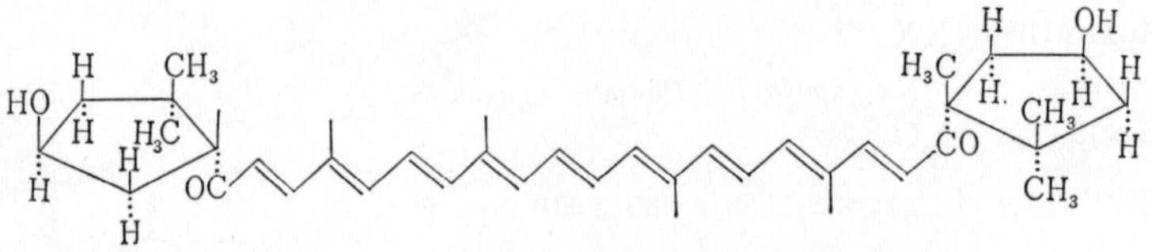

(XXIII)[1022]

≪Occurrence≫   *Capsicum annuum* L.,   mp. 201°, C$_{40}$H$_{60}$O$_4$

(8) Torularhodin (XXIX)[1393]

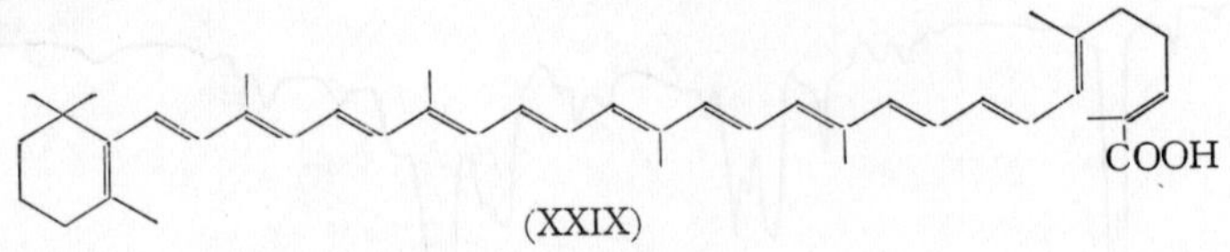

(XXIX)

≪Occurrence≫   *Torula rubra*[1394], *Rhodotorula mucilaginosa*[1393],   mp. 187~188°, C$_{41}$H$_{52}$O$_2$

UV $\lambda_{max}^{Petr.\ ether}$ mμ (E$_{1cm}^{1\%}$): 473(2720), 502(2120), 455 sh.

(9) Canthaxanthin (4, 4′-Diketo-β-carotene) (XXX)[1395]

≪Occurrence≫   *Cantharellus cinnabarinus*

Synth. all-*trans* (XXX) mp. 216~217°, C$_{40}$H$_{52}$O$_2$

UV $\lambda_{max}^{petrolether}$ mμ (E$_{1cm}^{1\%}$): 465~467(2200)

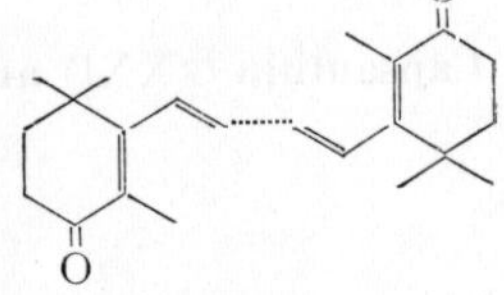

(XXX)

(10) Rhodoxanthin (XXXI)[1396]

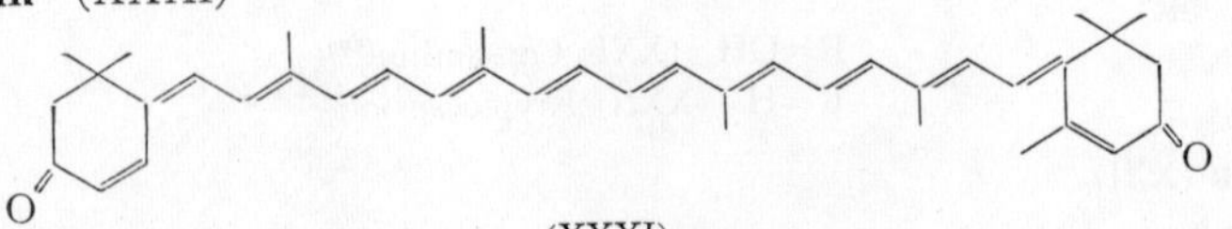

(XXXI)

1393) Isler, O. *et al*: *Helv. Chim. Acta*, **42**, 864 (1959).

1394) Karrer, P. *et al*: *ibid.*, **26**, 2109 (1943), **28**, 795 (1945). **29**, 355 (1946).

1395) Isler, O. *et al*: *ibid.*, **42**, 841 (1959).

1396) Karrer, P. *et al*: *Helv. Chim. Acta*, **42**, 466 (1959).

≪Occurrence≫   *Potamogeton natans, Taxus baccata, Cryptomeria japonica*

Synth.   mp. 215°, $C_{40}H_{50}O_2$

UV λ_{max}^{Hexane} m$\mu(\varepsilon)$: 490(140200), 516(104900)

(11) Bacterial Carotenoids[1397]

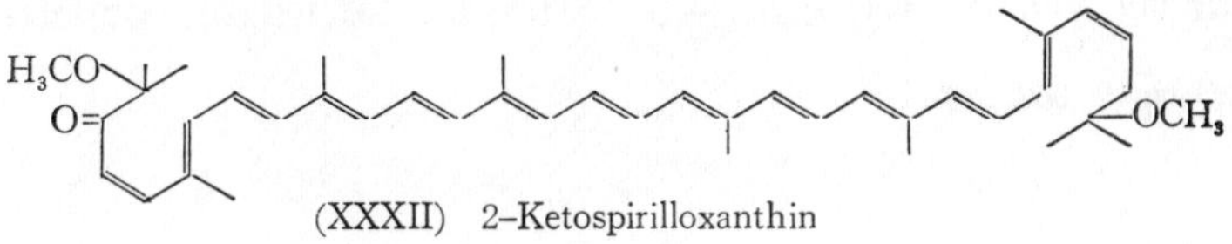

(XXXII) 2–Ketospirilloxanthin

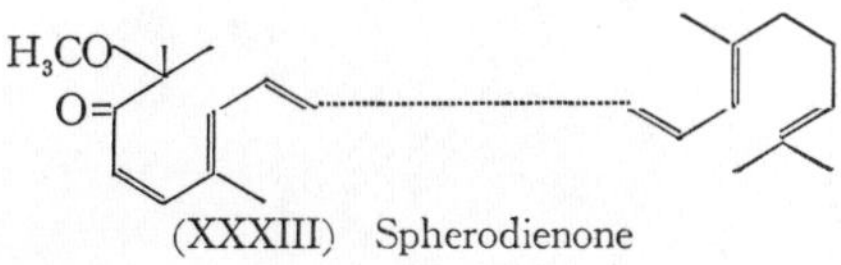

(XXXIII) Spherodienone

≪Occurrence≫   *Rhodopseudomonas*

(12) Arylic Carotenoids[1398) 1399]

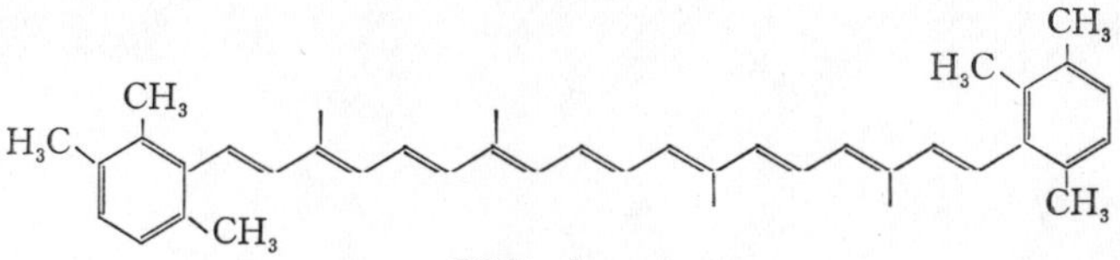

(VIII) Isorenieratene

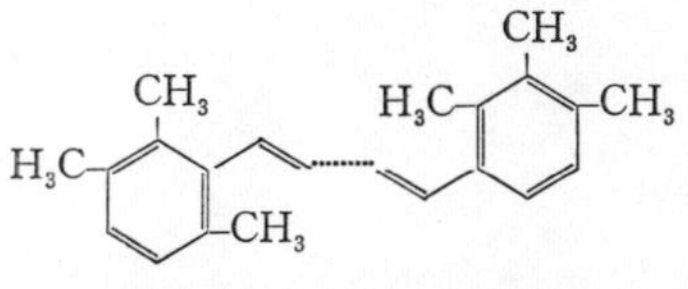

(XXXIV) Renieratene

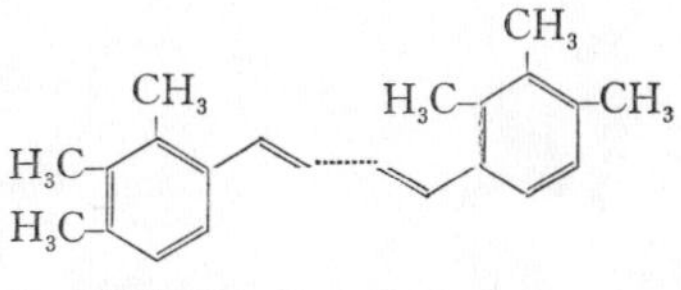

(XXXV) Renierapurpurin

≪Occurrence≫   *Reniera japonica* (海綿)

(a) **Isorenieratene** (VIII)

mp. 230° (uncorr.), Synth. 237~238°(corr), $C_{40}H_{48}$

UV λ_{max}^{C6H6} m$\mu(\varepsilon)$: 493(106000), 465(123000), 443 inf. (95000); λ_{max}^{CS2} mμ: 520, 484, 452

IR ν_{max} cm^{-1}: 964 (*trans*–CH=CH–), 815, 810 (two adjacent ring $\overset{H\ H}{C=C}$)

NMR   in CDCl$_3$, 60 Mc, TMS, τ: 8.02, 7.93, 7.81, 7.72 (1: 1: 1: 2) ($C-CH_3$)

(b) **Renieratene** (XXXIV)

mp. 191~192°[1398], 184~185°[1399], $C_{40}H_{48}$

1397)   Jensen, S. L: *Acta Chem. Scand.*, **17**, 303, 489 (1963).

1398)   Cooper, R.D.G. *et al*: *Soc.*, **1963**, 5637.

1399)   Yamaguchi, M: *Bull. Chem. Soc. Japan*, **30**, 111, 979 (1957), **31**, 51, 739 (1958), **32, 1171 (1959), 33,** 1560, (1960).

UV λ_{max}^{C6H6} mμ: 507, 476, 457; λ_{max}^{CS2} mμ (ϵ): 532(103000), 497(113000), 467(84000)

IR ν_{max}^{nujol} cm^{-1}: 926, 802

(c) **Renierapurpurin** (XXXV)

mp. 164°, $C_{40}H_{52}$

UV λ_{max}^{C6H6} mμ: 519, 487, 464; λ_{max}^{CS2} mμ (ϵ): 544(8800), 504(109000), 477(83000)

IR ν_{max}^{nujol} cm^{-1}: 962, 802

20 Alkaloids

Principal IR absorptions of alkaloids such as, streching vibration of (1) hydroxyl (ν_{O-H}), (2) imino group (ν_{N-H}), (3) carbonyl group ($\nu_{C=O}$) and other are described[1400~1404].

References on biosynthesis of alkaloids are summarized in [1406a)~c)], some of them are also indicated in each items of following description.

[A] Exocyclic Amines

(1) (+)-**Muscarine** (I)[1405]

≪Occurrence≫ *Amanita muscarina* (ベニテングダケ)

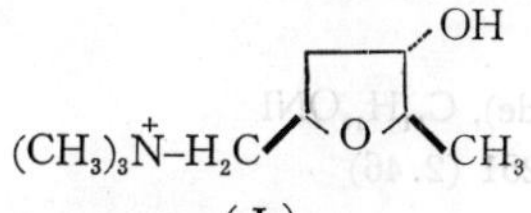

(I)

Chloride　mp. 181~182°, $[\alpha]_D$ +6.7° (H_2O), $C_9H_{20}O_2N\cdot Cl$

IR　See Fig. 202

1400) Marion, Ramsay, Jones: *J. A. C. S.*, **73**, 305 (1951).
1401) Pelletier, S. W., Jacobs, W. A: *J. A. C. S.*, **78**, 1914 (1956).
1402) Nishikawa, Perkin, Robinson: *Soc.*, **125**, 657 (1924).
1403) Wieland, Hsing: *Ann.*, **526** 188 (1936).
1404) Pleat, G. B., Harley, J. H., Wiberley, S. E: *J. Am. Pharm. Assoc. Sci. Ed.*, **40**, 107 (1951).
1405) Frydman, B. *et al*: *Tetrahedron*, **18**, 1063 (1962).
1406) Mothes, K. *et al*: *Angew. Chem.*, **75**, 357 (1963).
1406a) Barton, D. H. R: *Proc. Chem. Soc.*, **1963**, 293.　　1406b) Battersby, A. R: *ibid.*, **1963**, 189

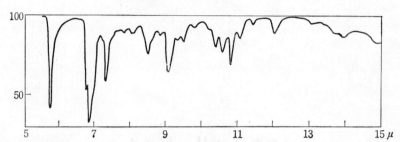

Fig. 202　IR Spectrum of Muscarine chloride (in nujol) (Eugster *et al.*)

(2)　(+)-Acanthoine (II) and **(+)-Acanthoidine** (III)[1406]

$$CH_3O-\text{〈〉}-CH=CH-CH-CH=CH-CH_2$$

（structure II with NH, CH, NH·HCl groups）

（structure III: $CH_3O-\text{〈〉}-CH_2-CH_2-CH-CH_2-CH_2-CH_2$ with NH, CH, NH·HCl groups）

(II)　　　　　　　　　　　(III)

≪Occurrence≫　　*Cardus acanthoides.*

(a)　(+)-**Acanthoine** (II)–2HCl　　mp. 192~193°, $[\alpha]_D$ +7.0°, $C_{16}H_{22}O_2N_4\cdot2HCl$

(b)　(+)-**Acanthoidine** (III)–2HCl　　mp. 249~251°, $[\alpha]_D$ +6.5°, $C_{16}H_{26}O_2N_4\cdot2HCl$

(3)　Taxine (IV)

　Taxine (IV)＝Taxinine (V)＋$HN(CH_3)_2$

　≪Occurrence≫　　*Taxus baccata* （イチイ）

　　　　　　　　mp. 105~111°, $[\alpha]_D$ +32~+35°

　　　　　　　　(EtOH),　$C_{37}H_{51}O_{10}N$

　　　　　　　　See the item (11) Diterpenoid [F], (14)

(V) Taxinine

(4)　Candicine (VI)[1407a]

　≪Occurrence≫　　*Phellodendron amurense* Rupl.

　　　　　　　　（オウバク）

　　　　　　　　mp. 228~229° (iodide), $C_{11}H_{18}ONI$

　UV λ_{max} mμ (log ε): 278 (3.28), 251 (2.46)

(VI)

(5)　Thalicthuberine (VII)[1407b]

　≪Occurrence≫　　*Thalictrum thunbergii* DC

　　　　　　　　（アキカラマツ）

　　　　　　　　mp. 126~127°, $[\alpha]_D$ ±0°,

　　　　　　　　$C_{21}H_{23}O_4N$

　UV λ_{max} mμ (log ε): 261 (4.84), 285 (4.50), 310

　　(4.32), 345 (3.50); λ_{min}: 230 (4.32), 278 (4.49),

　　300 (4.23), 340 (3.39);

　IR ν_{max}^{nujol} cm^{-1}: 930, 953, 1045 (–O–CH$_2$–O–)

(VII)

1407)　Tomita, T. *et al*: *Y. Z.*, **80**, 1300 (1960).

1407a)　Fujita, E. *et al*: *ibid.*, **79**, 1252 (1959).　　1408)　Sadtler Card No. 7161.

〔B〕 Imidazole Alkaloids

(1) Pilocarpine (IR)[1408]

(2) Chaksine (I)[1407]

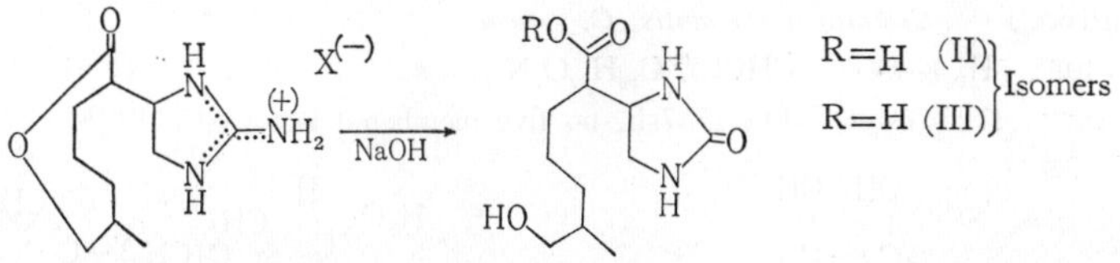

(I)

≪Occurrence≫ *Cassia absus* L.

(a) Chaksine (I)-iodide pKa 11, $C_{11}H_{20}O_2N_3I$

IR ν_{max}^{KBr} cm^{-1}: 1720, 1670, 1600, 1572

(b) Ureido–hydroxy acid (II)

mp. 122~123°, $C_{11}H_{20}O_4N_2$

Methylate   IR ν_{max}^{CCl4} cm^{-1}: 1740 (ester), 1710 (five membered cyclic urea)

(c) Ureido–hydroxy acid (III)

mp. 147°, $C_{11}H_{20}O_4N_2$

NMR of Methylate ppm: 4.40 (2NH), 6.01 m, 6.62 m (—CH$_2$–N), 6.31 s (OCH$_3$ overlap to —O–CH$_2$–), 7.52 (OH and —CH–COO–), 9.13 d (>CH–CH$_3$)

〔C〕 Oxazole Alkaloids

(1) Annuloline (I)[1410][1411]

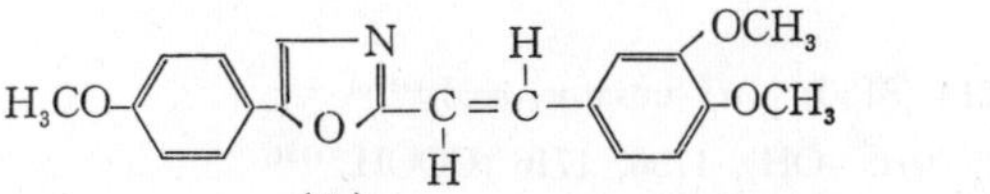

(I)

≪Occurrence≫ *Lolium multiflorum (Graminae)*,   mp. 105~106°, $C_{20}H_{19}O_4N$

Hydrochloride   mp. 174~177°, picrate   mp. 216~218°

UV $\lambda_{max}^{cyclohexane}$ mμ (log ε): 354 (4.48); λ_{min}: 285 (3.85).

IR $\lambda_{max} \mu$: absence of NH, OH and C=O, 10.35 (*trans*–disubstituted double bond)

NMR in CCl$_4$, 60Mc, TMS cps: 226 (9H), 389~453 (10 arom. H and/or olefinic H)

1409) Wiener, K. *et al*: *Chem & Ind.*, **1962**, 95; *J. A. C. S.*, **80**, 1521 (1958)
1410) Karimoto, R. S. *et al*: *Tetr. Letters* **1962**, 83.
1411) Axelrod, B. *et al*: *J. Org. Chem.*, **23**, 919 (1958).

[D] Pyrrolidine and Pyrrolizidine Alkaloids

(1) Hygrine, Cuscohygrine Spectra[1412]

(2) Pyrrolizidine Alkaloids of *Senecio* and *Crotalaria* Spp. [1413]~[1415]

(a) Monocrotaline (I) (Retronecine+Monocrotalic acid)[1413a]

《Occurrence》 *Crotalaria spetabilis, C. retusa*

m.p. 197~198°, $[\alpha]_D$ −54.7° (CHCl₃), $C_{16}H_{23}O_6N$

IR ν_{max} cm⁻¹: 1725 (ester C=O), 1737sh, no five membered lactone C=O[1413b]

(I) Monocrotaline   (II) Mikanoidine

(b) Mikanoidine (II) (Platynescine+Mikanecic acid)

《Occurrence》 *Senecio mikanoides*, $C_{18}H_{23}O_4N$[1413c], revised from former formula $C_{21}H_{29}O_6N$

Mikanecic acid $C_{10}H_{12}O_4$[1413c]

UV λ_{max}^{EtOH} mμ (ε): 216 (8540)   (conjtd. ⌐⌐)

IR ν_{max}^{nujol} cm⁻¹: 923 (α-methylenic acid), 900 (>C=CH₂), 950 (CH₂-C=CH-COOH), 960 (carboxyl OH), 835 (>C=C<H), 1690, 1680 (2-COOH), 1650, 1638sh (C=C), 2510, 2600 (carboxylic O-H)

(c) Seneciphylline (III) (Retronecine+Seneciphyllic acid)[1413d]

《Occurrence》 *Senecio platyphyllus*

mp. 232~233° $[\alpha]_D$ −56° (CHCl₃), $C_{18}H_{23}O_5N$

Seneciphyllic acid

mp. 115°, $C_{10}H_{14}O_5$

UV λ_{max}^{EtOH} mμ (ε): 214 (8130)   ($\alpha\beta$-unsatur. acid)[1413e]

IR ν_{max} cm⁻¹: 3452, (*tert.* -OH), 1736, 1716 (COOH)[1413f]

(III) Seneciphylline

1412) Willits, *el al*: *Anal. Chem.*, **22**, 430 (1950).
1413) Leonard, N. J., Manske R. H. F: The Alkaloids, Chemistry and Physiology VI, p. 35~121 (Academic Press, 1960).
1413a) Leonard, N. J. *et al*: *J. A. C. S.*, **71** 1760 (1949).   1413b) Adams, R. *et al*: *ibid.*, **74**, 5612 (1952).
1413c) Adams, R. *et al*: *ibid.*, **79**, 166 (1957).
1413d) Prelog, V. *et al*: *Helv. Chim. Acta*, **36**, 308 (1953).   1413e) Heilbron, I. *et al*: *Soc.*, **1947.**, 1586.
1413f) Adams, R. *et al*: *J. A. C. S.*, **74**, 700 (1952).

(d) Riddelline (IV) (Retronecine+Riddellic acid)

《Occurrence》 *Senecio riddelli*, etc., *Crotalaria juncea* L.

mp. 195~196°, $[\alpha]_D$ —109° (CHCl₃), $C_{18}H_{23}O_6N$

Riddellic acid $C_{10}H_{14}O_6$[1413g]

UV λ_{max}^{EtOH} mμ (ε): 215 (8300) (similar to seneciphyllic acid)

IR ν_{max} cm⁻¹: 1695, 1720 (conj. and uncon. C=O), 1637 (C=C)

(IV) Riddelline

(e) Rosmarinine (V) (Rosmarinecine+Senecic acid)[1413j]

《Occurrence》 *Senecio adnatus* DC., *S. brachypodus* DC.

mp. 209°, $[\alpha]_D$ —120° (CHCl₃), $C_{18}H_{27}O_6N$

UV λ_{max}^{EtOH} mμ (ε)=218 (6100)

Senecic acid $C_{10}H_{16}O^5$

Senecic acid → Lactone $C_{10}H_{14}O_4$ → Integerrinecic acid mp. 156°

UV λ_{max}^{H2O} mμ (ε): 215 (6195)[1413h] ; λ_{max}^{H2O} mμ (ε): 215 (4140)[1413i]

(V) Rosmarinine

(f) Integerrimine (VI) (Retronecine+Integerrinecic acid)[1413k]

《Occurrence》 *Crotalaria incana* L.,

mp. 172~172.5° $[\alpha]_D$ +4.3° (CHCl₃), $C_{18}H_{25}O_5N$

UV λ_{max}^{EtOH} mμ (ε): 212 (10900)

Intererrinecic acid $C_{10}H_{16}O_5$

UV λ_{max}^{EtOH} mμ (ε): 214 (9021)[1413h] ; λ_{max}^{H2O} mμ (ε): 218 (9333)

(VI) Integerrimine

(VII) Trichodesmine

(VIII) Retrorsine

(g) Trichodesmine (VII) (Retronecine+Trichodesmic acid)[1413l]

《Occurrence》 *Crotalaria juncea* L.

mp. 160~161°, $[\alpha]_D$ +38° (EtOH), $C_{18}H_{27}O_6N$

IR ν_{max}^{nujol} cm⁻¹: 1735 (normal ester C=O)

1413g) Adams, R. *et al*: *ibid.*, **75**, 4638 (1953). 1413h) Adams, R. *et al*: *ibid.*, **75**, 4631 (1953).
1413i) Kropman, M. *et al*: *Soc.*, **1949**, 2852. 1413j) Cram, D. J. *et al*: *J. A. C. S.*, **74**, 5828 (1952).
1413k) Leonard, N. J. *et al*: *J.A.C.S.*, **69**, 690 (1947). 1413l) Adams, R. *et al*: *ibid.*, **78**, 1922 (1956).

(h) **Retrorsine** (VIII) (Retronecine+Isatinecic acid)

≪Occurrence≫ *Senecio* spp., mp. 216~216.5°, 207~208°, $[\alpha]_D$ −48.6° (CHCl$_3$), C$_{18}$H$_{25}$O$_6$N

UV λ_{max}^{H2O} mμ (ε): 217.5 (7100)

Isatinecic acid[1413m] C$_{10}$H$_{16}$O$_6$

UV λ_{max}^{H2O} mμ (ε): 218 (4720) (*cis*-configuration)[1413n] ; 218 (9400) (Retronecic acid, *trans*-isomer)

(See UV of intererrinecic acid)

(i) **Junceine** (IX) (Retronecine+Junceic acid)

≪Occurrence≫ *Crotalaria juncea* L., mp. 191~192°, $[\alpha]_D$ −3° (pyridine), C$_{18}$H$_{27}$O$_7$N

Junceic acid C$_{10}$H$_{16}$O$_6$

IR ν_{max}^{nujol} cm^{-1}: 3340, 3480, 1710, 1725, 1740[1413o]

(j) **Biosynthesis of Necic Acids**[1414]

(k) **Pyrrolizidine Alkaloid of *Crotalaria* spp.** (X)~(XII)[1415)1416]

(IX) Junceine

(X) 1-Methoxymethyl-1,2-epoxypyrrolizidine
(*C. trifoliastrum* WILLD.)

(XI) Fulvine (XII) Crispatine
(*C. fulva* ROXB. (*C. crispata*))

(3) **Dendrobine** (XIII)[1417)1418)1418a]

≪Occurrence≫ *Dendrobium* spp. (金石斛)

mp. 134.5~136°, $[\alpha]_D$ −50.4° (EtOH), pKa' 7.80, C$_{16}$H$_{25}$O$_2$N

IR $\nu_{max}^{C=O}$ cm^{-1}: 1767 (KBr), 1763 (CHCl$_3$), 1760 (nujol), no hydroxyl

NMR (1) in CDCl$_3$, 60Mc, τ: 7.48 (≫N-CH$_3$), 5.15 (≫CHOCO),

8.59s (≫C-CH$_3$), 8.94d, 9.05d (-CH≪CH$_3$/CH$_3$), (2) in CDCl$_3$, 60Mc, τ:

9.02d (6H), 8.58s (3H), 7.48s (≫N-CH$_3$), 7.32t (1H), 7.21d (1H),

6.81t (1H), 5.16q (1H)

Mass m/e, 263, 235, 220, 210, 184, 96

(XIII)[1418a]

1413m) Areshkina, L. Ya: *Biokhimiya*, **16**, 461 (1951).   1413n) Christie. S. M. H. *et al*: *Soc.*, **1949**, 1700

1413o) Adams, R *et al*: *J. A. C. S.*, **78**, 1926 (1956).   1414) Hughes, C. *et al*: *Soc.*, **1962**, 34

1415) Culvenor, C. C. J. *et al*: *Australian J. Chem.*, **16**, 131 (1963); *Index Chem.*, **9** (3), 27058 (1963).

1416) Schoental, R: *ibid.*, **16**, 233, 239 (1963); *Index Chem.* **9**, (3), 28589, 28590 (1963).

1417) Okamoto, T. *et al*: *Chem. Pharm. Bull.*, **12**, 506 (1964); *7th Symposium on the Chemistry of Natural Products*, p. 62 (Fukuoka, 1963).

1418) Hirata, Y. *et al*: *ibid.*, p. 66 (Fukuoka, 1963).

1418a) Inubuse, Y. *et al*: *Y. Z.*, **83**, 1184 (1963).

〔E〕 Pyridine, Piperidine and Quinolizidine Alkaloids

(1) **Nicotine** (UV)[1412], **Piperidine, Anabasine** (IR)[1419], and **Aphylline** (IR)[1420]

(2) **Tobacco Alkaloids**[1421]

(I) Nicotine　　　　(II) Nicotyrine　　　　(III) Nornicotine　　　　(IV) Myosmine

≪Occurrence≫　　*Nicotiana tabacum* & other *N.* spp.

〔Note〕 Myosmine was isolated from the smoke.

(a) *l*-Nicotine (I)

bp. 246°, $[\alpha]_D$ −169.3°, d 1.0099, n_D 1.5282,　　$C_{10}H_{14}N_2$, picrate mp. 224°

UV λ_{max}^{EtOH} mμ ($E_{1cm}^{1\%}$): 262 (17.9); λ_{min}: 232 (6.48); $\lambda_{max}^{Acid-EtOH}$ mμ ($E_{1cm}^{1\%}$): 260 (29.7); λ_{min}: 231 (5.72)

(b) **Nicotyrine** (II)

bp. 280°, d 1.124, $C_{10}H_{10}N_2$, picrate mp. 171°

UV λ_{max}^{EtOH} mμ ($E_{1cm}^{1\%}$) 288 (62.1); λ_{min}: 251 (26.6); $\lambda_{max}^{Acid-EtOH}$ mμ ($E_{1cm}^{1\%}$): 310 (69.1), 244 (54.5);

λ_{min}: 264 (12.9), 226 (18.1)

(c) **Nornicotine** (III)

bp. 267°, $[\alpha]_D$ (*l*−) −88.8°, (*d*−)+88.8°, d 1.0737 (*l*), $_D$ 1.5378 (*l*), $C_9H_{12}N_2$, picrate mp. 192° (*l*)

UV λ_{max}^{EtOH} mμ ($E_{1cm}^{1\%}$): 262 (19.7); λ_{min}: 233 (7.13); $\lambda_{max}^{Acid-EtOH}$ mμ ($E_{1cm}^{1\%}$): 260 (33.7); λ_{min}: 231 (6.19)

(d) **Myosmine** (IV)

mp. 45°, $C_9H_{10}N_2$, picrate mp. 85°

UV λ_{max}^{H2O} mμ ($E_{1cm}^{1\%}$): 266 (26.6), 234 (77.7); λ_{min}: 261 (26.2) 214 (42.2); $\lambda_{max}^{Acid-EtOH}$ mμ ($E_{1cm}^{1\%}$): 262

(36.1), 226 (42.7); λ_{min}: 244 (32.6)

(3) **N-Methylhalfordinium chloride** (V)[1422]

$$\left[\text{(pyridyl-oxazole)} \quad \text{O—CH}_2\text{—CH·OH—C(OH)}=\text{(CH}_3)_2 \right]^{\oplus} \text{Cl}^{\ominus} \quad (V)$$

≪Occurrence≫　　*Halfordia scleroxyla*,　　mp. 210° (decomp.), $C_{20}H_{23}O_4N_2^+Cl^-$, picrate mp. 143~198°

IR ν_{max}^{CHCl3} cm⁻¹: 3602, 3587 (OH), no N–H, C=O absorption

1419) Marison, Ramsay, Jones: *J. A. C. S.*, **73**, 305 (1951).

1420) Bohlmann, *et al*: *Ber.*, **90**, 653 (1957).

1421) Swain, M. L. *et al*: *J. A. C. S.*, **71**, 1341 (1949).

1422) Crow, W. D. *et al*: *Tetr. Letters*, **1963**, 85.

1422a) Lavie, D. *et al*: *Chem. & Ind.*, **1963**, 781.　　1422b) Govindachari, T. R. *et al*: *Soc.*, **1957**, 551.

1422c) Proskurnia, N. F. *et al*: *J. Gen. Chem.* (*U. S. S. R.*), **14**, 1148 (1944); *C. A.*, **40**, 7213 (1946).

1422d) Shibata. S. *et al*: *Y. Z.*, **77**, 116 (1957).

(4) Gentianine (VI)[1422a) b)]

《Occurrence》 *Gentiana krilowi*[1422c)], *Swertia japonica* (センブ
リ)[1422d)] *Anthocleista procera*[1422a)] *Enicostemma littorale*[1422b)],　mp. 81~83°, [α]±0°, $C_{10}H_9O_2N$

UV λ_{max}^{EtOH} mμ (log ε)[1422b)] : 220 (4.38), 245 inf. (3.9), 280 inf. (3.2)

IR ν_{max}^{KBr} cm^{-1}: 1728 (γ–lactone), 1626 (C=C), 1592, 1574, 1475 (related
to C=N vibration), 1129, 1045 (pyridine ring)[1422a)], no bands in
1300~1400 (no C–Me)[1422b)]

NMR τ: 5.33t, 6.76t (J=5.9 resp.) (2–CH$_2$– in lactone ring), 4.23d, J=10.7, 4.05d, J=17.9
$\left(_H{>}C{=}CH_2\right)$, 2.92br, q, J=10.7, 17.9, $\left(^H_{>}C{=}C{<}\right)$, 0.83br$\left(_H{-}C{<}_N{>}C{-}_H\right)$

(VI)

(5) (—)-*trans*-4-Hydroxy-L-pipecolic acid (VII)[1423)]

《Occurrence》 *Acacia oswaldii*,　mp. 294° (decomp.), [α]$_D$ −13° (H$_2$O), $C_6H_{11}O_3N$

(VII)　(VIII)

(6) Carpaine (VIII)[1424)~1426)]

《Occurrence》 *Carica papaya* L. (パパイヤ),　mp. 121°, [α]$_D$ +21.9° (EtOH), $C_{14}H_{25}O_2N$

(7) Alkaloids of *Lobelia syphilitica* L.[1427)]

2,6–*cis*

R=H, R′=$<^H_{OH}$　Alkaloid C–1 (IX)

R=H, R′=O　Alkaloid D–1 (X)

R=CH$_3$, R′=$<^H_{OH}$　Alkaloid D–2b (XI)

H_5C_2–CH(OH)–H_2C$\diagdown$N$\diagup$CH$_2$–CH(OH)–C_2H_5
CH$_3$　2,6–*trans*
Alkaloid C–3
(XII)

H_3C—CH(OH)—H_2C$\diagdown$N$\diagup$CH$_2$—CH(OH)—C_2H_5
CH$_3$
Alkaloid D–3
(XIII)

(a) *cis*–8,10–Diethyl-nor-lobelionol (Alkaloid C–1) (IX)
　　Hydrochloride mp. 183~184°, $C_{13}H_{25}O_2N \cdot HCl$
　　IR ν_{max} cm^{-1}: 3410, 3390, 1700

(b) 8,10–Diethyl-nor-lobelidone (Alkaloid D–1) (X)
　　Hydrochloride mp. 183~184°, $C_{13}H_{23}O_2N \cdot HCl$
　　IR ν_{max} cm^{-1}: 1700

(c) (—)-*cis*–8,10–Diethyl-lobelionol (Alkaloid D–2b) (XI)
　　Hydrochloride mp. 120~121°, $C_{14}H_{27}O_2N \cdot HCl$

1423) Clarr–Lewis, J. W. *et al*: *Soc.*, **1961**, 189.
1424) Rapoport, H. *et al*: *J. A. C. S.*, **75**, 5290 (1953).
1425) Sicher, J. *et al*: *Coll. Czech. Chem. Comm.*, **24**, 950 (1959).
1426) Tichy, M. *et al*: *Tetr. Letters*, **1962**, 511.　　1427) Tschesche, R. *et al*: *Ber.*, **94**, 3327 (1961).

IR ν_{max} cm⁻¹: 3300, 1710

(d) (−)-**3-Dehydro-*trans*-8,10-diethyl-lobelidiol** (Alkaloid C-3) (XII)

Hydrochloride mp. 128°, $[\alpha]_D$ −114° (EtOH), $C_{14}H_{27}O_2N \cdot HCl$

UV λ_{max} mμ: 208, IR ν_{max} cm⁻¹: 3400, 3150 (OH), no carbonyl, 1650, 725 (C=C)

(e) (−)-**3-or-4-Dehydro-*trans*-8-methyl-10-ethyl-lobelidiol** (Alkaloid D-3) (XIII)

Hydrochloride mp. 120°, $[\alpha]_D$ −110° (EtOH), $C_{13}H_{25}O_2N \cdot HCl$

IR $\nu_{max}^{CHCl_3}$ cm⁻¹: 3300 (OH), 1650 (C=C)

(8) Alkaloids of *Sedum acre* L. [1428]

  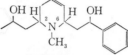

(XIV) Sedridine   (XV) Sedamine   (XVI) Sedinine

(a) (+)-**Sedridine** (XIV)

mp. 84°, $[\alpha]_D$ +29.5° (MeOH), $C_8H_{17}ON$

IR $\nu_{max}^{CCl_4}$ cm⁻¹ (Synth. (±)-sedridine): 3310, 2920, 2845, 1447, 1372, 1326, 1265, 1145, 1114, 1092, 1052, 1019, 985, 918, 897, 887, 864

(b) (±)-**Sedamine** (XV)

mp. 91°, $[\alpha]_D$ ±0°, $C_{14}H_{21}ON$, HCl-salt mp. 185°

UV λ_{max}^{MeOH} mμ (ε): 213 (5070), 252 (156), 258 (193), 264 (146)

IR ν_{max}^{KBr} cm⁻¹: 3170, 2950, 2790, 1470, 1380, 1344, 1204, 1128, 1107, 1081, 1050, 1026, 882, 769, 751, 703

(c) (−)-**Sedamine** (XV)[1428] [1429] $[\alpha]_D$ −75.4° (MeOH)

(d) (−)-**Sedinine** (XVI)[1430]

mp. 120°, $[\alpha]_D$ −105° (MeOH), $C_{17}H_{25}O_2N$

UV λ_{max}^{MeOH} mμ: (ε): 212 (5340), 252 (148), 258 (188), 264 (145)

IR ν_{max}^{KBr} cm⁻¹: 3390, 2890, 1494, 1454, 1421, 1318, 1186, 1137, 1064, 958, 927, 827, 792, 757, 741, 709, 699

NMR in $CDCl_3$, 60 Mc, TMS, ppm[1430]: 1.2d (>CHCH₃), 3.95m (—CH(OH)—CH₃), 4.78q

(—CH(OH)—C₆H₅), 3.16m (C₂-H, C₆-H), 5.38 (OH), 5.60 (H H / C=C \)

(9) (+)-**Nor-allosedamine** ((+)-8-Phenylnorlobelol-1) (XVII)[1431]

《Occurrence》 *Lobelia inflata* L., mp. 112~113°, $C_{13}H_{19}ON$

Hydrochloride mp. 153~154° (Synth.)

(XVII)

(10) Cassine (XVIII)

《Occurrence》 *Cassia excelsa* SHRAD.

mp. 57~58°, $[\alpha]_D$ −0.6° (EtOH), $C_{19}H_{37}O_2N$

IR $\nu_{max}^{CCl_4}$ cm⁻¹: 3530, 2930, 2860, 2810sh, 1720, 1360, 690

NMR 60 Mc, τ: no aldehydic or olefinic proton signal in low
field, 6.55 (>CH-OH), 7.95s (CO-CH₃), 8.98d (>CH-CH₃)

(XVIII)

1428) Franck, B: *Ber.*, **91**, 2803 (1958).   1429) Schöpf, C. *et al*: *Ann.*, **626**, 134 (1959).

1430) Franck, B: *Ber.*, **93**, 2360 (1960).   1431) Schöpf, C. *et al*: *Ann.*, **628**, 101 (1960).

(11) Tecomanine (XIX)[1433]

≪Occurrence≫   *Tecoma stans* JUSS.,

bp. 125°/0.1mm, $[\alpha]_D$ −175° (CHCl$_3$), C$_{11}$H$_{17}$ON

UV λ_{max}^{EtOH} mμ (log ε): 226 (4.10); $\lambda_{max}^{acid-EtOH}$ mμ (log ε): 223 (4.13)

IR ν_{max}^{CHCl3} cm^{-1}: 1700, 1620 ($\alpha\beta$–unsatur. cyclopentanone)

NMR δ: 5.95 (olef. proton), 2.75 (≻N–CH$_3$), 1.12d, 1.07d (2 ≻CH–CH$_3$)

(XIX)

(12) Julocrotine (XX)[1433]

(XX)

≪Occurrence≫   *Julocroton montevidensis* KLOTZSCH. (*Euphorbiaceae*)

mp. 108~109°, $[\alpha]_D$ −9° (CHCl$_3$), C$_{18}$H$_{24}$O$_3$N$_2$

UV λ_{max}^{EtOH} mμ (log ε): 252 (2.43), 258 (2.44), 264 (2.30), 268 (2.18); λ_{min}: 250 (2.41), 256 (2.37),

263 (2.27), 267 (2.16)

IR λ_{max}^{CHCl3} μ: 2.93 (sharp), 5.76 (w), 5.93 (s), 6.65 (s)

(13) Actinidine (XXI)

≪Occurrence≫   *Actinidia polygama* MIQ, (マタタビ), *Iridomyrex* (Argentine ant)

See the item **13** Isoprenoids and Non-terpenic Essential Oils, [I], (2)

(XXI)

(14) Alkaloids of *Skytanthus acutus* MEYER[1434]

(XXII) Skytanthine   (XXII–a)   (XXII–b)

Dehydroskytanthine

(XXIII)   (XXIV)

Alkaloid D

(a) Skytanthine (XXII)[1435]

Volatile base, C$_{11}$H$_{19}$N, picrate mp. 127°

NMR in CDCl$_3$, 60Mc, δ: 1.50 (3) (C=C–CH$_3$)

1432) Jones, G *et al* : *Thir. Letters*, **1963**, 397.   1433) Djerass C. *et al* : *J. Org. Chem.*, **26**, 1184 (1961).
1434) Casinori, C. G. *et al* : *Chem. & Ind.*, **1963**, 984.   1435) Djerassi, C : *Tetrahedron*, **18**, 183 (1962).

(b) Alkaloid D (XXIII) or (XXIV)

Non volatile, mp. 93°, $C_{11}H_{21}ON$

IR $\nu_{max}^{CHCl_3}$ cm^{-1}: 3550 (OH)

NMR 60Mc, TMS, ppm, in CDCl$_3$: 0.82d, $J=6.5$ ($>$CH-CH$_3$), in CHCl$_3$: 1.24s ($>$CH–CH$_3$), 2.3s ($>$N–CH$_3$)

(15) Perloline (XXV)[1436]

≪Occurrence≫ *Lolium perenne* L.

$C_{20}H_{17}O_4N_2$, perchlorate mp. 280° (decomp.)

IR ν_{max} cm^{-1}: 1695 (C=N$^+$)

(XXV)

(16) Dioscorine (XXVI)[1437)1438]

≪Occurrence≫ *Dioscorea hispida*.

mp. 54~55°, $[\alpha]_D$ −35.0° (CHCl$_3$), $C_{13}H_{14}O_2N$

UV λ_{max} mμ (ε)[1438]: 217 (16160) (conj. lactone)

IR ν_{max} cm^{-1} [1438]: 1712 (C=O)

(XXVI) (Revised Str.)[1437]

(17) Securinega Alkaloids

(XXVII) Securinine[1439)1440)1441] ≈ antipode (XXIX) Virosecurinine[1442)1443]

(XXVIII) Allosecurinine[1442)1443] (XXX) Norsecurinine[1446]

1436) Jeffreys, J. A. D. *et al*: *Proc. Chem. Soc.*, **1963**, 171.
1437) Morris, I. G. *et al*: *Soc.*, **1963**, 1841.
1438) Pinder, A. R.: *ibid.*, **1956**, 1577, **1953**, 1825, **1952**, 2236.
1439) Horii, Z. *et al*: *Chem. Pharm. Bull.*, **11**, 817 (1963).
1440) Horii, Z. *et al*: *7th Symposium on the Chemistry of Natural Products*, p.56 (Fukuoka, 1963).
1441) Saito, S. *et al*: *Chem. & Ind.*, **1962**, 1652, **1963**, 689.
1442) Nakano, T. *et al*: *Chem. & Ind.*, **1962**, 1651; *Tetr. Letters*, **1963**, 665.
1443) Satoda, I. *et al*: *ibid.*, **1962**, 1199.

(a) Securinine (XXVII)[1444)1445)1447)]

≪Occurrence≫   *Securinega suffruticosa* REHD. (*Euphorbiaceae*) (ヒトツツバキ)

mp. 143~144°, [α]_D −1042° (EtOH), $C_{13}H_{15}O_2N$

UV λ_{max}^{EtOH} mμ (log ε): 256 (4.27), 330 (3.30)

IR $\nu_{max}^{CCl_4}$ cm⁻¹: 1840, 1760 (lactone), 1640 (C=C)

NMR 1) in CDCl₃, 60Mc, ppm[1441)]: 1.79d (7-CH₂-), 2.05q (8-CH₂-), 1.58m (9-CH₂-), 2.58q
(10-CH₂-), 3.02t (10a-H), 3.86t (5a-H), 1.60m (11-CH₂-), 6.54m (5-H), 6.67q (4-H), 5.56s (3-H)

2) See Fig. 203[1440)]

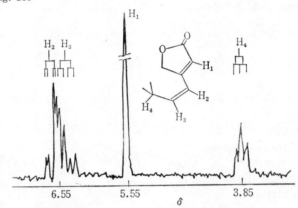

Fig. 203 NMR Spectrum of Securinine (Horii *et al.*)

(b) Dihydrosecurinine (XXVIII-a)[1447)1448)]

≪Occurrence≫   *Securinega suffruticosa* REHD. (ヒトツバハギ)

Securinine (XXVII) $\xrightarrow{\text{Pt-C,H}_2}$ (XXXIII-a)

bp. 158~164°/0.1mm, mp. 58~60°, $C_{13}H_{17}O_2N$

IR $\nu_{max}^{CCl_4}$ cm⁻¹: 1815, 1770, 1652

Hydrochloride   mp. 256~258°, $C_{13}H_{18}O_2NCl$

(XXXII)

(c) Virosecurinine (XXIX)[1442)1447)]

≪Occurrence≫   *Securinega virosa* PAX. et HOFFM. (タイワンヒトツバハギ)

mp. 139~140°[1442)], 144~146°[1447)], [α]_D +1148° (CHCl₃)[1442)], +1050° (antipode of securinine, [α]_D −1042°)
ORD: negative single Cotton effect, $C_{13}H_{15}O_2N$

UV λ_{max}^{EtOH} mμ (log ε)[1442)]: 256 (4.26); IR $\lambda_{max}^{CCl_4}$ μ: 5.56, 5.65, 6.09

NMR[1442)] in CDCl₃, TMS, ppm: 5.53s (C₃-H), 8 peaks at 6.27~6.78 (AB part of an ABX system by
the protons, attached to C₄, C₅ and C₅ₐ)

(d) Norsecurinine (XXX)[1446)]

≪Occurrence≫   *Securinega virosa* BAILL.,

mp. 81~82°, [α]_D −19.5° (EtOH), pKa 6.85, $C_{12}H_{13}O_2N$

UV λ_{max}^{EtOH} mμ (ε): 255.5 (22000)

IR $\nu_{max}^{CCl_4}$ cm⁻¹: 1802, 1770 (αβ-unsatur. lactone), 1640 (C=C)

1444) Muraveva, V. I. *et al*: *Doklady Acad. Nauk S. S. S. R.*, **110**, 998 (1956); *C. A.*, **51**, 8121 (1957).
1445) Turova *et al*: *Med. Prom.*, **10**, 27 (1956), **11**, 54 (1957); *C. A.*, **50**, 17336, 17201 (1957).
1446) Mathieson, D. W. *et al*: *J. Ph. Ph.*, **15**, 810 (1963).
1447) Horii, Z. *et al*: *Y. Z.*, **83**, 602 (1963).   1448) Horii Z. *et al*: *ibid.*, **83**, 800 (1963).

NMR τ: 4.33s (Ha), 6 peaks at 3~3.6 (AB part of an ABX system by Hb, Hc, Hd), **6.37t** (Hd, **X** part, coupled for Hc and Hf), 7.50t (Hf) $J_{fe} \simeq 11$, 8.1m (protons, j, k, l, m), 8.25d (He)

(18) Erythrina Alkaloids

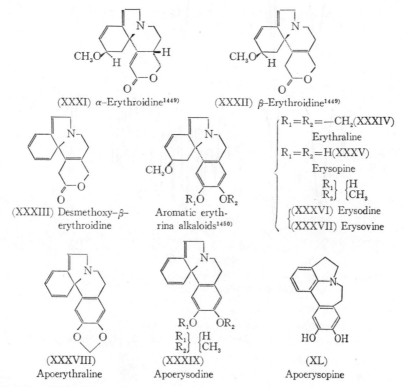

(XXXI) α–Erythroidine[1449]   (XXXII) β–Erythroidine[1449]

(XXXIII) Desmethoxy-β-erythroidine

Aromatic erythrina alkaloids[1450]

$\begin{cases} R_1=R_2=-CH_2 (XXXIV) \\ \quad Erythraline \\ R_1=R_2=H(XXXV) \\ \quad Erysopine \\ \left.\begin{matrix} R_1 \\ R_2 \end{matrix}\right\} \begin{cases} H \\ CH_3 \end{cases} \\ (XXXVI)\ Erysodine \\ (XXXVII)\ Erysovine \end{cases}$

(XXXVIII) Apoerythraline

(XXXIX) Apoerysodine

$\left.\begin{matrix} R_1 \\ R_2 \end{matrix}\right\} \begin{cases} H \\ CH_3 \end{cases}$

(XL) Apoerysopine

(a) α-Erythroidine (XXXI)[1451]

≪Occurrence≫   *Erythrina berteroana*

Hydrochloride mp. 226~228°, $[\alpha]_D$ +118°, pKa 7.79, $C_{16}H_{20}O_3NCl$

UV λ_{max}^{EtOH} mμ (log ε): 224 (4.5),   IR See Fig. 204

(b) β-Erythroidine (XXXII)[1451]~[1453]

≪Occurrence≫   *Erythrina berteroana, E. americana, E. peoppiginana, E. tholloniana*

mp. 99.5~100°, $[\alpha]_D$ 88.8°, pKa 7.80, $C_{16}H_{19}O_3N$

UV λ_{max}^{EtOH} mμ (log ε): (HCl salt) 238 (4.4),   IR See Fig. 204

(c) Desmethoxy-β-erythroidine (XXXIII)[1454]

≪Occurrence≫ β-Erythroidine (XXXII) $\xrightarrow{HF}$ (XXXIII),   mp. 108~109°, pKa 7.71, $C_{15}H_{15}O_2N$

UV λ_{max}^{EtOH} mμ (log ε)[1452]: 312 (3.6)

1449) Wenzinger, G.R. *et al*: *Proc. Chem. Soc.*, **1963**, 53.
1450) Boekelheide, V. *et al*: *J. Org. Chem.*, **29**, 1303 (1964).
1451) Boekelheide, V. *et al*: *J.A.C.S.*, **75**, 2563 (1953), **77**, 3342 (**1955**), **77** 3342 (1955).
1452) Boekelheide, V. *et al*: *ibid.*, **75**, 2550, 2558 (1953).
1453) Folkers, K. *et al*: *ibid.*, **59**, 1580 (1937).   1454) Boekelheide, V. *et al*: *ibid.*, **72**, 2062 (1950).
1455) Folkers, K. *et al*: *ibid.*, **62**, 436 (1940).
1456) Prelog, V. *et al*: *Helv. Chim. Acta*, **32**, 453 (1949).

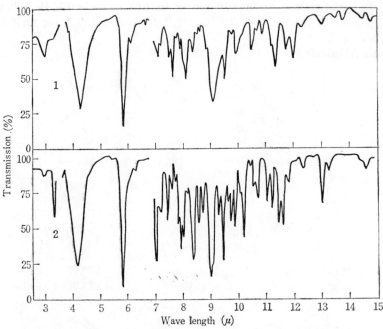

Fig. 204 IR Spectra of (1) α–Erythroidine–HCl and (2) β–Erythroidine–HCl in nujol (Boekelheide *et al*)

(d) Erythraline (XXXIV)[1455)1456)]

《Occurrence》 *Erythrina fusca* Lour., *E. glauca* Willd.

mp. 106~107°[1455)], 120°[1456)], [α]$_D$ +211.8°[1455)], +228° (EtOH)[1456)], pKa 5.97, $C_{18}H_{19}O_3N$

UV λ_{max}^{EtOH} mμ (log ε)[1456)] : 292 (3.6), 238 (4.3)

(e) Erysopine (XXXV)[1457)]

《Occurrence》 *Erythrina glauca* Willd.

mp. 241~242°, [α]$_D$ +265.6° (EtOH+glycerol), pKa 6.60, $C_{17}H_{19}O_3N$

UV λ_{max} mμ (log ε): 292 (3.6), 240 (4.3)[1458)]

(f) Erysodine (XXXVI)[1457)]

《Occurrence》 *Erythrina sandwicensis* Deg., *E. americana* Mill.

mp. 204~205°, [α]$_D$ +248°, 267° (EtOH)[1456)], pKa 6.29, $C_{18}H_{21}O_3N$

UV λ_{max} mμ (log ε): 285 (3.5), 235 (4.3)

(g) Erysovine (XXXVII)[1457)]

《Occurrence》 *Erythrina sandwicensis* Deg., *E. berteroana* Vrb.

mp. 178°, [α]$_D$ +252.2° (EtOH), pKa 6.45, $C_{18}H_{21}O_3N$

UV λ_{max} mμ (log ε): 285 (3.5), 239 (3.8)[1458)]

(h) Apoerythraline (XXXVIII)[1459)]

$$\text{Erythraline (XXXIV)–HBr} \xrightarrow{\text{HBr}} \text{(XXXVIII)}$$

Hydrobromide mp. 272~276°, [α]$_D$ −795° (H_2O), pKa 6.59, $C_{17}H_{16}O_2NBr$

UV λ_{max}^{H2O} mμ (log ε): 315 inf. (3.7), 295 (3.8), 240 inf. (3.8)

(i) Apoerysodine (XXXIX)[1459)]

1457) Folkers, K. *et al*: *J.A.C.S.*, **62**, 1677 (1940).   1458) Prelog, V: *Angew. Chem.*, **69**, 33 (1957).
1459) Prelog, V. *et al*: *Helv. Chim. Acta*, **34**, 1601 (1951).

Erysodine (XXXVI) $\xrightarrow{\text{HCl}}$ (XXXIX)

mp. 206~207°, $[\alpha]_D$ —965° (EtOH), pKa 6.54, $C_{17}H_{17}O_2N$

UV λ_{max}^{EtOH} mμ (log ε): 315 (3.7), 290 (3.7), 239 (4.0)

(j) Apoerysopine (XL)[1459]

Erysodine (XXXVI) $\xrightarrow{\text{HBr}}$ (XL)

mp. 172~173°, $[\alpha]_D$ 0° (EtOH), pKa 3.50, $C_{16}H_{15}O_2N$

$\lambda_{max}^{0.01NHCl-EtOH}$ mμ (log ε): 301 (4.0), 268 (4.0), 235 (4.3)

(19) Nuphar Alkaloids

(XLII) Desoxynupharidine[1463]~[1467][1470]
N–Oxide ((XLI) Nupharidine)

R=OH (XLIII) (−)-Nupharamine
R=H (XLIV) (−)-Desoxynupharamine[1468]

(XLV) Thiobinupharidine[1469]

(a) Nupharidine (XLI)[1463]

《Occurrence》 *Nuphar japonicum* DC. (センコツ)

mp. 220~221°(decomp.), $[\alpha]_D$ +17.2° (EtOH), $C_{15}H_{23}O_2N$

IR (Hydrochloride): See Fig. 205

(b) Desoxynupharidine (XLII)

bp. 77~79°/0.001mm (synth. *dl*-)[1463], $[\alpha]_D$ —109.9° (EtOH), $C_{15}H_{23}ON$

UV λ_{max}^{EtOH} mμ (log ε): 212 (3.9)[1467]

IR ν_{max}^{liq} cm^{-1}: 3110, 1505, 875 (furan), 2760, 2790 (*trans*-quinolizidine) (Fig. 205).

1460) Robinson, R. *et al*: *Chem. & Ind.*, **1954**, 783.
1461) Hamor, T. A. *et al*: *Proc. Chem. Soc.*, **1960**, 78.
1462) Woodward, R. B. *et al*: *ibid.*, **1960**, 76.
1463) Kotake, M. *et al*: *Bull. Chem. Soc. Japan.*, **35**, 1494 (1962); *Ann.*, **636**, 158 (1960); *6th Symposium on the Chemistry of Natural Products*, p.19 (Sapporo, 1962).
1464) Bohlmann, F. *et al*: *Ber.*, **94**, 3151 (1961).
1465) Arata, Y. *et al*: *Chem. Pharm. Bull.*, **10**, 675 (1962); *Y. Z.*, **82**, 326 (1962), **77**, 236 (1957).
1466) Kawasaki: *J. Chem. Soc. Japan.*, **81**, 156 (1960). 1467) Kusumoto, S; *ibid.*, **78**, 488 (1957).
1468) Arata, Y. *et al*: *Y. Z.*, **77**, 792 (1957), **79**, 127 (1959), **83**, 79 (1963).
1469) Achmatowicz, O. *et al*: *Tetr. Letters*, **1962**, 1121.

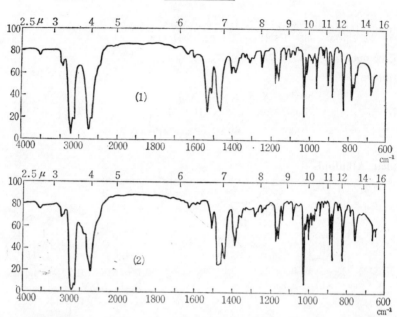

Fig. 205 IR Spectra of (1) Nupharidine hydrochloride (natural) and (2) Des-oxynupharidine hydrochloride (natural) in nujol (Kotake *et al.*)

Hydrochloride mp. 261~263°, $[\alpha]_D$ −21.4° (*N*–HCl)

  IR See Fig. 205

  Synthesis of *dl*–desoxynupharidine[1463) ~1466)1470)]

(c) (−)-**Nupharamine** (XLIII)[1468)]

  ≪Occurrence≫   *Nuphar japonicum* DC.

      bp. 130~134°/1 mm, $[\alpha]_D$ −35.4°, $C_{15}H_{25}O_2N$, picrolonate mp. 167.5~168°

  IR $\lambda_{max}^{liq.}$ μ: 3.18, 6.26, 6.36, 6.67, 11.44, 12.59, 13.04, 13.75

(d) (−)-**Desoxynupharamine** (XLIV)[1468)]

$$(-)\text{-Nupharamine (XLIII)} \xrightarrow{\text{Pd-C,H}_2} \text{(XLIV)}$$

  Synthesis

  bp. 129~133°/5 mm, $[\alpha]_D$ −66.67° (EtOH), picrolonate mp. 176°

(e) **Thiobinupharidine** (XLV)[1469)]

  ≪Occurrence≫   *Nuphar luteum* (L) Sm.,   mp. 129~130°, $[\alpha]_D$ +49.8° (diperchlorate in H_2O),
          $C_{30}H_{40}O_2N_2S$

  UV λ_{max}^{EtOH} mμ (ε) (diperchlorate): 298 (1115)

  IR ν_{max} cm^{-1}: 1660, 1425, 878 (furan), 2750 (CH–N), 3419 (C–H)

(f) **Minor alkaloids** of *Nuphar luteum* (L) Sm.

  Allothiobinupharidine $C_{30}H_{42}O_2N_2S$ (XLVI)–2HClO$_4$ mp. 320~325°

  Pseudothiobinupharidine (XLVII)–2HClO$_4$ mp. 173~175°

  Thiobidesoxynupharidine (XLVII–a)–2HClO$_4$ mp. 225~226°

1470) Bohlmann, F. *et al*: *Ber.*, **94**, 3151 (1961).

(20) Lupin Alkaloids

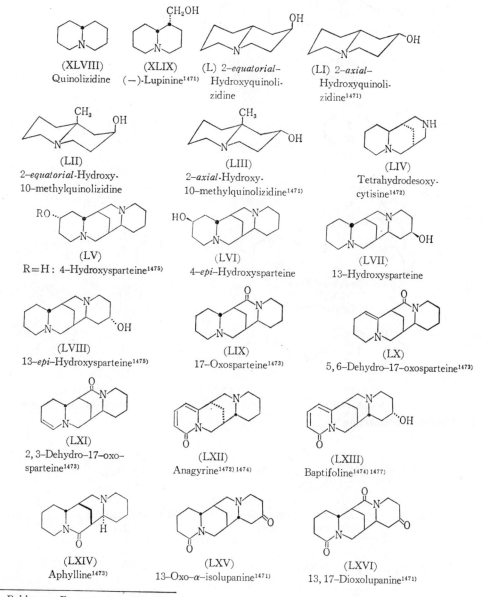

(XLVIII)
Quinolizidine

(XLIX)
(−)-Lupinine[1471]

(L) 2-*equatorial*-
Hydroxyquinoli-
zidine

(LI) 2-*axial*-
Hydroxyquinoli-
zidine[1471]

(LII)
2-*equatorial*-Hydroxy-
10–methylquinolizidine

(LIII)
2-*axial*-Hydroxy-
10–methylquinolizidine[1471]

(LIV)
Tetrahydrodesoxy-
cytisine[1472]

(LV)
R=H : 4–Hydroxysparteine[1475]

(LVI)
4-*epi*-Hydroxysparteine

(LVII)
13–Hydroxysparteine

(LVIII)
13-*epi*-Hydroxysparteine[1475]

(LIX)
17–Oxosparteine[1473]

(LX)
5, 6–Dehydro–17–oxosparteine[1473]

(LXI)
2, 3–Dehydro–17–oxo-
sparteine[1473]

(LXII)
Anagyrine[1472] [1474]

(LXIII)
Baptifoline[1474] [1477]

(LXIV)
Aphylline[1473]

(LXV)
13–Oxo-α–isolupanine[1471]

(LXVI)
13, 17–Dioxolupanine[1471]

1471) Bohlmann, F. *et al* : *Ber.*, **94**, 1767 (1961).
1472) Okuda, S. *et al* : *5th Symposium on the Chemistry of Natural Products*, p.27–1, (Sendai, 1961).
1473) Boblmann, F. *et al* : *Ber.*, **91**, 2157 (1958). 1474) Bohlmann, F. *et al* : *ibid.*, **91**, 2189 (1958).
1475) Bohlmann, F. *et al* : *ibid.*, **91**, 2194 (1958).
1476) Lloyd, H. A. : *J. Org. Chem.*, **26**, 2143 (1961).
1477) Bohlmann, F. *et al* : *Ber.*, **95**, 944 (1962). 1478) Bohlmann, F. *et al* : *ibid.*, **93**, 1956 (1960).
1479) Lloyd, H. A. *et al* : *J. Org. Chem.*, **25**, 1959 (1960).
1480) Bohlmann, F. *et al* : *Ber.*, **95**, 2365 (1962). 1481) Fraga, F. *et al* : *Tetrahedron*, **11**, 78 (1960).
1482) Goosen, A. : *Soc.*, **1963**, 3067. 1483) Gerrans, G. G. *et al* : *Chem. & Ind.*, **1963**, 1280.

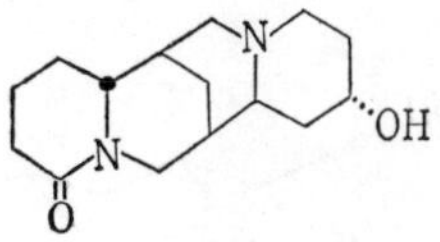

(LXVII)
13-*epi*-Hydroxy-α-
isolupanine

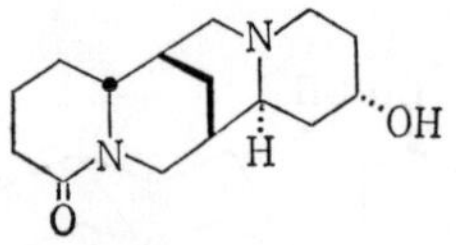

(LXVIII)
13-*epi*-Hydroxylupanine
(Jamaidine[1476])

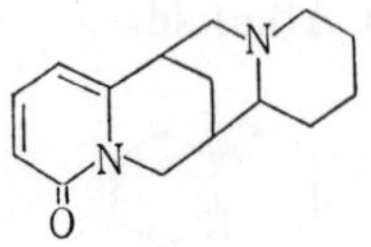

(LXIX)
(+)-Thermopsine[1477]

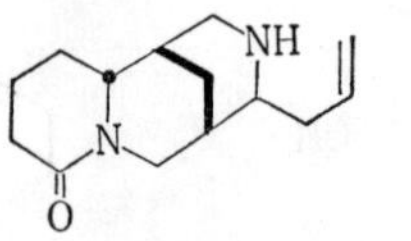

(LXX)
Angustifoline[1477)1478]

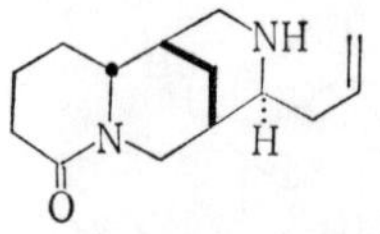

(LXXI)
Jamsaicencine[1476)1479]

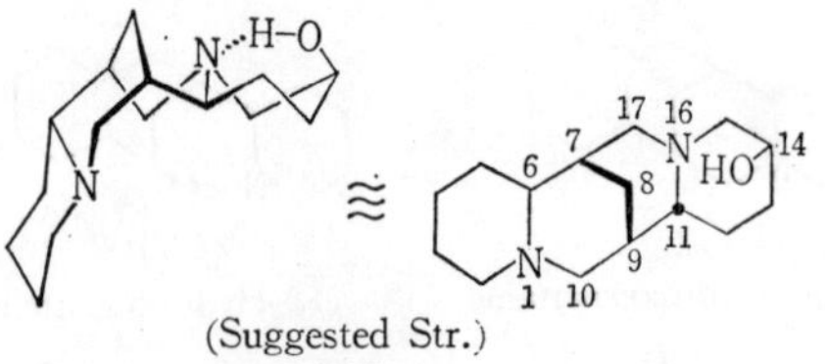

(Suggested Str.)
(LXXII)
Retamine[1480)1481]

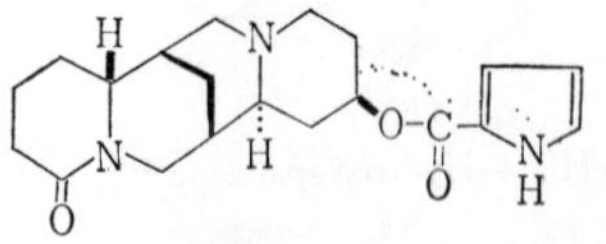

(LXXIII)
Calpurnine[1482]

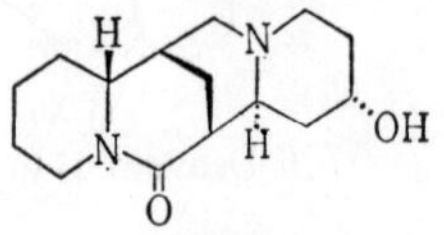

(LXXIV)
Virgiline[1483]

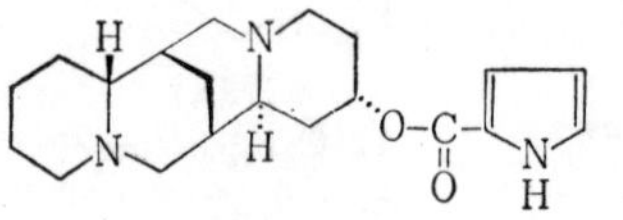

(LXXIV–a)
Oroboidine[1483]

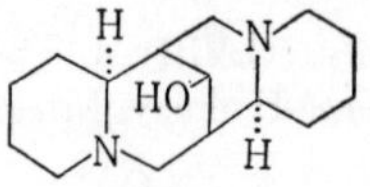

(LXXIV–b)
8–Hydroxyspartalupine[1484]

(a) **Absolute Configuration**[1472]

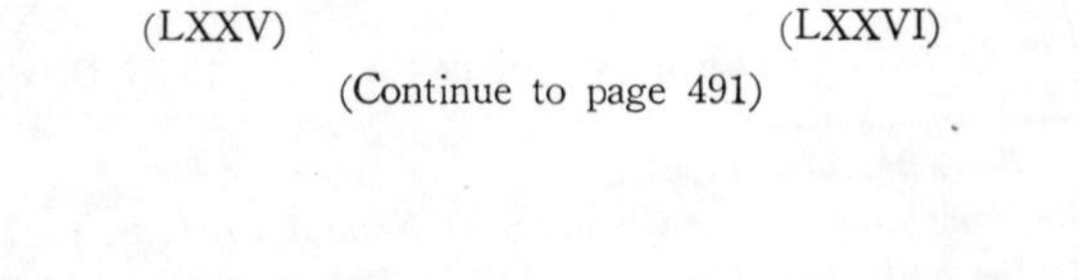

(LXIX) (−)-Thermopsine
(7R: 9R: 11S: 16S)

(LXII) (−)-Anagyrine
(7R: 9R: 11R: 16S)

(LXXV)

(LXXVI)

(Continue to page 491)

1484) Wiewiorowski *et al* : *Angew. Chem.*, **78**, 947 (1962).

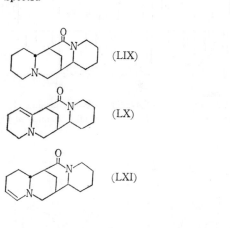

(LXXV) (−)-d-Isolupanine
(1S: 6S: 7R: 9R: 11S: 16S)

(LXXVI) (−)-Lupanine
(1S: 6S: 7R: 9R: 11R: 16S)

(LXXVII) (+)-α-Isosparteine   (LXXVIII) (+)-Sparteine   (LXXIX) (−)-β-Isosparteine
(1S: 6S: 7R: 9R: 11S: 16S)   (1S: 6S: 7R: 9R: 11R: 16S)   (1S: 6R: 7R: 9R: 11R: 16S)

(b) UV Spectra

(LIX)

(LX)

(LXI)

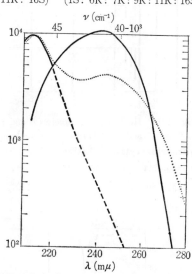

Fig. 206 UV Spectra of 17-oxosparteine - - - (LIX), 5,6-Dehydro-17-oxosparteine······(LX)
and 2,3-Dehydro-17-oxosparteine────(LXI) in ether   (Bohlmann)[1473]

(c) IR Spectra

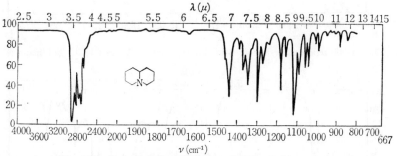

Fig. 207 Quinolizidine (XLVIII) in CHCl₃ (Bohlmann)[1473]

(−)-**Lupinine** (XLIX)[1472]

《Occurrence》 *Lupinus luteus* L.,   mp. 68°, [α]_D −20.0° (H₂O)

IR $\nu_{max}^{CHCl_3}$ cm⁻¹: 3260, 2820, 2765

CH₂OH

(XLIX)

(+)-Epilupinine (XLIX–a)[1472]

 mp. 77°, [α]$_D$ +37.0° (EtOH)

 IR ν$_{max}^{CHCl_3}$ cm^{-1}: 3625, 2820, 2760

(XLIX–a)

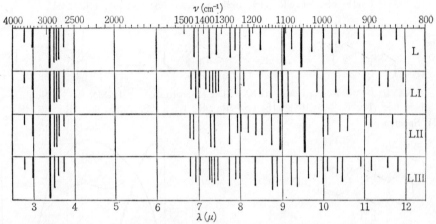

Fig. 207a IR Spectra of *equatorial* and *axial* Stereoisomers of 2–Hydroxyquinolizidine (L), (LI) and of 2–Hydroxy–10–methylquinolizidine (LII), (LIII) in CCl$_4$ (Bohlmann *et al.*)[1471]

Tetrahydrodesoxycytisine (LIV)[1472]

Tosylate mp. 144°, [α]$_D$ +9.8° (C$_6$H$_6$)

 IR ν$_{max}^{CHCl_3}$ cm^{-1}: 2810, 2765, 2740, 1597, 1330, 1162

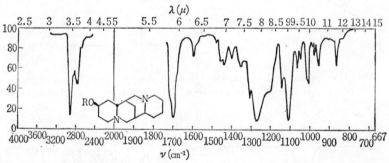

Fig. 208 Azobenzene carboxylate of 4–Hydroxysparteine (LV) in CHCl$_3$ (Bohlmann *et al.*)[1475]

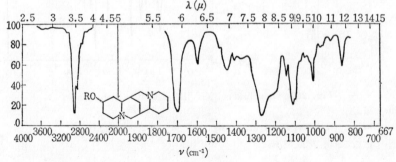

Fig. 209 Azobenzene carboxylate of 4–*epi*–Hydroxysparteine (LVI) in CHCl$_3$ (Bohlmann *et al.*)[1475]

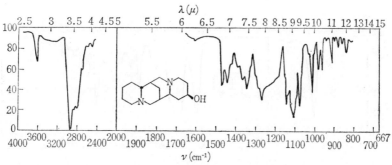

Fig. 210   13–Hydroxysparteine (LVII) in CHCl₃ (Bohlmann *et al.*)[1475]

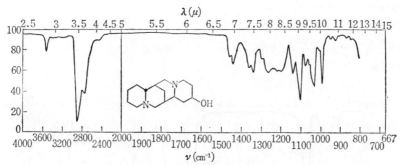

Fig. 211   13–*epi*–Hydroxysparteine (LVIII) in CHCl₃ (Bohlmann *et al.*)[1475]

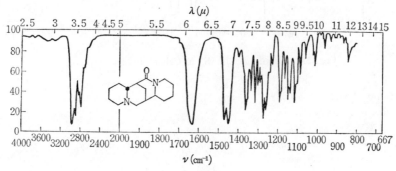

Fig. 212   17–Oxosparteine (LIX) in CCl₄ (Bohlmann)[1473]

Anagyrine   (LXII)

《Occurrence》   *Sophora flavescens*,

bp. 180°/0.01 mm, [α]_D −180° (MeOH), $C_{15}H_{20}ON_2$, perchlorate   mp. 315°

UV λ_{max}^{MeOH} mμ (ε): 309 (7800), 233 (6700)

IR   See Fig. 213

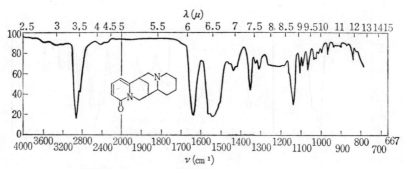

Fig. 213 Anagyrine (LXII) in CHCl₃ (Bohlmann *et al.*)[1474]

Baptifoline (LXIII)[1474)1477]

≪Occurrence≫ *Sophora flavescens,* mp. 210°, $[\alpha]_D$ −140° (EtOH), $C_{15}H_{20}O_2N$

UV $\lambda_{max}^{MeOH}m\mu$ (ε): 309 (8200), 233 (7000)

IR See Fig. 214

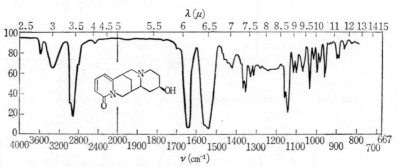

Fig. 214 Baptifoline (LXIII) in CHCl₃ (Bohlmann *et al.*)[1474]

Synthesis[1477]

epi-Baptifoline (LXIII) (13–α–OH)[1477] mp. 215°

IR: identical to Fig. 214

Aphylline (LXIV)[1473]

≪Occurrence≫ *Anabasis aphylla,* mp. 57° $[\alpha]_D$ +10.3°, $C_{15}H_{24}ON_2$

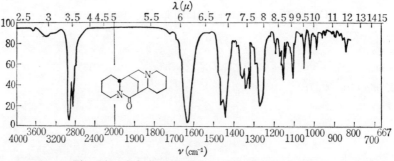

Fig. 215 Aphylline (LXIV) in CCl₄ (Bohlmann)[1473]

13–Oxo–α–isolupanine (LXV)[1471]

Hydroxylupanine ⟶ (LXV), mp. 115°, $C_{15}H_{22}O_2N_2$

IR $\nu_{max}^{CCl_4}$ cm^{-1}: 2790, 2745 (*trans*-quinolizidine), 1710 ($>$C=O), 1610 (lactam)

13, 17–Dioxolupanine (LXVI)[1471]

  13–Hydroxylupanine $\longrightarrow$ (LXVI) mp. 265°, $[\alpha]_D$ +105° (CHCl^3), C$_{15}$H$_{20}$O$_3$N

IR $\nu_{max}^{CCl_4}$ cm^{-1}: 1715 ($>$C=O), 1610 (lactam)

13–*epi*-Hydroxy–α–isolupanine (LXVII)[1471]

  13–Oxo–α–isolupanine (LXV) $\longrightarrow$ (LXVII) mp. 185°, C$_{15}$H$_{24}$O$_2$N$_2$

IR $\nu_{max}^{CCl_4}$ cm^{-1}: 3600, 3400 (OH), 2790, 2740 (*trans*-quinolizidine), 1600 (lactam)

(d) Jamaidine (13–*epi*-Hydroxylupanine) (LXVIII)[1476] [1485]

  《Occurrence》 *Ormosia jamaicensis, O. panamensis,* mp. 194.5~195°, $[\alpha]_D$ +63.7° (EtOH),
    C$_{15}$H$_{24}$O$_2$N$_2$, perchlorate mp. 175~177°

(e) (+)–Thermopsine (LXIX)[1477]

  《Occurrence》 *Lupinus* spp., mp. 207°, $[\alpha]_D$ +149° (EtOH), C$_{15}$H$_{20}$ON$_2$

IR ν_{max} cm^{-1}: 1653 (α–pyridone C=O) Synthesis[1477]

(f) Angustifoline (LXX)[1477]~[1479]

  《Occurrence》 *Lupinus polyphillus* L., mp. 80°, $[\alpha]_D$ −8.0 (EtOH), C$_{14}$H$_{22}$ON$_2$

IR See Fig. 216

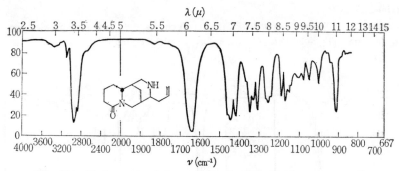

Fig. 216 IR Spectrum of Angustifoline in CCl$_4$ (Bohlmann *et al.*)[1478]

IR $\nu_{max}^{CCl_4}$ cm^{-1}: 1620 (—CO–N$<$), 3070, 915 (—CH=CH$_2$)

(g) Jamaicencine (LXXI)[1476] [1479]

  《Occurrence》 *Ormosia jamaicensis, O. panamensis,* mp. 80.5~81°, $[\alpha]_D$ +5.2 (EtOH), C$_{14}$H$_{22}$ON$_2$
  UV: no absorption above 220 mμ

IR ν_{max} cm^{-1}: 1625 (C=O of α–piperidone), 919, 998, 3078 (RCH=CH$_2$)

(h) Retamine (LXXII)[1480] [1481]

  《Occurrence》 *Retama sphaerocarpa* Boiss., mp. 168°, $[\alpha]_D$ +43.2° (EtOH), C$_{15}$H$_{26}$ON$_2$
  Isoretamine[1481] $[\alpha]_D$ −14.2, C$_{15}$H$_{26}$ON$_2$

(i) Calpurnine (LXXIII)[1482]

  《Occurrence》 *Calpurnia decandra,* mp. 152~154°, $[\alpha]_D$ +59° (CHCl$_3$), C$_{20}$H$_{27}$ON$_3$

IR $\nu_{max}^{CHCl_3}$ cm^{-1}: 3480 ($>$NH), 1710, 1695, 1615

1485) Lloyd, H. A. *et al*: *J. Org. Chem.*, **25**, 1959 (1960).

(21) **Matrine** and **Related Alkaloids**[1474)1486)1487)]

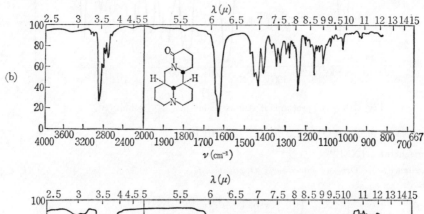

R=O (LXXX) Matrine   R=O (LXXXII) Allomatrine
R=H₂ (LXXXI) Matridine   R=H₂ (LXXXIII) Allomatridine

(LXXXIV) Sophoranol

(a) **Matrine** (LXXX)

≪Occurrence≫   *Sophora angustifolia* SIEB. et ZUCC. (クララ) and other *Sophora* spp.
mp. 76° (α–form), [α]$_D$ +39.1° (H_2O), $C_{15}H_{24}ON_2$, methiodide mp. 211°
IR See Fig. 217

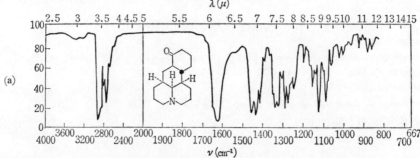

Fig. 217 IR Spectra of (a) Matrine and (b) Allomatrine in CCl_4 (Bohlmann *et al.*)[1487)]

1486) Tsuda, K. *et al*: *Chem. Pharm. Bull.*, **5**, 285 (1957).
1487) Bohlmann, F. *et al*: *Ber.*, **91**, 2176 (1958).

(b) Matridine (LXXXI)

《Occurrence》　*Sophora angustifolia* SIEB. et ZUCC. and other *Sophora* spp.

mp. 59~60°, bp. 161~163°/7 mm, $[\alpha]_D$ −11.6° (EtOH), $C_{15}H_{24}ON_2$, methiodide mp. 233~234°

IR　See Fig. 218

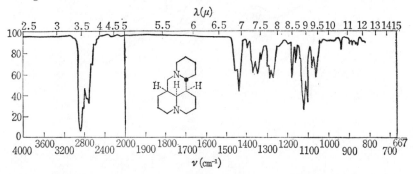

Fig. 218　IR Spectrum of Matridine in CCl_4 (Bohlmann *et al.*)[1109]

(c) Allomatrine (LXXXII)

《Occurrence》　*Sophora angustifolia* SIEB. et ZUCC. and other *Sophora* spp.

mp. 103~105°, $[\alpha]_D$ +77.9° (EtOH), $C_{15}H_{24}ON_2$, methiodide mp. 315°

IR　See Fig. 217

(d) Allomatridine (LXXXIII)[1474]

《Occurrence》　*Sophora angustifolia* SIEB. et ZUCC. and other *Sophora* spp.

mp. 76°, bp. 153~154°/5 mm, $[\alpha]_D$ +28.8° (EtOH), $C_{15}H_{24}ON_2$, methiodide mp. 265° (decomp.)

(e) Sophoranol (5–Hydroxymatrine) (LXXXIV)

《Occurrence》　*Sophora flavescens*,　　mp. 171°, $[\alpha]_D$ +66° (H_2O), $C_{15}H_{24}O_2N_2$

IR　See Fig. 219

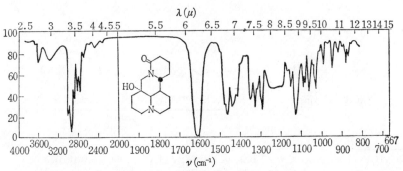

Fig. 219　IR Spectrum of Sophoranol in $CHCl_3$ (Bohlmann *et al.*)[1474]

Total synthesis of *dl*–matrine[1488], *dl*–matridine[1489] and allomatridine[1486]

Dipolemoments and the Conformation[1490]

1488)　Mandell, L. *et al* : *J. A. C. S.*, **85**, 2682 (1963).

1489)　Mandell, L. *et al* : *ibid.*, **83**, 1766 (1961).

1490)　Tsuda, K. *et al* : *ibid.*, **80**, 2426 (1958).

(22)　Lycopodium Alkaloids

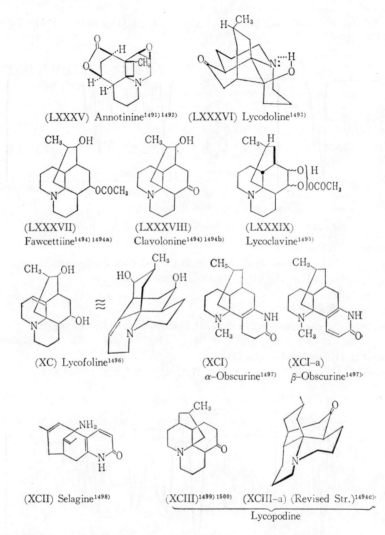

(LXXXV)　Annotinine[1491][1492]　　(LXXXVI)　Lycodoline[1493]

(LXXXVII)
Fawcettiine[1494][1494a]

(LXXXVIII)
Clavolonine[1494][1494b]

(LXXXIX)
Lycoclavine[1495]

(XC)　Lycofoline[1496]

(XCI)
α–Obscurine[1497]

(XCI–a)
β–Obscurine[1497]

(XCII)　Selagine[1498]

(XCIII)[1499][1500]　(XCIII–a)　(Revised Str.)[1494c]
Lycopodine

1491)　Manske, H. F. *et al*: *Can. J. Res.*, **B21**, 92 (1943); *J. A. C. S.*, **69**, 2126 (1947).
1492)　Wiesner, K. *et al*: *J. A. C. S.*, **78**, 2867 (1956); *Tetrahedron*, **4**, 87 (1958).
1493)　Ayer, W. A. *et al*: *Tetr. Letters*, **1962**, 87.
1494)　Taylor, D. R. *et al*: *Tetrahedron*, **15**, 173 (1961).
1494a)　Burnell, R. H.: *Soc.*, **1959**, 3091.
1494b)　Burnell, R. H. *et al*: *Can. J. Chem.*, in press (1961).
1494c)　Mc Lean, D. B.: *Chem. & Ind.*, **1960**, 261.
1495)　Ayer, W. A. *et al*: *Can. J. Chem.*, **40**, 2088 (1962); *Index Chem.*, **8**, 24559 (1963).
1496)　Burnell, R. H. *et al*: *Tetrahedron*, **18**, 1467 (1962).
1497)　Ayer, W. A. *et al*: *ibid.*, **18**, 567 (1962).
1498)　Valenta, Z. *et al*: *Tetr. Letters*, **1960**, 26.
1499)　Inubuse, Y. *et al*: *Y. Z.*, **82**, 1537 (1962).
1500)　Mac Lean, D. B. *et al*: *Chem. & Ind.*, **1960**, 261.

(a) **Annotinine** (LXXXV)[1491) 1492)]

≪Occurrence≫ *Lycopodium annotinum,* mp. 232°, $C_{16}H_{21}O_3N$, perchlorate mp. 267°

(b) **Lycodoline** (LXXXVI)[1493)]

≪Occurrence≫ *Lycopodium annotinum,* $C_{16}H_{25}O_2N$

IR ν_{max} cm^{-1}: 3545 (OH)

NMR τ: 9.14 d (〉CH–Me), no olefinic H

(c) **Fawcettiine** (Base C) (LXXXVII)[1494) 1494a)]

≪Occurrence≫ *Lycopodium fawcettii,* $C_{18}H_{29}O_3N$, perchlorate —— HClO$_4$·H$_2$O,

mp. 272~275° (decomp.), methiodide mp. 279~280°

Followed alkaloids are also isolated from *Lycopodium fawcettii*

Base A	$C_{16}H_{27}O_2N \cdot HClO_4$ (perchlorate)	mp. 221~222°
Base B	$C_{16}H_{25}O_2N$ (base)	mp. 178~179°
Base E	$C_{17}H_{27}O_2N \cdot HClO_4$ (perchlorate)	mp. 267~269° (decomp.)
Base F	$C_{16}H_{23}ON \cdot C_6H_3O_7N_3$ (picrate)	mp. 222~223°
Base G	$C_{18}H_{27}O_3N \cdot HClO_4$ (perchlorate)	mp. 198~200°

(d) **Clavolonine** (LXXXVIII)[1494)]

≪Occurrence≫ *Lycopodium clavatum*[1494b)], $C_{16}H_{25}O_2N$

(e) **Lycoclavine** (LXXXIX)[1495)]

≪Occurrence≫ *Lycopodium clavatum* var. *Megastachyon.*

(f) **Lycofoline** (XC)[1496)]

≪Occurrence≫ *Lycopodium annotinum, L. fawcettii,* $C_{16}H_{25}O_2N$

(g) **α–Obscurine** (XCI)[1497)]

≪Occurrence≫ *Lycopodium obscurum* L., mp. 282~283°, $C_{17}H_{26}ON_2$

UV λ_{max} mμ (log ε): 255 (3.73) IR $\nu_{max}^{CCl_4}$ cm^{-1}: 1700 m, 1675 s

NMR τ: 1.99 br (〉NH of pyridone), 7.55 (〉N–CH$_3$), 9.14 d, J=5 c/s (〉CH–CH$_3$)

≪Occurrence≫ *Lycopodium obscurum* L., mp. 317~318°, $C_{17}H_{24}ON_2$

(g)′ **β–Obscurine**

UV λ_{max} mμ: 230

NMR τ: 2.21 d, 3.63 d (AB quartet, J_{AB}~10 c/s) (3, 4–, 3, 6–or 5, 6–disubst.–2–pyridone),

7.46 (〉N–CH$_3$), 9.17 br. s (〉CH–CH$_3$)

(h) **Selagine** (XCII)[1498)]

≪Occurrence≫ *Lycopodium selago,* mp. 224~226°, $[\alpha]_D$ −99° (MeOH), pKa 7.18, $C_{15}H_{18}ON_2$

UV λ_{max}^{EtOH} mμ (ε): 231 (10700), 313 (8500) (α–pyridone)

IR $\nu_{max}^{CHCl_3}$ cm^{-1}: 1653, 1620, 1533

(i) **Lycopodine**[1494c) 1499) 1500)] (XCIII–a)

≪Occurrence≫ *Lycopodium clavatum* L. (ヒカゲノカズラ)

mp. 115~116°, $C_{16}H_{25}ON$, perchlorate, mp. 273~276° (decomp.)

(j) **Base L$_{30}$** (L$_8$)[1500) 1501)] (XCIV)

≪Occurrence≫ *Lycopodium annotinum* L., —— var. *acrifolium* FERN.[1501)], *L. clavatum* L.[1500)]

mp. 155~162°, $C_{16}H_{25}O_2N$

(k) **Serratinine** (XCV)[1502)]

≪Occurrence≫ *Lycopodium serratum* var. *Thunbergii* MAKINO (ホソバトウゲシバ)

mp. 244~245°, $[\alpha]_D$ −27.7° (EtOH), $C_{16}H_{25}O_3N$

IR ν_{max}^{nujol} cm^{-1}: 3472, 3436, 3185 br, (OH), 1724 (C=O), 1427 (δ_{C-H})

1501) Marion, L. *et al*: *J. A. C. S.*, **69**, 2126 (1947); *Can. J. Chem.*, **31**, 272 (1953).

1502) Inubuse, Y. *et al*: *Y. Z.*, **82**, 1339 (1962).

［F］　Amaryllidaceous Alkaloids

(1)　UV Spectra

Lycorine, dihydrolycorine and lycorine, anhydrohydromethine[1503]

(2)　Lycorane Alkaloids

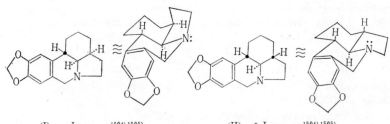

(I)　α-Lycorane[1504) 1505)]　　　　　(II)　β-Lycorane[1504) 1505)]

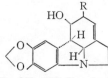　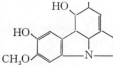

R=OH　(III) Lycorine　　(V) Pseudolycorine　　P=H (VI) Norpluviine
R=H　(IV) Caranine　　　　　　　　　　　　R=CH₃ (VII) Pluviine

(a)　Lycorine　(III)[1506)~1508),1528)]

≪Occurrence≫　*Amaryllis belladonna* & other spp., *Boöphone* spp., *Calostemma purpureum*, *Chlidanthus fragrans*, *Cliva* spp., *Crinum asiaticum* & other spp., *Elisena longipetala, Galanthus woronowii* & other spp., *Haemanthus* spp., *Hippeastrum* spp., *Hymenocallis* spp., *Leucojum* spp., *Lycoris radiata* & other spp., *Narcissus* spp., *Nerine bowdenii* & other spp., *Pancratium* spp., *Sternbergi* spp., *Ungernia* spp., *Vallota purpurea, Zephyranthes carinata* & other spp.

mp. 270° (decomp.), $[\alpha]_D$　$-83.8°$ (EtOH), $C_{16}H_{17}O_4N$

UV　See Fig. 220

$\lambda_{max}^{dil-HCl}$ mμ (E $_{1cm}^{1\%}$)[1508)] : 280 (114.2), 288(149.8), 295 (124.5)

λ_{max}^{EtOH} mμ (log ε)[1528)] : 235 (3.59), 291 (3.69)

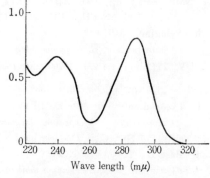

Fig. 220　UV Spectrum of Lycorine in 10% HCl (Kori *et al.*)

1503)　Kondo, Katsura : *Ber.*, **73**, 1424 (1940).　　1504)　Kotera, K : *Tetrahedron*, **12**, 240, 248 (1961).
1505)　Hill, R. K. *et al* : *J. A. C. S.*, **84**, 4951 (1962).　　1506)　Cook, J. W. *et al* : *Soc.*, **1954**, 4176.
1507)　Takeda, K. *et al* : *Chem. Pharm. Bull.*, **5**, 234 (1957).
1508)　Kori, S. *et al* : *Y. Z.*, **81**, 1042 (1961).

IR ν_{max}^{nujol} cm^{-1} [1528] : 3350, 3050, 1508, 1045, 944

Biosynthesis of lycorine and related alkaloids[1511] [1512]

(b) Caranine (IV)[1513]

≪Occurrence≫　Amaryllis belladonna, Crinum spp.

mp. 178~180°, $[\alpha]_D$ −196.6° (CHCl₃), $C_{16}H_{17}O_3N$ Perchlorate mp. 270°

UV λ_{max}^{EtOH} mμ (log ε): 233~236 sh (3. 47), 292~296 (3.68)

(c) Pseudolycorine (V)[1510]

≪Occurrence≫　Lycoris radiata (ヒガンバナ)

mp. 247~248° (decomp.), $[\alpha]_D$ −62° (EtOH), $C_{16}H_{19}O_4N\cdot4H_2O$

O–Acetate　　mp. 204~205°

IR λ_{max} μ: 5.73, 5.76

(d) Norpluviine (VI)[1510]

≪Occurrence≫　Lycoris radiata,　　mp. 274~275° (decomp.), $[\alpha]_D$ −160° (MeOH), $C_{16}H_{19}O_3N$

(e) Pluviine (VII)[1514]

≪Occurrence≫　Lycoris radiata, Narcissus pseudonarcissus.

mp. 225°, $[\alpha]_D$ −140° (CHCl₃), −181° (EtOH), $C_{17}H_{21}O_3N$

(f) Parkamine (VIII)[1515]

≪Occurrence≫　Amaryllis parkeri

mp. 251~253° (decomp.), $[\alpha]_D$ +69° (CHCl₃), $C_{18}H_{21}O_5N$, perchlorate
mp. 245° (decomp.)

(VIII)

(g) Falcatine (IX)[1516]~[1518]

≪Occurrence≫　Nerine falcata, N. laticoma

mp. 127~128°, $[\alpha]_D$ −197.8° (CHCl₃), $C_{17}H_{19}O_4N$,
methiodide mp. 250~255° (decomp.)

(IX) ⟶ Parkamine (VIII)[1516]

(IX)[1516]

(h) Zephyranthine (X)[1519]

≪Occurrence≫　Zephyranthes candida HERB. (タマスダレ)

mp. 201~202° (decomp.), $[\alpha]_D$ −43.17° (CHCl₃), pKa' 9.20,
$C_{16}H_{19}O_4N\cdot1/2H_2O$

UV λ_{max}^{EtOH} mμ (log ε): 291 (3.71)

IR ν_{max}^{KBr} cm^{-1}: 3470 (OH), 1040, 945 (–O–CH₂–O–)

(X)

1509)　Misukami, S: Tetrahedron, 11, 89 (1960).
1510)　Uyeo, S. et al: Soc., 1959, 172.
1511)　Battersby, A.R. et al: Proc. Chem Soc., 1960, 410, 1961, 243; Soc., 1964, 1595.
1512)　Wildman, W.C. et al: Proc. Chem. Soc., 1962, 180.
1513)　Wildman, W.C. et al: J.A.C.S., 77, 1253 (1955).
1514)　Boit, H.G. et al: Ber., 89, 163 (1956), 90, 363 (1957).
1515)　Boit, H.G. et al: Ber., 92, 2578 (1959).
1516)　Wildman, W.C. et al: J.A.C.S, 77, 4807 (1955).
1517)　Benington, F: J. Org. Chem., 27, 142 (1962).
1518)　Garbutt, D.F.C. et al: Soc., 1962, 5010.
1519)　Ozeki, S: Chem. Pharm. Bull., 12, 253 (1964).

(3) 5,10b-Ethanophenanthridines
(a) Stereochemical Correlation[1520)

(XI) (−)-Crinane   (XII) Crinine   (XIII) Buphanisine[1141) 1147)

(XIV) Powellane   7-CH$_3$O[1521) or 10-CH$_3$O[1522)   (XVI) Buphanidrine
                      (XV) Powelline

(XVII) Buphanamine[1523)   (XVIII) Crinamidine[1522)   (XIX) Undulatine[1520)

(XX) Flexinine[1523)   7-CH$_3$O[1522), 10-CH$_3$O[1524)   (XXII) Bupanitine
                        (XXI) Nerbowdine   = Nerbowdine[1525)

(XXIII) Ambelline[1526) 1537)   (XXIV) (−)-Epicrinine[1526) 1527)   (XXV) (+)-Crinane[1520)

1520) Wildman, W.C. et al: J.A.C.S., 82, 1472, 3368 (1960).
1521) Lloyd, H.A. et al: J. Org. Chem., 27, 373 (1962).
1522) Wildman, W.C. et al: ibid., 26, 181 (1961).
1523) Wildman, W.C. et al: ibid., 26, 881 (1951).
1524) Hauth, H. et al: Helv. Chim. Acta, 46, 810 (1963).
1525) Goosen, A. et al: Soc., 1961, 4038, 1960, 1097.
1526) Hauth, H. et al: Helv. Chim. Acta, 45, 1307 (1962).
1527) Wildman, W.C: J.A.C.S., 80, 2567 (1958), 82, 2620 (1960).

(XXVI) (+)-*epi* Crinine[1526)~1528)]   (XXVII) Haemultine

R=H, R'=OH (XXVIII) Haemanthamine[1520) 1529) ~1536)]
R=OH, R'=H (XXIX) *epi*-Haemanthamine[1529)]

(XXX) Haemanthidine   (XXXI) Crinamine[1529) 1532)]

(b) **Crinine** (Crinidine) (XII)[1520) 1527) 1528)]

《Occurrence》   *Crinum moorei, C. powellii, Nerine bowdenii*

mp. 209~210°, [α]$_D$ −11.1° (CHCl$_3$), C$_{16}$H$_{17}$O$_3$N, picrate mp. 237~239°

UV λ_{max}^{EtOH} mμ (log ε): 238 (3.53), 294 (3.71)

IR ν_{max}^{nujol} cm^{-1}: 3150, 1510, 1040, 938

(c) **Buphanisine** (XIII)[1520) 1526) 1528)]

《Occurrence》   *Ammocharis coranica*

mp. 124~126°, [α]$_D$ −25° (EtOH), C$_{17}$H$_{19}$O$_3$N

UV λ_{max}^{EtOH} mμ (log ε): 240 (3.57), 295 (3.75).

IR ν_{max} cm^{-1}: 1510, 1038, 932

(d) **Powelline** (XV)[1521) 1522)]

《Occurrence》   *Crinum moorei, C. powellii*

mp. 197~198°, [α]$_D$ 0° (CHCl$_3$), C$_{17}$H$_{19}$O$_4$N, picrate mp. 223~224°.

(e) **Buphanidrine** (XVI)

《Occurrence》   *Boöphone fischeri*[1527) 1528)]

mp. 88~89°, 90~92°, [α]$_D$ −6.93° (CHCl$_3$)[1527)], +2° (EtOH)[1528)], C$_{18}$H$_{21}$O$_4$N,

perchlorate mp. 240~242°.

UV λ_{max}^{EtOH} mμ (log ε): 281 (3.14).

1528) Hauth, H *et al*: *Helv. Chim. Acta*, **44**, 491 (1961).
1529) Fales, H, M *et al*: *J.A.C.S.*, **82**, 197 (1960).
1530) Boit, H.G *et al*: *Ber.*, **91**, 1965 (1958).   1531) Goosen, J *et al*: *Soc.*, **1960**, 1088.
1532) Wildman, W.C *et al*: *J. Org. Chem.*, **26**, 1617 (1961).
1533) Warren, F. L *et al*: *Soc.*, **1958**, 4696.   1534) Jeffs, P. W: *Proc. Chem. Soc.*, **1962**, 80.
1535) Suhadolnik, R. J *et al*: *ibid.*, **1963**, 132.   1536) Barton, D. H. R *et al*: *ibid.*, **1961**, 254.

(f) Buphanamine (XVII)[1523) 1528)]

≪Occurrence≫ *Boöphone disticha* HERB. (*Haemanthus toxicarius* HERB.)

mp. 183~185°, $[\alpha]_D$ −195° (CHCl₃), $C_{17}H_{19}O_4N$, perchlorate·H_2O mp. 180°.

UV λ_{max}^{EtOH} mμ (log ε): 285 (1480).

(g) Crinamidine (XVIII)[1522)]

≪Occurrence≫ *Crinum moorei, Nerine bowdenii* and other *Nerine* spp.

mp. 235~236° (decomp.) $[\alpha]_D$ −24° (CHCl₃), $C_{17}H_{19}O_5N$, methiodide mp. 265° (decomp.)

O–Acetylcrinamidine mp. 137~139°, $[\alpha]_D$ +15° (CHCl₃),

UV λ_{max}^{EtOH} mμ (ε): 287 (1600).

IR $\nu_{max}^{CHCl_3}$ cm⁻¹: no absorption at 5000~3000, 1733 (O–COCH₃)

(h) Undulatine (XIX)[1520) 1528)]

≪Occurrence≫ *Nerine undulata, N. bowdenii, N. flexuosa, Amaryllis belladonna.*

mp. 152~154°, $[\alpha]_D$ −33° (CHCl₃), $C_{18}H_{21}O_5N$, methiodide mp. 267~268° (decomp.)

UV λ_{max}^{EtOH} mμ (log ε): 286 (3.17), IR $\lambda_{max}^{CHCl_3}$ μ: 6.18, 9.58, 10.66.

(i) Flexinine (XX)[1522)]

≪Occurrence≫ *Nerine flexuosa*

mp. 232~234°, $[\alpha]_D$ −12.7° (EtOH), $C_{16}H_{17}O_4N$

(j) Nerbowdine (XXI)[1528)]

≪Occurrence≫ *Nerine bowdenii, Buphane disticha*

mp. 244~245° (decomp,) $[\alpha]_D$ −108.8 (CHCl₃), $C_{17}H_{21}O_5N$

UV λ_{max}^{EtOH} mμ (log ε): 284 (3.22)

IR ν_{max} cm⁻¹: 3450, 3200~2400, 1620, 1038, 935

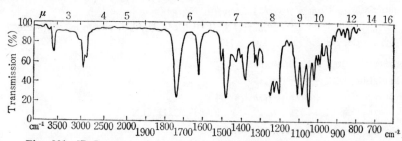

Fig. 221 IR Spectrum of Acetylnerbowdine in CH₂Cl₂ (Hauth *et al.*)[1528)]

(k) Acetylnerbowdine (XXI) **–diacetate**[1528)]

≪Occurrence≫ *Buphane disticha*

mp. 207~209°, $[\alpha]_D$ −116° (CHCl₃), $C_{19}H_{23}O_6N$

UV λ_{max}^{EtOH} mμ (log ε): 285 (3.22)

IR See Fig. 221.

NMR 60Mc, TMS, δ[1524)]: 1.95 (1–OAc), 2.03 (3–OAc), 6.19 (10–H), 5.79, (–O–CH₂–O).

(l) Ambelline (XXIII) [1522) 1526)]

≪Occurrence≫ *Nerine bowdenii*[1522)], *Ammocharis coranica* HERB.[1526)]

mp. 252~255° (decomp.), $[\alpha]_D$ +37° (EtOH), $C_{18}H_{21}O_5N$

UV λ_{max}^{EtOH} mμ (log ε): 283 (3.20), 240sh (3.68).

IR ν_{max}^{nujol} cm⁻¹: 3090, 1620, 1040, 938.

NMR 60Mc, TMS δ[1537)]: 6.58 (arom. H), 5.88 (O–CH₂–O), 6.52 d, 6.18 d AB–system, J=10

1537) Naegeli, P *et al*: *J. Org. Chem.*, **28**, 206 (1953).

$$\left(\overset{H}{\diagdown}C=C\overset{/H}{\diagup}\right)$$

(m) (−)-*epi*-**Crinine** (XXIV)[1527]

Oxocrinine $\xrightarrow{\text{LiAlH}_4}$ (XXIV)

mp. 209~209.5°, $[\alpha]_D$ −142° (CHCl₃), $C_{16}H_{17}O_3N$

UV λ_{max}^{EtOH} mμ (log ε): 237inf. (3.31), 295 (3.47)

(n) (+)-*epi*-**Crinine** (XXVI)[1526]

≪Occurrence≫ *Ammocharis coranica* HERD.

Perchlorate mp. 245~246°(decomp.), $[\alpha]_D$ +102° (EtOH), $C_{16}H_{17}O_3N\cdot HClO_4$

UV λ_{max}^{EtOH} mμ (log ε): 241 (3.54), 293 (3.70)

Base mp. 182~183°, $[\alpha]_D$ +139° (EtOH), $C_{16}H_{17}O_3N$

UV λ_{max}^{EtOH} mμ (log ε): 240 (3.65), 294 (3.87)

IR ν_{max}^{nujol} cm⁻¹: 3200~2100, 1515, 1045, 945

(o) **Haemultine** (XXVII)[1530][1532]

≪Occurrence≫ *Haemanthus multiflorus* MARTYN.

mp. 174~175°, $[\alpha]_D$ +174°, $C_{16}H_{17}O_3N$

(p) **Haemanthamine** (Natalensine) (XXVIII)[1529][1533]

≪Occurrence≫ Many spp. of *Haemanthus*, *Narcissus* and *Zephyranthes*.

mp. 204.5°, $[\alpha]_D$ +19.7° (MeOH), $C_{17}H_{19}O_4N$, picrate mp. 222°

UV λ_{max}^{EtOH} mμ (ε): 297 (5100) (methylenedioxyphenyl)

IR ν_{max}^{CHCl3} cm⁻¹: 3598,(O–H) Biosynthesis[1534]~[1536][1538]

(q) *epi*-**Haemanthamine** (XXIX)[1533][1544a]

mp. 216~217°, $[\alpha]_D$ −24.3° (CHCl₃), $C_{17}H_{19}O_4N$

(r) **Haemanthidine** (XXX)[1545]

≪Occurrence≫ *Haemanthus natalensis*, hybride " King Albert "

mp. 189~190°, $[\alpha]_D$ −41° (CHCl₃), $C_{17}H_{19}O_5N$

epi-Haemanthidine (XXIX)–*epi*–OH–isomer[1531]

≪Occurrence≫ *Haemanthus natalensis*, mp. 211°, $[\alpha]_D$ +44°, $C_{17}H_{19}O_5N$

(s) **Crinamine** (XXXI)[1526]

≪Occurrence≫ *Crinum asiaticum* var. *japonicum*, *Ammocharis coranica* HERD.[1526]

mp. 197~198°, $[\alpha]_D$ +156.6° (CHCl₃), $C_{17}H_{19}O_4N$

UV λ_{max}^{EtOH} mμ (log ε): 293 (3.71), 241sh (3.48).

IR ν_{max}^{nujol} cm⁻¹: 3200, 1515, 1040, 938.

1538) Battersby, A. R *et al*: *J.A.C.S.*, **83**, 4098 (1961).

(4) Acetal or Lactone Alkaloids[1539) 1549)]

R=H (XXXII)[1509) 1546) ~1549)]
Lycorenine

R=OCH₃ (XXXII-a)[1539)]
Nerinine

R=CH₃ (XXXIII)[1546) ~1549)]
Homolycorine

R=H (XXXIV)[1510)]
Desmethylhomolycorine

R=CH₃ (XXXV)[1518)]
Krigeine

R=H (XXXVI)[1518)]
Krigenamine

(XXXVII) Albomaculine[1539) 1550)]

(XXXVIII) Clivonine[1539) 1550)]

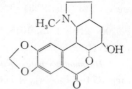

(XXXIX) Tazettine[1540) 1541)]

(a) **Lycorenine** (XXXII)[1546) ~1549)]

≪Occurrence≫ *Cooperanthes hortensis, Haemanthus albomaculatus, Lycoris radiata, Narcissus* spp., *Zephyranthes citrina*

mp. 198~200°, [α]D +180° (CHCl₃), C₁₈H₂₃O₄N, methiodide. mp. 260° (decomp.)

Lycorenine $\xrightarrow{\text{CrO}_3}$ Homolycorine (XXXIII)[1546)]

Acetyllycorenine

UV See Fig. 222, λ_{max} mμ (log ε)[1546)] : 235 (4.31), 281 (3.95), 3.13 (3.81).

1539) Boit, H. G *et al*: *Ber.*, **89**, 163, 1129 (1956), **90**, 57 (1957).
1540) Uyeo, S *et al*: *Soc.*, **1961**, 2485, **1959**, 1446, **1956**, 4749.
1541) Boit, H. G: *Ber.*, **87**, 1448, 1704 (1954). 1541a) Volpp, G. P *et al*: *ibid.*, **97**, 563 (1964).
1542) Barton, D.H.R *et al*: *Soc.*, **1962**, 806 ; *Proc. Chem. Soc.*, **1960**, 392.
1543) Uyeo, S: *4 th Symposium on the Chemistry of Natural Products*, p.28 (1960).
1544) Wildman, W. C *et al*: *J. Org. Chem.*, **25**, 2153 (1960).
1544a) Wildman, W. C *et al*: *Chem. & Ind.*, **1958**, 561. 1545) Boit, H. G: *Ber.*, **87**, 1339 (1954).
1546) Kitagawa, T *et al*: *Soc.*, **1955**, 1066.
1547) Boit, H. G. *et al*: *Ber.*, **87**, 681 (1954), **88**, 133 (1955).
1548) Wildman, W. C. *et al*: *J.A.C.S.*, **77**, 4399 (1955).
1549) Robinson, R. *et al*: *Chem. & Ind.*, **1955**, 1086.

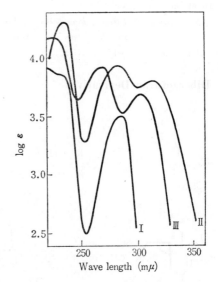

Fig. 222 UV Spectra of (I) Lycorenine, (II) Acetyllycorenine and (III) Homolycorine in EtOH (Kitagawa *et al.*)[1546]

(b) Nerinine (XXXII–a)[1539]

《Occurrence》 *Nerine sarniensis, Zephyranthes Candida*

mp. 209~210°, $[\alpha]_D$ +155° (CHCl$_3$), C$_{19}$H$_{25}$O$_5$N

(c) Homolycorine (XXXIII)[1546]~[1549]

《Occurrence》 *Hippeastrum* spp. *Lycoris albiflora, Narcissus* spp.

mp. 175°, $[\alpha]_D$ +85° (EtOH), C$_{18}$H$_{21}$O$_4$N

UV λ_{max}^{EtOH} mμ (log ε): 225 (4.18), 267 (3.91), 305 (3.69), see Fig. 222

IR $\nu_{C=O}$ cm^{-1}: 1712

(d) Desmethylhomolycorine (XXXIV)[1510]

《Occurrence》 *Lycoris radiata* mp. 213~214°, $[\alpha]_D$ +96.4° (CHCl$_3$) C$_{17}$H$_{19}$O$_4$N

UV λ_{max}^{EtOH} mμ (log ε): 229 (4.32), 269 (3.88), 307 (3.73)

IR λ_{max}^{nujol} μ: 3.05 (OH), 5.88 (C=O)

(e) Krigeine (XXXV)[1518]

《Occurrence》 *Nerine krigeii,* mp. 209~210° (decomp), $[\alpha]_D$ +245° (EtOH), C$_{18}$H$_{21}$O$_6$N

UV λ_{max}^{EtOH} mμ (ε): 285 (990).

(f) Krigenamine (XXXVI)[1518]

《Occurrence》 *Nerine krigeii* mp. 210~211°, $[\alpha]_D$ +210° (CHCl$_3$), C$_{18}$H$_{21}$O$_5$N

UV λ_{max}^{EtOH} mμ (ε): 280~285 (1345), IR λ_{max}^{KBr} μ: 6.18, 9.55, 10.66.

Krigenamine $\longrightarrow$ Falcatine (IX) methiodide

(g) Albomaculine (XXXVII)[1550]

《Occurrence》 *Haemanthus albomaculatus,* mp. 180~181°, $[\alpha]_D$ +71.1° (CHCl$_3$), C$_{19}$H$_{23}$O$_5$N.

UV λ_{max}^{EtOH} mμ (log ε): 224 (4.44), 266 (4.03), 295sh (3.41).

(h) Clivonine (XXXVIII)[1550]

《Occurrence》 *Clivia miniata,* mp. 199~200°, $[\alpha]_D$ +41.24° (CHCl$_3$), C$_{17}$H$_{19}$O$_5$N

UV λ_{max}^{EtOH} mμ (log ε): 226 (4.33), 268 (3.77), 308 (3.76)

1550) Wildman, W. C. *et al*: *J.A.C.S.*, **78**, 2899 (1956).

(i)   **Tazettine**   (XXXIX)[1540) 1541) 1551) 1552)]

≪Occurrence≫   *Elisena longipenta, Galanthus* spp., *Haemanthus* spp., *Hymenocallis* spp., *Lycoris radiata, Narcissus* spp., *Nerine* spp., *Ungernia* spp., *Zephyranthes* spp.

mp. 208~210°, $[\alpha]_D$ +160.4° (CHCl$_3$), +121.6° (EtOH), $C_{18}H_{21}O_5N$

(5)   Dibenzofuran Derivatives

R=OH, R'=H (XL) Galanthamine[1541a) 1542) 1553) 1554)]

R=H, R'=OH (XLI) (−)-*epi*-Galanthamine[1541a)]

(XLII)   Lycoramine[1543)]

(XLIII)   Chlidanthine[1539)]

(XLIV)   Narcissamine[1539)]

(XLV)   Narwedine[1542)]

(XLVI)   Nivalidine[1553)]

(a)   **Galanthamine** (Lycorenine)   (XL)[1541a) 1542) 1553) ~1555)]

≪Occurrence≫   *Cooperanthes hortensis, Crinum defixum, C. yemense, Eustephia yuyuensis, Galanthus erwesii, G. woronowii, Haemanthus rutilum, Hymenocallis* spp., *Leucojum* spp., *Lycoris radiata* and other spp., *Narcissus* spp., *Pancratium illyricum, Sternbergia fischeriana, Ungernia victoris, Vallota purpurea, Zephyranthes* spp.

mp. 127~129°, $[\alpha]_D$ −121.4° (EtOH), $C_{17}H_{21}O_3N$, hydrobromide mp. 246~247° methiodide 279°(decomp.)

UV: See Fig. 223
IR: See Fig. 224

1551)   Ikeda, T *et al*: *Chem. & Ind.*, **1956**, 411.   1552)   Wildman, W. C. *et al*: *ibid.*, **1955**, 1159.
1553)   Bubewa-Iwanowa, L: *Ber.*, **95**, 1348 (1962).
1554)   Uyeo, S: Congress Handbook, *Intern. Congr. Pure and Appl. Chem.* XVI th Congr., Paris, **1957**, 209.
1555)   Barton, D. H. R. *et al*: *Proc. Chem. Soc.*, **1961**, 254, **1962**, 179; *Soc.*, **1963**, 4545.

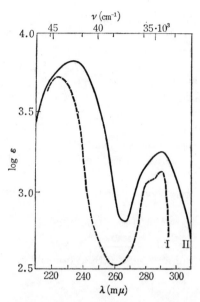

Fig. 223 UV Spectra of (I) Galanthamine and (II) Nivaldine in EtOH (Bubewa-Iwanowa)[1553]

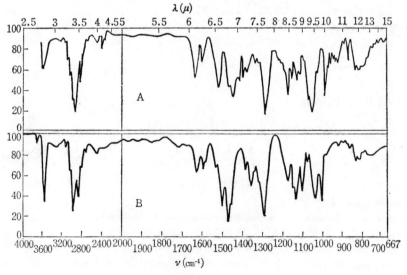

Fig. 224 IR Spectra of (A) Galanthamine and (B) Nivaldine
in CHCl₃ (Bubewa-Iwanowa)[1553]

IR $\nu_{max}^{CHCl_3}$ cm⁻¹ [1554] : 3575, 3700w, 3620w (OH)

Synthesis [1542], Biogenesis[1555]

(b) (−)-*epi*-**Galanthamine** (XLI)[1541]

≪Occurrence≫   *Lycoris squamigera*[1556]

Narwedine (XLV) $\xrightarrow{\text{Red}}$ (XLI)[1557]

1556) Wildman, W. C. *et al*: *J.A.C.S.*, **78**, 4145 (1956).

1557) Boit, H. G. *et al*: *Ber.*, **90**, 2197 (1957).

Galanthamine (XL) $\xrightarrow{\text{HCl}}$ (XLI)[1541]

mp. 188.5～189.5°, $[\alpha]_D$ −248°, $C_{17}H_{21}O_3N$

NMR　in $CDCl_3$, 60Mc, TMS, ppm: 2.35 (OCH_3), 3.85, ($>N-CH_3$), 3.62d, 4.10d, $J=15\,^c/_s$ (two arom.
protons), 4.89q (X part of, ABXsystem, $J_{AX}=11$, $J_{BX}=6\,^c/_s$) ($C_1-H\sim C_6-2$ H)

(c) Lycoramine (XLII)[1543) 1558) 1559)]

　≪Occurrence≫　*Lycoris radiata*[1558]

(XLII)=Dihydrogalanthamine[1559]

mp. 120～121°, −97.4° (EtOH), $C_{17}H_{23}O_3N$

(d) Chlidanthine (6-O-Methylapogalanthamine) (XLIII)[1539]

　≪Occurrence≫　*Chlidanthus fragrans*,　　mp. 238～239°, $[\alpha]_D$ −140° (EtOH), $C_{17}H_{21}O_3N$

(e) Narcissamine (XLIV)[1539]

　≪Occurrence≫　*Narcissus* spp.,　　mp. 193～199°, $[\alpha]_D$ −9.8° (EtOH), $C_{16}H_{19}O_3N$

(f) Narwedine (XLV)[1542]

　≪Occurrence≫　*Narcissus* spp.[1560] " Texas " Dafodils[1542]

(−)-Galanthamine (XL) $\xrightarrow{\text{MnO}_2}$ (XLV)[1542]

mp. 188～190°, $[\alpha]_D$ +19.3° (CHCl$_3$)[1560], +52° (CHCl$_3$) (natural)[1542], +40.5° (CHCl$_3$) (synthetic)[1542]
$C_{17}H_{49}O_3N$

(g) Nivalidine (XLVI)[1553]

　≪Occurrence≫　*Galanthus nivalis* L. var. *gracillus*

Galanthamine (XL) $\xrightarrow{\text{HCl}}$ (XLVI)

mp. 206～208°, $C_{17}H_{19}O_2N$

UV　See Fig. 223, $\lambda_{max}^{EtOH}\,m\mu$: 230～234, 290

IR　See Fig. 224, $\nu_{max}^{CHCl_3}\,cm^{-1}$: 3540 (OH), 1290 (CH$_3$O)

(6)　A novel Ring System

(XLVII) Montanine[1544]　　　(XLVIII) Coccinine[1544]　　　(XLIX) Manthine[1544]

(a) Montanine (XLVII)[1539) 1561)]

　≪Occurrence≫　*Haemanthus amarylloides, H. tigrinus, H. montanus.*

mp. 57～60°, $[\alpha]_D$ −87.6° (CHCl$_3$), $C_{17}H_{19}O_4N \cdot C_3H_6O$ (fr. acetone).

Anhydrous base $C_{17}H_{19}O_4N$, $[\alpha]_D$ −97.9° (CHCl$_3$), hydrate $C_{17}H_{19}O_4N \cdot H_2O$　mp. 88～89°,
perchlorate mp. 249～250° (decomp.), methiodide　mp. 269～272° (decomp.)

(b) Coccinine (XLVIII)[1539) 1561)]

　≪Occurrence≫　*Haemanthus amarylloides, H. tigrinus, H. coccineus*

mp. 162～163°, $[\alpha]_D$ −188.8° (EtOH), $C_{17}H_{19}O_4N$,　　perchlorate mp. 254～255° (decomp.)

UV　$\lambda_{max}^{EtOH}\,m\mu$ (log ε): 213 (4.25) 244 (3.58) 296 (3.64)

(XLVIII) $\xrightarrow{\text{K+}tert\text{-butylalcohol}}$ Dehydrococcinine (L)

1558)　Kondo, H *et al*: *Y.Z.*, **52**, 51 (1932).

1559)　Uyeo, S *et al*: *Pharm. Bull.*, **1**, 139 (1953).　　1560)　Boit, H.G. *et al*: *Ber.*, **90**, 2197 (1957).

1561)　Wildman, W.C. *et al*: *J.A.C.S.*, **77**, 1248 (1955).

Dehydrococcinine (L)

(L)

mp. 191~193°, $[\alpha]_D$ +40° (CHCl$_3$)

UV λ_{max}^{EtOH} mμ (log ε): 290 (7050), 232 inf. (11900) (two isolated benzenoid chromophores, not biphenyl)

IR ν_{max}^{nujol} cm^{-1}: 3257 (>NH), 2469 (H bonded phenol OH), 1592, 1034, 927

(c) **Manthine** (XLIX)[1544]

≪Occurrence≫ *Haemanthus amarylloides, H. coccineus.*

mp. 114~116°, $[\alpha]_D$ −71.3° (CHCl$_3$), C$_{18}$H$_{21}$O$_4$N

UV λ_{max}^{EtOH} mμ (log ε): 243 (3.64), 294 (3.71)

(7) Alkaloids of Unknown Structure

(a) **Manthidine** (LI)[1561]

≪Occurrence≫ *Haemanthus amarylloides.*

mp. 269~270°, $[\alpha]_D$ −26.6° (CHCl$_3$), C$_{18}$H$_{21}$O$_4$N

UV λ_{max}^{EtOH} mμ (ε): 240 (3.63), 293 (3.72)

(b) **Burnsvigine** (LII)[1539]

≪Occurrence≫ *Burnsvigia radulosa.*

mp. 242~244°, $[\alpha]_D$ −106.5° (CHCl$_3$), C$_{16}$H$_{17}$O$_4$N

(c) **Hippandrine** (LIII)[1562]

≪Occurrence≫ *Hippeastrum brachyandrum, H. rutilum.*

mp. 194°, $[\alpha]_D$ −110° (CHCl$_3$), C$_{17}$H$_{14}$O$_4$N

(d) **Macronine** (LIV)[1563]

≪Occurrence≫ *Crinum macrantherum.*

mp. 203~205°, $[\alpha]_D$ +413° (CHCl$_3$) pK$_{MCS}$ 6.31, C$_{18}$H$_{19}$O$_5$N

UV λ_{max}^{EtOH} mμ (log ε): 228.5 (4.49), 268 (3.74), 308 (3.78)

(e) **Macranthine** (LV)[1563]

≪Occurrence≫ *Crinum macrantherum*

mp. 238~240°, $[\alpha]_D$ −19° (CHCl$_3$), pK$_{MCS}$ 7.23, C$_{16}$H$_{19}$O$_5$N

〔G〕 Quinazolone Alkaloids

(1) Febrifugine, 2-methyl-4-quinazolone, 3-allyl-4-quinazolone (UV)[1564], and Chiang Chan alkaloid (diosdichroine B) (IR)[1565]

1562) Boit, H. G: *Ber.*, **92**, 2582 (1989).   1563) Hanth, H. *et al*: *Helv. Chim. Acta*, **47**, 185 (1964).
1564) Koepfli *et al*: *J.A.C.S.*, **71.**, 1048 (1949).
1565) Williams, H. *et al*: *J.Org. Chem.*, **17**, 14, 20 (1952).

〈2〉 Alkaloids of *Glycosmis arborea* (Roxb.) DC.

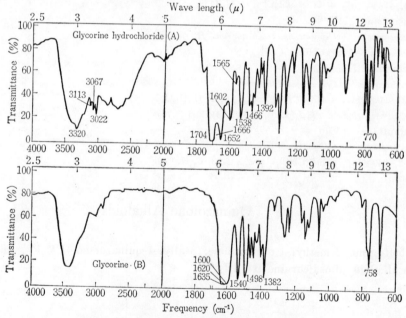

(I) Glycosmicine[1566]　　(II) Glycorine　　(III) Glycosminine[1567]　　(IV) Arborine

〈a〉 Glycosmicine (I)[1566]

mp. 270~271°, $C_9H_8O_2N_2$

UV λ_{max}^{EtOH} mμ (log ε): 219 (4.70), 244 (3.97), 311 (3.64); $\lambda_{max}^{0.01N-NaOH}$ mμ (log ε): 223 (4.70), 313 (3.75)

IR ν_{max}^{KBr} cm^{-1}: 1701, 1661 (2 C=O at 4 and 2 positions), 1605, 1484 (quinazoline dione system), 1315 (N–CH$_3$)

NMR in CDCl$_3$, 60Mc, TMS, δ: 3.58 (N–CH$_5$), 8.26~7.13 (arom. protons), 8.22dd, (1–H), 7.72t (other ring protons), 8.59m (>NH)

〈b〉 Glycorine (II)[1566]

mp. 145~147°, $C_9H_8ON_2$　——HCl mp. 242° decomp.

UV λ_{max}^{EtOH} mμ (log ε): 269 (3.59), 278 (3.66), 306 (3.91), 317 (3.84) (pract. same to 1–methyl–4–quinazolone); $\lambda_{max}^{0.01N-HCl}$ mμ (log ε): 282 (3.68), 295 (3.73), 304 (3.69)

IR ν_{max}^{KBr} cm^{-1}: 3313, 3067, 3022, 1392 (bonded NH$_4$ salt), 3320 (?), 1704, 1652, 1602, 1538 (4–quinazolone system), (1652: pyridylbetaine type) (see Fig. 225)

Fig. 225 IR Spectra of (A) Glycorine–HCl and (B) Glycorine in KBr (Bhattacharyya *et al.*)[1566]

1566) Bhattacharyya, J. *et al*: *Tetrahedron*, **19**, 1011 (1963).
1567) Chatterjee, A. *et al*: *J.A.C.S.*, **76**, 2459 (1954).

NMR of Hydrochloride (V) in dimethylsulfoxide, 60MC, TMS δ: 8.08

(four arom. protons), 9.68s ($\overset{\diagdown N}{\underset{N}{\bigcirc}}_{H}$), 4.08s ($\geqslant \overset{\oplus}{N}-CH_3$, normal shift

by assuming pyridyl methyl grp. accounted for 3.83), 6.25 (H_2O)

(c) **Glycosminine** (III) [1566)1567]

mp. 225~227°, $C_{15}H_{12}O_2N$

UV λ_{max}^{EtOH} mμ (log ε): 225 (4.44), 265 (3.95), 303 (3.66), 312 (3.57)

IR ν_{max}^{KBr} cm^{-1}: 3356 (NH or OH), 1676, 1613 (2 or 3 or both subst. 4-quinazolone), 770 (O-di-

subst. benzene ring), 748, 713 (monosubst. benzene ring)

NMR in CDCl$_3$, 60MC, TMS, δ: 10.25 (bonded $\geqslant$NH), 8.26d (1-H), 4.08 (-CH$_2$-), 7.5 (remaining

protons)

(d) **Arborine** (IV) [1568]

mp. 155~156°, $C_{16}H_{14}ON_2 \cdot C_6H_6$ (fr. benzene)

IR ν_{max} cm^{-1}: (CHCl$_3$) 3663, 3425, (KBr) 3534, 3414. ν_{max}^{CHCl3} cm^{-1}: 1730, 1642, 1610, 1603, 1531,

1497, 1466, 1437

Characteristic IR Absorptions of Functional Groups ν_{max} cm^{-1}

(i) Free Group

O–H: 3636~3610, N–H: 3500~3400, –CO–NH–: 3330~3280, 3100~3060

(ii) Bonding Type

OH······O–: 3500~3200, OH······O=C$\langle$: 3300~3240, $\geqslant$NH······N$\langle$: 3300~3150

Ar.–$\overset{O}{\overset{\|}{C}}$–N=C$\langle$: 1531, –$\overset{O}{\overset{\|}{C}}$–N=C$\langle$: 1608, 1451~1468, 1453~1443, Ar.–$\overset{O}{\overset{\|}{C}}$–N=C$\langle$: 1642 ($\nu_{C=O}$ of (IV))

Ar.–$\overset{O}{\overset{\|}{C}}$–NH–CH$\langle$: 1667 ($\nu_{C=O}$ of (V)), C=C: 1610, 1603, 1715~1724

NMR in CDCl$_3$, TMS, cps: 218.2 (3) (N–CH$_3$), 256.2 (2)

(–CH$_2$–⬡), 438.4, >438.4 complex (9) (protons

of benzyl, phenyl)

(IVa) Arbosine, tautomeric
form of (IV)

(3) **Alkaloids of *Adhatoda vasica* NEES.** [1569]

(VI) Vasicine　　　(VII) Vasicinone

(a) **Vasicine** (VI)

≪Other Occurrence≫ *Peganum harmala*

mp. 211~212°, [α]$_D$ −208~−254°, $C_{11}H_{12}ON_2$

(VI) $\xrightarrow{Oxyd.}$ Vasicinone (VII)

1568) Chakravarti, D. *et al*: *Tetrahedron.*, **16**, 224 (1962).

1569) Metha, D. R *et al*: *J. Org. Chem.*, **28**, 445 (1963).

(b) Vasicinone (VIII)

<Other occurrence> *Peganum harmala*

mp. 201∼202°, $[\alpha]_D$ −74° (mix. of *l*- and *dl*-), $C_{11}H_{10}O_2N_2$, —HCl mp. 232∼234°

UV λ_{max}^{H2O} mμ: 227, 272, 302, 315

IR λ_{max} μ: 3.21, 3.42, 6.03, 6.875, 7.22, 7.516, 7.74, 8.275, 8.48, 9.06, 9.265, 9.73, 10.13, 10.31, 11.155, 11.33, 11.54, 11.71, 12.92, 13.35, 13.89, 14.41

[H] Tropane Alkaloids

(1) Spectra

UV Atropine sulfate, tropine, homatropine, and tropinone[1570], cocaine hydrochloride, benzoylecgonine, and ecgonine[1570].

IR *l*–Hyosciamine, tropine[1400], *l*-hyosciamine[1400], atropine[1404], methoscopolamine, atropine sulfate[1571], homatropine hydrobromide[1572], homatropine methylbromide[1573].

(2) Valeroidine (I)[1574]

<Occurrence> *Duboisia myoporoides*, mp. 85°, $[\alpha]_D$ −9.1° (EtOH), $C_{13}H_{23}O_3N$

(I)

(3) Biosynthesis of Tropane alkaloids[1575]

[I] Quinoline Alkaloids

(1) Spectra

UV Quinine sulfate, apoquinine, cinchonine[1578], cinchonidine, quinine and quinidine[1579]

IR Quinine[1400], cinchonidine[1400]

1570) Castille *et al*: *Bull. Soc. Chim. biol.*, **10**, 623 (1928).　　1571) Sadtler Card No. 7160

1572) Sadtler Card No. 7162.　　1573) Sadtler Card No. 83–B.

1574) Fodor, G., *et al*: *Soc.*, **1961**, 3219, **1957**, 1347.

1575) Mothese K., *et al*: *Biochim. biophys. Acta*, **46**, 588 (1961); *Angew. Chem.*, **73**, 251 (1961).

1576) Younken H. W., *et al*: *J. P. S.*, **51**, 121 (1962).

1577) Leete E: *J. A. C. S.*, **82**, 612 (1960).

1578) Landolt–Börnstein: Zahlenwerte und Funktionen 6 Auf. I band, 3 Teil, Molekulen II. s 302, Abb 244 (Springer–Verlag 1951).　　1579) Grant, Jones: *Anal. Chem.*, **22**, 679 (1950).

(2) IR Spectra of Cinchona Alkaloids[1580]

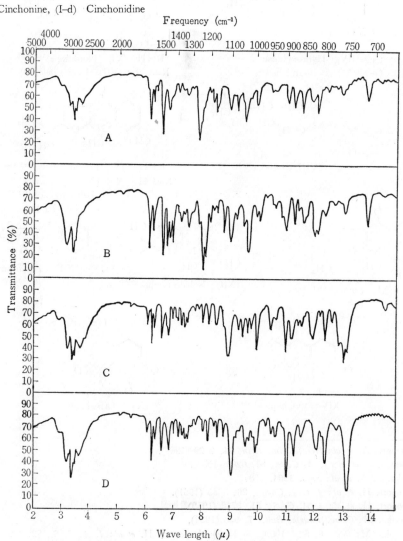

CH₂-CH-CH-CH=CH₂

Q ;

R=-OCH₃
(I-a) Quinine, (I-b) Quinidine
R=H
(I-c) Cinchonine, (I-d) Cinchonidine

(I-a) Quinine (I-b) Quinidine
(I-d) Cinchonidine (I-c) Cinchonine

Fig. 225 a IR Spectra of (A) Quinidine, (B) Quinine, (C) Cinchonine and (D) Cinchonidine in KBr (Hayden *et al.*)[1580]

1580) Hayden, A. L. *et al*: *J. A. P. A.*, **49**, 497 (1960).

(3) **Rutaceous Quinoline Alkaloids** (I)~(XXVIII)

R=H: (I) Lunidine[1580) 1580a)]
R=OH: (II) Hydroxylunidine[1580) 1581)]

(II–a) Lunidonine[1580)]

(III) Lunacridine[1580) 1584)]

(+) (IV) Hydroxylunacridine[1580)]
(−) (V) Balfourolone[1583)]

R=H (VI) Nororixine[1585)]
R=CH₃ (VII) Isoorixine[1585)]

(VIII) Orixine[1585) 1586)]

R=H (IX) Lunacrine[1580) ~1582) 1584) 1588)]
R=OH (X) Hydroxylunacrine[1581)]
=Balfourodine[1584) 1587)]

R=H (XI) Lunine[1581) 1588)]

(XII) Isobalfourodine[1587)]

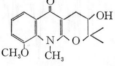

(XIII) Platidesmine[1594)]

(XIV) Orixidinine[1590)]

(XV) Orixidine[1590)]

1580a) Rüegger, A. *et al*: *Helv. Chim. Acta*, **44**, 2329 (1963).
1581) Goodwin, S. *et al*: *J. A. C. S.*, **81**, 6209 (1959).
1582) Clarke, E. A. *et al*: *Soc.*, **1964**, 438.
1583) Rapoport, H. *et al*: *J. Org. Chem.*, **26**, 3585 (1961).
1584) Rapoport, H. *et al*: *J. A. C. S.*, **81**, 3738 (1959).
1585) Terasaka, M: *Chem. Pharm. Bull.*, **7**, 946 (1959),
1586) Terasaka, M: *ibid.*, **8**, 523 (1960). 1587) Rapoport, H. *et al*: *J. A. C. S.*, **82**, 4395 (1960).
1588) Goodwin, S. *et al*: *ibid.*, **81**, 1908, 3065 (1959).
1589) Taylor, W. C. *et al*: *Australian J. Chem.*, **16**, 480 (1963); *Index Chem.*, **10**, 30804 (1963).
1590) Narahashi, K: *Chem. Pharm. Bull.*, **10**, 792 (1962).

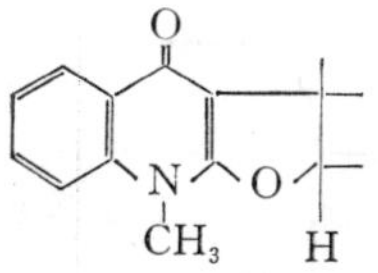

(XVI) Ifflaiamine[1589]

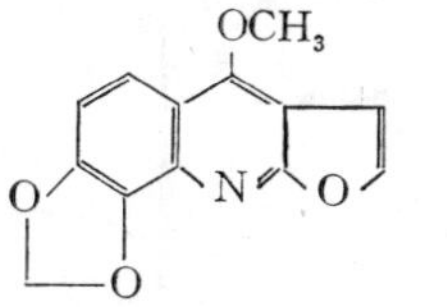

(XVII) Kokusagine[1581)1595)]

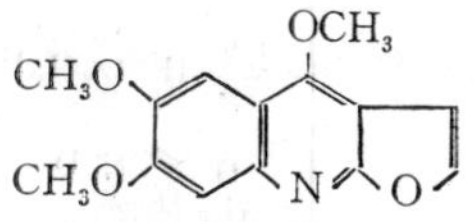

(XVIII) Kokusaginine[1597]

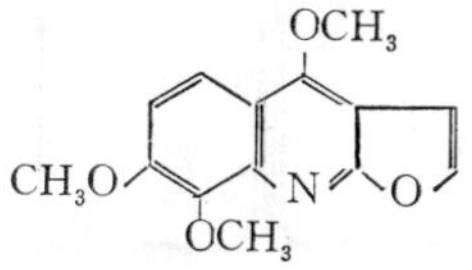

(XIX) Skimmianine[1581)1591)1595)]

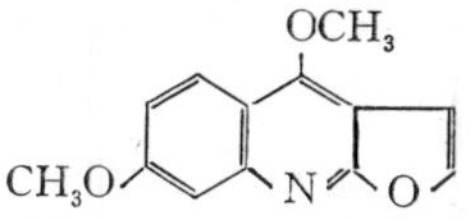

(XX) Evolitrine[1592)~1594)]

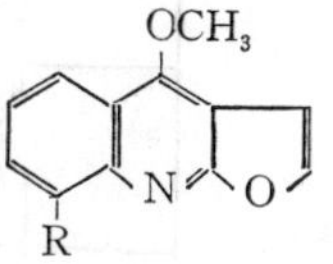

R=H (XXI) Dictamnine[1595a)]
R=OCH₃ (XXII) γ–Fagarine[1595a)]

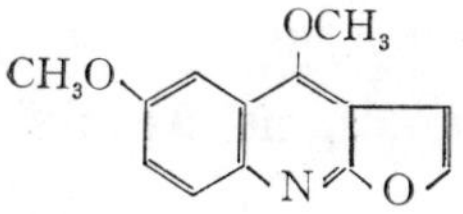

(XXIII) 6–Methoxy–
dictamnine[1594)]

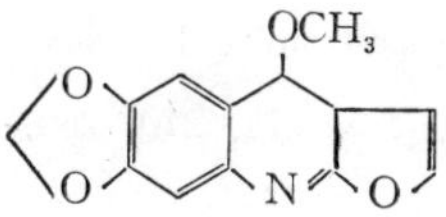

(XXIV) Maculine[1596)]

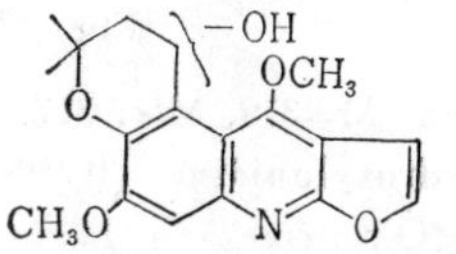

(XXV) Shuazine[1597)]

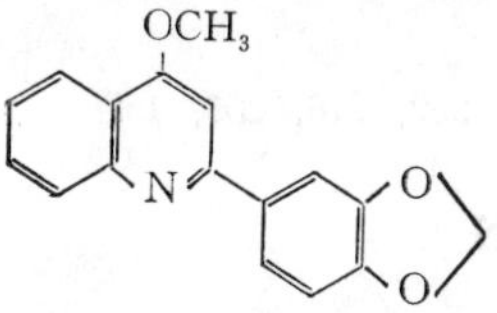

(XXVI) Graveolinine[1598)]

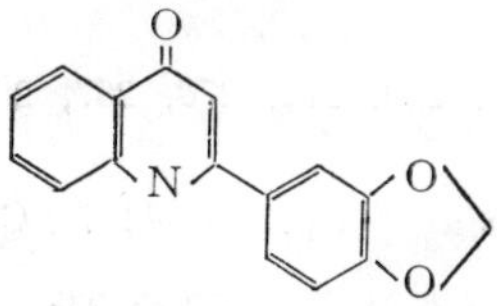

(XXVII) Graveoline[1599)]

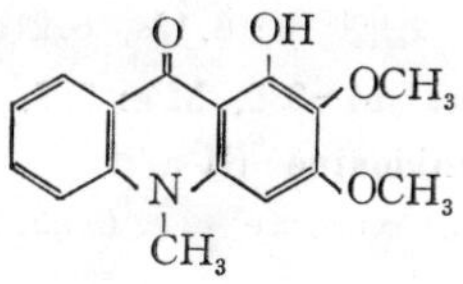

(XXVIII) Arborinine[1600)]

(a) Lunidine (I)[1580a)]

《Occurrence》 *Lunasia quercifolia* mp. 65~66.5°, $[\alpha]_D$ +28° (EtOH), $C_{17}H_{21}O_5N$

UV λ_{max}^{EtOH} mμ (log ε): 226 (4.42), 266 (4.35), 317 (3.93)

NMR in CDCl₃, 60Mc, TMS, ppm: See Fig. 226

1591) Ohta, T. *et el*: *ibid.*, **8**, 377 (1960).   1592) Rapoport, H. *et al*: *J. Org. Chem.*, **25**, 2251 (1960).
1593) Ohta, T. *et al*: *Bull. Chem. Soc. Japan*, **31**, 161 (1958).
1594) Werny, F. *et al*: *Tetrahedron*, **19**, 1293 (1963).
1595) Ishii: *Y. Z.*, **81**, 1633 (1961).
1595a) Grundon, M. F *et al*: *Soc.*, **1957**, 2177.
1596) Ohta, T. *et al*: *Y. Z.*, **82**, 549 (1962).
1597) Kuzorkov, A. D. *et al*: *Zhur. Obskch. Khim.*, **33**, 121 (1963); *Index Chem.*, **9**, 27673 (1963).
1598) Chatterjee, A. *et al*: *Chem. & Ind.*, **1962**, 1982.
1599) Arthur, H. R. *et al*: *Soc.*, **1961**, 4360.
1600) Chakravarti, D. *et al*: *Tetrahedron*, **16**, 251 (1962).

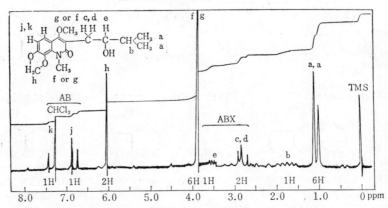

Fig. 226 NMR Spectrum of Lunidine (Rüegger *et al.*)[1580]

Mass M=319, M/e: 317, 302, 299, 285, 274, 270, 260, 258, 246, 232, 216, 203, 188

(b) Hydroxylunidine (II)[1580] [1581]

«Occurrence»   *Lunasia quercifolia*   mp. 124~125°, [α]$_D$ +27.6° (EtOH), C$_{17}$H$_{21}$O$_6$N

UV   λ$_{max}^{EtOH}$ mμ (log ε): 239 (4.39), 257 (4.39), 284 (3.93), 293 (3.90), 332 (3.55), 345sh (3.41)

IR   λ$_{max}^{CHCl3}$ μ: 6.13s, 6.21m, 6.33s, 9.23s, 10.57w

Mass M=335, M/e: 317, 301, 299, 289, 274, 270, 267, 258, 246, 232, 216, 203, 188

(c) Lunidonine (II-a)[1580]

«Occurrence»   *Lunasia quercifolia*   mp. 116~117°, C$_{17}$H$_{19}$O$_5$N

UV   λ$_{max}^{EtOH}$ mμ (log ε): 226 (4.41), 266 (4.35), 312 (3.93)

IR   ν$_{max}^{CH2Cl2}$ cm^{-1}: 1710, 1640, 1603, 1593, 1365, 1360, 1060

NMR Difference is observed between NMR spectra of (I) (Fig. 226) and (II-a) (Fig. 227) on the shifts of protons, b, c, d and e when measured at the same condition in the item (a).

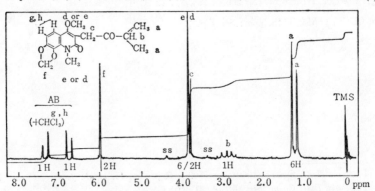

Fig. 227 NMR Spectrum of Lunidonine (Rüegger *et al.*)[1590]

Mass M=317, m/e: 317, 302, 299, 285, 274, 270, 260, 258, 246, 232, 216, 203, 188

(d) Lunacridine (III)[1580] [1584] [1588]

«Occurrence»   *Lunasia quercifolia*   mp. 86~87°, [α]$_D$ +28.1° (EtOH), C$_{17}$H$_{23}$O$_4$N

UV   See Fig. 228, λ$_{max}^{EtOH}$ mμ (log ε). 239 (4.38), 257 (4.41), 285 (3.94), 294 (3.91), 333 (3.55)

IR   λ$_{max}^{CHCl3}$ μ: 3.0 br, 6.11s, 6.19w, 6.30s

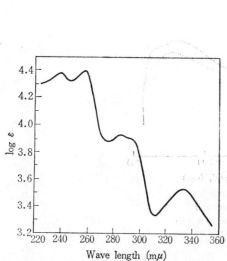

Fig. 228 UV Spectrum of Lunacridine
in EtOH (Goodwin *et al.*)[1588]

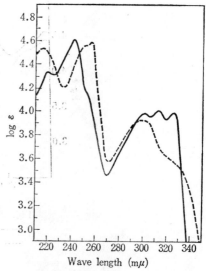

Fig. 229 UV Spectra of Lunacrine
—— in EtOH,
······ in 0.1 N–HCl–EtOH

(e) Hydroxylunacridine (IV)[1580] [1581]

《Occurrence》 *Lunasia quercifolia*, mp. 100~102°, $[\alpha]_D$ +32° (EtOH), $C_{17}H_{23}O_5N$

UV λ_{max}^{EtOH} mμ (log ε): 239 (4.39), 257 (4.39), 284 (3.93), 293 (3.90), 332 (3.55), 345sh (3.41), no

change by acidification, IR $\lambda_{max}^{CHCl_3}$ μ: 6.13s, 6.21m, 6.33s, 9.32s, 10.57w

(f) Balfourolone (V)[1584]

《Occurrence》 *Balfourodendron riedelianum*, mp. 99~100°, $[\alpha]_D$ −36°, $C_{17}H_{23}O_5N$

UV λ_{max}^{MeOH} mμ (log ε): 212 (24000), 239 (25000), 258 (27000), 285 (8200), 293 (7800), 331 (3500);

IR $\lambda_{max}^{CHCl_3}$ μ: 6.15s, 6.22s, 6.31s, 6.80s, 6.92m, 7.09w, 7.31s, 7.71w, 8.05s, 8.59m, 8.89m,

9.30s, 10.11m

(g) Nororixine (VI) and **Isoorixine** (VII)[1585]

《Occurrence》 Orixine (VI) $\xrightarrow{HCl}$ (VI) $\xrightarrow{CH_2N}$ (VII)

Nororixine (VI)

mp. 196°, $C_{16}H_{19}O_6N$

UV See Fig. 230

IR ν_{max}^{KBr} cm^{-1}: 3459 (OH), 3155, 3080 (NH), 1660, 1640 (NH–CO)

Isoorixine (VII)

mp. 127°, $C_{17}H_{21}O_6N$

UV See Fig. 230

UV λ_{max}^{EtOH} mμ (log ε): (VI) 224 (4.53), 258 (4.47), 265 (4.49), 320 (4.12), (VII) 228 (4.47), 260

(4.36), 267 (4.40), 320 (4.00), (2–quinolone type)

IR ν_{max}^{KBr} cm^{-1}: 3440 (OH), 1640, 1620 (R_2N–CO)

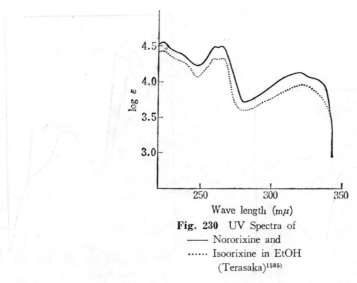

Wave length (mμ)

Fig. 230 UV Spectra of
—— Nororixine and
⋯⋯ Isoorixine in EtOH
(Terasaka)[1585]

(h) Orixine (VIII)[1585) 1586]

≪Occurrence≫ *Orixa japonica* THUNB. (コクサギ)

mp. 152.5°, $C_{17}H_{21}O_6N$

UV λ_{max}^{EtOH} mμ (log ε): 254 (4.66), 318 (3.51); λ_{min}: 237 (4.22), 270 (3.28)

IR See Fig. 231, ν_{max}^{nujol} cm^{-1}: 3450, 3350 (OH); ν_{max}^{KBr} cm^{-1}: 3440, 3350 (OH), 1477, 1275, 1047, 927 (–O–CH$_2$–O)

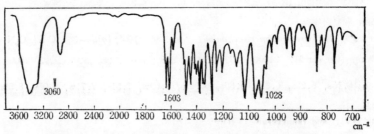

Fig. 231 IR Spectrum of Orixine in KBr (Terasaka)[1586]

(i) Lunacrine (IX)[1580) ~1582) 1584) 1588]

≪Occurrence≫ *Lunasia amara* BLANCO, —— var *repanda*

mp. 117~119°, $[\alpha]_D$ −50° (EtOH), $C_{16}H_{19}O_3N$

UV[1588] See Fig. 229, λ_{max}^{EtOH} mμ (log ε): 220 (4.34), 242 (4.61), 300 (4.00), 313 (4.01), 326 (3.99)

IR $\lambda_{max}^{CHCl_3}\mu$[1588]: 6.18s, 6.29s, 6.45s, 6.64s, 6.84m, 6.96w, 9.37s, 9.88v.w, 10.61s;

$\lambda_{max}^{CCl_4}\mu$: 3.24vw, 3.31sh, 3.36s, 3.46m, 3.51m.

NMR Comparation of lunacrine (IX) and lunine (XI)[1588]

See Fig. 232 and Tab. 96, in CDCl$_3$, 60Mc, cps relative to C_6H_6

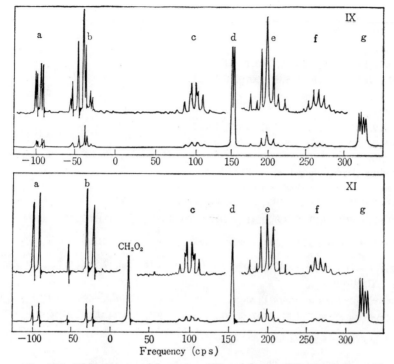

Fig. 232 NMR Spectra of Lunacrine (IX) and Lunine (XI) (Goodwin *et al.*)

(IX) Lunacrine   (XI) Lunine

Table 96 NMR Spectra of Lunacrine and Lunine (in CDCl$_3$, 60Mc, cps) (Goodwin)[1588]

Pattern	cps to C$_6$H$_6$	Intensity ratio	System
(IX) Lunacrine			
a	−100, −98, −93, −91	1/2	ABX $\begin{cases} J_{AX}= 7 \\ J_{AB}= 7 \\ J_{BX}= 2 \\ \delta_{AB}= 9 \end{cases}$
b	−58.2, −54.4, −47.6, −41, −39, −33.9		
c	86.4, 93.1, 94.7, 101.8, 104.3, 111.4	1/2	ABX $\begin{cases} J_{AB}=15.3 \\ J_{AX}= 9.9 \\ J_{BX}= 8.7 \\ \delta_{AB}=16.7 \end{cases}$
e	174.9, 184.6, 190.2, 198.0, 199.8, 206.7, 214		
d	151, 153	6	2s
f	246.8, 253.5, 260.2, 267, 273.6, 280.2	1	–CH(CH$_3$)$_2$
g	318.9, 322, 325.9, 329	6	

(XI) Lunine

a	−100, −91.8	} 1/1	d
b	−30.5, −21.8		d
c	86.1, 93.1, 94.7, 101.8, 103.9, 110.9	} 1/2	} ABX
e	176, 185.4, 190.9, 198.8, 199.9, 207, 213.7		
O–CH₂–O	ca 23	2	s
d	156	1	s
f	247.9, 254.4, 260.4, 267.4, 273.7, 280,7	6	} –CH(CH₃)₂
g	318.2, 321.2, 325, 328	3	

(j) **Balfourodine** (Hydroxylunacrine) (X)[1581) 1584) 1587)]

≪Occurrence≫　*Balfourodendron riedelianum, Lunasia amara*[1581)]

mp. 188∼190° [1584)], 201∼203° [1581)], $[\alpha]_D$ +49° (EtOH), $C_{16}H_{19}O_4N$

UV λ_{max}^{MeOH} mμ (ε)[1584)] : 219 (22000), 241 (43500), 299 (10600), 312 (11100), 325 (9600)

$\lambda_{max}^{0.1N-HCl-MeOH}$ mμ (ε): 214 (28000), 257 (38000), 299 (8700), 315sh (4300)

IR[1584)] λ_{max}^{CHCl3}: 6.17s, 6.26s, 6.42s, 6.59s, 6.69s, 6.96m, 6.72m, 7.90s, 9.31s, 12.03m

(k) **Lunine** (XI)[1581) 1588)]

≪Occurrence≫　*Lunasia amara* BLANCO,　mp. 228∼229°, $[\alpha]_D$ −38.5° (CHCl₃), $C_{16}H_{17}O_4N$

UV λ_{max}^{EtOH} mμ (log ε): 222 (4.31), 247 (4.60), 268sh (3.97), 314 (4.02), 325 (4.01);

$\lambda_{max}^{0.1N-HCl}$ mμ (log ε): 257 (4.61), 332 (3.86)

IR λ_{max}^{CHCl3} μ: 6.09s, 6.30s, 6.39vs, 6.59m, 9.42s, 9.56m, 10.67s

NMR See Fig. 232 and Tab. 96

(l) **Isobalfourodine** (XII)[1587)]

≪Occurrence≫　*Balfourodendron riedelianum*,　mp. 204∼205°, $[\alpha]_D$ +15° (EtOH), $C_{16}H_{19}O_4N$

UV λ_{max}^{MeOH} mμ (ε): 216 (17500), 242 (46600), 298 (7200), 320 (10400), 330 (8900)

$\lambda_{max}^{0.2N-HCl-MeOH}$ mμ (ε): 215 (28600), 253 (44200), 304 (7400), 326sh (4200)

IR λ_{max}^{CHCl3} μ: 6.20, 6.25, 6.32, 6.45, 6.64, 6.82

(m) **Platidesmine** (XIII)[1594)]

≪Occurrence≫　*Platydesma campanulata*,　mp. 137∼138°, $C_{15}H_{17}O_3N$

UV λ_{max}^{EtOH} mμ (log ε): 320.5 (3.55), 307.2 (3.49), 294.5sh (3.24), 283 (3.65), 272 (3.73), 262.3

(3.65), 253.5 (3.50), 238.1 (4.43), 229.3 (4.57)

IR λ_{max}^{CHCl3} μ: 2.79w, 3.36s, 6.11s, 6.29s, 6.59s, 6.81m, 7.00s, 7.16s, 7.31s, 7.49m, 7.64m,

7.72m, 8.08∼8.31m, 8.47m, 8.58m, 8.72w, 8.91s, 9.08s, 9.82s, 10.04m, 10.50m

(n) **Orixidinine** (XIV) and **Orixidine** (XV)[1590)]

Orixine (VIII) $\xrightarrow{\text{heating with } 10\sim20\% HCl}$ (XIV) + (XV)

Orixidinine (XIV)

mp. 209∼210°, $C_{15}H_{15}O_5N$

UV λ_{max}^{EtOH} mμ (log ε): 226.5 (4.49), 235sh (4.46), 248 (4.41), 265sh (4.01), 314 (3.97), 322 (3.91),

328 (3.95)

IR ν_{max}^{CHCl3} cm⁻¹: 1681, 1645, 1510, 1466

Orixidine (XV)

mp. 191°, $C_{15}H_{13}O_4N$

UV λ_{max}^{EtOH} mμ (log ε): 243 (4.40), 258.5 (4.42), 268 (4.46), 307 (3.95), 321 (4.01), 336 (4.12)

(o) Ifflaiamine (XVI)[1589]

《Occurrence》 Australian *Flindersia* spp., $C_{15}H_{17}O_2N$

(p) Kokusagine (XVII)[1581) 1601) 1602]

《Occurrence》 *Orixa japonica* THUNB. (コクサギ), *Evodia xanthoxyloides* F. MUELL., *E. littoralis*
ENDL., *Flindersia maculosa* LINDL., *Lunasia amara*, mp. 199~201°, $C_{13}H_9O_4N$

IR λ_{max}^{CHCl3} μ[1581] : 6.06m, 6.13s, 6.33m, 6.53s, 9.36s, 9.56s, 10.25s, 10.86w

(q) Kokusaginine (XVIII)[1603]

《Occurrence》 *Orixa japonica* THUNB. (コクサギ), mp. 169°, $C_{14}H_{13}O_4N$

(r) Skimmianine (XIX)[1581) 1591) 1595]

《Occurrence》 *Ruta graveolens* L., *Xanthoxylum schinifolium* S. et Z. (イヌザンショウ), *Lunasia amara*
BLANCO, mp. 181~183°, $C_{14}H_{13}O_3N$

IR λ_{max}^{CHCl3} μ[1581] : 6.12s, 6.17s, 6.31s, 6.43w, 9.50s, 10.06s, 10.21m, 10.50m

(s) Evolitrine (XX)[1592)~1594]

《Occurrence》 *Cusparia macrocarpa, Evodia littoralis, Platydesma campanulata*
mp. 114~115°, $C_{13}H_{11}O_3N$, picrate mp. 201~202°
UV λ_{max} mμ: 246, 308, 319, 333

(t) Dictamnine (XXI)[1595a]

《Occurrence》 *Casimiroa edulis, Evodia littoralis, Dictamnus albus, Skimmia repens*
mp. 132~133°, $C_{12}H_{11}O_2N$

UV λ_{max} mμ (log ε) of synth.[1595a] : 308 (3.88), 312 (3.86), 328 (3.82); λ_{max}^{EtOH} mμ (ε): 235 (58900),
312 (8700)

(u) γ-Fagarine (XXII)

《Occurrence》 *Casimiroa edulis, Fagara coco*, mp. 142°, $C_{13}H_{11}O_3N$

UV λ_{max} mμ (log ε) of synth.[1595a] : 308 (3.87), 323 (3.85), 334 (3.79); λ_{max}^{EtOH} mμ (ε): 242 (60300),
308 (6600)

(v) 6-Methoxydictamnine (XXIII)[1594]

《Occurrence》 *Platydesma campanulata*, mp. 134~135°, $C_{13}H_{11}O_3N$

UV λ_{max}^{EtOH} mμ (log ε): 350 (4.68), 333 (4.75), 307.3 (5.04), 295.5 (4.99), 284sh (4.81), 260.7(4.91),

248.8 (6.10)

IR λ_{max}^{CHCl3} μ: 3.40m, 6.18s, 6.33s, 6.49m, 6.64s, 6.83s, 7.07m, 7.33s, 7.67s, 7.92m, 8.11s, 8.23s,
8.67s, 9.01s, 9.15s, 9.68m, 10.20m, 11.78w, 12.06m, 14.30w

(w) Maculine (XXIV)[1596) 1604]

《Occurrence》 *Flindersia maculosa* LINDL., mp. 195~196° (synth), $C_{13}H_9O_4N$
IR Fig. 233.

1601) Terasaka, M; *Pharm. Bull.*, **2**, 159 (1954); *Y. Z.*, **73**, 773 (1953).
1602) Hughes, G. K. *et al*: *Australian J. Sci. Res.* **A5**, 412 (1952).
1603) Terasaka, M. *et al*: *Y. Z.*, **75**, 1040 (1955).
1604) Brown, R. F. C. *et al*: *Australian J. Chem.*, **7**, 181 (1954).

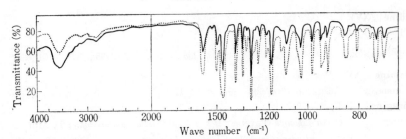

Fig. 233 IR Spectra of Maculine ······natural, ——synth.

(Ohta *et al.*)[1596]

(x) **Shuazine** (XXV)[1597]

《Occurrence》 *Choisya ternata* H.B.& K., $C_{18}H_{19}O_5N$

(y) **Graveolinine** (XXVI)[1598]

《Occurrence》 *Ruta graveolens* L., mp. 115~116°, $C_{17}H_{13}O_3N$

UV $\lambda_{max}^{EtOH} m\mu$ (log ε): 236 (4.55), 275 (4.01), 310~312 (3.99)

IR $\lambda_{max} \mu$: 6.18, 6.26 (arom.), 6.41 (arom. conjtd. C=C), 6.66 (arom.), 9.61, 10.69

(z) **Graveoline** (XXVII)[1599]

《Occurrence》 *Ruta graveolens* of Hong Kong[1605]

mp. 204~205°, $C_{17}H_{13}O_3N$

=Synth. 1–methyl–2–(3,4–methylenedioxyphenyl–4–quinolone)[1606]

(z′) **Arbrinine** (XXVIII)[1600]

《Occurrence》 *Glycosmis arborea*, mp. 175~176°, $C_{16}H_{15}O_4N$

IR $\nu_{max}^{CHCl_3}$ cm^{-1}: 3510~3470 (vw), 3050, 2913, 2817, 2644, 1640, 1595, 1563, 1500, 1457, 1351, 1318, 1280, 1142, 1108, 1053, 988, 933, 850

(4) **Perloline** (XXIX)[1607]

《Occurrence》 *Lolium perenne* L. (perenial rye-grass)

$C_{20}H_{18}O_4N_2$

Perchlorate mp. 280° (decomp.) IR ν_{max} cm^{-1}: 1695 (C=N$^+$)

N–Acetate mp. 232° (decomp.) IR ν_{max} cm^{-1}: 1650br (C=O)

O–Acetate mp. 296° (decomp.) IR ν_{max} cm^{-1}: 1766, 1650 (C=O)

(XXIX)

(5) **Perlolidine** (XXX)[1607]

《Occurrence》 *Lolium perenne* L.

Perloline (XXIX) ⟶ (XXX)

IR ν_{max} cm^{-1}: 3148 (>N–H), 1665 (C=O)

(XXX)

1605) Arthur, H. R. *et al*: *Austral. J. Chem.*, **13**, 510 (1960).
1606) Goodwin, S. *et al*: *J.A.C.S.*, **79**, 2239 (1957).
1607) Jeffreys, J. A. *et al*: *Proc. Chem. Soc.*, **1963**, 171.

(6) Calycanthine (XXXII)

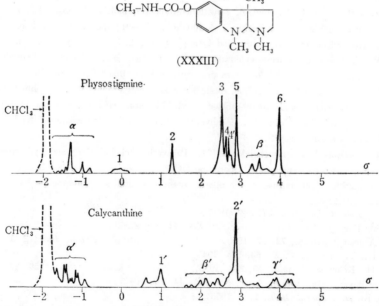

(Former Str. [1608]))   (Revised Str. [1609])~[1611]))

(XXXI)   (XXXII)

Calycanthine occurs in the seeds of *Calycanthus* species accompanying with two indole alkaloids, chimonanthine and folicanthine. It had been once regarded also as an indol alkaloid. But it's structure has been revised to (XXXII) from the results of NMR spectrum by Woodward[1609], mass spectrum by Clayton[1610] and X-ray analysis of dihydrobromide dihydrate $C_{22}H_{28}N_4Br_2 \cdot 2 H_2O$ by Hamor[1611]. In spite of revision of the structure, calycanthine has close correlation with other two alkaloids, described in the item 〔K〕, Indole Alkaloids, (17).

≪Occurrence≫   *Calycanthus glaucus* WILLD., *C. floridus* L.

mp. 245°, $C_{22}H_{26}N_4$, dibenzoate mp. 235°

UV $\lambda_{max}^{CH3CN} m\mu$ (log ε)[1609] : 252 (4.26), 310 (3.80)

NMR[1609]   Comparation with physostigmine (XXXIII) in $CHCl_3$ 40Mc, cyclohexane as an internal standard δ +3.9: See Fig. 234.

$$CH_3–NH–CO–O$$

(XXXIII)

Physostigmine

Calycanthine

Fig. 234 NMR Spectra of Physostigmine and Calycanthine (Woodward *et al.*)[1609]

1608) Robinson, R. *et al*: *Chem. & Ind.*, **1954**, 783.

1609) Woodward, R. B. *et al*: *Proc. Chem. Soc.*, **1960**, 76.

1610) Clayton, E. *et al*: *Tetrahedron*, **18**, 1495 (1962).

1611) Hamor, T. A. *et al*: *Proc. Chem. Soc.*, **1960**, 78.

Physostigmine (XXXIII) α (arom. protons), 1 (NH–CO), 2 $\left(\underset{|}{\searrow}N\diagup\overset{|}{\underset{H}{C}}\diagdown N\underset{|}{\diagup}\right)$, 3$\left(CH_3-\overset{|}{N}-\overset{|}{C}=C\right)$,

4, 4′(CH$_3$–NH–CO overlap. to $>$N–CH$_2$–CH$_2$–), 5(CH$_3$–N$<$), β(–CH$_2$–CH$_2$–C$\diagdown\!\!\diagup$), 6(CH$_3$–C$\diagdown\!\!\diagup$)

Calycanthine (XXXII) α'(arom. protons),1′$\left(\underset{|}{\searrow}N\diagup\overset{|}{\underset{H}{C}}\diagdown N\underset{|}{\diagup}\right)$, $\beta'(>$N–CH$_2$–CH$_2$–), 2′(CH$_3$–N$<$ + $>$NH?)

$\gamma'\left(CH_2-CH_2-C\diagdown\!\!\diagup\right)$

Remarkable difference was observed between the NMR spectra of calycanthine (XXXII) and of physostigmine (XXXIII) on the shifts, due to cyclic methylene (–CH$_2$–CH$_2$–C$\diagdown\!\!\diagup$). In Fig. 234, (XXXIII) shows the signal β as rather simple triplets, but (XXXII) reveals complicated multiplets β' and γ' which are interpreteted that assymmetricity of two methylenes in six membered ring of (XXXII) being larger than those in five membered ring of (XXXIII).

Mass[1610] m/e 346 (M), 288, 245, 232, 231, 172.

[J]   Isoquinoline Alkaloids

(1)   Spectra

UV : Papaverine[1612)1612a], 1–(β–picolyl)–6, 7–methylenedioxyisoquinoline, 1–nicotinyl–6, 7–methylenedioxy-isoquinoline, papaveraldine, 1–(β–picolyl)–3, 4–dihydro–6, 7–methylenedioxyisoquinoline, 1–benzyl–3, 4–dihydro–6, 7–methylenedioxyisoquinoline and 1–(α–picolyl)–1, 2, 3, 4–tetrahydro–6, 7–methylenedioxyiso-quinoline[1613] Tetrandrine, desmethoxytetrandrine[1614], trilobine, isotrilobine and anhydrodemethyltetrand-rine[1614]., Oxyacanthine, epistephanine, tetrandrine[1615], cepharanthine[1616] and cotarnine[1617]. Berberine[1617], narcoine[1618] and narceine[1618]. Laurifoline, dicentrine and menispermine[1619]. Morphine[1620], codeine[1620], thebaine[1621], tuduranine and morphothebaine[1621]. Morphine–HCl[1622], codeine–HCl[1623], thebaine–HCl[1623], phenolic dihydrothebaine[1624], β–dihydrothebaine[1625], dihydrothebainol–6–methylether[1625] and N–methylemetine dimethiodide[1626].

IR : α–Methylmorphimethine and chelidonine[1400]. Papaverine, chellidonine and hydrastine[1400]. Hydra-stine, narceine, narcotine, morphine, morphine–HCl[1627], codeine, thebaine, hydrastine–HCl[1627] and bebeerine sulfate[1628]. Tetrandrine (++), phaeanthine (−−) and isotetrandrine (−+)[1629].

1612)   Hofmann: *Helv. Chim. Acta*, **37**, 849 (1954).
1612a)   Dyer, Mc Bay: *J. Am. pharm. Assoc. Sci. Ed.*, **44**, 156 (1955).
1613)   Noller, Azima: *J.A.C.S.*, **72**, 17 (1950).   1614) Inubushi, Y: *Pharm. Bull.*, **2**, 1 (1954).
1615)   Kondo, H., Tanaka, K: *Y. Z.*, **63**, 273 (1943).
1616)   Kondo. H., Keimatsu, I: *ibid.*, **55**, 894 (1935).   1617) Skinner, B: *Soc.*, **1950**, 823.
1618)   Dyer, Mc Bay: *J. Am. pharm. Assoc. Sci. Ed.*, **44**, 156 (1955).
1619)   Tomita, Kusuda: *Pharm. Bull.*, **1**, 5 (1953).
1620)   Clark, McBay: *J. Am. pharm. Assoc. Sci. Ed.*, **43**, 39 (1954).
1621)   Hofmann: *Helv. Chim. Acta*, **37**, 849 (1954).   1622) Goto K: *Ann.*, **521**, 177 (1936).
1623)   Elvidge, Quart: *J. Pharm. Pharmacol.*, **13**, 219 (1940).
1624)   Steiner: *Bull. Soc. Chim. Biol.*, **6**, 231 (1924).   1625) Small *et al*: *J. Org. Chem.*, **12**, 839 (1947).
1626)   Schmid, Karrer: *Helv. Chim. Acta*, **33**, 863 (1950), **34**, 1948 (1951).
1627)   Battersby *et al*: *Soc.*, **1949**, 3207, S 59.
1628)   Sadtler Card No. 7163.   1628) Sadtler Card No. 7159.
1629)   Tomita, Inubushi: *Pharm. Bull.*, **4**, 413 (1956).

(2) Tetrahydroalkylisoquinolines
(a) Lophocerine (I) and Pilocereine (II-a)[1630]

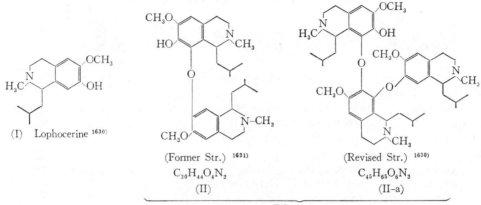

(I) Lophocerine [1630]

(Former Str.) [1631]
$C_{30}H_{44}O_4N_2$
(II)

(Revised Str.) [1630]
$C_{45}H_{65}O_6N_3$
(II-a)

Pilocereine

≪Occurrence≫ *Lophocereus chottii*
Methyl ether   mp. 153~155°[1631]

Lophocerine (1–Isobutyl–2–methyl–6–methoxyl–7–hydroxy–1, 2, 3, 4–tetrahydroisoquinoline) (I)[1631]
Picrate $C_{23}H_{30}O_9N_4$ mp. 151.5~152.5°

(3) Benzylisoquinolines and Tetrahydrobenzylisoquinolines

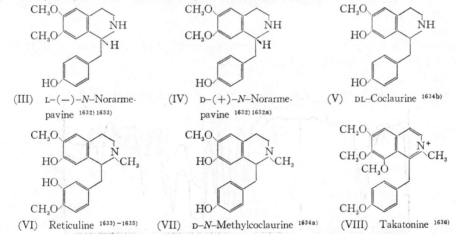

(III) L–(–)–N–Norarme-
pavine [1632) 1633]

(IV) D–(+)–N–Norarme-
pavine [1632) 1632a]

(V) DL–Coclaurine [1634b]

(VI) Reticuline [1633)~1635]

(VII) D–N–Methylcoclaurine [1634a]

(VIII) Takatonine [1636]

1630) Tomita, M. *et al*: *Tetr. Letters*, **1963**, 127.
1631) Djerassi, C. *et al*: *J.A.C.S.*, **79**, 2203 (1957).
1632) Tomita, M. *et al*: *Y. Z.*, **83**, 15 (1963).
1632a) Yang, T–H *et al*: *ibid.*, **83**, 22 (1963).   1633) Tomita, M. *et al*: *ibid.*, **84** 362 (1964).
1633a) Tomita, M. *et al*: *ibid.*, **84**, 365 (1964).   1634) Kunitomo, J: *ibid.*, **81**, 1253, 1257 (1961).
1634a) Kunitomo, J: *ibid.*, **81**, 1261 (1961).   1634b) Ln, S–T: *ibid.*, **83**, 19 (1963).
1635) Gopinath, K. W. *et al*: *Ber.*, **92**, 776, 1657 (1959).
1636) Fujita, E. *et al*: *Y. Z.*, **79**, 1082 (1959).

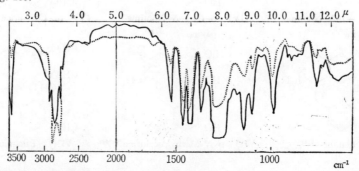

(IX)　Laudanosine　　　　　　(X)　Armepavine　　　　　(XI)　Cularine [1637) 1638]

(a)　L–(−)–*N*-**Norarmepavine**　(III)[1632) 1633]

<<Occurrence>>　*Machilus kusanoi* HAYATA (オオバタブ), *M. thunbergii* S. et Z. (タブノキ)

mp. 157∼158°, $[\alpha]_D$ −30.46° (CHCl$_3$), C$_{18}$H$_{21}$O$_3$N

UV　λ_{max}^{EtOH} mμ (log ε): 282.5 (3.78); λ_{min}: 253 (2.84)

(b)　D–(+)–*N*-**Norarmepavine**　(IV)[1632a]

<<Occurrence>>　*Magnolia kachirachiai* DANDY

mp. 157∼158°, $[\alpha]_D$ +31.49° (CHCl$_3$), C$_{18}$H$_{21}$O$_3$N

UV　λ_{max}^{EtOH} mμ (lgo ε): 282.5 (3.80); λ_{min}: 253 (3.02)

(c)　DL–**Coclaurine**　(V)[1634b]

<<Occurrence>>　*Machilus kusanoi* HAYATA (オオバタブ)

mp. 219∼220°, $[\alpha]_D$ ±0° (MeOH), C$_{17}$H$_{19}$O$_3$N

UV　λ_{max}^{EtOH} mμ (log ε): 287 (3.85); λ_{min}: 254 (2.97)

(d)　**Reticuline**　(VI)[1633) ∼1635]

<<Occurrence>>　*Anona reticulata* L.

Perchlorate　mp. 203∼204°, $[\alpha]_D$ +87.54° (EtOH) C$_{19}$H$_{23}$O$_4$N·HClO$_4$

UV　λ_{max}^{EtOH} mμ: 285[1635]

Synth. DL–Reticuline[1634]

Perchlorate　mp. 144∼145° (──HClO$_4$·2 H$_2$O)[1635], 138.5∼139.5° (decomp.) (──HClO$_4$·H$_2$O)[1634]

UV　λ_{max}^{EtOH} mμ (log ε): 286 (3.75); λ_{min}: 256 (2.92)

IR　See Fig. 235.

3.0　　4.0　　5.0　　　6.0　7.0　8.0　　9.0 10.0 11.0 12.0 μ

3500　3000　2500　　2000　　　1500　　　　　　1000　　　　cm^{-1}

Fig. 235　IR Spectra of Reticuline, ──synth. DL, ……nat. D (Kunitomo)[1634]

(e)　D–*N*-**Methylcoclaurine** (Coclanoline A)　(VII)[1634a]

<<Occurrence>>　*Cocculus laurifolius* DC. (コウシュウヤク)

Oxalate　mp. 143.5∼144.5° (decomp.), $[\alpha]_D$ +100.14° (MeOH), C$_{18}$H$_{21}$O$_3$N·(COOH)$_2$·$^1/_2$H$_2$O

UV　λ_{max}^{EtOH} mμ (log) of base: 285 (3.75); λ_{min}: 255 (2.92)

(f)　**Takatonine**　(VIII)[1636]

<<Occurrence>>　*Thalictrum minus* (タカトグサ)

1637)　Kametani, T *et al*: *Soc.*, **1963**, 4289; *7 th Symposium on the Chemistry of Natural Products*, p. 48 (Fukuoka, 1963).
1638)　Manske, R.H.F: *Can. J. Research*, **16 B**, 81 (1939), **18 B**, 97 (1940); *J.A.C.S.*, **72**, 55 (1950).

Iodide mp. 192~193°, $[\alpha]_D$ ±0°, $C_{21}H_{24}O_4N \cdot I$

UV λ_{max}^{MeOH} mμ (log ε): 220 (4.54), 265 (4.54), 315~320 (3.75); λ_{min}: 242 (4.31), 296 (3.71).

(g) Laudanosine (IX)

　　《Occurrence》 　Opium, 　mp. 89°, $[\alpha]_D$ +103.23° (EtOH), $C_{21}H_{27}O_4N$

　　NMR[1647] in $CHCl_3$ 40 Me, τ: +2.75 for $CHCl_3$; (±)-Laudanosine——6.18 (3 and 6–OCH_3), 6.22
　　　　(4–OCH_3), 6.45 (5–OCH_3), 7.48 (N–CH_3)

(h) Armepavine (X)

　　《Occurrence》 　Papaver caucasicum BIEB., Nelumbo nucifera[1663]

　　mp. 92° (DL–), $C_{20}H_{25}O_3N$

　　NMR[1647] in $CHCl_3$, 40 Mc, τ: +2.75 for $CHCl_3$: 6.20, 6.28 (3 and/or 6–OCH_3), 6.58 (5–OCH_3)
　　　　of O–Methylether.

(i) Cularine (XI)[1637][1638]

　　《Occurrence》 　Corydalis claviculata, Dicentra cucullaria, D. eximia, D. formorsa, D. oregana

　　mp. 115°, $[\alpha]_D$ +285° (MeOH), $C_{20}H_{23}O_4N$

　　DL–Cularine (synth.)[1637] 　mp. 113~114°

UV λ_{max}^{EtOH} mμ (log ε): 226 (4.38), 285 (3.92)

(j) Biosynthesis of Benzylisoquinoline Alkaloids

Papaverine[1640], narcotine[1641], hydrastine[1642][1643]

（4） N-bridged Tetrahydrodibenzocyclooctanes

(a) Argemonine (XII)[1644][1645]

　　《Occurrence》 　Argemone hispida, A. munita DUR. &
　　　　HILG. subsp. rotundata G. B. OWNB.

(XII)[1645]

(XIII) Papaverine 　　　　　　　　(XIV) Pavine

N–Methylpavine (XII)[1646] =Argemonine[1644]

　　Argemonine = O–Dimethylrotundine[1644]

　　mp. 151~151.5°, $C_{21}H_{25}O_4N$

Methiodide

　　mp. 273~274° (decomp.), 　UV See Fig. 236

1639) Taylor, W.I.: Tetrahedron, **14**, 42 (1961). 　　1640) Battersby, A.R. et al: Soc., **1962**, 3526.

1641) Battersby, A.R. et al: Proc. Chem. Soc., **1962**, 365.

1642) Spencer, J.D. et al: J.A.C.S., **84**, 1059 (1962).

1643) Gear, J. R. et al: Nature, **191**, 1393 (1962).

1644) Soine, T. O. et al: J.P.S., **49**, 187 (1960), **50**, 321 (1961), **51, 1196** (1962).

1645) Stermitz, F. R. et al: J.A.C.S., **85**, 1551 (1963).

1646) Battersby, A. R. et al: Soc., **1955**, 2888.

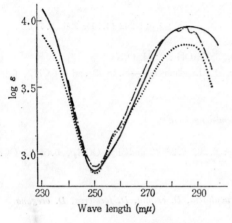

Fig. 236　UV Spectra of Argemone Alkaloids
(Soine *et al.*)[1644]
−·−·− Argemonine
······ Norargemonine
—— Rotundine

UV　λ_{max}^{EtOH} mμ: 280.2 sh, 287, 291.2 sh; λ_{min}: 250
IR　(Methiodide)　See Fig. 237.

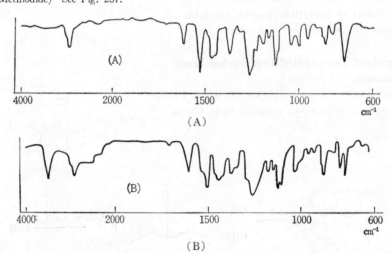

Fig. 237　IR Spectra of (A) Argemonine (Dimethylrotundine) methiodide and (B) Rotundine
in KBr (Soine *et al.*)[1644]

(b)　Norargemonine　$C_{20}H_{23}O_4N$　(XII–a)

《Occurrence》　*Argemone hispida*

$$(XII\text{–}a) \xrightarrow{CH_2N_2} Argemonine \quad (XII)$$

UV　λ_{max}^{EtOH} mμ: 287; λ_{min}: 250 (Fig. 236)

(c)　Rotundine　$C_{19}H_{21}O_4N$　(XII–b)

《Occurrence》　*Argemone munita* spp. *rotundata*
mp. 245~245.5°, $[\alpha]_D$ −265.8° (MeOH)
UV　λ_{max}^{EtOH} mμ: 288; λ_{min}: 250 (Fig. 236)
IR　See Fig. 237.

(5)　Aporphine Alkaloids

(a)　NMR Spectra[1647] in $CHCl_3$, 40 Mc, τ: +2.75 for $CHCl_3$

1647)　Bick, I.R.C *et al*: Soc., **1961**, 1896.

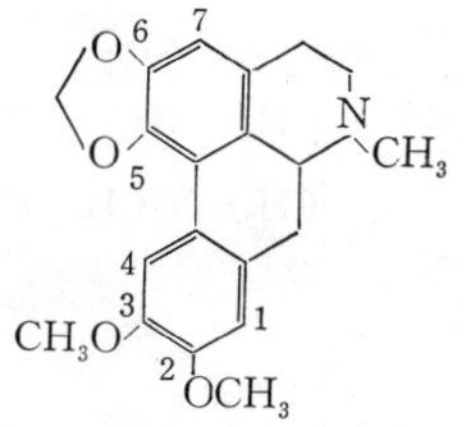

(XIII) Dicentrine

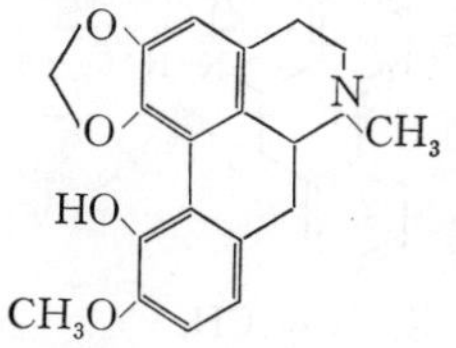

(XIV) Bulbocapnine

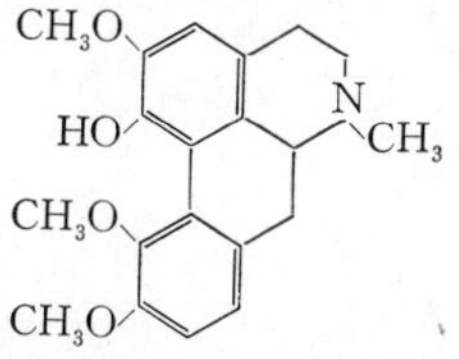

(XV) Corydine

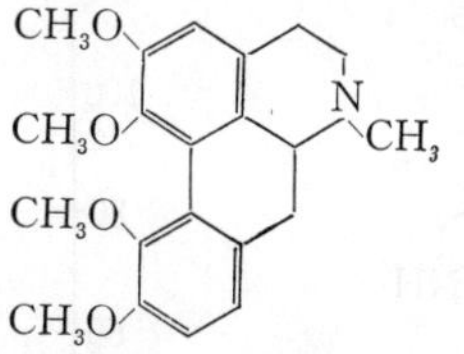

(XVI) Glaucine

Dicentrine (XIII) See the item (h) (XLIX)

τ: 6.20 (2-and 3-OCH$_3$), 7.62 (N-CH$_3$), 4.05~4.20 (-O-CH$_2$-O-)

Bulbocapnine (XIV) See the item (h) (LIII)

τ: 6.20 (3-OCH$_3$), 7.65 (N-CH$_3$), 3.97~4.15 (-O-CH$_2$-O-)

Corydine (XV) See the item (h) (LI)

τ: 6.18 (3-6-OCH$_3$), 6.35 (4-OCH$_3$), 7.55 (N-CH$_3$)

Glaucine (XVI) See the item (d) (XXXII)

τ: 6.20 (2-6-OCH$_3$), 6.25 (3-OCH$_3$), 6.45 (5-CH$_3$), 7.57 (N-CH$_3$)

(b) Mass Spectra[1648]

(i) *Aporphine Type*

Fragmentation

M-1 peak (a')

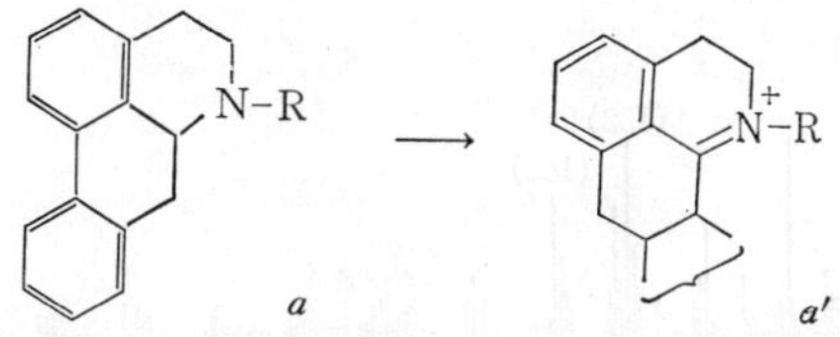

a   a'

M-15(CH$_3$) peak (b) and M-31 (OCH$_3$) peak (c)

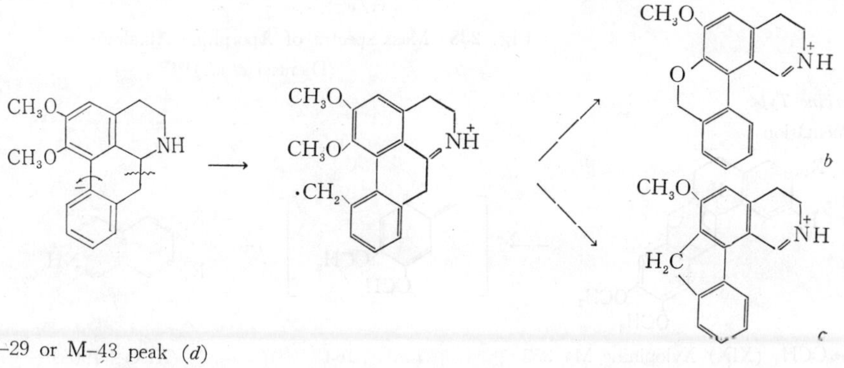

b

c

M-29 or M-43 peak (d)

1648) Djerassi, C et al: J.A.C.S., 85, 2807 (1963).

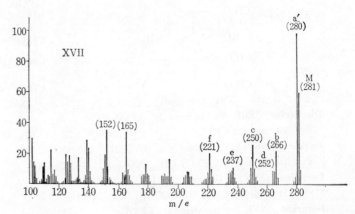

R=H or CH₃

d–15 (CH₃) peak (e) and d–31 (OCH₃) (f), m/e 165 and 152

Spectra

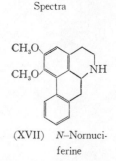

(XVII) N–Nornuci-
ferine

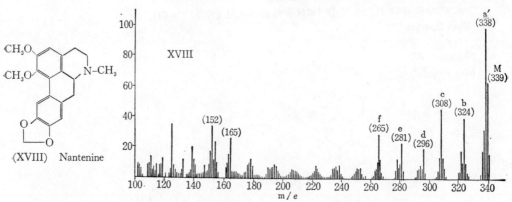

Fig. 238 Mass spectra of Aporphine Alkaloids
(Djerassi et al.)[1648]

(ii) Berbine Type
Fragmentation

R₁=R₂=OCH₃ (XIX) Xylopinine M⁺ 355   i   164   (XIX) 190
R₁, R₂=-O-CH₂-O- (XX) Tetrahydroberberine   i-CH₃   149   (XX) 174
M⁺ 339

— 532 —

Spectra

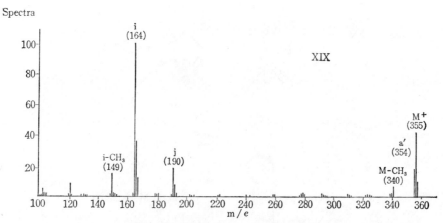

Fig. 239 Mass Spectrum of Xylopinine

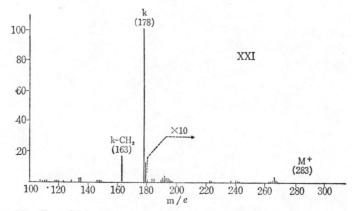

Fig. 240 Typical Mass Spectrum of Benzylisoquinoline (XXI) (Djerassi *et al.*)[1648]

(iii) *Benzylisoquinoline Type*

Fragmentation

		k	*k*–CH₃
R=H (XXI) 1–(2′–Hydroxybenzyl)–6–methoxy–7–hydroxy– 1,2,3,4–tetrahydroisoquinoline	(XXI)	178	163
R=CH₃ (XXII) 1–(2′–Hydroxybenzyl)–6,7–dimethoxy– 1,2,3,4–tetrahydroisoquinoline	(XXII)	192	178

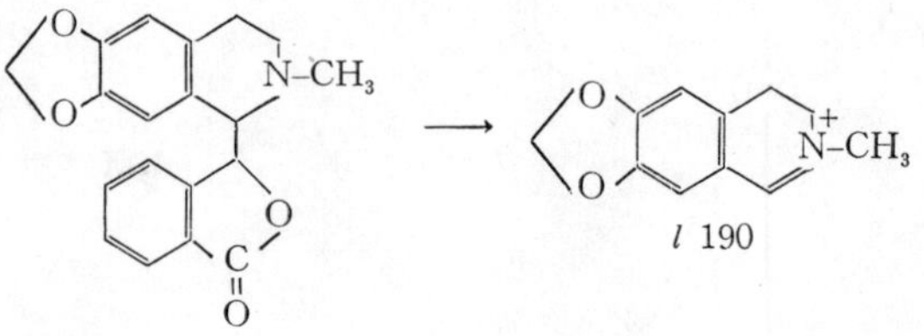

(XXIII) Hydrastine

Exception:

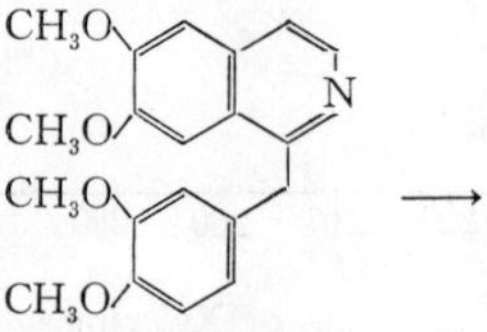

$\longrightarrow$ m/e M–1, M–15, M–31

(XXIV) Papaverine

Spectra: See Fig. 240

(c) Lauraceous Alkaloids

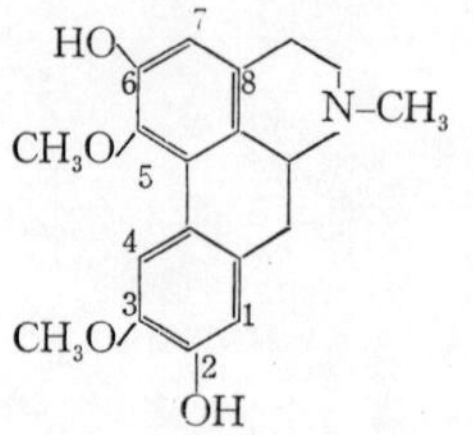

(XXV) Boldine [1649]

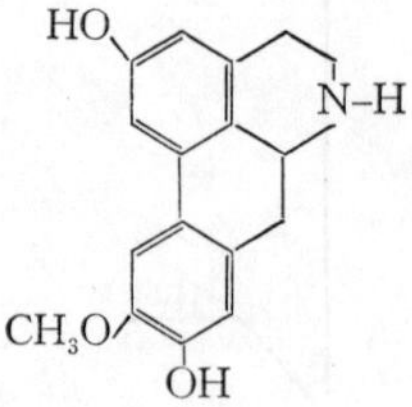

(XXVI) Laurolitsine [1633a] [1650]

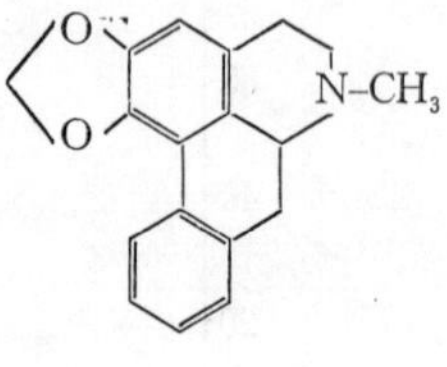

(XXVII) Litsericine [1649]

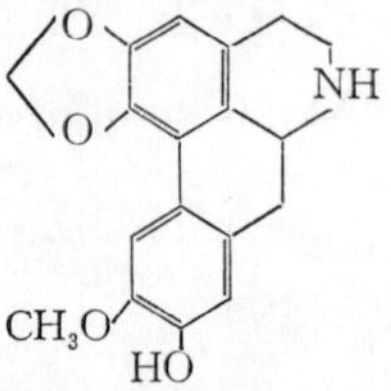

(XXVIII) Actinodaphnine

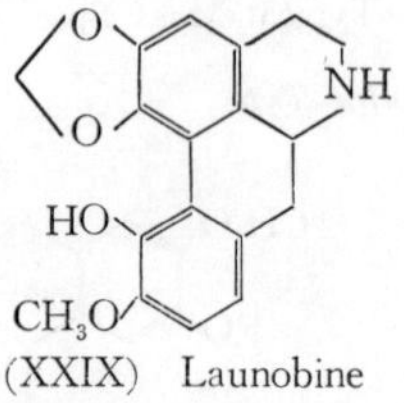

(XXIX) Launobine

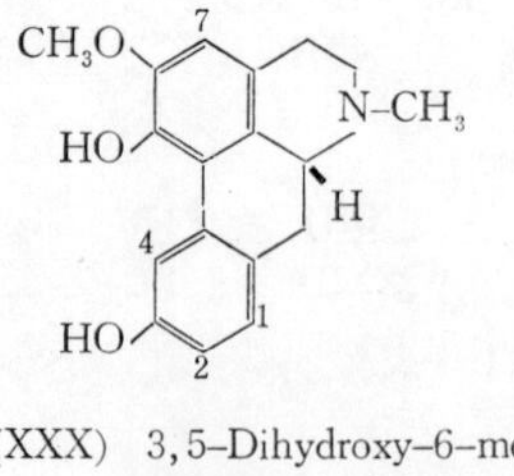

(XXX) 3,5-Dihydroxy-6-methoxy
aporphine

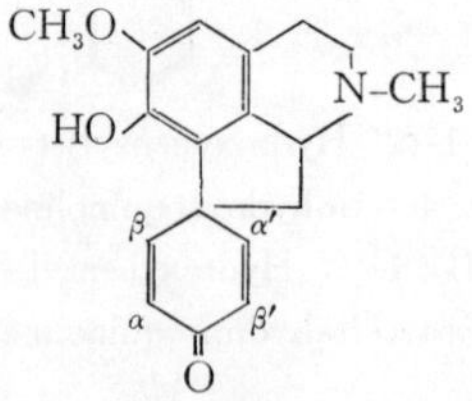

(XXXI) Glaziovine

1649) Nakasato, T. *et al*: *Y.Z.*, **79**, 1267 (1959).
1650) Nakasato, T. *et al*: *Chem. Pharm. Bull.*, **7**, 780 (1959).

Boldine (XXV) [1649]

≪Occurrence≫　　*Neolitsea sericea* KOIDZ. (シロダモ), *Peunmus boldus* MOLINA

mp. 161~162°, $[\alpha]_D$ +114.3° (EtOH), $C_{19}H_{21}O_4N$

UV　λ_{max}^{EtOH} mμ (log ε): 282(4.15), 304(4.15)

Laurolitsine (XXVI) [1633a) 1649) 1650]

≪Occurrence≫　　*Neolitsea sericea* KOIDZ. (シロダモ),　　mp. 138~140°, $[\alpha]_D$ + 102.5° (EtOH),

$C_{18}H_{19}O_4N$,　　picrolonate mp. 239° (decomp.)

O,O–Dimethylether–HClO$_4$ mp. 212° (decomp.)

UV　λ_{max}^{EtOH} mμ (log ε): 283 (4.16), 304 (4.16), see Fig. 241.

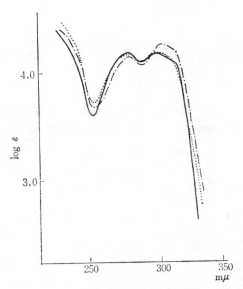 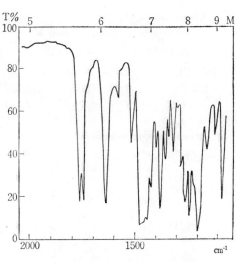

Fig. 241　UV Spectra of——Boldine and

⋯⋯*O, O*–Dimethyllaurolitsine–HClO$_4$

(Nakasato *et al.*)

Fig. 242　IR Spectrum of

O, O, N-Triacetyllaurolitsine in

nujol　　(Nakasato *et al.*)

O, O, N–Triacetate　　mp. 152~154°, $[\alpha]_D$ +293° (EtOH)

IR　See Fig. 242, ν_{max} cm^{-1}: 1767, 1749 (ester C=O), 1641 (>N–COCH$_3$), 1207 (phenolic acetate)

Litsericine (XXVII) [1649]

≪Occurrence≫　　*Neolitsea sericea* KOIDZ. (シロダモ)

mp. 157~158°, $[\alpha]_D$ +74.9° (EtOH), $C_{17}H_{21}O_3N$

UV　λ_{max}^{EtOH} mμ (log ε): 291 (3.54), 235~245 sh.

Actinodaphnine (XXVIII) [1651]

≪Occurrence≫　　*Laurus nobilis* L. (ゲッケイジュ),　　mp. 196°, $[\alpha]_D$ +32.48° (EtOH), $C_{18}H_{17}O_4N$

UV　λ_{max}^{EtOH} mμ (log ε): 221 (4.47), 285 (4.14), 308 (4.19)

IR　ν_{max} cm^{-1}: 938 (C–O), 1048 (C–O–C)

Launobine (XXIX) [1651]

≪Occurrence≫　　*Laurus nobilis* L. (ゲッケイジュ)

1651)　Tomita, M. *et al* : *Y. Z.*, **83**, 763 (1963).

mp. 214~215°, $[\alpha]_D$ +192.69° (EtOH), +227° (CHCl$_3$)

UV λ_{max}^{EtOH} mμ (log ε): 223 (4.37), 270 (4.06), 309 (3.72)

IR ν_{max} cm^{-1}: 953 (C–O), 1054 (C–O–C)

3,5-Dihydroxy-6-methoxyaporphine (XXX) [1651a]

≪Occurrence≫　　Octea glaziovii MEZ.,　　mp. 149~152° (decomp.), $[\alpha]_D$ −35° (CHCl$_3$),
C$_{18}$H$_{19}$O$_3$N·H$_2$O

UV λ_{max}^{EtOH} mμ (ε): 218 (38100), 266 (10200), 275 (13400), 307 (9040); $\lambda_{max}^{EtOH-NaOH}$ mμ (ε): 340 (9590)

IR ν_{max}^{nujol} cm^{-1}: 3460 w, 3300 m, 1600 m, 1302 m, 1136 m

NMR　in D$_2$O–NaOD, 60 Mc, C$_6$H$_6$ as an external standard δ, corrected to TMS, δ : 0, C$_6$H$_6$: 295 cps :
　　2.57 s (N–CH$_3$), 3.92 (6–OCH$_3$), 7.18 d, J=7.8 c/s (1–H), 6.62 dd, J=7.8, 2.5 (2–H), 8.08 d,
　　J=2.5 (4–H), 6.77 s (7–H)

Glaziovine (XXXI) [1651a]

≪Occurrence≫　　Octea glaziorii MEZ.,　　mp. 235~237° (decomp.) $[\alpha]_D$ +7° (CHCl$_3$), C$_{18}$H$_{19}$O$_3$N

UV λ_{max}^{EtOH} mμ (ε): 288 (3710); $\lambda_{max}^{EtOH-KOH}$ mμ (ε): 308 (5440)

IR $\nu_{max}^{CHCl_3}$ cm^{-1}: 1657 s, 1619 s

NMR　in CDCl$_3$, 60 Mc, TMS, δ=0 : 2.4 s (N–CH$_3$), 3.85 (OCH$_3$), 6.0 br. s (OH), 6.65 s (arom. H),
　　6.25~6.6 (α, α'), 6.7~7.3 (β, β'), two AB–type quartets with fine structure (vinyl protons α, α',
　　β, β')

(d) Magnoliaceous Alkaloids [1652~1664]

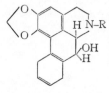

(XXXII)=(XVI) Glaucine　　　(XXXIII) Magnoflorine　　　R=CH$_3$: (XXXIV)
Ushinsunine
R=H: (XXXV)
Michelalbine

(XXXVI) Oxoushin-　　　(XXXVII) Roemerine　　　(XXXVIII) Michepressine [1658]
sunine (Liriodenine)

1651a)　Rapoport, H. et al : J.A.C.S., **86**, 694 (1964).
1652)　Yang, T-H. et al : Y. Z., **82**, 816 (1962).
1653)　Tomita, M. et al : ibid., **82**, 1199 (1962).　　1654)　Tomimatsu, et al : ibid., **82**, 1560 (1962).
1655)　Yang, T-H et al : ibid., **82**, 216 (1963).　　1656)　Yang, T-H et al : ibid., **82**, 794 (1962).
1657)　Ito, K : ibid. **80**, 705 (1960).　　1658)　Ito, K : ibid., **81**, 703 (1960).
1659)　Tomita, M. et al : ibid., **82**, 925 (1962).　　1660)　Tomita, M. et al : ibid., **82**, 616 (1962).

Glaucine (XXXII)

≪Occurrence≫ *Liriodendron tulipifera* L. (ユリノキ), *Corydalis tuberosa, C. ternata, Glaucium flavum, G. serpieri, Dicentra formosa (Papaveraceae)*

mp. 120°, $[\alpha]_D$ +113.3° (EtOH), $C_{21}H_{25}O_4N$

Hydrobromide   mp. 233~234° (decomp.)

UV λ_{max}^{EtOH} mμ (log ε): 238 (4.11), 304 (4.12); λ_{min}: 255 (3.60), 292 (4.03)[1653] ; λ_{max}^{EtOH} mμ (log ε): 305 (4.18), 283 (4.13)[1652]

NMR See the itm (5), (a)

Magnoflorine (XXXIII)

≪Occurrence≫ *Magnolia grandiflora.* L., *M. kobus* DC.[1652], *M. coco* DC., *M. kachirachirai*[1652], *Liriodendron tulipifera*[1653], *Michelia champaca* L.[1655], *M. compressa* MAXIM[1657], *Aristolochia debilis* SIEB. et ZUCC., *Menispermum dauricum* DC., *Cryptocarya* spp., *Thalictrum thunbergii* DC.,[1654] *Nandina domestica* THUNB[1655].

Iodide[1654]   mp. 252° (decomp.), $[\alpha]_D$ +203.2° (MeOH), $C_{20}H_{24}O_4NI$

UV λ_{max}^{MeOH} mμ (log ε): 226 (4.64), 271 (3.88), 314 (3.81)

Ushinsunine (XXXIV)

≪Occurrence≫ *Michelia champaca* L. (キンコウボク)[1655] *M. compressa* MAX. var. *Formosana* (タイワンオガタマノキ)[1656], *M. Compressa* MAX. (オガタマノキ)[1659]

mp. 180~181°[1656], 122~123°[1659] (bismorph.), $[\alpha]_D$ −113.6° (EtOH), $C_{18}H_{17}O_3N$

UV λ_{max}^{EtOH} mμ (log ε)[1656]: 274 (4.20), 321 (3.51); λ_{min}: 255 (3.94), 301 (3.18)

IR $\nu_{max}^{CHCl_3}$ cm^{-1} [1656]: 1055, 945 (–O–CH$_2$–O–), 3510 (O–H), see Fig. 243

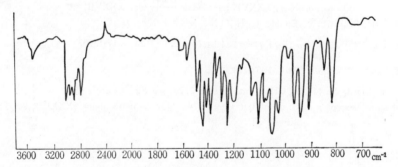

3600 3200 2800 2400 2000 1800 1600 1400 1200 1100 1000 900 800 700 cm^{-1}

Fig. 243 IR Spectrum of Ushinsunine in CHCl$_3$

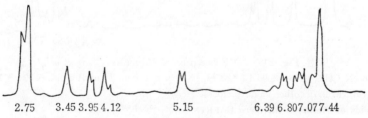

2.75   3.45 3.95 4.12   5.15   6.39 6.80 7.07 7.44

Fig. 244 NMR Spectrum of Ushinsunine (Yang, *et al.*)

Determination of configuration at C_A and C_B by NMR[1656]

In CHCl$_3$, 56.44 Mc, τ=2.75 for CHCl$_3$ as an internal reference, see Fig. 244.

τ 5.15 d, J: 2.5 cps was assigned for spin-spin coupling of C_A–H$\sim$$C_B$–H. The configuration of ushinsunine was determined as type (A), because the coupling constants, calculated by Karplus's equation for (A) is 2$\sim$4 cps and for (B) is 7$\sim$9 cps and thereby, the structure of ushinsunine should be (XXXIV) or it's antipode. Other signals of ushinsunine are almost identical with those of rodemerine (XXXVII).

(A) (B)

Michelalbine (XXXV)[1656]

≪Occurrence≫ *Michelia alba* DC. (ギンコウボク), mp. 205$\sim$207°, $[\alpha]_D$ $-105.25°$ (CHCl$_3$), $C_{17}H_{15}O_3N$

UV λ_{max}^{EtOH} mμ (log ε): 275 (4.23), 328 (3.49); λ_{min}: 255 (3.98), 304 (3.17)

IR ν_{max}^{nujol} cm^{-1}: 3240 (OH), 1050, 960 (–O–CH$_2$–O–)

Oxoushinsunine (Liriodenine) (XXXVI)

≪Occurrence≫ *Liriodendron tulipifera* L.[1653) 1660)], *Michelia champaca* L.[1655)] *M. compressa* MAXIM[1657)],
—— var. *Formosana*[1656)]

mp. 289°[1639)], 278$\sim$280° (decomp.)[1655)], 276$\sim$279° (decomp.)[1659)], $C_{17}H_9O_3N$

UV λ_{max}^{EtOH} mμ (log ε)[1659)]: 274.5 (4.04), 273 (4.25), 320.5 (3.58); λ_{min}: 244 (4.03), 254.5 (4.01), 300 (3.32)

Roemerine (XXXVII)[1656) 1664)]

≪Occurrence≫ *Roemeria refracta* DC., *Cryptocarya angulata*, *Nelumbo nucifera* GAERTN[1664)]

Oxoushinsunine (XXXVI) methiodide $\xrightarrow{\text{Zn+HCl}}$ (XXXVII)[1656)]

mp. 100$\sim$101°, $[\alpha]_D$ $-72.5°$ (EtOH), $C_{18}H_{17}O_2N$

UV See Fig. 245, λ_{max}^{EtOH} mμ (log ε)[1664)]: 234 (4.15), 274 (4.24), 317 (3.57)

IR λ_{max}^{CHCl3} cm^{-1} [1664)]: 1401, 1361, 1053, 942

NMR[1656)] in CHCl$_3$ 56.44 Mc, CHCl$_3$ int. ref. τ: 2.75, see Fig. 245.

Chemical shifts, except τ 5.15 are almost identical to those of ushinsunine (XXXIV), see Fig. 244.

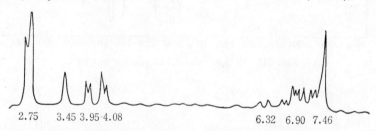

2.75   3.45 3.95·4.08   6.32 6.90 7.46

Fig. 245 NMR Spectrum of Roemerine (Yang *et al.*)

τ: 7.46 (N–CH$_3$), 4.08$\sim$3.95 d (–O–CH$_2$–O–), 3.45 (arom. H of A ring), 6.32 $\leqq$7.46 (3–CH$_2$–of B, C rings), ca. 2.75 (other arom. protons)

(e) Alkaloids of *Nelumbo nucifera* GAERTN (ハス)[1661) ~1664)]

1661) Tomita, M. *et al*: *Y. Z.*, **81** 942 (1961).
1662) Tomita, M. *et al*: *ibid.*, **81**, 1202 (1961).   1663) Tomita, M. *et al*: *ibid.*, **82**, 1458 (1962).
1664) Tomita, M. *et al*: *ibid.*, **81**, 469 (1961).

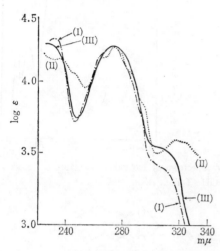

(XXXIX) Nuciferine   (XL) Nornuciferine   (XXXVII) Roemerine

Nuciferine (XXXIX)[1664]

Fig. 246 UV Spectra of
—·—·— Nuciferine,
······ Roemerine
——— Nornuciferine in EtOH
(Tomita *et al.*)

mp. 162.5~163.5°, $[\alpha]_D$ −215° (EtOH), $C_{19}H_{21}O_2N$,   UV: See Fig. 246

N-Nornuciferine (XVII) See the item (b)

Nornuciferine (XL)   mp. 195~196°, $[\alpha]_D$ −265° (CHCl₃), $C_{18}H_{19}O_2N$,   UV: See Fig. 246

Roemerine (XXXVII) See the item (d)

(f) Berberidaceous Alkaloids

(XLI) Domesticine   (XLII) *O*-Methyldomesticine[1665]   (XLIII) Isoboldine[1666]

Domesticine (XLI)[1667]

≪Occurrence≫   *Nandina domestica* Thunb. (ナンテン)[Note]

mp. 108°, 115~117°, $[\alpha]_D$ +60.5°, $C_{19}H_{19}O_4N$

UV λ_{max}^{EtOH} mμ: 283, 311, λ_{min}: 256, 291[1667]

[Note] The plant contains also *O*-methyldomesticine, protopine, berberine, jateorrhizine, magnoflorine, menisperine and nandinine.

1665) Tomita, M. *et al*: *Y.Z.*, **76**, 751 (1956).
1666) Tomita, M. *et al*: *J. Chem. Soc. Japan*, **82**, 1708 (1961).
1667) Tomita, M. *et al*: *Y.Z.*, **81**, 1090 (1961).

Cleavage reaction by Na in liquid ammonia[1668]

O–Methyldomesticine (Nantenine＝Domestine) (XLII)[1667]

≪Occurrence≫   *Nandina domestica* THUNB. (ナンテン)

mp. 138.5° $[\alpha]_D$ +111°, $C_{20}H_{21}O_4N$

Cleavage reaction by Na or Li in liquid ammonia[1669]

Isoboldine (XLIII)[1666]

≪Occurrence≫   *Nandina domestica* THUNB.

mp. 180°, $[\alpha]_D$ +83° (CHCl₃), $C_{19}H_{21}O_4N$

(XLIII) ⟶ Laurifoline chloride (XLVII)

   Triacetate

   UV: See Fig. 247

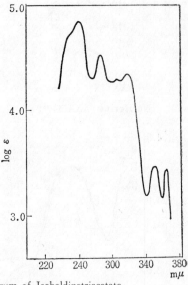

Fig. 247 UV Spectrum of Isoboldinetriacetate
in EtOH (Tomita *et al.*)[1666].

(g) Menispermaceous Alkaloids

(XLIV) Tuduranine   (XLV) Stephanine   (XLVI) Crebanine

(XLVII) Laurifoline   (XLVIII) Cocsarmine   (XLVIII–a) Protostephanine

Tuduranine (XLIV)[1670]

≪Occurrence≫   *Sinomenium acutum* R. & W.,   mp. 125°, $C_{18}H_{19}O_3N$

Hydrochloride   mp. 286°(decomp.), $[\alpha]_D$ −148°

Stephanine (XLV)[1671][1671a][1735]

≪Occurrence≫   *Stephania capitata* SPRENG, *S. japonica* MIERS[1735]

mp. 155°, $[\alpha]_D$ −88.3° (CHCl₃), $C_{19}H_{19}O_3N$

1668) Kitamura, T: *Y.Z.*, **80**, 1104 (1960).
1669) Kitamura, T. *et al*: *ibid.*, **80**, 219 (1960), **81**, 254 (1961).
1670) Goto, K. *et al*: *Ann.*, **521**, 135 (1936), **539**, 262 (1939).
1671) Tomita, M *et al*: *Y.Z.*, **77**, 290 (1957).   1671a) Shirai, H. *et al*: *ibid.*, **76**, 1287 (1956).

Crebanine (XLVI)[1672]

≪Occurrence≫   *Stephania sasakii, S. Capitala*   mp. 126°, $C_{20}H_{21}O_4N$ (Synth.)   mp. 123~123.5°

UV λ_{max}^{EtOH}: 280 (4.29)

Laurifoline (XLVII)[1673]

≪Occurrence≫   *Cocculus laurifolius* DC. (コウシュウウヤク)

Chloride   mp. 253 (decomp.) $[\alpha]_D + 26.32°$ (H_2O), $C_{20}H_{24}O_4NCl$, picrate mp. 222° (decomp.)

Cossarmine (XLVIII)[1674]

≪Occurrence≫   *Cocculus sarmentosus* DIELS. (ホウザンツヅラフジ)

Iodide   mp. 205~207° (decomp.), $[\alpha]_D + 27.9°$ (EtOH), $C_{21}H_{26}O_4NI\cdot 3H_2O$

UV λ_{max}^{EtOH} mμ (log ε): 282 (4.11), 308 (4.22); $\lambda_{max}^{EtOH-NaOH}$ mμ: 272, 345

***O*-Methylcocsarmine iodide** (Glaucine (XXXII) methiodide=*O, O*-Dimethyllaurifolineiodide)

mp. 219~221°, $C_{22}H_{28}O_4NI$

Protostephanine (XLVIII–a)[1672a]

≪Occurrence≫   *Stephania japonica* MIERS (ハスノハカズラ)

mp. 95°, $C_{21}H_{27}O_4N$

UV   See Fig. 247 a.

(h) Papaveraceous Alkaloids

(XLIX)=(XIII)   Dicentrine

$R_1=R_4=CH_3$, $R_2=R_3=H$
(L) Corytuberine
$R_1=R_3=R_4=CH_3$, $R_2=H$
(LI)=(XV) Corydine
$R_1=R_2=R_4=CH_3$, $R_3=H$
(LII) Isocorydine

(LIII)=(XIV)   Bulbocapnine

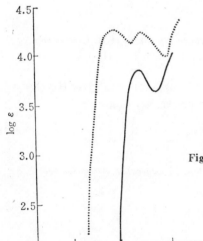

Fig. 247 a UV Spectra of
——Protostephanine and
······Dicentrine
(Kondo *et al.*)[1672a]

1672) Govindachari, T. R. *et al*: *Soc.*, **1958**, 983.
1672a) Kondo, H. *et al*: *The Annual Report of ITSUU Lab.*, **1**, 1, 5, 12 (1950), **4**, 6 (1953), **5**, 1 (1954), **7**, 30 (1956), **9**, 24, 33 (1958).
1673) Tomita, M. *et al*: *Pharm. Bull.*, **1**, 1, 5 (1953).
1674) Tomita, M. *et al*: *Y. Z.*, **83**, 190 (1963).

Dicentrine (XLIX)

≪Occurrence≫　*Dicentra pusilla* S. & Z., *D. formosa, Stephania capitata (Menispermaceae)*
mp. 169°, $[\alpha]_D +62°$ (CHCl$_3$), C$_{20}$H$_{21}$O$_4$N, methine mp. 158～159°
UV　See Fig. 247 a,　　NMR　See the item (a)

Corytuberine (L)

≪Occurrence≫　*Corydalis tuberosa, C. nobilis,*　mp. 240°(decomp.), $[\alpha]_D +282°$ (EtOH), C$_{19}$H$_{21}$O$_4$N

Corydine (LI)

≪Occurrence≫　*Corydalis tuberosa, C. ternata, Dicentra* spp.
mp. 149°, $[\alpha]_D +205°$ (CHCl$_3$), C$_{20}$H$_{23}$O$_4$N,　　NMR　See the item (a)

Isocorydine (LII)

≪Occurrence≫　*Corydalis lutea, C. platicarpa, Glaucium* spp., *Dicentra canadensis*
mp. 186°, $[\alpha]_D +195°$ (CHCl$_3$), C$_{20}$H$_{23}$O$_4$N

Bulbocapnine (LIII)

≪Occurrence≫　*Corydalis tuberosa* and other spp.,　mp. 199°, 202°, $[\alpha]_D +237°$ (CHCl$_3$), C$_{19}$H$_{19}$O$_4$N
Synth. *dl*-Bulbocapnine[1675]　mp. 213～214°

UV　λ_{max}^{EtOH} mμ (log ε): 223 (4.51), 270 (4.20), 306 (3.74)

IR　$\nu_{max}^{CHCl_3}$ cm^{-1}: 3580 (OH): 2804 (NCH$_3$)　　NMR　See the item (a)

(i)　Alkaloids occur in other Families

Anolobine (LIV)

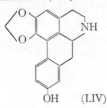

≪Occurrence≫　*Asimina triloba (Anonaceae)*
mp. 262° (decomp.), $[\alpha]_D -22.5°$ (CHCl$_3$), C$_{17}$H$_{15}$O$_3$N

Xylopine (*O*–Methylanolobine) (LIV–a)[1676]

≪Occurrence≫　*Xylopia discreta* S. et H. *(Anonaceae)*
mp. 182～186°, $[\alpha]_D -23.4°$ (MeOH), C$_{18}$H$_{17}$O$_3$N
UV: See Fig. 248
N–Acetate　mp. 213～214°, $[\alpha]_D -194°$ (CHCl$_3$), C$_{20}$H$_{19}$O$_4$N
UV　λ_{max}^{EtOH} mμ (ε): 217 (35400), 284 (20400)

Tylophorine (LIX–b)[1677]

≪Occurrence≫　*Tylophora asthmatica (Asclepiadaceae)*

1675)　Tachibana, I: *Y. Z.*, **79**, 1244 (1959).　　1676)　Schmutz, J: *Helv. Chim. Acta*, **42**, 335 (1959).
1677)　Govindachari, T.R. *et al*: *Tetrahedron*, **9**, 53 (1960).

(6) Tetrahydroberberine-type Tertiary Bases (Berberine-type Bases)

R_1	R_2	R_3	R_4	R_5	R_6	R_7	
MeO	MeO	H	MeO	MeO	H	H	(LV) Xylopinine
MeO	MeO	MeO	MeO	H	H	H	(LVI) Tetrahydropalmatine
O–CH$_2$–O		MeO	MeO	H	HO	H	(LVII) Ophiocarpine
MeO	MeO	MeO	HO	H	H	H	(LVIII) Corydalmine
O–CH$_2$–O		O–CH$_2$–O		H	H	H	(LIX) Tetrahydrocoptisine
HO	MeO	MeO	MeO	H	H	H	(LX) Corypalmine
O–CH$_2$–O		HO	MeO	H	H	H	(LXI) Nandinine
MeO	HO	MeO	MeO	H	H	H	(LXII)Tetrahydrocolumbamine
O–CH$_2$–O		MeO	MeO	H	H	H	(LXIII) Tetrahydroberberine
O–CH$_2$–O		MeO	MeO	H	H	O	(LXIII–a) Oxyberberine

(a) **Xylopinine (Norcoralydine)** (LV)[1676]

   《Occurrence》 *Xylopia discreta* S. et H. (*Ano-naceae*)

mp. 182~183°, $[\alpha]_D$ −297° (CHCl$_3$), C$_{21}$H$_{25}$O$_4$N

UV See Fig. 248

Aporphine type-Xylopine (LIV–a)

UV λ_{max}^{EtOH} mμ (ε): 270~310 (12000~25000)

Tetrahydroprotoberberine type–Xylopinine (LV)

UV λ_{max}^{EtOH} mμ (ε): 280~290 (5000~8000)

(b) **Tetrahydropalmatine** (LVI)[1683a]

   《Occurrence》 *Corydalis tuberosa* and other spp.

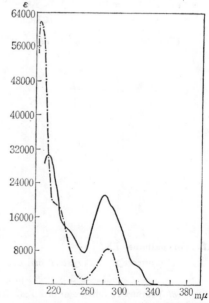

Fig. 248 UV Spectra of ——Xylopine
and –·–·– Xylopinine in EtOH
(Schmutz)[1676]

mp. (*d*–) 142°, (*dl*–) 151°, [α]$_D$ +291° (EtOH) (*d*–), $C_{21}H_{27}O_5N$

UV　$λ_{max}^{MeOH}$ mμ (log ε)[1679] : 282 (3.833)

NMR[1709] in $CHCl_3$, 40 Mc, H_2O as ext. reference, cps: (+76.0, +80.2, +83.7) t (4)(arom. protons), −34.2 s (12) (4CH_3O–)

Discretinine (Monodesmethyl–(−)–tetrahydropalmatine)[1676]

≪Occurrence≫　*Xylopia discreta* S. et H.,　　mp. 212~214° (decomp.), [α]$_D$ −371° (pyridine), $C_{20}H_{23}O_4N$

Discretamine (Bisdesmethyl–(−)–tetrahydropalmatine)[1676]

≪Occurrence≫　*Xylopia discreta* S. et H.,　　mp. 221~224° (decomp.), [α]$_D$ −368° (pyridine), $C_{19}H_{21}O_4N$

(c)　**Ophiocarpine** (LVII)[1678]

≪Occurrence≫　*Corydalis ophiocarpa*,　　mp. 188°, [α]$_D$ −283° (CHCl$_3$), pKa 5.57, $C_{20}H_{21}O_5N$

UV　$λ_{max}^{EtOH}$ mμ (log ε): 291 (3.74)

IR　$ν_{max}^{CCl_4}$ cm^{-1}: 3526, 2807, 2771, 2756

(d)　**Corydalmine** (LVIII)[1679]

≪Occurrence≫　Chinese corydalis,　　mp. 238~239°, [α]$_D$ +337.4°, $C_{20}H_{23}O_4N$

UV　$λ_{max}^{MeOH}$ mμ (log ε): 284 (3.559)

(e)　**Tetrahydrocoptisine** (Stylopine) (LIX)

≪Occurrence≫　Chinese corydalis[1679], *Chelidonium majus* L.

Coptisine (LXXVII) $\xrightarrow{NaBH_4}$ (LIX), confirmed[1691a]

mp. (*dl*–) 218°[1693], (*l*)–197~200°[1679], 222~224°[1680], 213~215°[1691a], [α]$_D$ −322.5° (CHCl$_3$) (*l*–), $C_{19}H_{17}O_4N$

UV　$λ_{max}^{MeOH}$ mμ (log ε)[1679] (*dl*–): 290 (3.664), see Fig. 249

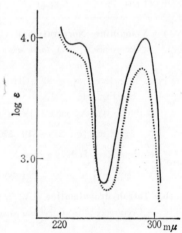

Fig. 249　UV Spectra of
──Tetrahydrocoptisine and
‥‥‥‥Tetrahydrocorysamine
(Tani)[1691a]

(f)　**Corypalmine** (LX)

≪Occurrence≫　*Corydalis* and other spp., *Dicentra oregana*,　　mp. (*l*–) 246°, (*d*–) 241°, (*dl*–) 223°, 212~213.5°[1683c], [α]$_D$ +280° (CHCl$_3$) (*d*–), $C_{20}H_{23}O_4N$

(*dl*–) form=Tetrahydrojatrorrhizine (see LXXV)

(g)　**Nandinine** (LXI)

≪Occurrence≫　*Nandina domestica* (ナンテン),　　mp. 78°, [α]$_D$ +63.2° (EtOH), $C_{19}H_{19}O_4N$

(h)　***l*–Tetrahydrocolumbamine** (LXII) [1679]

≪Occurrence≫　Chinese corydalis,　　mp. 240~241°, [α]$_D$ −352° (CHCl$_3$), $C_{20}H_{23}O_4N$

1678)　Ohta, H. *et al*: *Tetr. Letters*, **1963**, 859.　　1679)　Imaseki, J. *et al*: *Y. Z.*, **82**, 1214 (1962).

UV λ_{max}^{MeOH} mμ (log ε): 284 (3.760)

(i) dl–Tetrahydroberberine (LXIII) [1683c]

[Preparation] Berberine (LXXIII)–iodide $\xrightarrow{+AgOAc \ in \ MeOH, \ H_2, \ Ni\text{-alloy catalyst}}$ (LXIII)[1683d], iodide $\xrightarrow{Zn+H_2SO_4}$ (LXIII)[1683c]

mp. 168°, 171~173°, $C_{20}H_{21}O_4N$

NMR[1709] in $CHCl_3$, 40 Mc, H_2O as ext. reference, cps: (+73.2, +78.5, +82.0) t (4) (arom. protons), +45.8s (2) (–O–CH$_2$–O–), −34.8s (6) (2CH$_3$O–)

(j) Oxyberberine (LXIII–a)[1683e]

≪Occurrence≫ *Berberis mingetsensis* HAYATA (ウスバヘビノボラズ), mp. 199~201°, $C_{20}H_{17}O_5N$

R$_1$	R$_2$	R$_3$	R$_4$		
O–CH$_2$–O		MeO	MeO	(XLIV)	d–Thalictricavine
MeO	MeO	MeO	MeO	(LXV)	d–Corydaline
MeO	HO	MeO	MeO	(LXVI)	d–Corybulbine
HO	MeO	MeO	MeO	(LXVII)	d–Isocorybulbine

(k) Thalictricavine (XLIV)[1692]

≪Occurrence≫ *Corydalis tuberosa*, mp. 149~175°, $[\alpha]_D$ +292° (CHCl$_3$), $C_{21}H_{23}O_4N$

dl–form mp. 209°

IR: See Fig. 250

1680) Bandelin, J. *et al*: *J.A.P.A.*, **45**, 702 (1956).

1681) Tomita, M. *et al*: *Y.Z.*, **77**, 274 (1957).

1682) Tomita, M. *et al*: *ibid.*, **77**, 69, 73, 79 (1957).

1683) Tomita, M. *et al*: *Chem. Pharm. Bull.*, **5**, 10 (1957); *Y.Z.*, **80**, 880, 885 (1960).

1683a) Kunitomo, J: *ibid.*, **82**, 611 (1962).

1683b) Tomita, M. *et al*: *Y.Z.*, **80**, 1238 (1960).

1683c) Imaseki, I. *et al*: *ibid.*, **81**, 1281 (1961).

1683d) Tani, Ch *et al*: *ibid.*, **77**, 805 (1957).   1683e) Yang, T.H. *et al*: *ibid.*, **80**, 847, 849 (1960).

1684) Spenser, I.D. *et al*: *Proc. Chem. Soc.*, **1962**, 228.

1685) Barton, D.H.R. *et al*: *ibid.*, **1963**, 267.   1686) Battersby, A.R. *et al*: *ibid.*, **1963**, 268.

1687) Imaseki, I. *et al*: *Y.Z.*, **80**, 1802 (1960).

1688) Yang, T–H: *J. ibid.* **80**, 1304 (1960).

1688a) Kitasato, Z: *J. Pharm. Soc. Japan*, **542**, 315 (1927).

1688b) Yamaguchi, K: Shokubutsu Seibun Bunsekihō (Analytical Methods in Phytochemistry) Vol I p. 555 (Nankodo Pub. Inc. Tokyo, 1963)

1689) Arthur, H.R. *et al*: *Soc.*, **1959**, 1840, 4010, 4012.

1690) Arthur, H.R. *et al*: *ibid.*, **1959**, 4007.

1691) Tani, Ch *et al*: *Y.Z.*, **82**, 594 (1962).

1691a) Tani, Ch *et al*: *ibid.*, **82**, 598 (1962).   1692) Kondo, Y: *ibid.*, **83**, 1017 (1963).

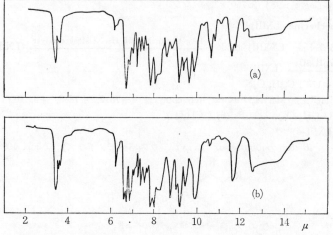

Fig. 250 IR Spectra of (a) Thalictricavine and (b) Corydaline
in CHCl$_3$ (Kondō)[1692]

IR $\nu_{max}^{CHCl_3}$ cm^{-1}: 2766 (characteristic absorption for *trans*–quinolizidine)[1697]

NMR in CHCl$_3$, 60 Mc, TMS, δ: See Fig. 251

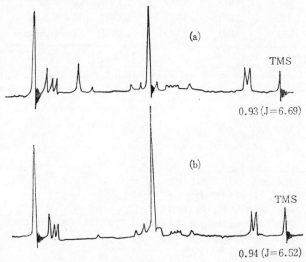

0.93 (J=6.69)

0.94 (J=6.52)

Fig. 251 NMR Spectra of (a) Thalictricavine and (b) Corydaline
(Kondō)[1692]

$\delta_{Me}=0.93d$, $J=6.7$cps

(LXVIII) 1 (*a*)–Methylquinolizidine   (LXIX) 1 (*e*)–Methylquinolizidine

$\delta_{Me(a)}$ $<$ $\delta_{Me(e)}$[1698]

0.1~0.4ppm

↓   ↓

Acetate   Acetate

$\delta_{Me(a)}$ 0.90ppm $\delta_{Me(e)}$ 0.92ppm

ORD[1692] : Comparation of *d*–thalictricavine (LXIV), *d*–corydaline (LXV) and *l*–tetrahydropalmatine (LVI) (Fig. 252)

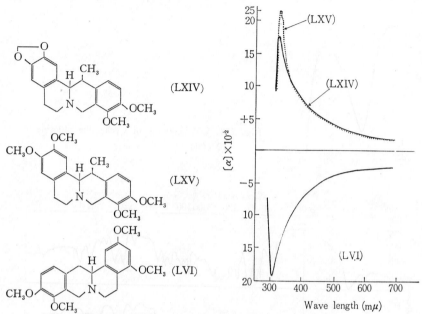

Fig. 252 ORD of Protoberberine Alkaloids (Kondō)[1692]

(l) ***d*–Corydaline (LXV)[1692]**

≪Occurrence≫ *Corydalis tuberosa* and other spp., mp. 136°, $[\alpha]_D$ +300° (CHCl$_3$), C$_{22}$H$_{27}$O$_4$N

IR: See Fig. 250, NMR: See Fig. 251, ORD: See Fig. 252

(m) ***d*–Corybulbine (LXVI)**

≪Occurrence≫ *Corydalis tuberosa* and other spp., mp. 242°, $[\alpha]_D$ +303.3° (CHCl$_3$), C$_{21}$H$_{25}$O$_4$N

(n) ***d*–Isocorybulbine (LXVII)**

≪Occurrence≫ *Corydalis tuberosa*, mp. 179~180°, $[\alpha]_D$ +299.8° (CHCl$_3$), C$_{21}$H$_{25}$O$_4$N

(7) Tetrahydroberberinemethine-type Bases

R$_1$	R$_2$	R$_3$	R$_4$		
CH$_3$O	HO	HO	CH$_3$O	(LXX)	Cyclanoline
HO	CH$_3$O	HO	CH$_3$O	(LXXI)	Steponine
CH$_3$O	HO	CH$_3$O	CH$_3$O	(LXXII)	Phellodendrine

(a) Cyclanoline (LXX)[1682]

≪Occurrence≫ *Cyclea insularis* DIELS.

Chloride mp. 214~215° (decomp.), $[\alpha]_D$ −115.8° (MeOH), C$_{20}$H$_{24}$O$_4$NCl·H$_2$O

UV See Fig. 253, λ_{max}^{MeOH} mμ (log ε): 233 (4.13), 286 (3.88); λ_{min}: 255 (2.79)

IR See Fig. 254[1681]

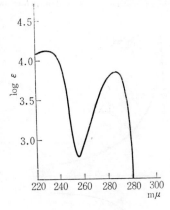

Fig. 253 UV Spectrum of Cyclanoline chloride in MeOH (Tomita *et al.*)

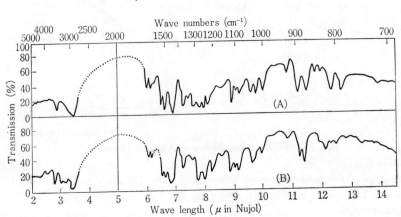

Fig. 254 IR Spectra of (A) Steponine and (B) Cyclanoline in nujol
(Tomita *et al.*)

(b) **Steponine** (LXXI)[1681]

«Occurrence» *Stephania japonica* MIERS. (ハスノハカズラ)

Chloride mp. 235° (decomp.), $[\alpha]_D$ −129.89° (H_2O), $C_{20}H_{24}O_4NCl\cdot H_2O$

UV λ_{max}^{MeOH} mμ (log ε): 234 (4.12), 282 (3.85); λ_{min}: 256 (2.76)

IR See Fig. 254

(c) **Phellodendrine** (LXXII)[1683)1683a]

«Occurrence» *Phellodendron amurense* RUPR. (オウバク)

Iodide mp. 259°, $[\alpha]_D$ −147° (MeOH), $C_{20}H_{24}O_4NI$

UV λ_{max}^{MeOH} mμ (log ε): 287 (4.03)

Synthesis of *dl*-form[1683b]

(8) **Berberinium Bases**

R$_1$	R$_2$	R$_3$	R$_4$	R$_5$	
O–CH$_2$–O		CH$_3$O	CH$_3$O	H	(LXXIII) Berberine
CH$_3$O	CH$_3$O	CH$_3$O	CH$_3$O	H	(LXXIV) Palmatine
HO	CH$_3$O	CH$_3$O	CH$_3$O	H	(LXXV) Jatrorrhizine
O–CH$_2$–O		O–CH$_2$–O		Me	(LXXVI) Worenine, Corysamine
O–CH$_2$–O		O–CH$_2$–O		H	(LXXVII) Coptisine

(a) Berberine (LXXIII)

≪Occurrence≫ *Berberis vulgaris* and other spp., *Mahonia acanthifolia* and other spp., *Nandina domestica*, *Xylopia polycarpa* (*Berberidaceae*), *Archangelisia flava*, *Coscinium fenestratum* (*Menispermaceae*), *Coptis japonica* and other spp., *Hydrastis canadensis*, *Thalictrum folilosum*, *T. thunbergii* (*Ranunculaceae*), *Argemone mexicana*, *A. alba*, *Chelidonium majus*, *Corydalis cheilantheifolia* and other spp. (*Papaveraceae*), *Evodia meliaefolia*, *Phellodendron amurense*, *Toddaiia aculeata* (*Rutaceae*)

Iodide C$_{20}$H$_{18}$O$_4$NI, mp. 263~265° (decomp.)[1683e], C$_{20}$H$_{18}$O$_4$NI·^{1}/$_2$H$_2$O mp. 255° (decomp.)[1684e].

UV λ_{max}^{MeOH} mμ (log ε): 225 (4.64), 270 (4.02), 331 (3.88)

Chloride[1683c] mp. 205° (decomp.), C$_{20}$H$_{18}$O$_4$NCl·2H$_2$O

Biosynthesis[1684]~[1687]

***dl*–Tetrahydroberberine** (LXIII) See the item (6), (i)

(b) Palmatine (LXXIV)

≪Occurrence≫ *Jateorhiza palmata*, *Coptis japonica*, *Berberis vulgaris* and other spp., *Mahonia philippinensis*, *Coscinium blumeanum*, *Fibraurea tinctoria*, *Phellodendron amurense*, *Tinospora bakis*, *Cocculus leaeba*.

Chloride[1683c] mp. 206° (pecomp.), C$_{20}$H$_{18}$O$_4$NCl

Tetrahydropalmatine (LVI) See the item (6), (b)

(c) Jatrorrhizine (LXXV)

≪Occurrence≫ *Jateorhiza columba*, *J. palmata*, *Berberis vlgaris* and other spp., *Coscinium blumeanum*, *Fibraurea chloleuca*, *F. tinctoria*, *Mahonia philippinensis*.

Chloride mp. 217~216°[1683e]

Picrate[1688] mp. 217~219°, C$_{20}$H$_{20}$O$_4$N·C$_6$H$_2$O$_7$N$_3$

Iodide $\xrightarrow{\text{Zn+dil.-H}_2\text{SO}_4}$ Tetrahydrojatro–rrhizine (LX)

(d) Worenine (LXXVI)[1688a]

≪Occurrence≫ *Coptis japonica* (オウレン).   mp. 212~213°, C$_{20}$H$_{19}$O$_4$N

$\xrightarrow{\text{Red}}$ Tetrahydroworenine

1693) Tani, Ch *et al*: *Y. Z.*, **82**, 748 (1962).   1694) Tani, Ch *et al*: *ibid.*, **82**, 751 (1962).
1695) Tani, Ch *et al*: *ibid.*, **82**, 755 (1962).   1696) Takao, N: *Chem. Pharm. Bull.*, **11**, 1312 (1963).

(e) Corysamine (LXXVI)[1691a) 1693) 1694) 1696)]

≪Occurrence≫ *Corydalis insica* (ムラサキケマン)

Chloride mp. 230° (decomp.), $C_{20}H_{16}O_4NCl \cdot 3^1/_2H_2O$

UV See Fig. 255

(LXXVI) $\xrightarrow{\text{NaBH}_4}$ Tetrahydrocorysamine[1693)]

Tetrahydrocorysamine mp. 202~203°, $C_{20}H_{19}O_4N$|

UV See Fig. 249

IR See Fig. 256

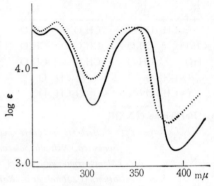

Fig. 255 UV Spectra of
—— Coptisine chloride and
······Corysamine chloride in
MeOH (Tani *et al.*) [1691a)]

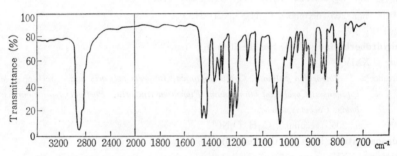

Fig. 256 IR Spectrum of Tetrahydrocorysamine in nujol (Tani *et al.*) [1693)]

(f) Coptisine (LXXVII)[1691a) 1693)]

≪Occurrence≫ *Corydalis insica* (ムラサキケマン) and other spp., *Chelidonium majus* L. (クサノオウ),
Coptis japonica (オウレン), mp. >290°, $C_{19}H_{15}O_4N$

Chloride

UV See Fig. 225

(LXXVII) $\xrightarrow{\text{NaBH}_4}$ Tetrahydrocoptisine (LIX), see the item (6) (e)

(9) Protopine Alkaloids[1688b)]

(10) Benzophenanthridine Alkaloids

(a) Alkaloids of *Zanthoxylum* spp.

R=O (LXXVIII) Oxynitidine
R=H₂ (LXXIX) Dihydronitidine

(LXXX) Nitidine

R=O (LXXXI) Oxyavicine
R=H$_2$ (LXXXII) Dihydroavicine

(LXXXIII) Avicine

Oxynitidine (LXXVIII)[1689]

≪Occurrence≫　*Zanthoxylum nitidum* DC. (*Fagara nitida*),　　mp. 284~285°, [α]$_D$ 0° (CHCl$_3$), C$_{21}$H$_{17}$O$_5$N

UV　λ$_{max}^{EtOH}$ mμ (log ε): 367 (36.3), 333 (4.18), 320 (4.20), 288 (4.81), 277 (4.72), 251 (4.59)

Dihydronitidine (LXXIX)[1689]

≪Occurrence≫　*Zanthoxylum nitidum* DC. (*Fagara nitida*),　　mp. 221~223°, [α]$_D$ 0° (CHCl$_3$), C$_{21}$H$_{19}$O$_4$N

UV　λ$_{max}^{EtOH}$ mμ (log ε): 311 (4.29), 278 (4.54), 228 (4.61)

(LXXIX) $\xrightarrow{\text{HCl}}$ Nitidine chloride (LXXX) C$_{21}$H$_{18}$O$_4$NCl·2H$_2$O

(LXXIX) $\xrightarrow{\text{KCN}}$ ── ψ–cyanide C$_{22}$H$_{18}$O$_4$N$_2$, mp. 216° (decomp.)

(LXXIX) $\xrightarrow{\text{50\% HOAc, Hg(OAc)}_2,\ \text{H}_2\text{S}}$ ── Acetate C$_{23}$H$_{21}$O$_6$N·4H$_2$O, mp. 255~260°

Synthesis[1689]

Oxyavicine (LXXXI)[1690]

≪Occurrence≫　*Zanthoxylum avicennae* DC.,　　mp. 257~259°, 275~277°, C$_{20}$H$_{13}$O$_5$N

UV　λ$_{max}^{EtOH}$ mμ (log ε): 332 (4.19), 322 (4.21), 289 (4.76), 278 (4.70), 248 (4.50)

Dihydroavicine (LXXXII)[1690]

≪Occurrence≫　*Zanthoxylum avicennae* DC.,　　mp. 211~215.5°, C$_{20}$H$_{15}$O$_4$N

UV　λ$_{max}^{EtOH}$ mμ (log ε): 322 (4.33), 278 (4.50), 232 (4.60)

(LXXXII) $\xrightarrow{\text{50\% HOAc, Hg(OAc)}_2,\ \text{H}_2\text{S}}$ Avicine (LXXXIII) acetate C$_{22}$H$_{17}$O$_6$N·2H$_2$O, mp. 160°(decomp.)

(b) Papaveraceous Alkaloids

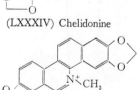

(LXXXIV) Chelidonine　　　　(LXXXV) Methoxychelidonine　　　　(LXXXVI) Corynoline

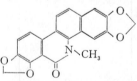

(LXXXVII) Sanguinarine　　　　(LXXXVIII) Oxysanguinarine　　　　(LXXXIX) Chelerythrine

Chelidonine (LXXXIV)

≪Occurrence≫　*Chelidonium majus* L. (クサノオウ),　　mp. 136°, [α]$_D$ +118° (CHCl$_3$), C$_{20}$H$_{19}$O$_5$N

IR　See Fig. 258

Methoxychelidonine (LXXXV)

≪Occurrence≫　*Chelidonium majus* L.,　　[α]$_D$ +115.8°, C$_{21}$H$_{21}$O$_6$N

Corynoline (LXXXVI)[1691][1696][1698]

1696a)　Takao, N: *Chem, Pharm. Bull.*, **11**, 1306 (1963).

≪Occurrence≫   *Corydalis incisa.* (ムラサキケマン),   mp. 216～217°, $[\alpha]_D$ ±0°, $C_{21}H_{21}O_5N$

UV See Fig. 257, λ_{max}^{MeOH} mμ (log ε): 242 (3.94), 290 (3.85); λ_{min}: 227 (3.82), 265 (3.10)

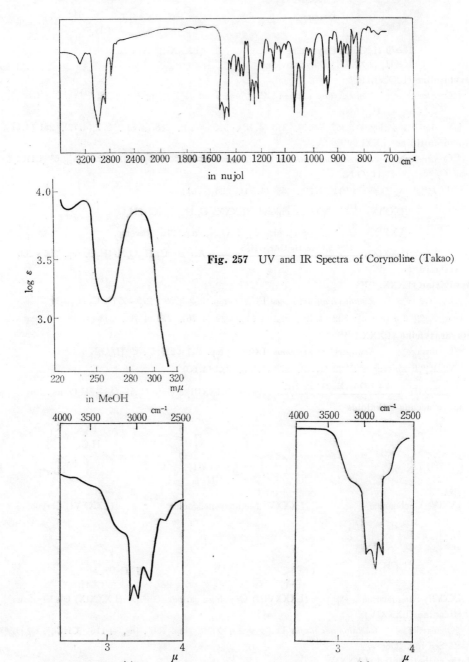

Fig. 257 UV and IR Spectra of Corynoline (Takao)

Fig. 258 IR Spectra of (a) Corynoline and (b) Chelidonine in CHCl$_3$ (Takao)[1696]

IR See Fig. 257 and 258.

The position of hydroxyl group at C_{11} and the stereochemical structure of corynoline (LXXXVI) have been established by the characteristic frequency of ν_{O-H} as follows:

IR ν_{max}^{nujol} cm^{-1}: 3240 (alcoholic OH) (see Fig. 257); IR $\nu_{max}^{CHCl_3}$ cm^{-1}: 3000~3100 shoulder

The frequency is assigned for hydroxyl group, bonded with N–atom, which also appears in the spectrum of chelidonine (LXXXIV). Nonchelated hydroxyl absorption does not appear in the both spectra at 3400~3600 cm^{-1} region. This fact suggests that the hydroxyl group is attached to C_{11}–position but not to C_{12} to form the chelation with N–atom as shown in (LXXXVI–a). Further frequencies are assigned as: ν_{max} cm^{-1}: 804, 848, 869, 883 (1, 2, 3, 4–*tetra.* subst. benzene (A) and 1, 2, 4, 5–*tetra.* subst. benzene (B).)

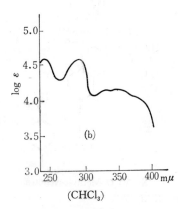

NMR: H$_2$O as external standard, ppm 3.5s $\left(\!\!>\!\!C_3\!-\!Me\right)$

(LXXXVI–a) Corynoline

Sanguinarine (LXXXVII)[1691)1695]

≪Occurrence≫ *Bocconia cordata* (*Macleya cordata*) (タケニグサ), *Corydalis insica* (ムラサキケマン)

(LXXXVII) $\xrightarrow{\text{NaBH}_4}$ Dihydrosanguinarine

Dihydrosanguinarine (XC)

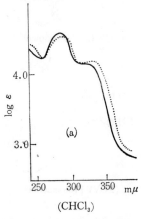

mp. 188~189°, C$_{20}$H$_{15}$O$_4$N
UV See Fig. 259

(XC) Dihydrosanguinarine

Oxysanguinarine (LXXXVIII)[1695]

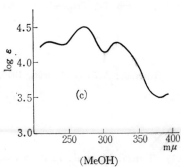

(a)

(CHCl$_3$)

(b)

(CHCl$_3$)

(c)

(MeOH)

Fig. 259 UV Spectra of
(a) ······ Dihydrosanguinaline,
────Dihydrochelerythrine
(b) Oxysanguinarine and
(c) Chelerythrine chloride
(Tani *et al.*)[1695]

≪Occurrence≫   *Bocconia cordata* (*Macleya cordata*) (タケニグサ),   mp. 360°, $C_{20}H_{13}O_5N$

UV See Fig. 259

Chelerythrine (LXXXIX)[1695]

≪Occurrence≫   *Bocconia cordata* (*Macleya cordata*), *Chelidonium majus* (クサノオウ) etc.

Chloride $C_{20}H_{17}O_4N \cdot Cl$

UV See Fig. 259

(LXXXIX) $\xrightarrow{\text{NaBH}_4}$ Dihydrochelerythrine (XCI)

Dihydrochelerythrine

mp. 165~166°, $C_{21}H_{19}O_4N$
UV See Fig. 259

(XCI) Dihydrochelerythrine

(11) Ipecac Alkaloids

(XCII) Emetine[1699]

(XCIII) Emetamine[1700]

(XCIV) *O*–Methylpsychotrine[1700]

(XCV) Protoemetine[1700]

(a) Emetine (XCII)[1699][1701]~[1707]

≪Occurrence≫   *Uragoga ipecacuanha* BAILL (トコン),   mp. 74°, $[\alpha]_D$ −25.8~32.7° (EtOH),

$C_{29}H_{40}O_4N_2$

1697) Bohlmann, F: *Ber.*, **91**, 2157 (1958).
1698) Kotake, M. *et al*: *Bull. Chem. Soc. Japan*, **35**, 1335 (1962).
1699) Battersby, A.R. *et al*: *Soc.*, **1961**, 3899.   1700) Battersby, A.R. *et al*: *ibid.*, **1959**, 1744, **1748.**
1701) Clark, D.E. *et al*: *ibid.*, **1962**, 2479, 2490.
1702) Van Tamelen, E.E. *et al*: *J.A.C.S*, **81**, 6214 (1959).
1703) Battersby, A.R. *et al*: *Soc.*, **1960**, 3474.
1704) Evstigneeva, R.P. *et al*: *Tetrahedron*, **4**, 223 (1958).
1705) Brossi, A. *et al*: *Helv. Chim. Acta*, **42**, 1515 (1959).
1706) Grüssner, A. *et al*: *ibid.*, **42**, 2431 (1959).   1707) Tomita, M. *et al*: *Y. Z.*, **82**, 734, 741 (1962).

IR ν_{max}^{Film} cm^{-1}: 1610 (at 1550~1650)

Synthesis[1699) 1701) 1705) ~1707)]

(b) Emetamine (XCIII)[1699) 1700)]

≪Occurrence≫ *Uragoga ipecacuanha* BAILL., mp. 155~156°, $C_{29}H_{36}O_4N_2$

UV See Fig. 259 a

Emetine (XCII) $\xrightarrow{\text{Pd-C, 180~190°}}$ (XCIII)

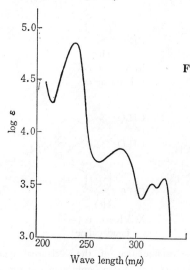

Fig. 259 a UV Spectrum of Emetamine in EtOH (Battersby *et al.*)[1699)]

(c) O–Methylpsychotrine (XCIV)[1700)]

≪Occurrence≫ *Uragoga ipecacuanha* BAILL., mp. 121~123°, $C_{29}H_{34}O_4N_2$

IR ν_{max}^{nujol} cm^{-1}: 1618, 1610, 1576 (at 1550~1650)

Hydrogeneoxalate mp. 151~155°, $[\alpha]_D+45.9°$ (H_2O)

(d) Protoemetine (XCV)[1700)]

≪Occurrence≫ *Uragoga ipecacuanha* BAILL.

Perchlorate

monohydrate mp. 140~142° $C_{19}H_{27}O_3N \cdot HClO_4 \cdot H_2O$

Anhydrous mp. 193~195°, $[\alpha]_D-10.9°$ (EtOH), $C_{19}H_{27}O_3N$

UV λ_{max}^{EtOH} mμ (log ε) (anhydrous): 232 (3.92), 283 (3.61); λ_{min}: 220 (3.85), 254 (2.68)

IR ν_{max} cm^{-1} (monohydrate): 2810 (aldehyde C–H), 1728 (aldehyde C=O), 1622 (arom. ring)

(e) Ipecac Alkaloid A[1700)]

≪Occurrence≫ *Uragoga ipecacuanha* BAILL., mp. 149~150°

Picrate mp. 195~196° (decomp.) Oxalate mp. 182~190° (decomp.)

(12) Biscoclaurine Alkaloids

(a) Absolute Configuration

L-(+)-Laudanosine (XCVI), with the established absolute configuration by Hardegger *et al*[1708)], was submitted to cleavage reaction with Na in liquid ammonia by Tomita *et al*[1707)] and was derived to *d*–N, O,O–trimethylcoclaurine (XCVIII) via L-(+)-laudanidine (XCVII). This proved that those alkaloids

---

1708) Hardegger, E. *et al*: *Helv. Chim. Acta*, **39**, 889 (1956).

have the same configuration and are L–type.   The same examination was carried on D–(−)–laudanosine (XCVI–a), whose absolute configuration has already been established as the antipode of (XCVI), was derived to *l–N, O,O*–trimethylcoclaurine (XCVIII–a).   The result proved D–series configuration of the latter bases.

L–Series :

| (XCVI) L–(+)–
Laudanosine | (XCVII) L–(+)–
Laudanidine | (XCVIII) *d–N, O, O–*
Trimethylcoclaurine |

D–Series :

| (XCVI–a) D–(−)–
Laudanosine | (XCVII–a) D–(−)–
Laudanidine | (XCVIII–a) *l–N, O, O–*
Trimethylcoclaurine |

As the results, the absolute configulation of the coclaurine type bases, obtained by the cleavage reaction of biscoclaurine type alkaloids with Na in liquid ammonia was established, thus the absolute configuration of the two assymmetric centers in biscoclaurine–type bases being established simultaneously.   The absolute configurations at two assymmetric centers A and B in various biscoclaurine–type bases are indicated as follows :

(i) *Dauricine Type*

(XCIX)   (CIII)

Str. No.	Compound	General Formula	R$_1$	R$_2$	Absolute Configuration	
					A	B
C	Dauricine	XCIX	CH$_3$	H	D–(−)	D–(−)
CI	Magnoline	XCIX	H	H	L–(+)	D–(−)
CII	Berbamunine	XCIX	H	H	D–(−)	L–(+)
CIII	Magnolamine				L–(+)	L–(+)

(ii) *Oxyacanthine-berbamine Type*

(CIV)

(CXII)

Str. No.	Compound	General Formula	R_1	R_2	Absolute Configuration	
					A	B
CV	Isotetrandrine	CIV	CH_3	CH_3	D–(–)	L–(+)
CVI	Tetrandrine	CIV	CH_3	CH_3	L–(+)	L–(+)
CVII	Phaeanthine	CIV	CH_3	CH_3	D–(–)	D–(–)
CVIII	Berbamine	CIV	CH_3	H	D–(–)	L–(+)
CIX	Pycnamine	CIV	CH_3	H	D–(–)	D–(–)
CX	Obamegine	CIV	H	H	D–(–)	L–(+)
CXI	Fangchinoline	CIV	H	CH_3	L–(+)	L–(+)

Str. No.	Compound	General Formula	R_1	R_2	R_3	R_4	Absolute Configuration	
							A	B
CXIII	O–Methylrepandine	CXII	CH_3	CH_3	CH_3	CH_3	L–(+)	L–(+)
CXIV	Obaberine	CXII	CH_3	CH_3	CH_3	CH_3	L–(+)	D–(–)
CXV	Oxyacanthine	CXII	CH_3	CH_3	CH_3	H	L–(+)	D–(–)
CXVI	Repandine	CXII	CH_3	CH_3	CH_3	H	L–(+)	L–(+)
CXVII	Aromoline	CXII	CH_3	CH_3	H	H	L–(+)	D–(–)
CXVIII	Daphnandrine	CXII	CH_3	H	H	CH_3	L–(+)	D–(–)
CXIX	Trilobamine	CXII	CH_5	H	H	H	L–(+)	D–(–)
CXX	Sepeerine	CXII	CH_3	H	CH_3	H	L–(+)	D–(–)

(CXXI)

(CXXII)

Str. No	Compound	General Formula	R	Absolute Configuration	
				A	B
CXXI	Cepharanthine	——		?–(+)	D–(–)
CXXIII	Epistephanine	CXXII	CH_3	——	D–(–)
CXXIV	Hypoepistephanine	CXXII	H	——	D–(–)

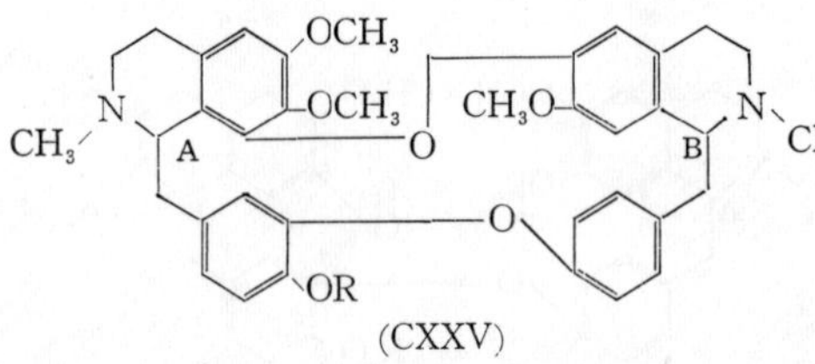

(CXXV)

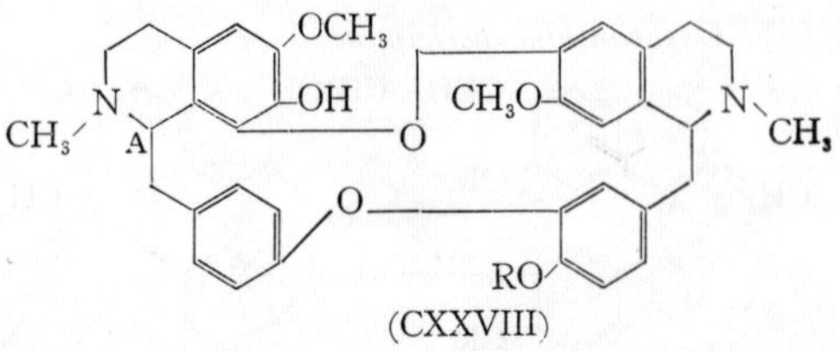

(CXXVIII)

Str. No.	Compound	General Formula	R	Absolute Configuration	
				A	B
CXXVI	Thalicberine	CXXV	H	L-(+)	L-(+)
CXXVII	O-Methylthalicberine	CXXV	CH$_3$	L-(+)	L-(+)
CXXIX	Thalicrine	CXXVIII	H	L-(+)	D-(−)
CXXX	Homothalicrine	CXXVIII	CH$_3$	L-(+)	D-(−)

(iii) *Isochondodendrine-bebeerine Type*

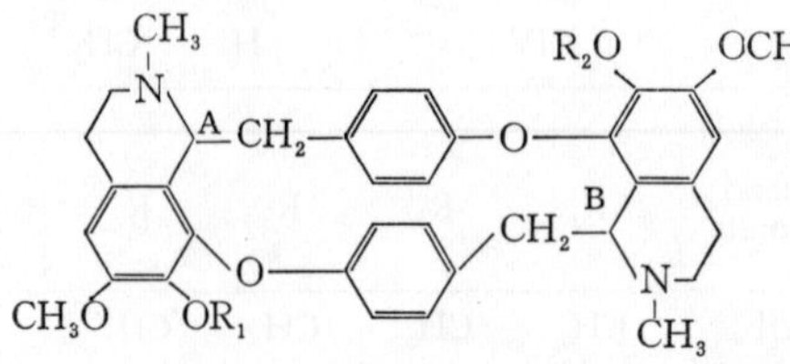

(CXXXI)

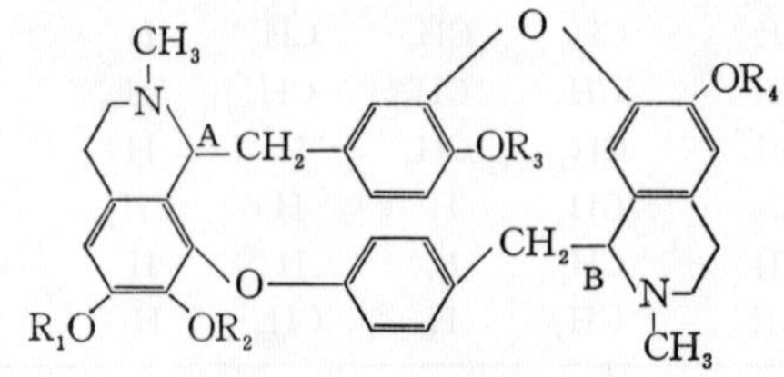

(CXXXV)

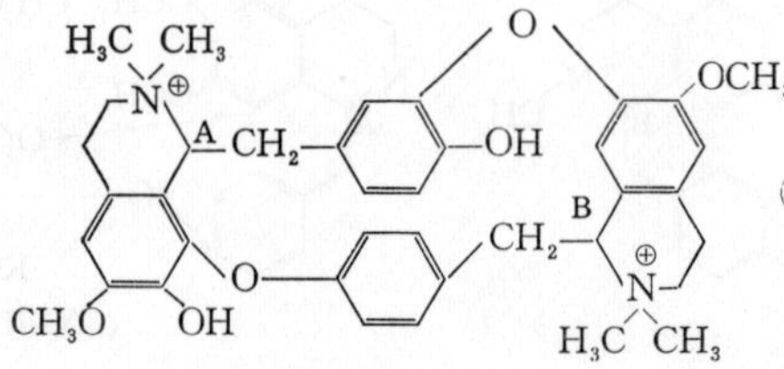

(CXXXIX)

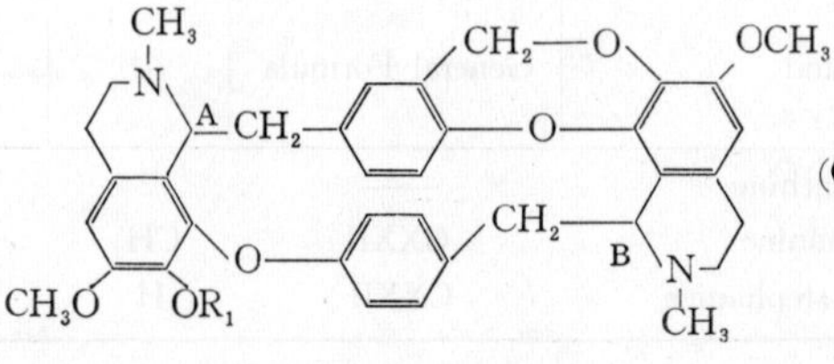

(CXL)

Str. No	Compound	General Formula	R_1	R_2	R_3	R_4	Absolute Configuration	
							A	B
(CXXXII)	Cycleanine	(CXXXI)	CH$_3$	CH$_3$	—	—	D-(−)	D-(−)
(CXXXIII)	Norcycleanine	(CXXXI)	CH$_3$	H	—	—	D-(−)	D-(−)
(CXXXIV)	Isochondrodendrine	(CXXXI)	H	H	—	—	D-(−)	D-(−)
(CXXXVI)	Bebeerine	(CXXXV)	CH$_3$	H	H	CH$_3$	L-(+)	L-(+)
(CXXXVII)	Curine	(CXXXV)	CH$_3$	H	H	CH$_3$	D-(−)	D-(−)
(CXXXVIII)	d-Chondocurine	(CXXXV)	H	CH$_3$	H	CH$_3$	D-(−)	L-(+)
(CXXXIX)	Tubocurarine	——	—	—	—	—	D-(−)	L-(+)
(CXLI)	Insularine	(CXL)	CH$_3$	—	—	—	?-(−)	D-(−)
(CXLII)	Insulanoline	(CXL)	H	—	—	—	?-(−)	D-(−)

(b) NMR Spectra[1647][1709]

in CHCl$_3$, 40 Mc, τ: (+2.75 for CHCl$_3$)[1647], or in CHCl$_3$, 40 Mc, H$_2$O as reference, cps[1709]

Table 97 Chemical Shifts (τ) of Biscoclaurine Alkaloids in CHCl$_3$, 40 Mc

Str. No.	Compound	Type	OCH$_3$				Ring CH$_2$	NCH$_3$	
			4''	6	6'	7		2'	2
C	Dauricine	i		6.20	6.20	6.45 / (7') 6.45	7.23	7.50	7.50
CV	Isotetrandrine	ii *(cps)	−30.3 s	−37.2 s	−42.3 s	−61.0 s	+50.1~82.4m	−85.1 s	−97.9 s
CVI	Tetrandrine	ii	6.10	6.27	6.65	6.82	7.00	7.41	7.70
CVII	Phaeanthine	ii	6.13	6.28	6.68	6.80	6.92	7.40	7.70
CVIII	Berbamine	ii	—	6.18	6.35	6.85	7.1	7.35	7.75
CIX	Pycnamine	ii	—	6.27	6.82	6.82	—	7.37	7.65
CXI	Fangchinoline	ii	6.03	6.23	6.60	—	6.94	7.38	7.62
CXIII	O-Methylrepandine	ii	6.05	6.25	6.60	6.95	7.22	7.45	7.45
CXV	Oxyacanthine	ii	—	6.27	6.44	6.85	7.12	7.52	7.52
CXVI	Repandine	ii	—	6.27	6.62	6.98	7.20	7.50	7.50
CXVII	Aromoline	ii	—	6.23	6.44	—	—	7.51	7.51
CXVIII	Daphandrine	ii	6.12	6.25	6.40	—	7.12	—	7.50
CXXI	Cepharanthine	ii	6.10	—	6.30	—	7.05	7.35	7.42
"		*(cps)	6,7-O-CH$_2$-O- / +30.4 s		6',4''-2 OCH$_3$ / −34.3 s −40.8 s		+61.9m ~79.9m	−101.4 s	
CXXXII	Cycleanine	iii *(cps)	−34.9 s (6),		−49.4 s (6)		+38.3~ +90.6m	−84.5 s	

(Bick *et al.*)[1647], *(Sasaki)[1709]

1709) Sasaki, Y: *Y. Z.*, **80**, 241 (1960).

(c) Dauricine Type Alkaloids

Dauricine (C)[1710]

≪Occurrence≫   *Menispermum dauricum* DC. (コウモリカズラ)

mp. 115° (amorph.), $[\alpha]_D$ −139° (MeOH), $C_{38}H_{44}O_6N_2$

O–Methyldauricine[1711]

UV λ_{max}^{EtOH} mμ (log ε): 285 (3.98); λ_{min}: 257 (3.21)

NMR   See Table 97

Magnoline (CI)

≪Occurrence≫   *Magnolia fuscata* (*Michelia fuscata*) (トウオガタマ)

mp. 178〜179°, $[\alpha]_D$ −9.6° (pyridine), $C_{36}H_{40}O_6N_2$

Berbamunine (CII)[1711]

≪Occurrence≫   *Berberis amurensis* RUPR. var *japonica* (アカジクヘビノボラズ)

mp. 190〜191°, $[\alpha]_D$ +87° (MeOH), $C_{36}H_{40}O_6N_2$

UV $\lambda_{max}^{0.1NHCl}$ mμ (log ε): 282 (3.95); λ_{min}: 257 (3.38)

Magnolamine (CIII)[1712]

≪Occurrence≫   *Magnolia fuscata* (トウオガタマ)

mp. 117〜119°, $[\alpha]_D$ +111.6° (EtOH), $C_{36}H_{40}O_7N_2$

Tetramethylmagnolamine   mp. 148〜149°

IR   See Fig. 260

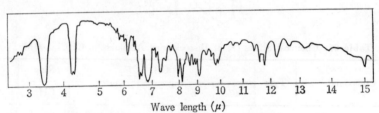

Fig. 260 IR Spectrum of Tetramethylmagnolamine in nujol

(Tomita, *et al.*)[1712]

Wave length (μ)

Thalicarpine (CXLI)[1713]

≪Occurrence≫   *Thalictrum dasycarpum* F. & L.

(CXLI)

mp. 160〜161°, $[\alpha]_D$+133° (MeOH), +89° (CHCl$_3$), $C_{41}H_{48}O_8N_2$

UV λ_{max}^{MeOH} mμ (ε): 282 (17000), 302 (13000)

NMR   in CDCl$_3$, τ: 7.55, 7.52 (6 H, N–Me), 6.40, 6.29, 6.21, 6.19, 6.17, 6.09 6.05 (21 H, OMe),
3.79, 3.47, 3.40, 3.37, 3.32, 1.77 (7 H arom. protons)

1710) Inubushi, Y. *et al*: *Y. Z.*, **72**, 762 (1952).
1711) Tomita, M. *et al*: *ibid.*, **77**, 1075, 1079 (1957).
1712) Tomita, M. *et al*: *ibid.*, **78**, 103 (1958), **79**, 325 (1959).
1713) Kupchan, S. M. *et al*: *J.A.C.S.* **86**, 2177 (1964); *J.P.S.*, **52**, 984 (1963).

(d) Oxyacanthine–berbamine Type Alkaloids

Isotetrandrine (CV)[1714)~1716) 1716a)]

≪Occurrence≫ *Mahonia japonica, Berberis thunbergii,* mp. 182°, $[\alpha]_D$ +146° (CHCl$_3$), C$_{38}$H$_{42}$O$_6$N$_2$

NMR See Table 97

Tetrandrine (CVI)[1714)]

≪Occurrence≫ *Menispermum dauricum* (コウモリカズラ) *Stephania bernardifolia*[1717)]
mp. (*d*–) 217°, (*dl*–) 257~258°, $[\alpha]_D$ +263° (CHCl$_3$), C$_{38}$H$_{42}$O$_6$N$_2$.

NMR See Table 97

Synthesis[1718)]

Phaeanthine (CVII)[1719)]

≪Occurrence≫ *Gyrocarpus americanus, Phaeanthus ebracteolatus*
mp. 210°, $[\alpha]_D$ −278° (CHCl$_3$), C$_{38}$H$_{42}$O$_6$N$_2$

NMR See Table 97

Berbamine (CVIII)[1716a) 1720)]

≪Occurrence≫ *Berberis thunbergii* DC. *Mahonia japonica, Stephania cepharantha, S. sasakii*
mp. 170~172°, $[\alpha]_D$ +103.1° (CHCl$_3$), C$_{37}$H$_{40}$O$_6$N$_2$

NMR See Table 97

Pycnamine (CIX)[1721)]
mp. 186~187°, $[\alpha]_D$ −283°, C$_{37}$H$_{40}$O$_6$N$_2$

NMR See Table 97

Obamegine (CX)[1722) 1723)]

≪Occurrence≫ *Berberis tschonoskiana* (オオバメギ)
mp. 164~166°, $[\alpha]_D$ +98.9° (MeOH), C$_{36}$H$_{38}$O$_6$N$_2$

UV λ_{max}^{EtOH} mμ (log ε): 283 (3.85); λ_{min}: 258 (3.28)

IR See Fig. 261

Stepholine (CXLII)[1723)]

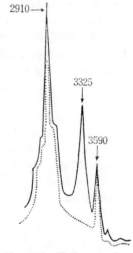

(CXLII)

Fig. 261 IR Spectra of
—— Stepholine and
‥‥‥ Obamegine
in CHCl$_3$ (Tomita *et al*)[1723)].

1714) Tomita, M. *et al*: *Y. Z.*, **71**, 226, 1035 (1951).
1715) Tomita, M. *et al*: *ibid.*, **71**, 1039 (1951).
1716) Tomita, M. *et al*: *ibid.*, **72**, 211 (1952); *Pharm. Bull.*, **2**, 372 (1954).
1716a) Tomita, M. *et al*: *Y. Z.*, **80**, 845 (1960).
1717) Kupchan, S. M. *et al*: *J.P.S.*, **50**, 819 (1961).
1718) Tomita, M. *et al*: *Y. Z.*, **82**, 1148 (1962).
1719) Kidd, D. D. A. *et al*: *Soc.*, **1954**, 669.
1720) Inubusi, Y: *Y. Z.*, **72**, 220 (1952).
1721) Bruchhausen, F. V. *et al*: *Arch. Pharm.*, **293**, 785 (1960).
1722) Kugo, T: *Y. Z.*, **79**, 322 (1959).
1723) Tomita, M. *et al*: *ibid.*, **83**, 940 (1963).

≪Occurrence≫   *Stephania japonica* MIERS. (ハスノハカズラ)

Benzene Adduct

mp. 171~173°, $[\alpha]_D + 273°$ (CHCl₃), $C_{36}H_{38}O_6N_2 \cdot 1^2/_3 \; C_6H_6$

UV λ_{max}^{EtOH} mμ (log ε): 284 (4.03)

IR $\nu_{max}^{CHCl_3}$ cm^{-1}: 3590, 3325 (OH)

O, O–Dimethylstepholine (Isotetrandline) (CV)

Fangchinoline (CXI)[1717) 1724)]

≪Occurrence≫   *Stephania bernandifolia, S. tetrandra*

mp. 237~238°, $[\alpha]_D + 203~204°$ (CHCl₃), $C_{37}H_{40}O_6N_2$

NMR See Table 97

O–**Methylrepandine** (CXIII)[1726)]

≪Occurrence≫   *Daphnandra dielsii, D. repandula*

Repandine (CXVI) $\xrightarrow{\text{CH}_2\text{N}_2}$ (CXIII)

mp. 210~213°, $[\alpha]_D - 80.4°$ (CHCl₃), $C_{38}H_{42}O_6N_2$

NMR See Table 97

Obaberine (*O*–Methyloxyacanthine) (CXIV)[1725)]

≪Occurrence≫   *Berberis Tschonoskiana* (オオバメギ)

mp. 139~140°, $[\alpha]_D + 302°$ (CHCl₃), $C_{38}H_{42}O_6N_2$

UV λ_{max}^{EtOH} mμ (log ε): 283 (3.86); λ_{min}: 260 (3.40),   NMR See Table 97

Oxyacanthine (CXV)[1727)]

≪Occurrence≫   *Berberis thunbergii* DC. (メギ), *B. aquifolium, Mahonia japonica*

mp. 216~217°, $[\alpha]_D + 279°$ (CHCl₃), $C_{37}H_{40}O_6N_2$

NMR See Table 97

Repandine (CXVI)[1726)]

≪Occurrence≫   *Daphnandra repandula*

mp. 255°, $[\alpha]_D - 106°$ (CHCl₃), $C_{37}H_{40}O_6N_2$

NMR See Table 97

Aromoline (CXVII)[1728)]

≪Occurrence≫   *Daphnandra aromatica, D. micrantha*

mp. 174~175°, $[\alpha]_D + 327°$, $C_{36}H_{38}O_6N_2$

NMR See Table 97

1724) Chang, K. Ch. *et al*: *Ber.*, **72**, 519 (1939).
1725) Kugo, T. *et al*: *Y. Z.*, **80**, 1304, 1425 (1960).
1726) Fujita, E. *et al*: *ibid.*, **72**, 1232 (1952).
1727) Fujita, E. *et al*: *ibid.*, **72**, 213, 217 (1952).
1728) Bick, I.R.C. *et al*: *Soc.*, **1960**, 4928, **1949** 2767.

Daphnandrine (CXVIII)[1728]

≪Occurrence≫　　*Daphnandra aromatica, D. micrantha*

mp. 283° (chloroform adduct), $[\alpha]_D$ +474° (CHCl$_3$), C$_{36}$H$_{38}$O$_6$N$_2$

NMR　See Table 97

Daphnoline (CXLIII)[1728]

≪Occurrence≫　　*Daphnandra aromatica, D. micrantha*　(CXLIII)=(CXII): R$_1$=Me, R$_2$=R$_3$=R$_4$=H

mp. 195°, C$_{35}$H$_{36}$O$_6$N$_2$

Trilobamine (Daphnoline) (CXIX)[1729) 1730]

≪Occurrence≫　　*Cocculus trilobus* DC. (アオツヅラフジ), *Daphnandra aromatica, D. micrantha*

CHCl$_3$–Adduct　　mp. 195°, $[\alpha]_D$ +356° (CHCl$_3$), C$_{35}$H$_{36}$O$_6$N$_2$·CHCl$_3$

O, O–Dimethylether hydrochloride

mp. 258~260° (decomp.), $[\alpha]_D$ +272.2° (H$_2$O), C$_{37}$H$_{40}$O$_6$N$_2$·2 HCl·2 H$_2$O

Sepeerine (CXX)[1731]

≪Occurrence≫　　*Nectandra rodiei* R.SCHOMB.,　　mp. 197~199°, $[\alpha]_D$ +391°, C$_{36}$H$_{38}$O$_6$N$_2$·2H$_2$O

anhydrous base　　mp. 164~166°

UV　λ_{max} mμ (ε): 284 (6150)

IR　ν_{max}^{KBr} cm^{-1}: 3570 (OH), 3440~3260 (OH and NH), 1613, 1590

N, O–Diacetate　　mp. 156~158°

IR　ν_{max}^{KBr} cm^{-1}: 1767, 1195 (phenolic OAc), 1645 (〉N–Ac)

Cepharanthine (CXXI)[1732) 1733]

≪Occurrence≫　　*Stephania cepharantha* HAYATA (タマサキツヅラフジ)

Benzene Adduct　　mp. 103°, $[\alpha]_D$ +300° (CHCl$_3$), C$_{37}$H$_{38}$O$_6$N$_2$·C$_6$H$_6$

Base　　mp. 155°, C$_{37}$H$_{38}$O$_6$N$_2$

NMR　See Table 97

Epistephanine (CXXIII)[1734~1736]

≪Occurrence≫　　*Stephania japonica* MIERS (ハスノハカズラ)

mp. 203~204°, $[\alpha]_D$ +183.5° (CHCl$_3$), C$_{37}$H$_{38}$O$_6$N$_2$

Hypoepistephanine (CXXIV)[1734) ~1736]

≪Occurrence≫　　*Stephania japonica* MIERS,　　mp. 256~257°, $[\alpha]_D$ +183.8° (CHCl$_3$), C$_{36}$H$_{36}$O$_6$N$_2$

Trilobine (CXLIV) and **Isotrilobine** (CXLV)[1737) 1737a]

≪Occurrence≫　　*Cocculus laurifolius* DC., *C. sarmentossu* DIELS,

　　　　　　C. trilobus DC.

R=H (CXLIV)　Trilobine
R=CH$_3$ (CXLV) Isotrilobine

1729) Inubushi, Y: *Pharm. Bull.*, **3**, 384 (1955).　　1730) Bick, I.R.C. *et al*: *Soc.*, **1949**, 2767.

1731) Grundon, M.F. *et al*: *ibid.*, **1960**, 2739.

1732) Tomita, M. *et al*: *Pharm. Bull.*, **1**, 101 (1953), **2**, 89, 375 (1954).

1733) Kunitomo, J: *Y. Z.*, **82**, 981 (1962).

1734) Tomita, M. *et al*: *Pharm. Bull.*, **2**, 378 (1954).

1735) Tomita, Y. *et al*: *Y. Z.*, **83**, 996 (1963).　　1736) Watanabe, Y: *ibid.*, **80**, 166 (1960).

1737) Tomita, M. *et al*: *Y. Z.*, **83**, 282, 288, 676, 760, (1963).

1737a) Inubushi, Y: *Tetr. Letters*, **1962**, 1133.

Trilobine (CXLĪV)

mp. 237°, $[\alpha]_D + 307.9°$ (CHCl$_3$), C$_{35}$H$_{34}$O$_5$N$_2$

IR λ_{max} μ: 3.025 ($>$NH)

N–Methyltrilobine (Isotrilobine) (CXLV)

Isotrilobine (CXLV)

≪Occurrence≫　　The same plants as above and *Stephania hernandifolia*[1739a]

mp. 215°, $[\alpha]_D$ $+312.6°$ (CHCl$_3$), C$_{36}$H$_{36}$O$_5$N$_2$

Epistephanine (CXXIII) $\longrightarrow$ (CXLV) antipode

Thalicberine (CXXVI)[1738]

≪Occurrence≫　　*Thalictrum thunbergii* DC. (アキカラマツ)

mp. 161°, $[\alpha]_D$ $+231.2°$ (CHCl$_3$), C$_{37}$H$_{40}$O$_6$N$_2$·H$_2$O

UV λ_{max}^{EtOH} mμ (log ε): 240 (4.17), 280 (3.92); λ_{min}: 260 (3.45)

O–Methylthalicberine (CXXVII)[1738]

≪Occurrence≫　　*Thalictrum thunbergii* DC.

mp. 186~187°, $[\alpha]_D$ $+265.9°$ (CHCl$_3$), C$_{38}$H$_{42}$O$_6$N$_2$

UV　almost identical to that of (CXXVI)

Thalicrine (CXXIX)[1739]

≪Occurrence≫　　*Thalictrum thunbergii* DC.

mp. 221~222°, $[\alpha]_D$ $+341.2°$ (CHCl$_3$), C$_{36}$H$_{38}$O$_6$N$_2$·H$_2$O

UV　See Fig. 262

UV λ_{max}^{MeOH} mμ (log ε): 284 (3.90); λ_{min}: 260 (3.43)

Homotharictrine (CXXX)[1739]

≪Occurrence≫　　*Thalictrum thunbergii* DC.

mp. 235~236° (decomp.) $[\alpha]_D$ $+425.3°$ (CHCl$_3$),

C$_{37}$H$_{40}$O$_6$N$_2$

UV　See Fig. 262 λ_{max}^{MeOH} mμ (log ε): 284 (3.93);

λ_{min}: 260 (3.53)

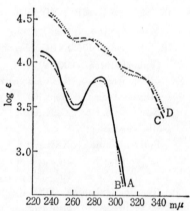

Fig. 262　UV Spectra of (A) Thalicrine,
(B) Homothalicrine, (C) Homothalicrine methylmethine and
(D) O–Ethylthalicrine methylmethine
in MeOH

(Tomimatsu *et al.*)[1739]

1738)　Tomimatsu, T. *et al*: *Y. Z.*, **79**, 1256, 1260, 1386 (1959), **80**, 1137 (1960), **83**, 153, 159 (1963).

1739)　Tomimatsu, T. *et al*: *ibid.*, **82**, 311, 315, 320 (1962).

1739a)　Tomita, M. *et al*: *ibid.*, **79**, 977 (1959).

Phaeantharine (CXLVI)[1740]

≪Occurrence≫   *Phaeanthus ebracteolatus*
Chloride

$C_{38}H_{36}O_6N_2Cl_2 \cdot 5H_2O$

Perchlorate   mp. 180~184°

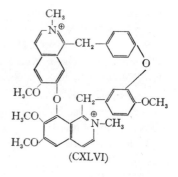

(CXLVI)

(e) Isochondodendrine-bebeerine Type Alkaloids

Cycleanine (CXXXII)[1741]

≪Occurrence≫   *Stephania capitata* SPRENG.

mp. 273~274°, $[\alpha]_D$ −15.08° (CHCl₃), $C_{38}H_{42}O_6N_2$

UV   See Fig. 263,

NMR   See Table 97

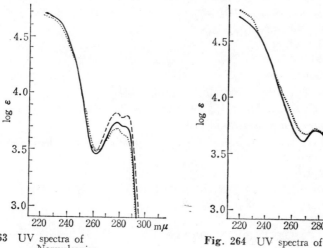

Fig. 263   UV spectra of
——Norcycleanine,
······Cycleanine and
------Isochondrodendrine in EtOH

Fig. 264   UV spectra of
——Insulanoline and
······Insularine in MeOH

(Kikuchi *et al.*)[1742]

Norcycleanine (CXXXIII)[1742]

≪Occurrence≫   *Cyclea insularis* DIELS. (ミヤコジマツヅラフジ)

mp. 249~251° (decomp.), $[\alpha]_D$ −26.54° (MeOH), $C_{37}H_{40}O_6N_2 \cdot 1/2 H_2O$

UV λ_{max}^{EtOH} mμ (log ε): 278 (3.742), 284 (3.711); λ_{min}: 262 (3.471), 283 (3.709), see Fig. 263

Isochondrodendrine (CXXXIV)[1717)1744a)1745]

≪Occurrence≫   *Stephania bernandifolia, Chondrodendron tomentosum* and other spp., *Cyclea
insularis, Cissampelos pareira* L., Pareira Bravae Radix

mp. 309~310°, 318~319°, $[\alpha]_D$ +59° (pyridine), +120° (0.1N−HCl), $C_{36}H_{38}O_6N_2$

UV   See Fig. 263

1740) Knabe, J: *Ber.*, **91**, 1613 (1958).   1741) Fujita, E. *et al*: *Y. Z.*, **71**, 1043 (1951).
1742) Kikuchi, T. *et al*: *ibid*, **78**, 1408 (1958), **79**, 262 (1959).
1742a) King, H: *Soc.*, **1940**, 737.   1743) Kondo, H. *et al*: *Annual Rept. ITSUU Lab.*, **4**, 6, 12 (1953).

d-**Bebeerine** (CXXXVI)[1742a]

mp. 215°, $C_{36}H_{38}O_6N_2$

l-**Curine** (l-Bebeerine) (CXXXVII)[1744) 1744a) 1745)]

≪Occurrence≫ *Chondrodendron tomentosum, Cissampelos pareira*, Pareira Bravae Radix

mp. 217~219°[1744a)], 165~167°[1745)], $[\alpha]_D$ −190° (CHCl$_3$), −280° (0.1N HCl), $C_{36}H_{38}O_6N_2$

UV $\lambda_{max}^{EtOH} m\mu$ (ε): 285 (9200)

——2HCl mp. 265~266° (decomp.)[1745)]

d-**Chondrocurine** (Condrocurine) (CXXXVIII)[1743) 1745)]

≪Occurrence≫ *Chondrodendron tomentosum*[1745)], "Bebeerine Hydrochloride" (Gehe)[1743)].

mp. 232~234°, $[\alpha]_D$ +186° (N–HCl), +200° (0.1 N–HCl)[1745)], $C_{36}H_{38}O_6N_2$

d-**Tubocurarine** (CXXXIX)[1745)~1747)]

≪Occurrence≫ *Chondrodendron tomentosum*

mp. 274~275° (decomp.), $[\alpha]_D$ +215° (0.01 N–HCl), $C_{38}H_{44}O_6N_2Cl_2$

Total Synthesis[1748)]

Insularine (CXLI)[1749) 1751)]

≪Occurrence≫ *Cyclea insularis* DIELS. *Stephania japonica* MIELS[1735)]

Oxalate mp. 188~190° (decomp.), $[\alpha]_D$ +71.6° (MeOH), $C_{38}H_{40}O_6N_2 \cdot 2(COOH)_2 \cdot 2H_2O$

Dipicrate mp. 231~232° (decomp.), $[\alpha]_D$ +27.95° (CHCl$_3$)

Base UV See Fig. 264.

Insulanoline (CXLII)[1742) 1750) 1751)]

≪Occurrence≫ *Cyclea insularis* DIELS

mp. 195° (decomp.), $[\alpha]_D$ +48.6° (MeOH), $C_{37}H_{38}O_6N_2 \cdot H_2O$

UV $\lambda_{max}^{MeOH} m\mu$ (log ε): 278 (3.707); λ_{min}: 269 (3.602), see Fig. 264

(f) Alkaloids of *Ocotea rodiai* (Greenheart Bark)[1752)]

The structures of all alkaloids are undetermined. Physical constants and the formulae of their hydrochloride are described.

Rodiasine (CXLVII)

mp. 292°, $[\alpha]_D$ +74° (H$_2$O), $C_{38}H_{44}O_6N_2 \cdot 2HCl \cdot 2\frac{1}{2}H_2O$

IR ν_{max} cm^{-1}: 3385 (phenolic OH)

Norrodiasine (CXLVIII)

mp. 292°, $[\alpha]_D$ +74° (H$_2$O), $C_{37}H_{42}O_6N_2 \cdot 2HCl \cdot 2H_2O$

IR λ_{max} cm^{-1}: 3365 (phenolic OH)

Dirosine (CXLIX)

mp. 303°, $[\alpha]_D$ +97° (H$_2$O), $C_{37}H_{42}O_6N_2 \cdot 2HCl \cdot 1\frac{1}{2}H_2O$

IR λ_{max} cm^{-1}: 3360 (phenolic OH)

Octeamine (CL)

mp. 290°, $[\alpha]_D$ +250° (H$_2$O), $C_{36}H_{38}O_6N_2 \cdot 2HCl \cdot H_2O$

Otocamine (CLI)

mp. 281°, $[\alpha]_D$ +268° (H$_2$O), $C_{37}H_{40}O_6N_2 \cdot 2HCl \cdot H_2O$

1744) Hultin, E: *Acta Chem. Scand.*, **16**, 559 (1962), **17**, 753 (1962).

1744a) Kupchan, S. M. *et al*: *J. A. P. A.*, **49**, 727 (1960).

1745) Dutcher, J. D: *J. A. C. S.*, **68**, 419 (1946), **74**, 2221 (1952).

1746) Bick, I. R. C. *et al*: *Soc.*, **1953**, 3893.   1747) King, H: *Soc.*, **1935**, 1381.

1748) Hellmann, H. *et al*: *Ann.*, **639**, 77 (1961).

1749) Tomita, M. *et al*: *Y. Z.*, **77**, 69, 997 (1957).

1750) Kikuchi, T. *et al*: *ibid.*, **78**, 1413 (1958).   1751) Kunitomo, J: *ibid.*, **82**, 1152 (1962).

1752) Hearst, P. J: *J. Org. Chem.*, **29**, 466 (1964).

IR λ_{max} cm^{-1}: no phenolic OH peak

Demerarine (CLII)

mp. 278°, $[\alpha]_D$ −181° (H$_2$O), C$_{36}$H$_{38}$O$_6$N$_2$·2HCl·H$_2$O

IR ν_{max} cm^{-1}: 3545 (phenolic OH)

Ocodemerine (CLIII)

mp. 275°, $[\alpha]_D$ −170° (H$_2$O), C$_{37}$H$_{40}$O$_6$N$_2$·2HCl·1$^1/_2$ H$_2$O

IR λ_{max} cm^{-1}: no phenolic OH peak

(13) Morphinane Alkaloids[1753]=[1772]

(a) Menispermaceous Alkaloids

(CLIV) Sinomenine   (CLV) Isosinomenine

R$_1$=R$_2$=CH$_3$ (CLVI) Hasubanonine[1735]   (CLVIII) Metaphanine

R$_1$=H, R$_2$=CH$_3$

or   (CLVII) Homostephanoline

R$_1$=CH$_3$, R$_2$=H

Sinomenine (CLIV)[1753]

≪Occurrence≫   *Sinomenium acutum* R. et W. (オオツヅラフジ)

mp. 161~162°, $[\alpha]_D$ −70.8° (EtOH), C$_{19}$H$_{23}$O$_4$N

UV λ_{max} mμ (log ε): 262 (3.83); λ_{min} 249 (3.74)

IR λ_{max}^{CS2} μ: 2.83 (OH), 5.88 (conjtd C=O), 12.67 (1, 2, 3, 4 tetrasubst. benzene)

NMR cps: +77.1 s. (ring protons), −36.1, −50.0 (2 OCH$_3$), −91.2 (NCH$_3$), +27.0 d $\left(>\!\!=\!\!<^H\right)$

Isosinomenine (CLV)[1753]

≪Occurrence≫   *Sinomenium acutum* R. et W.   mp. 210~212°, $[\alpha]_D$ +53.8°, C$_{19}$H$_{23}$O$_4$N

UV λ_{max} mμ (log ε): 270 (3.96); λ_{min}: 250 (3.80)

IR λ_{max}^{CS2} μ: 2.83 (OH), 5.89 (conjtd. C=O), 12.64 (1, 2, 3, 4 tetrasubst. benzene)

NMR cps: +85.1~+81.3 d (ring protons), −34.4 (2 CH$_3$O), −92.5 (NCH$_3$), +28.3 s $\left(>\!\!=\!\!<^H\right)$

Hasubanonine (CLVI)[1735][1753a][1754a]

≪Occurrence≫   *Stephania japonica* MIERS (ハスノハカズラ)

1753) Sasaki, Y: *4 th Symposium on the Chemistry of Natural Products*, p. 7 (1960).

1753a) Kondo, H. *et al*: *The Annual Report of ITSUU Lab.*, **2**, 1 (1951), **3**, 1 (1952).

1754) Tomita, M. *et al*: *Y. Z.*, **76**, 856 (1956).

1754a) Watanabe, Y. *et al*: *ibid.*, **83**, 991 (1963).

mp. 116, $C_{21}H_{27}O_5N$

Homostephanoline (CLVII)[1754]

≪Occurrence≫ *Stephania japonica* MIERS

mp. 233°, $[\alpha]_D$ −247.8° (CHCl$_3$), $C_{20}H_{25}O_5N$

O–Methylhomostephanoline (Hasubanonine) (CLVI)

mp. 116°

Metaphanine (CLVIII)[1735)1755]

≪Occurrence≫ *Stephania japonica* MIERS

mp. 232～234°, $[\alpha]_D$ −41.1° (CHCl$_3$), $C_{19}H_{23}O_5N$

IR ν_{max}^{CHCl3} cm^{-1}: 3480 (OH), 1730 (C=O)

(b) Euphorbiaceous Alkaloids

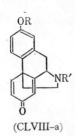

R=R′=H (CLVIII–a) Crotonosine (former str.)[1771] it's revised structure (CLXXIV) is shown in the item (d).

R=R′=Me (CLVIII–b) *N*, *O*–Dimethylcrotonosine ("Base A"[1771]) [Note]

(CLVIII–a)

Crotonosine Described in the item (d) *N*, *O*–Dimethylcrotonosine ("Base A") (CLVIII–b)[1771]

≪Occurrence≫ *Croton linearis* JACQ.

mp. 127～128°, $C_{19}H_{21}O_3N$

UV λ_{max}^{EtOH} mμ (log ε): 227 (4.40), 230 (4.41), 280 (3.55), 285 (3.55)

IR λ_{max}^{nujol} cm^{-1}: 1658 (C=O in cross conjugated dienone), 1618 (arom. OCH$_3$), 1605, 1486 (arom. C=C)

Linearisine (CLVIII–c)[1771]

≪Occurrence≫ *Croton linearis* JACG.

mp. 219～222° (decomp.), $C_{18}H_{21}O_3N$

UV λ_{max}^{EtOH} mμ (log ε): 228 (4.30), 282 (3.19), 288 (3.22)

IR ν_{max}^{nujol} cm^{-1}: 1664 ($\alpha\beta$–unsatur. C=O), 1608 (C=C), 1600, 1500 (arom. C=C), 865, 685 (arom. substitution).

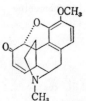

(CLVIII–c) Linearisine

(c) Morphine Alkaloids

(CLIX) Codeine   (CLX) Acetylcodeine   (CLXI) Dihydrocodeine   (CLXII) Codeinone

---

1755) Takeda, K: *The Annual Report of ITSUU Lab.*, **11**, 61 (1960).
[Note] The structure of crotonosine has been revised to (CLXXIV).

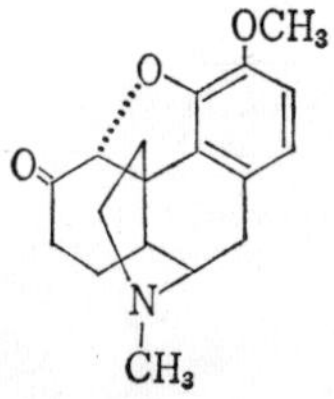

(CLXIII) Dihydrocodeinone 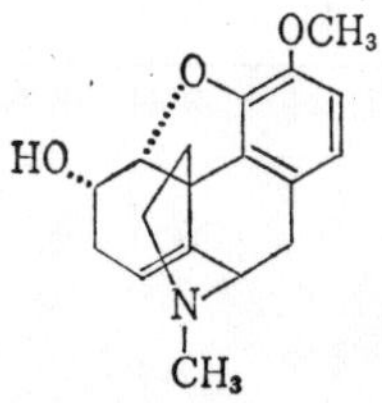　(CLXIV) Neopine 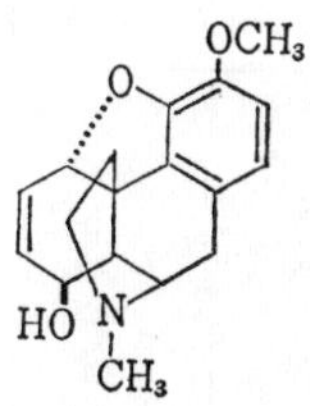　(CLXV) Pseudocodeine 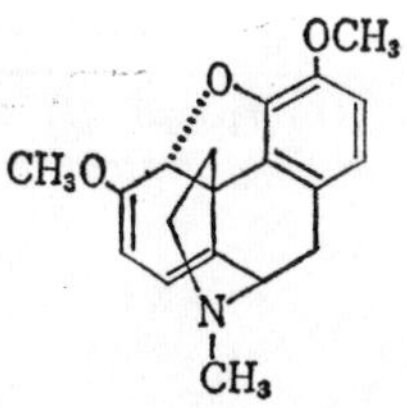　(CLXVI) Thebaine

NMR Spectra

See Fig. 265, and Table 98

in CHCl$_3$, 60 Mc, $\tau_{\text{cyclohexane}}$=8.564, τ:

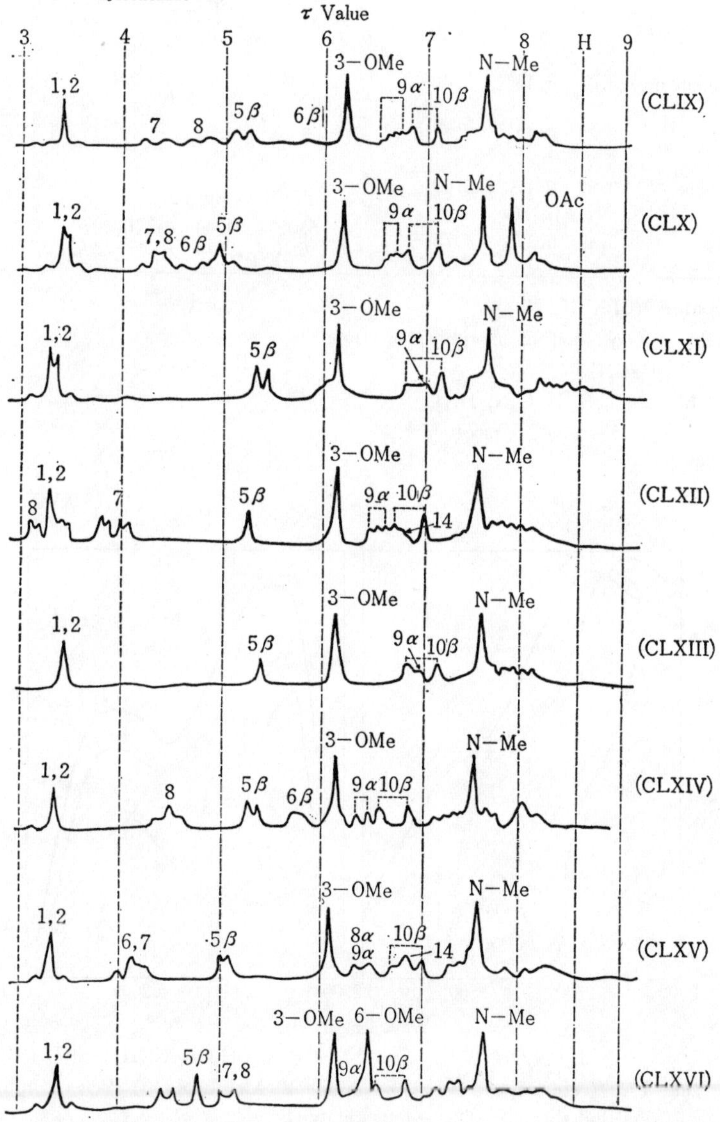

Fig. 265　NMR Spectra of Morphine Alkaloids (Okuda *et al.*)[1756]

Table 98 NMR Spectra of Morphine Alkaloids (Okuda *et al.*)[1756]

Str. No.	Compound	1, 2-H	3-OCH₃	5β–H	6–H	7–H	8–H	9α–H	10β–H	N–CH₃	etc.
CLIX	Codeine	3.40	6.16	5.15 $J_{5.6}$ 6.4	5.84(β)	4.29 J 9.6	4.74 J 9.6	6.63 $J_{9\alpha,14}$3.2 $J_{9\alpha,10\alpha}$5.6	6.92 $J_{9\alpha,10\alpha}$18.4	7.56	
CLX	Acetylcodeine	3.33	6.15	4.98 $J_{5.6}$ 6.7	ca. 4.9(β)	4.38 J 9.9	4.64 J 9.9	6.64 $J_{9\alpha,14}$3.3 $J_{9\alpha,10\alpha}$5.7	6.92 $J_{9\alpha,10\alpha}$18.6	7.57	6α–OAc 7.85
CLXI	Dihydrocodeine	3.32	6.12	5.42 $J_{5.6}$ 5.1	ca. 6.0(β)			6.95 $J_{9\alpha,14}$2.6 $J_{9\alpha,10\alpha}$5.8	7.01 $J_{9\alpha,10\alpha}$18.1	7.61	
CLXII	Codeinone	3.35	6.15	5.31		3.94 $J_{9\alpha,14}$2.3 $J_{7.8}$10.4	3.31 $J_{9\alpha,14}$1.7 $J_{7.8}$10.4	6.57 $J_{9\alpha,14}$3.3 $J_{9\alpha,10\alpha}$5.4	6.85 $J_{9\alpha,10\alpha}$18.0	7.54	14β–H ca. 6.8
CLXIII	Dihydrocodeinone	3.46	6.15	5.42				6.90 J ?	7.00 $J_{9\alpha,10\alpha}$18.0	7.51	
CLXIV	Neopine	3.39	6.14	5.36 $J_{5.6}$ 4.6	5.78(β)		4.52	6.43 $J_{9\alpha,10\alpha}$6.1	6.73 $J_{9\alpha,10\alpha}$17.9	7.55	
CLXV	Pseudocodeine	3.37	6.17	5.09	4.18	4.18	ca. 6.4(α)	6.52 J ?	6.91 $J_{9\alpha,10\alpha}$18.0	7.62	14β–H 6.81
CLXVI	Thebaine	3.40	6.17	4.75		5.02 $J_{7.8}$6.6	4.49 $J_{7.8}$6.6	ca. 6.4 J ?	6.67 $J_{9\alpha,10\alpha}$17.7	7.54	6–OMe 6.41

Pseudomorphine (CLXVII)[1757]

≪Occurrence≫ Oxydation of morphine by fungi or K₃[Fe(CN)₆]

mp. ca. 330° (decomp.), $C_{34}H_{36}O_6N_2 \cdot H_2O$

Dimethylether mp. 155~156°, tetraacetate
mp. 294~296°

UV See Fig. 266

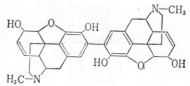

(CLXVII)

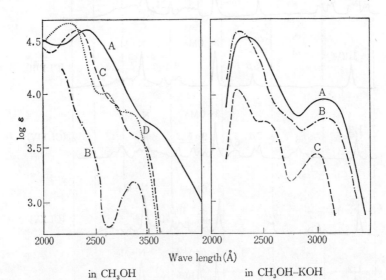

in CH₃OH   in CH₃OH–KOH

Fig. 266 UV spectra of (A) Pseudomorphine, (B) Morphine, (C) Pseudomorphine dimethylether and (D)——tetraacetate (Bentley *et al.*)

Biosynthesis

1–Benzylisoquinoline as precursor[1758], biosynthesis of morphine alkaloids[1759]~[1761], interconversion by moulds[1762][1763]

(d) Miscellaneous Phenanthrene Alkaloids

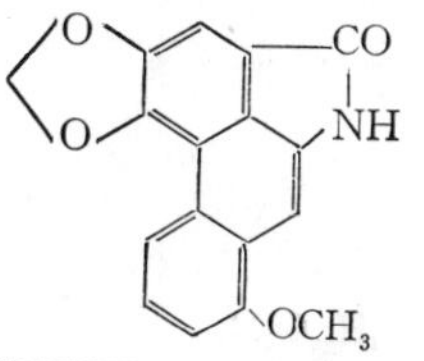

(CLXVIII) Aristololactam

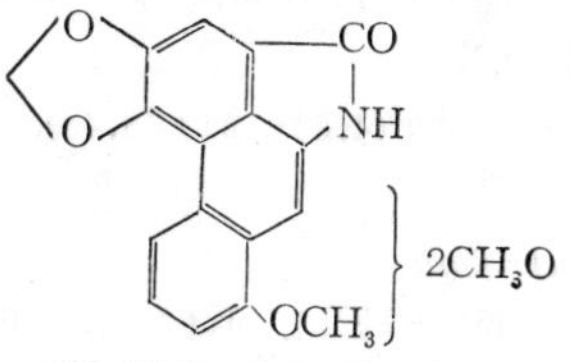

(CLXIX) Aristolored

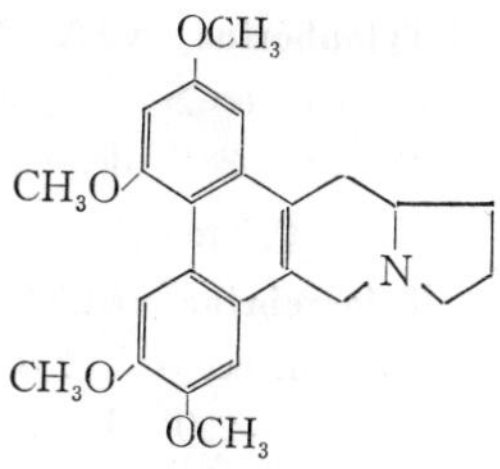

(CLXX) Tylophorine

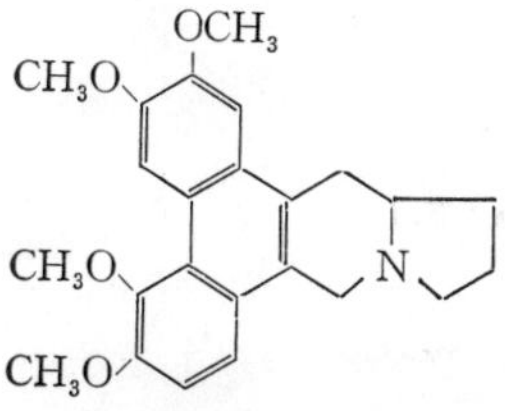

(CLXXI) Tylocrebrine

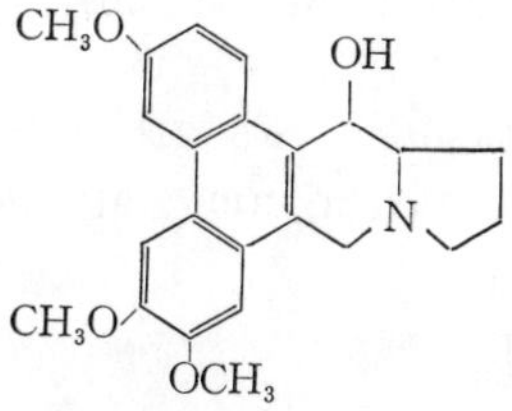

(CLXXII) Tyloprinine

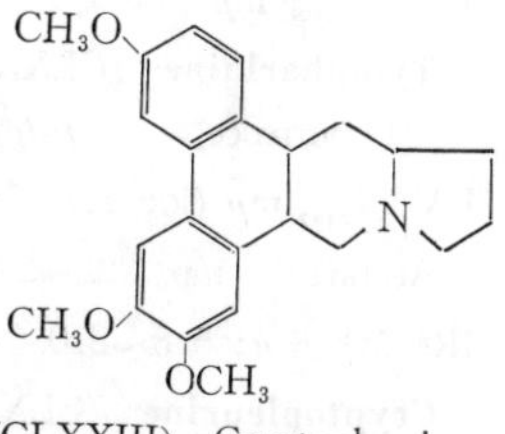

(CLXXIII) Cryptopleurine

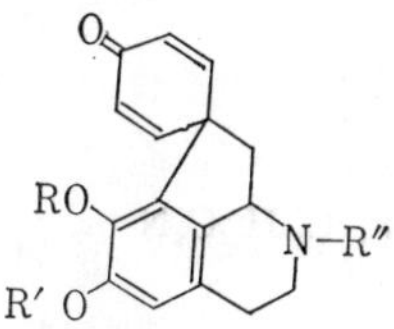

$R=R''=H,$ $R'=CH_3$ or
$R'=R''=H$ $R=CH_3$
(CLXXIV) Crotonosine (Revised Str.)[1772]

Aristololactam   (CLXVIII)[1764]

≪Occurrence≫   *Aristolochia fangchi* Wu, *A. debilis* Sieb. et Zucc. (ウマノスズクサ)
Reduction product of aristolochic acid (see the item 1, 〔D〕, (2))

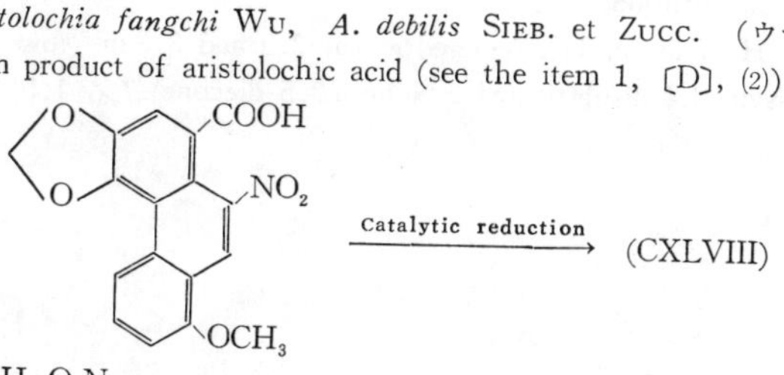

$\xrightarrow{\text{Catalytic reduction}}$ (CXLVIII)

mp. 305° (decomp.), $C_{17}H_{11}O_4N$

UV λ_{max} mμ (log ε): 241 (4.50), 250 (4.87), 259 (4.56), 291 (4.16), 300 (4.15)

IR λ_{max}^{nujol} μ: 3.15 (NH), 5.91 (γ lactam)

1756) Okuda, S. *et al*: *Chem. Pharm. Bull.*, **12**, 104 (1964).
1757) Benthley, K. W. *et al*: *Soc.*, **1959**, 2574.   1758) Battersby, A. R. *et al*: *Proc. Chem. Soc.*, **1963**, 203
1759) Battersby, A. R. *et al*: *ibid.*, **1960**, 287, 360, **1962**, 3534.
1760) Rapoport, H. *et al*: *J. A. C. S.*, **82**, 2765 (1960), **83**, 4045 (1961).
1761) Rapoport, H. *et al*; *Nature*, **189**, 310 (1961).
1762) Tsuda, K. *et al*: *Chem. Pharm. Bull.*, **8**, 1056 (1960).
1763) Tsuda, K. *et al*: *ibid.*, **10**, 67 (1962).

Aristolored   (CLXIX)[1764)~1766)]

≪Occurrence≫   *Aristolochia reticulata, A. sepentaria,*   mp. 286.5°, $C_{19}H_{15}O_6N$

UV $\lambda_{max}^{EtOH} m\mu$ (ε): 253 (42400), 265 (31500), 294 (19350), 300 (19100), 305 (18800).

335 inf. (5850), 352 inf. (5000), 395 (8200)

Tylophorine   (CLXX)[1677)1767)]

≪Occurrence≫   *Tylophora asthmatica, T. crebriflora   (Asclepiadaceae)*

mp. 282~284° (decomp.), $C_{24}H_{27}O_4N$

UV $\lambda_{max}^{EtOH} m\mu$ (log ε): 257 (4.82), 290 (4.51), 340 (3.43), 355 (2.96)

Tylocrebrine   (CLXXI)[1767)]

≪Occurrence≫   *Tylophora crebriflora*

mp. 218~220° (decomp.), $[\alpha]_D$ −45° (CHCl₃), pKa (50% EtOH) 6.7, $C_{24}H_{27}O_4N$,

hydrochloride   mp. 214~217° (decomp.)

UV $\lambda_{max}^{EtOH} m\mu$ (log ε): 263 (4.81), 342 (3.25), 360 (3.09)

Tylopharinine   (CLXXII)[1768)]

≪Occurrence≫   *Tylophora asthmatica*   mp. 248~249°, $C_{23}H_{25}O_4N$

UV $\lambda_{max} m\mu$ (log ε): 258 (4.61), 287 (4.34), 340 (2.91), IR $\lambda_{max}^{nujol} \mu$: 3.1

Acetate   mp. 222~223° (decomp.)

IR $\lambda_{max}^{CHCl_3} \mu$: 5.8~8.0 (—OCOCH₃)

Cryptopleurine   (CLXXIII)[1769)]

≪Occurrence≫   *Cryptocarya pleurosperma*[1770)]   mp. 197~198°, $C_{24}H_{27}O_3N$

Synth. *dl*-form mp. 199~200°, methiodide mp. 272~274°

UV $\lambda_{max}^{EtOH} m\mu$ (log ε): 258 (4.76), 284 (4.51), 358 (2.75)

Crotonosine   (CLXXIV)[1771)1772)][Note]

≪Occurrence≫   *Croton linearis* JACQ.,   mp. 197° (softened), 300°, $C_{17}H_{17}O_3N$

UV $\lambda_{max}^{EtOH} m\mu$ (log ε): 226 (4.30), 235 (4.33), 282 (3.37), 290 (3.41)

IR $\lambda_{max}^{nujol} cm^{-1}$: 3320 (NH), 2600 (bonded OH), 1664 (C=O in cross-conj. dienone), 1622 (C=C),

858 (C=C arom. substitution)

NMR τ: 3.42 (1 H of arom. H), centered at ca. 2.9 and 3.8 m (tow AB quartets of the β and α

protons of assym. 4,4–disubstituted cyclohexa 2,5–dienone) $J_{\alpha\alpha'}$ 1.5, $J_{\beta\beta'}$ 2.5 (*trans* annular coupling

of the α and β protons)

1764)   Tomita, M. *et al*: *Y. Z.* **79**, 973, 1470 (1959).

1765)   Coutts, R. T. *et al*: *J. Ph. Ph.*, **11**, 607 (1959).

1766)   Coutts, R. T. *et al*: *Soc.*, **1959**, 4120.   1767)   Govindachari, T. R. *et al*: *ibid.*, **1962**, 1008 (1962).

1768)   Govindachari, T. R. *et al*: *Tetrahedron.*, **14**, 288 (1961).

1769)   Bradsher, C. K. *et al*: *J. A. C. S.*, **80**, 930 (1958).

1770)   Webb, L. J: *Australian J. Sci.*, **11**, 26 (1948).

1771)   Haynes, L. J. *et al*: *Soc.*, **1963**, 1784, 1789.

1772)   Haynes, L. J. *et al*: *Proc. Chem. Soc.*, **1963**, 280.

[Note]   cf. (b) (CLVIII–a)

〔K〕 Indole Alkaloids

(1) Spectra

UV Lysergic acid, dihydrolysergic acid[1773], isolysergic acid, setoclavine, isosetoclavine, penniclavine, isopenniclavine and chanoclavine[1774]. Norharman, yohimbine δ–yohimbine[1775], py–tetrahydroalstonine [1776], serpentine, py–tetrahydroserpentine and py–tetrahydroalstonine[1777]. Corynantheic acid, corynantheine[1778], semperivirine, yobyrine, yobyrone, tetrahydroyobyrine[1779], reserpine, reserpic acid, and 3, 4, 5–trimethoxybenzoic acid[1780].

Rauwolfia alkaloids[1781]~[1787], methylreserpate[1788], ajmalidine and vomalidine[1787]. Gelsemine, strychnine[1789], strychnolic acid, dihydrostrychnolone, strychronic acid, strychnolone, strychnic acid, strychnidine[1790], uleine, and dihydrouleine[1791].

IR Gramine, evodiamine, isoevodiamine, harman, rutaecarpine, N–acetylharmaline, physostigmine, gelsemine, harmaline, and N–acetylharmaline.

Yohimbine, δ–yohimbine, alkaloid C and isorauhimbine[1792]. py–Tetrahydroalstonine, py–tetrahydroserpentine, alstonine–2, 4–D.P.H.–HCl and the related model compounds[1793]. Serpentine, py–tetrahydroserpentine[1794][1795], tetrahydroserpentine, tetraphylline, raumitorine, tetrahydroalstonine, reserpinine, acrine, reserpiline, and isoreserpinine[1796]. Melinonine F chloride, melinonine G iodide, sempervirine hydroiodide, and merinonine E iodide[1797]. Reserpine, 11–desmethoxyreserpine (raunormine), reserpic acid, raunormic acid[1798], rescinnamine[1799], vomalidine[1800], sandwicine, and ajmaline[1801]. Penniclavine, isopenniclavine, setoclavine, chanoclavine[1802], ergometrine and ergotinine.

Strychnine[1803], brucine[1804], strychinone[1805] uleine and dihydroureine[1806].

1773) Jacobs, Craig: *Science*, **83**, 166 (1936).
1774) Hofmann *et al*: *Helv. Chim. Acta*, **40**, 1358 (1957).
1775) Hofmann: *ibid.*, **37**, 849 (1954).   1776) Bader: *ibid.*, **36**, 215 (1953).
1777) Bader, Schwarz: *ibid.*, **35**, 1594 (1952).   1778) Chatterjee, Karrer: *ibid.*, **33**, 802 (1950).
1779) Prelog: *ibid.*, **31** 588 (1948).   1780) McMullen *et al*: *J. Am. pharm. Assoc. Sci Ed.*, **44**, 446 (1955).
1781) Klohs: *J. A. C. S.*, **76**, 2843 (1954).
1782) Schlittler *at al*: *Experimentia* **11**, 64 (1955); *C. A.*, **50**, 2622 (1956).
1783) Harrisson *et al*: *J. Am. pharm Assoc. Sci Ed.*, **44**, 688 (1955).
1784) Szalkowski, Mader: *J. Am. pharm. Assoc. Sci Ed.*, **45**, 613 (1956).
1785) *Drug Standards*, **25** No. 261 (1957).
1786) Haycock, Mader: *J. Am. pharm. Assoc. Sci. Ed.*, **46**, 744 (1957).
1787) Hofmann A., Frey A. J: *Helv. Chim. Acta*, **40**, 1866 (1957).
1788) Hofmann: *ibid.*, **37**, 849 (1954).
1789) Kebrle, Schmid, Waser, Karrer: *Helv. Chim. Acta*, **36**, 102 (1953).
1790) Prelog, Szpifogel: *Helv. Chim. Acta*, **28**, 1669 (1945).
1791) Schmutz, Hunziker, Hirt: *ibid.*, **40**, 1189 (1957).
1791a) Marion, Ramsay, Jones: *J. A. C. S.*, **73**, 305 (1951).
1792) Hofmann, A: *Helv. Chim. Acta*, **37**, 849 (1954).   1793) Bader: *ibid.*, **36**, 215 (1953).
1794) Bader, Schwarz: *ibid.*, **35**, 1594 (1952).   1795) Kloks *et al*: *J.A.C.S.*, **76**, 1332 (1954).
1796) Neuss, N., Boaz, E: *J. Org. Chem.*, **22**, 1001 (1957).
1797) Büchli, Vamvacas, Schmid, Karrer: *Helv. Chim. Acta*, **40**, 1167 (1955).
1798) Harrisson, J.W.E *et al*: *J. Am. pharm. Assoc. Sci Ed.*, **44**, 688 (1955).
1799) Klohs, M. W *et al*: *J.A.C.S.*, **77**, 2241 (1955).
1800) Hofmann, A., Frey, A. J: *Helv. Chim. Acta*, **40**, 1866 (1957).
1801) Gormhn, Neuss, Djerassi, Kutney, Seheuer: *Tetrahedron*, **1**, 328 (1957).
1802) Hofmann, *et al*: *Helv. Chim. Acta*, **40**, 1358 (1957).

(2) **NMR** 60 Mc, TMS τ: [1806a]

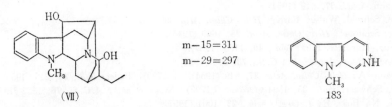

(I) Indole (CCl$_4$) (II) N–Methylindole (CDCl$_3$) (III) 2–Methylindole (CDCl$_3$)

H←—3.62 t
H←—3.32 t
arom. H 3.10 m

H←—3.52 d
H←—3.18 d
arom. H 2.83 m
CH$_3$ ←— 6.63 s

H←—3.87 s
CH$_3$←~7.80 s
arom. H 2.92 m

CH$_3$←—7.70 s
H←—3.20 d
arom. H 2.82 m

(IV) 3–Methylindole (CDCl$_3$)

7.05 t
CH$_2$·CH$_2$·NH$_2$ ←—8.72 s
H←—3.08 d
1.33

(V) Tryptamine (CDCl$_3$)

(3) **Characteristic Mass Number**[1807]

(a) **Aspidosperma Alkaloids**
   m/e: 124

(b) **Indoline alkaloids**
   m/e: 130

(c) **Indole** and **Indoline Alkaloids**
   m/e 143, 144, 156, 168, 169, 170

(d) **Physostigmine** (VI)
   m/e 275

(VI)   218   203

174   161   160

(e) **Ajmaline** (VII)
   m/e 326

m—15=311
m—29=297

(VII)   183

1803) Sadtler Card No. 3801.   1804) Sadtler Card No. 2252.   1805) Sadtler Card No. 3801.
1806) Schmutz, Hunziker, Hirt: *Helv. Chim. Acta*, **40**, 1189 (1957).
1806a) Cohen, L. A. *et al*: *J.A.C.S.*, **82**, 2184 (1960).
1807) Spiteller, G: *Z. Anal. Chem.*, **197**, 1 (1963); *Tetr. Letters*, **1963**, 147, 153.

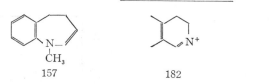

157   182   144

(4) Reviews and Biosynthesis

Ph–N–C–N System alkaloids[1808], bis-indole alkaloids[1809], biosynthesis[1810) 1811)

(5) Exocyclic Bases and Pyrrole Derivatives

(a) **Gramine** (VIII)

≪Occurrence≫   *Hordeum vulgare* L. (オオムギ), *Acer saccharinum*
(サトウカエデ)[1813)

mp. 134°, $[\alpha]_D \pm 0°$, $C_{11}H_{14}O_2$, Perchlorate mp. 180~181° (decomp.)   (VIII)

(b) **Serotonine** (5–Hydroxytryptamine) (IX)

≪Occurrence≫   *Lycopersicum esculentum* (トマト)[1812)

Hydrochloride   mp. 167~168° $C_{10}H_{12}ON_2$. HCl

(IX)

(c) **Bufotenine** (X) and **N, N-Dimethyl-5-methoxytryptamine** (XI)

Bufotenine (X)

≪Occurrence≫   *Piptadenia peregrina* (*Leguminosae*), *Amanita mappa*, secretion of *Bufo vulgaris* (ガマ)

(Pharmacology) Hallucinate action.

Methiodide   mp. 213~214°,

Oxalate mp. 82~84°, $C_{12}H_{16}ON_2$ (base)

N, N-Dimethyl-5-methoxytryptamine (XI)[1813)

≪Occurrence≫   *Dictyoloma incanescens*, $C_{13}H_{18}ON_2$

R=H (X) Bufotenine
R=CH₃ (IX) *N, N*–Dimethyl–5–methoxytryptamine

1808) Robinson, B: *Chem. & Ind.*, **1963**, 218.
1809) Chatterjee, A: *Sci. & Ind. Res.* (*India*), **23**, 178 (1964).
1810) Battersby, A. R. *et al*: *Proc. Chem. Soc.*, **1963**, 369.
1811) Wenkert, E. *et al*: *J.A.C.S.*, **81**, 1474 (1959).
1812) West, G. B: *J. Ph. Ph.*, **11**, Suppl. 275 T (1959).
1813) Pachter, I. R. *et al*: *J. Org. Chem.*, **24**, 1285 (1959).

(d)　Psilocybine (XII) and **Psilocine** (XIII)[1814]

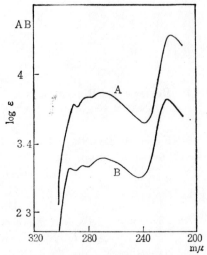

Fig. 267　UV Spectra of (A) Psilocybine
and (B) Psilocine in MeOH

≪Occurrence≫　　*Psilocybe mexicana* HEIM., and other
spp., *Stropharia cubensis* EARLE (Fungi)

(Pharmacology) Hallucinate action

Psilocybine (XII)

mp. 220~228°, $[\alpha]_D \pm 0°$,　　$C_{12}H_{17}O_4N_2P$

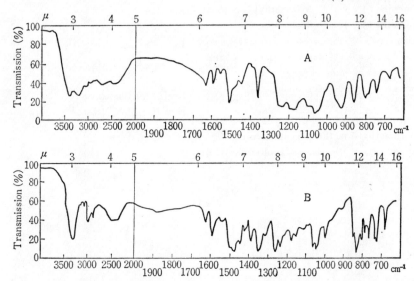

Fig. 268　IR Spectra of (A) Psilocybine and (B) Psilocine in KBr
(Hofmann *et al.*)[1814]

UV　See Fig. 267, λ_{max}^{MeOH} mμ (log ε): 220 (4.6), 267 (3.8), 280 sh (3.7), 290 (3.6)

Psilocine (XIII)　　mp. 166~167°,　　$C_{12}H_{16}ON_2$

UV　See Fig. 267

IR　See Fig. 268

(e)　Betanin (Phytolaccanin) (XIV) and **Betanidin** (Phytolaccanidin) (XV)[1815]

≪Occurrence≫　　Purple pigments from *Beta vulgaris* var. *rubra*, and *Phytolacca decandra* L.

1814)　Hofmann, A *et al*: *Helv. Chim. Acta*, **42**, 1557 (1959).
1815)　Whyler, H *et al*: *Helv. Chim. Acta*, **42**, 1696, 1699 (1959), **44**, 249 (1961), **45**, 638, 640 (1962), **46**, 1745 (1963).

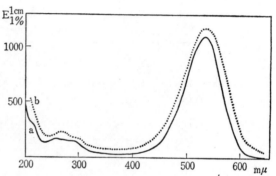

(1) Methanolysis→
    Trimethyl ester
(2) Ac₂O+Pyridine
    or (1) CH₂N₂

$R_1=H,\ R_2=$glucose
or $R_1=$glucose, $R_2=H$ }(XIV) Betanin

$R_1=R_2=H$ (XV) Betanidin

$R_1=R_2=H$ (XVI)
Neobetanidin

Betanin (XIV) $C_{24}H_{26}O_{13}N_2$

Pyridinium salt

$C_{24}H_{26}O_{13}N_2 \cdot (C_5H_5N)_{0.5}(H_2O)_{1.5}$

UV   See Fig. 269, $\lambda_{max}^{H_2O}$ mμ ($E_{1cm}^{1\%}$): 536~
538 (1100)

IR   See Fig. 270

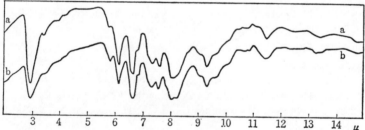

Fig. 269 UV Spectra of (a) Phytolaccanin pyridinium
salt and (b) Phytolaccanidin hydrochloride

Fig. 270 IR Spectra of (a) Betanin pyridinium salt and (b) Phytolaccanin
pyridinium salt in KBr (Wyler *et al.*)[1815]

Betanidin (XV) $C_{18}H_{16}O_8N$

UV   See Fig. 269, λ_{max} mμ (ε): 542~546 (51000), 295 sh. (7170), 271~272 (8530)
IR   See Fig. 271

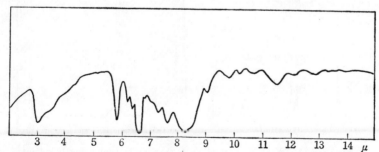

Fig. 271 IR Spectrum of Phytolaccanidin hydrochloride in KBr (Wyler *et al.*)[1815]

NMR: in F_3CCOOH, 60 Mc, τ: 1.33 d, $J=13$ (C_{11}–H), 2.63 s, 2.93 s (C_7–H, C_4–H), 3.32 s (C_{18}–H), 3.60 d, $J=13$ (C_{12}–H), 4.45 m (C_2–H), 5.30 m (C_{15}–H), 6.35 m (C_3–H, C_3–H, C_{14}–H, C_{14}–H), 4.00 (?) (see Fig. 272)

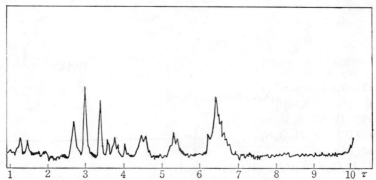

Fig. 272　NMR Spectrum of Betanidin–HCl (Wyler *et al.*)

Di-*O*-acetylneobetanidin-trimethylester (XVII, XVII-a)

(XVII) in CDCl₃　　　　　　　　(XVII–a) in F_3CCOOH

mp. 200.5°, $[\alpha]_D$ −151°, $C_{25}H_{24}O_{10}N_2$.

UV　λ_{max}^{MeOH} mμ: 383 ; $\lambda_{max}^{H^+ \cdot MeOH}$ mμ: 483

NMR　60 Mc, see Fig. 273

　　　(XVII) in CDCl₃ τ: 5.16 dd, J_{cis} 10 c/s, J_{trans} 5 c/s (C_2–H), 6.35 dd, $J=5$, 17 c/s, (C_3–H, *cis* to C_2–H), 6.80 dd, $J=10$, 17 c/s (C_3–H, *trans* to C_2–H), (ABX-system by C_3–H, C_3–H, C_2–H), 1.90 s (C_{14}–H and C_{18}–H), 4.54 d, 2.32 d, $J=14$ c/s (AB-system by C_{11}–H~C_{12}–H (*trans*), 7.72 s, 7.68 s (2 CH₃CO), 5.97 s (2 CH₃O), 6.20 (CH₃O at C_{10}))

　　　(XVII–a) in $F_3C \cdot COOH$: Signals generally shift 0.2~0.5 ppm to lower region from those of (XVII). The sift of the signal 1.30, due to C_{11}–H is most remarkable.

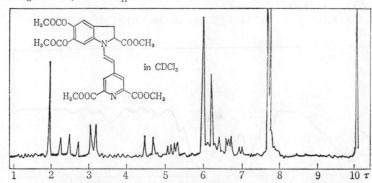

Fig. 273 (1)　NMR Spectra of Di-*O*-acetylneobetanidintrimethylester (XVII) in CDCl₃ (Wyler *et al.*)

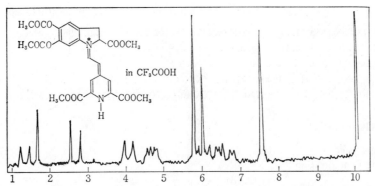

Fig. 273 (2) NMR Spectra of Di–*O*–acetylneobetanidintrimethylester
(XVII–a) in CF₃COOH (Wyler *et al.*)

(f) Erythroskyrine (XVIII)[1816]

≪Occurrence≫ *Penicillium islandicum*

(XVIII) (Suggested Str.)

mp. 130～133°, $C_{26}H_{33}O_6N$

UV See Fig. 274, λ_{max}^{EtOH} mμ: 409, 260; $\lambda_{max}^{0.01N-NaOH-EtOH}$

mμ: 392, 260 $\left(\begin{matrix} O \\ \parallel \\ C \end{matrix} \diagdown (CH=CH)_5-CH{\diagup}^O{\diagdown} \right)$

IR See Fig. 275, ν_{max}^{KBr} cm⁻¹; 1010, 1550, 1564, 1584
(Charact.)

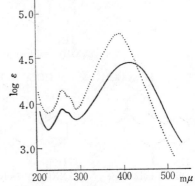

Fig. 274 UV Spectra of Erythroskyrine
——— in EtOH
┅┅┅ in 0.01 *N*-NaOH-EtOH

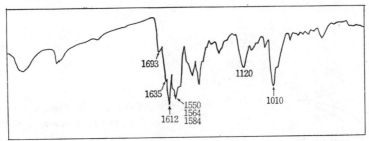

Fig. 275 IR Spectrum of Erythroskyrine in KBr (Shibata, *et al.*)[1816]

Decahydroerythroskyrine (XIX)

Liquid. The UV and IR Spectra are resemble to those of
tenuazoic acid (XX), isolated from *Alternaria tenuis*[1817]

(XX)

1816) Shibata, S *et al* : *7 th Symposium on the Chemistry of Natural Products*, p 138 (Fukuoka, 1963).
1817) Hofmann, A *et al* : *Helv. Chim. Acta*, **45**, 2005 (1962).

UV	λ_{max}^{EtOH} mμ (log ε)	$\lambda_{max}^{0.1N-NaOH}$ mμ (log ε)	λ_{max}^{EtOH} mμ (Cu-Complex)
(XIX)	225 (3.855), 284 (4.04)	246 (4.145), 288 (4.145)	230, 296
(XX)	217 (3.715), 277 (4.11)	239 (3.98), 279 (4.08)	225, 292

IR ν_{max} (1500~1700 cm^{-1})

(XIX) 1610, 1640~1660 br., 1690~1710 br., (XIX–Cu) 1510, 1583, 1695 (sharp)

(XX) 1616   1660   1710   (XX–Cu) 1500, 1600, 1680

(6) Ergot Alkaloids

(i) *Lysergic acids*

(a) **Absolute Configuration**[1817) 1818)]

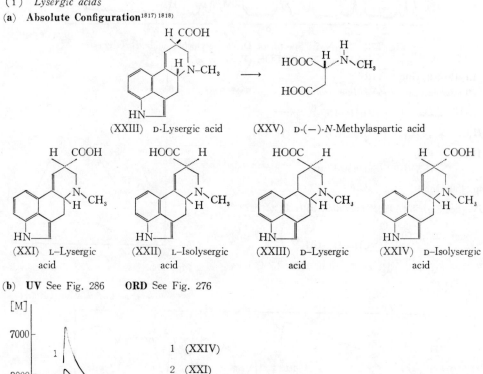

(XXIII) D-Lysergic acid   (XXV) D-(−)-N-Methylaspartic acid

(XXI) L–Lysergic acid   (XXII) L–Isolysergic acid   (XXIII) D–Lysergic acid   (XXIV) D–Isolysergic acid

(b) **UV** See Fig. 286   **ORD** See Fig. 276

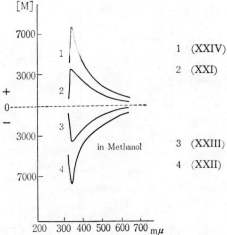

1 (XXIV)

2 (XXI)

3 (XXIII)

4 (XXII)

in Methanol

Fig. 276 ORD–Curves of Lysergic acids and Isolysergic acids in MeOH (Leemann *et al.*)[1818)]

1818) Leemann, H. G. *et al*: *Helv. Chim. Acta*, **42**, 2696 (1959).

(c) NMR of D-**Lysergic acid diethylamide** (LSD) (XXV)[1772a]

in D_2O, 60 Mc, external

C_6H_6 reference (τ: 3.59):

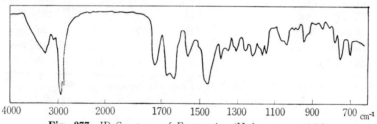

(XXV)

(ii) *Peptide Alkaloids*

≪Occurrence≫　　*Claviceps purpurea* (バッカク)

(5R : 8R : 2′R : 5′S : 11′S : 12′S)

(XXV)　Ergotamine[1819]　　　　(XXVI)　Pyroergotamine[1820]

(a) Ergotamine (XXV)

mp. 241~243°, $[\alpha]_D$ +385° (CHCl$_3$),　　$C_{33}H_{35}O_5N_5$

Synthesis[1819], IR　See Fig. 277

Fig. 277　IR Spectrum of Ergotamine (Hofmann *et al.*)[1819]

(b) Pyroergotamine (XXVI)[1820]

Ergotamine (XXV) ⟶ Dihydroergotamine $\xrightarrow[\text{thermal decomposition}]{}$ (XXVI)

mp. 185°,　　$C_{17}H_{18}O_4N_2$

NMR in CDCl$_3$, 60 Mc, TMS, τ: 2.75 m (5) (arom. protons), 4.29 t　(1 : 2 : 1)　J=4.5 c/s　(C$_5$-H),

　　~6.7 br. m. (4), (–CH$_2$–C$_6$H$_5$), 7.54 s (3) (–COCH$_3$), ~8.0 br. m. (C$_2$ and C$_3$ methylenes) Methyl

signal of 2–methyl–2–ethyl–1,3–dioxolane (XXVI–a) shifts to τ 8.80

(XXVI–a)

[Note]　τ 7.54 s is regarded as typical for CH$_3$CO–(diacetyl 7.70, pyruvic acid 7.47)

1819)　Hofmann, A *et al*: *Helv. Chim. Acta*, **46**, 2306 (1963).

1820)　Green, M *et al*: *ibid.*, **44**, 1417 (1961).

(c) IR Spectra of Peptide Alkaloids[1821]

(XXIII) Lysergic acid
(Lys.)

(XXIV) Isolysergic acid
(Isolys.)

CH$_3$

Lys. –NH–CH–CH$_2$OH

(XXVII) Ergometrine

(Maleate mp. 167°)

Peptide alkaloids

Peptide alkaloids			R	R′	mp.
Ergotamine	(XXV)	Lys.	H	CH$_2$·C$_6$H$_5$	203° (decomp.) (tartrate)
Ergotaminine	(XXVIII)	Isolys.	H	CH$_2$·C$_6$H$_5$	241∼243° (decomp.)
Ergocristine	(XXIX)	Lys.	CH$_3$	CH$_2$·C$_6$H$_5$	170∼190° (decomp.)
Ergocristinine	(XXX)	Isolys.	CH$_3$	CH$_2$·C$_6$H$_5$	266° (decomp.)
Ergocornine	(XXXI)	Lys.	CH$_3$	CH(CH$_3$)$_2$	182∼184° (decomp.)
Ergocorninine	(XXXII)	Isolys.	CH$_3$	CH(CH$_3$)$_2$	228° (decomp.)
Ergokryptine	(XXXIII)	Lys.	CH$_3$	CH$_2$–CH(CH$_3$)$_2$	212° (decomp.)
Ergokryptinine	(XXXIV)	Isolys.	CH$_3$	CH$_2$–CH(CH$_3$)$_2$	240∼242° (decomp.)

Fig. 278∼282 IR Spectra of Ergot Alkaloids in KBr (Yamaguchi *et al.*)[1821]

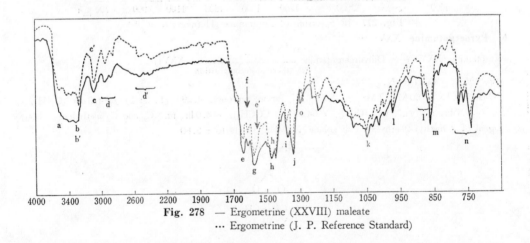

Fig. 278 — Ergometrine (XXVIII) maleate
··· Ergometrine (J. P. Reference Standard)

1821)   Yamaguchi, K *et al*: *Bull. of National Inst. Hyg. Sci.* (Tokyo), **80**, 22 (1962).

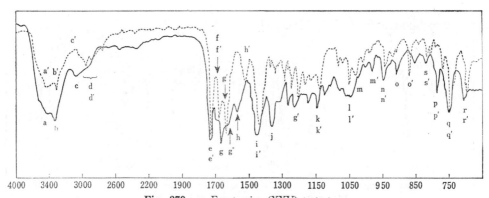

Fig. 279 — Ergotamine (XXV) tartrate
··· Ergotaminine (XXVIII)

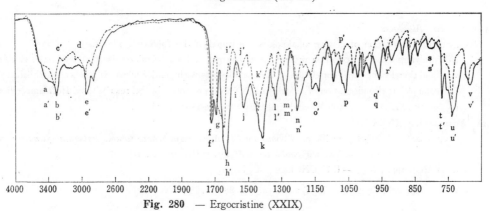

Fig. 280 — Ergocristine (XXIX)
··· Ergocristinine (XXX)

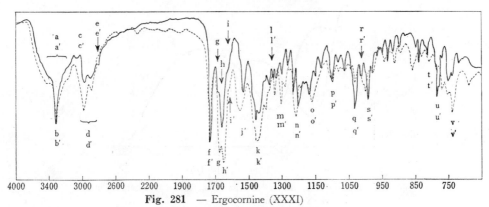

Fig. 281 — Ergocornine (XXXI)
··· Ergocorninine (XXXII)

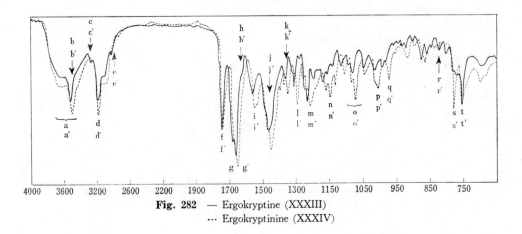

Fig. 282 — Ergokryptine (XXXIII)
··· Ergokryptinine (XXXIV)

(iii) *Clavine Alkaloids*

Structural correlation of clavine alkaloids is illustrated in the following schema.

Dihydrolysergol I, the reduction product of D–lysergic acid was obtained also by reduction of elymoclavine and thereby the steric correlation of clavine alkaloids with peptide alkaloids has been established[1822].
For example, the configuration of agloclavine (XXXV), presented by Schreier[1823] and Hofmann[1823a] was revised to (XXXVI) by Stoll[1822] and Spilsbury[1824] (see next page[1822]).

(a) **Agroclavine** (XXXVI)

≪Occurrence≫   Ergot parasitic on *Elymus mollis, Agropyrum semicostatum, Imperata cylindrica* var.
koengi, Pennisetum typhoideum and *Aspergillus fumigatus*

mp. 206° (decomp.), $[\alpha]_D$ —151° (CHCl₃), $C_{16}H_{18}N_2$

UV See Fig. 283, λ_{max}^{EtOH} mμ (log ε): 225 (3.88), 284 (3.88), 293 (3.81)[1823)1823a]

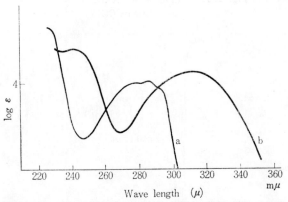

Fig. 283 UV Spectra of (a) Agroclavine and (b) Lysergine in
EtOH (Yamatodani)[1825]

1822) Stoll, A: *Kagaku no Ryoiki* (Nankodo-Pub. Inc.,Tokyo), **14**, 13 (1960)
1823) Schreier, E: *Helv. Chim. Acta*, **41**, 1984 (1958).
1823a) Hofmann, A *et al*: *ibid.*, **40**, 1358 (1957).
1824) Spilsbury, J. F *et al*: *Soc.*, **1961**, 2085.
1825) Yamatodani, S: *Ann. Rept. Takeda Lab.*, **19**, 8, 15 (1960).

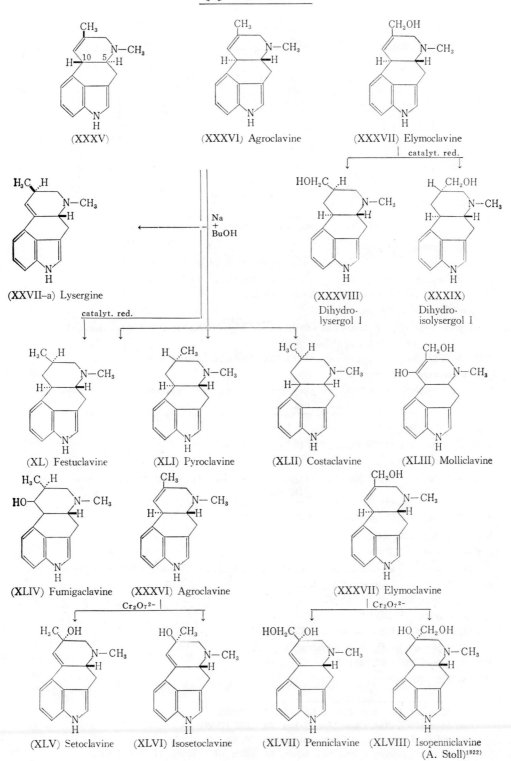

(XXXV)

(XXXVI) Agroclavine

(XXXVII) Elymoclavine

catalyt. red.

(XXVII-a) Lysergine

(XXXVIII) Dihydro-lysergol I

(XXXIX) Dihydro-isolysergol I

Na + BuOH

catalyt. red.

(XL) Festuclavine

(XLI) Pyroclavine

(XLII) Costaclavine

(XLIII) Molliclavine

(XLIV) Fumigaclavine

(XXXVI) Agroclavine

(XXXVII) Elymoclavine

$Cr_2O_7^{2-}$

$Cr_2O_7^{2-}$

(XLV) Setoclavine

(XLVI) Isosetoclavine

(XLVII) Penniclavine

(XLVIII) Isopenniclavine (A. Stoll)[1822]

— 585 —

IR See Fig. 284

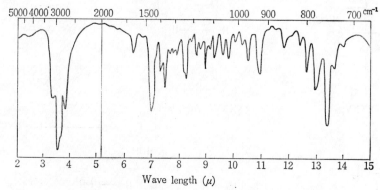

5000 4000 3000   2000   1500   1000 900   800   700 cm⁻¹

2   3   4   5   6   7   8   9   10   11   12   13   14   15

Wave length (μ)

Fig. 284   IR Spectrum of Agroclavine in nujol (Yamatodani)[1826]

(b) **Elymoclavine** (XXXVII)

≪Occurrence≫   Ergot parasitic on *Elymus mollis*, *Pennisetum typhodeum*

mp. 249° (decomp.), [α]$_D$ −109° (EtOH), −152° (pyridine), $C_{16}H_{18}ON_2$

(XXXVII) ⟶ α and β–Dihydrolysergol[1827]

(c) **Lysergine** (XXXVII–a)

≪Occurrence≫   Agroclavine (XXXVI) $\xrightarrow{\text{Na+BuOH}}$ (XXXVII–a) + Festuclavine (XL)

+ Costaclavine (XLII)[1828]

UV   See Fig. 283, λ_{max} mμ (log ε): 226 (4.23), 239 (4.20), 310 (3.49)[1824]

(d) **Festuclavine** (XL)[1824) 1829]

≪Occurrence≫   *Aspergillus fumigatus*, ergot parasitic on *Elymus mollis* TRIN. (テンキグサ)

Agroclavine (XXXVI) $\xrightarrow{\text{Red.}}$ (XL)[1828]

mp. 238∼239°, [α]$_D$ −128° (pyridine), $C_{16}H_{20}ON_2$

UV λ_{max}^{EtOH} mμ (log ε): 224 (4.53), 276 (3.81), 281 (3.84)[1824]

(e) **Pyroclavine** (XLI)

≪Occurrence≫   Ergot parasitic on *Agropyrum semicostatum* (カモジグサ), *Aspergillus fumigatus*

Agroclavine (XXXVI) $\xrightarrow{\text{Na+BuOH}}$ (XLI)

mp. 204°, [α]$_D$ −105° (pyridine), $C_{16}H_{20}N_2$

UV λ_{max}^{EtOH} mμ: 225, 275, 282, 292

(f) **Costaclavine** (XLII)[1824) 1828]

≪Occurrence≫   Ergot parasitic on *Agropyrum semicostatum* (カモジグサ), *Aspergillus fumigatus*

mp. 182°, [α]$_D$ +59° (pyridine), $C_{16}H_{20}N_2$

UV λ_{max} mμ: 225, 275, 282, 292

(g) **Molliclavine** (XLIII)[1829]

≪Occurrence≫   Ergot parasitic on *Ellymus mollis* (テンキグサ)

mp. 253° (decomp.), [α]$_D$ +30° (pyridine), $C_{16}H_{18}O_2N_2$

UV λ_{max} mμ: 226, 287, 294

IR See Fig. 285

1826)   Abe, M *et al*: *Ann. Rept Takeda Lab.*, **22**, 116 (1963).
1827)   Abe, M *et al*: *J. Agr. Chem. Soc. Japan*, **33**, 1036 (1959).
1828)   Yamatodani, S: *Bull. Agr. Chem. Soc. Japan*, **20**, 59, 95 (1956).
1829)   Abe, M *et al*: *J. Agr. Chem. Soc. Japan*, **33**, 1031, 1039 (1959), **34**, 249 (1960).

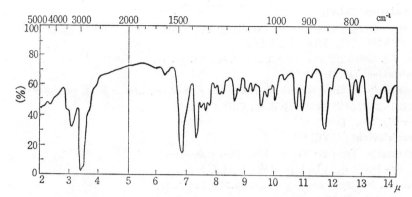

Fig. 285 IR Spectrum of Molliclavine in nujol (Abe *et al.*)[1829]

(h) Fumigaclavine B (XLIV)[1824) 1830]

≪Occurrence≫   *Aspergillus Fumigatus*,   mp. 244~245°, [α]$_D$ −113° (pyridine), $C_{16}H_{20}ON_2$

UV λ$_{max}^{EtOH}$ mμ (log ε): 225 (4.49), 275 (3.79), 282 (3.82), 293 (3.72)

(i) Setoclavine (XLV)[1823a]

≪Occurrence≫   Ergot parasitic on *Pennisetum typhoideum*

mp. 229~234° (decomp.), [α]$_D$ +174° (pyridine), $C_{16}H_{18}ON_2$

UV See Fig. 286, λ$_{max}^{EtOH}$ mμ (log ε):243 (4.38), 313 (4.04)

IR [1830a]

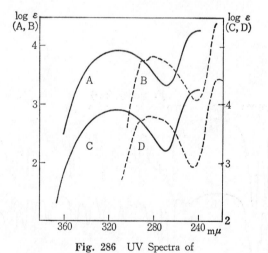

Fig. 286 UV Spectra of
(A) Lysergic acid and Isolysergic acid
(B) Dihydrolysergic acid
(C) Setoclavine, Isosetoclavine, Penniclavine and
    Isopenniclavine
(D) Chanoclavine
        in EtOH (Hofmann *et al.*)[1823a]

1830) Abe, M *et al*: *Ann. Rept. Takeda Lab.*, **21**, 95 (1962).
1830a) Hofmann *et al*: *Helv. Chim. Acta*, **40**, 1358 (1957).

(j) Isosetoclavine (XLVI)

≪Occurrence≫ Ergot parasitic on *Pennisetum typhoideum*,

mp. 234~237°, [α]$_D$ +107°, C$_{16}$H$_{18}$ON$_2$

UV See Fig. 286, λ$_{max}^{EtOH}$ mμ (log ε): 242 (4.42), 317 (4.10),   IR[1830a]

(k) Penniclavine (XLVII)[1823a)1829]

≪Occurrence≫   Ergot parasitic on *Pennisetum typhoideum*, *Elymus mollis* (テンキグサ)

mp. 222~225°, [α]$_D$ +153° (pyridine), C$_{16}$H$_{18}$O$_2$N$_2$

UV See Fig. 286. IR[1830a]

(l) Isopenniclavine (XLVIII)[1823a]

≪Occurrence≫ Ergot parasitic on *Pennisetum typhoideum*

mp. 163~165°, [α]$_D$ +146° (pyridine), C$_{16}$H$_{18}$O$_2$N$_2$

UV See Fig. 286, λ$_{max}^{EtOH}$ mμ (log ε): 242 (4.31), 313 (3.94)

(m) Chanoclavine (XLIX)[1823a]

≪Occurrence≫   Ergot parasitic on *Pennisetum typhoideum*

mp. 220~222°, [α]$_D$ −240° (pyridine), C$_{16}$H$_{20}$ON$_2$

$$(XLIX) \xrightarrow{\text{red.}} \text{Festuclavine} \quad (XL)$$

UV See Fig. 286, λ$_{max}^{EtOH}$ mμ (log ε): 225 (4.44), 284 (3.82), 293 (3.76)
IR[1830a]

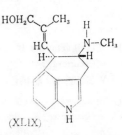

(XLIX)

(n) Triseclavine (L)[1829)1831]

≪Occurrence≫ Ergot parasitic on *Elymus mollis* (テンキグサ)

mp. 232°, [α]$_D$ +174° (pyridine), C$_{16}$H$_{18}$ON$_2$

IR See Fig. 287

(L) Suggested Str.[1829]

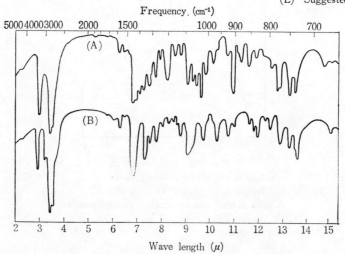

Frequency. (cm⁻¹)

(A)

(B)

Wave length (μ)

Fig. 287 IR Spectra of (A) Penniclavine and
(B) Triseclavine in nujol (Abe *et al.*)[1831]

1831) Abe, M *et al*: *Bull. Agr. Chem. Soc. Japan*, **19**, 92 (1955); *J. Agr. Chem. Japan*, **34**, 248 (1960).

(o) 6-Methyl-$\Delta^{8,9}$-ergolene-8-carboxilic acid (LI)[1832]

≪Occurrence≫ *Claviceps paspali* S. et H.

mp. 245～247°, $[\alpha]_D$ −208° (0.1N–NaOH), $C_{16}H_{16}O_2N_2$,
Hydrochloride.

mp. 257～258° (decomp.), $C_{16}H_{17}O_2N_2Cl$

UV See Fig. 288

IR See Fig. 289

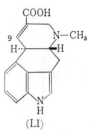

(LI)

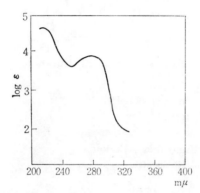

Fig. 288 UV Spectrum of (LI) (Kobel *et al.*)

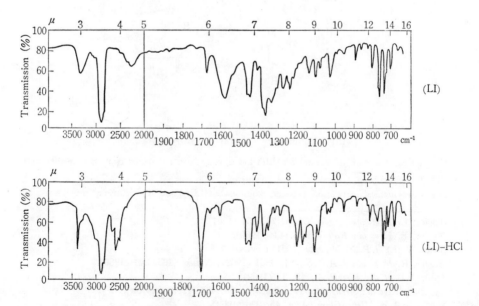

Fig. 289 IR Spectra of (LI) and (LI)–HCl in nujol (Kobel *et al.*)[1832]

NMR (LI)–HCl in $CDCl_3$, 60Mc, TMS, δ (ppm): 7.0～7.35m (4 arom. protons), 7.68s (C_9–H of
α,β–unsatur. carboxilic acid) (lysergic acid XXIII): 6.67 (C_9–H of arom. system) (see Fig. 290)

1832) Kobel, H *et al* : *Helv. Chim. Acta*, **47**, 1052 (1964).

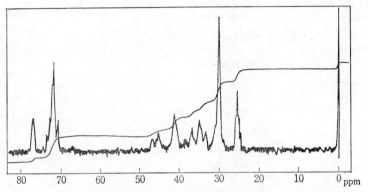

Fig. 290 NMR Spectrum of (LI)–HCl (Kobel *et al.*)[1832)

(iv) *Biosynthesis*

　　Biosynthesis of lysergic acid and ergot alkaloids[1833) ~ 1837)

　　Biosynthesis of clavine alkaloids[1826) 1838) ~ 1842)

(7) Yohimbe, Alstonia and Rauwolfia Alkaloids
(a) Absolute Configuration and IR Spectra

　　　　(LII) Yohimbine　　　　(LIII) (syn–*trans*) Yohimbane

　　Absolute configuration of yohimbine (LII) has already been suggested from the results of ORD[1843) 1844) and MRD[1845).

　　The stereochemical correlation between yohimbane (LIII) and it's isomers, pseudo–yohimbane (LIV),

1833) Taylor, E. H *et al*: *Angew. Chem.* **73**, 251 (1961).
1834) Taylor, E. H *et al*: *Nature*, **188**, 494 (1960).
1835) Tyler, V. E: *J.P.S.*, **50**, 629 (1961) (review).
1836) Baxter, R. M *et al*: *J.A.C.S.*, **84**, 4351 (1962); *Nature*, **195**, 241 (1960).
1837) Paul, A.G *et al*: *J.A.P.A.*, **49**, 14 (1959).
1838) Ramstad, E *et al*: *Arch. Biochem. Biophys.*, **98**, 457 (1962) (review).
1839) Brack, A *et al*: *Helv. Chim. Acta*, **45**, 276 (1962).
1840) Plieninger, H *et al*: *Ann.*, **642**, 214 (1961).
1841) Taylor, E. H *et al*: *J.P.S.*, **50**, 681 (1961).
1842) Abe, M *et al*: *Ann. Rept. Takeda Lab.*, **21**, 88 (1962).
1843) Klyne, W: *Chem. & Ind.*, **1953**, 1032, **1954**, 1198.
1844) Aldrich *et al*: *J.A.C.S.*, **81**, 2481 (1959).
1845) Djerassi, C *et al*: *ibid.*, **78**, 6362 (1959).

alloyohimbane (LV) and *epi*–alloyohimbane (LVI) have also been established[1846) 1847)].

(LIV) (*anti–trans*)
Pseudo–yohimbane

(LV) (*syn–cis*)
Allo–yohimbane
(Isoreserpine type)

(LVI) (*anti–cis*)
epi–Alloyohimbane
(Reserpine type)

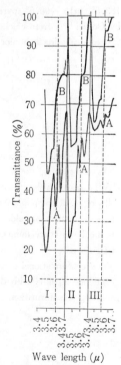

IR The C–H streching vibration (3.4~3.7μ) in chloroform solution can be used for identification of the configuration of the C_3–H of the alkaloids or their derivatives (Wenkert's rule)[1847)]. Normal and allo–type Alkaloids (C_3–α–H): Exhibit two or more distinct and characteristic peaks of medium intensity on the high wave length side of the major 3.46μ band (see Fig. 291).

(LIII), (LV)–Type—Yohimbine, alloyohimbine, ajmalicine, yohimbone, yohimbane, β–yohimbine, corynanthine, alloyohimbane, rauwolscine, methylisoreserpate, corynantheine, tetrahydroalstonine, aricine, tetraphylline, reserpinine, mayumbine, isoreserpiline, corynantheidine.

Pseudo–or *epi*–allo type alkaloides (C_3–β–H) show slight shoulders on the longer wave length side of the main peak (3.46μ) (see Fig. 291)

(LIV), (LXI)–Type—Reserpine, rescinnamine, deserpidine, methyl reserpate, 3–*epi*–α–yohimbine, pseudo–yohimbine, pseudoyohimbane, isoajmalicine, *epi*-alloyohimbane, isorauhimbine, raunescine, isoraunescine, raujemidine, pseudo-reserpine, isoreserpine, raumitorine, reserpiline, akuammigine.

The absolute configuration at C_{17} of yohimbine (LII) and reserpine (LVII) has been established by Ban[1848)], using the method of Prelog. (LII) gave L–(+)–atrolactic acid (mp. 89~90°, $[\alpha]_D$ +1.50° (EtOH)) (LVIII), while (LVII) gave D–(−)–atrolactic acid (mp. 88~90°, $[\alpha]_D$ −9.78° (EtOH) (LIX)).

Fig. 291 IR Spectra of (A) normal and (B) pseudo type Alkaloids(Wenkert *et al.*)[1847)]

1846) Janot, M.M *et al*: *Bull. Soc. Chim. France*, **19**, 1085 (1952).

1847) Wenkert, E *et al*: *J.A.C.S.*, **78**, 6417 (1956).

1848) Ban, Y: *18 th Nippon Yakugaku-taikai Koenyôshi Shû*, p. 191 (Tokyo, 1963); *Symposium on the Chemistry of Natural Products*, **5 th**, p. 25–1 (1961), **6 th**, p. 1 (1962).

(LII) Yohimbine $\longrightarrow$ [structure] $\longrightarrow$ [structure] $\longrightarrow$ [structure] (LVIII)

[structure with CH_3O indole] $\longrightarrow$ [structure] $\longrightarrow$ [structure] (LIX)

$R= $ —OC— [trimethoxybenzene with OCH_3, OCH_3, OCH_3] (LVII) Reserpine

R= H (LVII–a) Methyl reserpate

Correlation with cinchona alkaloids: Yohimbine (LII) was derived to dihydrocorynantheane (mp. 182.5~ 183.5°, $[\alpha]_D$ −96.6° (pyridine) (LX)) via eight steps by Ban[1848]. The latter compound was also obtained by Ochiai[1849], from cinchonine, the absolute configuration of which has already been established via 2′-oxohexahydrocinchonine (LXI).

[structure with O·COC₆H₅] $\xrightarrow{\text{5 steps}}$ [indole structure] $\xleftarrow{\text{8 steps}}$ (LII) Yohimbine

(LXI) 2′-Oxohexahydrocinchonine   (LX) Dihdrocorynantheane

As the results, the absolute configurations of yohimbine and reserpine were confirmed as (LII) and (LVII), respectively. Rosen[1850] modified Wenkert's rule[1847] for IR absorption of indole alkaloids as follwing: All compounds possessing a conformationally stable *axial*-H at C_3 position exhibit 2 or more peaks or distinct shoulders between 2700~2900 cm^{-1} region at least one of which appears below 2800 cn^{-1} on the lower wave number side of the major band (ca. 2900 cm^{-1}), whereas those containing *equatorial*-H at C_3 do not.

The conformational correlation of ajmalicine (δ–yohimbine=tetrahydroserpentine) (LXI), a typical hetero-yohimbine alkaloid with dihydrocorynantheane (LX) has also heen established by Wenkert[1851] by the following transformation.

[indole structure with CH_3OOC] $\xrightarrow{\text{5 steps}}$ [indole structure]

(LXI) Ajmalicine   (LX)

1849) Ochiai, E. *et al*: *Chem. Pharm. Bull.*, **7**, 386 (1959).
1850) Rosen, W: *Tetr. Letters*, **1961**, 481.
1851) Wenkert, E *et al*: *J.A.C.S.*, **80**, 3484 (1958).

ORD See the reports of Djerassi et al.[1852], Jeffereys[1853], Aldrich et al.[1854] and Klyne[1855].

(b) Correlation of Configuration with the Spectra in Hetero-Yohimbine Alkaloids[1854]

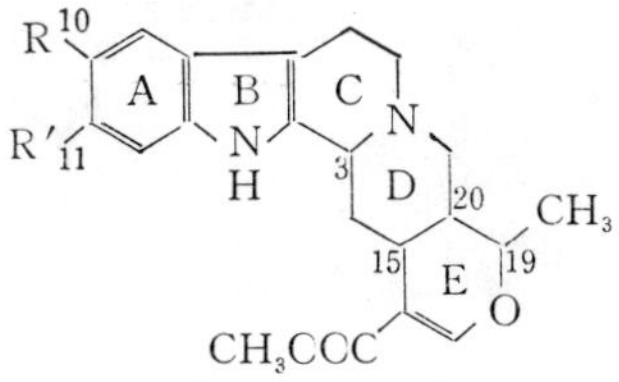

(LXI-a) Heteroyohimbine Alkaloids

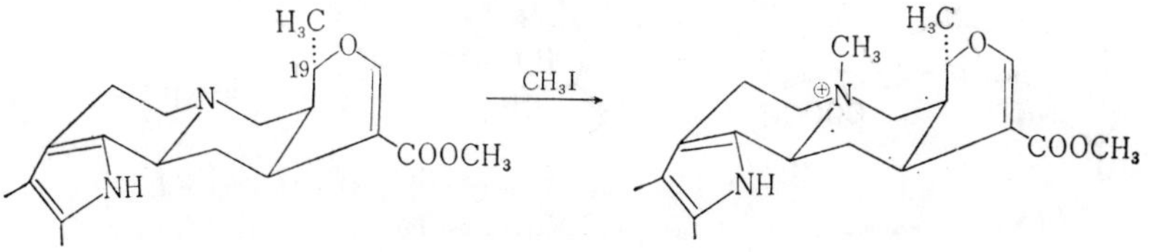

(LXII) *allo* (LXII-a)

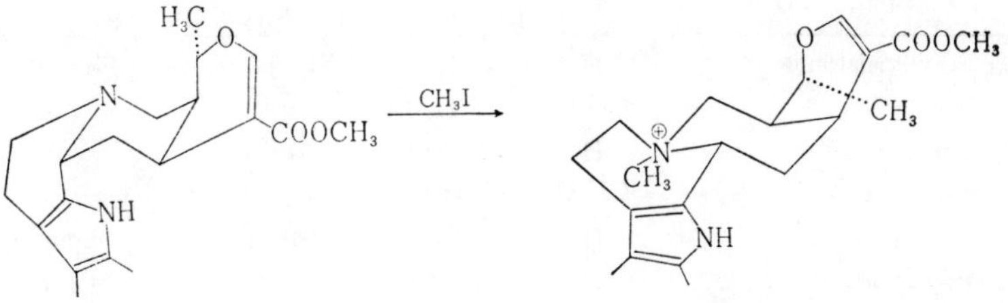

(LXIII) *epiallo* (LXIII-a)

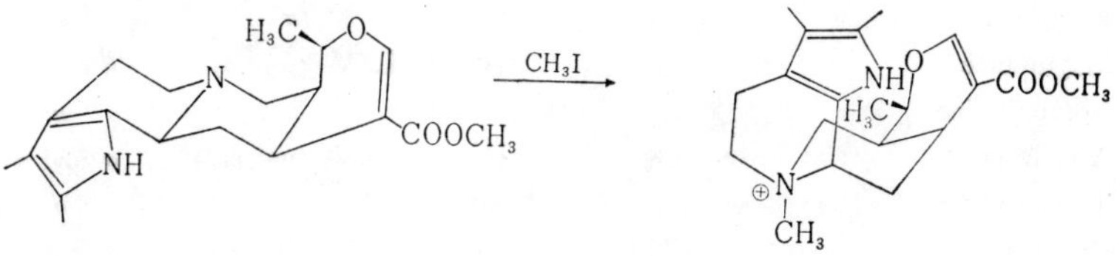

(LXIV) *allo* (LXIV-a)

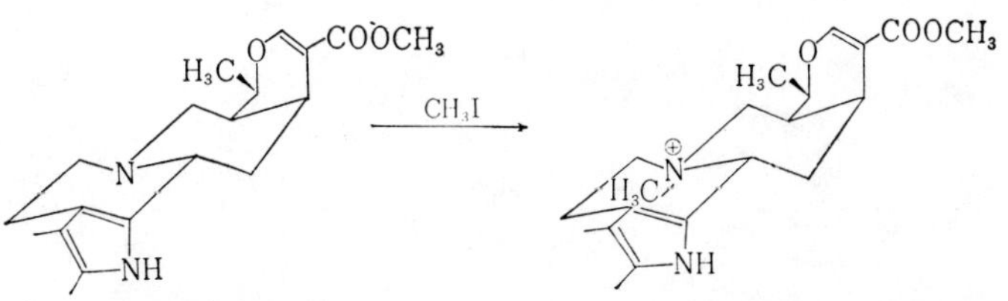

(LXV) *epiallo* (LXV-a)

1852) Djerassi, C et al: *J. A. C. S.*, **78**, 6371 (1956).
1853) Jeffreys, J.A.D. *Soc.*, **1959**, 3077.
1854) Aldrich, P.E et al: *J.A.C.S.*, **81**, 2481 (1959).
1855) Klyne, W: *Chem. & Ind.*, **1953**, 1032, **1954**, 1198.

(LXVI) *normal* CH$_3$I → (LXVI–a)

(LXVII) *normal* (LXVIII) *pseudo* (LXIX) *pseudo*

Table. 99 Configuration of Heteroyohimbine Alkaloids (Shamma *et al.*)[1856]

Str. No.	Alkaloids (LXI–a)	R	R′	Group and Stereo-chemistry	19–CH$_3$	C/D fusion
(LXX)	Tetrahydroalstonine	H	H	LXII (*allo*)	α and e	*trans*
(LXXI)	Aricine	CH$_3$O	H	"	"	"
(LXXII)	Reserpinine	H	CH$_3$O	"	"	"
(LXXIII)	Isoreserpiline	CH$_3$O	CH$_3$O	"	"	"
(LXXIV)	Akuammigine	H	H	LXIII (*epiallo*)	α and e	*cis*
(LXXV)	Isoreserpinine	H	CH$_3$O	"	"	"
(LXXVI)	Reserpiline	CH$_3$O	CH$_3$O	"	"	"
(LXXVII)	Rauniticine			LXIV (*allo*)	β and a	*trans*
(LXXVIII)	Raunitidine			"	"	"
(LXXIX)	Mayumbine	H	H	LXV (*epiallo*)	β and e	*trans*
(LXXX)	Isoraunitidine			"	"	"
(LXXXI)	Ajmalicine	H	H	LXVI (*normal*)	α and a	*trans*
(LXXXII)	Tetraphylline	H	CH$_3$O	"	"	"
(LXXXIII)	Raumitorine	CH$_3$O	H	LXVII (*normal*)	β and e	*trans*

1856) Shamma, M *et al*: *J.A.C.S.*, **85**, 2507 (1963).

Table. 100 Physical Constants and Occurrence of Heteroyohimbine Alkaloids[1857]~[1866]

Str. No.	Alkaloids	Formula	mp.	$[\alpha]_D$	pKa	λ_{max} mμ	λ_{max} μ (ν_{max} cm^{-1})	Ref.
(LXX)	Tetrahydroalstonine	$C_{21}H_{24}O_3N_2$	230~232°	−98°(CHCl₃)	5.83	250	see Fig. 293 5.81, 6.11	[1857] [1866]
(LXXI)	Aricine	$C_{22}H_{26}O_4N_2$	190° (decomp.)	−63° (pyridine)	5.75	see Fig. 292 225, 280	see Fig. 293 5.9, 6.1	[1859] [1860] [1866]
(LXXII)	Reserpinine	$C_{22}H_{26}O_4N_2$	238~239°	−117° (CHCl₃)	6.01	230, 299	see Fig. 293 5.88, 6.21 12.1	[1858] [1866]
(LXXIII)	Isoreserpiline	$C_{23}H_{28}O_5N_2$	211~212°	−82° (pyridine)	6.07	228,300,304	see Fig. 293 (LXXV)	[1860] [1866]
(LXXIV)	Akuammigine	$C_{21}H_{24}O_3N_2$ H₂O	113°	−42°(EtOH)	—	see Fig. 292-a	(3546, 1683, 1629, 741)	[1865]
(LXXV)	Isoreserpinine	$C_{22}H_{26}O_4N_2$	225~226° (decomp.)	−5°(pyridine)	6.49	see Fig. 292 ident.w(LXXIII)	see Fig. 293	[1860] [1866]
(LXXVI)	Reserpiline	$C_{23}H_{28}O_5N_2$	amorph	−38°(EtOH)	6.20	see Fig. 292 229, 304	see Fig. 293 5.99, 6.20	[1860] [1866]
(LXXVII)	Rauniticeine	$C_{21}H_{24}O_3N$	233~235°	−38.4° (CHCl₃)	6.24	228, 282	3.4(α-OH), 5.88, 6.13	[1861]
(LXXVIII)	Raunitidine	$C_{22}H_{26}O_4N_2$	276~278°	−69.5° (CHCl₃)	6.20	229, 298	ca. same to (LXXVII)	[1861]
(LXXIX)	Mayumbine	$C_{21}H_{24}O_3N_2$	216°	−68° (pyridine)	5.85	ident. w (LXXXI)	5.9, 6.2	[1862]
(LXXX)	Isoraunitidine	$C_{22}H_{26}O_4N_2$	259~261°	+131° (pyridine)	6.42	228, 298	——	[1861]
(LXXXI)	Ajmalicine	$C_{21}H_{24}O_3N_2$	250~252° (decomp.)	−58.1° (CHCl₃)	6.31	225, 283	see Fig. 293 6.9, (820~830)	[1858] [1866]
(LXXXII)	Tetraphylline	$C_{22}H_{26}O_4N_2$	220~223° (decomp.)	−73°(CHCl₃)	6.39	229, 298	see Fig. 293 2.9,5.96,6.14 6.22	[1863] [1866]
(LXXXIII)	Raumitorine	$C_{21}H_{26}O_4N_2$	138°	+60°(CHCl₃)	—	ca. ident. w (LXXI)	see Fig. 293 6.34~3.7	[1864] [1866]

≪Occurrence≫   (LXX) *Rauwolfia sellowii*, (LXXI) *R. canescens, R. heterophylla, R. sellowii*, (LXXII) *R. Canescens, R. serpentina*, (LXXIII) *R. canescens, R. schueli, R. vomitoria*, (LXXIV) *Picralima nitida*, (LXXV) *R. canescens*, (LXXVI) *R. serpentina, R. canescens, R. schueli, R. micrantha, R. vomitoria*, (LXXVII) *R. nitida*, (LXXVIII) *R. nitida*, (LXXIX) *Pseudocinchona mayumbensis*,   (LXXX) Raunitidine (LXXVIII) $\xrightarrow{Ac_2O}$ (LXXX), (LXXXI) yohimbe bark, *Vinca rosea, Rauwolfia serpentina, R. canescens, R. micrantha, R. heterophylla, R. verticillata, R. sellowii*, (LXXXII) *R. tetraphylla, R. degeneri*, (LXXXIII) *R. vomitoria*

As illustrated in Fig, 293, alkaloids are separated to three groups from their spectral behavior.

1st group—Rather simple ester band at 8.45μ: ajmalicine (LXXXI) tetraphylline. (LXXXII).

2nd group—Three distinct absorptions at 8.15, 8.32 and 8.45μ; tetrahydroalstonine, (LXX) reserpinine (LXXII).

3rd group—Broad band centered at 8.26μ; reserpiline (LXXVI), isoreserpinine (LXXV).

1857) Bader, F.E: *Helv. Chim. Acta*, **36**, 215 (1953).
1858) Hofmann, A: *ibid.*, **37**, 314, 849 (1954).   1859) Prelog, V *et al*: *ibid.*, **37**, 1805 (1954).
1860) Stoll, A *et al*: *ibid.*, **38**, 270 (1955).   1861) Salkin, R *et al*: *J.P.S.*, **50**, 1038 (1961).
1862) Janot, M.M *et al*: *Compt. rend.*, **234**, 850 (1952).
1863) Djerrasi, C *et al*: *J.A.C.S.*, **79**, 1217 (1957).
1864) Goutarel, R *et al*: *Compt. rend.*, **239**, 302 (1954); *Bull. Soc. Chim. France.*, **1954**, 1481.
1865) Robinson, R *et al*: *Soc.*, **1954**, 3479.

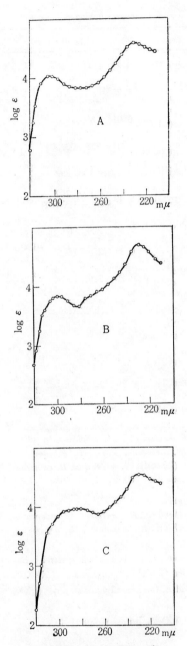

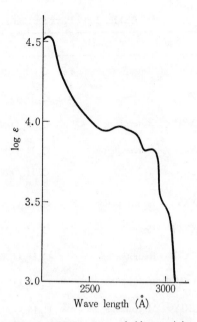

Fig. 292-a UV Spectrum of Akuammigine
(Robinson *et al.*)[1865]

Fig. 292 UV Spectra of (A) Reserpiline
(B) Isoreserpinine and
(C) Aricine in EtOH
(Stoll *et al.*)[1860]

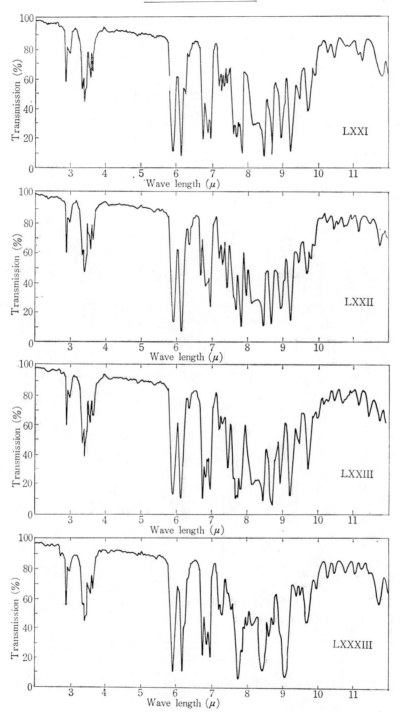

Fig. 293 IR Spectra of (LXXI) Aricine, (LXXII) Reserpinine, (LXXIII) Isoreserpiline, and (LXXXIII) Raumitorine in CS$_2$ (Neuss)[1866]

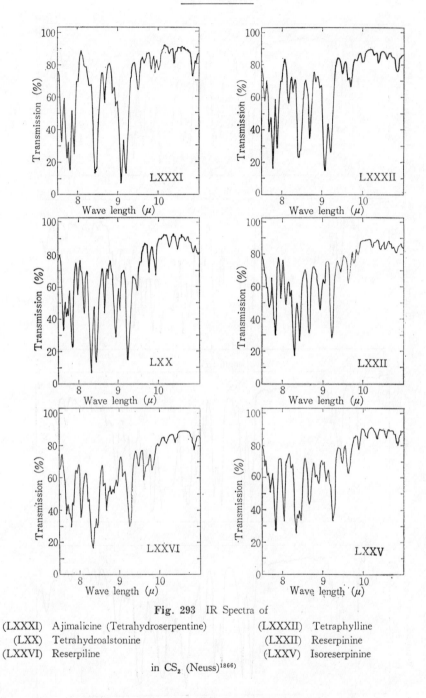

Fig. 293 IR Spectra of

(LXXXI)　Ajimalicine (Tetrahydroserpentine)　　(LXXXII)　Tetraphylline
(LXX)　Tetrahydroalstonine　　　　　　　　　　　(LXXII)　Reserpinine
(LXXVI)　Reserpiline　　　　　　　　　　　　　　(LXXV)　Isoreserpinine

in CS₂ (Neuss)[1866]

Table 101 NMR of Heteroyohimbine Alkaloids (see Table 99) (Shamma *et al.*)[1856]

in CDCl$_3$, 40 Mc, TMS: ppm=0

Str. No.	Alkaloids	Free Bases	Methiodide Salts		Δ ppm C$_{19}$–Me (base)–(salt)
		C$_{19}$–Me chem. shift (*J*cps)	⊕N–CH$_3$ chem. shift	C$_{19}$–Me chem. shift (*J*cps)	
(LXX)	Tetrahydroalstonine	1.38(6.1)	3.42	1.46(5.8)	−0.08
(LXXI)	Aricine	1.37(6.3)	3.43	1.47(5.4)	−0.10
(LXXII)	Reserpinine	1.38(6.1)	3.39	1.48(5.4)	−0.10
(LXXIII)	Isoreserpiline	1.39(6.2)	3.41	—	[Note 1]
(LXXIV)	Akuammigine	— —	—	—	—
(LXXV)	Isoreserpinine	1.32(6.5)	3.32	1.39(6.2)	−0.07
(LXXVI)	Reserpiline	1.32(6.3)	3.35	1.39(5.7)	−0.07
(LXXVII)	Rauniticine	1.42(6.7)	3.50	1.40(6.5)	+0.02
(LXXVIII)	Raunitidine	1.42(7.1)	3.50	1.41(6.1)	+0.01
(LXXIX)	Mayumbine	1.35(6.3)	—	—	—
(LXXX)	Isoraunitidine	1.35(6.8)	3.31	1.43(6.0)	−0.08
(LXXXI)	Ajmalicine	1.16(6.7)	3.35	1.26(6.2)	−0.10
(LXXXII)	Tetraphylline	1.16(6.5)	3.36	1.26(6.2)	−0.10
(LXXXIII)	Raumitorine	— —	—	—	—

[Note 1] See the item (h)

(c) Mass Spectra[1867]

(i) *Stereoisomers of Yohimbine*

General mass peaks for yohimbine, α–yohimbine, 3–*epi*–β–yohimbine, pseudoyohimbine, *allo*–yohimbine, corynanthine and 3–*epi*–corynanthine are as following: M=354, M–1=353 (very strong), M–18 (H$_2$O), M–59 (–COOCH$_3$) (See Fig. 294)

Those mass peaks are useful for indentification of yohimbine-type alkaloids, but not applicable for criterion of the isomers.

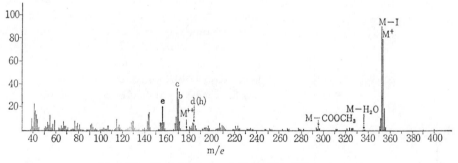

Fig. 294 Mass Spectrum of Yohimbine (Djerassi *et al.*)

1867) Djerassi, C. *et al*: *J.A.C.S.*, **84**, 2161 (1962).

Fragmentation

a (M-1)

b (170)

e (156)   d (184)   c (169)

g   f   h (184)

j (197)   i (225)

N–Methylyohimbine; d, h=198, b=184, e=170

(ii) *Heteroyohimbine Alkaloids*

(LXXXIV)

In mass spectrum of ajmalicine (LXXXIV, R=R′=H, 3 α, 15 α, 20 β), peak *e* (m/e 156) appears stronger than that of yohimbine. This fact is interpreteted that the peak *e* arises (wave line in (LXXXIV)) from cleavage (*f→g→e*) of three allylically activated centers, the 14–15 bond in *g* now being especially labilized because of the additional $\varDelta_{16-17}$ in ring E. The peak *h* (m/e 184) also appears stronger than the

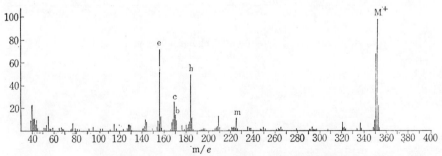

Fig. 295 Mass Spectrum of Ajmalicine (Djerassi *et al.*)

corresponding peak of yohimbine and may, therefore, 14–15 linkage be promoted by Δ_{16-17}. There exists a significant difference in the relative intensities of the e (m/e 156), c (m/e 169) and h (m/e 184) in the spectrum of ajmalicine as compared to those of *tetra* hydroalstonine (LXXXIV, R=R′=H, 3α, 15 α, 20 α) and this can evidently be used as a criterion for a D/E *trans vs. cis* ring juncture (see Fig. 295 and 296).[1867]

(d) Synthesis

Total synthesis of yohimbine (LII)[1868], *dl–allo*-yohimbane (LV)[1869], isorauhimbine (LXXXV) → α–Yohimbine (LXXXVI)[1870]

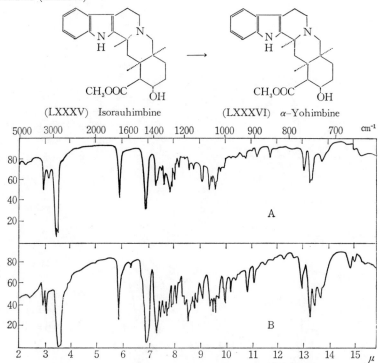

(LXXXV) Isorauhimbine   (LXXXVI) α–Yohimbine

Fig. 296 IR Spectra of (A) Isorauhimbine and (B) α–Yohimbine in nujol
(Hofmann *et al.*)

(e) 17-Ketoyohimbine (LXXXVII)[1871]

≪Occurrence≫ Yohimbine (LII) $\xrightarrow{\text{Al-phenoxide}}$ (LXXXVII)

mp. 254~255°, [α]$_D$ +15.8° (pyridine) $C_{21}H_{24}O_3N_2 \cdot CH_3OH$

UV λ_{max}^{EtOH} mμ (log ε): 232 (4.05), 283 (3.96)

IR λ_{max}^{nujol} μ: 5.78, 5.88

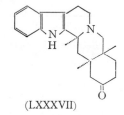

(LXXXVII)

(f) Serpentine (LXXXVIII)[1872) 1873]

≪Occurrence≫   *Rauwolfia canescens, R. heterophylla, R. sellowii, R. serpentina, R. tetraphylla*

1868) Van Tamelen, E. E. *et al*: *J.A.C.S.*, **80**, 5006 (1958).
1869) Rapala, R. T. *et al*: *ibid.*, **79**, 3770 (1957).
1870) Hofmann, A. *et al*: *Helv. Chim. Acta*, **40**, 156 (1957).
1871) Kimoto, S. *et al*: *Chem. Pharm. Bull.*, **7**, 650 (1959).
1872) Fritz, H.: *Ann.*, **655**, 148 (1962).   1873) Bader, F. E. *et al*: *Helv. Chim. Acta*, **35**, 1594 (1952).

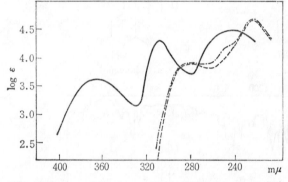

(LXXXVIII)

mp. 153~154°, C$_{20}$H$_{26}$O$_3$N$_2$·5 H$_2$O, methiodide mp. 271~272° (decomp.)
UV See Fig. 297

IR ν$_{max}^{CH3Cl3}$ cm^{-1}: 1698, 1610 (H$_3$C·O$_2$C–C=C–C–O—), see Fig. 298.

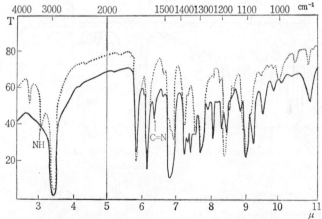

Fig. 297 UV Spectra of
―――― Serpentine
------ py–Tetrahydroserpentine
–·–·–·– py–Tetrahydroalstonine
in EtOH (Bader *et al.*)[1873]

Fig. 298 IR Spectra of
―― Serpentine
······ py–Tetrahydroserpentine in nujol (Bader *et al.*)[1873]

(g) Holeinine (LXXXIX)[1874]

(LXXXIX)

≪Occurrence≫ *Ochrosia sandwicensis* A. GRAY
Chloride mp. 283~285° (decomp.)
[α]$_D$ −134.5 (MeOH), C$_{24}$H$_{31}$O$_5$N$_2$Cl, perchlorate 228~**231°**
UV λ$_{max}$ mμ: 221.5, 296.2, 302, 307.
IR ν$_{max}^{KBr}$ cm^{-1}: 3750, 3395, 2900, 1695, 1634, 1480, **1460**, 1440, 1390, 1310, 1210, 1178, 1158, 1125, 1100, **1025**, 855, 773.

(h) Isoreserpiline (LXXIII)

1874) Scheuer, P. J. *et al*: *J. Org. Chem.*, **26**, 3069 (1961).

See Table 100 and 101

NMR[1875] in CDCl$_3$, 60, 100 Mc, TMS, δ 0 (ppm). 1.37 d, $J=6$ (3) (CH$_3$–C̣H–O—), 4.44 octet (1)

(CH$_3$—C̣H–O—), 3.72 s (3) (CH$_3$CO$_2$–), 3.83 s, 3.87 s (6) (CH$_3$O), 4.2–4.7 q (1) (allylic H),
6.77 s, 6.90 s (2) (aromatic H), 7.57 s (1) (olefinic H), 7.95 s (1) (>NH)

(i) Alstonidine (XC)[1876]

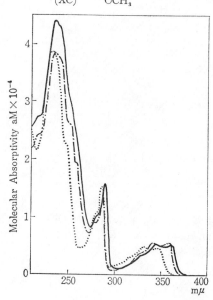

(XC)

≪Occurrence≫   *Alstonia constricta*
mp. 186~188°, C$_{22}$H$_{24}$O$_4$N$_2$
UV See Fig. 299
IR $\lambda_{max}^{CHCl_3}$ μ: 5.89, 6.14 (correspond to tetrahydroalstonine),
3.18 (bonded OH), 6.16, 6.39, 6.74, 7.13 (indole–*N*–methyl-
harman), see Fig. 300

Fig. 299 UV Spectra of
——— Alstonidine
—·—·—· Indole–*N*–methylharman
············ Harman
in 0.05 M KOH–CH$_3$OH

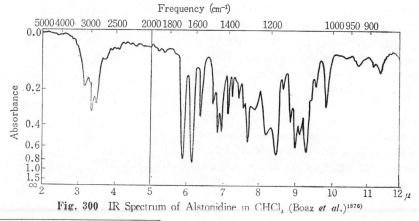

Fig. 300 IR Spectrum of Alstonidine in CHCl$_3$ (Boaz *et al.*)[1876]

1875) Djerassi, C *et al*: *J.A.C.S.*, **85**, 1523 (1963).
1876) Gordon, H. Rose, H. A. Boaz, H. *et al*: *J.A.P.A.*, **46**, 508, 509, 510 (1957).

(j) Dihydrocorynantheol (XCI)[1877]

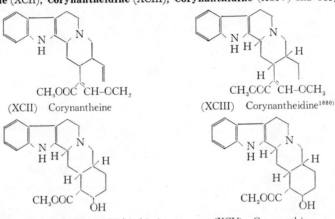

≪Occurrence≫ *Aspidosperma marcaravianum*
　　mp. 181~183°, $[\alpha]_D$ −19° (CHCl₃), C₁₉H₂₆ON₂
　　UV λ_{max}^{EtOH} mμ (log ε): 226 (4.56), 281 (3.87), 290 sh (3.80)
　　IR λ_{max}^{film} μ: 3.1 s, 6.88 s, 6.93 s

(XCI)

(k) Melinonines See the item (15) Strychnos and Curare Alkaloids

(l) Corynantheine (XCII), **Corynantheidine** (XCIII), **Corynanthidine** (XCIV) and **Corynanthine** (XCV)

CH₃OOC CH–OCH₃
　　(XCII)   Corynantheine

CH₃CC CH–OCH₃
　　(XCIII)   Corynantheidine[1880]

CH₃OOC   OH
　　(XCIV)   Corynanthidine (α–Yohimbine)

CH₃OOC   OH
　　(XCV)   Corynanthine

Corynantheine (XCII)
　　≪Occurrence≫   *Pseudocinchona africana*,   mp. 117° & 169°, $[\alpha]_D$ +27.7°, C₂₂H₂₆O₃N₂
UV λ_{max} mμ: 225, 285 (typical unconj. indole)[1878], 250 inf. (unsatur. enolester)
IR λ_{max} μ: 5.89, 6.2 (approx. equal intensity) (unsatur. enolester)[1878]

Corynantheidine (XCIII)
　　≪Occurrence≫   *Pseudocinchona africana*
　　mp. 83°, $[\alpha]_D$ −147° (MeOH), C₂₂H₂₈O₃N₂·CH₃COCH₃,
　　mp. 125°, $[\alpha]_D$ −171° (MeOH), C₂₂H₂₈O₃N₂
　　UV λ_{max} mμ: 225, 280, 250 inf. (very similar to corynantheine)[1879]
　　IR λ_{max} μ: 5.9, 6.2 (unsatur. ester), 9 (ether)[1879]

Corynanthidine (α–Yohimbine=Rauwolscine) (XCIV)
　　≪Occurrence≫   *Rauwolfia canescens, R. serpentina*
　　　　mp. 243~244°, $[\alpha]_D$ −11.5°, C₂₁H₂₆O₃N₂
IR See Fig. 296[1880]

Corynanthine (Rauhimbine) (XCV)
　　≪Occurrence≫   *Rauwolfia serpentina*, Yohimbe bark.
　　mp. 218~225°, $[\alpha]_D$ −82° (pyridine), C₂₁H₂₆O₃N₂
　　UV See Fig. 301, λ_{max}^{EtOH} mμ (log ε): 226 (4.56), 283 (3.87), 290(3.79)[1858]

1877) Gilbert, B *et al*: *J. Org. Chem.*, **27**, 4702 (1962).
1878) Janot, M. M *et al*: *Bull. Soc. Chim. France*, **1951**, 588, **1953**, 1033.

IR   normal or allo type See the item (a)[1847].

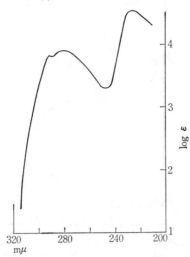

Fig. 301 UV Spectrum of Rauhimbine (Corynanthine) in EtOH   (Hofmann)[1858]

Serpentinine (XCVI)[1879]

(XCVI)

≪Occurrence≫   *Rauwolfia serpentina* BENTH.

mp. 264~265° (decomp.), $[\alpha]_D$ +81° (EtOH) pKa' 10.4, 6.9, $C_{42}H_{44}O_6N_4$

UV   See Fig. 302, λ_{max}^{EtOH} mμ (log ε): 227 (4.74), 257 (4.56), 282~283 (4.15), 291 (4.14), 308 (4.22), 371 (3.78) (tetraammonium carboline and indole)

IR   See Fig. 303, $\nu_{max}^{CHCl_3}$ cm^{-1}: 1712, 1623, 3340 (NH or OH) (see the item (g) Serpentine (LXXXVIII));

$\nu_{max}^{CHCl_3}$ cm^{-1}: 1698, 1610, ν_{max}^{KBr} cm^{-1}: 1720, 1709 (conj. and non-conj. C=O); ν_{max}^{nujol} cm^{-1}: 1730, 1701.

NMR   in CHCl$_3$, 40 Mc, relative to H$_2$O, cps (Fig. 304): +39.0, +52.0 (2-COOCH$_3$), +113.9 d, +142.7 d (2 >CHCH$_3$)

1879) Kaneko, S: *Y. Z.*, **80**, 1357~1382, 1493 (1960).

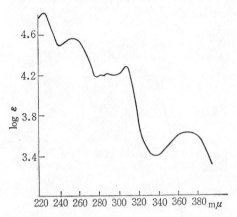

Fig. 302   UV Spectrum of Serpentinine in EtOH (Kaneko)

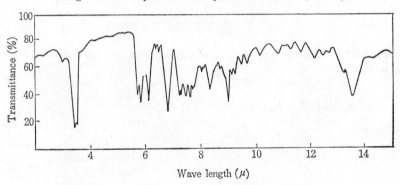

Fig. 303   IR Spectrum of Serpentinine in nujol (Kaneko)

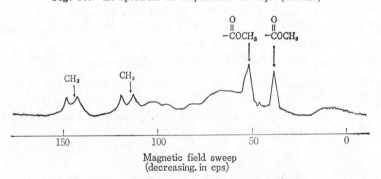

Fig. 304   NMR Spectrum of Serpentinine (Kaneko)

(n)   Absolute Configurations of Ajmaline-type Alkaloids and Sarpagine[1880]

(XCVII) Ajmaline   ≈   (XCVIII) Sandwicine

1880)   Bartlett, M. F. *et al*: *J.A.C.S.*, **84**, 622 (1962).

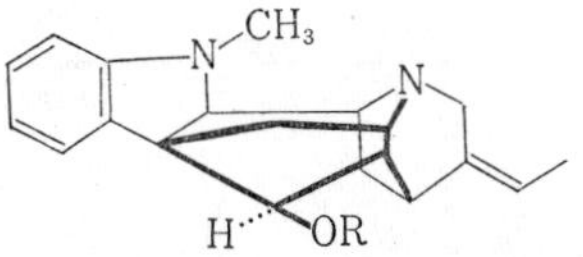

R=H: (XCIX) Tetraphyllicine
R=—COPh(OMe)₃: (C) Rauvomitine

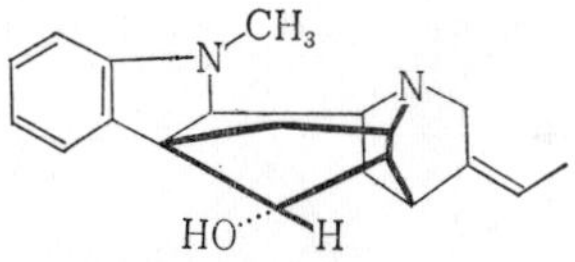

(CI) Mauiensine

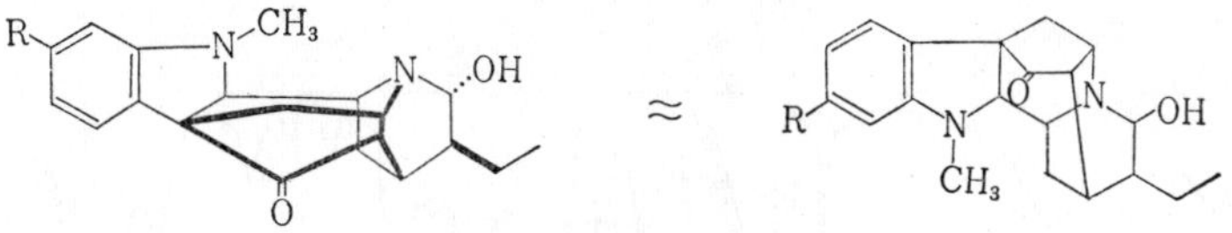

R=H: (CII) Ajmalidine
R=OCH₃: (CIII) Vomalidine

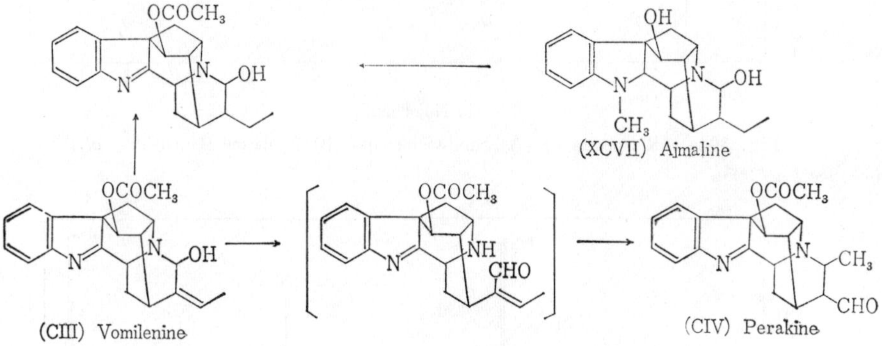

(XCVII) Ajmaline

(CIII) Vomilenine

(CIV) Perakine

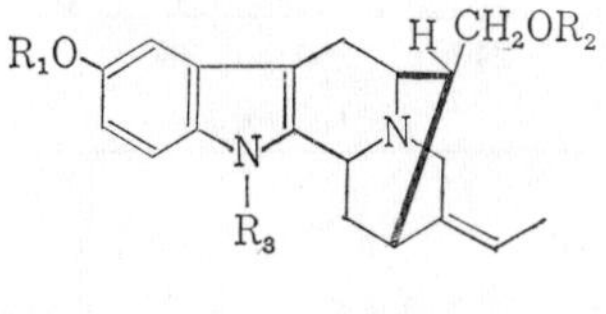

R₁=R₂=R₃=H: (CV) Sarpagine
R₁=CH₃, R₂=R₃=H: (CVI)
O-Methylsarpagine = Lochnerine

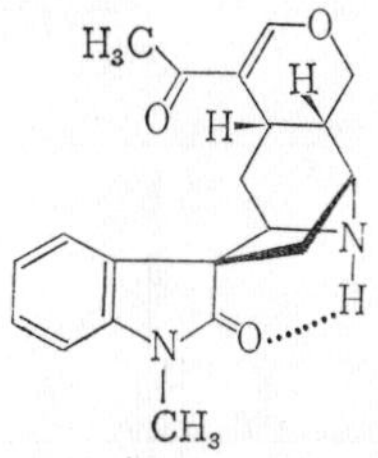

(CVII) Alkaloid C

(o) **Ajmaline** (XCVII)

    ≪Occurrence≫   *Rauwolfia serpentina, R. canescens, R. micrantha, R. heterophylla, R. vomitoria, R. degeneri, R. sellowii, R. caffra,* Yohimbe bark

    mp. 158~160°, $[\alpha]_D$ +128° (CHCl₃), $C_{20}H_{26}O_2N_2 \cdot CH_3OH$

    mp. 205~207°, $[\alpha]_D$ +144° (CHCl₃), $C_{20}H_{26}O_2N_2$

  UV λ_{max}^{EtOH} mμ (ε): 247 (8730), 295 (3070)[1887]

  IR λ_{max} μ: 3.00, 3.17 (OH), no absorptions bands at 5.82, 7.24 (=C–Me) or 9.0 (ether bridge)[1881], see Fig. 304a

1881) Robinson, R. *et al*: *Soc.*, **1954**, 1242.

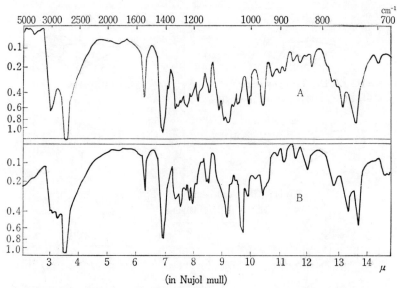

(in Nujol mull)

Fig. 304a　IR Spectra of (A) Sandwicine and (B) Ajmaline (Gorman *et al.*)[1882]

Mass : See Fig. 305

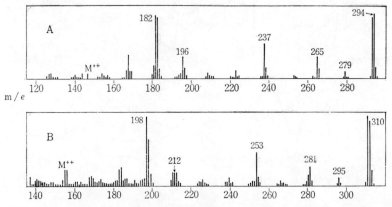

Fig. 305　Mass Spectra of (A) compound (CVIII) derived from Ajmaline (XCVII) and of
(B) compound (CIX) derived from Sarpagine (CV)　　　　(Biemann)[1883]

1882)　Gorman, M. *et al* : *Tetrahedron*, **1**, 328 (1957).
1883)　Biemann, K : *J.A.C.S.*, **83**, 4801 (1961).

Fragmentation:

(XCVII)   (CVIII) 294   265

238

237 CH_3

182   183

(p) **Sandwicine** (XCVIII)[1882)]

    ≪Occurrence≫   *Rauwolfia sandwicensis, R. degeneri, R. mauiensis*

       Amorphous, $[\alpha]_D$ +171° (MeOH), $C_{20}H_{26}O_2N_2$, —— 2 HI   mp. 238~240°, $[\alpha]_D$ +84° (MeOH)

  UV λ_{max}^{EtOH} mμ (log ε): 246 (3.91), 292 (3.46) (typical dihydroindole)

  IR   See Fig. 304a

(q) **Tetraphyllicine** (XCIX)[1884)]

    ≪Occurrence≫   *Rauwolfia tetraphylla, R. sellowii*

    mp. 320~322°, $[\alpha]_D$ +21° (pyridine), pKa′ 8.5, $C_{20}H_{24}ON_2$

  UV λ_{max}^{EtOH} mμ (log ε): 250 (4.07), 294 (3.60); λ_{min}: 229 (3.75), 262 (3.32)

  IR λ_{max}^{nujol} μ: 6.24, 6.80, OH absorption is hardly noticeable

  Rauvomitine (C)[1884) 1885)]

    ≪Occurrence≫   *Rauwolfia vomitoria*,   mp. 115~117°, $[\alpha]_D$ −173.4° (CHCl₃), $C_{30}H_{34}O_5N_2$

(r) **Mauiensine** (CI)[1882) 1886)]

    ≪Occurrence≫   *Rauwolfia mauiensis*

       Tetraphyllicine (XCIX) $\xrightarrow[\text{reflux}]{\text{K-}t\text{-butoxide}}$ (CI)

  mp. 237~238°, $C_{20}H_{24}ON_2$

  IR λ_{max} μ: 10.05, 10.9, 12.2 (trisubst. double bond)

  Dihydromauiensine   mp. 198~201°, 215~217°, $[\alpha]_D$ +229° (MeOH), $C_{20}H_{26}ON_2$

(s) **Ajmalidine** (CII)[1884) 1887) 1888)]

    ≪Occurrence≫   *Rauwolfia sellowii*

  mp. 241~242°, pKa 6.3 (80% DMF), $C_{20}H_{24}O_2N_2$

    UV   See Fig. 306, λ_{max}^{EtOH} mμ (ε): 247 (9150), 295 (3020)

    IR $\lambda_{max}^{CHCl_3}$ μ: 5.77

1884) Djerassi, C. *et al*: *J.A.C.S.*, **79**, 1217 (1957), **78**, 1259 (1956).

1885) Janot, M. M. *et al*: *Compt. rend.*, **241**, 1840 (1955).

1886) Scheuer, P. J. *et al*: *J. Org. Chem.*, **28**, 2641 (1963).

1887) Neuss, N. *et al*: *J.A.C.S.*, **77**, 6687 (1955).

1888) Hofmann, A. *et al*: *Helv. Chim. Acta*, **40**, 1866 (1957).

(t) Vomalidine (CIII)[1888]

《Occurrence》   *Rauwolfia vomitoria*

mp. 242~243°, $[\alpha]_D$ +210° (pyridine), pKa 6.5 (in 80% methylcellosolve), $C_{21}H_{26}O_3N_2$

UV See Fig. 306 (typical pattern of indoline alkaloids)

IR See Fig. 307 $\lambda^{CHCl_3}_{max}$ μ: 2.86 (OH), 5.74 (C=O in five membered ring), 6.22, 6.26, 11.8 (1, 2, 4–trisubst. benzene)

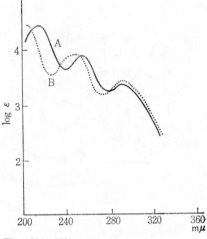

Fig. 306 UV Spectra
(A) Vomalidine and
(B) Ajmalidine
(Hofmann *et al.*)[1888]

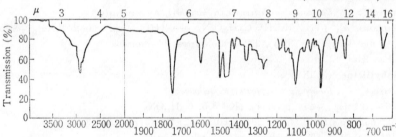

Fig. 307 IR Spectrum of Vomalidine in CHCl₃   (Hofmann *et al.*)[1888]

(u) Vomilenine (CIII)[1888] [1889]

《Occurrence》   *Rauwolfia vomitoria*

mp. 207°, $[\alpha]_D$ −72°, (pyridine), pK 4.65, $C_{21}H_{22}O_3N_2$

UV See Fig. 308[1889]

IR See Fig. 309[1889]

NMR See Fig. 310

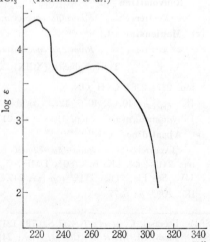

Fig. 308 UV Spectrum of Vomilenine in MeOH[1889]

1889) Hofmann, A. *et al*: *Helv. Chim. Acta*, **40**, 1866 (1957).

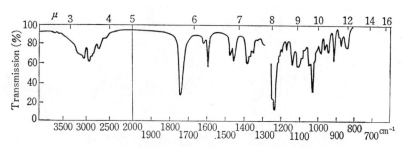

Fig. 309 IR Spectrum of Vomilenine in CH_2Cl_2[1889)

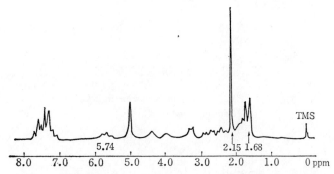

Fig. 310 NMR Spectrum of Vomilenine in $CDCl_3$, 60 Mc, TMS=0 ppm
(Hofmann *et al.*)[1889)

(v) Perakine (CIV)[1889) 1890)

《Occurrence》 *Rauwolfia perakensis, R. vomitoria.*

mp. 185°, $[\alpha]_D$ +112° ($CHCl_3$) pKa' 5.01, $C_{21}H_{22}O_3N_2$

$$\text{Vomilenine (CIII)} \xrightarrow{\text{HOAc}} \text{(CIV)}$$

UV $\lambda_{max}^{neutral\ medium}$ mμ: 219, 257; λ_{min}: 236; $\lambda_{max}^{Acidic\ med.}$ mμ: 266~267; λ_{min}: 238 (indolenine)

IR λ_{max}^{nujol} cm^{-1}: 1739, 1713 (satur. C=O)

NMR δ: 6.91~7.60 m (4 arom. H), 4.21 dd, J=4.1, 8.4 (C_3–H),
　　4.97 d, J=2.1 (C_{17}–H), 2.17 ($COCH_3$), 1.42 d, J=6.3 (>$CHCH_3$).

(w) Sarpagine (CV)[1883) 1891)

《Occurrence》 *Rauwolfia serpentina*

mp. >320°, $[\alpha]_D$ +54° (pyridine), $C_{19}H_{22}O_2N_2$

UV See Fig. 311

Mass See Fig. 305[1883)

1890) Ulshafer, P. R. *et al*: *Tetr. Letters*, **1961**, 363.
1891) Hofmann, A. *et al*: *Helv. Chim. Acta*, **36**, 1143 (1953), **40**, 508 (1957).

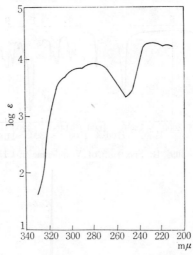

Fig. 311 UV Spectrum of Sarpagine in EtOH
(Hofmann *et al.*)

Fragmentation :

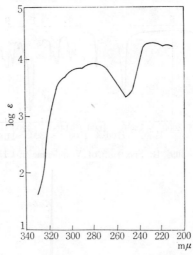

(CV)

(CIX) 310

253

254

198

199

O-**Methylsarpagine** (Lochnerine=C–Alkaloid T) (CVI)[1880) 1892)

《Occurrence》 *Vinca* (*Lochnera*) *rosea* var. *alba*

mp. 202~202.5°, [α]_D +56° (pyridine), $C_{20}H_{24}ON_2$

UV See Fig. 312[1444)

IR See Fig. 312a

1892) Karrer, P. *et al*: *Helv. Chim. Acta*, **40**, 705 (1957).

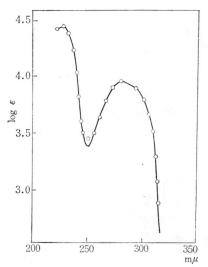

Fig. 312 UV Spectrum of C–Alkaloid T in EtOH

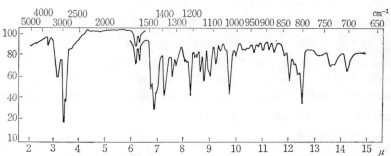

Fig. 3i2a IR Spectrum of C–Alkaloid T in nujol (Karrer *et al.*)[1892]

(**x**) **Alkaloid C** (CVII)[1893]

　　≪Occurrence≫　*Alstonia muelleriana*

　　mp. 168~169°, $[\alpha]_D$ +200° (EtOH), $C_{20}H_{22}O_3N_2$, X–ray analysis (CVII)

(**y**) **Echitamine**　See the item (10) Aspidosperma and Pleiocarpa Alkaloids, *Type (IX)*.

(**z**) **Biosynthesis**　Rauwolfia alkaloids[1894], reserpine[1895], ajmaline[1896].

(8)　Rutaceous Indole Alkaloids

　　≪Occurrence≫　*Evodia rutaecarpa* HOOKER fil. et THOMSON (ゴシュユ)[1897]

(CX)　Rutaecarpine	(CXI)　Evodiamine[1897) 1899]
	(CXI–a)　Rhetsine[1898]

1893) Nordman, C. E. *et al*: *J.A.C.S.*, **85**, 353 (1963).　　1894) Leete, E: *Tetrahedron*, **14**, 35 (1961).

1895) Peets, E. A. *et al*: *J.A.P.A.*, **47**, 280 (1958).

1896) Leete, E. *et al*: *J.A.C.S.*, **82**, 6338 (1960), **84**, 1068 (1962).

1897) Nakasato, T. *et al*: *Y. Z.*, **82**, 619 (1962).

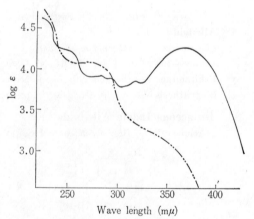

(CXI–a) ←[NaBH₄ 1898)]— ... ↔ ... —[KMnO₄ 1897)]→ (CXI)

(CXII)   Hydroxyevodiamine (Rhetsinine)

(CXII) ⇄ [cryst. fr. C₆H₆ / cryst. fr. (CH₃)₂CO 1897)] ... ↔ ...

(CXIII)   Dehydroevodiamine[1897)]

(a) Rutaecarpine (CX)[1528)]

≪Other occurrence≫   *Hortia arborea* ENGL.[1900)]

mp. 256°, [α]$_D$ ±0° (CHCl₃), C₁₈H₁₃ON₃

UV λ$_{max}^{EtOH}$ mμ (log ε): 235 sh, 278 (3.83), 290 (3.88), 332 (4.49), 345 (4.54), 364 (4.44);

IR λ$_{max}^{nujol}$ μ: 3.02 (>NH), 6.12 (*tert.* amide), 6.20 (benzenoid)

(b) Evodiamine[1897)1899)] **and Rhetsine**[1898)]   (CXI)

≪Other occurrence≫   *Zanthoxylum rhesta* DC (DL–form)

D–Evodiamine[1897)] from *E. rutaecarpa*,   mp. 270~272°, [α]$_D$ +440° (CHCl₃), C₁₉H₁₇ON₃

UV λ$_{max}^{EtOH}$ mμ (log ε): 268 (4.08), See Fig. 313

IR λ$_{max}^{nujol}$ μ: 3.10 (NH), 6.15 (C=O)

DL–Evodiamine[1899)] from *Z. rhesta*

mp. 270~271° (decomp.), C₁₉H₁₇ON₃

UV λ$_{max}$ mμ (log ε): 266 (4.08), 312 (4.08),
    324 (3.29)

(c) Hydroxyevodiamine (Rhetsinine) (CXII)

≪Other occurrence≫

Zanthoxylum rhesta DC[1898)1899)]

mp. 188~190°[1897)], 192° (decomp.)[1898)],
    206~207°[1899)], [α]$_D$ ±0°[1898)], C₁₉H₁₇O₂N₃

UV λ$_{mat}^{EtOH}$ mμ (log ε)[1897)]: 285 (3.92), 292
    (3.87), 315(3.74), 390 (3.90)

IR λ$_{max}^{nujol}$ cm⁻¹[1898)]: 3448 (O–H), 3330(N–H),
    1695 (amide)

Fig. 313   UV Spectra
—··—Evodiamine and
———Dehydroevodiamine
in EtOH (Nakasato *et al.*)[1897)]

1898)   Chatterjee, A. *et al*: *Tetrahedron*, **7**, 257 (1959).
1899)   Gopinath, K. W. *et al*: *ibid.*, **8**, 293 (1960).
1900)   Pachter, I. J *et al*: *J.A.C.S.*, **82**, 5187 (1960), **83**, 635 (1961).

(d) Dehydroevodiamine (CXIII)[1897]

mp. 189~190°, $[\alpha]_D \pm 0°$ (CHCl$_3$), C$_{19}$H$_{15}$ON$_3$

UV λ_{max}^{EtOH} (log ε): 282 (3.90), 292 (3.88), 315 (3.83), 365 (4.26), see Fig. 313;

$\lambda_{max}^{0.015N-KOH-EtOH}$ mμ (log ε): 307 (4.19), 380~395 sh, IR λ_{max}^{nujol} μ: 5.96.

(e) Hortiamine, 6–Methoxyrhetsinine, and Hortiacine[1900]

≪Occurrence≫ *Hortia arborea* ENGL., *H. braziliana* VEL. (*Rutaceae*).

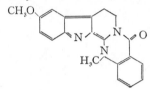

(CXIV) Hortiamine

(CXV) Isohortiamine

(CXVI) 6–Methoxyrhetsinine

(CXVII) Hortiacine

Hortiamine (CXIV)

Red orange needles, mp. 209° (decomp.), C$_{20}$H$_{17}$O$_2$N$_3$

UV λ_{max}^{CH3CN} mμ (ε): 411 (47300) (see Fig.314)[1900]

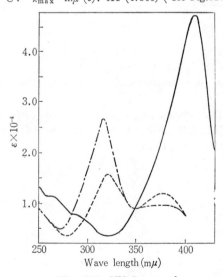

Fig. 314 UV Spectra of

Hortianine ———————— in acetonitril

------------ in EtOH

6–Methoxyrhetsinine —·—·— in acetonitril

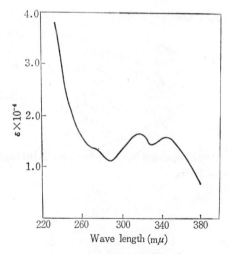

Fig. 315 UV Spectrum of
Isohortiamine in EtOH
(Pachter *et al.*)[1900]

Isohortiamine (CXV)

Pale yellow needles, mp. 340~343° (decomp.), C$_{20}$H$_{17}$O$_2$N$_3$

UV λ_{max}^{EtOH} mμ (ε): 227 (42300), 318 (16900), 346 (16000) (see Fig. 315)

6–Methoxyrhetsinine (CXVI)

$$(\text{CXVI}) \xrightarrow{\text{heating}} (\text{CXIV}) \quad \text{Hortiamine}$$

Hortiamine hydrate

Yellow prisms, mp. turns to red 209° (decomp.) (same to hortiamine), $C_{20}H_{19}O_3N_3$

UV λ_{max}^{CH3CN} mμ (ε): 318 (26000), 375 (9700), see Fig. 314

IR λ_{max}^{nujol} μ: 5.98, 6.05 (C=O), 2.95, 3.10 (N–H)

Hortiacine (CXVII)

Pale yellow needles, mp. 252~252.5°, $C_{19}H_{15}O_2N_3$

(9) Eserine and Related Alkaloids

(a) **Eserine** (Physostigmine) (CXVIII)

≪Occurrence≫ *Physostigma venenosum* BALF. (Calabar bean)

mp. 105~106°, $[\alpha]_D$ −75.8°, $C_{15}H_{21}O_3N_2$

NMR See the item (6) Calycanthine, Fig. 234[1609]

in CCl$_4$, 60Mc, external C_6H_6 reference, τ[1772a]

(CXVIII)

(b) **Sporidesmine** (CXIX)[1901]

(CXIX)

≪Occurrence≫ *Pithomyces chartarum* (mould)

$C_{18}H_{20}O_6N_3S_2Cl$

IR ν_{max}^{CHCl3} cm^{-1}: no absorption in 3600~3200, 1701, 1670

(C=O)

NMR τ: 2.92 (arom. proton), 4.70 (C–H), 5.16 (OH), 5.42

(CH–OH), 6.13 (OCH$_3$), 6.18 (OCH$_3$), 6.60 (OH), 6.70

(N–CH$_3$), 6.93 (N–CH$_3$), 7.97 (C–CH$_3$)

(10) Aspidosperma and Pleiocarpa Alkaloids

Classification[1902] [1902a]

[*Type I*]

(CXX)

(CXXI) Aspidospermine[1903]~[1905]

1901) Hodges, R *et al*: *Chem. & Ind.*, **1963**, 42.

1902) Djerassi, C. *et al*: *Helv. Chim. Acta*, **46**, 742 (1963).　　1902a) Schmid, H. *et al*: *Ann.*, **668**, 97 (1963).

1903) Biemann, K. *et al*: *J.A.C.S.*, **85**, 631 (1963); *Tetr. Letters*, **1961**, 485.

1904) Smith, G. F. *et al*: *Soc.*, **1960**, 1463.　　1905) Djerassi, C. *et al*: *Tetrahedron*, **16**, 212 (1962).

Alkaloids [Type I]	(CXX)	R_1	R_2	R_3	R_4	R_5	R_6
Aspidospermine	(CXXI)	H	OMe	Ac	H	H	Me
(−)–Pyrifolidine	(CXXII)	OMe	OMe	Ac	H	H	Me
Spegazzinine	(CXXIII)	H	OH	Ac	α–OH	H	Me
Spegazzinidine	(CXXIV)	OH	OH	Ac	OH	H	Me
Vindoline	(CXXV)	H	OMe	Me	OH COOMe	OAc	=CH$_2$
Limaspermine	(CXXVI)	H	OH	COEt	H	H	CH$_2$OH
Demethoxypalosine	(CXXVII)	H	H	COEt	H	H	Me
Aspidospermidine	(CXXVIII)	H	H	H	H	H	Me
Aspidolimine	(CXXVIII–a)	OMe	OH	COEt	H	H	Me
Aspidocarpine	(CXXVIII–b)	OMe	H	Ac	H	H	Me
3′–Methoxylimaspermine	(CXXVIII–c)	OMe	OH	COEt	H	H	CH$_2$OH
Limapodine	(CXXVIII–d)	H	OH	Ac	H	H	CH$_2$OH
3′–Methoxylimapodine	(CXXVIII–e)	OMe	OH	Ac	H	H	CH$_2$OH

[*Type II*]

(CXXIX) Quebrachamine

[*Type III*]

(CXXX)

Alkaloids [Type III]	(CXXX)	R_1	R_2	R_3	R_4
Aspidolimidine	(CXXXI)	H	OMe	OH	Me
Aspidoalbine	(CXXXII)	OMe	OMe	OH	Et
Haplocine	(CXXXII–a)	H	H	OH	Et
Haplocidine	(CXXXII–b)	H	H	OH	Me

[Type *IV*]

(CXXXIII)

Alkaloids [*Type IV*]	(CXXXIII)	R
Vincadifformine	(CXXXIV)	H$_2$
Tabersonine	(CXXXV)	H$_2$, $\Delta_{6,7}$
Minovincine	(CXXXVI)	O
Minovincinine	(CXXXVII)	H OH

1906) Conroy, H. *et al*: *J.A.C.S.*, **80**, 5178 (1958).

[*Type V*]

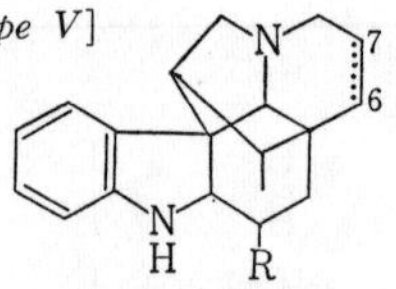

(CXXXVIII)

Alkaloids	[*Type V*]	(CXXXVIII)	R
Tuboxenine		(CXXXIX)	H
Vindolinine		(CXL)	COOMe, $\Delta_{6,7}$

[*Type VI*]

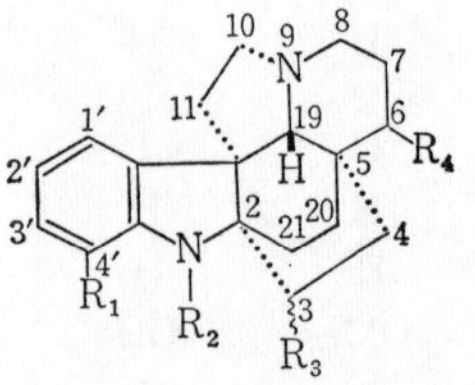

(CXLI)

Alkaloids	[*Type VI*]	(CXLI)	R_1	R_2	R_3	R_4
Pyrifoline		(CXLII)	OMe	Ac	H	OMe
Refractidine		(CXLIII)	H	CHO	H	OMe
Refractine		(CXLIV)	OMe	CHO	3α–CO_2Me	H
Aspidofractine		(CXLV)	H	CHO	3α–CO_2Me	H
Pleiocarpine		(CXLVI)	H	COOMe	3α–CO_2Me	H
Pleiocarpinine		(CXLVII)	H	Me	3α–CO_2Me	H
Kopsinine		(CXLVIII)	H	H	3α–CO_2Me	H

[*Type VII*]

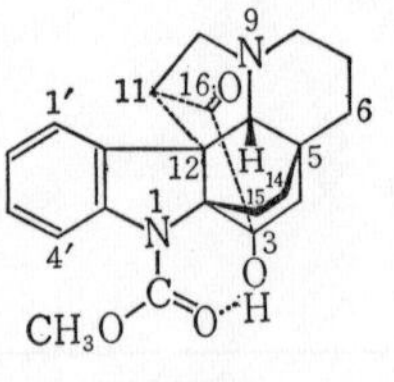

(CXLIX)

(CXLIX)   Kopsine

(CL)   Fruticosamine an isomer of (CXLIX)

(CLI)   Fruticosine an isomer of (CXLIX)

[*Type VIII*]

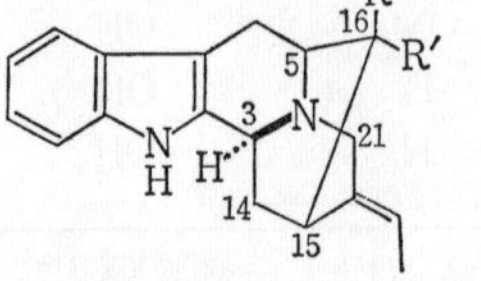

(CLII)

Alkaloids	[*Type VIII*]	(CLII)	R	R'
Polyneuridine		(CLIII)	CH_2OH	CO_2Me
Akuammidine		(CLIV)	CO_2Me	CH_2Me

[*Type IX*]

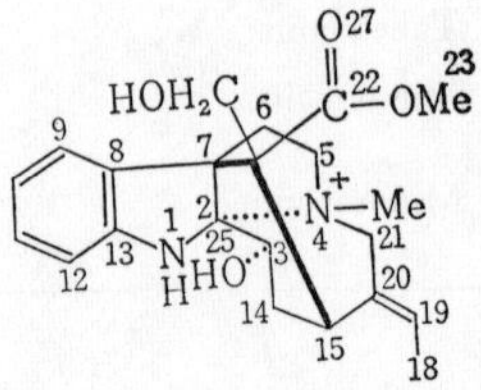

(CLV)   Echitamine

— 618 —

[*Type X*]

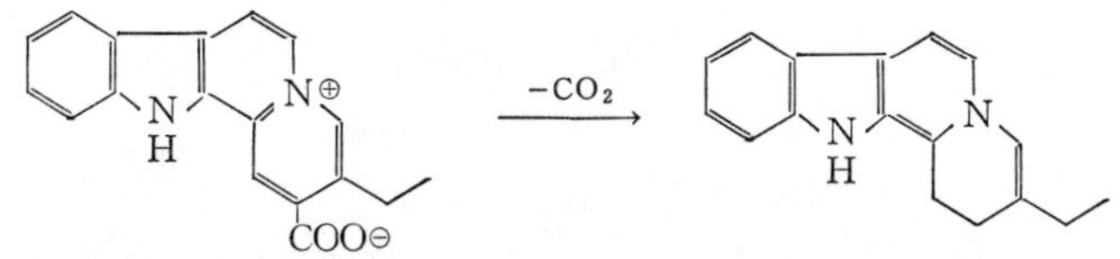

$$-CO_2 \longrightarrow$$

(CLVI) Flavocarpine

(CLVII) Flavopereirine

[*Type XI*]

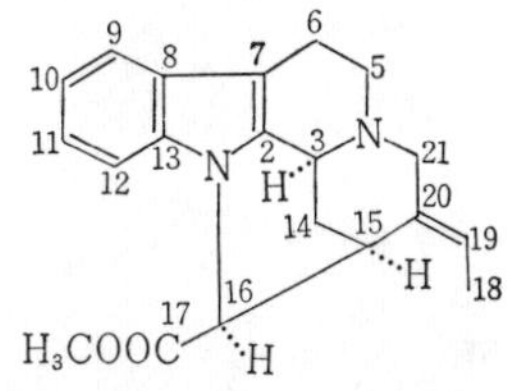

[*Type XII*]

(CLVIII) Pleiocarpamine

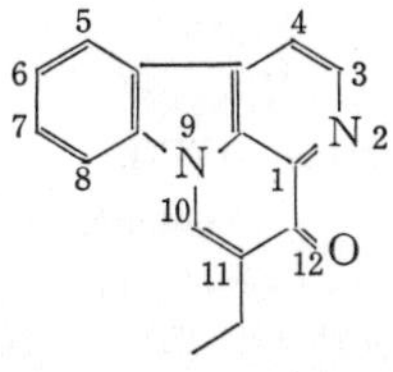

(CLIX) Tuboflavine

[*Type XIII*]

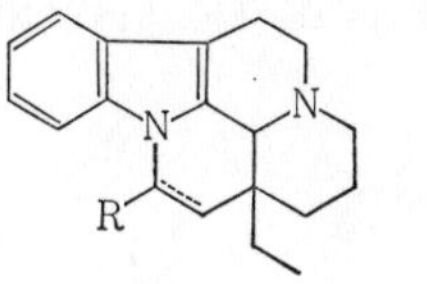

R=O   (CLX)   Eburnamonine
R=OH   (CLXI)   Eburnamine
R=H, $\Delta_{10,11}$ (CLXII) Eburnamenine

[*Type XIV*]

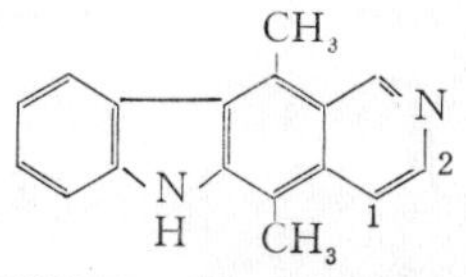

(CLXIII)   Ellipticine
(CLXIV)   1, 2–Dihydro-
          ellipticine

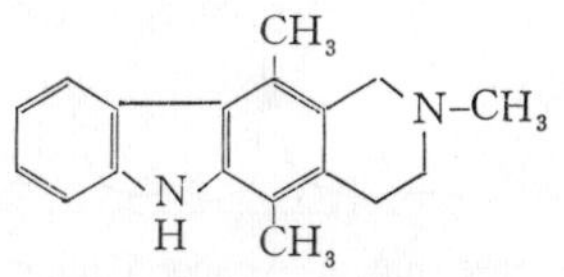

(CLXV)   *N*–Methyltetra-
        hydroellipticine

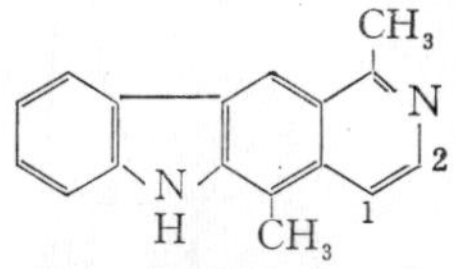

(CLXVI)   Olivacine
(CLXVII)   1, 2–Dihydro-
          olivacine

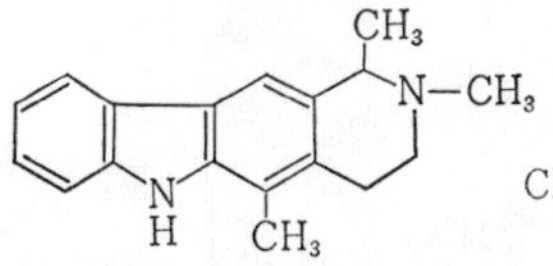

(CXVII–a)   Guatambuine

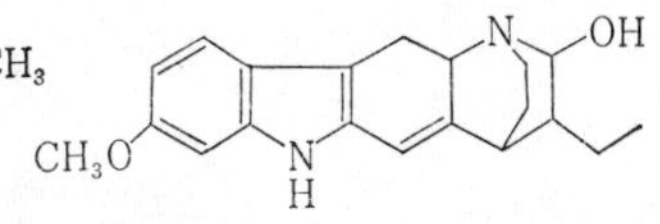

(CLXVIII)   Elliptinine

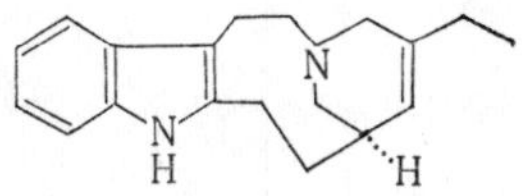

(CLXIX)   Cleavamine

[*Type XV*]

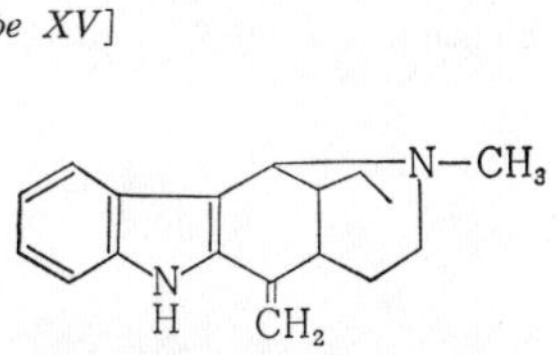

(CLXX)   Uleine

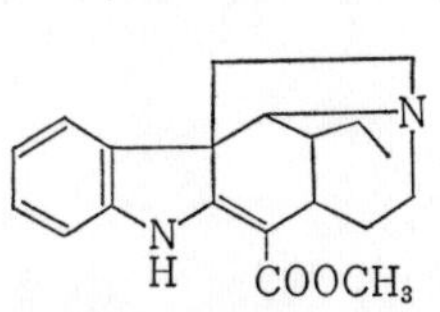

(CLXXI)   Tubotaiwine

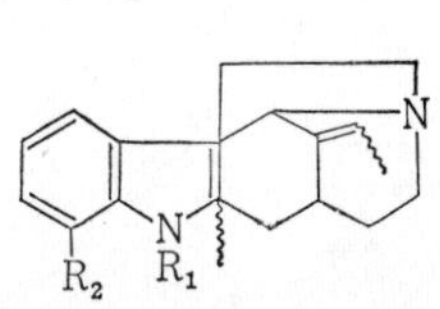

$R_1=R_2=H$: (CLXXII)
     Aspidospermatidine
$R_1=COCH_3$, $R_2=OCH_3$: (CLXXIII)
     Aspidospermatine

[*Type XVI*]

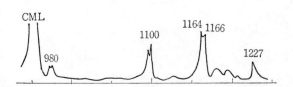

R₁=R₂=H: (CLXXIV)   Quebrachidine
R₁=CH₃, R₂=COCH₃: (CLXXV)   Vincamedine

(i)   *Type 1* (CXX)

(a) **Aspidospermine** (CXXI)

≪Occurrence≫   *Aspidosperma quebracho blanco, Vallesia glabra*

mp. 208∼209°, $C_{22}H_{30}O_2N_2$

UV λ_{max}^{EtOH} mμ (log ε): 218 (4.53), 255 (4.07), 290 (3.45).

IR $\nu_{max}^{CHCl_3}$ cm⁻¹: 2915s, 2780m, 1630s, 1592m, 1486m, 1460s, 1385s, 1335m, 1319m, 1290m, 1269m, 1171w, 1117w, 1103w, 1073w, 909w[1906]

NMR[1906]   in CHCl₃, 40 Mc, toluene; aromatic resonance as 1000 cps see Fig. 316: 1097 $\left(\begin{smallmatrix}C\\C\end{smallmatrix}\!\!>\!\!CH\text{--}N\right)$, 1100 (OCH₃), 1164

   (N–CH₃), 1167 (COCH₃), 1227 (CH–CH₃)

Mass[1903]   See Fig. 317

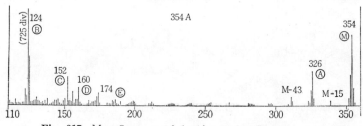

Fig. 316   NMR Spectrum of Aspidospermine (Conroy *et al.*)[1906]

Fig. 317   Mass Spectrum of Aspidospermine (Biemann *et al.*)[1903]

Fragmentation:

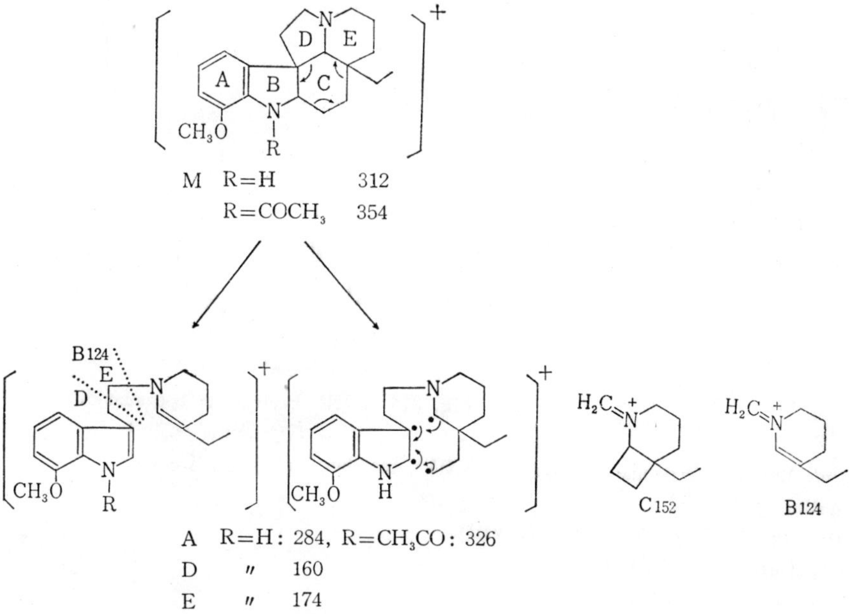

M R=H   312
   R=COCH₃   354

A R=H: 284, R=CH₃CO: 326
D   〃   160
E   〃   174

Total synthesis[1917]

(b) **(−)-Pyrifolidine** (CXXII)[1903)~1905)]

  ≪Occurrence≫   *Aspidosperma quebracho blanco, A. pyrifolium*

  mp. 148~150°, $[\alpha]_D$ +93° (CHCl₃), $C_{23}H_{32}O_2N_2$

  UV λ_{max}^{EtOH} mμ (log ε): 223 (4.45), 252 (4.0), 286 (3.35)

  IR λ_{max}^{KBr} μ: 3.59m, 6.02s, 6.24m, 6.69s.

(c) **Spegazzinine** (CXXIII)[1907) 1908)]

  ≪Occurrence≫   *Aspidosperma chakensis*

  mp. 104.5~106°, $[\alpha]_D$ +175.6° (CHCl₃), pKa 6.0, 13.0, $C_{21}H_{28}O_3N_2$

  UV shows *N*-acyldihydroindole

  IR λ_{max} μ: 6.14 (amide)

(d) **Spegazzinidine** (CXXIV)[1907)]

  ≪Occurrence≫   *Aspidosperma chakensis*

  mp. 237~238°, $[\alpha]_D$ +186° (CHCl₃), pKa 2.9, 6.4, 10.7 (D.F.A.), $C_{21}H_{28}O_4N_2$

  UV λ_{max}^{EtOH} mμ (log ε): 225 (4.30), 260 (3.89), 295inf.; $\lambda_{max}^{EtOH-NaOH}$ mμ (log ε): 216 (4.29), 237 (4.24),
    304 (3.44)

  IR $\lambda_{max}^{CHCl_3}$ μ: 2.85, 6.13, 6.33

  NMR in CDCl₃, 60Mc, TMS, δ (ppm), see Fig. 318

1907) Djerassi, C. *et al*: *J.A.C.S.*, **84**, 3480 (1962).
1908) Djerassi, C. *et al*: *J. Org. Chem.*, **21**, 979 (1956).

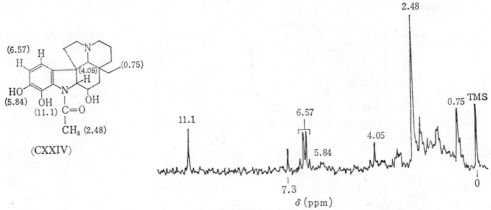

Fig. 318 NMR Spectrum of Spegazzinidine
(Djerassi *et al.*)[1907]

Mass   m/e 124 (aspidospermine skeleton in (CXXIV), see the item (i), (a) (CXXI)

(e) Vindoline (CXXV)

See the item (13) Vinca Alkaloids (CXCVIII)

(f) Limaspermine (CXXVI)[1909]

≪Occurrence≫   *Aspidosperma limae*

mp. 175~175.5°, $[\alpha]_D$ +108° (CHCl$_3$), C$_{20}$H$_{30}$O$_3$N$_2$

UV λ_{max}^{EtOH} mμ (log ε): 221 (4.40), 261 (3.92), 292 (3.55), 258inf. (3.80); λ_{min}: 283 (3.22)

IR $\nu_{max}^{CHCl_3}$ cm^{-1}: no OH absptn., 1764 (arylacetate), 1734 (alkylacetate), 1664 (non chelated amide), 1634 (arom. band).

NMR   in CDCl$_3$ 60Mc, TMS, ppm: see Fig. 319.

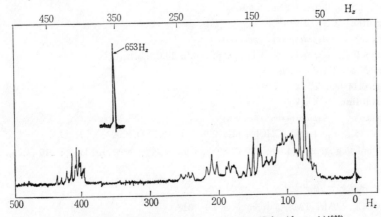

Fig. 319 NMR Spectrum of Limaspermine (Schmid *et al.*)[1909]

653s (1) (chelated O–H), 415m (3) (arom. H), 247q (1) (C$_2$-H), 211t, *J*=7 (2) (–CH$_2$-$\overset{15}{\text{CH}_3}$), ca. 180 (2) (charact. absptn. for aspidospermine type), 154q, *J*=7.5 (—N–CO–CH$_2$–CH$_3$)

Mass: m=370, m/e a=342, b=324, c=140, d=146

1909) Schmid, H *et al*: *Helv. Chim. Acta*, **45**, 2260 (1962).

1909a) Biemann, K *et al*: *Tetr. Letters*, **1961**, 485.

Fragmentation :

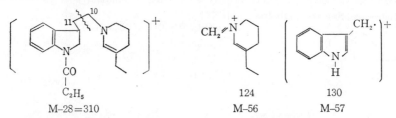

M=370

a

b

c

d

(g) Demethoxypalosine (CXXVII)[1911)

《Occurrence》 *Aspidosperma limae* WOODSON

mp. 117~120°, $[\alpha]_D$ −20° (CHCl$_3$), C$_{22}$H$_{30}$ON$_2$

UV λ_{max}^{EtOH} mμ (log ε): 253 (4.15), 280 (3.64), 289 (3.59); λ_{min}: 226 (3.67), 275 (3.63), 287 (3.57); λ_{max} μ: 6.04 (amide).

NMR δ: 8.14 (pair of doublets J=8, 2), 1.23t, 2.35q (J=7 resp.) (*N*-propionyl)

Mass M=338, M−28, 130, 124

M−28=310   124 M−56   130 M−57

(h) Aspidospermidine (CXXVIII)[1903) 1909a)

《Occurrence》 *Aspidosperma quebracho blanco*, mp. 110~112°, C$_{19}$H$_{26}$N$_2$

Mass M=282 (see the item (b) Aspidospermine, M=312, ΔM=30)

A 254, C 152, E 144, D 130, B 124

(i) Aspidolimine and Aspidocarpine [1902) 1910) 1911)

R=C$_2$H$_5$: (CXXVIII–a) Aspidolimine

R=CH$_3$: (CXXVIII–b) Aspidocarpine

Aspidolimine (CXXVIII–a)

《Occurrence》 *Aspidosperma limae*

mp. 150~151°, $[\alpha]_D$ +133° (CHCl$_3$), C$_{23}$H$_{32}$O$_3$N$_2$

UV λ_{max}^{EtOH} mμ (log ε): 228 (4.43), 264 (3.94); λ_{min}: 250 (3.80); $\lambda_{max}^{0.05N-KOH-EtOH}$ mμ (log ε): 224 (4.42), 307 (3.73)

IR ν_{max}^{nujol} cm^{-1}: 3597, 1631, 1603, 1580; ν_{max}^{CHCl3} cm^{-1}: 1634, 1605, 1582

NMR in CDCl$_3$, 60Mc, TMS, c/s: 659s (chelated OH), 399 AB-q, J=8 (2) (arom. H), 247q (C$_2$–H),

1910) Schmid, H. *et al*: *Helv. Chim. Acta*, **45**, 1283 (1962).

1911) Gilbert. B. *et al*: *Chem. & Ind.*, **1962**, 1949.

ca. 180 (charact. absptn. for aspidospermine type), 233s (3)(OCH$_3$), 154q, $J=7$ ($>$N–CO–CH$_2$–CH$_3$), 75t, $J=7$ ($>$N–CO–CH$_2$–CH$_3$), 37t, $J=6$ ($>$C–CH$_2$–CH$_3$)

Aspidocarpine (CXXVIII–b)

≪Occurrence≫   *Aspidosperma megalocarpa, A. limae*[1911]

mp. 167.5~168.5°, [α]$_D$ +140° (CHCl$_3$), C$_{22}$H$_{30}$O$_3$N$_2$

(j) 3′-Methoxylimaspermine (CXXVIII–c)[1902a]

≪Occurrence≫   *Aspidosperma limae*

mp. 174~175°, [α]$_D$ +118° (CHCl$_3$), C$_{23}$H$_{32}$O$_4$N$_2$

UV λ_{max}^{EtOH} mμ (log ε): 224 (4.42), 234 (4.38), 307 (3.73); λ_{min}: 230 (4.38), 284 (3.40)

IR $\nu_{max}^{CHCl_3}$ cm^{-1}: 3425 br. (OH) 1634 (chelated amide) 1603, 1582 (arom.)

NMR in CDCl$_3$, 60Mc, TMS, cps: 657 s (1) (chelated phenolic OH), 390~415 q, $J=8$ (2) (arom. H), 246 q (C$_2$–H), 232 s (OCH$_3$), 213 t, (2), $J=7$ (—$\overset{15}{C}$H$_2$—), ca. 185 (2) (aspidospermine type), 150 s (C$_{13}$–H), 154 q, $J=7.5$ (N–CO–CH$_2$–CH$_3$), 76 t, $J=7.5$ (N–CO–CH$_2$–CH$_3$)

(k) Limapodine (CXXVIII–d)[1902]

≪Occurrence≫   *Aspidosperma limae*,   mp. 177~178°, [α]$_D$ +110° (CHCl$_3$), C$_{21}$H$_{28}$O$_3$N$_2$

IR $\nu_{max}^{CHCl_3}$ cm^{-1}: 1629

(l) 3′-Methoxylimapodine (CXXVIII–e)[1902a]

≪Occurrence≫   *Aspidosperma limae*,   Amorphous, [α]$_D$ +584° (CHCl$_3$), C$_{20}$H$_{24}$O$_2$N$_2$

NMR (cps): Almost identical to that of 3′–methoxylimaspermine (CXXVIII–c) except the signal at 139 s (N–CO–CH$_3$) instead of 76 t (N–CO–CH$_2$–CH$_3$) of the latter

(m) Cylindrocarpidine and **Cylindrocarpine**[1905]

R=COCH$_3$, R′=CH$_3$ (CXXVIII–f)
Cylindrocarpidine
R=C$_6$H$_5$CH=CH–CO–, R′=CH$_3$
(CXXVIII–g) Cylindrocarpine

≪Occurrence≫   *Aspidosperma* spp.

Cylindrocarpidine (CXXVIII–f)

mp. 118~118.5°, [α]$_D$ −122° (CHCl$_3$).

UV ≒ aspidospermine (CXXI)   IR $\lambda_{max}^{CHCl_3}$ μ: 5.81, 6.13, 6.23, 6.29

Cylindrocapine (CXXVIII–g)

mp. 168~169°, [α]$_D$ −181° (CHCl$_3$)

UV λ_{max}^{EtOH} mμ (log ε): 216 (4.53), 248 (4.05);   IR $\lambda_{max}^{CHCl_3}$ μ: 5.79, 6.06, 6.23

(n) Tabersonine   See the item (14) Iboga Alkaloids, (r), (CCXLVII)

(o) Alkaloids of *Aspidosperma obscurinervium*[1912]

R	R′	$\Delta_{6,7}$	
OMe	Et	−	(CXXVIII–h) Dihydroobscurinervine
OMe	Et	+	(CXXVIII–i) Obscurinervine
OMe	Me	−	(CXXVIII–j) Dihydroobscurinervidine
OMe	Me	+	(CXXVIII–k) Obscurinervidine

1912) Djerassi, C. *et al*: *J.A.C.S.*, **86**, 2451 (1964).

Dihydroobscurinervine (CXXVIII–h)

  mp. 184°~185°, $[\alpha]_D$ −61° (CHCl$_3$), C$_{25}$H$_{32}$O$_5$N$_2$

UV λ_{max}^{EtOH} mμ (log ε): 220 (4.46), 256 (3.81), 313 (3.43); λ_{min}: 247 (3.76), 283 (2.91)

IR $\nu_{max}^{CHCl_3}$ cm^{-1}: 2780 m, 1750 vs, 1610 m

NMR in CDCl$_3$, 60 Mc, δ: 6.32 s (1) (arom. H), 4.52 q, J=5.2 (1) (CH$_2$CHOCO), 4.28 d−4.12 d (2)
   (CH$_2$ of cyclic ether), 3.87 s, 3.80 s (6) (2 arom. –OCH$_3$), 2.62 d, 1.78 d J=19 (2)
   (rigid C–CH$_2$–CO), 0.95 t, J=7 (3) (C–CH$_2$–CH$_3$)

  Mass m/e 440 (M$^+$), 425, 411, 244, 206 ((M−CO)$^{2+}$)

Obscurinervine (CXXVIII–i)

  mp. 203~204° (decomp.), $[\alpha]_D$ −54° (CHCl$_3$), C$_{25}$H$_{30}$O$_5$N$_2$

UV λ_{max}^{EtOH} mμ (log ε): 220 (4.49), 253 (3.81), 312 (3.43); λ_{min}^{EtOH}: 247 (3.78), 282 (2.99)

IR $\nu_{max}^{CHCl_3}$ cm^{-1}: 2780 m, 1750 vs, 1605 m

NMR δ: 6.32 s, 5.71 m (2) (vinyl protons), 4.52 q, 4.28 d~4.13 d, 3.87 s, 3.80 s, 2.57 d, 2.02 d,
   J=18 (2) (rigid C–CH$_2$–CO), 0.97 t (see (CXXIII–h))

  Mass m/e 438 (M$^+$), 423, 409, 244, 205

Dihydroobscurinervidine (CXXVIII–j)

  mp. 189~190° (decomp.), $[\alpha]_D$ −44° (CHCl$_3$), C$_{24}$H$_{30}$O$_5$N$_2$

UV λ_{max}^{EtOH} mμ (log ε): 219 (4.52), 255 (3.80), 312 (3.43); λ_{min}: 247 (3.77)

IR $\nu_{max}^{CHCl_3}$ cm^{-1}: 2780 m, 1750 vs, 1610 m

NMR δ: 6.35 s (1) (arom.), 4.53 q, J=5.2 (1) (CH$_2$–CHOCO), 4.18 m (2) (CH$_2$ of cyclic ether),
   3.88 s, 3.82 s (6) (2–OCH$_3$), 2.62 d, 1.78 d, J=19 (2) (rigid C–CH$_2$–CO), 1.07 d, J=7 (3)
   (N–CH–CH$_3$) (see (CXXVIII–h))

  Mass m/e 426 (M$^+$), 411, 246, 244, 199

Obscurinervidine (CXXVIII–k)

  mp. 206~207° (decomp.), $[\alpha]_D$ −39° (CHCl$_3$), C$_{24}$H$_{28}$O$_5$N$_2$

UV λ_{max}^{EtOH} mμ (log ε): 219 (4.50), 253 (3.77), 310 (3.42); λ_{min}: 247 (3.74), 291 (2.91)

IR $\nu_{max}^{CHCl_3}$ cm^{-1}: 2780 m, 1760 vs, 1605 m

NMR δ, see Fig. 320: 6.35 s (1) (arom. H), 5.73 m (2) (vinyl protons), 4.55 br. (1) (CH$_2$CHOCO),
   4.18 m (2) (CH$_2$ of cyclic ether), 3.88 s, 3.82 s (6) (arom.–OCH$_3$), 2.57 d, 2.02 d, J=18 (2)
   (rigid C–CH$_2$–CO), 1.07 d, J=6 (3) (N–CH–CH$_3$)

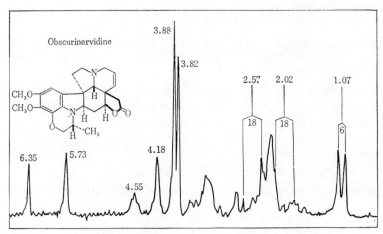

Fig. 320 NMR Spectrum of Obscurinervidine (Djerassi *et al.*)[1912]

Mass m/e 424 (M⁺), 409 (M–CH₃), 260 (g), 246 (c), 244 (a), 212 (M²⁺), 198 (M–CO)²⁺

(a) 244   (c) 246   (g) 260

(ii) *Type (II)*

(a) **Quebrachamine** (CXXIX)[1913~1917]

≪Occurrence≫   *Aspidosperma quebracho blanco, A. polyneuron* M. Arg.

mp. 146~147°, [α]_D −108° (CHCl₃), C₁₉H₂₆N₂

UV   λ_max^EtOH mμ (ε): 230 (35200), 287 (7170), 293 (6860); λ_min: 256 (2200)

IR   See Fig. 321
NMR   in CDCl₃, 60 Mc, TMS, τ: See Fig. 322

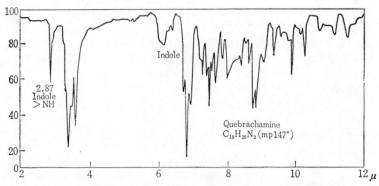

Fig. 321 IR Spectrum Quebrachamine in CHCl₃ (Witkop)[1913]

1913) Witkop, B: *J.A.C.S.*, **79**, 3193 (1957).   1914) Witkop. B *et al*: *ibid.*, **82**, 2184 (1960).
1915) Biemann, K. *et al*: *ibid.*, **84**, 4578 (1962); *Tetr. Letters*, **1961**, 299.
1916) Schmutz, J. *et al*: *Helv. Chim. Acta*, **42**, 874 (1959).
1917) Stork, G. *et al*: *J.A.C.S.*, **85**, 2872 (1963).

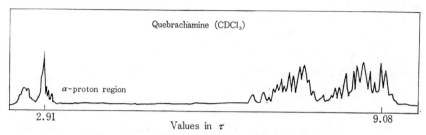

Fig. 322 NMR Spectrum of Quebrachamine (Witkop *et al.*)[1913]

Mass: See Fig. 323

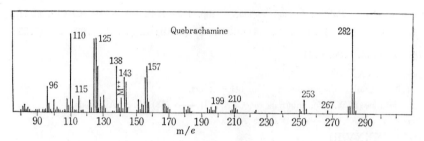

Fig. 323 Mass Spectrum of Quebrachamine (Biemann *et al.*)[1915]

Fragmentation:

Synthesis[1917]

(iii) *Type-III* (CXXX)

(a) **Aspidolimidine** (CXXXI)[1911]

≪Occurrnece≫ *Aspidosperma limae,* mp. 196~199°, $[\alpha]_D$ +239° (CHCl$_3$), C$_{22}$H$_{28}$O$_4$N$_2$

NMR δ: 10.78 s (H–bonded phenolic OH), 3.88 s (arom. –OCH$_3$), 2.23 s (N–CO–CH$_3$), 6.73 d~7.09 d,

J=8 (*ortho* arom. protons), 2.5~3.2 m$\left(\diagdown CH_2 \diagup N \diagdown CH_2 \diagup \right)$

(b)   Aspidoalbine (CXXXII)[1918]

≪Occurrence≫   *Aspidosperma album* R. Bent.

mp. 170~172°, $[\alpha]_D$+159° (MeOH), $C_{24}H_{32}O_5N$

UV λ_{max}^{EtOH} mμ (log ε): 227 (4.16), 267 (3.88)

IR λ_{max}^{EtOH} μ: 6.17, 6.31

(c)   Haplocine (CXXXII–a)[1919]

≪Occurrence≫   *Haplophyton cimicidum*,   mp. 186~187°, $[\alpha]_D$+196° (CHCl₃), $C_{22}H_{28}O_3N_2$

UV λ_{max}^{EtOH} mμ (log ε): 219.5 (4.44), 258 (3.95), 292 (3.57); λ_{min}: 241 (3.74), 282 (3.54)

IR λ_{max}^{KBr} μ: 6.13, 6.24, 6.37

(d)   Haplocidine (CXXXII–b)[1919]

≪Occurrence≫   *Haplophyton cimicidum*,   mp. 183~184°, $[\alpha]_D$+231° (CHCl₃), $C_{21}H_{26}O_3N_2$

UV λ_{max}^{EtOH} mμ (log ε): 218.5 (4.40), 258 (3.91), 291 (3.60); λ_{min}: 241 (3.07), 282 (3.55)

IR λ_{max}^{KBr} μ: 6.16, 6.26, 6.41

(iv)   *Type-IV* (CXXXIII)

The following alkaloids are described in other items.

Vincadifformine (CXXXIV), minovincine (CXXXVI) ⟶ (13) Vinca Alkaloids.

Tabersonine (CXXXV) ⟶ (14) Iboga Alkaloids.

(v)   *Type-V* (CXXXVIII)

(a)   Tuboxenine (CXXXIX)[1920]

≪Occurrence≫   *Pleiocarpa tubiciua*,   mp. 139~140°, $[\alpha]_D$+5.4 (CHCl₃), $C_{19}H_{24}N_2$

UV λ_{max}^{EtOH} mμ (log ε): 206 (4.39), 244 (3.81), 295 (3.44),; λ_{min}: 225 (3.58), 270 (2.96)

IR ν_{max}^{KBr} cm⁻¹: 3180 (>NH), 1605 (indoline), 739 (*ortho* disubst. benzene)

NMR   in CDCl₃, 100 Mc, TMS, ppm=0: 6.5~7.5 m (4 arom. protons), ~3.9 (N–H), 3.2 m (5)

$\left(\begin{smallmatrix} H_2 & H_2 \\ C-N-C \\ C \\ H \end{smallmatrix}\right)$, 1.1~2.9 (11) (4–CH₂–, 3–CH<), 0.85 d, J=7 (3) (>CH–CH₃), 1.7 (>C–CH–C< CH₃)

Mass: M=280 (a), 171 (b), 156 (d), 136 (e), 124 (f), 123 (g), 122 (h), 110 (i)

M=280 (a)   171 (b)   156 (d)   136 (e)

124 (f)   123 (g)   122 (h)   110 (i)

1918) Djerassi, C. *et al*: *Tetr. Letters*, **1962**, 1001.
1919) Cava, M.P. *et al*: *Chem. & Ind.*, **1963**, 1242.
1920) Schmid, H. *et al*: *Helv. Chim. Acta*, **47**, 358 (1964).

(b) **Vindolinine** (CXL) See the item (13) Vinca Alkaloids

(vi) *Type-VI* (CXLI)

(a) **Pyrifoline** (CXLII)[1921]

    ≪Occurrence≫ *Apsidosperma pyrifolium* MART., $C_{22}H_{30}O_3N_2$

    NMR in $CDCl_3$, 60 Mc, TMS, ppm: 3.81 (arom.–OCH_3), 2.12 (N–$COCH_3$)

    Mass: M–28, 173, 174 (d), 186, 187, 188 (e), 109(a), 131 (b), 154 (c)

109 (a)   131 (b)

M–28

(b) **Refractidine** (CXLIII)[1921]

    ≪Occurrence≫   *Aspidosperma pyrifolium* MART.

    mp. 158~160°, $[\alpha]_D$ −140° ($CHCl_3$), pKa′ 6.5, $C_{21}H_{26}O_2N_2$

UV λ_{max}^{EtOH} mμ (log ε): 208 (4.36), 253 (4.16), 278 (3.72), 288 (3.67) (unsat. *N*–acyldihydroindole)

NMR N–CHO, no signals corresponding to olefinic protons, arom.–OCH_3, C–CH_3 or C_2–H

    Mass See the item (e) Pleiocarpine

(c) **Refractine** (CXLIV)[1922]

    ≪Occurrence≫   *Aspidosperma refractum*

    mp. 157.5~159° & 191~192°, $[\alpha]_D$ −23° ($CHCl_3$), $C_{23}H_{28}O_4N_2$

    NMR in $CDCl_3$, 60 Mc, TMS, ppm: arom. protons, 3.85 (arom.–OCH_3), 3.73 ($COOCH_3$), 9.47

          (N–CHO), no signals corresponding to olefinic protons, C–Me or C_2–H

    Mass[1902] M–28, 124, 109, see the item (e) Pleiocarpine

(d) **Aspidofractine** (CXLV)[1922]

    ≪Occurrence≫   *Aspidosperma refractum*,   mp. 193~193.5°, $[\alpha]_D$−142° ($CHCl_3$), $C_{22}H_{26}O_3N_2$

    Mass See the item (e) Pleiocarpine

(e) **Pleiocarpine** (CXLVI)[1923]

    ≪Occurrence≫   *Pleiocarpa munica* BENTH., *Hunteria eburnea* PICHON.

    mp. 141~142°, $[\alpha]_D$ −145° ($CHCl_3$), pKMCS 6.19, $C_{23}H_{28}O_4N_2$

UV λ_{max}^{EtOH} mμ (log ε): 206.5 (4.49), 246 (4.20), 282.5 (3.51), 295 (3.48); λ_{min}: 226 (3.81), 267 (3.24),

    288 (3.47) (see Fig. 324)

IR λ_{max}^{CCl4} λ: 5.74, 5.87 (C=O), no absptn. of OH and NH

NMR in $CDCl_3$, 60 Mc, cps: 463 d, $J=7$ (C'_4–H), 430 m (3) (arom. protons), 229 s (3), 223 s (3)

    (2 $COOCH_3$), 180 (charact. to aspidospermine type), no signals of C–CH_3, olefinic proton or C_2–H

1921) Gilbert, B: *Tetr. Letters*, **1962**, 59.
1922) Djerassi, C. *et al*: *J.A.C.S.*, **84**, 1499 (1962).

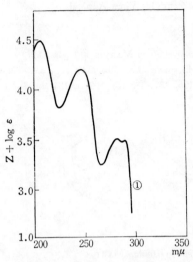

Fig. 324 UV Spectrum of Pleiocarpine
in EtOH (Schmid)

Mass Spectra of Pleiocarpine-type Alkaloids[1902]

Fragmentation: Pleiocarpine, refractine and aspidofractinine

a M-28

$+$ CH$_2$=CH$_2$

R$_1$=R$_2$=H

a' b 109 d

$+$ CH$_2$=CH$_2$

M $\longrightarrow$ f

e 124

(CXLI-a) M$^+$=280

g h

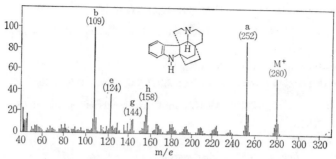

Fig. 325 Mass Spectrum of (CXLI–a) (Djerassi *et al.*)[1902]

(f) Pleiocarpinine (CXLVII)[1923]

≪Occurrence≫ *Pleiocarpa mutica* BENTH.

mp. 127° & 135~136°, $[\alpha]_D$ −124° (CHCl$_3$), pKMCS 6.94, C$_{22}$H$_{28}$O$_2$N$_2$

UV λ_{max}^{EtOH} mμ (log ε): 206 (4.42), 254 (3.97), 300 (3.52), λ_{min}: 231 (3.55): 276 (3.12)

IR $\lambda_{max}^{CCl_4}$ μ: 5.78 (C=O), no absptn. of OH or NH

(g) Kopsinine (CXLVIII)[1923]

≪Occurrence≫ *Pleiocarpa mutica* BENTH., mp. 105°, $[\alpha]_D$ −69° (CHCl$_3$), C$_{21}$H$_{26}$O$_2$N$_2$

UV λ_{max}^{EtOH} mμ (log ε): 205 (4.43), 246 (3.83), 295 (3.45); λ_{min}: 227 (3.63), 270 (2.97)

IR $\lambda_{max}^{CCl_4}$ μ: 3.0 (NH), 5.78 (–COOCH$_3$)

Mass See the item (e) Pleiocarpine

(h) Other Pleiocarpa Alkaloids[1923]

≪Occurrence≫ *Pleiocarpa mutica* BENTH.

Pleiomutine (CXLVIII–a)

mp. 230° (decomp.) (amorph), $[\alpha]_D$ −97° (CHCl$_3$), C$_{42~43}$H$_{52~56}$O$_2$N$_4$

UV λ_{max}^{EtOH} mμ (log ε): 226 (4.52), 247 (4.02), 281 (3.97), 290 (3.98)

Pleiomutinine (CXLVIII–b)

mp. >220°, C$_{40}$H$_{46~48}$O$_2$N$_4$

UV λ_{max}^{EtOH} mμ (log ε): 209 (4.72), 255~257 (4.20), 310~313 (3.66), 330 (3.64)

Eburnamenine (CLXII) See *Type–(XIII)*
Pleiocarpamine (CLVIII) See *Type–(XI)*

(i) Pleiocarpinilam and Kopsinilam[1924]

R=CH$_3$ (CXLVIII–c)
Pleiocarpinilam
R=H (CXLVIII–d)
Kopsinilam

≪Occurrence≫ *Pleiocarpa tubicina* STAPP., *P. mutica* BENTH., *Hunteria eburnea* PICHON.
Pleiocarpinilam (CXLVIII–c)

mp. 250°, $[\alpha]_D$ −53.4° (CHCl$_3$), C$_{22}$H$_{26}$O$_3$N$_2$

1923) Schmid, H. *et al*: *Helv. Chim. Acta*, **44**, 1503 (1961), **45**, 854 (1962).
1924) Schmid, H. *et al*: *ibid*, **45**, 1090 (1962).

UV λ_{max}^{EtOH} mμ (log ε): 253 (4.01), 300 (3.49); λ_{min}: 228 (3.50), 275 (3.13)

IR ν_{max}^{KBr} cm^{-1}: 1736 (COOCH$_3$), 1696 (five membered ring lactam), 1605 (indoline)

Kopsinilam (CXLVIII-d)

mp. 254~254.5°, $[\alpha]_D$−13° (CHCl$_3$), C$_{21}$H$_{24}$O$_3$N$_2$

UV λ_{max}^{EtOH} mμ (log ε): 246 (3.89), 295 (3.50); λ_{min}: 227 (3.62), 269 (3.10)

IR ν_{max}^{KBr} cm^{-1}: 3286 (NH), 1742 (COOCH$_3$), 1684 (five membered lactam), 1608 (indoline)

(vii) *Type-VII* (CXLIX)

(a) **Kopsine** (CXLIX)[1925)1926]

≪Occurrence≫ *Kopsia fruticosa* (*Apocynaceae*)

mp. 210~214°, $[\alpha]_D$−17.5° (CHCl$_3$), pKa′ 4.28, C$_{22}$H$_{24}$O$_4$N$_2$

UV $\lambda_{max}^{EtOH-H2O}$ mμ (log ε)[1926]: 241 (4.09), 278 (3.35), 285 (3.33)

IR ν_{max}^{CHCl3} cm^{-1} [1925]: 3268 (intramolecular chelated OH), 1679 (chelated N–COOCH$_3$, non chelated

N–COOCH$_3$: 1704~1710)

NMR in CDCl$_3$, 60 Mc, TMS, cps: 415~470 m (arom. protons), 236 (3) (>N–COOCH$_3$), no signals
due to C–CH$_3$, N–CH$_3$, —CHO, C=C–H or C$_2$–H

(b) **Fruticosamine**, an isomer of (CXLIX)[1926]

≪Occurrence≫ *Kopisia fruticosa*, mp. 177~181°, $[\alpha]_D$+43° (CHCl$_3$), pKa′ 4.04, C$_{22}$H$_{24}$O$_4$N$_2$

UV $\lambda_{max}^{EtOH-H2O}$ mμ (log ε): 243 (4.17), 278 (3.40), 286 (3.37)

(c) **Fruticosine**, an isomer of (CXLIX)[1926]

≪Occurrence≫ *Kopsia fruticosa*, mp. 225~226°, $[\alpha]_D$−19° (CHCl$_3$), pKa′ 4.62, C$_{22}$H$_{24}$O$_4$N$_2$

UV $\lambda_{max}^{EtOH-H2O}$ mμ (log ε): 245 (4.18), 280 (3.42), 286 (3.38)

(viii) *Type-VIII* (CLII)

(a) **Polyneuridine** (CLIII)[1877]

≪Occurrence≫ *Aspidosperma polyneuron* MÜLL. ARG.

mp. 245~247.5°, (EtOH–solvate), $[\alpha]_D$−68° (EtOH–solvate in pyridine), pKa′ 6.60 (D.F.A),
C$_{21}$H$_{24}$O$_3$N$_2$·C$_2$H$_5$OH

UV λ_{max}^{EtOH} mμ (log ε): 228 (4.51), 281 (3.82); λ_{min}: 249 (3.33); $\lambda_{max}^{EtOH-HCl}$ mμ (log ε): 221 (4.60),
273 (3.82); λ_{min}: 242 (3.31)

IR λ_{max}^{nujol} μ: 2.99, 5.81

NMR (CLIII) $\xrightarrow{\text{LiAlH}_4,\ \text{Ac}_2\text{O}}$ (CLIII–a) [Note]

Polyneuridinoldiacetate (CLIII–a)=((CLII), R=R′=CH$_2$OCOCH$_3$)

In CDCl$_3$, 60 Mc, TMS, δ: 6.9~7.6 (4) (arom. H), 8.43 (indole NH), 5.1~5.6 q $\left(>C=C<_H^H\right)$,

1.45~1.70 d (3) (=CH–CH$_3$), 3.9~4.2 (4) (2-CH$_2$–OCOCH$_3$), 3.65~3.87 (C$_3$–H), 3.4~3.6
(−CH$_2$−)[21], 2.8~3.2 (C$_5$–H), 2.8~3.2 (−CH$_2$−)[6], 2.55~2.8 (C$_{15}$–H), 1.45~1.85 (−CH$_2$−)[14]

Mass M$^+$=352, M−CH$_3$<M−H$_2$O, M−CH$_2$OH, M−CO$_2$CH$_3$, M−(H$_2$O+CO$_2$CH$_3$), 249 (q),
182 (n), 169(c)

1925) Schmid, H. *et al*: *Helv. Chim. Acta*, **45**, 1146 (1962), **46**, 572 (1963).
1926) Battersby, A.R *et al*: *Soc.*, **1963**, 22.
[Note] Because (CLIII) does not resolve in CDCl$_3$

M$^+$=352

(q) 249

(n) 182

(c) 169

(b) Akuammidine (CLIV)

Streoisomer of polyneuridine (CLIII), see the item (12) Picralima Alkaloids, (b)

(ix) Type-IX (CLV)

(a) Echitamine (CLV)[1927)～1930)]

≪Occurrence≫ *Alstonia scholaris* R. BR

Chloride $C_{22}H_{29}O_4N_2Cl$

UV λ_{max}^{EtOH} mμ (log ε): 235 (3.93), 295 (3.55)

IR $\lambda_{max}^{Hexachlorobutadiene}$ μ: no N–CH$_3$ at 7.2, 3.04 (OH), 3.17 (NH), 5.79 (–CCOMe), 13.2 (*ortho*

disubstituted benzene), no $\overset{+}{\rightarrow}$NH at 4.15

Absolute configuration by X–ray analysis[1928) 1929)]

Echitamine (see the item (15) Strychnos and Curare Alkaloids, (iii), (e), (CCLXXVIII))

(x) Type-X

(a) Flavocarpine (CLVI)[1931)]

≪Occurrence≫ *Pleiocarapa mutica* BENTH.,   mp. 307°, [α]$_D$ 0°, $C_{18}H_{14}O_2N_2$

UV λ_{max}^{EtOH} m$\mu(\varepsilon)$: 223 (40800), 242 (48600), 250 (51200), 291 (23000), 351 (23500), 389 (22000);

$\lambda_{max}^{0.01N-HCl-EtOH}$ m$\mu(\varepsilon)$: 230 (29900), 246 (32300), 298 (16900), 333 (11600), 371 (13600), 400 (12700)

IR ν_{max}^{KBr} cm^{-1}: 1630, 1596, 1525, 1480, 1405, 1375, 1360, 1220, 1190, 1085, 800, 745

(b) Flavopereirine (CLVII) See the item (11) Geissospermum Alkaloids

(xi) Type-XI

(a) Pleiocarpamine (CLVIII)[1923) 1932)]

≪Occurrence≫ *Pleiocarpa mutica* BENTH, *Hunteria eburnea* PICHON.

mp. 159°, [α]$_D$+136° (MeOH), +123° (CHCl$_3$), pK 6.91 (CMS), $C_{20}H_{22}O_2N_2$

UV λ_{max}^{EtOH} mμ (log ε): 230 (4.47), 285 (3.91)

IR ν_{max}^{CCl4} cm^{-1}: 1736, 1770; ν_{max}^{nujol} cm^{-1}: 1730

1927) Govindachari, T.R. *et al*: *Tetarhedron*, **15**, 132 (1961).
1928) Monohar, H. *et al*: *Tetr. Letters*, **1961**, 814.
1929) Hamilton, J.A. *et al*: *Soc.*, **1962**., 506.
1930) Conroy, H. *et al*: *Tetr. Letters*, **1960**, No.6, 1.
1931) Büchi, G *et al*: *J.A.C.S.*, **84**, 3393 (1962).
1932) Karrer, P. *et al*: *Helv. Chim. Acta*, **47**, 878 (1964).

NMR   in deuteroacetone, 100 Mc, TMS, ppm: (See Fig. 326) 5.22 dq, $J_1=7$, $J_2=2\left(=C_{19}{<}{}^{H}\right)$, 1.48 dd,

$J_1=7, J_2=2\left(=C_{19}{<}_{CH_3}\right)$, 1.68 d, $J=13(C_{21}-H)$, 2.53 d $(C_{21}-H)$, 5.26 d, $J=4$ $(C_{16}-H)$, 3.54 q $(C_{15}-H)$

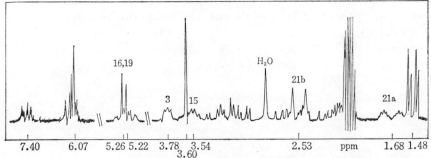

R=COOCH₃

Fig. 326   NMR Spectrum of Pleiocarpamine (Karrer *et al.*)[1932]

Mass   M⁺=322, M−15, M−59 (COOCH₃)=263, 180 (a)

180 (a)

(b)   C–Mavacurine   (CLVIII–a)

    See the item (15) Strychnos and Curare Alkaloids

(CLVIII–a)

(xii)   *Type-XII*   (CLIX)

  (a)   Tuboflavine   (CLIX)[1933]

    《Occurrence》   *Pleiocarpa tubicina* STAPF.,   mp. 207~208°, $[\alpha]_D$ 0° (pyridine), $C_{16}H_{12}ON_2$

  UV λ_{max}^{EtOH} mμ (log ε): 401, (3.97), 323 (3.71), 289 (4.38), 264 (4.37), 215 (4.59);   UV $\lambda_{max}^{0.05N-HCl-EtOH}$

    mμ (log ε): 460 (3.52), 345 (4.01), 277 (4.57)

  IR ν_{max}^{KBr} cm⁻¹: 1616 (C=O of pyridone), no absorptn. of >NH

  NMR   in CDCl₃, 60 Mc, TMS, ppm: 2.80 q, $J=7.5$ & 1.32 t, $J=7.5\left(>C-H_2-CH_3\right)$, in CDCl₃, 100 Mc:

    9.00 d, $J=4.5$ (C_3-H), 7.70 (C_4-H), ca. 7.70 (C_8-H), 8.08 m (3) $(C_6-H, C_7-H, C_{10}-H)$, 7.44 q

    (C_5-H)

1933)   Schmid, H. *et al*: *Helv. Chim. Acta*, **46**, 498 (1963).

(xiii) *Type-XIII*

≪Occurrence≫ *Hunteria eburnea*[1933a)], *Haplophyton cimicidum*[1919)], *Rhazya stricta*[1933b)]
See the item (15) Strychnos and Curare Alkaloids, (ii), (a)[2026) 2079)]

(a) **Eburnamonine** (CLX)[1919) 1933a~c)]

mp. 183°, $[\alpha]_D +89°$ (CHCl$_3$), C$_{19}$H$_{22}$ON$_2$

Mass[1933b)] : 294 (M), 265 (a), 237 (d), 224 (b), 209 (c) (see mass spectrum of Eburnamenine (CLXII))
Synthesis[1933c)]

(b) **Eburnamine** (CLXI)

mp. 181°, $[\alpha]_D -93°$ (CHCl$_3$), C$_{19}$H$_{24}$ON$_2$

(c) **Eburnamenine** (CLXII)

mp. (picrate, 196°), $[\alpha]_D +183°$ (CHCl$_3$), C$_{19}$H$_{22}$N$_2$
Mass[1933b)] : 278 (M), 249 (a), 208 (b), 193 (c)

Eburnamonine (CLX)

(a)   (b)

(xiv) *Type-XIV*

(a) **Ellipticine** (CLXIII)[1934) 1937)]

≪Occurrence≫ *Aspidosperma subincanum* MART. *Ochrosia elliptica*

mp. 311~315° (decomp.), C$_{17}$H$_{14}$N$_2$

UV λ_{max}^{EtOH} mμ (log ε): 227~234 sh (4.32), 238 (4.36), 245 sh (4.31), 276 (4.74), 287 (4.90), 295 (4.88),

318~322 sh (3.52), 333 (3.71), 343~347 sh (3.47), 384 (3.61), 401 (3.58); $\lambda_{max}^{HCl-EtOH}$ mμ : 241,

249, 271 sh, 307

Methonitrate (CLXIII–a)   mp. 293~304° (decomp.),
C$_{17}$H$_{14}$N$_2$·CH$_3$·NO$_3$

UV λ_{max}^{MeOH} mμ (log ε): 423 (3.68), 356 (3.72), 307 (4.86),
249 (4.36), 241 (4.38)

IR ν_{max}^{KBr} cm^{-1}: 1645, 1605, 1588, 1500, 1460, 1415, 1385,
1320, 1245, 1190, 1115, 825, 805, 755, 715

(CLXIII–a)

1933a) Taylor, W. I *et al*: *Compt. rend.*, **249**, 1259 (1959).
1933b) Biemann, K *et al*: *Tetr. Letters*, **22**, 993 (1962).
1933c) Taylor, W. I *et al*: *Tetr. Letters*, **1959**, No. 2, 20.
1934) Büchi, G *et al*: *Tetrahedron*, **15**, 167 (1961).
1935) Schmutz, J *et al*: *Helv. Chim. Acta*, **44**, 444 (1961).

(b) 1, 2-Dihydroellipticine (CLXIV)[1934) 1935)

Methonitrate (LXIV–a)

mp. 301~303° (decomp.), $C_{17}H_{16}N_2 \cdot CH_3 \cdot NO_3$

UV λ_{max}^{MeOH} mμ (log ε): 382 (4.43), 313 (4.30), 302 (4.18),
281 (4.64), 271 sh (4.46), 244 sh (4.30), 236, (4.45)

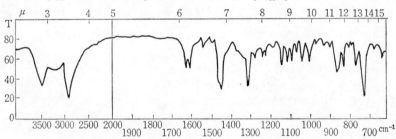

(CLXIV–a)

IR ν_{max}^{KBr} cm^{-1}: 1645, 1595, 1575, 1500, 1442, 1410, 1395, 1330, 1250, 1210, 822, 778, 750

(c) Olivacine (CLXVI)[1934) 1936)

≪Occurrence≫   *Aspidosperma subincanum, A. australe*

mp. 308~314°, $[\alpha]_D$ +0° (pyridine), $C_{17}H_{14}N_2$

UV λ_{max} mμ (ε): 224 (24300), 238 (21300), 276 (50600), 287 (71400), 292 (67000), 314 (4600),
329 (6250), 375 (4600)

(d) 1, 2-Dihydroolivacine (CLXVII)[1934) 1935)

≪Occurrence≫   *Aspidosperma subincanum, A. ulei,*   mp. 317~319° (decomp.), $C_{17}H_{16}N_2$

(e) Guatambuine (*N*–Methyltetrahydroolivacine) (LXVIII–a)[1936)

≪Occurrence≫   *Aspidosperma australe*

(+)-**Guatambuine** (μ–Alkaloid C)

mp. 245~248°, $[\alpha]_D$ +112° (pyridine), $C_{18}H_{20}N_2$

UV λ_{max} mμ (ε): 240 (41300), 250 (31300), 262 (22700), 299 (19300), 330 (4300)

(−)-**Guatambuine**

mp. 247~248°, $[\alpha]_D$ −106° (pyridine).

(f) Elliptinine (CLXVIII)[1937)

≪Occurrence≫   *Ochrosia elliptica (Apocynaceae)*,   mp. 231~233°, $[\alpha]_{589}$ −255°, $C_{20}H_{24}O_2N_2$

UV λ_{max}^{EtOH} mμ (log ε): 222 sh (end absptn. 4.37), 311 (4.27); $\lambda_{max}^{HCl-EtOH}$ mμ (log ε): 217 (4.45),
308 (4.24)

IR $\lambda_{max}^{CHCl_3}$ μ: 2.80, 2.88, 6.16, 6.30

(g) Cleavamine (CLXIX) $C_{19}H_{24}N_2$[1938)

(xv) *Type-XV*

(a) Uleine (CLXX)[1936) 1936a, b)

≪Occurrence≫   *Aspidosperma ulei* Mfg., *A. australe.*

mp. 76~118°, $[\alpha]_D$ +18.50° (CHCl$_3$), pKa 8.23 (MCS), $C_{18}H_{22}N_2$

Fig. 327 IR Spectrum of Uleine in KBr (Schmutz *et al.*)

1936) Ondetti, M. A *et al*: *Tetrahedron*, **15**, 161 (1961).
1936a) Schmutz, J *et al*: *Helv. Chim. Acta*, **40**, 1189 (1957).
1936b) Schmutz, J *et al*: *ibid.*, **41**, 288 (1958).
1937) Goodwin, S *et al*: *J.A.C.S.*, **81**, 1903 (1959).
1938) Kutney, J. P *et al*: *Chem. & Ind.*, **1963**, 648.

UV λ_{max}^{EtOH} mμ (ε)[1936a]: 209 (24000), 309 (19900); λ_{max} mμ (ε)[1936]: 307~308 (17000) 316 (17000)

IR See Fig. 327,

IR ν_{max}^{KBr} cm^{-1}: 1625 (conj. |═), 1630, 1615 (conj. indole ring), 840 (〉C=CH—), 740 (*ortho* subst. benzene).

(b) **Minor Alkaloids** *of Aspidosperma ulei* MFG.[1936b]

　　u–Alkaloid B (CLXX–a)

　　mp. 215~218° (decomp.), [α]$_D$ 0° (CHCl$_3$), C$_{18}$H$_{20}$N$_2$

　　u–Alkaloid C (CLXX–b)

　　mp. 249~252° (decomp.), [α]$_D$ +112° (pyridine), C$_{18}$H$_{20}$N$_2$

　　u–Alkaloid D (CLXX–c)

　　mp. 308~312° (decomp.), [α]$_D$ 0° (pyridine), C$_{17}$H$_{16}$N$_2$

(c) **Tubotaiwine** (CLXXI)[1902a]

　　≪Occurrence≫ 　*Aspidosperma limae* WOODSON, *Pleiocarpa tubicina*

　　Amorphous, [α]$_D$ +584° (CHCl$_3$), C$_{20}$H$_{24}$O$_2$N$_2$

　　Picrate 　mp. 168~170° (decomp.)

　　See the item (12) Picralima Alkaloids (CXCIV)

(d) **Aspidospermatidine** (CLXXII)[1903]

　　≪Occurrence≫ 　*Aspidosperma quebracho* BLANCO, 　　mp. 184~186°, C$_{18}$H$_{22}$N$_2$

UV λ_{max}^{EtOH} mμ (log ε): 242 (3.83), 296 (3.49)

Mass See Fig. 328

(e) **Aspidospermatine** (CLXXIII)[1903]

　　≪Occurrence≫ 　*Aspidosperma quebracho* BLANCO, 　　mp. 157~159°, [α]$_D$ −73° (EtOH), C$_{21}$H$_{26}$O$_2$N$_2$

UV λ_{max}^{EtOH} mμ (log ε): 219 (4.54), 255 (4.10), 290 (3.62)

Mass See Fig. 328

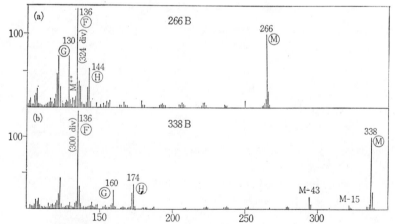

Fig. 328 Mass Spectra of (a) Aspidospermatidine and (b) Aspidospermatine (Biemann *et al.*)[1903]

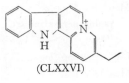

(CLXXII)
M=266

Ⓗ=144, Ⓕ=136, Ⓖ=130

(xvi) *Type-XVI*

(a) **Quebrachidine** (CLXXIV)[1939]

≪Occurrence≫ *Aspidosperma quebracho* Blanco

mp. 276~278°, [α]$_D$ +54° (CHCl$_3$) pKa′ 6.7 (sec. basic N), C$_{21}$H$_{24}$O$_3$N$_2$

UV λ$_{max}^{EtOH}$ mμ (ε): 242 (6390), 290 (2920) (dihydroindole)

IR ν$_{max}$ cm^{-1}: 3590 (N–H), 3370 (OH), 1722, 1235 (ester)

NMR δ: 6.6~7.2 m (4) (arom. protons), 3.6 (carbomethoxyl), 1.5 d (ethylidene), 5.1 q, *J*=6.5 (vinyl proton)

Mass *m/e* 130, 143 (indole nucleus w. one and two carbons resp.)

Biosynthesis of Aspidosperma Alkaloids[1940]

(11) Geissospermum Alkaloids

(a) **Flavopereirine** (CLXXVI)[1931][1941]~[1944][1947]

≪Occurrence≫ *Geissospermum lavea, G. vellosii, Strychnos melinoniana*

mp. 233~235°, C$_{17}$H$_{14}$N$_2$[1947]

Perchlorate   mp. 308°, C$_{14}$H$_{15}$O$_4$N$_2$·Cl

(CLXXVI)

UV λ$_{max}^{0.015N-HCl-EtOH}$ mμ (log ε): 238 (4.57), 294 (4.22), 350 (4.31), 389 (4.21); λ$_{min}$: 273 (4.04),

307 (4.08), 380 (4.17); λ$_{max}^{0.015N-KOH-EtOH}$ mμ (log ε): 231 (4.39), 236 (4.38), 289 (4.51), 319 (4.09),

365 (4.36), 450 (3.73); λ$_{min}$: 233 (4.39), 264 (4.14), 310 (4.06), 328 (4.03), 418 (3.68)

Synthesis[1931][1943][1944]

(b) **Pereirine** (CLXXVII)[1945]

≪Occurrence≫ *Geissospermum vellosii*

mp. 142.5~143°, C$_{19}$H$_{26}$ON

Hydrate   mp. 98~100°, C$_{19}$H$_{26}$ON·H$_2$O

UV λ$_{max}^{EtOH}$ mμ (log ε): 245 (3.93), 300 (3.47); λ$_{min}$: 222.5

(3.47), 270 (2.81) (as free base)

IR See Fig. 329, ν$_{max}^{KBr}$ cm^{-1}: 3460 (OH), 3380 (NH), 1610

(arom.), 1500~1450, 745 (*ortho*-disubst. benzene)

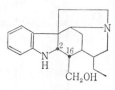

(CLXXVII)

1939) Gorman, M: *Tetr. Letters*, **1963**, No. 1, 39.

1940) Wenkert, E: *J.A.C.S.*, **84**, 98 (1962).

1941) Janot, M.M *et al*: *Compt. rend.*, **244**, 2066 (1957).

1942) Rapoport, H *et al*: *J.A.C.S.*, **80**, 1604 (1958).

1943) Ban, Y *et al*: *Tetrahedron*, **16**, 5 (1962).

1944) Wenkert, E *et al*: *J.A.C.S.*, **84**, 3732 (1962); *J. Org. Chem.*, **27**, 2283 (1962).

1945) Bertho, A *et al*: *Ber.*, **91**, 2581 (1958), **94**, 2737 (1961).

1946) Chatterjee, A *et al*: *J. Sci. & Ind. Res.* **23**, 178 (1964).

1947) Rapoport, H *et al*: *J.A.C.S.*, **82**, 4404 (1961), **81**, 3166 (1959), **80**, 1601 (1958).

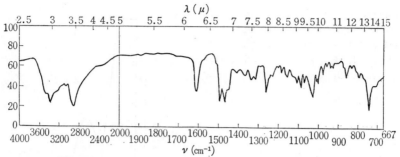

$\lambda\,(\mu)$

ν (cm^{-1})

Fig. 329 IR Spectrum of Pereirine in KBr (Bertho *et al.*)

(**c**) **Geissospermine and Related Alkaloids**[1945]~[1948]

《Occurrence》 *Geissospermum vellosii*

Geissospermine (CLXXVIII) + H$_2$O $\xrightleftharpoons[10\%\mathrm{AcOH}]{2N\mathrm{HCl}}$ Geissoshizine (CLXXIX)+Geissoshizoline (CLXXX)

C$_{40}$H$_{48}$O$_3$N$_4$ C$_{21}$H$_{24}$O$_3$N   C$_{19}$H$_{26}$ON

Geissoshizine (CLXXIX) $\xrightleftharpoons[\mathrm{dilHCl}]{\mathrm{conc.\ HCl\ or\ H_3PO_4}}$ Apogeissoshizine   +H$_2$O

C$_{21}$H$_{24}$O$_3$N   (CLXXXI) C$_{21}$H$_{22}$O$_2$N$_2$

H$_3$COOC

(CLXXVIII) Geissospermine

(CLXXIX) Geissoshizine

CH$_2$OH

(CLXXX) Geissoshizoline

COOCH$_3$

(CLXXXI) Apogeissoshizine

1948) Janot, M.M *et al*: *Tetrahedron*, **14**, 113 (1961).

Geissospermine (CLXXVIII)

mp. 213~214°, $[\alpha]_D$ −101° (EtOH), $C_{40}H_{48}O_3N_4$

UV See Fig. 330, λ_{max} mμ: 251, 289, 292.5[1946]

IR See Fig. 331, λ_{max} μ: 2.79, 2.93 (OH and N–H), 5.76 (C=O)

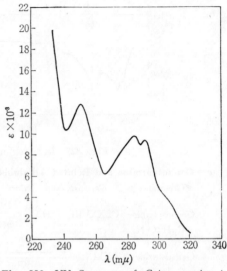

Fig. 330 UV Spectrum of Geissospermine in EtOH (Rapoport *et al.*)[1947]

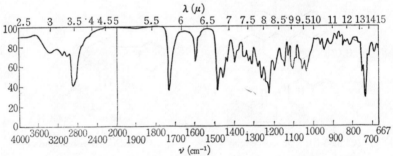

Fig. 331 IR Spectrum of Geissospermine in KBr (Bertho *et al.*)[1945]

Hydrate   mp. 145~147°, $C_{40}H_{48}O_3N_4 \cdot 2\,H_2O$

Geissoshizine (CLXXIX)

mp. 180~182°, $C_{21}H_{24}O_3N_2$

Geissoshizoline (CLXXX)

Chloroformate   mp. 105~108°, $C_{14}H_{26}ON_2 \cdot CHCl_3$

Apogeissoshizine (CLXXXI)[1947]

Hydrochloride   mp. 139~142°, $C_{21}H_{22}O_2N_2 \cdot HCl$

(d) **Vellosimine, Vellosiminol and Geissolosimine**[1949]

《Occurrence》   *Geissospermum vellosii*

1949) Rapoport, H *et al*: *J. Org. Chem.*, **27**, 2981 (1962).

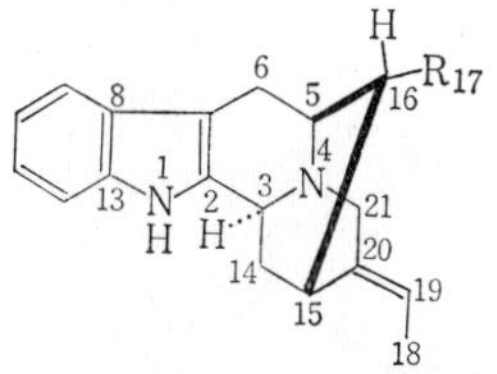

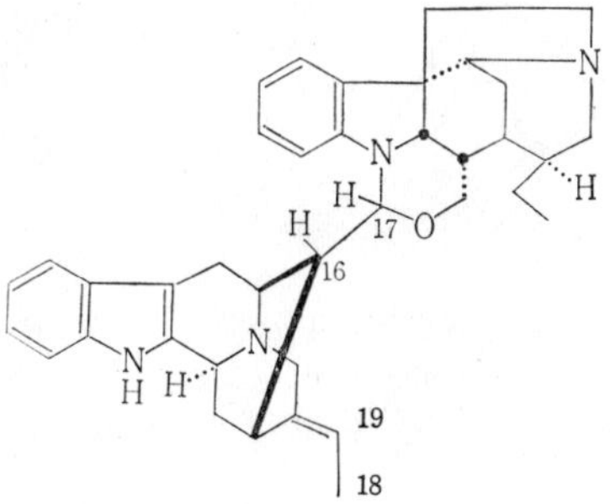

R=CHO: (CLXXXII) Vellosimine
R=CH₂OH: (CLXXXIII) Vellosiminol   (CLXXXIV) Geissolosimine

Vellosimine (CLXXXII)

mp. 305~306°, $[\alpha]_D$ +48°, $C_{19}H_{20}ON_2$

UV λ_{max}^{EtOH} mμ (ε): 280 (8000), 289 (6430) (typical indole)

IR ν_{max} cm⁻¹: 3473 (indole NH), 1720, 2720, 2840 (unconj.–CHO)

NMR in liquid SO₂, δ: 7.1~7.7 (4) (arom. protons), 6.2~6.9 (neg!) (2, 3–disubstit. indole), 9.58 (–CHO),

8.66 centr. br. (indole N–H), 5.57 centr. q (1: 3: 3: 1) and 1.72 d, J=7.0 $\left(\_\!\stackrel{\displaystyle\frown\!\!\frown}{}\!\!+\!\!\stackrel{\displaystyle\frown\!\!\frown}{}\_\right)$ 1.72

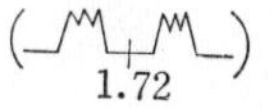

(C=CH and CH₃ in C=C$\langle^H_{CH_3}$), 4.45 dd $\left(-\stackrel{H}{\underset{H}{C_3}}-\stackrel{H}{\underset{H}{C_{14}}}-\right)$

Vellosiminol (CLXXXIII)

mp. 242~243°, $[\alpha]_D$ +37.7°, $C_{19}H_{22}ON_2$

Geissolosimine (CLXXXIV)

mp. 140°, $[\alpha]_D$ +70.4°, $C_{36}H_{44}ON_4 \cdot 1/2\ H_2O$

UV λ_{max}^{EtOH} mμ (ε): 250 (13600), 284 (6700), 292 (7000)

NMR in CDCl₃, δ: 5.21 d, J=10 $\left(-\stackrel{H}{\underset{|}{C_{16}}}-\stackrel{H}{\underset{|}{C_{17}}}\langle^N_O\right)$ 5.55 q (1: 3: 3: 1) (=C₁₉$\langle^H$), 1.71 d, J=7

(18–CH₃), 9.42 br. (indole NH)

(12) Picralima Alkaloids

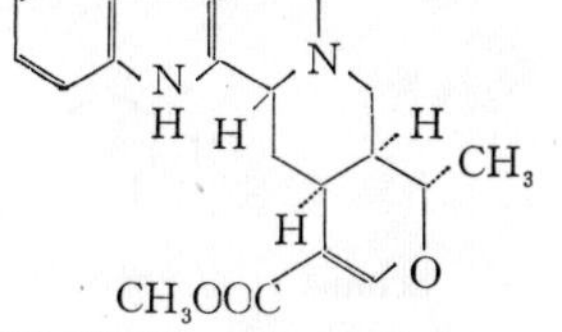

(CLXXXV) Akuammigine 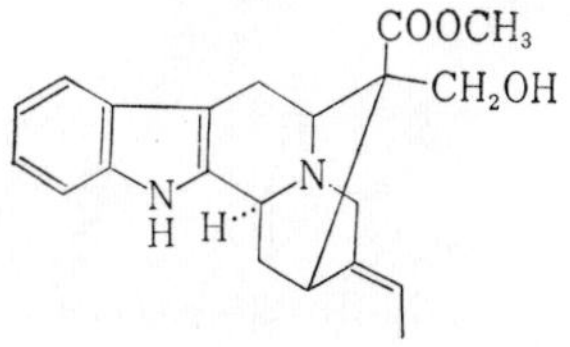   (CLXXXVI) Akuammidine

R=H (CLXXXVII) Pseudo-
akuammigine   Britten[1958]   Oliver[1956]

R=OH (CLXXXVII–a) Akuammine   (CLXXXVIII) Picraline

(CLXXXIX) Akuammicine   (CXCIII) Condylocarpine

(CXC) 19,20–Dihydroakuammicine   (CXCIV) 19,20–Dihydrocondylocarpine (Tubotaiwine)

(CXCI) Tubifoline   (CXCV) Condyfoline

(CXCII) 20–epi–Condifoline   (Schmid et al.)[1965]

(CXCVI) Mossambine   (CXCVII) Norfluorocurarine

(a) **Akuammigine** (CLXXXV)

≪Occurrence≫　　*Picralima nitida*

See the item (7) Yohimbe, Alstonia and Rauwolfia Alkaloids (b), (LXXIV), Tab. 99, 100

(b) **Akuammidine** (CLXXXVI)[1877)1950)1951)]

≪Occurrence≫　　*Picralima nitida*,　　mp. 248.5°, $[\alpha]_D$ +21° (EtOH), $C_{21}H_{24}O_3N_2 \cdot H_2O$

C_{16}-Stereoisomer of polyneuridine　See the item (10) Aspidosperma and Pleiocarpa Alkaloids, *Type–VIII*
(a) (CLIII)

Mass[1877)] : M$^+$=352, M–CH$_3$, M–H$_2$O, M–CH$_2$OH, M–COOCH$_3$, 249 (q), 182 (n), 169 (c) (see mass
spectrum of polyneuridonone)

(c) **Pseudoakuammigine** (CLXXXVII)[1952)1953)1957)]

≪Occurrence≫　　*Picralima klaineana*,　　mp. 162~164°, pKa 7.15 (EtOH·H$_2$O 1:1), $C_{20}H_{26}O_3N_2$

UV　λ_{max}^{EtOH} mμ (ε): 245 (11000), 291 (4000) ; $\lambda_{max}^{0.2N-HCl-EtOH}$ mμ (ε): 242 (11000), 288 (4000)

IR　$\nu_{max}^{CCl_4}$ cm^{-1}: 1737 (C=O)

NMR　τ: 4.61 (olefinic protons), 6.23 (ester OCH$_3$), 7.18 (arom.–NCH$_3$), 8.50 (=CH–CH$_3$)

(d) **Akuammine** (Vincamajoridine)[1955)] (CLXXXVII–a)[1952)1954)1956)]

≪Occurrence≫　　*Picralima klaineana, Vinca major*

mp. 254~259° (decomp.), $[\alpha]_D$ −66.7° (EtOH), $C_{22}H_{26}O_4N_2$

IR　λ_{max} μ: 12.33 (1, 2, 4-trisubstit. benzene)

Hydrochloride　　mp. 215~218° (decomp.),　　$C_{22}H_{26}O_4N_2 \cdot HCl$

O–Methylether　　mp. 242~243°

UV　λ_{max}^{EtOH} mμ (ε): 244 (11500), 308 (4200)

(e) **Picraline** (CLXXXVIII)[1956)1958)]

≪Occurrence≫　　*Picralima klaineana*

mp. 160~162°, pKa 5.65 (50% EtOH), $C_{23}H_{26}O_5N_2$

UV　λ_{max}^{EtOH} mμ (ε): 237 (7400), 289 (3200) ; $\lambda_{max}^{70\%HClO_4}$ mμ (ε): 241 (5500), 246 (5400), 310 (6200)

IR　$\nu_{max}^{CHCl_3}$ cm^{-1}: 3400, 1737 (NH and C=O)

Mass　m/e 351 (–CH$_3$COO), 337 (–CH$_3$COOCH$_2$), 239 (–CH$_3$COOCH$_2$–Ċ–CO$_2$CH$_3$)

(f) **Akuammicine** (CLXXXIX)[1959)~1966)]

≪Occurrence≫　　*Picralima nitida*,　　mp. 182°, $[\alpha]_D$ −745° (EtOH), pKa 7.45, $C_{20}H_{22}O_2N_2$[1959)]

UV　Dihydroindole type[1960)1961)], λ_{max}^{EtOH} mμ (log ε): 234 (4.10), 299 (4.07), 329 (4.21) ; λ_{min} : 262, 307[1967)]

(CLXXXIX)　$\xrightarrow{\text{HCl}}$　Base (CLXXXIX–a)[1962)]

1950) Hamet, R: *Compt. rend.*, **221**, 699 (1945), **236**, 319 (1953) ; *Bull. Soc. Pharm. Bordeaux*, **90**, 178 (1952) ; *C.A.*, **48**, 8794 (1954).　　1951) Janot, M.M *et al* : *Compt. rend.*, **253**, 131 (1961).

1952) Joule, J.A *et al* : *Soc.*, **1962**, 312.　　1953) Janot, M.M *et al* : *Bull. Soc. Chim. France*, **1961**, 1659.

1954) Robinson, R *et al* : *Experientia*, **9**, 89 (1953).

1955) Janot, M.M *et al* : *Compt. rend.*, **238**, 2550 (1954), **240**, 909 (1955).

1956) Oliver, L *et al* : *Bull. Soc. Chim. France*, **1963**, 646.

1957) Britten, A.Z *et al* : *Chem. & Ind.*, **1963**, 1120.

1958) Britten, A.Z *et al* : *ibid.*, **1963**, 1492, *Soc.*, **1963**, 3850.

1959) Robinson, R *et al* : *Soc.*, **1955**, 2049.　　1959a) Edwards, P.N. *et al* : *Proc. Chem. Soc.*, **1960**, 215.

1960) Robinson, R *et al* : *Experientia*, **9**, 89 (1953).　　1961) Hamet, R: *Compt. rend.*, **233**, 560 (1951).

1962) Smith, G.F *et al* : *Soc.*, **1960**, 792.　　1963) Karrer, P *et al* : *Helv. Chim. Acta*, **43**, 717 (1960).

(CLXXXIX–a)   mp. 80～84°

UV λ_{max}^{EtOH} mμ (ε): 283 (5320), 224 (11200) ; $\lambda_{max}^{HCl-EtOH}$ mμ (ε):

276 (3950), 223 (8800)

Characteristic UV absorption of Indolenium salt (CLXXXIX–d)[1963]

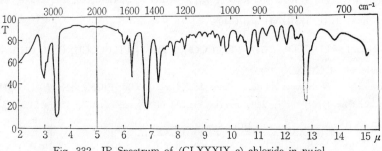

(CLXXXIX–e) C–Fluorocurarine [Note]   (CLXXXIX–d)   2HCl$^{\ominus}$

UV $\lambda_{max}^{N-HCl-MeOH}$ mμ (log ε): 236 (3.94), 296 (3.75) ; $\lambda_{max}^{10N-HCl-MeOH}$ mμ (log ε): 238 (3.88)–243 (3.85)

(double maxima), 265 inf. (3.05) $\xrightarrow[\text{Standing}]{}$ 303 (3.74)

IR λ_{max} μ: 6.03, 13.39 (1, 2–disubstit. benzene)[1959)1960]

Characteristic IR absorption of Indoleine–type Alkaloid (CLXXXIX–c)[1963]

IR λ_{max}^{nujol} μ: 6.33 (See Fig. 332)

Fig. 332 IR Spectrum of (CLXXXIX–c)–chloride in nujol

(Karrer *et al.*)[1963]

Pseudoakuammicine (CLXXXIX–f)[1959)1959a]

≪Occurrence≫   *Picralima nitida*, $C_{20}H_{22}O_2N_2$

2, 16–Dihydroakuammicine (CLXXXIX–g)[1964]

mp. 143～145°, $C_{20}H_{24}O_2N_2$

UV λ_{max}^{MeOH} mμ (ε): 299 (3900), 246 (8700),   IR ν_{max}^{CCl4} cm^{-1}: 3400 w,1747 s, 1729 s.

2, 16, 19, 20–Tetrahydroakuammicine (CLXXXIX–h)[1964]

mp. 135～137°, $C_{20}H_{26}O_2N_2$

1964) Smith, G. F *et al* : *Soc.*, **1961**, 152, 1458.
1965) Schmid, H *et al* : *Helv. Chim. Acta*, **46**, 1996 (1963).
[Note] See the item (15) Strychnos and Curare Alkaloids (d), (CCLXXVII).

UV λ_{max}^{MeOH} mμ (ε): 299 (3600), 240 (8300),　　IR ν_{max}^{CCl4} cm^{-1}: 3400 w, 1746 s, 1726 s,　　Mass[1966]

19, 20–Dihydroakuammicine (CXC)[1967]

mp. 173～175°, [α]$_D$ +636.6°, (MeOH), $C_{20}H_{24}O_2N_2$

UV λ_{max}^{MeOH} mμ (log ε): 233 (4.02), 296 (3.98), 327 (4.10); λ_{min}: 215, 261, 305

IR λ_{max}^{CHCl3} μ: 3.01, 6.01, 6.28, 6.84, 6.98

NMR　TMS　cps: 223 s (OCH$_3$), 4 arom. protons, 228 br.s (NH)

(g) Condylocarpine (CXCIII)[1965][1968]

《Occurrence》　*Diplorrhynchuchus condylocarpon* spp. *mossambicensis*

mp. 159～162°, [α]$_D$ +900° (CHCl$_3$), $C_{20}H_{22}O_2N_2$

UV λ_{max}^{MeOH} mμ (log ε): 228 (4.04), 295 (4.01), 328 (4.17),　　IR　See Fig. 333

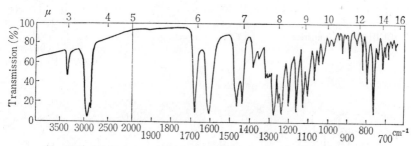

Fig. 333　IR Spectrum of Condylocarpine in nujol (Stauffacher)[1968]

(h) 19, 20–Dihydrocondylocarpine (Tubotaiwine) (CXCIV)[1965]

See the item (10) Aspidosperma and Pleiocarpa Alkaloids (c) (CLXXI)

(i) Condyfoline (CXCV)[1968]

《Occurrence》　*Diplorrhynchus condylocarpon* spp. *mossambicensis*

mp. 76～80°, [α]$_D$ +348° (AcOEt), $C_{18}H_{22}N_2$,　　UV λ_{max}^{Ether} mμ: 250

(j) Mossambine (CXCVI)[1966][1968]

《Occurrence》　*Diplorrhynchus condylocarpon* spp. *mossambicensis*

mp. 238～242°, [α]$_D$ −470° (CHCl$_3$), $C_{20}H_{22}O_3N_2$

UV λ_{max}^{MeOH} mμ (log ε): 228 (4.13), 296 (4.06), 329 (4.21)

(k) Norflurocurarine (CXCVII)[1967][1968]

《Occurrence》　*Diplorrhynchus condylocarpon* spp. *mossambicensis*

See the item (15) Strychnos and Curare Alkaloids, (iii), (c), (CCLXXVI)

(13) Vinca Alkaloids

(a) Structures

(CXCVIII)　Vindoline　　　　　　　　　　(CXCIX)　Vincadiformine

1966) Budzikiewicz, H *et al*: *Tetrahedron*, **19**, 1265 (1963).

1967) Karer, P *et al*: *Helv. Chim. Acta*, **44**, 1877 (1961).

1968) Stauffacher, D: *ibid*, **44**, 2006 (1961).

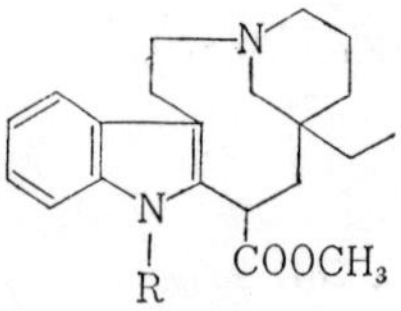

R=H (CC) Vincadine
R=CH₃ (CCI) Vincaminoreine

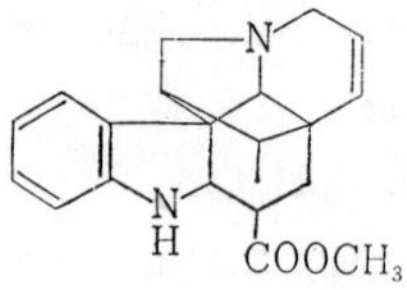

or

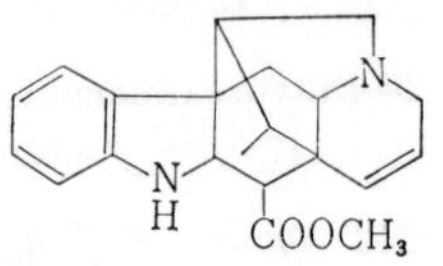

(CCII) Vin dolinine

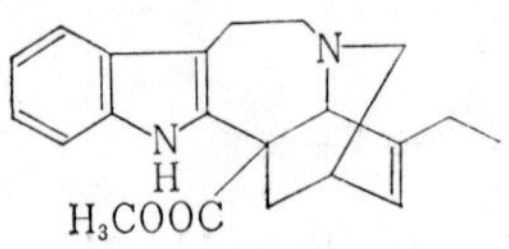

(CCIII) Catharanthine

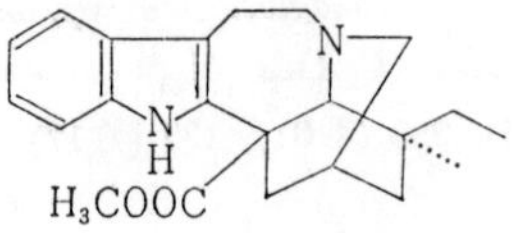

(CCIV) Coronaridine

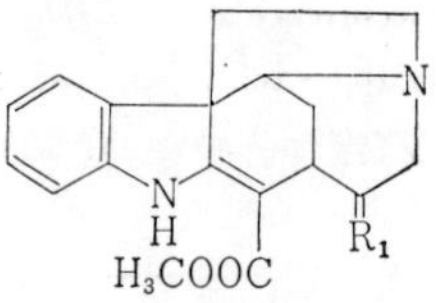

$R_1 = \begin{smallmatrix} C_2H_5 \\ OH \end{smallmatrix}$

(CCV) Lochneridine

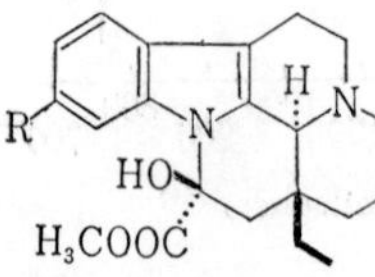

R=H: (CCVI) Vincamine
R=OCH₃: (CCVII) Vincine

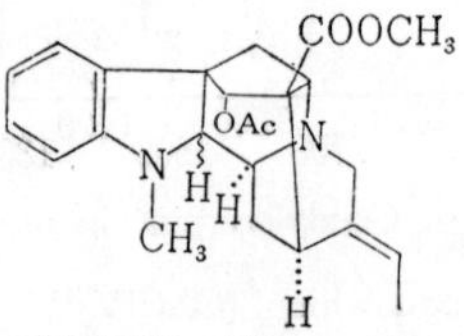

(CCVIII) Vincamedine

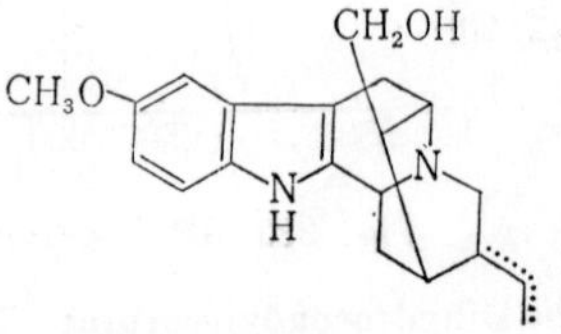

(CCVIII–a) O–Methylsarpagine

$R_1=COOCH_3$, $R_2=CH_3$, $R_3=OCH_3$, $R_4=COCH_3$:
(CCIX) Vinblastine (Vincaleucoblastine (VLB))
$R_1=COOCH_3$, $R_2=CHO$, $R_3=OCH_3$, $R_4=COCH_3$:
(CCX) Leurocristine (LCR)

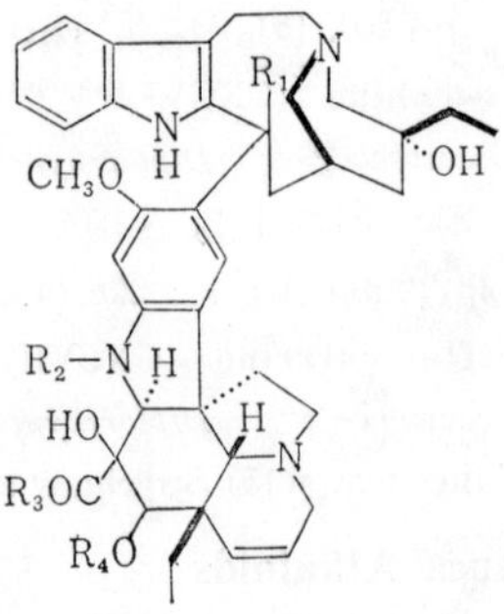

Systematic solvent-extraction, recrystallisation, partition paper chromatography and thin-layer chromatography[1969]~[1971]

1969) Svoboda, H: *J.A.P.A.*, **47**, 834 (1958); Gordon, H *et al*: *J.A.P.A.*, **48**, 659 (1959); *J. Pharm. Sci.*, **50**, 409 (1961), **51**, 217, 519 (1962).
1969a) Cone, N. J *et al*: *ibid.*, **52**, 688 (1963).   1970) Gordon, H *et al*: *ibid.*, **51**, 217, 409, 519, 707 (1962).
1971) Gorman, M. *et al*: *J.A.C.S.*, **84**, 1058 (1962).

(b) **Physical Constants**

Table 102 Physical Constants of Vinca Alkaloids　(Gordon *et al.*[1970] and others)

Str. No.	Alkaloids	Formula	mp.	$[\alpha]_D$ (CHCl$_3$)	pK'$_a$ (DMF)	Occurr-ence	Ref.
			Part 1.　Structures have been confirmed				
(CXCVIII)	Vindoline	$C_{25}H_{32}O_6N_2$	154~155°	−18	5.5	*V.r.,Vma.*	[1970)1971)]
(CXCIX)	Vincadiformine	$C_{21}H_{26}O_2N_2$	120~126° (*dl−*)	*dl*　0 *d* +402 *l* −540 (EtOH)	—	*V.d., R.s.*	[1972)1973) 1973a)]
(CC)	Vincadine	$C_{21}H_{28}O_2N_2$	—	—	—	*V.mi.*	[1975)]
(CCI)	Vincaminoreine	$C_{22}H_{30}O_2N_2$	138~139°	+26.5	—	*V.mi.*	[1975)]
(CCII)	Vindolinine	$C_{21}H_{24}O_2N_2$	−2HCl 210~212°(d.)	−8(H$_2$O)	3.3, 7.1	*V.r.*	[1969)1970)]
(CCIII)	Catharanthine	$C_{21}H_{24}O_2N_2$	126~128°	+29.8	6.8	*V.r.*	[1969)1970)1974)]
(CCIV)	Coronaridine	$C_{21}H_{26}O_2N_2$	See (14) Tabernaemontama & Voacauga Alkaloids				
(CCV)	Lochneridine	$C_{20}H_{24}O_3N_2$	211~214°(d.)	+607.5	5.5	*V.r.*	[1970)1976)]
(CCVI)	Vincamine	$C_{21}H_{26}O_3N_2$	—	—	—	*V.mi., V.d.*	[1978)1979)]
(CCVII)	Vincine	$C_{22}H_{28}O_4N_2$	Methiodide 215~217°(d.)	Methiodide −21(MeOH)	—	*V.mi.*	[1977)]
(CCVIII)	Vincamedine	$C_{24}H_{28}O_4N_2$				*V.ma.*	[1980)]
(CCVIII-a)	*O*-Methylsarpa-gine	$C_{20}H_{24}ON_2$	200~202.5°	+5.6°(Py.)		*V.r.*	[1892) [Note] 1969)1970)1981)]
(CCIX)	Vinblastine	$C_{46}H_{56}O_9N_4$	211~216°(d.)	+42	5.4, 7.4	*V.r.*	[1983)]
(CCX)	Leurocristine	$C_{46}H_{54}O_{10}N_4$	218~220°(d.)	+17.0 (CH$_2$Cl$_2$)	5.0, 7.4	*V.r.*	[2000)1982)1983)]

[Note] See　(7) Yohimbe, Alstonia and Rauwolfia Alkaloids, (w), (CVI)　and　(15) Strychnos　and Curare Alkaloids, (i), (p), (CCLXVIII).

Part 2.　Compounds of unknown structures at 1962

Str. No.	Alkaloids	Formula	mp.	$[\alpha]_D$ (CHCl$_3$)	pK'$_a$ (DMF)	Occurr-ence	Ref.
(CCXI)	Perivine	$C_{20}H_{24}O_3N_2$	180~181°	−124.0	7.5	*V.r.*	[1969)]
(CCXII)	Lochnericine	$C_{21}H_{24}O_2N_2$	190~193° (d.)	−432	4.2	*V.r.*	[1970)1984)]
(CCXIII)	Minovicine	$C_{21}H_{24}O_3N_2$	picrate 213~216° (d.)	−534(MeOH)		*V.mi.*	[1985)]
(CCXIV)	Sitsirikine	$C_{21}H_{26}O_3N_2$	1/2·H$_2$SO$_4$ 239~241° (d.)	Base + 23	7.6	*V.r.*	[1970)]
(CCXV)	Minovine	$C_{22}H_{28}O_2N_2$	80~ 82° (d.)	0·		*V.mi.*	[1985)]
(CCXVI)	Virosine	$C_{22}H_{26}O_4N_2$	258~264° (d.)	−160.5	5.85	*V.r.*	[1969)1970)]
(CCXVII)	Vindolicine	$C_{25}H_{32}O_6N_2$	248~251°, 265~267° (d.)	− 48.4	5.4	*V.r.*	[1969)1970)]

1972)　Djerassi, C *et al*: *Helv. Chim. Acta*, **46**, 742 (1963).
1973)　Smith, G. F *et al*: *Soc.*, **1963**, 4002.　　1974)　Neuss, N *et al*: *Tetr. Letters*, **1961**, 206.
1975)　Mokry, J *et al*: *ibid.*, **1962**, 1185.　　1976)　Djerassi, C: *Chem. & Ind.*, **1962**, 1986.
1977)　Strouf, O *et al*: *ibid.*, **1962**, 2037.　　1978)　Clauder, O *et al*: *Tetr. Letters*, **1962**, 1147.
1979)　Trojaneak, J *et al*: *ibid.*, **1961**, 702.
1980)　Trojanek, J: *Coll. Czech. Chem. Comm.*, **27**, 2981 (1962); *Index Chem.* 8, 26111 (1963).
1981)　Neuss, N *et al*: *J.A.C.S.*, **81**, 4754 (1959).
1982)　Neuss, N *et al*: *J.A.C.S.*, **86**, 1441 (1964).　　1983)　Neuss, N *et al*: *ibid.*, **84**, 1509 (1962).
1984)　Nair, C.P.N *et al*: *Tetrahedron*, **6**, 89 (1959).
1985)　Zachystalova, D *et al*: *Chem. & Ind.*, **1963**, 610.

Str. No.	Alkaloids	Formula	mp.	$[\alpha]_D$ (CHCl$_3$)	pK'a$_{(DMF)}$	Occurr-ence	Ref.
(CCXVIII)	Vincarodine	C$_{44}$H$_{52}$O$_{10}$N$_4$	253~256° (d.)	−197.4	5.8	V. r.	1970)
(CCXIX)	Leurosine	C$_{46}$H$_{58}$O$_9$N$_4$	202~205° (d.)	+ 72	5.5, 7.5	V. r.	1969) 1970) 1981)
(CCXX)	Isoleurosine	C$_{46}$H$_{60}$O$_9$N$_4$	202~206° (d.)	+ 61.2	4.8, 7.3	V. r.	1970)
(CCXXI)	Catharine	C$_{46}$H$_{52}$O$_9$N$_4$·CH$_3$OH	271~275° (d.)	− 54.2	5.34	V. r.	1970)
(CCXXII)	Carosine	C$_{46}$H$_{56}$O$_{10}$N$_5$	214~218°	+ 6.0	4.4, 5.5	V. r.	1970)
(CCXXIII)	Pleurosine	C$_{46}$H$_{56}$O$_{10}$N$_4$	191~194°	+ 61.0	4.4, 5.55	V. r.	1970)
(CCXXIV)	Neoleurocristine	C$_{46}$H$_{56}$O$_{12}$N$_4$	188~196° (d.)	− 57.87	4.68	V. r.	1970)
(CCXXV)	Neoleurosidine	C$_{48}$H$_{62}$O$_{11}$N$_4$	219~225° (d.)	+ 41.6	5.1	V. r.	1970)
(CCXXVI)	Vindolidine	C$_{48}$H$_{64}$O$_{10}$N$_4$	244~250° (d.)	−113.2	5.3	V. r.	1970)
(CCXXVII)	Vincamicine	Dimeric	224~228° (d.)	+418	4.80, 5.85	V. r.	1970)
(CCXXVIII)	Leurosidine	Dimeric	208~211° (d.)	+ 55.8	5.0, 8.8	V. r.	1970)
(CCXXIX)	Carosidine	Dimeric	263~278 (d.) 283 d	− 89.8	—	V. r.	1970)

V. ma.=Vinca major, V. mi.=V. minor, V. r.=V. rosea, V. d.=V. difformis, R. s.=Rhazya stricta

(c) UV and IR Spectra

Table 103 UV and IR Spectra of Vinca Alkaloids (Gordon *et al.*[1970] and others)

Str. No.	Alkaloids	UV		IR	
		λ_{max}^{EtOH} mμ	a) log ε, b) ε, c) log E$_{1cm}^{1\%}$ d) log a$_M$	a) ν_{max}cm^{-1}, b) $\lambda_{max}\mu$	
(CXCVIII)	Vindoline	212, 250, 304	d) 4.49, 3.74, 3.57	Fig. 334	
(CXCIX)	Vincadiformine	263, 222	b) 7050, 31500	a) lig : 1579 (C=N)	
(CC)	Vincadine	—	—	a) KBr, CCl$_4$: 3450, 1730, 720	
(CCI)	Vincaminoreine	230, 288, 296	a) 4.59, 3.93, 3.91	a) KBr : 1740, 741	
(CCII)	Vindolinine	245, 300			
(CCIII)	Catharanthine	226, 284, 292		Fig. 335	
(CCV)	Lochneridine	230, 293, 328	d) 4.04, 3.94, 4.07	Fig. 336	
(CCVI)	Vincamine			a) nujol : 1756, 1074, 747, 727	
(CCVII)	Vincine	(0.5N–HCl–MeOH) 223, 269, 291	a) 4.44, 3.84, 3.67	a) nujol : 3320	
(CCIX)	Vinblastine	214, 259		Fig. 337	
(CCX)	Leurocristine	220, 255, 296	a) 4.65, 4.21, 4.18	Fig. 338, b) CHCl$_3$ 5.94	
(CCXI)	Perivine	226sh, 240sh, 314	c) 2.67 (314mμ)	[Note]–CCXI	
(CCXII)	Lochnericine	227, 299, 328	a) 4.10, 4.15, 4.32	[Note]–CCXII	
(CCXIII)	Minovicine	(MeOH) 228, 302sh, 328	a) 4.01, 3.96, 4.15	[Note]–CCXIII	
(CCXIV)	Stisirikine	224, 282, 288		Fig. 339	

[Note] **IR Spectra** of ;

  (CCXI) Perivine $\lambda_{max}\mu$: 2.90, 3.40, 5.79, 6.07, 6.34, 6.50, 6.89, 6.96, 7.52, 7.68, 8.58, 8.80, 8.95, 9.08, 9.46, 9.82, 10.63

  (CCXII) Lochnericine $\lambda_{max}\mu$: 2.96, 3.39, (>NH, OH?), 3.57 (C–H), 5.98, 6.21 (ester), 6.21, 6.80 (indole)

  (CCXIII) Minovicine ν_{max}^{CHCl3} cm^{-1}: 1612, 1680 (carbomethoxymethylene indole), 1702 (additional C=O), 3290~3440 (>NH); ν_{max}^{nujol} cm^{-1}: 745 (arom.)

Sir. No.	Alkaloids	UV		IR
		λ_{max}^{EtOH} mμ	a) log ε, b) ε, c)log $E_{1cm}^{1\%}$ d) log a_M	a) ν_{max}cm^{-1}, b) $\lambda_{max}\mu$
(CCXVI)	Virosine	226, 270	c) 2.28, 2.33	Fig. 340, [Note]–CCXVI
(CCXVII)	Vindolicine	212, 257, 303	d) 4.45, 3.96, 3.81	Fig. 341
(CCXVIII)	Vincarodine	230, 272, 298	c) 2.82, 2.30, 2.04	Fig. 342
(CCXIX)	Leurosine	213, 261sh, 287, 310	c) 2.26(213mμ), 2.24(261mμ)	Fig. 343, [Note]–CCXIX
(CCXX)	Isoleurosine	214, 261, 287	d) 4.70, 4.21, 4.09	Fig. 344
(CCXXI)	Catharine	222, 265, 292	d) 4.74, 4.17, 4.08	Fig. 345
(CCXXII)	Carosine	255, 294	c) 2.21, 2.19	Fig. 346
(CCXXIII)	Pleurosine	267, 308	c) 2.32, 1.89	Fig. 347
(CCXXIV)	Neoleurocristine	220, 257, 298	c) 2.74, 2.23, 2.29	Fig. 348
(CCXXV)	Neoleurosidine	214, 268	c) 2.77, 2.29	Fig. 349
(CCXXVI)	Vindolidine	261, 311	c) 2.33, 2.12	Fig. 350
(CCXXVII)	Vincamicine	214, 264, 315, 341	c) 2.88, 2.39, 2.18 2.22	Fig. 351
(CCXXVIII)	Leurosidine	214, 265		
(CCXXIX)	Carosidine	212, 254, 303		Fig. 352

Fig. 334～352 IR Spectra of Vinca Alkaloids (Gordon *et al.*)[1969)1970]

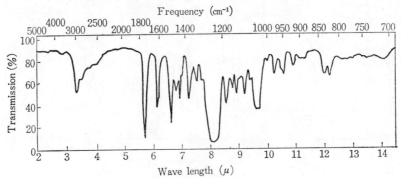

Fig. 334 Vindoline·2HCl in mineral oil mull

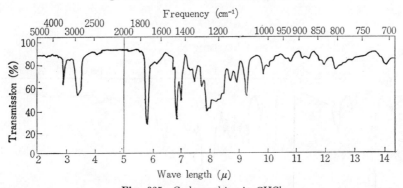

Fig. 335 Catharanthine in CHCl$_3$

[Note] **IR Spectra** of; (CCXVI) Virosine λ_{max}^{CHCl3} μ: 2.81, 3.03, 3.43, 5.67, 6.88, 7.93

(CCXIX) Leurosine λ_{max}^{CHCl3} μ: 2.80, 2.90 (>NH), 3.33, 3.39, 5.72 (ester), 6.16, 6.63, 6.83, 6.95, 7.26, 7.49, 8.1 (ester), 8.64, 8.71, 8.82, 8.91, 9.02, 9.15, 9.39, 9.59, 10.75, 11.10

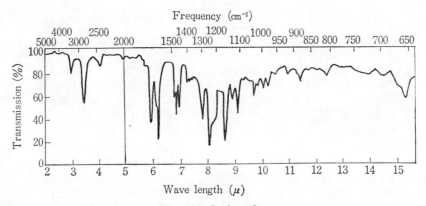

Fig. 336 Lochneridine

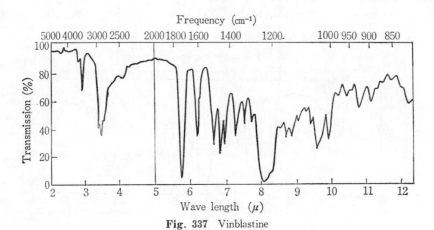

Fig. 337 Vinblastine

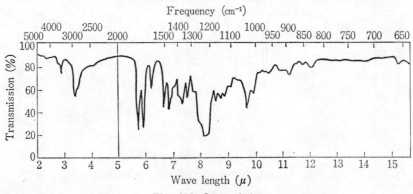

Fig. 338 Leurocristine

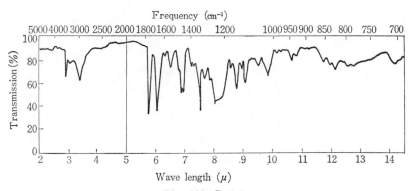

Fig. 339 Perivine

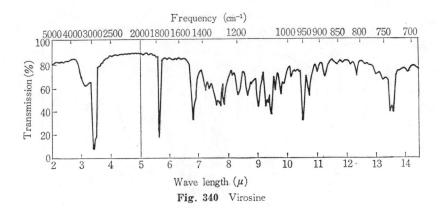

Fig. 340 Virosine

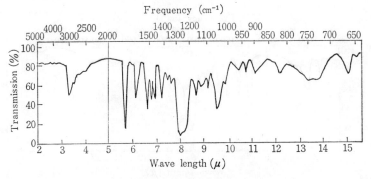

Fig. 341 Vindolicine

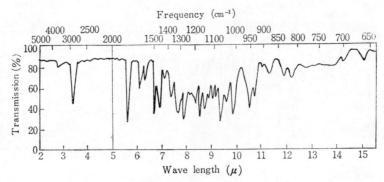

Fig. 342   Vincarodine

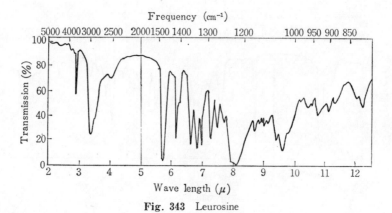

Fig. 343   Leurosine

Fig. 344   Isoleurosine

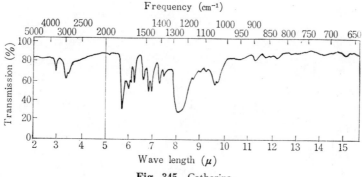

Fig. 345 Catharine

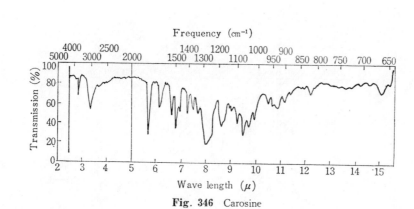

Fig. 346 Carosine

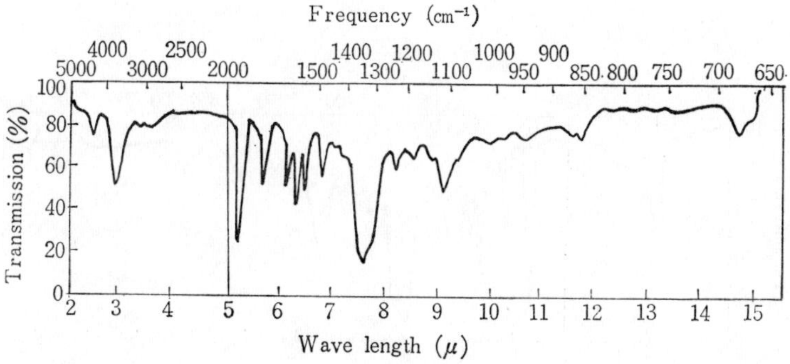

Fig. 347 Pleurosine

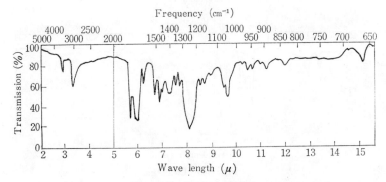

Fig. 348 Neoleurocristine

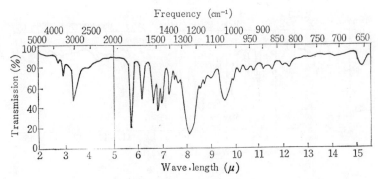

Fig. 349 Neoleurosidine

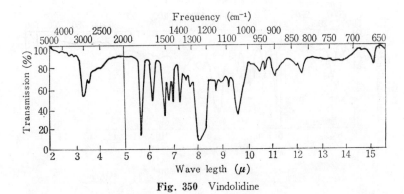

Fig. 350 Vindolidine

Frequency (cm⁻¹)

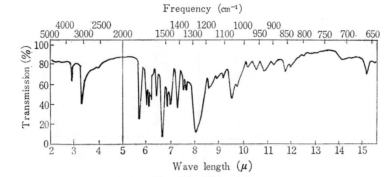

Wave length (μ)

Fig. 351 Vincamicine

Frequency (cm⁻¹)

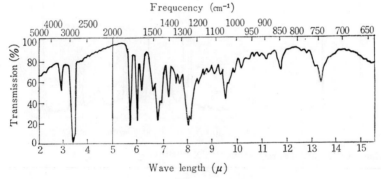

Wave length (μ)

Fig. 352 Carosidine

(d) NMR

Vindoline (CXCVIII)[1970) 1971)]

in CDCl₃, 60 Mc, TMS, δ (ppm):

See Fig. 353 [Note]

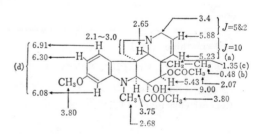

[Note] a) two *cis*-vinyl protons,   b) triplet, $J=7.5$ c/s,   c) 12 bands multiplet, $J=7.5$,

　　　　 d) typical 1, 2, 4-aromatic protons pattern, Jortho$=8$, Jmeta $=2.5$

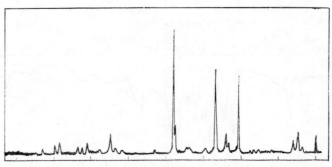

Fig. 353 NMR Spectrum of Vindoline (Gordon *et al.*)[1970]

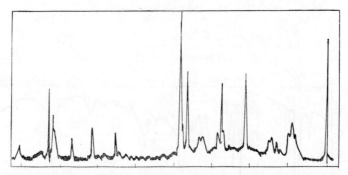

Fig. 354 NMR Spectrum of Vinblastine (Gordon *et al.*).[1970]

Vinblastine (CCIX) and **Leurocristine** (CCX)[1970) 1983]

δ due to:	N–CH$_3$	arom. OCH$_3$ &–COOCH$_3$	–COOCH$_3$	–OCOCH$_3$	$>$NH (indole)	CHO, OH
(CCIX)	2.73	3.8	3.63	2.12	8.09	9.8
(CCX)	—					9.5, 8.9

Catharanthine (CCIII)

τ: 8.90 t, $J=7$ (–CH$_2$–CH$_3$), 4.10 $\left(C=C<^H\right)$

(14) Iboga Alkaloids

(a) **General Formulae** (CCXXX) and structural correlation[1986]

$$R_1\overset{12}{}\quad R_2\overset{13}{}$$

(CCXXX)[1987]

1986) Taylor, W. I *et al*: *J.A.C.S.*, **79**, 3298 (1957), **80**, 123, 126 (1958).

1987) Renner, U *et al*: *Helv. Chim. Acta*, **42**, 1572 (1959).

Str. No.	Alkaloids	R_1	R_2	R_3	R_4
(CCXXXI)	Ibogamine	H	H	H	CH_3
(CCXXXII)	Ibogaine	OCH_3	H	H	CH_3
(CCXXXIII)	Tabernanthine	H	OCH_3	H	CH_3
(CCXXXIV)	Coronaridine	H	H	$COOCH_3$	CH_3
(CCXXXV)	Voacangine	OCH_3	H	$COOCH_3$	CH_3
(CCXXXVI)	Isovoacangine	H	OCH_3	$COOCH_3$	CH_3
(CCXXXVII)	Conopharingine	OCH_3	OCH_3	$COOCH_3$	CH_3
(CCXXXVIII)	Ibogaline	OCH_3	OCH_3	H	CH_3
(CCXXXIX)	Voacangarine	OCH_3	H	$COOCH_3$	CH_2OH

≪Occurrence≫　　See Table 104

Table 104　Occurrence of Iboga Alkaloids

(Renner *et al.*[1987] etc.)

Species \ Alkaloids	(CCXXXI)	(CCXXXII)	(CCXXXIII)	(CCXXXIV)	(CCXXXV)	(CCXXXVI)	(CCXXXVII)	(CCXXXVIII)	(CCXXXIX)
Tabernanthe iboga	○	○	○		○			○[1988]	
Tabernaemontana oppositifolia	○			○	○				
T. psychotrifolia				○	○				
T. australis				○					
Conopharyngia durissima						○	○		
Ervatamia Coronaria				○					
E. divaricata				○					
Stemmadenia galeottiana	○								
S. donnell–smithii			○		○	○			
Voacanga africana					○				○[1989]
V. thouarsii var. *obtusa*					○				
V. dregei					○				
Cabunea eglandulosa					○				

(c) Ibogamine (CCXXXI)

　　mp. 162∼164°, $[\alpha]_D$ −54° (EtOH), $C_{16}H_{24}N_2$

UV　See Fig. 355

IR　See Fig. 356[1990]

Mass　See the item (m)

(d) Ibogaine (CCXXXII)

　　mp. 151∼152°, $[\alpha]_D$ −53° (EtOH), $C_{20}H_{26}ON_2$

UV　Fig. 335

IR　See Fig. 357

Mass　See the item (m)

1988)　Neuss, N: *J. Org. Chem.* **24**, 2047 (1959).
1989)　Seebeck, E *et al*: *Helv. Chim. Acta.*, **41**, 169 (1958).
1990)　Prelog, V *et al*: *ibid.*, **39**, 742 (1956).

(e) Ibolutein (CCXXXII–a)

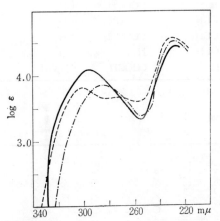

(CCXXXII) Ibogaine

(CCXXXII–a) Iboluteine

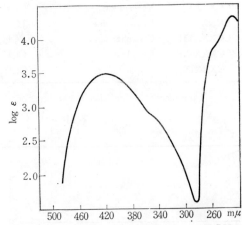

Fig. 355 UV Spectra of —— Ibogaine,
—·—·— Ibogamine and ······ Taber,
nanthine in EtOH

(Schlittler *et al.*)[1991]

Fig. 358 UV Spectrum of Iboluteine in EtOH

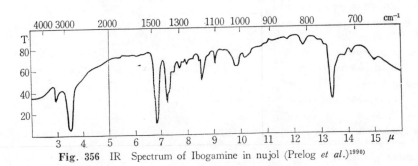

Fig. 356 IR Spectrum of Ibogamine in nujol (Prelog *et al.*)[1990]

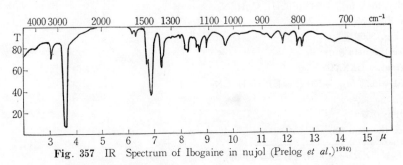

Fig. 357 IR Spectrum of Ibogaine in nujol (Prelog *et al.*)[1990]

1991) Schlittler, E *et al*: *ibid.*, **36**, 1337 (1953).

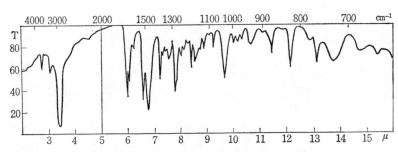

Fig. 359 IR Spectrum of Iboluteine in nujol

(Prelog *et al.*)[1990]

≪Occurrence≫   *Tabernanthe iboga*

mp. 142°, $[\alpha]_D$ −114° (CHCl₃), C₂₀H₂₆O₂N₂

UV See Fig. 358,   IR See Fig. 359

(f) Tabernanthine (CCXXXIII)[1986]

mp. 207∼211°, C₂₀H₂₆ON₂

Hydrochloride   mp. 275∼277°, $[\alpha]_D$−66° (MeOH)

UV See Fig. 355

Mass See the item (m)

(g) Coronaridine (CCXXXIV)[1992)1993]

mp. 92∼93°, $[\alpha]_D$ −34° (CHCl₃), C₂₁H₂₆O₂N₂

Hydrochloride mp. 235°, $[\alpha]_D$ −8.5° (MeOH)

UV λ_{max}^{EtOH} mμ (ε) (Base): 226, (3480), 286 (8300), 294 (7800)

IR (Base)λ_{max}^{CS2} μ: 2.82, 2.86, 3.33, 3.45, 5.82, 6.17, 7.04, 8.00, 8.51, 8.77, 9.26, 9.80

(h) Voacangine (CCXXXV)[1986]

mp. 136∼137°, $[\alpha]_D$ −34° (CHCl₃), pKa′ 7.4 (40% MeOH), C₂₂H₂₈O₃N₂

UV λ_{max} mμ (ε): 226 (28400), 288∼293 (9140), 300sh (8600)

(i) Isovoacangine (CCXXXVI)[1987]

mp. 155∼156°, $[\alpha]_D$ −51° (CHCl₃), pK_MCS 5.65, C₂₂H₂₈O₃N₂

UV λ_{max}^{MeOH} mμ (log ε): 227.5 (4.57), 270sh (3.79), 299 (3.90)

IR See Fig. 360

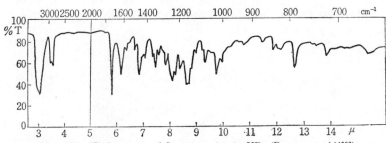

Fig. 360 IR Spectrum of Isovoacangine in KBr (Renner *et al.*)[1987]

(j) Conopharingine (CCXXXVII)[1987)1994]

mp. 141∼143°, $[\alpha]_D$ −40.5° (CHCl₃), pK_MCS 5.61, C₂₃H₃₀O₄N₂

1992) Gorman, M *et al*: *J.A.C.S.*, **82**, 1142 (1960).

1993) Kupchan, S, M *et al*: *J. Pharm. Sci.*, **52**, 598 (1963).

1994) Thomas, J: *J. Ph. Ph.*, **15**, 487 (1963).

UV λ_{max}^{MeOH} mμ (log ε): 224.5 (4.47), 304 (4.05)

IR See Fig. 361, ν_{max}^{nujol} cm^{-1}: 3350 (NH), 1725 (ester C=O)[1994]

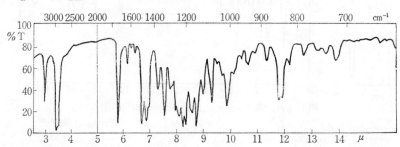

Fig. 361 IR Spectrum of Conopharingine in nujol (Renner *et al.*)[1987]

(k) Ibogaline (CCXXXVIII)[1988]

    mp. 141~143°, [α]$_D$ −42.9° (CHCl$_3$), C$_{21}$H$_{28}$O$_2$N$_2$

UV λ_{max}^{EtOH} mμ (log ε): 227 (4.40), 302 (3.92) (typical of 2,3–diethyl–5,6–dimethoxyindole);

IR $\lambda_{max}^{CHCl_3}$ μ: 2.91 (indole NH), 6.13[a], 6.28[a], 6.39[a], 6.74[a], 7.65, 8.64, 8.80, 9.79, 11.96[a] [Note]

Mass See the item (m)

(l) Voacangarine (CCXXXIX)[1989]

    mp. 166~167°, [α]$_D$ −29° (CHCl$_3$), C$_{22}$H$_{28}$O$_4$N$_2$

UV λ_{max}^{EtOH} mμ (log ε): 226 (4.45), 286 (3.99), 302 inf, (3.91), 314 (3.65); λ_{min}: 302 (3.91)

IR See Fig. 362, ν_{max}^{nujol} cm^{-1}: 3350, 3150 (OH & NH), 1732, 1625, 1590 (ester C=O)

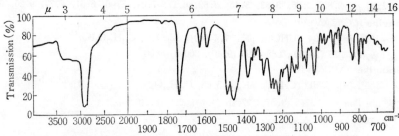

Fig. 362 IR Spectrum of Voacangarine in nujol (Seebeck *et al.*)[1989]

[Note] Absorption bands **a** appear only in the spectrum of 2,3–diethyl–5,6–dimethoxyindole but not in that of ibogamine (CCXXXI).

(m) **Mass Spectra**[1995)]

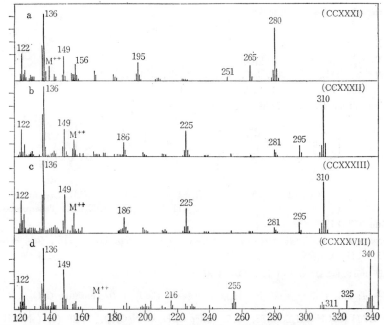

Fig. 363 Mass Spectra of (a) Ibogamine, (b) Ibogaine, (c) Tabernanthine
and (d) Ibogaline Intensity m/e 136: m/e 310=1.7: 1 (Biemann *et al.*)[1995)]

M : Ibogamine (CCXXXI) (280), Ibogaine (CCXXXII) (310), Tabernanthine (CCXXXIII) (310), Ibogaline
(CCXXXVIII) (340)

Fragmentation of Ibogaine (CCXXXII)

1995) Biemann, K *et al*: *J.A.C.S.*, **83**, 4805 (1961).

(n)   Voachalotine (CCXL)[1996]~[1998]

≪Occurrence≫   *Voacanga chalotiana*

Retroaldolization of akuammidine (see the item (12) Picralima Alkaloids (b)
(CLXXXVI))[1997]

mp. 231°, $[\alpha]_D$ +21° (CHCl$_3$), C$_{22}$H$_{26}$O$_3$N$_2$

Mass[1996] : 366 (M), 365, 364, 336, 335, 264, 263, 250, 196,
184, 183, 182, 181, 170, 169, 168, 157

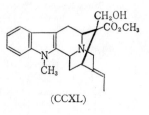

(CCXL)

(o)   Vobasine, Dregamine and **Tabernaemontanine**[1999]

(CCXLI) Vobasine

+ 2 H
[1999]

R$_1$=C$_2$H$_5$, R$_2$=H : (CCXLII) Dregamine
R$_1$=H, R$_2$=C$_2$H$_5$: (CCXLIII) Tabernaemontanine

Vobasine (CCXLI)[1999][2000]

≪Occurrence≫   *Voacanga africana*, *Peschiera affinis*,   mp. 112.5~115°, $[\alpha]_D$ −158° (CHCl$_3$),
pK 6.13, C$_{21}$H$_{24}$O$_3$N$_2$

UV λ_{max}^{MeOH} mμ: (log ε): 239 (4.19), 315 (4.27)

IR λ_{max}^{CH2Cl2} μ: 2.91 (NH), 3.60 (N–CH$_3$), 5.79 (ester C=O), 6.07 (unsatur. C=O)

NMR δ: 1.70d (C=CH–CH$_3$), 2.58, 2.65 (ester–OCH$_3$ & N–CH$_3$), 3.50s (2) (–N–CH$_2$–C=), 5.43q
(=CH–CH$_3$), ~7.3 (arom. protons), 9.37 (>NH···OC)

Mass   352 (M), 293 (M–59), 194 (D), 180 (A), 158 (C), 122 (B)

(CCXLI)

COOCH$_3$

180 (A)   122 (B)

158 (C)   194 (D)   M−59

Dregamine (CCXLII)[1991][2001]

≪Occurrence≫   *Voacanga dregei* E. M.,   mp. 137~140°, $[\alpha]_D$ −89.6° (CHCl$_3$), C$_{21}$H$_{26}$O$_3$N$_2$
Methine   mp. 97~100°, & 134~136°, $[\alpha]_D$ −24.1° (CHCl$_3$), pK$_{MCS}$ 7.13, C$_{22}$H$_{28}$O$_3$N$_2$

1996)   Clayton, E *et al*: *Tetrahedron*, **18**, 1449 (1962).
1997)   Janot, M *et al*: *Bull. Soc. Chim. France*, **1962**, 1079; *C.A.*, **57**, 9900e (1962).
1998)   Defay, N *et al*: *Bull. Soc. Chim. Belg.*, **70**, 475 (1961); *C.A.*, **57**, 11254c (1962).
1999)   Biemann, K *et al*: *Helv. Chim. Acta*, **46**, 2186 (1963).
2000)   Weisbach, J.A. *et al*: *J. Pharm. Sci.*, **52**, 350 (1963).
2001)   Neuss, N *et al*: *Experientia*, **15**, 414 (1959).

UV λ_{max}^{EtOH} mμ (log ε): 227 (4.24), 251.5 (4.36), 320 (4.14)

IR λ_{max}^{CH2Cl2} μ; 2.90 (NH), 3.55~3.60 (–N(CH$_3$)$_2$), 5.77 (ester C=O), 6.10 (unsatur. ketone)

Tabernaemontanine (CCXLIII)[1999) 2001) ~2003]

≪Occurrence≫ *Tabernaemontana* spp., mp. 215~216°, $[\alpha]_D$ −58° (CHCl$_3$), C$_{21}$H$_{26}$O$_3$N$_2$

(p) Affinine and Affinisine[2000]

≪Occurrence≫ *Peschiera affinis*

Affinine (CCXLIV)

mp. 256° (decomp.), C$_{20}$H$_{24}$O$_2$N$_2$

UV λ_{max}^{EtOH} mμ (log ε): 238 (4.18), 318 (4.34) (α–ketoindole)

IR See Fig. 363a, λ_{max}^{nujol} μ: 6.06~6.08 (ketoindole), 5.79~5.80 (ester)

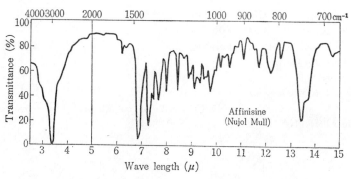

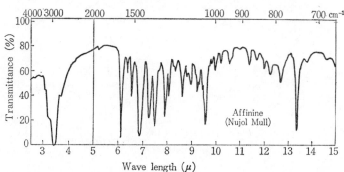

Fig. 363a IR Spectra of Affinisine and Affinine (Weisbach *et al.*)[2000]

Affinisine (CCXLV) mp. 115~118°

Hydrochloride mp. 287° (decomp.), $[\alpha]_D$ +40.3° (MeOH), C$_{19}$H$_{24}$ON$_2$·HCl

UV λ_{max}^{EtOH} mμ (log ε): 224 (4.65), 282 (3.92), 292 (3.82) (nonconj. indole)

IR See Fig. 363a, λ_{max}^{nujol} μ: 5.5~6.5 (C=O)

(q) Olivacine (CCXLVI)[2004]

≪Occurrence≫ *Tabernaemontana psychotrifolia*

See the item (10) Aspidosperma and Pleiocarpa Alkaloids, *TypeXIV* (c), (CLXVI)

(CCXLVI)

2002) Ratnagiriswaran, A. N. *et al*: *Quart. J. Pharm. Pharmacol*, **12**, 174 (1939).
2003) Renner, U *et al*: *Experientia*, **17**, 209 (1961).   2004) Wenkert, E: *J.A.C.S.*, **84**, 98 (1962).

(r) **Tabersonine** (CCXLVII)[1902) 2005) 2006)]

   ≪Occurrence≫   *Amsonia tabernaemontana* WALT.   Oily base,

C$_{21}$H$_{24}$O$_2$N$_2$, m/e 336

NMR δ: 5.75 (2H, fairly sharp), two olefinic protons, C–C$_2$H$_5$,

     COOCH$_3$, four arom. protons, 1⟩NH

     (CCXLVII) $\xrightarrow{\text{catalyt. red.}}$ (CCXLVIII)

     (CCXLVII) $\xrightarrow{\text{LiAlH}_4}$ (CCXLIX) $\xrightarrow{\text{Ac}_2\text{O}}$ (CCL)

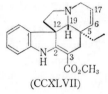

(CCXLVII)

6,7-Dihydrotabersonine (CCXLVIII)

   [α]$_D$ −540° (EtOH), C$_{21}$H$_{26}$O$_2$N$_2$

UV λ$_{max}$ mμ (log ε): 230 (4.05), 295 (4.17)

Tabersonol (CCXLIX)

   mp. 186°, [α]$_D$ +81.5° (EtOH), C$_{19}$H$_{26}$ON$_2$

UV λ$_{max}$ mμ (log ε): 245 (3.84), 300 (3.5)

IR λ$_{max}$ μ: 3.03 (charact. for dihydroindole)

Tabersonoldiacetate (CCL)

   mp. 196°, C$_{23}$H$_{30}$O$_3$N$_2$

UV λ$_{max}$ mμ (log ε): 205 (4.07), 280sh (3.48)

IR λ$_{max}$ μ: 5.8 (ester), 6.02 (–C(=O)–NRR′), 8.1 (Ac),no 3.1 (analog to strychnine)

(s) **Bis-indole Alkaloids**[2007)]

R$_1$=CO$_2$CH$_3$,

R$_2$=H : (CCLI)

Voacamine

(CCLII) Voacorine

Voacamine (CCLI)[2008) ~2010)]

   ≪Occurrence≫   *Voacanga africana, V. obtusa.*

     mp. 223°, [α]$_D$ −53° (CHCl$_3$), pKa 5.45, 7.14, C$_{43}$H$_{52}$O$_5$N$_4$

UV λ$_{max}$ mμ (log ε): 225 (4.45), 295 (4.16)

IR λ$_{max}^{nujol}$ μ: 5.8~5.9 (two ester grps.)

NMR δ: 1.66d, J=7cps, 5.20q, J=7 (C=CH–CH$_3$), 2.58s (COOCH$_3$), 2.44s (N–CH$_3$), 7.48, 7.70

   (two indole ⟩NH)

2005)   Janot, M.M *et al*: *Tetr. Letters*, **1962**, 271.

2006)   Janot, M.M *et al*: *Compt. rend.*, **248**, 3005 (1959); *C.A.*, **53**, 22729g (1959).

2007)   Chatterjee, A *et al*: *J. Sci. & Ind. Res.*, **23**, 178 (1964).

2008)   Janot, M.M *et al*: *C. R. Acad. Sci. Paris.*, **240**, 1719 (1955).

2009)   La Barre *et al*: *Bull. Acad. roy. Med. Belg.*, **20**, 194 (1955).

2010)   Büchi, G *et al*: *J.A.C.S.*, **85**, 1893 (1963).

Voacorine (CCLII)[2011)~2013)]

≪Occurrence≫   *Voacanga bracteata*,   mp. 273°, $[\alpha]_D$ −42° (CHCl₃), pKa 6.39, $C_{43}H_{52}O_6N_4$

UV λ_{max} mμ: 225, 295 (almost ident. with that of voacamine (CCLI), 5–methoxyindole)

IR Very similar to that of voacamine (CCLI), except possessing only one C=O band at 5.83μ

Vobtusine (CCLIII)[2008) 2014)~2016)]

≪Occurrence≫ *Voacanga africana, V. thoursii,* —— var. *obtusa, V. schweinfurthii, Callichilia subsessilis,*
        C. stenocepala, C. barteri,

mp. 305~306° (decomp.)[2015)], 312° (decomp.)[2016)], $[\alpha]_D$ −321°[2015)], −320°[2016)], (CHCl₃), pKa 6.95,
    $C_{42}H_{48}O_6N_4$

UV λ_{max} mμ: 220, 267, 300, 325

IR ν_{max} cm⁻¹: 3335 (bonded N–H), 1681 (amide), 1608 (arom.)[2015)], λ_{max} μ: 5.95, 6.17, 6.22[2008)]

Voacamidine (CCLIV)[2017)]

≪Occurrence≫   *Voacanga* spp.,   mp. 128~130°, $[\alpha]_D$ −174.5° (CHCl₃), $C_{45}H_{56}O_6N_4$

UV λ_{max} mμ: 227.5, 292.5, IR OH, NH, C=O, C=C

(15) Strychnos and Curare Alkaloids

(i) *Melinonines and related Alkaloids*

(CCLV) Melinonine A   (CCLVI) Melinonine B   (CCLVII) Melinonine E

(CCLVIII) Melinonine F

(CCLIX) Melinonine G

(CCLX) Yohimbol methine

N_b–β–CH₃: (CCLXI) Hunterburnine β–methine   (CCLXIII) Dihydrocorynantheolmethine
N_b–α–CH₃: (CCLXII) Hunterburnine α–methine

2011) Janot, M.M *et al*: *Compt. rend.*, **242**, 2981 (1956).
2012) Janot, M.M *et al*: *ibid.*, **244**, 1955 (1957).
2013) Djerassi, C *et al*: *J. Bull. Soc. Chim. France*, **1963**, 1899.
2014) Seebeck, E *et al*: *Helv. Chim. Acta*, **41**, 169 (1958).
2015) Schuler, B.O.G *et al*: *Soc.*, **1958**, 4776.   2016) Taylor, D.A.H. *et al*: *ibid.*, **1961**, 2587.
2017) Renner, U: *Experientia*, **13**, 468 (1957).

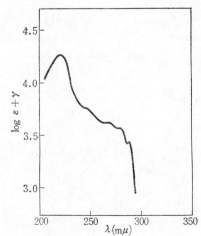

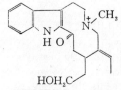

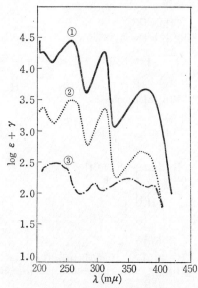

(CCLXIV) Hunterbrinemethine

(CCLXV) Burnamicine

(CCLXVI) Macusine **A**

(CCLVII) Macusine B

(CCLXVII–a) Tombozine

(CCLXVIII) C–Alkaloid T

(a) Melinonine A (CCLV)[2018]

≪Occurrence≫   *Strychnos melinoniana*, **Baillon**
Chloride mp. 255°, $[\alpha]_D$ −125° (H_2O), $C_{22}H_{27}O_3N_2 \cdot Cl$
UV See Fig. 364

Fig. 364 UV Spectrum of
Melinonine A chloride in
EtOH–H_2O (1 : 1) $\gamma = -0.4$
(Karrer *et al.*)[2018]

Fig. 365 UV Spectra of ① Melinonine
F chloride ($\gamma = 0$), ② Melinonine
E perchlorate ($\gamma = -1.0$) and
Melinonine G chloride
③ ($\gamma = -2.0$) in EtOH

2018) Karrer, P *et al*: *Helv. Chim. Acta*, **40**, 1167, 1793 (1957).

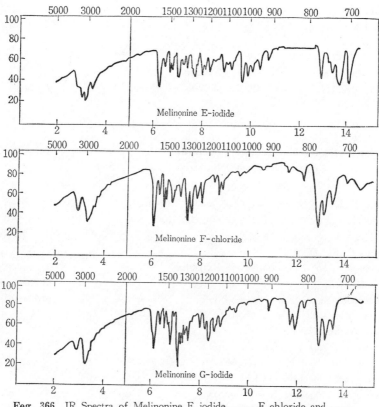

Feg. 366 IR Spectra of Melinonine E iodide, ――― F chloride and ―――
G iodide in KBr (Karrer *et al.*)[2018]

(b) Melinonine B (CCLVI)[2018]

≪Occurrence≫ *Strychnos melinoniana* BAILLON

Chloride mp. 311° (decomp.), $[\alpha]_D$ −14.8 (27.5% MeOH), $C_{20}H_{27}ON_2Cl$

IR See Fig. 367, λ_{max}^{nujol} μ: 3.05 (OH), 3.20 (indole NH), 13.27 (*ortho*–disubst. benzene), 6.81,

6.09 (w) (arom.); λ_{max}^{KRr} μ: 3.09, 3.22, 6.83, 13.28

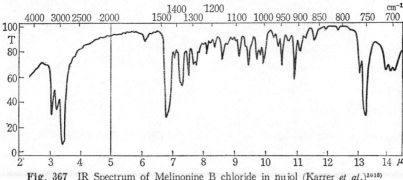

Fig. 367 IR Spectrum of Melinonine B chloride in nujol (Karrer *et al.*)[2018]

N,O–Diacetylmelinonine B chloride

    mp. 136~142°

UV See Fig. 368

(c) Melinonine E (CCLVII)[2018]

    ≪Occurrence≫   *Strychnos melinoniana* BAILLON

    Perchlorate mp. 187.5° & 220~222°, $C_{20}H_{23}ON_2 \cdot ClO_4 \cdot H_2O$

UV See Fig. 365

    Iodide mp. 234~238° (decomp.), $C_{20}H_{23}ON_2 \cdot I \cdot 1^1/_2H_2O$

IR See Fig. 366

(d) Melinonine F (Harmanchloromethylate) (CCLVIII)[2018]

    ≪Occurrence≫   *Strychnos melinoniana* BAILLON

    Chloride mp. 288° (decomp.), $C_{13}H_{13}N_2 \cdot Cl$

UV See Fig. 365 λ_{max}^{EtOH} mμ (log ε): 253 (4.46), 3.08 (4.27), 377 (3.67); λ_{min}: 225 (4.08), 279 (3.60), 328 (3.06)

IR See Fig. 366

(e) Melinonine G (CCLIX)[2018]

    ≪Occurrence≫   *Strychnos melinoniana* BAILLON

    Iodide $C_{17}H_{15}N_2 \cdot I$

UV See Fig. 365 (chloride), IR See Fig. 366

    Picrate mp. 229.5~230.5°

(f) Yohimbol methine (CCLX)[2021]

    ≪Occurrence≫   *Hunteria eburnea* PICHON

    Chloride mp. 264~265°, $[\alpha]_D$ +53° (MeOH), $C_{20}H_{27}ON_2 \cdot Cl$

(g) Hunterburnine β–methine (CCLXI)[2019)~2021]

    ≪Occurrence≫   *Hunteria eburnea* PICHON

    Chloride mp. 307~308°, $[\alpha]_D$ +105° (27.5% MeOH), $C_{20}H_{27}O_2N_2 \cdot Cl$

    Iodide mp. 277~280°

(h) Hunterburnine α–methine (CCLXII)[2019)~2021]

    ≪Occurreuce≫   *Hunteria eburnea* PICHON

    Chloride mp. 335°, $C_{20}H_{27}O_2N_2 \cdot Cl$

UV λ_{max}^{EtOH} mμ (ε): 273 (8700), 300 (4300), 311sh (3700); λ_{min}: 244 (6500), 294 (3700)

(i) Dihydrocorynantheol methine (CCLXIII)[2021]

    ≪Occurrence≫   *Hunteria eburnea* PICHON

    Chloride mp. 272~273° (decomp.)[2018], 296~297°[2021], $[\alpha]_D$ +101° (MeOH), $C_{20}H_{29}ON_2 \cdot Cl$

(j) Huntrabrine methine (CCLXIV)[2021]

    ≪Occurrence≫   *Hunteria eburnea* PICHON

    Chloride mp. 285~287°, $[\alpha]_D$ +54° (H$_2$O), $C_{20}H_{27}O_2N_2 \cdot Cl$

UV λ_{max}^{EtOH} mμ (ε): 271 (8800), 300 (4300), 310sh (3800)

IR ν_{max}^{nujol} cm^{-1}: 3120, 1220, 1135, 1031, 923, 913, 839, 814

(k) Hunteracine (CCLXIV–a)[2021]

    ≪Occurrence≫   *Hunteria eburnea* PICHON

    Chloride mp. 343~344° (decomp.), $[\alpha]_D$ −91° (H$_2$O–MeOH), $C_{20}H_{25}ON_2 \cdot Cl$

UV λ_{max}^{EtOH} mμ (ε): 234 (7900), 291 (2200)

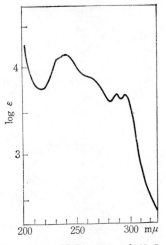

Fig. 368 UV Spectrum of N,O–Diacetylmelinonine B chloride in H$_2$O (Karrer *et al.*)[2018]

2019) Asher, J.D.M *et al*: *Proc. Chem. Soc.*, **1962**, 72.

2020) Scott, C.C *et al*: *ibid.*, **1962**, 355.   2021) Taylor, W.I *et al*: *J. Org. Chem.*, **28**, 1445 (1963).

(l) Hunteramine (CCLXIV–b)[2021]

   ≪Occurrence≫   *Hunteria eburnea* PICHON.,   mp. 206∼208°, pKₐ′ 4.6, $C_{26}H_{34}O_{10}N_2$ (possibly)

UV λ_{max}^{EtOH} mμ $\left(E_{1cm}^{1\%}\right)$: 221 (800), 271 (150), 278 (150), 282sh (140), 289 (120).

   IR ν_{max}^{nujol} cm⁻¹: 3350∼3170, 1160, 1075, 745

(m) Burnamicine (CCLXV)[2022][2027]

   ≪Occurrence≫   *Hunteria eburnea* PICHON.   mp. 198∼200°, $[\alpha]_D$ −281°, pKₐ 8.9 (50% MeOH), $C_{20}H_{26}O_2N_2$

UV λ_{max} mμ (ε): 309∼312 (14600) $\underset{OH^-}{\overset{H^+ \text{ or } NaBH_4}{\rightleftarrows}}$ typical indole absorption,

$\lambda_{max}^{neutral\ or\ base}$ mμ (ε): 311 (14600), 236sh (13500); λ_{min}: 265 (5200); λ_{max}^{acid} mμ (ε): 269 (7280), 280 (7000), 290 (5000); λ_{min}: 238 (1800), 277 (6700), 289 (4900)

   IR ν_{max}^{nujol} cm⁻¹: 3425, 3195 (OH and NH)[2027], ν_{max}^{KBr}cm⁻¹: 1630 (C=O)[2022]

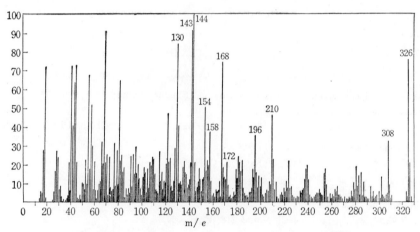

Fig. 369 Mass Spectrum of Burnamicine (Taylor *et al.*)[2022]

   NMR four arom. protons, ethylidene group., N–CH₃

   Mass See Fig. 369, M=326

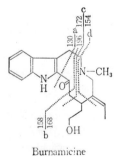

Burnamicine

(n) Macusine A (CCLXVI)[2023][2024][2024a]

   ≪Occurrence≫   *Strychnos toxifera* SCHOMB.

   Chloride   mp. 252° (decomp.), $[\alpha]_D$ −57.5° (H₂O)

   Iodide   mp. 274° (decomp.), $C_{22}H_{27}O_3N_2I$

UV $\lambda_{max}^{H_2O}$ mμ (log ε): 222 (4.73), 273 (3.84), 277 (3.84), 288 (3.71);

   λ_{min}: 251 (3.58), 275 (3.83), 286 (3.70) (characteristic of 2,3–disubst. indole, almost ident. with that of yohimbine methochloride)

   IR ν_{max} cm⁻¹: 3320 (OH), 3120 (NH), 1729 (COOR)

(o) Macusine B (CCLXVII)[2023][2024][2024a]

   ≪Occurrence≫   *Strychnos toxifera* SCHOMB.

   Chloride   mp. 248∼249° (decomp.), $[\alpha]_D$ +15.6° (H₂O), $C_{20}H_{27}ON_2Cl \cdot H_2O$

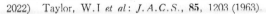

2022) Taylor, W. I *et al*: *J. A. C. S.*, **85**, 1203 (1963).
2023) Mc Phail *et al*: *Proc. Chem. Soc.*, **1961**, 223.
2024) Battersby, A. R *et al*: *ibid.*, **1961**, 17.   2024a) Battersby, A. R *et al*: *Soc.*, **1960**, 1848.

UV λ_{max}^{H2O} mμ (log ε): 222 (4.61), 273 (3.84), 280 (3.82), 291 (3.74); λ_{min}: 241 (3.22), 278 (3.81), 287 (3.60)

IR ν_{max} cm^{-1}: 3310 (OH), 3210 (NH), no ester band

Tombozine (CCLXVII–a)[2030]

≪Occurrence≫   *Diplorrhynchus condylocarpon* PICHON. spp. *mossambicensis* DUVIGN.

mp. 270~272°, $[\alpha]_D$ +38° (EtOH), $C_{19}H_{22}ON_2$

UV λ_{max}^{MeOH} mμ (log ε): 222 (4.60), 280 (3.87), 289 (3.77), 272 sh (3.85)

(p) C–Alkaloid T (Sarpagine methylether) (CCLXVIII)[2025]

≪Occurrence≫   " Calebassencurare " from *Strychnos* spp., *Vinca rosea* var. *alba*

See the item (7) Yohimbe, Alstonia and Rauwolfia Alkaloids, (w), (CVI)

(ii) *Eburnamine and related Alkaloids*

R=H, R₁=OH   (CCLXIX)   Eburnamine

R=OH, R₁=H   (CCLXX)   Isoeburnamine

R, R₁=O   (CCLXXI)   Eburnamonine

R=H, Δ_{10-11} (CCLXXII)   Eburnamenine

(a) Eburnamine, Isoeburnamine, Eburnamonine and Eburnamenine[1933a~c) 2019) 2026) 2027]

≪Occurrence≫   *Hunteria eburnea* PICHON

See the item (10) Aspidosperma and Pleiocarpa Alkaloids, *Type XIII*, (CLX), (CLXI), (CLXII)

Eburnamine (CCLXIX)[2027]

mp. 186~187°, $[\alpha]_D$ −93° (CHCl₃), pKa′ 7.7, $C_{19}H_{24}ON_2$

UV λ_{max}^{EtOH} mμ (ε): 229 (33800), 282 (8100), 276 sh (8400); λ_{min}: 249 (2500)

Isoeburnamine (CCLXX)[2027]

mp. 217~220°, $[\alpha]_D$ +111, (CHCl₃), pKa′ 7.8, $C_{19}H_{24}ON_2$

UV λ_{mxa}^{EtOH} mμ (ε): 229 (34600), 282 (7900), 277 sh (7800), 291 (6800); λ_{min}: 248 (2100)

Eburnamonine (CCLXXI)

mp. 183°, $[\alpha]_D$ +89° (CHCl₃), pKa′ 6.1, $C_{19}H_{22}ON_2$

UV λ_{max}^{EtOH} (ε): 241 (19800), 268 (10200), 295 (4800), 303 (4800); λ_{min}: 220 (8400), 258 (9400), 288 (4200), 299 (4600)

IR ν_{max}^{nujol} cm^{-1}: 1700 (C=O)

Mass See (CLX)

(b) Other Alkaloids from *Hunteria eburnea*[2027]

Eburnamenine (CCLXXII)

mp. 196°, $[\alpha]_D$ +183° (CHCl₃), pKa′ 6.45, $C_{19}H_{22}N_2$

UV λ_{max}^{EtOH} mμ (ε): 223(23200), 258 (29200), 301 (7600), 311 (8600); λ_{min}: 236 (12000), 285 (5900), 305 (6300).

Mass See (CLXII)

2025) Karrer, P *et al*: *Helv. Chim. Acta*, **40**, 705 (1957).

2026) Taylor, W. *J et al*: *J.A.C.S.*, **82**, 5941 (1960).

2027) Taylor, W. *I et al*: *J. Org. Chem.*, **28**, 2197 (1963).

Burnamine (CCLXXII–a)

mp. 197~198°, $[\alpha]_D$ −131° (CHCl$_3$), pKa′, 6.3, C$_{21}$H$_{26}$O$_4$N$_2$

UV λ_{max}^{EtOH} mμ (ε): 234 (6800), 288 (3100)

Neburnamine (CCLXXII–b)

mp. 285~290°, $[\alpha]_D$ −199° (MeOH), pKa′ 9.9, 7.7

UV λ_{max}^{EtOH} mμ $\left(E_{1cm}^{1\%}\right)$: 293 (140), 230 sh (460), 284 (130)

IR ν_{max}^{nujol} cm^{-1}: 3600, 3460, 3250 (OH and NH), 1741 (C=O)

Pleiocarpine (CXLVI)

Pleiocarpamine (CLVIII)

See the item (10) Aspidosperma and Pleiocarpa Alkaloids, *Type VI*, (e), (CXLVI) and *Type XI*, (a),(CLVIII)

(c) Corymine (CCLXXIII)[2028]

≪Occurrence≫　*Hunteria corymbosa*

mp. 189~192°, $[\alpha]_D$ +27.3° (CHCl$_3$),

pKa 7.86, C$_{22}$H$_{26}$O$_4$N$_2$

UV λ_{max}^{EtOH} mμ (log ε): 258 (4.99), 314 (4.50);

$\lambda_{max}^{0.2N-HCl-EtOH}$ mμ (log ε): 245 (5.00), 299 (4.51)

IR ν_{max} cm^{-1}: 3150 (OH), 1725 (C=O)

NMR in CDCl$_3$, τ: 6.18 (CH$_3$O), 7.28 (N–CH$_3$),

8.35 d, J=6.5 c/s and 4.62 q, J=6.5 $\left(\!\!\diagdown\!\!C=CH\cdot CH_3\right)$

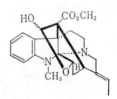

(CCLXXIII)

(iii) *Fluorocuraine, Strychnine and related Alkaloids*

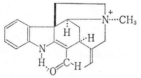

$\beta \rightarrow X$ or $\beta \rightarrow Y$ (CCLXXIV)[2029]
Stemmadenine

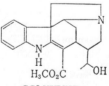

(CCLXXV) Condylocarpine

(CCLXXVI)
Norfluorocurarine

(CCLXXVII)
Fluorocurarine

(CCLXXVIII)
Echitamidine

2028) Smith, G. F *et al*: *Proc. Chem. Soc.*, **1962**, 298.
2029) Sandoral, A *et al*: *Tetr. Letters*, **1962**, 409.

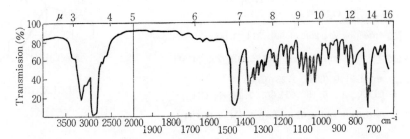

R₁=R₂=H (CCLXXIX) Strychnine
R₁=OCH₃, R₂=H (CCLXXX) β–Colubrine
R₁=R₂=OCH₃ (CCLXXXI) Brucine

R=H (CCLXXXII) Wieland Gumlich
    aldehyde[2038]
R=Ac (CCLXXXIII) Diaboline

(a) **Stemmadenine** (CCLXXIV)[2029)2030)]

  ≪Occurrence≫ *Stemmadenia* spp., *Diplorrhynchus condylocarpon* spp., *mossambicensis*
mp. 189~191°, $[\alpha]_D$ +329° (pyridine), $C_{21}H_{26}O_3N_2$

UV λ_{max}^{MeOH} mμ (log ε): 226 (4.57), 284 (3.89), 292 (3.86), 275 sh (3.81), 305 sh (3.33)

IR See Fig. 369–a

Fig. 369-a IR Spectrum of Stemmadenine in nujol (Stauffacher)[2030)]

NMR δ: 9.5 (indole NH), 7.35 (four arom. protons), 3.79 (COOCH₃), 5.4 q $\left(\rangle\!\!=\!\!<^H_{CH_3} \right)$, 1.7 d $\left(\rangle\!\!=\!\!<^H_{CH_3} \right)$

Mass 354 (M), 324 (M—HCHO), 336 (M— H₂O), 123 (a)

(CCLXXIV) − HCl $\xrightarrow[2031)]{KMnO_4}$ Condylccarpine (CCLXXV) + HCHO

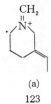

(a)
123

(b) **Condylocarpine** (CCLXXV)[2029~2031)]

  ≪Occurrence≫ *Diplorrhynchus condylocarpon*, spp. *mossambicensis*
mp. 159~162°, $[\alpha]_D$ +900° (CHCl₃), $C_{20}H_{22}O_2N_2$

UV λ_{max}^{MeOH} mμ (log ε): 228 (4.04), 295 (4.01), 328 (4.17)

IR Fig. 370

2030) Stauffacher, D: *Helv. Chim. Acta.*, **44**, 2006 (1961).
2031) Biemann, K *et al*: *Tetr. Letters*, **1962**, 527.

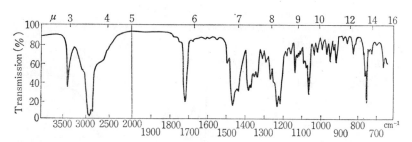

Fig. 370 IR Spectrum of Condylocarpine in nujol (Stauffacher)[2030]

NMR 60 & 100 Mc, δ: 1.58 d (a), 5.32 q (f), 3.78 s (3) (c), 2.95 (?), 3.92 (d), 4.12 (e), 5.3~7.0 m (g).

Mass :

$$M = 326$$

(c) Norfluorocurarine (CCLXXVI)[1967)1968)2030]

≪Occurrence≫ *Diplorrhynchus condylocarpon* PICHON spp., *mossambicensis* DUVIGN. mp. 184~186°, $[\alpha]_D$ −123° (CHCl₃), $C_{19}H_{20}ON_2$

UV λ_{max}^{MeOH} mμ (log ε): 242 (3.98), 290 sh (3.51), 299 (3.57), 360 (4.25)

IR See Fig. 371, $\lambda_{max}^{CHCl_3}$ μ: 6.07, 6.18, 6.47, 6.82

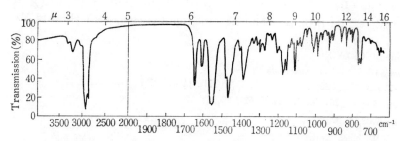

Fig. 371 IR Spectrum of Norfluorocurarine in nujol (Stauffacher)[2030]

Wieland Gumlich aldehyde (CCLXXXII) ⟶ C–Curarine I

C–Curarine I $\xrightarrow{\text{mineral aids}}$

Norfluorocurarine (CCLXXVI) $\xrightarrow{\text{CH}_3\text{I}}$ C–Curarine Ⅲ (CCLXXVII)[2036]

(d) Fluorocurarine (C–Fluorocurarine, C–Curarine III) (CCLXXVII)[2032)2033)]

≪Occurrnce≫ "Calabash curare", *Strychnos macrophylla, S. mitscherlichii, S. tomentosa, S. trinervis*

Salts $C_{20}H_{23}ON_2X$: X=Cl, I, ClO_4

UV See Fig. 372, see the item (11) Geissospermum Alkaloids, (f) Indolenium Salt.

IR λ_{max}^{KBr} μ: 6.0~6.25 (chelated

α, β–unsatur. aminoaldehyde),
3.54 (CaF$_2$–prism) (ν_{C-H} of
—CHO)

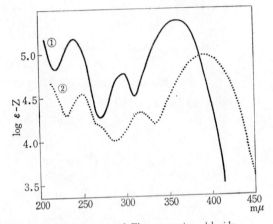

Fig. 372 UV Spectra of Fluorocurarine chloride
① in EtOH, Z=−1.13
② in 0.01 N–NaOH–EtOH, Z=−0.63
(Karrer *et al.*)

Tetrahydrofluorocurarine

Picrate mp. 153°

Chloride $[\alpha]_D$ +60.5° (MeOH) $C_{20}H_{27}ON_2Cl$

UV λ_{max}^{EtOH} mμ (log ε): 246 (3.95), 302 (3.50) (typical indoline absptn.); $\lambda_{max}^{2N-HCl-EtOH}$ mμ: 261, 26
(typical indolinium absorption).

IR $\lambda_{max}^{KBr\,or\,nujol}$ μ: 2.98, 3.17 (OH & NH), 6.21 (indoline)

(e) Echitamidine (CCLXXVIII)[2034)]

≪Occurrence≫ *Alstonia scholaris* R. BR.

mp. 244°, $[\alpha]_D$ −515°, $C_{20}H_{24}O_3N_2$

NMR δ: 3.89 (COOCH$_3$), 6.7~7.4 (arom. protons), 8.68 (NH), 1.16 d, J=6 c/s (3) $\left(\!\!\!\begin{array}{c}\rangle CH-CH_3\end{array}\!\!\!\right)$

Mass 340 (M), 322 (M–H$_2$O), see echitamine (item (10) Aspidosperma and Pleiocarpa Alkaloids, *Type IX*, (a), (CLV)

(f) Strychnine (CCLXXIX)

≪Occurrnce≫ *Strychnos nux vomica*,

mp. 278~279.5°, 286~288°, $[\alpha]_D$ −109.9° (EtOH), pKa 7.37 $C_{21}H_{22}O_2N_2$

UV λ_{max} mμ (log ε): 257 (4.2), 281 (3.62), 290 (3.53)

IR $\lambda_{max}^{CHCl_3}$ μ: 6.03, 6.27

Stereospecific total synthesis[2035)]

2032) Karrer, P *et al*: *Helv. Chim. Acta.*, **42**, 461 (1959).
2033) Karrer, P *et al*: *ibid.*, **41**, 1257 (1958).
2034) Djerassi, C *et al*: *Tetr. Letters*, **1962**, 653.
2035) Woodward, R.B *et al*: *Tetrahedron*, **19**, 247 (1963).
2036) Fritz, H *et al*: *Ann.* **663**, 150 (1963).

(g) *β*–**Colubrine** (CCLXXX)[2039]

   ≪Occurrence≫   *Strychnos nux vomica*,   mp. 222°, $[\alpha]_D$ −156° (CHCl₃), $C_{22}H_{24}O_3N_2$

(h) **Brucine** (CCLXXXI)

   ≪Occurrence≫   *Strychnos nux vomica*,   mp. 175~177°, $[\alpha]_D$ −127° (EtOH), $C_{23}H_{36}O_4N_2$

(i) **Wieland Gumlich aldehyde**[2042] (Caracurine VII) (CCLXXXII)[2037)2038)2040)~2043]

   ≪Occurrence≫   *Strychnos toxifera*[2041], derived from strychnine[2043] and toxiferine via alkaloid A8 [2040]

   mp. 213~214°, (decomp.), $[\alpha]_D$ −134.5°, (MeOH), $C_{19}H_{22}O_2N_2$[2042] [Note]

  UV See Fig. 373

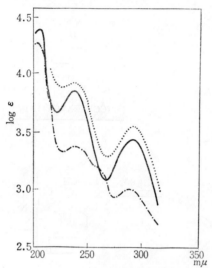

Fig. 373 UV Spectra of Caracurine VII—HCl
———— in H₂O, ·········· in 0.05 *N*
H₂SO₄, —·—·— in 0.05 *N*–NaOH
(Karrer *et al.*)[2041]

(j) **Alkaloid A 8** (Hemitoxiferine I) (CCLXXXII–a)[2038)2040]

   ≪Occurrence≫   *Strychnos toxifera*, derived from toxiferine I, from Wieland Gumlich aldehyde:

      (CCLXXXII) $\xrightarrow{\text{CH}_3\text{I}}$ (CCLXXXII–a)

Chloride $C_{20}H_{25}O_2N_2Cl$, $[\alpha]_D$ −43° (H₂O)

Picrate mp. 233~235°, $C_{20}H_{25}O_2N_2 \cdot C_6H_2O_7N_3$

(CCLXXXII–a)

(k) **Diaboline** (CCLXXIII)[2037)2038]

   ≪Occurrence≫   *Strychnos diaboli*

      Wieland Gumlich aldehyde (CCLXXXII) $\xrightarrow{\text{Ac}_2\text{O}}$ (CCLXXXIII)

  $[\alpha]_D$ +42.0° (CHCl₃)

Hydrochloride mp. >300°, $C_{21}H_{24}O_3N_2 \cdot HCl \cdot H_2O$

  IR λ_{max}^{KBr} *μ*: 6.03 (N–Ac), 6.24 (indoline)

2037) Battersby, A.B *et al*: *Proc. Chem. Soc.*, **1959**, 126.
2038) Karrer, P *et al*: *Helv. Chim. Acta*, **45**, 2266, (1962).
2039) Rosemund, P: *Ber.*, **95**, 2639 (1962).   2040) Battersby, A.R. *et al*: *Soc.*, **1960**, 736.
2041) Karrer, P *et al*: *Helv. Chim. Acta*, **37**, 1983, 1993 (1954).
2042) Karrer, P *et al*: *ibid.*, **41**, 1405 (1958).   2042a) Karrer, P *et al*: *ibid.*, **42**, 201 (1959).
[Note] The formula $C_{20}H_{23}O_2N_2Cl$ (hydrochloride)[2041] has been revised to $C_{19}H_{22}O_2N_2$ (base)[2042].

(iv) **Bis–indole Alkaloids**[2007) 2040) ~2044) 2047)]

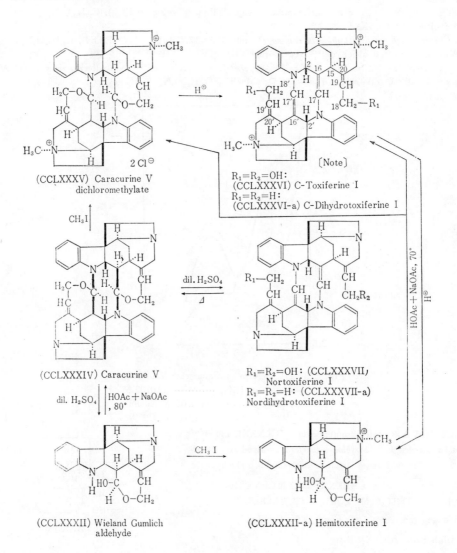

(CCLXXXV) Caracurine V
dichloromethylate

$R_1=R_2=OH$:
(CCLXXXVI) C-Toxiferine I
$R_1=R_2=H$:
(CCLXXXVI-a) C-Dihydrotoxiferine I

(CCLXXXIV) Caracurine V

$R_1=R_2=OH$: (CCLXXXVII)
Nortoxiferine I
$R_1=R_2=H$: (CCLXXXVII-a)
Nordihydrotoxiferine I

(CCLXXXII) Wieland Gumlich
aldehyde

(CCLXXXII-a) Hemitoxiferine I

(CCLXXXVI–b)

2043) Wieland *et al*: *Ann.*, **494**, 191 (1932), **506**, 60 (1933);
Robinson, R *et al*: *Soc.*, **1955**, 2253.
2043a) Karrer, P *et al*: *Helv. Chim. Acta*, **38**, 166 (1955).
2044) Karrer, P *et al*: *Helv. Chim. Acta*, **44**, 34 (1961).
[Note] The former structure of toxiferine I (CCLXXXVI–b)[2040) ~2042)]
has been revised to (CCLXXXVI)[2047)].

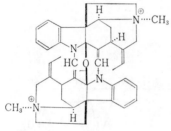

(CCLXXXVIII) Caracurine II

(CCLXXXIX) C—Alkaloid D

Caracurine V (CCLXXXIV) $\underset{\Delta}{\overset{H^+}{\rightleftarrows}}$ Caracurine Va $\longrightarrow$ Caracurine VII

$\longrightarrow$ Caracurine II (CCLXXXVIII)[2041)2044)~2046)]

C—Toxiferine I (CCLXXXVI) $\overset{H^+}{\longrightarrow}$ Hemitoxiferine I (CCLXXXII–a)[2045)]

$\longrightarrow$ Caracurine II (CCLXXXVIII)–methochloride[2045)]

(CCXC) C–Curarine

$R_1=OH$, $R_2=R_3=H$: (CCXCI) Calebassine
$R_1=R_3=OH$, $R_2=H$: (CCXCII) C–Alkaloid A

(a)   or   (b)[2051)]

(a) **Caracurine V** (CCLXXXIV)[2042)2042a)]

《Occurrence》   *Strychnos toxifera*
mp. >300°, $C_{38}H_{40}O_2N_4$
UV See Fig. 374
IR See Fig. 375

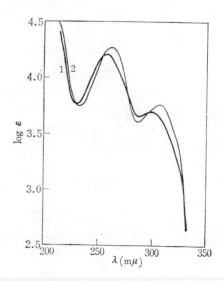

Fig. 374 UV Spectra of Caracurine V
–2HCl in ① H_2O and
② MeOH
(Karrer *et al.*)[2042)]

2045) Battersby, A.R *et al*: *Proc. Chem. Soc.*, **1961**, 412, 413.
2046) Sim, G.A *et al*: *ibid.*, **1961**, 416.

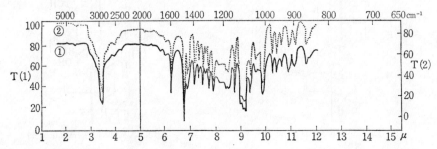

Fig. 375 IR Spectrum of Caracurine V in CHCl₃ (Karrer *et al.*)²⁰⁴²⁾

NMR²⁰⁴²⁾ in CDCl₃, 56.4 Mc, τ: 2.7~3.7 (8) (arom. protons), 4.15 (2) $(C_{19,19'})$, 5.31 (2) $(C_{17,17'})$, 5.5~6.6 m (10) (protons attach to $C_{2,2'}$, $_{18,18',21,21'}$)

(b) C–Toxiferine I (CCLXXXVI)²⁰⁴⁰⁾²⁰⁴²⁾²⁰⁴⁷⁾

≪Occurrence≫   *Strychnos toxifera*, Calabash curare,   mp. 270°, $C_{40}H_{46}O_2N_4{}^{2+}$
Dichloride $C_{40}H_{46}O_2N_4Cl_2\cdot 4H_2O$, $[\alpha]_D$ —529° (H₂O)

UV λ_{max}^{EtOH} mμ (log ε): 293 (4.61); λ_{max}: 237 (3.75)

IR See Fig. 376

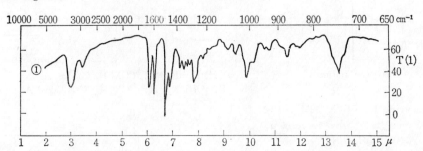

Fig. 376 IR Spectrum of C–Toxiferine I–dichloride in KBr (Karrer *et al.*)²⁰⁴²⁾

NMR in CDCl₃, 60 Mc, TMS, c/s=0, external standerd D₂O²⁰⁴⁷⁾: 375 t, J=4.5 $(C_{19}$–H, $C_{19'}$–H), 356 s (2) $(C_{17}$–H, $C_{19'}$–H), 284 d, J=4.5 $(C_{18}$–, $C_{18'}$–methylene protons), 412 (arom. protons)

(c) C–Dihydrotoxiferine I (CCLXXXVI–a)²⁰⁴²⁾²⁰⁴²ᵃ⁾²⁰⁴³ᵃ⁾²⁰⁴⁷⁾

≪Occurrence≫   *Strychnos toxifera*, $[\alpha]_D$ —599~—611° (50 % EtOH), $C_{40}H_{46}N_4{}^{2+}$
Picrate mp. 182~185°

IR See Fig. 377

2047) Karrer, P *et al*: *Helv. Chim. Acta*, **44**, 620 (1961).

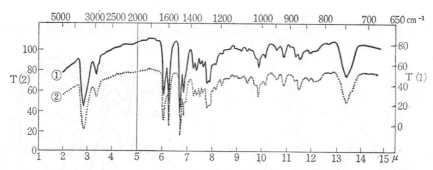

Fig. 377 IR Spectrum of C–Dihydrotoxiferine–dichloride in KBr (Karrer *et al.*)[2042]

NMR c/s (condition was same to that of (b)[2047]): 373 q, (C_{19}–H, $C_{19'}$–H), 359 s (2) (C_{17}–H, $C_{17'}$–H), 416 (arom. protons)

(d) Nordihydrotoxiferine (CCLXXXVII–a)[2042] [2043a]

≪Occurrence≫ *Strychnos toxifera*, $[\alpha]_D$ −567° (MeOH), $C_{38}H_{40}N_4$

IR See Fig. 378

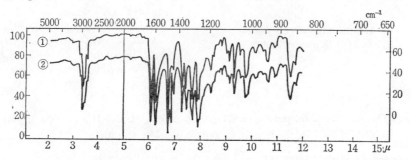

Fig. 378 IR Spectrum of Nordihydrotoxiferine in CCl_4 (Karrer *et al.*)[2042]

(e) Caracurine II (CCLXXXVIII)[2041] [2044] ~[2046]

≪Occurrence≫ *Strychnos toxifera*, mp. 248~249°, $[\alpha]_D$ −232° (CHCl$_3$), $C_{38}H_{38}O_2N_2$

UV See Fig. 379; $\lambda_{max}^{EtOH-H2O}$ mμ (log ε): 246 (4.22), 291 (3.73); λ_{max}^{H2SO4} mμ (log ε): 236 (3.87), 322 (4.35)

IR *ortho*–disubst. benzene, ether linkages, no OH, $>$NH, $>$C=N—, $>\overset{|}{C}=\overset{|}{C}$–N$<$, and C=O absorptions.
NMR in CDCl$_3$, 56.4 Mc, τ: 2.7~3.7 (8) (arom. protons), 4.76 (2) (C_{19}–H, $C_{19'}$), 5.14 (2) (C_{17}–H, $C_{17'}$–H), 5.5~6.6 m (10) (protons attached to C_2, $C_{2'}$, C_{18}, $C_{18'}$, C_{21}, $C_{21'}$), 580 s (2) (C_2–H, $C_{2'}$–H)

(f) C–Alkaloid D (CCLXXXIX)[2044] [2045] [2047a, b)]

≪Occurrence≫ Calabash Curare, $C_{40}H_{48}O_2N_2{}^{2+}$
Picrate mp. $>$270°

UV See Fig. 380, λ_{max}^{H2O} mμ (log ε): 238 (4.15), 284 (3.66); $\lambda_{max}^{0.1N-KOH}$ mμ (log ε): 244 (4.13), 288 (3.56); λ_{max}^{H2SO4} mμ (log ε): 236 (3.86), 322 (4.32)

2047a) Karrer, P. *et al*: *Helv. Ceim. Acta*, **35**, 1864 (1952).
2047b) Karrer, P. *et al*: *ibid.*, **36**, 102 (1953).

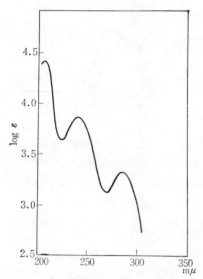

Fig. 379   UV Spectrum of Caracurine
II—HCl in H_2O (Karrer *et al.*)[2041]

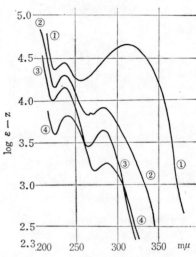

Fig. 380   UV Spectra of C–Alkaloid
D–dichloride in ① 12 *N*–HCl, ②
N–HCl, ③ H_2O and ④ 0.1 *N*–KOH
(① Z=−0.5, ② Z=−0.3, ③ Z=0,
④ Z=+0.3)

(Karrer *et al.*)[2044]

NMR in D_2O, 56.4 Mc, TMS, (ext.) cps: 377 q, *J*=6 (2) (C_{19}–H, $C_{19'}$–H), 316 s (2) (C_{17}–H,
  $C_{17'}$–H), 291s (2) (C_2—H, $C_{2'}$–H), 126 d, *J*=6(6) (C_{18}–Me, $C_{18'}$–Me)

(g) C–Curarine (CCXC)[2047b) 2048) 2049) 2052)]
  ≪Occurrence≫   Calabash curare, $C_{40}H_{44}ON_4^{2+}$
  Picrate mp. 306~307°
UV   See Fig 381, IR of derivatives[2048)], NMR[2049)]

(h) Calebassine (CCXCI)[2047b) 2048) ~2052)]
  ≪Occurrence≫   Calabash curare, $C_{40}H_{48}O_2N_4^{2+}$
  Picrate mp. 210~212°

UV   See Fig. 382 (indoline carbinolamine)
NMR   60 Mc, TMS, (ext.) cps[2051)]: Dichloride in D_2O — 416 d, *J*=8 (2), 455 br. (6) (arom. protons), 374 q,
  J=7 (C_{19}–H,$C_{19'}$–H), 291 s (2) (C_{17}–H, $C_{17'}$–H), 213 s

  $\left(\underset{bb'}{>}N\overset{\oplus}{-}CH_3\right)$, 179 s ($C_{16}$–H, $C_{16'}$–H), 138 d, *J*=7 (18–Me, 18'–Me)

2048)   Karrer, P *et al*: *Helv. Chim. Acta.*, **43**, 141 (1960).
2049)   Karrer, P *et al*: *Tetrahedron*, **14**, 138 (1961).
2050)   Karrer, P *et al*: *Helv. Chim. Acta*, **41**, 673 (1958).
2051)   Karrer, P *et al*: *ibid.*, **44**, 2211 (1961).
2052)   Karrer, P *et al*: *ibid.*, **42**, 2650 (1959), **29**, 1853 (1946).

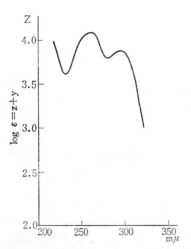

Fig. 381 UV Spectrum of C–Curarine-
chloride in water

(Karrer *et al.*)[2047b]

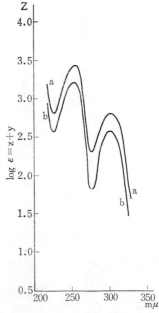

Fig. 382 UV Spectra of ⓐ C–Calebassine chloride in H_2O ($y=0.4$) and
ⓑ C–Alkaloid A chloride ($y=0.8$)
in H_2O

(Karrer *et al.*)[2047b]

(i) C–Alkaloid A (CCXCII)[2047b) 2048) ~2053)]

≪Occurrence≫ Calabash curare, $C_{40}H_{48}O_4N_4^{2+}$
Picrate mp. 228~229°

UV See Fig. 382 (indoline carbinolamine)

NMR 60 Mc, TMS (ext.), cps[2051] : Dichloride in D_2O — 423 d, $J=7.5$ (2), 455 br. (b) (arom. protons), 382 t,

$J=7$ (2) (C_{19}–H, $C_{19'}$–H), 291 s(2) (C_{17}–H, $C_{17'}$–H), 216 s $\left(\overset{\oplus}{\underset{b,b'}{\gt N}}-CH_3\right)$, 184 s (2) ($C_{16}$–H,

$C_{16'}$–H), 288 d, $J=7$ (4) (18–CH$_2$–, 18'–CH$_2$–)

2053) Swan, G. A : *Soc.*, **1958**, 2038.

(16) Gelsemium and Mitragyna Alkaloids

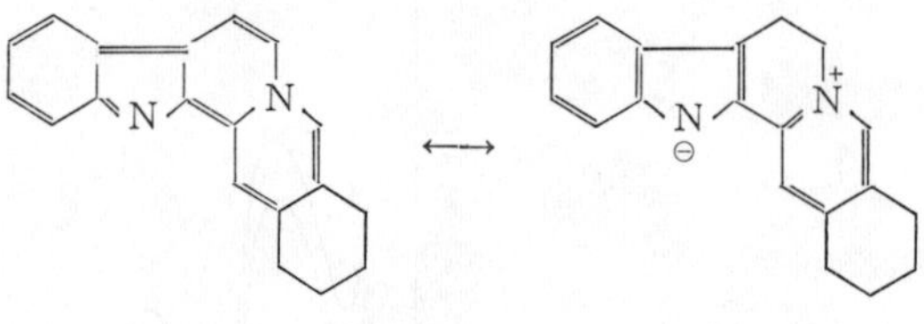

(CCXCIII) Sempervirine[2054]

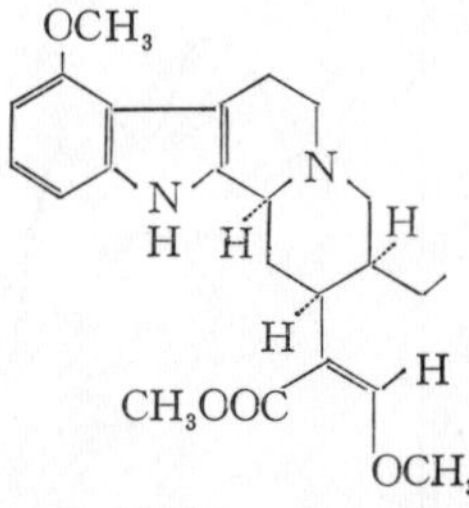

(CCXCIV) Mitragynine[2055]

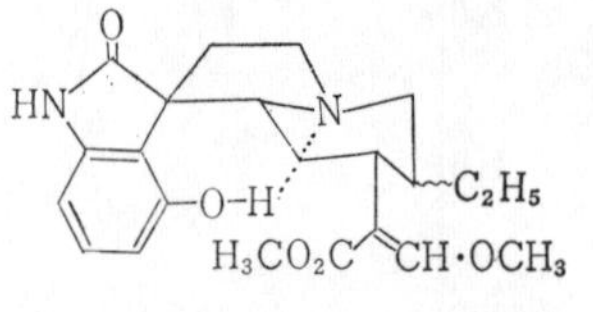

(CCXCV) Rotundifoline[2057) 2060)]

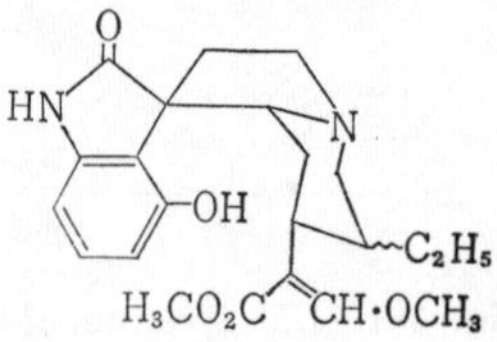

(CCXCV-a) Isorotundifoline[2055]

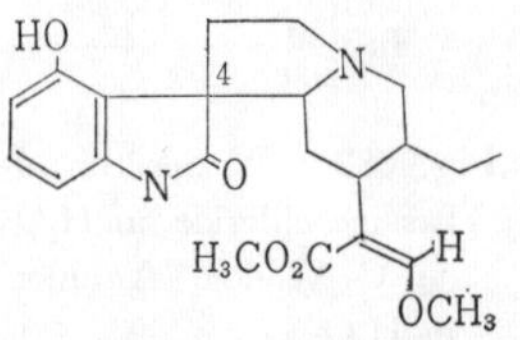

(CCXCV-b) Stipulatine[2061]

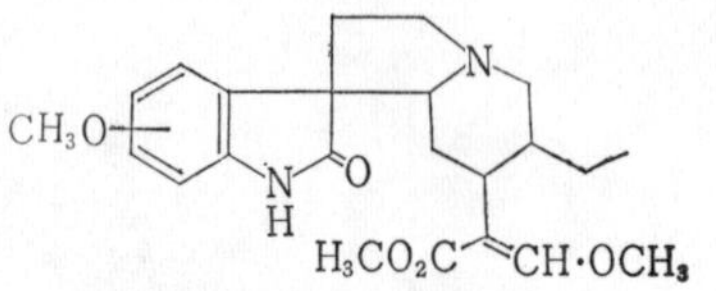

(CCXCVI) Rhynchociline
(CCXCVII) Ciliaphylline[2059]

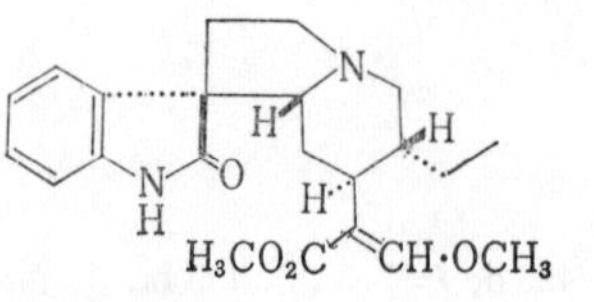

2062)

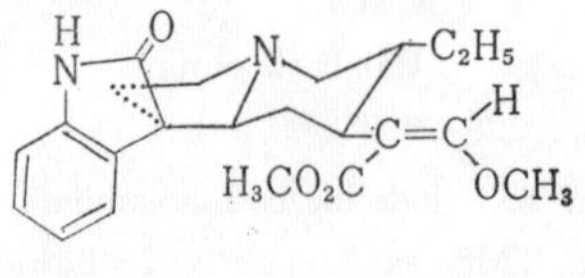

2062a)

(CCXCVII) Rhyncophylline

2053a) Marion. L. et al: J.A.C.S., 71, 1694 (1949).

2054) Woodward. P.B et al: ibid., 71, 379 (1949).

2055) Joshi, B.S. et al: Chem. & Ind., 1963, 573.

2056) Djerassi, C. et al: J.A.C.S., 85, 1523 (1963).

2057) Beckett, A.H. et al: Chem. & Ind., 1963, 1122.

2058) Badger, G.M: Proc. Chem. Soc., 1963, 206.

2059) Beckett, A.H. et al: J. Ph. Ph. 15, 158 T, 166 T, 267 T (1963).

2060) Hendrickson, J.B: Chem. & Ind., 1961, 713.   2060a) Badger, G.M. et al: Soc., 1950, 867.

2061) Hendrickson, J.B et al: Tetr. Letters, 1963, 929.

2062) Hendrickson, J.B: J.A.C.S., 84, 643, 650 (1962).

2062a) Taylor. W.L. et al: ibid., 84, 1318, 3871 (1962).

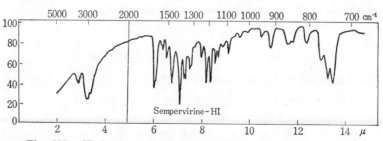

2062)

2062a)

(CCXCVIII) Isorhyncophylline

(CCXCIX) Corynoxine[2062b]

R_1, 8, 3, 2, 13, 4, 20, 21, 15, 19, 17

20–β–H, $R_1=R_2=$H
(CCC) Mitraphylline[2056]
20–α–H, $R_1=$OCH$_3$, $R_2=$H
(CCCI) Aricine Oxindole
20α–H, $R_1=R_2=$OCH$_3$
(CCCII) Carapanaubine

(CCCIII) Mitraginol[2060a],

(CCCIV) Uncarine A[2062]

(CCCV) Uncarine B[2062]

2064) (CCCVI) Gelsemine 2065) 2066)

R=OCH$_3$ (CCCVII) Gelsemicine
R=H (CCCVIII) Gelsedine

(a) **Sempervirine** (CCXCIII)[2018] 2053) 2053a) 2054)

≪Occurrence≫ *Gelsemium sempervirens* AIT. (*Loganiaceae*)

mp. 258∼260°, C$_{19}$H$_{16}$N$_2$·H$_2$O

Nitrate (synth.) mp. 267° (decomp.), C$_{19}$H$_{16}$N$_2$·HNO$_3$·0.75H$_2$O

UV λ_{max}^{EtOH} mμ (log ε): 243 (4.58), 248 (4.57), 297 (4.21), 346 (4.23), 386 (4.22); λ_{min}: 276 (4.04), 311 (4.08), 370 (4.16); $\lambda_{max}^{0.015 N-KOH-EtOH}$ mμ (log ε): 230 (4.44), 245 (4.46), 289 (4.46), 322 (4.14), 362 (4.29), 435 (3.72); λ_{min}: 237 (4.43), 268 (4.15), 310 (4.05), 333 (4.08), 425 (3.71)

IR See Fig. 382a

Sempervirine–HI

Fig. 382a IR Spectrum of Sempervirine–HI in KBr (Karrer *et al.*)[2018]

Methiodide

UV λ_{max}^{EtOH} mμ (log ε): 242 (4.53), 295 (4.21), 338 (4.29), 395 (4.22)

(b) **Mitragynine** (9 Methoxycorynantheidine) (CCXCIV)[2055]

    ≪Occurrence≫   *Mitragyna speciosa* KORTH., $C_{23}H_{30}O_4N_2$

    UV λ_{max} mμ ($\varepsilon \times 10^{-3}$): 226 (41.15), 249sh (13.13), 284sh (6.65), 292 (6.6)

    IR ν_{max} cm^{-1}: 3365 (NH), 1690, 1640 (CH$_3$OOC–$\overset{\shortmid}{C}$=$\overset{\shortmid}{C}$–O–) (closely resemble to corynantheidine)
    NMR: almost superimposable to corynantheidine at 0~4 δ region)

    See the item (7) Yohimbe, Alstonia and Rauwolfia Alkaloids (l), Corynantheidine (XCIII)

(c) **Rotundifoline** (CCXCV)[2057~2060]

    ≪Occurrence≫   *Mitragyna stipulosa, M. ciliata, M. inermis (Rubiaceae)*

    mp. 239~240°, [α]$_D$ +124.7° (CHCl$_3$), pKa 5.3 (H$_2$O), $C_{22}H_{28}O_5N_2$

    UV λ_{max}^{EtOH} mμ (log ε): 221 (4.39), 289 (3.42), 242.5sh (4.14); λ_{min}: 272 (3.21)

    IR ν_{max}^{nujol} cm^{-1}: 3240, 2450 br., 1700, 1625, 1275, 1250, 1107, 847, 780, 750, 730

    NMR in CDCl$_3$, TMS, ppm[2058]: Mitraphylloid (CCCIII) and Rhyncophylloid (CCXCV)

(CCCIII) Y=OCH$_3$ (CCXCV)

	>CH–CH$_3$	>CH·CH$_2$–CH$_3$	–2OCH$_3$	C$_{17}$–H	arom. H
(CCCIII)	0.7~0.9d	—	3.74s, 3.82s	7.40s	
(CCXCV)	—	0.7~0.9m	3.64s, 3.74s	7.39s	6.3~7.2m [Note]

(d) **Isorotundifoline** (CCXCV–a)[2059]

    ≪Occurrence≫   *Mitragyna stipulosa*

    mp. 130~132°, [α]$_D$ −7.7° (CHCl$_3$), pKa 7.4 (H$_2$O), $C_{22}H_{28}O_5N_2$

    UV λ_{max}^{EtOH} mμ (log ε): 218 (4.43), 289 (3.49), 242sh (4.13)

    IR ν_{max}^{nujol} cm^{-1}: 3250, 1695, 1685, 1625, 1605, 1230, 1140, 1100, 1095, 915, 900, 790, 770, 740

    Mass See the item (m)

(e) **Stipulatine** (CCXCV–b)[2061]

    ≪Occurrence≫   *Mitragyna speciosa*

    mp. 238~240°, [α]$_D$ +108° (CHCl$_3$), pKa 5.2 (50% EtOH), $C_{22}H_{28}O_5N_2$

(f) **Rhinchociline** (CCXCVI) and **Ciliaphylline** (CCXCVII)[2059]

    ≪Occurrence≫   *Mitragyna ciliata*

    Rhinchociline (CCXCVI)

    mp. 178~180°, [α]$_D$ +6.2° (CHCl$_3$), pKa 6.7 (80% MCS), $C_{23}H_{30}O_5N_2$

    UV λ_{max}^{EtOH} mμ (log ε): 225 (4.41), 242 (4.24), 286 (3.48)

    IR ν_{max}^{nujol} cm^{-1}: 3525w, 3100, 1685, 1605, 1270, 1240, 970 w, 780, 730

[Note] The signals (6.3~7.2 ppm) were examined for four *ar*–methoxyindoles and the presence of hydroxylgroup at C$_9$ position was confirmed by the comparative study.

Ciliaphylline (CCXCVII), an isomer of (CCXCVI)

mp. 222~223°, $[\alpha]_D$ −89.5° (CHCl$_3$), pKa 6.75, C$_{23}$H$_{30}$O$_5$N$_2$

UV λ_{max}^{EtOH} mμ (log ε): 222 (4.44), 244 (4.24), 287 (3.46)

IR ν_{max}^{nujol} cm^{-1}: 1728, 1705, 1640, 1620, 1500, 1402, 1380, 1335, 1300, 1180~1300, 1170, 1150, 780, 770, 740

(g) **Rhynchophylline** (CCXCVII)[2059) 2062) 2062a) 2062c]

≪Occurrence≫ *Mitragyna ciliata, M. stipulosa, M. inermis, M. rubrostipulata, Uncaria rhyncho-phylla* (カギカヅラ)

mp. 212~214°, $[\alpha]_D$ −14.4° (CHCl$_3$), pKa 6.8 (H$_2$O), C$_{22}$H$_{28}$O$_4$N$_2$

UV λ_{max}^{EtOH} mμ (log ε): 208.3 (4.45), 243.3 (4.21), 282.0 (2.93); λ_{min}: 222.2 (3.76), 277.8 (2.89)

IR ν_{max}^{nujol} cm^{-1}: 1725, 1700, 1640, 1280, 1250, 1180, 1125, 1100, 780, 745

Mass See the item (m)

(h) **Isorhynchophylline** (CCXCVIII)[2059) 2062) 2062a]

≪Occurrence≫ *Mitragyna rubrostipulata, Uncaria rhynchophylla*

mp. 144°, $[\alpha]_D$ +8.6° (CHCl$_3$), pKa 6.25 (H$_2$O), C$_{22}$H$_{28}$O$_4$N$_2$

UV λ_{max}^{EtOH} mμ (log ε): 204 (4.46), 239.8 (4.24), 277.8sh (3.25); λ_{min}: 220.8 (4.02)

IR ν_{max}^{nujol} cm^{-1}: 3200, 1695, 1610, 1600, 1105, 750

(i) **Corynoxine** (CCXCIX)[2062) 2062b]

≪Occurrence≫ *Pseudocinchona africana*, mp. 166~168°, $[\alpha]_D$ −14° (pyridine), −3° (CHCl$_3$), C$_{22}$H$_{28}$O$_4$N

(j) **Mitraphylline** (CCC)[2056) 2059) 2062]

≪Occurrence≫ *Mitragyna stipulosa*, mp. 267~268°, pKa 5.3, C$_{21}$H$_{24}$O$_4$N$_2$

UV λ_{max}^{EtOH} mμ (log ε): 208 (4.67), 241.5 (4.19), 281 (3.03); λ_{min}: 222 (4.0), 274 (2.99)

IR ν_{max}^{nujol} cm^{-1}: 3600, 3250, 1715, 1700, 1620, 1290, 1270, 1190, 1170, 1105, 775, 760

(k) **Aricine Oxindole** (CCCI)[2056]

≪Occurrence≫ Oxidation product of aricine (see the item (7) Yohimbe, Alstonia and Rauwolfia Alkaloids (b), (LXXI))

(LXXI)   (CCCI)

mp. 185~187°, $[\alpha]_D$ +140° (CHCl$_3$), C$_{22}$H$_{26}$O$_5$N$_2$·CH$_3$OH [Note]

UV λ_{max}^{EtOH} mμ (log ε): 227~230 (4.35), 291 (3.83)

IR ν_{max}^{CHCl3} cm^{-1}: 1745, 1700, 1626

[Note] The formula C$_{24}$H$_{28}$O$_5$N$_2$·CH$_3$OH in original report[2056] should be revised to C$_{22}$H$_{26}$O$_5$N$_2$·CH$_3$OH.

2062b) Janot, M.M. *et al*: *Bull. Soc. Chim. France*, **24**, 1292 (1957).
2062c) Ban, Y. *et al*: *Chem. Pharm. Bull.*, **11**, 441, 446, 451 (1963).

(l)   Carapanaubine   (CCCII)[2056]

≪Occurrence≫   *Aspidosperma carapanauba*

mp.   221~223°, $[\alpha]_D$ −101° (CHCl₃), $C_{23}H_{28}O_6N_2$

UV λ_{max}^{EtOH} mμ (log ε): 215 (4.57), 244 (4.23), 278 inf (3.80), 300 (3.66); λ_{min}: 238 (4.15);

$\lambda_{max}^{HCl-EtOH}$ mμ (log ε): 222 (4.56), 246 inf (4.15), 278 (3.79), 300 (3.61); λ_{min}: 263 (3.72)

IR λ_{max}^{nujol} μ: 3.09, 5.90, 5.99, 6.15

NMR   in CDCl₃, 60, 100 Mc, TMS, δ (ppm).

Compare to that of isoreserpiline (see the item (7) Yohimbe, Alstonia and Rauwolfia Alkaloids (h),
(LXXIII)) 1.40d, J=6 (3) (CH₃–ĊH–O–), 4.56 octet, J=6, (1) (CH₃–ĊH–O–), 3.61s (3) (–COOCH₃),
3.90s, 3.92s (6) (CH₃O), 4.2~4.7q (1) (allylic H), 6.55s, 6.74s (2) (arom. H), 7.44s (1) (olefinic H),
8.73s (1) (>NH)

(m)   Mass Spectra of Mitragyna Alkaloids[2056]

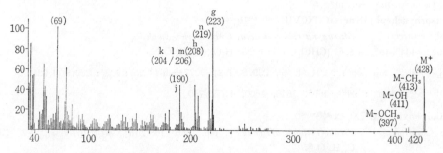

Fig. 383   Mass Spectrum of Carapanaubine (Djerassi *et al.*)[2056]

The mass spectra are quite different from those of heteroyohimbine alkaloids (see the item (7) Yohimbe,
Alstonia and Rauwolfia Alkaloids, (c), (ii)).

Rhyncophylline   (CCXCVII): 384 (M⁺), 369 (M−Me), 367 (M−OH), 355 (M−Et), 353 (M−OCH₃),
239 (g), 224 (g−Me), 210 (g−Et), 208 (g−OCH₃), 159 (n), 144~146 (k, l, m), 130 (j), 69 (i)

Mitraphylline   (CCC): 368 (M⁺), 353 (M−Me), 351 (M−OH, f), 337 (M−OCH₃), **223** (g), 208 (h),
159 (n), 144~146 (k, l, m), 130 (j), 69 (i)

Carapanaubine   (CCCII): 428 (M⁺), 413 (M−Me), 411 (M−OH), 397 (M−OCH₃), **223** (g), 219 (n),
208 (h), 204~206 (h, l, m), 190 (j), 69 (i)

Fragmentation :

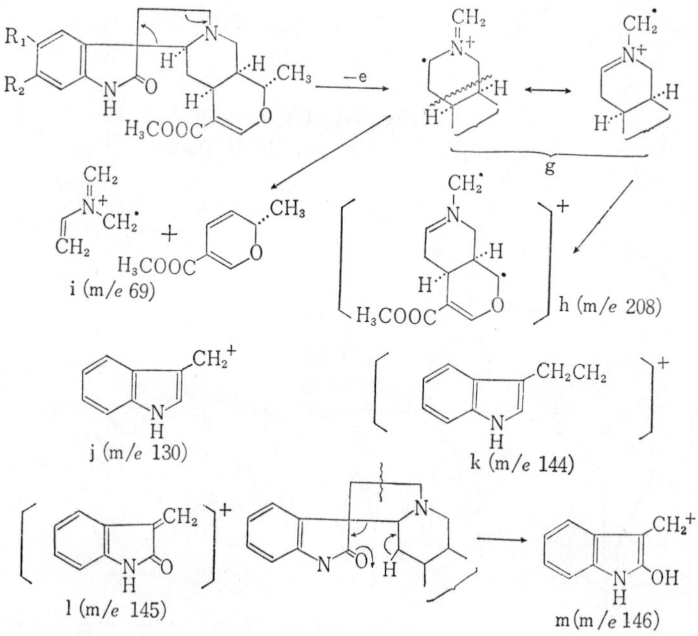

(n) Mitraginol (CCCIII)[2058)2060)2060a)]

≪Occurrence≫ *Mitragyna rotundifolia*

mp. ca. 130°, $[\alpha]_{5500A}$ −0.5° (c=4.24), −5.2° (c=2.12), −10.4° (c=1.06), $C_{22}H_{26}O_5N_2$ [Note]

NMR[2058)] See NMR of the item, (c) Rotundifoline (CCXCV)

(o) Uncarine A and **Uncarine B**[2062)2063)]

≪Occurrence≫ *Uncaria Kawakami*

Uncarine A (CCCIV)

Amorphous, pka 4.2, $C_{21}H_{24}O_4N_2$ Hydrochloride mp. 220° (decomp.), $[\alpha]_D$ +113.1°

UV See Fig. 384, IR See Fig. 385

Uncarine B (CCCV)

mp. 216~217°, pka 5.5, $C_{21}H_{24}O_4N_2$ Hydrochloride mp. 227~228° (decomp.), $[\alpha]_D$ +93.6°

UV See Fig. 384, IR See Fig. 385

2063) Nozoe: *Chem. Pharm. Bull.*, **6**, 300 (1958).

[Note] The molecular formula of Mitraginol in original report[2060a)] is $C_{21}H_{26}O_5N_2$ but it should be revised to $C_{22}H_{26}O_5N_2$ if the structure being (CCCIII).

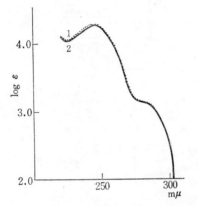

Fig. 384 UV Spectra of (1) Uncarine A and (2) Uncarine B in MeOH (Nozoe)[2063]

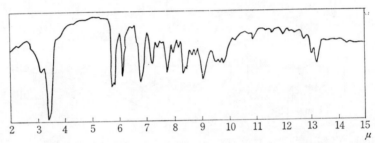

Fig. 385 IR Spectrum of Uncarine B in nujol (Nozoe)[2063]

(p) Gelsemine (CCCVI)[2064]~[2070]

《Occurrence》 *Gelsemium sempervirens,* mp. 176°, $[\alpha]_D$ +13° (CHCl$_3$), pKa 7.75, C$_{20}$H$_{22}$O$_2$N$_2$
Hydrochloride mp. 326°, $[\alpha]_D$ +5° (H$_2$O)

UV See Fig. 386, λ_{max} mμ (log ε): 252 (3.89), 282 (3.18)[2069]

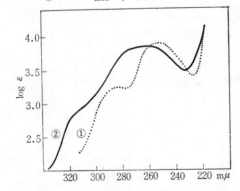

Fig. 386 UV Spectra of
① Gelsemine-HCl and
② Gelsemine in 0.1 *N*–NaOH
(Taylor *et al.*)[2067]

2064) Teuber, H.J. *et al*: *Ber.*, **93**, 3100 (1960).
2065) Conroy, H. *et al*: *Tetr. Letters.*, **1959**, No. 4, 6.
2066) Lovell, F.M. *et al*: *ibid.*, **1959**, No. 4, 1.
2067) Taylor, W.I. *et al*: *Helv. Chim. Acta*, **34**, 1139 (1951).
2068) Janot, M.M. *et al*: *ibid.*, **34**, 1962 (1951).
2069) Jones, G. *et al*: *Soc.*, **1953**, 2344.
2070) Marion, L. *et al*: *J.A.C.S.*, **78**, 5127 (1956).

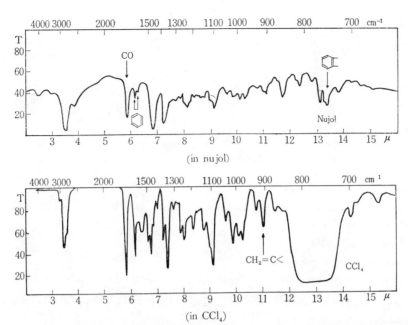

Fig. 387 IR Spectra of Gelsemine (Taylor *et al.*)[2067], (Janot *et al.*)[2068]

IR   See Fig. 387 in nujol[2067], in CCl₄[2068]

NMR   in CDCl₃, τ[2065] : 12 strong lines of the ABX system—three symmetrically split quartets centered at 3.72 (X), 4.95 (A), 5.10 (B) with J=11.3 c/s, 17.8 (X), 1.5, 11.3 (A) and 1.5, 17.8 (B)—

further splitting was not observed for X. ($>$C–C=CH₂)

X–Ray[2066]

(q) Gelsemicine (CCCVII)[2971) 2072]

 ≪Occurrence≫   *Gelsemium sempervirens*,   mp. 171~172°, $[\alpha]_D$ −142° (CHCl₃), C₂₀H₂₆O₄N₂

NMR   in CDCl₃, TMS, ppm: 7.26dd, J=1.3, 8.5 c/s (arom. C₉–H), 6.55dd J=3.0, 8.5 (arom. C₁₀–H), 6.48m (C₁₂–H, arom. peak pattern of 11–methoxyindole alkaloids).

(r) Gelsedine (CCCVIII)[2071) 2072]

 ≪Occurrence≫   *Gelsemium sempervirens*,   mp. 172.6~174°, $[\alpha]_D$ −158° (CHCl₃), C₁₉H₂₄O₃N₂

NMR   in CDCl₃, TMS, ppm: 6.81~7.44m (4) (arom.–unsubst. oxindole), 3.96s (MeO), 1.72q, J=7.5 c/s (2) ~1.00t, J=7.5 (3) (−CH₂–CH₃), neither 3.2 (charact. of *N*–Me–indoles) or N_b–Me

(s) Biosynthesis[2073]

2071)   Marion, L. *et al* : *ibid.*, **75**, 4372 (1953).

2072)   Wenkert, E. *et al* : *J. Org. Chem.*, **27**, 4123 (1962).

2073)   Hendrickson. J.B. *et al* : *J.A.C.S.*, **84**. 643 (1962).

(17) Calycanthaceous Alkaloids

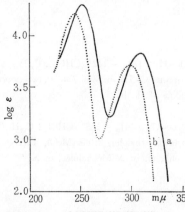

R=R′=H (CCCIX) Chimonanthine   (CCCXII) Hodgkinsine   (CCCXIII) Calycanthine
R=H, R′=Me (CCCX) Calycanthidine[2074]
R=R′=Me (CCCXI) Folicanthine

(a) **Chimonanthine** (CCCIX)[2074a~2076]

≪Occurrence≫   *Chimonanthus fragrans* LINDLE.   mp. 188~189°, $[\alpha]_D$ −329° (EtOH), $C_{22}H_{26}N_4$

UV λ_{max} mμ: 246, 304 (ph–N–C–N)

IR $\nu_{max}^{CCl_4}$ cm^{-1}: 3440 (NH)

(b) **Calycanthidine** (CCCX)[2074) 2074a) 2077]

≪Occurrence≫   *Calycanthus floridus*,   mp. 142°, $[\alpha]_D$ −317° (EtOH), pKa 7.37, 6.04, $C_{23}H_{28}N_4$

UV $\lambda_{max}^{neutral}$ mμ (ε): 250 (15200), 308 (6050) ; $\lambda_{max}^{dil-HCl}$ mμ (ε): 243 (15050), 298 (5100) (ph–N–C–N)

(c) **Folicanthine** (CCCXI)[2074a) 2076) 2078]

≪Occurrence≫   *Calycanthus floridus, C. occidentalis, Chimonanthus fragrans* (*Calycanthaceae*)
mp. 118~120°, $[\alpha]_D$ −364° (EtOH), $C_{24}H_{30}N_4$

UV See Fig. 388, λ_{max}^{EtOH} mμ (log ε): 254 (4.25), 311 (3.83)

Fig. 388 UV Spectra of Folicanthine (a) in EtOH,
(b) in HCl–EtOH (Hodson *et al.*)[2078]

(d) **Hodgkinsine** (CCCXII)[2007) 2079) 2080]

≪Occurrence≫   *Hodgkinsonia frutescens* (*Rubiaceae*)
mp. 128°, $[\alpha]_D$ +60° (0.3 *N*–HCl), $C_{22}H_{26}N_4$
UV λ_{max} mμ: 246, 305 (ph.–N–C–N)

2074) Saxton. J.E. *et al*: *Proc. Chem. Soc.*, **1962**. 148.
2074a) Grant. I.J. *et al*: *ibid.*, **1962**, 148.
2075) Hendrickson, J.B. *et al*: *ibid.*, **1962**, 283.
2076) Hodson, H.F. *et al*: *ibid.*, **1961**, 465.
2077) Barger, G. *et al*: *J. Rec. trav. Chim. Pays-Bas.*, **57**, 548 (1938).
2078) Hodson, H.F. *et al*: *Soc.*, **1957**, 1877.

(e) **Calycanthine** (CCCXIII)

Former Structure (CCCXIII–a)

≪Occurrence≫ *Calycanthus glaucus, C. floridus*

The former structure (CCCXIII–a) of calycanthine had been belonged to indole alkaloid which has been revised to (CCCXIII). See the item 〔I〕 Quinoline alkaloids, (6), (XXXII).

(18) Other Indole Alkaloids

(a) **Adifoline** (CCCXIV)[2081]

(CCCXIV)

≪Occurrence≫ *Adina cordifolia* HOOK (*Rubiaceae*)

mp. 195～196° (decomp.), $C_{22}H_{20}O_6N_2 \cdot 3H_2O$

UV λ_{max}^{EtOH} mμ (log ε): 235 (4.59), 285 (4.38), 365 (3.67);

$\lambda_{max}^{0 \cdot 01N-HCl-EtOH}$ mμ (log ε): 248 (4.47), 299 (4.44), 418 (3.79);

$\lambda_{max}^{0 \cdot 01N-NaOH-EtOH}$ mμ (log ε): 244 (4.65), 269 (4.56), 390 (3.72)

IR ν_{max}^{nujol} cm^{-1}: 3510, 3300, 3075, 3570～2530 (one or more CO_2H), 1709s, 1706s, 1689m, 1678s, 1664m, 1658m, 1638s, 1607s, 1577, 1220s, 1124s, (weak or medium) 1515, 1312, 1250, 1241, 1193, 1081, 1024, 812, 790, 769, 719

(b) **Haplophytine** (CCCXV)[2082]

(CCCXV)
(Partial Str.)

≪Occurrence≫ *Haplophyton cimicidum,* $C_{27}H_{31}O_5N_3$

UV λ_{max}^{EtOH} mμ (log ε): 265 (4.02), 305 (3.52)

O–Methylether mp. 288～291° (decomp.) $[\alpha]_D$ +12° (CHCl$_3$), $C_{28}H_{33}O_5N_3$

IR $\nu_{max}^{mineral\ oil}$ cm^{-1}: 1750, 1715, 1603

(c) **Alkaloids of *Decodon verticillatus* ELL.** (*Lythraceae*)[2083]

Table. 105 Alkaloids of *Decodon verticillatus* (Ferris)

Str. No.	Alkaloids	Formula	mp.	$[\alpha]_D^{CHCl_3}$	λ_{max}^{MeOH} mμ	λ_{max}^{Basic} mμ	$\nu_{max}^{CHCl_3}$ cm^{-1}
(CCCXIV)	Decaline	$C_{26}H_{31}O_5N$	80～81 102～118	−136	293, 280	293, 280	1720
(CCCXV)	Vertaline	$C_{26}H_{31}O_5N$	194	−170	293, 280	293, 280	1720
(CCCXVI)	Decinine	$C_{26}H_{31}O_5N$	222	−142	294	293, 313	3500, 1724
(CCCXVII)	Decamine	$C_{26}H_{31}O_5N$	222	−145	294	293, 313	3480, 1728
(CCCXVIII)	Vertine	$C_{26}H_{29}O_5N$	245	+ 39	285, 260, 308	318, 298	3530, 1700
(CCCXIX)	Decodine	$C_{25}H_{29}O_5N$	193	− 97	287, 312	310	3500, 1723
(CCCXX)	Verticillatine	$C_{25}H_{27}O_5N$	312	+119	293	300	——

2079) Anet, E.F.F.L. *et al*: *Aust. J. Chem.*, **14**, 173 (1961).
2080) Robinson, B: *Chem. & Ind.*, **1963**, 218.
2081) King, T.J. *et al*: *Soc.*, **1961**. 2714.
2082) Snyder, H.R. *et al*: *J.A.C.S.*, **80**, 3708 (1958).
2083) Ferris, J.P: *J. Org. Chem.*, **27**, 2985 (1962).

[L] Steroid Alkaloids

(1) Holarrhena Alkaloids

≪Cccurrence≫   *Holarrhena antidysenterica* and other *Holarrhena.* spp.

(a) Classification[2084)~2086)]

(i) *Conarrhimine Type* (I)

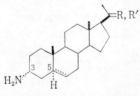

(I)

OCOC$_5$H$_9$=

OCO–CH$_2$–CH=C$\big\langle$CH$_3$ / CH$_3$

R	R'	R''	R'''			
H	H	H	H	II	(II)	Conarrhimine
Me	H	H	H	III	(III)	Conimine
H	H	Me	H		(IV)	Conamine
Me	Me	H	H		(V)	Conessimine
Me	H	Me	H		(VI)	Isoconessimine
Me	Me	Me	H	VII	(VII)	Conessine
Me	Me	Me	OH	VIII	(VIII)	Holarrhenine
Me	Me	H	CCCC$_5$H$_9$		(IX)	Holafrine
Me	Me	Me	CCCC$_5$H$_9$		(X)	Holarrhetine

(ii) *Conkurchine Type* (XI)

(XI)

R	R'	R''		
H	H	H	(XII)	Conkurchine
Me	H	H	(XIII)	Conessidine
Me	Me	Me	(XIV)	Trimethylconkurchine

(iii) *Kurchamine Type* (XV)

(XV)

R	R'		
H	Me	(XVI)	Kurchamine
Me	H	(XVII)	Kurchimine

(iv) *3–Aminopregnanes*

=R, R'

3	5	R, R'		
α–NH$_2$	α	=O	(XVIII)	Funtumine
α–NH$_2$	α	R=α–OH R'=H	(XIX)	Funtimidine
α–NH$_2$	Δ$_{5-6}$	=O	(XX)	Holamine

2084) Seebeck, E. *et al*: *Helv. Chim. Acta*, **41**, 11 (1958).
2085) Tschsche, R. *et al*: *Ber.*, **95**. 1144 (1962).
2086) Alauddin, M. *et al*: *J. Ph. Ph.* **14**, 469 (1962).
2087) Bhattacharyya, P.K *et al*: *Tetrahedron*, **18**, 1457 (1962).
2088) Tschesche, R. *et al*: *Ber.*, **91**, 1504 (1958).

	3 R	
	β–NH$_2$	(XXI) Holaphyllamine
	β–NH·CH$_3$	(XXII) Holaphylline

R=H (XXIII) Paravallarine
R=OH (XXIII–a) Paravallaridine

CH$_3$HN

(v) 20–*Aminopregnanes*

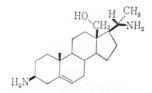

R	R'	R''	5		
β–OH	H	H	α	(XXIV)	Funtuphyllamine A
β–OH	Me	H	α	(XXV)	Funtuphyllamine B
β–OH	Me	Me	α	(XXVI)	Funtuphyllamine C
=O	Me	H	α	(XXVII)	Funtumafrine B
=O	Me	Me	α	(XXVIII)	Funtumafrine C
β–OH	H	H	$\Delta_{5\sim6}$	(XXIX)	Holafebrine

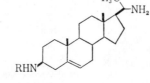

R=H (XXXI) Irehediamine A
R=Me (XXXII) Irehediamine B

(XXX) Holarrhimine

(b) **Occurrence and Physical Constants**

Table. 106 (Alauddin *et al.*)[2086]

Str. No.	Alkaloids	Formula	mp.	$[\alpha]_D^{CHCl_3}$	Occurrence [Note]	Ref.
		(i) *Conarrhimine Type* (I)				
(II)	Conarrhimine	C$_{21}$H$_{34}$N$_2$	Impure preparation	——	H.a.	2091)
(III)	Conimine	C$_{22}$H$_{36}$N$_2$	134	−30 (EtOH)	H.a.	2092) 2093)
(IV)	Conamine	C$_{22}$H$_{36}$N$_2$	130	−21	H.a.	2091) 2093)
(V)	Conessimine	C$_{23}$H$_{38}$N$_2$	100	−22.3	H.a.	2094)

2089) Lábler, L. *et al*: *Coll. Czech. Chem. Comm.*, **24**, 370 (1959); *J. Ph. Ph.* **11**, 565 (1959).
2090) Minh, T. H. *et al*: *Bull. Soc. Chim. France*, **1963**, 594.
2091) Siddiqui, S: *Proc. Indian Acad. Sci.*, **3 A**, 249 (1936).
2092) Siddiqui, S. *et al*: *J. Indian Chem. Soc.*, **11**, 787 (1934).
2093) Tschesche, R. *et al*: *Ber.*, **89**, 1288 (1956). 2094) Tschesche, R. *et al*: *ibid.*, **87**, 1719 (1954).
2095) Siddiqui, S. *et al*: *Proc. Indian Chem. Soc.*, **2 A**, 426 (1935).
2096) Bertho, A. *et al*: *Ann.*, **619**, 96 (1958). 2097) Fabre, H. *et al*: *Soc.*, **1953**, 1115.
2098) Uffer, A: *Helv. Chim.*, *Acta*, **39**, 1834 (1956).

Str. No.	Alkaloids	Formula	mp.	$[\alpha]_D^{CHCl_3}$	Occurrence [Note]	Ref.
(i) *Conarrhimine Type* (1)						
(VI)	Isoconessimine	$C_{23}H_{38}N_2$	92	+30 (EtOH)	*H. a.*	2095) 2089) 2096)
(VII)	Conessine	$C_{24}H_{40}N_2$	125	− 2	*H.* sps.	2097)
(VIII)	Holarrhenine	$C_{24}H_{40}ON_2$	198	− 7	*H. c.*	2084) 2098)
(IX)	Holafrine	$C_{29}H_{46}O_2N_2$	116~117	−19.1	*H. af.*	IR 2084)
(X)	Holarrhetine	$C_{30}H_{48}O_2N_2$	74~ 75	−14.9	*H. af.*	IR 2084)
(ii) *Conkurchine Type* (XI)						
(XII)	Conkurchine	$C_{21}H_{32}N_2$	152	−67.4	*H. a.*	2099)
(XIII)	Conessidine	$C_{22}H_{34}N_2$	123	−52.2	*H. a.*	2093) 2100)
(XIV)	Trimethylconkurchine	$C_{24}H_{38}N_2$	125~127	+12.0	*H. a.*	2093)
(iii) *Kurchamine Type* (XV)[2085]						
(XVI)	Kurchamine	$C_{22}H_{36}N_2$	115~117	−16	*H. a.*	2088)
(XVII)	Kurchimine	$C_{22}H_{36}N_2$	104~106	−21	*H. a.*	2085)
(iv) *3-Aminopregnanes*						
(XVIII)	Funtumine	$C_{21}H_{35}ON$	126	+95	*F. l.*	2101)
(XIX)	Funtimidine	$C_{21}H_{37}ON$	182	+10	*F. l.*	2101)
(XX)	Holamine	$C_{21}H_{33}ON$	135	+23	*H. f.*	2102)
(XXI)	Holaphyllamine	$C_{21}H_{33}ON$	−HCl 260	−HCl +33	*H. f.*	2102) 2103)
(XXII)	Holaphylline	$C_{22}H_{35}ON$	128	+23	*H. f.*	2102)
(XXIII)	Paravallarine	$C_{22}H_{33}O_2N$	181	−52	*P. m.*	2104)
(XXIII–a)	Paravallaridine	$C_{22}H_{33}O_3N$			*P. m.*	2091)
(v) *20-Aminopregnanes*						
(XXIV)	Funtuphyllamine A	$C_{21}H_{37}ON$	173	+13	*F. a.*	2105)
(XXV)	Funtuphyllamine B	$C_{22}H_{39}ON$	214	+24	*F. a., M. b.*	2105) 2106)
(XXVI)	Funtuphyllamine C	$C_{23}H_{41}ON$	172	+24	*F. a.*	2105)
(XXVII)	Funtumafrine B	$C_{22}H_{37}ON$	160	+43	*F. a.*	2105)
(XXVIII)	Funtumafrine C	$C_{23}H_{39}ON$	174.	+45	*F. a., M. b.*	2105)
(XXIX)	Holafebrine	$C_{21}H_{35}ON$	177	−61	*H. f., K. a.*	2107)
(XXX)	Holarrhimine	$C_{21}H_{36}ON_2$	183	−14.2	*H. a.*	2087) 2108)
(XXXI)	Irehdiamine A					2090)
(XXXII)	Irehdiamine B					2090)

2099) Bertho, A: *Ann.*, **573**, 210 (1951).   2100) Bertho, A: *ibid.*, **558**, 62 (1947).

2101) Janot, M. M. *et al*: *C. R. Acad. Sci., Paris*, **248**, 982 (1959).

2102) Janot, M. M. *et al*: *ibid.*, **251**, 559 (1960).

2103) Janot, M. M. *et al*: *Bull. Soc. Chim. France*, **1959**, 896.   2104) Le Men, J: *ibid.*, **1960**, 860.

2105) Janot, M. M. *et al*: *C. R. Acad. Sci., Paris*, **250**, 2445 (1960).

2106) Janot, M. M. *et al*: *Ann. Pharm. Franç.* **18**, 673 (1960).

2107) Janot, M. M. *et al*: *Bull. Soc. Chim. France*, **1962**, 285.

2108) Siddiqui, S. *et al*: *J. Indian Chem. Soc.*, **9**, 553 (1932).

(vi) *Alkaloids, incompletely characterised*

Str. No.	Alkaloids	Formuls	mp.	$[\alpha]_D^{CHCl_3}$	Occurrence [Note]	Ref.
(XXXIII)	Holarrhidine	$C_{21}H_{36}ON_2$	181~182	−23	*H. a.*	2085) 2089)
(XXXIV)	Conkuressine	$C_{24}H_{40}N_2$	86.5~87.5	+ 7	*H. a.*	2085)
(XXXV)	Kurcholessine	$C_{24}H_{40}O_2N_2$	145~146	− 4	*H. a.*	2085)
(XXXVI)	Conkurchinine	$C_{25}H_{36}N_2$	161	−47(EtOH)	*H. a.*	2100)
(XXXVII)	Holarrhessimine	$C_{22}H_{36}ON_2$	160~164	−30	*H. a.*	2094)
(XXXVIII)	Holarrhine	$C_{20}H_{38}O_3N_2$	240	−17(MeOH)	*H. a.*	2108)
(XXXIX)	α–Hydroxyconessine	$C_{24}H_{40}ON_2$	—	−9(EtOH)	*H. a.*	2096)
(XL)	Kurchenine	$C_{21}H_{32}O_2N_2$	335~336	$-78\left(\begin{array}{c}H_2SO_4\\ in\ H_2O\end{array}\right)$	*H. a.*	2109)
(XLI)	Kurchessine	$C_{24}H_{40}N_2$	132~133	−36	*H. a.*	2088)
(XLII)	Kurchine	$C_{23}H_{38}N_2$	75	− 7	*H. a.*	2100) 2110) 2111)
(XLIII)	Lettocine	$C_{17}H_{25}O_2N$	350~352	—	*H. a.*	2112)
(XLIV)	Malouphylline	$C_{24}H_{40}O_2N_2$	259	−10	*M. b.*	2113)

[Note] *F. a.=Funtumia africana, F. l.=F. latifolia, H. a.=Holarrhena antidysenterica, H. af.=H. africana, H. c.=H. congolensis, H. f.=H. febrifuga, K. a.=Kibatalia arborea, M. b.=Malouetia bequertiana, P. m.=Paravallaris microphylla*

(c) **IR Spectra** See Fig. 389 and 390

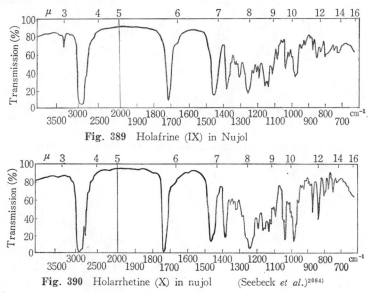

Fig. 389 Holafrine (IX) in Nujol

Fig. 390 Holarrhetine (X) in nujol   (Seebeck *et al.*)²⁰⁸⁴⁾

(d) **Synthesis**

Conessine (VII) 2114)~2116)

2109) Bertho, A *et al*: *Ber.*, **66**, 786 (1933).   2110) Ghosh, S *et al*: *Arch. Pharm.*, **270**, 100 (1932).

2111) Haworth, R. D: *Soc.*, **1932**, 631.   2112) Peacock, D. H *et al*: *ibid.*, **1935**, 734.

2113) Goutarel, R: *Tetrahedron*, **14**, 126 (1961).   2114) Stork, G *et al*: *J. A. C. S.*, **84**, 2018 (1962).

2115) Marshall, J. A *et al*: *ibid.*, **84**, 1485 (1962).

2116) Barton, D. H. R *et al*: *Proc. Chem. Soc.*, **1961**, 206.

(2) Salamander Alkaloids [2086) 2117) ~2125)]

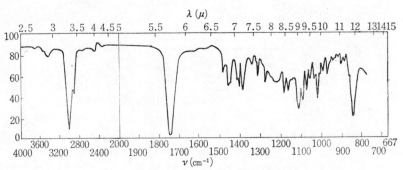

(XLV) Samandarone [2119)]

(XLVI) Samandarine [2118) 2119)]

(XLVII) Samandaridine [2120)]

(XLVIII) Cycloneosamandione [2121)]

《Occurrence》 *Salamander maculosa, S. atra* (Alpen Salamander)

(a) **Samandarone** (XLV) [2119) 2120) 2123)]

Samandarine (XLVI) $\xrightarrow{\text{CrO}_3}$ (XLV) [2124)]

mp. 191~192°, $C_{19}H_{29}O_2N$

IR See Fig. 391, X-Ray [2120)]

Fig. 391 IR Spectrum of Samandarone in $CHCl_3$ (Habermehl *et al.*) [2118)]

2117) Schöpf, C: *Experientia*, **17**, 285 (1961).   2118) Habermehl, G *et al*: *Ber.*, **94**, 2361 (1961).

2119) Habermehl, G: *ibid.*, **96**, 840 (1963).   2120) Habermehl, G: *ibid.*, **96**, 143 (1963).

2121) Habermehl, G *et al*: *Angew. Chem.*, **75**, 247 (1963), **74**, 154 (1962), *IUPAC Symposium, Kyoto*, p. 119 (1964).

2122) Schöpf, C *et al*: *Ann.*, **633**, 127 (1960).   2123) Schöpf, C *et al*: *ibid.*, **514**, 69 (1934).

2124) Schöpf, C *et al*: *Ber.*, **83**, 372 (1950).   2125) Schöpf, C *et al*: *Ann.*, **552**, 37 (1942).

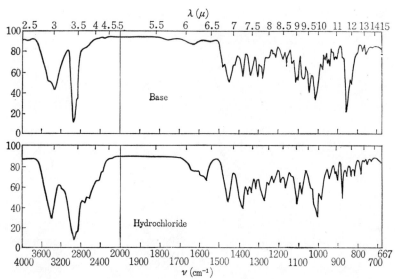

Fig. 392 IR Spectra of Samandarine in KBr (Habermehl *et al.*)[2118]

(b) Samandarine (XLVI) [2119) 2123]

Methanolate mp. 187~188°, $[\alpha]_D$ +43.7° (acetone), $C_{19}H_{31}O_2N \cdot MeOH$

Hydrochloride mp. 321~322°

IR : See Fig. 392

(c) Samandaridine (XLVII) [2119) 2120) 2125]

Samandarine (XLVI) $\xrightarrow{\text{CrO}_3.}$ Samandarone (XLV) $\xrightarrow{\text{OCH·COOH}}$

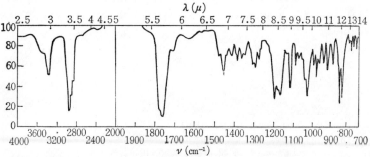

$\xrightarrow{\oplus}$ Samandaridine (XLVII) [2119]

mp. 289°, $[\alpha]_D$ +29.5° (2 N–HOAc), $C_{21}H_{31}O_3N$, Hydrochloride mp. 342~343° (decomp.)

IR **See** Fig. 393, 840 cm⁻¹ doublet (oxazolidine ring)

X–Ray [2120]

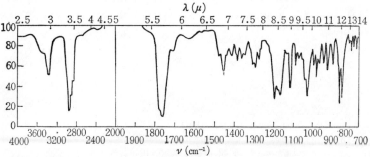

Fig. 393 IR Spectrum of Samandaridine in KBr (Habermehl *et al.*)[2121]

(d) Cycloneosamandione (XLVIII) [2121)2122)]

mp. 118~119°, $[\alpha]_D$ −207.6° (acetone), $C_{19}H_{29}O_2N$

——HCl. MeOH mp. 272~273°, Methiodide mp. 303~304°

X–Ray[2121)]

(3) Buxus Alkaloids [2126)]

≪Occurrence≫ *Buxus sempervirens* L. (*Buxaceae*)

(XLIX) Cyclobuxine

(L) Buxus–Alkaloid–L

	R$_1$	R$_2$		
	CH$_3$	H	(LI)	Buxamine
	H	H	(LII)	Norbuxamine
	CH$_3$	OH	(LIII)	Buxaminol

(a) Cyclobuxine (Buxus-Alkaloid A) (XLIX) [2127)2128)]

mp. 246°, $[\alpha]_D$ +100° (CHCl$_3$), $C_{25}H_{42}ON_2$

IR $\lambda_{max}^{CHCl_3}$ μ: 6.09, 11.20 (terminal methylene)

NMR in CDCl$_3$ or CCl$_4$, 60 Mc, τ: 5.20, 5.43 d, $J<1$ c/s (2) (terminal methylene), 5.92 octuplet, J's

3, 7, 9.5 (1) (CH$_2$–CHOH–CH), 7.53, 7.57 (6) (2 NCH$_3$), 8.87, 9.03 (6) $\left(\!\!>\!\!C\text{–}CH_3\right)$, 8.92 d, $J=6$

(3) $\left(\!\!>\!\!CH\text{–}CH_3\right)$, 9.72−9.95 d, $J=4$ (2) (cyclopropyl methylene)[2126)]

(b) Buxus-Alkaloid L (L) [2129)]

mp. 201°, $[\alpha]_D$ +70° (CHCl$_3$), $C_{27}H_{48}N_2$ [2126)]

(c) Buxamine (LI) [2126)]

Amorphous, $[\alpha]_D$ +32° (CHCl$_3$), $C_{26}H_{44}N_2$,

Oxalate mp. 263~267°

UV λ_{max}^{MeOH} mμ (log ε): 238 (4.42), 246 (4.45), 254 (4.24), 230 sh (4.28), 277 sh (2.39), 288 sh (2.24);

λ_{min}: 241, 252

IR $\nu_{max}^{CH_2Cl_2}$ cm^{-1}: 1580 (NH$_2$), 1372, 1360, 1325, 1150, 1090, 1021, 1002, 982, 860, 830

NMR See Tab. 107

Mass m/e 384

2126) Stauffacher, D: *Helv. Chim. Acta*, **47**, 968 (1964).

2127) Kupchan, S. M *et al*: *J. A. C. S.*, **84**, 4590, 4592 (1962).

2128) Schlittler, E *et al*: *Helv. Chim. Acta*, **32**, 2209, 2226 (1949).

2129) Schlittler, E *et al*: *ibid.*, **33**, 878 (1950).

Table 107 NMR Spectra of Buxus Alkaloids in CDCl$_3$, 60 Mc, TMS, δ (Stauffacher)[2126]

Assign to	(LI) Buxamine	(LII) Norbuxamine	(LIII) Buxaminol
4 $>$C–CH$_3$	0.70, 0.72, 0.74, 1.02 4 s (3)	0.68,0.70,0.75,1.01 4 s (3)	0.71,0.73,0.90,1.01 4 s (3)
$>$CH–CH$_3$	1.09 d, $J=6$ (3)	1.08 d, $J=6$ (3)	1.15 d, $J=6$ (3)
	1.2~1.95 m (ca. 12)	1.2~1.95 m (ca. 12)	1.2~1.95 m (ca. 11)
Allyl-H & others	1.95~2.2 m (ca. 7)	1.95~2.3 m (ca. 8)	1.95~2.3 m (ca. 11)
N–CH$_3$	2.29 s (6)	2.45 s (3)	2.28 s (6)
$>$CH–N$<$	2.3~3.2 m (ca. 2)	2.3~3.2 m (ca. 2)	2.3~3.2 m (ca. 2)
$>$CH–O–			4.0~4.35 m (1)
Vinyl–H	5.45~5.63 m (1)	5.48~5.66 m (1)	5.40~5.57 m (1)
Vinyl–H	5.95 br. s (1)	5.99 br. s (1)	5.90 br. s (1)

(d) Norbuxamine (LII) [2126]

Amorphous, $[\alpha]_D$ +20° (CHCl$_3$), C$_{25}$H$_{42}$N$_2$, Bishydrogentartrate mp. 300° (decomp.)

UV λ_{max}^{MeOH} mμ (log ε): 238 (4.44), 246 (4.47), 254 (4.26), 215 sh (3.84), 230 sh (4.27), 290 sh (2.18); λ_{min}: 242, 253

IR ν_{max}^{CH2Cl2} cm^{-1}: 1580 (NH$_2$), 1375, 1360, 1328, 1168, 1140, 1120, 1095, 1004, 972, 860, 830

NMR See Table 107, Mass: m/e 370

(e) Buxaminol (LIII) [2126]

mp. 199~200°, $[\alpha]_D$ +38° (CHCl$_3$), C$_{26}$H$_{44}$ON$_2$

UV λ_{max}^{MeOH} mμ (log ε): 238 (4.45), 245 (4.49), 254 (4.28), 277 (2.68), 287 (2.58), 224 sh (4.13), 230 sh (4.30); λ_{min}: 241, 252, 272, 285

IR ν_{max}^{CH2Cl2} cm^{-1}: 3400 (OH), 1630 w (Δ), 1582 (NH$_2$), 1380, 1360, 1330, 1155, 1105, 1090, 1040, 1027, 1018, 978, 855, 845, 836

(4) Solanaceous Steroid Alkaloids

(a) Classification

(i) *Solanidane Alkamine Type* (LIV) [2086][2130]

(LIV)

(LV) Solanidine

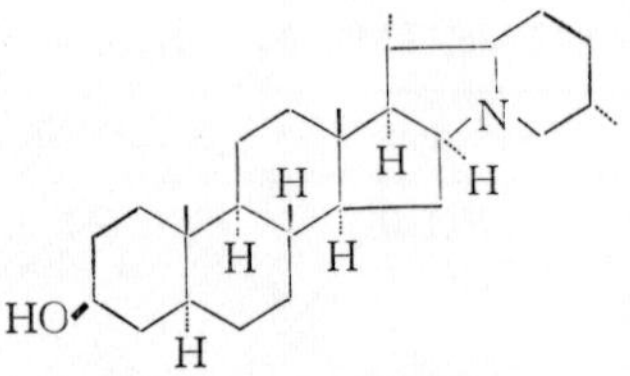

(LVI) Demissidine

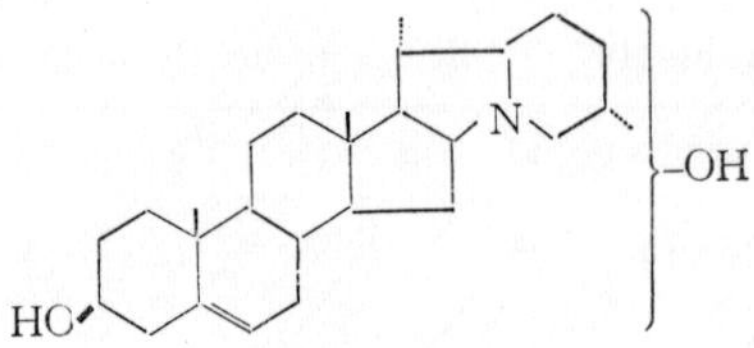

(LVII) Leptinidine

(LVIII) Soladulcamaridine $C_{27}H_{43}ON$

(ii) *Spirosolane Alkamines Type* (LIX) [2086) 2130)]

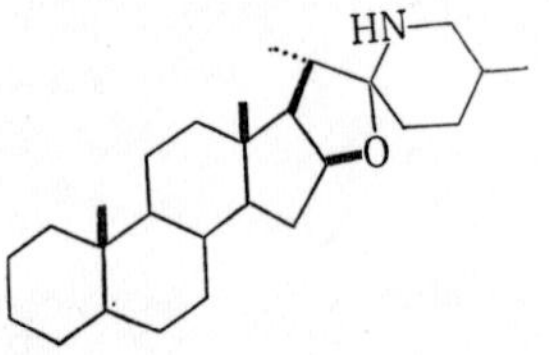

(LIX)

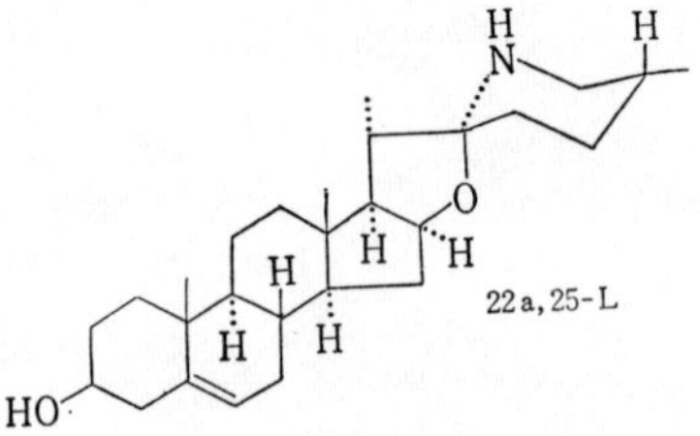

(LX) Solasodine

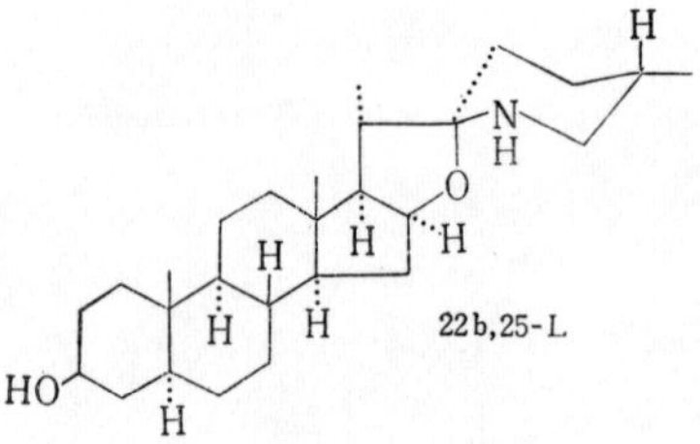

22 a, 25-L

22 b, 25-L

22 a, 25-D

(LXI) Tomatidine

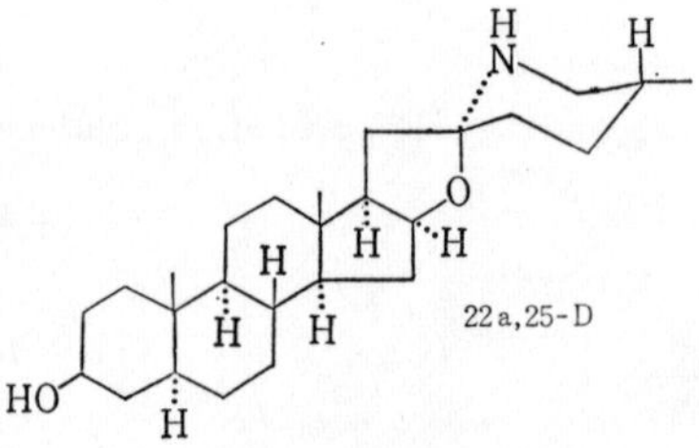

(LXII) Soladulcidine

(LXI–a) 5,6–Dehydrotomatidine

(LXIII) Solauricidine $C_{27}H_{43}O_2N$

(LXIV) Solangustidine $C_{27}H_{43}O_2N$

(iii) *Alkamine Type* (LXV) [2131)]

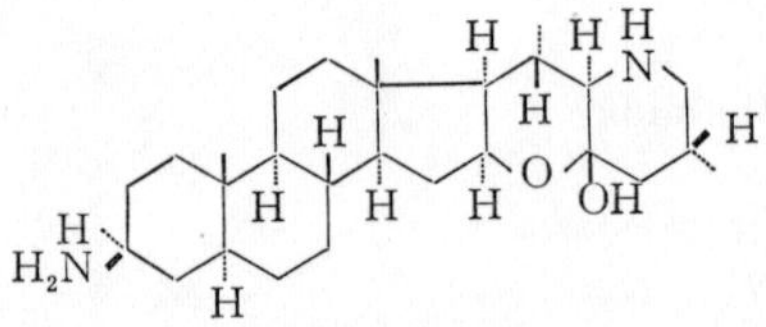

(LXV) Solanocapsine

2130) Manske, R. H. F: The Alkaloids, Vol VII, p. 343~361 (Academic Press, 1960).

2131) Schreiber, K et al: Ann., **655**, 114 (1962).

(iv) *Sugars* [2130]

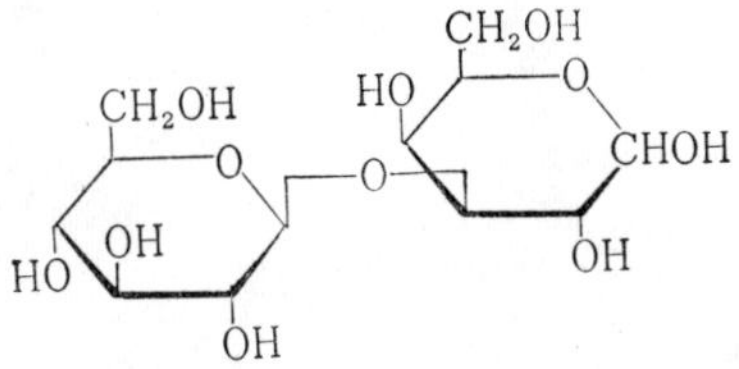

(LXVI) Solabiose

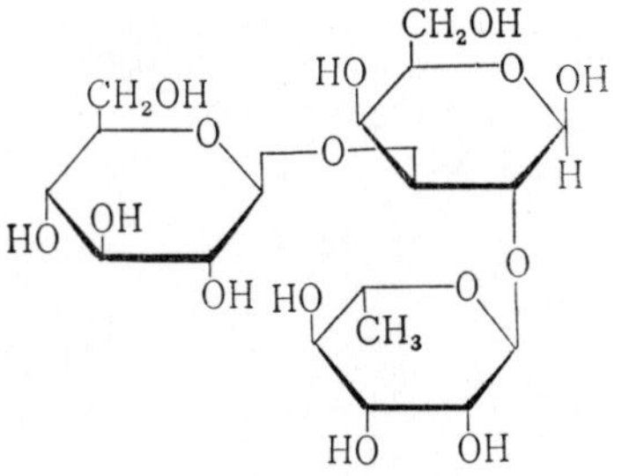

(LXVII) β-Solatriose

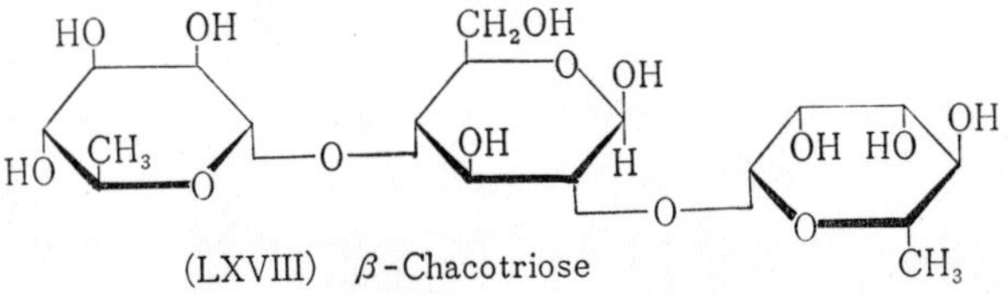

(LXVIII) β-Chacotriose

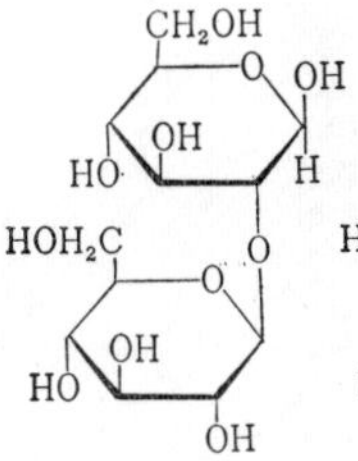

(LXIX) Sophorose 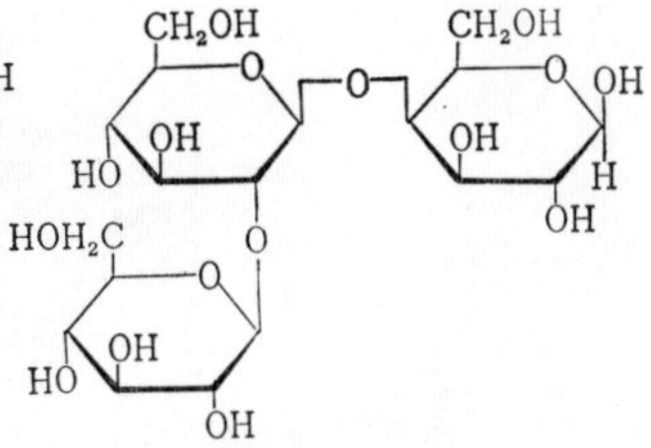   (LXX) Lycotriose 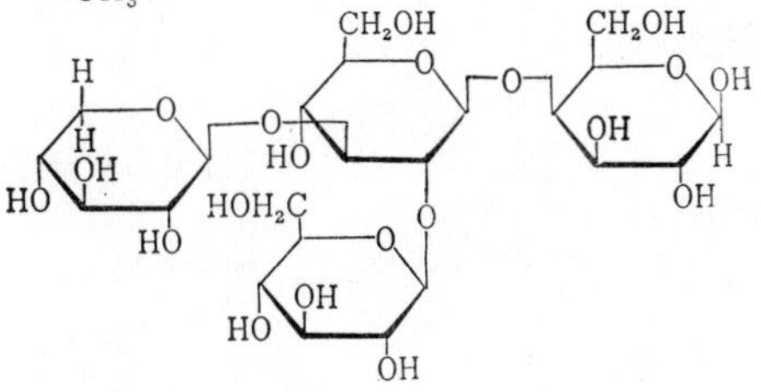   (LXXI) β-Lycotetraose

(b) Physical Constants and Occurrence of Alkamines and their Glycosides

Table 108 (Alauddin *et al.*)[2086]

Alkamines				Glycoalkaloids				Source	Ref.
Str. No.	Alkamines	mp.	$[\alpha]_D$ CHCl$_3$	Glycoside	mp.	$[\alpha]_D$ py.	Sugar		
(i) *Solanidane Alkamines* (LIV)									
(LV)	Solanidine C$_{27}$H$_{43}$ON	219	−27	α–Chaconine C$_{45}$H$_{73}$O$_{14}$N	242	−85	α–L–Rhampyr. (1→2 gluc.)–α–L–	S.t. S.n.	
							rhampyr. (1→4 gluc.)–D–gluc. (LXVIII)	S.dul. S.ch.	2132)
				β–Chaconine C$_{39}$H$_{63}$O$_{10}$N	255	−61.5	α–L–Rhampyr. (1→4) D–gluc.	"	2132)
				γ–Chaconine C$_{33}$H$_{55}$O$_6$N	243~244	−40	D–Gluc.	"	2132)
				α–Solanine (Solanine) C$_{45}$H$_{73}$O$_{15}$N	285	−59	α–L–Rhampyr. (1→2 galact.)–β–D –glucpyr. (1→3) galact. (LXVII)	"	2132) ~ 2134)
				β–Solanine C$_{39}$H$_{63}$O$_{11}$N	295	−13 (Me)	β–D–Glucpyr. (1→3) D–galact. (LXVI)	"	2132)

2132) Kuhn, R *et al*: *Ber.*, **88**, 1492, 1690 (1955).   2133) Schreiber, K: *ibid.*, **87**, 1007 (1954).
2134) Uhle, F. C *et al*: *J. Biol. Chem.*, **160**, 243 (1945).

Alkamines				Glycoalkaloids				Source	Ref.
Str. No.	Alkamines	mp.	$[\alpha]_D$ CHCl$_3$	Glycoside	mp.	$[\alpha]_D$ py.	Sugar		
				γ–Solanine C$_{33}$H$_{53}$O$_6$N	240~250	−26 (Me)	D–Galact.	"	2132)
				Solacauline C$_{43}$H$_{69}$O$_{14}$N	260~265	−30	Trisaccharide of 1 mol gluc. 2 mol D–xyl.	S. ac.	2133)
(LVI)	Demissidine C$_{27}$H$_{45}$ON	220~222	+21	Demissine C$_{50}$H$_{83}$O$_{20}$N	305~308	−20	Branched tetrasaccharide of 1 mol D–xyl. 1 mol D–galact. 2 mol D–gluc. (LXXI)	S. dem.	2133) 2135)
(LVII)	Leptinidine C$_{27}$H$_{43}$O$_2$N	247~248	−24	Leptinine I C$_{45}$H$_{73}$O$_{15}$N	230	−90	Trisaccharide of 1 mol D–gluc. 2 mol L–rham.	S. ch.	2136)
				Leptinine II C$_{45}$H$_{73}$O$_{16}$N	225	−62	Trisaccharide of 1 mol D–galact. 1 mol L–rham. 1 mol D–gluc.	"	2136)
(LVIII)	Soladulcamaridine C$_{27}$H$_{43}$ON	220~222	−78 (Me)	Soladulcamariine C$_{49}$H$_{79}$O$_{17}$N	193~197	—	Tetrasaccharide of 1 mol D–gluc. 1 mol L–rham. 2 mol L–arab.	S. dul.	2137)

(ii) *Spirosolane Alkamines* (LIX)

(LX)	Solasodine (Solanidine S) C$_{27}$H$_{43}$O$_2$N	202	−80 (Me)	Solasodamine C$_{51}$H$_{83}$O$_{20}$N	298~302	−72 (Me)	L–Rham.–L–rham.–D–galact.–D–gluc.	S. sod. S. aur. S. mar. S. av.	2138)
				Solasonine (Solanine S) C$_{45}$H$_{73}$O$_{16}$N	301~303	−88	L–Rham.–D–galact.–D–gluc.	S. av. S. sod. S. xan. S. nod. S. tor. S. lac.	2139)~ 2143)
				Solamargine C$_{45}$H$_{73}$O$_{15}$N	301~310	−114	L–Rham.–L–rham.–D–gluc.	S. mar. S. n.	
(LXI)	Tomatidine C$_{27}$H$_{45}$O$_2$N	210~211	−8 (Me)	Tomatine C$_{50}$H$_{83}$O$_{21}$N	263~267	−19	Branched tetrasaccharide of 1 mol xyl, 1 mol D–galact, 2 mol D–gluc. (LXXI)	L. pi. L. esc. L. h.	2138) 2139) 2141) 2144)
(LXI–a)	5, 6–Dehydrotomatidine C$_{27}$H$_{43}$O$_2$N	206	−45	—	—	—	—	S. t.	2145)
(LXII)	Soladulcidine C$_{27}$H$_{45}$O$_2$N	206.5	−52.6	α, β, γ–Soladulcine	—	—	—	S. dul.	2146)~ 2148)

2135) Kuhn, R *et al*: *Ber.*, **90**, 203 (1957).   2136) Kuhn, R *et al*: *ibid.*, **94**, 1088, 1096 (1961).
2137) Rasmussen, H. B *et al*: *Acta Chem. Scand.*, **12**, 802 (1958).
2138) Briggs, L. H *et al*: *Soc.*, **1958**, 1419.   2139) Boll, P. M: *Acta Chem. Scand.*, **12**, 358 (1958).
2140) Briggs, L. H *et al*: *Soc.*, **1958**, 1422.   2141) Kuhn, R *et al*: *Ber.*, **88**, 289 (1955).
2142) Uhle, F. C: *J. A. C. S.*, **76**, 4245 (1954).   2143) Taylor, D. A. H: *Soc.*, **1958**, 4216.
2144) Schreiber, K: *Der Züchter*, **27**, 289 (1957).   2145) Schreiber, K: *Angew. Chem.*, **69**, 483 (1957).
2146) Briggs, L. H *et al*: *Soc.*, **1952**, 1654.   2147) Schreiber, K: *Planta Med.*, **6**, 94 (1958).
2148) Tuzson, P *et al*: *Acta Chim Acad. Sci. Hungar.*, **12**, 31 (1957).

Alkamines				Glycoalkaloids				Source	Ref.
Str. No.	Alkamines	mp.	$[\alpha]_D$ CHCl$_3$	Glycoside	mp.	$[\alpha]_D$ py.	Sugar		
(LXIII)	Solauricidine $C_{27}H_{43}O_2N$	219	−90 (Me)	Solauricine $C_{45}H_{73}O_{16}N$	270	—	L–Rham.–D–galact.– D–gluc.–D–gluc.	S. aur.	2149)
(LXIV)	Solangustidine $C_{27}H_{43}O_2N$	amor-ph.	—	Solangustine $C_{33}H_{53}O_7N\cdot H_2O$	235	—	D–gluc.	S. ang	2150)

(iii) *Alkamine* (LXV)

(LXV)	Solanocapsine $C_{27}H_{46}O_2N_2$	222	+25.5	—	—	—	—	S. ps.	2151)

[Note] (Sugar) arab.=arabinose, galact.=galactose, gluc.=glucose, glucpyr.=glucopyranosyl, rham.=rham-
nose, rhampyr.=rhamnopyranosyl, xyl.=xylose.

(Source) *L. esc.=Lycopersicum esculentum, L. h.=L. hirsutum, L. pi.=L. peruvianum, S. ac.=
Solanum acaulia, S. ang.=S. angustifolium, S. aur.=S. auriculatum, S. av.=S. aviculare, S. ch.=
S. chacoense, S. dem.=S. demissum, S. dul.=S. dulcamara, S. lac.=S. laciniatum, S. mar.=S. ma-
rginatum, S. n.=S. nigrum, S. nod.=S. nodiflorum, S. ps=S. pseudocapsicum, S. sod.=S. sodomeum,
S. t.=S. tuberosum, S. tor.=S. torvum, S. xan.=S. xanthocarpum.*

(c) Gas Chromatographic Separation [2152) and **Synthesis** [2153)

(5) Veratrum Alkaloids
(a) Classification [2086) 2153a) 2154)

(i) *Alkamines of Glycoalkaloids*

(LXXII) Veratramine

(LXXIII) Jervine

(LXXIII–a) Isojervine

(LXXIV) Rubijervine

(LXXV) Isorubijervine

2149) Briggs, L. H *et al* : *Soc.*, **1942**, 17.   2150) Tutin, F *et al* : *Soc.*, **105**, 559 (1914).
2151) Boll P. M *et al* : *Farm. Tidende*, **67**, 198 (1957) ; *C. A.* **53**, 22265 h (1959) ; *Acta Chem. Scand.*, **13**, 2039 (1959).
2152) Sato, Y *et al* : *J. Org. Chem.*, **26**, 628 (1961)   2153) Uhle, F. C : *J.A.C.S.*, **83**, 1460 (1961).
2153a) Manske, R. H. F : The Alkaloids, Vol. VII, p.363~417 (Academic Press, 1960).
2154) Kupchan, S. M : *J.P.S.*, **50**, 273 (1961).

(ii) *Alkamines of Ester Alkaloids*

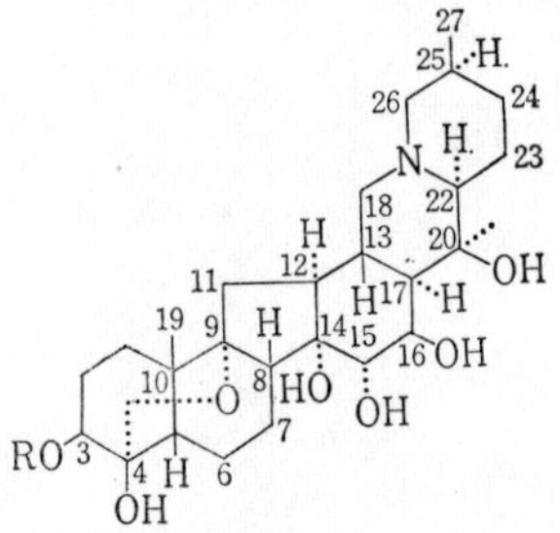

(LXXVI) R=H: Zygadenine

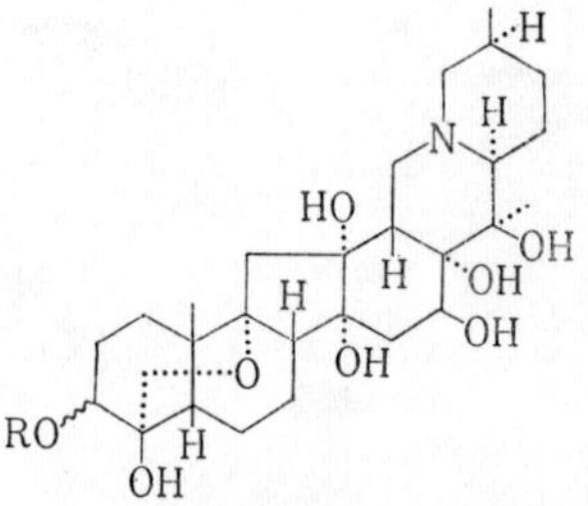

(LXXVII) 3β–OR,R=H: Veracevine
(LXXVII–a) 3α–OR,R=H: Cevine

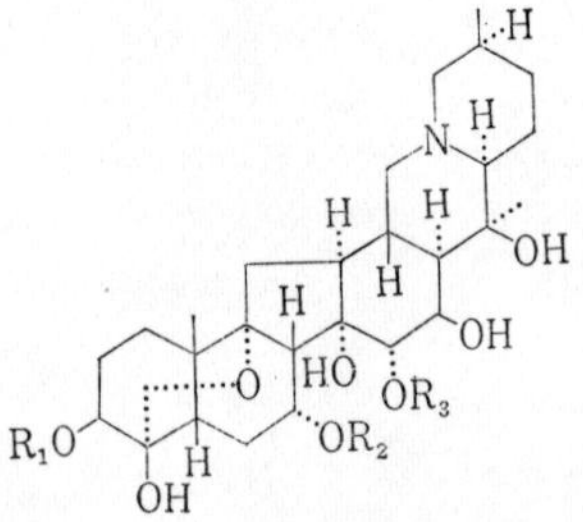

(LXXVIII) R₁=R₂=R₃=H:
Germine

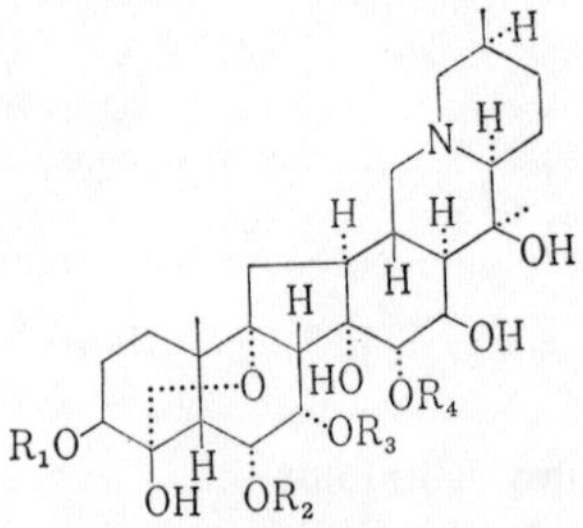

(LXXIX) R₁=R₂=R₃=R₄=H:
Protoverine

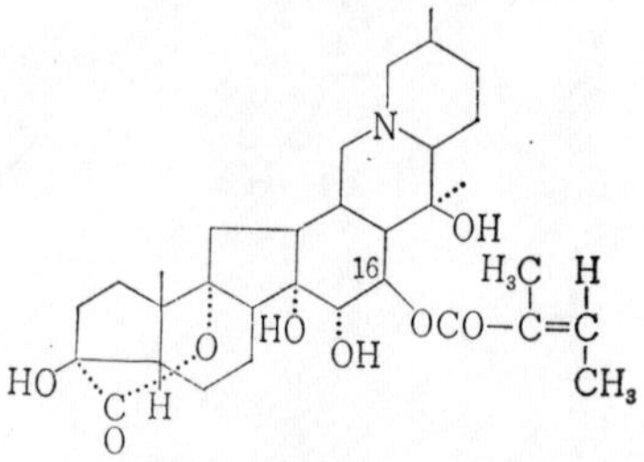

(LXXX) Zygadenilic acid
δ–lactone Angelate

$C_{27}H_{45\sim47}O_7N$

(LXXXI) Sabine

(iii) *Esterifying Acids* (R–)

(Ac) Acetic

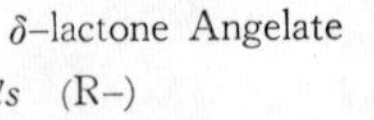

(An) Angelic

(Ti) Tiglic

(MB) (+) or (−)–2–Methyl
–butyric

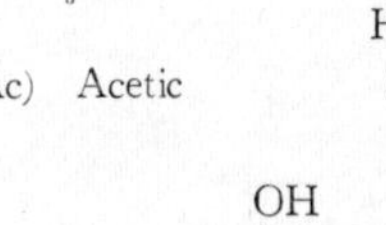

(HMB) 2–Hydroxy–
2–methylbutyric

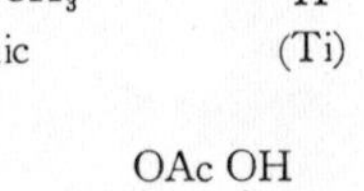

(HMAB) *erythro*–2–Hydroxy–
2–methyl–3–acetoxybutyric

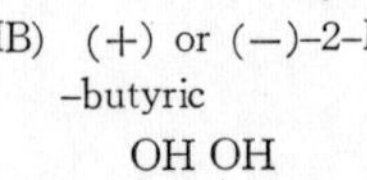

(DMB) (+)–*threo*–or (−)
erythro–2,3–Dihydroxy–
2–methylbutyric

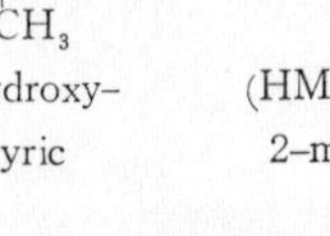

(Va) Vanillic

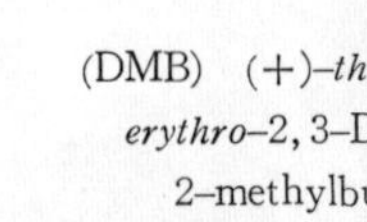

(Ve) Veratric

(b) Physical constants and Occurrence of Alkamines and derived Alkaloids

Table 109
(Alauddin *et al.*[2086] etc.)

(i) *Glyco–alkaloids*

Str. No.	Alkamine	mp.	$[\alpha]_D$ (CHCl$_3$)	Derived Alkaloids (3–glucosides)	mp.	$[\alpha]_D$ (CHCl$_3$)	Source 〔Note〕	Ref.
(LXXII)	Veratramine C$_{27}$H$_{39}$O$_2$N	204~207	−71	Veratrosine C$_{33}$H$_{49}$O$_7$N	242~243 (decomp.)	−55 (EtOH/ CHCl$_3$)	*V. a.*, *V. vi.*	[2155)~2158)]
(LXXIII)	Jervine C$_{27}$H$_{39}$O$_3$N	240~245	−167.5	Pseudojervine C$_{53}$H$_{49}$O$_8$N	300~301	−131	*V.a.*, *V.vi.*, *Sch. off.*	UV, NMR [2157)~2164)]
(LXXIII–a)	Isojervine C$_{27}$H$_{39}$O$_3$N	acetone compd. 105~112	″ −37 (EtOH)					UV, IR, NMR [2164)~2167)]
(LXXIV)	Rubijervine C$_{27}$H$_{43}$O$_2$N	242~243	+8				*V.a.*	[2168)2169)]
(LXXV)	Isorubijervine C$_{27}$H$_{43}$O$_2$N	235~237	+6.5 (EtOH)	Isorubijervosine C$_{33}$H$_{53}$O$_7$N	279~280	−20	*V.a.*, *V.vi.*	[2168)2170)2171)]

(ii)–(iii) *Ester–alkaloids*

Str. No.	Alkamine	Ester–alkaloids	Esterifying acids (R) position of ester ( )	mp.	$[\alpha]_D$ (pyridine)	Source 〔Note〕	Ref.
(LXXVI)	Zygadenine C$_{27}$H$_{43}$O$_2$N mp. 201~204° $[\alpha]_D$ −45(CHCl$_3$) [2172)]	Zygacine C$_{29}$H$_{45}$O$_8$N	Ac–(3)	Amorph	−24 (CHCl$_3$)	*Z. pa.*, *Z. ven.*	IR [2174)~2176)]
	Synthesis of ester [2173)]	Angeloylzygadenine C$_{44}$H$_{69}$O$_{11}$N	An–(3)	222~224	−35	*V. a.*	[2177)2178)] UV, IR
		Vanilloylzygadenine C$_{35}$H$_{49}$O$_{10}$N	Va–(3)	258~259	−27	*Z. pa.*, *Z. ven.*	[2179)]
		Veratroylzygadenine C$_{36}$H$_{51}$O$_{10}$N	Ve–(3)	270~271	−27	*V.a.*,*V.fim.*, *V.es.*,*V.n.*	[2177)2180)2181)] UV, IR
		C$_{32}$H$_{51}$O$_8$N·3H$_2$O	*l*-MB–(3)	175	−7.8	*V.st.*	[2182)]

2155) Jacobs, W. *et al*: *J. Biol. Chem.*, **181**, 55 (1949).
2156) Klohs, M.W *et al*: *J.A.C.S.*, **75**, 2133 (1953). 2156a) Tamm, C. *et al*: *ibid.*, **74**, 3842 (1952).
2157) Mitsuhashi, H. *et al*: *6th Symposium on the Chemistry of Natural Products*, p.178 (Sapporo, 1962).
2158) Bailey, D.M. *et al*: *Tetr. Letters*, **1963**, 555.
2159) Jacobs, W. *et al*: *J. Biol. Chem.*, **155**, 565 (1944).
2160) Klohs, M. W. *et al*: *J.A.C.S.*, **75**, 4925 (1953).
2161) Okuda, S. *et al*: *Chem. & Ind.*, **1961**, 512. 2162) Poethke, W: *Arch. Pharm.*, **276**, 170 (1938).
2163) Tsukamoto, T. *et al*: *Y. Z.*, **74**, 729 (1954).
2164) Takasugi *et al*: *6th Symposium on the Chemistry of Natural Products*, p.172 (Sapporo, 1962).
2165) Masamune *et al*: *7th ibid.*, p.29 (Fukuoka, 1963).
2166) Wintersteiner, O. *et al*: *Tetrahedron*, **18**, 795 (1962).
2167) Wintersteiner, O. *et al*: *J. Org. Chem.*, **29**, 262 (1964).
2168) Jacobs, W.A. *et al*: *J. Biol. Chem.*, **160**, 555 (1945).
2169) Pelletier, S.W. *et al*: *J.A.C.S.*, **79**, 4531 (1957).
2170) Klohs, M.W. *et al*: *ibid.*, **75**, 2133 (1953).
2171) Weisenborn, F.L. *et al*: *ibid.*, **75**, 259 (1953). 2172) Kupchan, S.M: *ibid.*, **81**, 1925 (1959).
2173) Shimizu, B: *Y. Z.*, **80**, 32 (1960). 2174) Shimizu, B: *ibid.*, **78**, 443 (1958).
2175) Kupchan, S.M: *J.A.C.S.*, **77**, 689 (1955), **78**, 3546 (1956).
2176) Shimizu, B. *et al*: *Y. Z.*, **79**, 609, 615, 619 (1959).
2177) Kupchan, S.M. *et al*: *J.A.P.A.*, **48**, 737 (1959).
2178) Suzuki, M. *et al*: *Y. Z.*, **77**, 1050 (1957). 2179) Kupchan, S.M. *et al*: *J.A.C.S.*, **74**, 2382 (1957).
2180) Klohs, M.W. *et al*: *ibid.*, **75**, 4925 (1953).
2181) Stoll, A. *et al*: *Helv. Chim. Acta*, **36**, 1570 (1953). 2182) Kawasaki, T. *et al*: *Y. Z.*, **82**, 210 (1962).

Str. No.	Alkamine	Ester–alkaloids	Esterifying acids (R) position of ester ( )	mp.	$[\alpha]_D$ (pyridine)	Source [Note]	Ref.
(LXXVII)	Veracevine $C_{27}H_{43}O_8N$ mp.181~183° $[\alpha]_D$ −33(CHCl₃)	Cevacine $C_{24}H_{45}O_2N$	Ac–(3)	205~207	−27	V.sab.	2183)~2186)
		Cevadine (Veratrine) $C_{32}H_{49}O_9N$	An–(3)	208~215	+6	V.sab.,V. vi.,Sch. off.	2187)2188)
		Veratridine $C_{36}H_{51}O_{11}N$	Ve–(3)	160~180	−19	V.a.,V.sab. V.vi.	2186)2187)
(LXXVII -a)	Cevine $C_{27}H_{43}O_8N$		Vigorous treatment of cevagenine, cevadine and veratridine with alkali				2184)
(LXXVIII)	Germine $C_{27}H_{43}O_8N$ mp.218~221° $[\alpha]_D$ +4(CHCl₃ EtOH)2190)	Germanitrine $C_{39}H_{59}O_{11}N$	Ac–(7), (−)–MB–(15), An–(3)	228~229	−61	V.fim.	2189)~2192)
		Germbudine $C_{37}H_{59}O_{12}N$	(−)–MB–(15), (+)–D MB(3)	160~164	−8	V.vi.	2193)2194)
		Neogermbudine $C_{37}H_{59}O_{12}N$	(−)–MB–(15), (−)–D MB(3)	149~152	−12	V.a.,V.vi.	2195)2196)
		Germerine $C_{37}H_{59}O_{11}N$	(−)–MB–(3), (+)–HM B–(15)	200~203	−14	V.a.,V.vi., V.n.	2197)~2201)
		Germidine $C_{34}H_{53}O_{10}N$	Ac–(3), (−)–MB–(15)	230~231	−11	V.vi., Z. ven.	2198)2201)2202)
		Isogermidine $C_{34}H_{53}O_{10}N$	Ac–(7), (−)–MB–(15)	221~223	−63	V.vi., Z. ven.	2198)~2201) 2203)
		Germinitrine $C_{39}H_{57}O_{12}N$	Ac, An, Ti	175~176	−36	V.fim.	2180)2191)
		Germitetrine $C_{41}H_{63}O_{14}N$	Ac–(7), (−)–MB–(15), HMAB–(3)	229~230	−74	V.a.	2195)2204)2205)
		Desacetylgermit- etrine $C_{39}H_{61}O_{13}N$	(−)–MB–(15), HMAB–(3)	143~149	−8	V.a.	2195)2196)
		Germitrine $C_{39}H_{61}O_{12}N$	Ac–(7), (−)–MB–(3), (+)–HMB–(15)	216~219	−69	V.vi.	2197)2198)
		Neogermitrine $C_{36}H_{55}O_{11}N$	2mol Ac–(3, 7), (−)–MB–(15)	234~235	−78	V.vi, V.fim. V.es,Z.ven.	2198)2202)2206) 2207)
		Protoveratridine $C_{32}H_{51}O_9N$	(−)–MB–(3)	272~273	−9	V.a, V.vi, Z. ven.	2198)2203)

2183) Barton, D.H.R. et al: Experientia, **10**, 81 (1954).

2184) Kupchan, S.M. et al: Tetrahedron, **7**, 47 (1959).

2185) Idem: J.A.C.S., **75**, 5519 (1953).

2186) Vejdelek, Z.L. et al: Coll. Trad. Chim. Tcke'cosl., **21**, 995 (1956).

2187) Ikawa, M. et al: J. Biol. Chim., **159**, 517 (1945).

2188) Kupchan, S.M. et al: J.A.P.A., **49**, 242 (1960).   2189) Blount, B.K.: Soc. **1935**, 122.

2190) Idem: J.A.C.S., **81**, 1913 (1959).

2191) Klohs, M.W. et al: J.A.C.S., **74**, 4473 (1952).

2192) Kupchan, S.M. et al: J.A.P.A., **48**, 731 (1959).

2193) Myers, G.S. et al: J.A.C.S., **77**, 3348 (1955).

2194) Myers, G.S. et al: J.A.P.A., **48**, 737 (1959).

2195) Myers, G.S. et al: ibid. 48, 440 (1959).

2196) Myers, G.S. et al: J.A.C.S., **78**, 1621 (1956).   2197) Fried, J. et al: ibid., **72**, 4621 (1950).

2198) Kupchan, S.M: ibid., **81**, 1921 (1959).   2199) Myers, G.S. et al: ibid., **74**, 3198 (1952).

2200) Poethke, W: Arch. Pharm., **275**, 571 (1937).

2201) Weisenborn, F.L. et al: J.A.C.S., **76**, 5543 (1954).

2202) Fried, J. et al: ibid., **74**, 3041 (1952).   2203) Kupchan, S.M. et al: ibid., **74**, 3202 (1952).

2204) Glen, W.L et al: Nature, **170**, 932 (1952).   2205) Nash, H.A et al: J.A.C.S., **75**, 1942 (1953).

2206) Kupchan, S.M. et al: ibid., **75**, 4671 (1953).   2207) Klohs, M.W. et al: ibid., **76**, 1152 (1954).

Str. No.	Alkamine	Ester–alkaloids	Esterifying acids (R) position of ester ( )	mp.	$[\alpha]_D$ (pyridine)	Source [Note]	Ref.
(LXXIX)	Protoverine $C_{27}H_{43}O_9N$ mp. 195~200° $[\alpha]_D$ −11 (EtOH) [2209]	Escholerine $C_{41}H_{61}O_9N$	2mol Ac–(6, 7), (−)– MB–(15), An–(3)	235~236	−30	V.esch.	2208)~2211)
		Deacetylproto- veratrine $C_{39}H_{61}O_{13}N$	Ac–(6), (−)–MB–(15), (+)–HMB–(3)	191~192	−15	V.a.	2196) 2212)
		Protoveratrine A $C_{41}H_{63}O_{14}N$	2mol Ac–(6, 7), (−)–M B–(15),(+)–HMB–(3)	267~269	−40	V.a.,V.v. Z.ven.	2199) 2205) 2206) 2210)
		Protoveratrine B $C_{41}H_{63}O_{15}N$	2mol Ac–(6, 7), (−)–M B–(15),(+)–DMB–(3)	268~270	−37	V.a.,V.vi. Z.ven.	2213)~2215)
(LXXX)	Zygadenilic acid–δ–lactone	—— angelate $C_{32}H_{47}O_8N\cdot H_2O$	An–(16)	235	−10.9	V.gr. IR	2216)
(LXXXI)	Sabine $C_{27}H_{45\sim47}O_7N$ mp. 173~176° —	Sabatine $C_{29}H_{47\sim49}O_8N$	Ac–	256~258	—	Sch. off.	2217)
		Sabadilline I, II, III	—	—	—	Sch. off.	2218)

〔Note〕 (Source) *Sch. off.* = *Schoenocaulon officinale, V.a.= Veratrum album, V.es.= V.escholtzii, V.fim.= V.fimbriatum, V.gr.= V.grandiflorum, V.n.=V.nigrum, V.pa.= V.paniculatus, V.sab.=V.sabadilla, V.st.= V.stamineum, V.vi.= V.viride, Z.ven.= Zygadenus venenosus*

(c) UV, IR and NMR Spectra

See Table 109.

UV   α–and β–Cevine[2219], α–, β– and iso–germine[2219], veratramine, dihydroveratramine[2220], jervine,

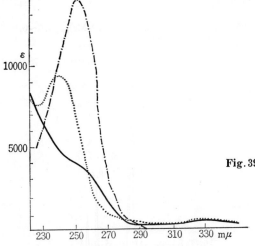

Fig. 394 UV Spectra of —·— Jervine, —— Isojervine and ······ Dihydro- isojervine (Takasugi *et al.*)[2164]

2208) Klohs, M.W. *et al*: *J.A.C.S.*, **74**, 1871 (1952).
2209) Kupchan, S.M. *et al*: *ibid.*, **81**, 4753 (1959), **82**, 2242 (1960).
2210) Kupchan, S.M. *et al*: *J. Pharm. Sci.*, **50**, 52, 396 (1961).
2211) *Idem*: *J.A.P.A.*, **48**, 735 (1959).
2212) *Idem*: *J.A.C.S.*, **82**, 2616 (1960).   2213) *Idem*: *ibid.*, **82**, 2252 (1960).
2214) Stoll, A. *et al*: *Helv. Chim. Acta*, **36**, 718 (1953).
2214a) Klohs, N.W. *et al*: *J.A.C.S.*, **74**, 5107 (1952).   2215) Kupchan, S.M: *ibid.*, **81**, 1009 (1959).
2216) Kawasaki, T. *et al*: *Chem. Pharm. Bull.*, **10**, 519 (1962).
2217) Mitchner, H. *et al*: *J.A.P.A.*, **48**, 303 (1959).
2218) Svoboda, G.R. *et al*: *J. Pharm. Sci.*, **52**, 772, 777 (1963).
2219) Jaffe, Jacobs: *J. Biol. Chem.*, **193**, 325 (1951).
2220) Jacobs, Sato: *ibid.*, **191**, 71 (1951).

isojervine[2221], protoverine, protoveratrine and isoprotoverine[2219]

IR   Cevine, cevagenine, cevadine[2222], jervine and protoveratrines[2223]

Jervine (LXXIII)[2164]

UV   See Fig. 394, λ_{max}^{EtOH} mμ (ε): 252 (14000), ΔM_D +748°(EtOH)

NMR of Diacetyljervine τ: 4.57 (C$_6$–H), 5.33br. (C$_3$–H), 7.75s (3) (18 Me), 7.90 (N–COCH$_3$), 7.98 (O–COCH$_3$), 8.97s (19–Me), 8.93d, J=6 (26–Me), 9.11d, J=7 (21–Me)

Triacetylated Product[2191]   C$_{33}$H$_{43}$O$_6$, mp. 239~240°, [α]$_D$ −29°

Jervine (LXXIII)–O,N–diacetate $\xrightarrow{\text{Ac}_2\text{O, H}_2\text{SO}_4}$

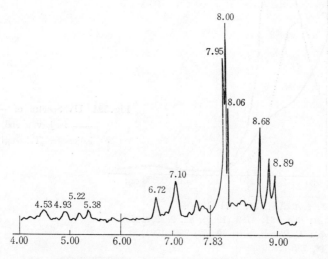

(LXXIII–b) Triacetate

UV λ_{max}^{EtOH} mμ (log ε): 251 (4.08), 300 (3.30)

IR λ_{max} μ: 5.75 (OAc), 5.88 (ketonic carbonyl), 6.09 (N–Ac)

Isojervine (LXXIII–a)[2164]

UV   See Fig. 394 λ_{max}^{EtOH} mμ (log ε): 250sh (3000), ΔM_D +669° (EtOH)

IR ν_{max} cm^{-1}: 1684, 1630 ($\alpha\beta$–unsatur. carbonyl)[2164]

Acetone compound; λ_{max}^{nujol} μ: 3.05m, 5.92s, 6.04, 6.11 (ident. to chloroform adduct)[2167]

N–Acetylisojervine   mp. 207~210°

IR λ_{max} μ: 3.00s, 5.92s, 6.14, 6.20[2167]

Triacetylisojervine

NMR[2164]   60Mc, τ: See Fig. 395, 8.06s (allyl–Me), 8.68s (19–Me), 4.0~6.0 (4), 4.53 (C$_6$–H), 6.00~7.83 (11), 8.89d, J=7 (21–Me)[2165]

Fig. 395 NMR Spectrum of Triacetylisojervine (Takasugi *et al.*)[2164]

Veratroylzygadenine (LXXVI), R=Ve

UV λ_{max}^{EtOH} mμ (log ε): 265 (4.12), 294 (3.90)

2221)   Jacobs, Huebner: *J.Biol. Chem.*, **170**, 635 (1947).

2222)   Stoll, A., Seebeck, E: *Helv. Chim. Acta*, **35**, 1270 (1952).

2223)   Grant, E.W., Kennedy, E.E: *J. A. P. A.* **44**, 129 (1955).

IR ν_{max} cm^{-1}: 1701 (ester C=O)[2178)

Angeloylzygadenine (LXXVI), R=An

IR λ_{max} μ: 5.88, 6.04 (α, β–unsatur. carboxylic acid)[2178)

Germine (LXXVIII)[2202)

IR λ_{max}^{nujol} μ: 3.04~3.26, 3.50, 6.80, 7.21, 8.24, 8.60, 8.82, 9.04, 9.34, 9.62, 9.97, 10.20, 10.54, 11.10, 11.62, 12.36

Germerine (LXXVIII) (−)–MB–(3), (+)–HMB–(15)[2202)

IR λ_{max}^{nujol} μ: 3.10, 3.50, 5.74, 5.81, 6.80, 7.21, 7.41, 7.98, 8.73, 8.91, 9.20, 9.46, 9.63, 9.82, 10.37, 10.58, 11.10, 12.13, 12.34, 12.81

Germidine (LXXVIII) Ac–(3), (−)–MB–(15)[2202)

IR λ_{max}^{nujol} μ: 2.92~3.15, 3.50, 5.76, 6.80, 7.21, 7.98, 8.60, 8.95, 9.25, 9.54, 10.16, 10.42, 10.84, 11.10, 11.55, 11.75, 12.01, 12.36, 12.68

Germitetrine (LXXVIII) Ac–(7), (−)–MB–(15), HMAB–(3)

IR See Fig. 396

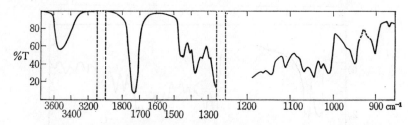

Fig. 396 IR Spectrum of Germitetrine in CHCl$_3$ (Myers *et al.*)[2196)

Germitrine (LXXVIII) Ac–(7), (−)–Me–(3), (+)–HMB–(15)[2202)

IR λ_{max}^{nujol} μ: 2.99, 3.50, 5.90, 6.80, 7.21, 7.52~7.93, 8.34, 8.65, 8.95, 9.30, 9.66, 9.90, 10.23, 10.58, 11.10, 11.41, 12.36, 12.52, 13.02

Neogermitrine (LXXVIII) 2 mol Ac–(3,7), (−)–MB–(15)[2202)

IR λ_{max}^{nujol} μ: 2.97, 3.50, 5.71, 5.82, 6.80, 7.21, 7.98, 8.65, 9.00, 9.30, 9.50, 9.82, 10.09, 10.23, 10.42, 11.10, 11.52, 12.10, 12.31, 12.55, 12.81

Protoveratrine (LXXIX) 2 mol Ac–(6,7), (−)–MB–(15), (+)–HMB–(3) (A) or (+)–DMB–(3)[2202)

IR λ_{max}^{nujol} μ: 2.87, 2.99, 3.50, 5.74, 6.78, 7.21, 7.46, 7.62, 7.98~8.82, 9.25, 9.46, 9.78, 10.20, 10.46, 10.95, 12.07, 12.55, 12.74, 13.20

Escholerine (LXXIX) 2 mol Ac–(6,7), (−)–MB–(15), An–(3)[2207)

IR λ_{max} μ: 5.7, 8.1 (C=O, C–O–C)

Protoverine (LXXIX) See Fig. 397

"Protoveratrine" (Klohs *et al.* 1952)[2214a)   mp. 255° (decomp.), $[\alpha]_D$ −8.3 (CHCl$_3$), $C_{39}H_{61}O_{13}N$

IR See Fig. 397

Neoprotoveratrine (LXXIX), 2 mol Ac–(6,7), (−)–MB–(15), (+)–DMB–(3) (Protoveratrine B)[2214a)

mp. 255.4~255.8°, $[\alpha]_D$ −39° (pyridine), $C_{41}H_{63}O_{15}N$

IR See Fig. 397

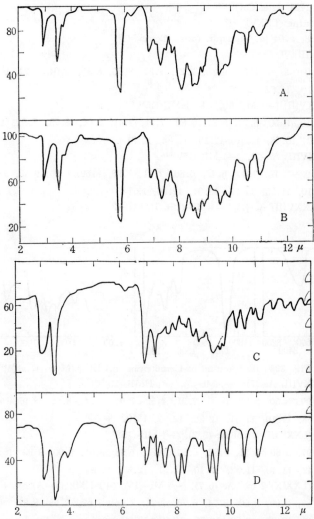

Fig. 397 IR Spectra of (A) Protoveratrine (in CHCl₃), (B) Neoprotoveratrine (in CHCl₃), (C) Protoverine (in nujol) and (D) α–Methyl–α, β–dihydroxy-butyric acid (DMB) (in nujol) (Klohs *et al.*)[2214a]

(d) Fritillaria Alkaloids

Table 110 Fritillaria Alkaloids (Alauddin *et al.*)[2186]

Str. No.	Alkaloid	Formula	mp.	$[\alpha]_D$(CHCl₃)	Source, Ref.
(LXXXII)	Alginine	$C_{23}H_{39}O_3N$	271~272	+108(EtOH)	*F. sewerzowii*[2224]
(LXXXIII)	Amianthine	$C_{27}H_{41}O_2N$	251~253	−87	*Amianthium muscae-toxicum*[2225]
(LXXXIV)	Base	$C_{27}H_{45}O_3N$	256	—	[2226]

2224) Yunusov, S.Y. *et al*: *J. Gen. Chem., U.R.S.S.*, **9**, 1911 (1939).
2225) Neuss, N: *J.A.C.S.*, **75**, 2772 (1953).
2226) Chu, T.T. *et al*: *Acta chim. Sinica.*, **22**, 210 (1956).

Str. No.	Alkaloid	Formula	mp.	$[\alpha]_D(CHCl_3)$	Source, Ref.
(LXXXV)	Beilpeimine	$C_{27}H_{43}O_3N$	155~157	−53(EtOH)	2226)
(LXXXVI)	Chinpeimine	$C_{27}H_{43}O_3N$	247~248	−21	2226)
(LXXXVII)	Fritiminine		258~260	—	2226)
(LXXXVIII)	Imperoline	$C_{27}H_{45}O_3N$	237~238	− 8	2227)
(LXXXIX)	Imperonine	$C_{27}H_{43}O_3N$	239	−65	2227)
(XC)	Peimidine	$C_{27}H_{45}O_2N$	222	−74(EtOH)	2228)
(XCI)	Peimine (Peimunine, Verticine, Apoverticine)	$C_{27}H_{45}O_3N$	223~224	−26(EtOH)	2229)~2232)
(XCII)	Peiminine (Peimiphine, Peimitidine, Verticilline, Fritillarine)	$C_{27}H_{43}O_3N$	212~213	−78	2228) 2232) 2233)
(XCIII)	Peimissine	$C_{27}H_{43}O_4N$	270	−51(EtOH)	2228)
(XCIV)	Sipeimine (Imperialine)	$C_{27}H_{43}O_3N$	267	−36	2234)~2236)
(XCV)	Sonpeimine	$C_{27}H_{43}O_4N$	256~258	—	2226)

Verticine (Peimine) (XCI)[2237)2238)]

≪Occurrence≫ *Fritillaria verticillata* WILD. var *Thunbergii* BAKER (バイモ)

mp. 223~224°, $[\alpha]_D$ −19.0° (EtOH), pKa′ 9.5, $C_{27}H_{45}O_3N$

IR ν_{max}^{nujol} cm^{-1}: 3333, 1113, 1058, 1037 (OH)

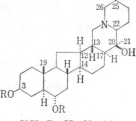

(XCI) R=H : Verticine

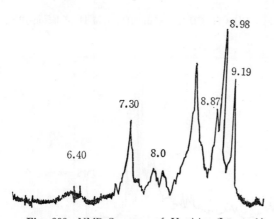

Fig. 398 NMR Spectrum of Verticine (Itō *et al.*)

NMR in CDCl$_3$, 60 Mc, TMS τ: See Fig. 398, 9.19 s (3) (19–Me), 8.98 s (3) (21–Me), 8.93 d (J=7 c/s)

(27–Me), 7.5~8.0 m (N–CH, N–CH$_2$—), 7.30 (shifts depend on concentration) (O–H), 6.40

(CH–OH)

2227) Boit, H.G. *et al*: *Ber.*, **91**, 1968 (1958).   2228) Chou, T.Q: *J.A.P.A.*, **36**, 215 (1947).

2229) Chou, T.Q. *et al*: *J.A.C.S.*, **63**, 2936 (1941).

2230) Chu, T.T. *et al*: *Acta chim. Sinica.*, **22**, 210, 361 (1956).

2231) Itō, S. *et al*: *Chem. Pharm. Bull.*, **9**, 253 (1961).

2232) Wu, Y.H: *J.A.C.S.*, **66**, 1778 (1944).   2233) Chi, Y.F. *et al*: *ibid.*, **62**, 2896 (1940).

2234) Bauer, S. *et al*: *Chem. Zvesti*, **12**, 584 (1958).   2235) Boit, H.G: *Ber.*, **87**, 472 (1954).

2236) Chu, T.T. *et al*: *Acta chim. Sinica.*,: **21**, 241 (1955).

2237) Itō, S. *et al*: *6th Symposium on the Chemistry of Natural Products* p. 166 (Sapporo, 1962).

Chemical shifts of verticine series[2238]

Peiminoside (*O*–β–D–Glucopyranosyl–(1→3)–peimine) $C_{33}H_{55}O_8N$[2239]

[M] Diterpenoid Alkaloids

(1) Tuberostemonine (V-a) or (V-b)[2251)2251a)2252)2252a]

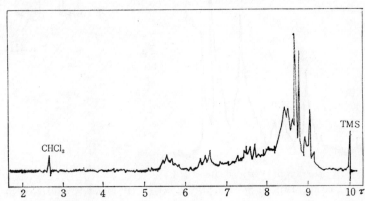

(a)   (b)

(V) Tuberostemonine[2250)2252]

≪Occurrence≫ *Stemona tuberosa* (タマビャクブ)

mp. 86~88°, pK 6.4 (60% EtOH), $C_{22}H_{33}O_4N$

IR ν_{max}^{KBr} cm^{-1}: 1765

NMR   ppm: 8.63, 8.77 (2 $-\overset{\overset{O}{\|}}{C}-CH_3$), 9.02 t (>CH–CH$_2$–CH$_3$), 5.62m (2H of, δ–H of lactone)[2252],

see Fig. 399[2250]

![NMR Spectrum]

Fig. 399 NMR Spectrum of Tuberostemonine (in CDCl$_3$, 60 Mc, TMS)
(Uyeo *et al.*)[2250]

Dehydrogenated Compound 'I' (VI-a) or (VI-b)[2250]

[Preparation] Tuberostemonine (V–a or b) $\xrightarrow[-H_2]{40\%Pd\text{-}Asbest}$ (VI–a or b)

2238) Nozoe, T. *et al*: *Chem. Pharm. Bull.*, **11**, 1327 (1963).
2239) Morimoto, H. *et al*: *ibid.*, **8**, 302, 871 (1960).
2250) Uyeo, S. *et al*: *Y. Z.*, **84**, 663 (1964).
2251) Uyeo, S. *et al*: *6th Symposium on the Chemistry of Natural Products*, p. 82 (Sapporo, 1962).
2251a) Suzuki, S. *et al*: *Y. Z.*, **54**, 573 (1934). **59**, 443 (1939).
2252) Götz, M. *et al*: *Tetr. Letters*, **1961**, 707.
2252a) Schild, H. *et al*: *Ber.*, **69**, 74 (1936).

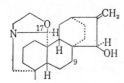

(a)　　　　　　　　(b)

(VI) Dehydrogenated Compound "I"

mp. 275°, $C_{22}H_{27}O_3N$

UV $\lambda_{max}^{EtOH}\,m\mu$ (log ε): 228 (4.60), 264 (3.81), 292 (3.76), 301 (3.77) (indole derivs.)

IR $\nu_{max}^{KBr}\,cm^{-1}$: 3425, 1740; $\nu_{max}^{CHCl3}\,cm^{-1}$: 3580, 1765 (OH and lactone)

NMR of Acetate in CDCl$_3$, 60 Mc, TMS, τ: 2.98 (2 arom. protons), 7.61 (COCH$_3$), 7.79 (arom-**Me**),

8.98 t (—CH$_2$–CH$_3$), 8.48 d (—$\overset{\overset{\text{O}}{\|}}{C}$–CH–CH$_3$), 5.13 t, J=5 c/s (1) (–CO–O–C–H)

(4) Atisine[2253][2254], Veatchine and related Alkaloids[2253]~[2271a]

(VII) Atisine[2256][2257][2266]

$R_1=R_2=H$ (VIII) Dihydroatisine[2258]
$R_1=H,\ R_2=OH$ (IX) Dihydroajaconine[2258]
$R_1=R_2=O$ (X) Atidine (Plane str.)[2254][2254a]

2253) Iwai, I. *et al*: *Chem. Pharm. Bull.*, **11**, 766, 770, 774, (1963); *6 th Symposium on the Chemistry of Natural Products*, p. 131 (Sapporo, 1962).
2254) Pelletier, S.W: *Tetrahedron*, **14**, 76 (1961).
2254a) *Idem.*: *Chem & Ind.*, **1956**, 1016, **1957**, 1670.
2254b) Craig, L.C. *et al*: *J. Biol. Chem.*, **143**, 611 (1942).
2255) Pelletier, S.W: *J.A.C.S.*, **82**, 2398 (1960).
2256) Whalley, W.B: *Tetrahedron*, **18**, 43 (1962).
2257) Bell, R.A. *et al*: *Tetr. Letters*, **1963**, 269.　　2257a) Cneto, J.F: *J.A.P.A.*, **35**, 204 (1946).
2258) Dvornik, D. *et al*: *Tetrahedron*, **14**, 54 (1961).
2258a) Jacobs, W.A. *et al*: *J. Biol. Chem.*, **170**, 515 (1947), **174**, 1001 (1948).
2259) Suzuki, A. *et al*: *Bull. Chem. Soc. Japan*, **34**, 455 (1961).
2259a) Majima, R. *et al*: *Ber.*, **65**, 599 (1932).
2259b) Suginome, H. *et al*: *J. Fac. Sci. Hokkaido Univ. Ser.* III, *Chem.*, **4**, 25 (1950); *Bull. Chem. Soc. Japan*, **32**, 352 (1959).
2260) Amiya, T: *ibid.*, **33**, 1175 (1960).

(XI) Ajaconine[2258]

R=COCH₃ (XII) Lucidusculine[2259) 2259a, b) 2260]
R=H (XII–a) Luciculine

(XIII) Isoatisine[2254) 2267]

R₁=OH, R₂=H (XIV) Garryfoline[2257]
R₁=H, R₂=OH (XV) Veatchine[2256) 2271a]

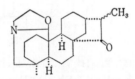

(a) The structure by Djerassi[2270]

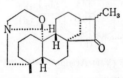

(b) Revised Structure by Whalley[2256]

(XVI) Cuauchichicine

(XVII) Garryine[2256) 2271]   (XVIII) Miyakonitine[2262]   (XIX) Miyakonitinone[2262]

2261) Katsui, N. *et al*: *Bull. Chem. Soc. Japan*, **33**, 1037 (1960).
2262) Kakimoto, S. *et al*: *ibid.*, **32**, 349, 352, 1153 (1959).
2263) Sugasawa, T.: *4 th Symposium on the Chemistry of Natural Products*, p 11 (1960).
2264) Ochiai, E. *et al*: *Chem. Pharm. Bull*, **7**, 542 (1959).
2265) Sugasawa, T.: *ibid.*, **9**, 889, 897 (1961).
2266) Nagata, W. *et al*: *J.A.C.S.*, **85**, 2342 (1963).
2266a) Pelletier, S.W. *et al*: *Tetr. Letters*, **1963**, 205.
2267) Leonard, N.J. *et al*: *J.A.C.S.*, **80**, 5185 (1958).
2268) Pelletier, S.W: *ibid.*, **82**, 2398 (1960).

(a) Atisine (VII)[2254) 2258) 2266) ~2268)]

≪Occurrence≫   *Aconitum heterophyllum, A. anthora,*

mp. 57~60°, pKa 12.2, $C_{22}H_{33}O_2N$,

Hydrochloride mp. 311~312°, $[\alpha]_D$ +28°

dl–**Atisine** (Synthetic)[2266)]

Enone (XX)[2266)]   mp. 160~168°

UV λ_{max}^{EtOH} mμ (ε): 208 (13100), 232 sh

IR $\nu_{max}^{CHCl_3}$ cm^{-1}: 1703, 1628, 942

$$(XX) \xrightarrow{\text{Red}} 8\beta\text{–Alcohol (and it's epimer)}^{2266a)} \longrightarrow dl\text{–Atisine}$$

Synthesis[2253) 2253a) 2266)], Biosynthesis[2266)]

(b) Dihydroatisine (VIII)[2258)]

〔Preparation〕

$$\text{Atisine (VII)} \xrightarrow[\text{treatment}^{2258a)}]{\text{alkaline}} \begin{cases} \text{(vigorous)} \longrightarrow \text{(VIII)} \\ \text{(mild)} \longrightarrow \text{(XIII) Isoatisine} \end{cases}$$

mp. 156~158°, $[\alpha]_D$ −45°, $C_{22}H_{35}O_2N$

(c) Dihydroajaconine (IX)[2258)]

〔Preparation〕

$$\text{Ajaconine (XI)} \xrightarrow{\text{NaBH}_4 \text{ in MeOH}} \text{(IX)} \cdot H_2O$$

Hydrate   mp. 99~101°, $[\alpha]_D$ −39°, $[M]_D$ −150°, pKa′ 7.7, $C_{22}H_{35}O_3N \cdot H_2O$

IR ν_{max} cm^{-1}: 3350, 1795, 1652, 895

Triacetate   mp. 133~135°, $[\alpha]_D$ −92°

IR ν_{max} cm^{-1}: 1738, 1728, 1656, 1246, 893

(d) Atidine (X)[2254) 2254a)]

≪Occurrence≫   *Aconitum heterophyllum*

mp. 183°, $[\alpha]_D$ −47°, pKa 7.5, $C_{22}H_{33}O_3N$

(e) Ajaconine (XI)[2254) 2254a) 2258)]

≪Occurrence≫   *Delphinium ajacis* L. (ヒエンソウ), *D. consolida*

mp. 167°, $[\alpha]_D$ −122°, $[M]_D$ −439°, pKa′ 11.8 (dil. MeOH), $C_{22}H_{33}O_3N$

IR ν_{max} cm^{-1}: 3380, 3300, 3100, 1665, 889

(f) Lucidusculine (XII)[2259) 2259a, b) 2260) 2288)]

≪Occurrence≫   *Aconitum lucidusculum* NAKAI

mp. 170~171°, $[\alpha]_D$ −95° (CHCl₃), $C_{24}H_{35}O_4N$

IR λ_{max} μ: 6.1, 11.3 (⟩C=CH₂)

(g) Isoatisine (XIII)[2254) 2258a) 2267)]

〔Preparation〕 See the item (b)

mp. 150~151°, $[\alpha]_D$ −16°, $C_{22}H_{32}O_2N$

(h) Garryfoline (Laurifoline[2270)])(XIV)[2257) 2269) 2270)]

≪Occurrence≫   *Garrya laurifolia* HARTW.

mp. 130~133°, $[\alpha]_D$ −60° (CHCl₃), pK 11.81, $C_{22}H_{33}O_2N$

2269) Mosettig, E. *et al*: *J.A.C.S.*, **83**, 3163 (1961).

2270) Djerassi, C. *et al*: *ibid.*, **77**, 4801, 6633 (1955).

IR See Fig. 400

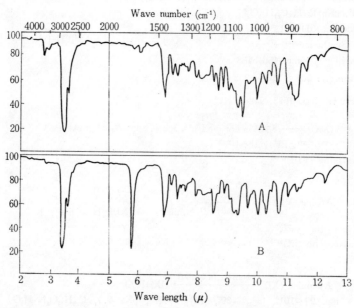

Wave number (cm⁻¹)

Wave length (μ)

Fig. 400 IR Spectra of (A) Garryfoline and (B) Cuauchichicine in CHCl₃ (Djerassi *et al.*)[2270]

$$(XIV) \xrightarrow{10\%HCl} Cuauchichicine \quad (XVI)$$

$$(XIV) \longrightarrow$$

(XIV-a) (−)-β-Dihydro-
kaurene

(XIV-b) Steviol[2269]

(i) Veatchine (XV)[2256) 2271) 2271a]

≪Occurrence≫ *Garrya veatchii*, mp. 119∼120°, [α]ᴅ −69°, pKa 11.5, C₂₂H₃₃O₂N

dl-**Dihydroveatchine** (Synthetic) mp. 138∼141°[2271]

IR $\nu_{max}^{CHCl_3}$ cm⁻¹: 3618, 3479, 1662, 905

Diacetate chloride (XV-a)[2255] mp. 254∼257°

IR ν_{max}^{nujol} cm⁻¹: 1739, 1238 (OAc), 1669 ($>$C$=$N$\overset{+}{-}$)

Total Synthesis[2271]

(j) Cuauchichicine (XVI)[2256) 2270]

≪Occurrence≫ *Garrya laurifolia* HARTW.

mp. 152∼155°, [α]ᴅ −71.4° (CHCl₃), pK 11.15, C₂₂H₃₃O₂N

UV no selective absorption IR see Fig. 400

(k) Garryine (XVII)[2256) 2271) 2271a]

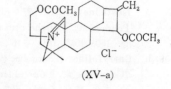

(XV-a)

2271) Nagata, W. *et al*: *J.A.C.S.*, **86**, 929 (1964).
2271a) Wiesner, K. *et al*: *Can. J. Chem.*, **30**, 608 (1952).

≪Occurrence≫　*Garrya veatchii*

Hydrate　　mp. 74〜82°, $[\alpha]_D$ −84°, pKa 8.7, $C_{22}H_{33}O_2N \cdot H_2O$

Total synthesis of *dl*-Garryine[2271]

(l) Miyaconitine (XVIII)[2262]

　≪Occurrence≫　*Aconitum miyabei*, 　mp. 218°, $[\alpha]_D$ −88° (CHCl$_3$), $C_{24}H_{31}O_6N$

UV　λ_{max}^{EtOH} mμ (log ε): 295 (1.6)

IR　ν_{max}^{KBr} cm^{-1}: 3410 (OH), 1728, 1715, 1676 (C=O), 3090, 1648, 877 ($>$C=CH$_2$)

(m) Miyaconitinone (XIX)[2262]

　≪Occurrence≫　*Aconitum miyabei*, 　mp. 285°, $[\alpha]_D$ −27.6° (AcOH), pKa′ 6.5　$C_{24}H_{31}O_6N$

UV　λ_{max}^{EtOH} mμ (log ε): 290 (2.6)

IR　ν_{max}^{KBr} cm^{-1}: 3410 (OH), 1728, 1715, 1676 (C=O), 3090, 1648, 833 ($>$C=CH$_2$)

(5) Songorine and related Alkaloids

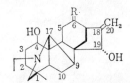

R=O (XXI) Songorine[2265]
(Napellonine[2254)2263)])

$R <^{\ H}_{\ OH}$ (XXII) Napelline[2254)2265)]

(a) Songorine (Napellonine=Shimoburobase I) (XXI)[2254),2263)~2265)]

　≪Occurrence≫　*Aconitum songoricum*, aconite root (Ochiai Nr. 13)

mp. 198〜202°, $[\alpha]_D$ −141.1° (MeOH), $C_{22}H_{31}O_3N_2$

UV　λ_{max} mμ (log ε): 290 (2.50)

IR　ν_{max} cm^{-1}: 3546, 3448 (OH), 1709 (six membered ring ketone), 1664 (double bond)

(b) Napelline (XXII)[2254)2254b)2265)]

　≪Occurrence≫　*Aconitum napellus*, 　mp. 166°, $C_{22}H_{33}O_3N$,

Hydrochloride　mp. 222°, $[\alpha]_D$ −94° (H$_2$O)

(6) Hetisine and related Alkaloids[2272)~2279)]

(XXIII)　Hetisine[2272)]

or

(XXIV)　Pseudokobusine[2274)]

or

(XXV)　Kobusine[2275)2276)2277)]

R=C$_6$H$_5$CO　(XXVI)　Ignavine[2279)]
R=H　(XXVII)　Anhydroignavinol

R=C₆H₅CO (XXVIII) Hypognavine[2278)]
R=H (XXIX) Hypognavinol

(a) Hetisine (XXIII)[2254) 2272)]

≪Occurrence≫   *Aconitum heterophyllum,*   mp. 253~256°, $[\alpha]_D$ +14° (EtOH), $C_{20}H_{27}O_3N$

(b) Pseudokobusine (XXIV)[2274)]

≪Occurrence≫   *Aconitum yesoensis* NAKAI, *A. lucidusculum* NAKAI

mp. 271 (decomp.), $[\alpha]_D$ +50° (CHCl₃), $C_{20}H_{27}O_3N$

N-Acetyl-seco-pseudokobusine (XXIV-a)

$$(XXIV) \xrightleftharpoons[20\%KOH]{Ac_2O+pyridine, OH^-} (XXIV\text{-a})$$

mp. 230~231°, $[\alpha]_D$ −81.4° (MeOH), $C_{22}H_{29}O_4N$

UV λ_{max} mμ (ε): 299 (33), neg. to carbonyl reagents

IR ν_{max} cm⁻¹: 1596 (Ac), 1700 (C=O in six membered ring)

NMR τ: 7.87 s ($-\overset{\overset{O}{\|}}{C}-CH_2-\overset{|}{\underset{|}{C}}-$)

(XXIV-a)

(c) Kobusine (XXV)[2275) 2276) 2277)]

≪Occurrence≫   *Aconitum sachalinense* FR. SCHMIDT

mp. 267~267.5°, $[\alpha]_D$ +104.4° (MeOH), $C_{20}H_{27}O_2N$

IR ν_{max} cm⁻¹: 1648, 888

Dihydrokobusine   mp. 229~231°, $[\alpha]_D$ +78.8° (MeOH)

IR ν_{max}^{nujol} cm⁻¹: 3465, 3125 (OH)

$$(XXIV) \xrightarrow{CrO_3, pyridine} (XXIV\text{-b})+(XXIV\text{-}c_c)$$

(XXIV-b)   (XXIV-c)

2272) Wiesner, K. *et al*: *Tetr. Letters*, **1962**, 621.
2273) Singh, N. *et al*: *J. Ph. Ph.*, **14**, 288 (1962).
2274) Natsume, M: *Chem. Pharm. Bull.*, **10**, 879 (1962), 8, 374 (1960).
2275) Natsume, M. *et al*: *ibid.*, **10**, 883 (1962).   2276) Okamoto, T: *ibid.*, **7**, 44 (1959).
2277) Natsume, M: *ibid.*, **7**, 539 (1959).   2278) Sakai, S. *et al*: *ibid.*, **7**, 50, 55 (1959), **1**, 152 (1953).
2279) Ochiai, E. *et al*: *ibid.*, **7**, 550, 556 (1959).

Kobusinone (XXIV–b)   mp. 273~275°, $C_{20}H_{25}O_2N$

UV λ_{max} mμ (ε): 225 (9000), 330 (25) ($\alpha\beta$–unsatur. ketone)

IR ν_{max}^{nujol} cm^{-1}: 3120 (OH), 1692, 1632 ($\alpha\beta$–unsatur. ketone)

Ketokobusinone (XXIV–c)

mp. 189~191°, 280~282°, $C_{20}H_{23}O_2N$

UV λ_{max} mμ (ε): 218 (7800), 307 (270) ($\alpha\beta$–unsat. ketone)

IR ν_{max}^{nujol} cm^{-1}: 1718 (six membered ring ketone), 1705, 1625

   (α, β–unsatur. ketone in six membered ring)

(d) Ignavine (XXVI) [2279) 2279a)]

≪Occurrence≫ *Aconitum sanyoense* var. *typicum* NAKAI

   (カツヤマブシ)

mp. 172~174° (from cold acetone), 228~230° (from hot

   acetone), $C_{27}H_{31}O_6N$,

Hydrochloride mp. 245~246° (decomp.) $[\alpha]_D$ +58.27° (H_2O)

UV See Fig. 401

(e) Anhydroignavinol (XXVII)[2279) 2279a)]

[Preparation]

Ignavine (XXVI) $\xrightarrow[\text{hydrolysis}]{\text{Alkaline}}$ (XXVII)

mp. 302~304°, (decomp.), $C_{20}H_{25}O_4N \cdot \frac{1}{2}H_2O$

UV See Fig. 401

IR See Fig. 402, λ_{max} μ: 3.10, 3.23 (OH), 5.08, (C=C),

8.45 (>C—OH), 9.12 (>CH—OH), 9.68 (CH_2OH), 11.22

(>C=CH$_2$) ; IR[2242)].

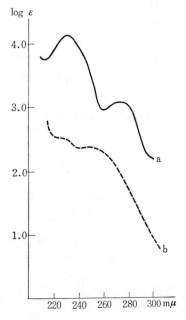

Fig. 401 UV Spectra of (a)
Ignavine and (b)
Anhydroignavinol
(Ochiai *et al.*)[2279a)]

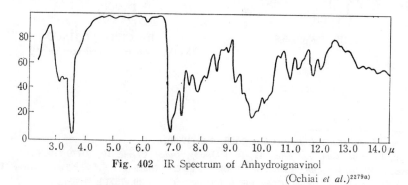

Fig. 402 IR Spectrum of Anhydroignavinol

(Ochiai *et al.*)[2279a)]

(f) Hypognavine (XXVIII) [2278)]

≪Occurrence≫ *Aconitum sanyoense* NAKAI,   mp. 239~241°, $C_{27}H_{31}O_5N$

(XXVIII) $\xrightarrow{\text{CrO}_3, \text{ pyridine}}$ Hypognavinone (XXVIII–a)

Hypognavinone (XXVIII–a)   mp. 305~306°, $[\alpha]_D$ +96.38° ($CHCl_3$), pKa′ 7.38 (50% MCS),

$C_{27}H_{29}O_5N$

UV λ_{max}^{EtOH} mμ (log ε): 231 (4.35), 273 (3.03)

IR ν_{max}^{nujol} cm^{-1}: 3500 (OH), 1709, 1705 (C=O), 1600, 1580 (C_6H_5), 1630 (C=CH$_2$)

2279a) Ochiai, E. *et al* : *Y. Z.*, **72**, 816 (1952).
2242) Sadtler Card No. 7536.

(g) Hypognavinol (XXIX)[2278]

[Preparation] Hypognavine (XXVIII) $\xrightarrow{\text{Hydrolysis}}$ (XXIX)

mp. 307~308°, $[\alpha]_D$ +68°, $C_{20}H_{27}O_4N$

Triacetate–HBr   mp. 283~284°(decomp.), $[\alpha]_D$+52.3° (MeOH), pKa′ 8.35, $C_{20}H_{24}O_4N$ $(COCH_3)_3 \cdot$ HBr

IR $\nu_{max}^{CHCl_3}$ cm^{-1}: 3572, 3297 (OH), 3020 (C=CH$_2$), 2483 $\left(\equiv\overset{+}{N}H\right)$, 1748 (C=O)

(7) Aconitine and related Alkaloids [2273)2279a)~2292]

R=Et: (XL) Aconitine[2242)2280]
R=Me: (XL–a) Mesaconitine

R=C_6H_5CO–(XLI): Indaconitine
R=$(CH_3O)_2$: C_6H_3–CO–
(XLII): Pseudoaconitine

(XLIII) Delphinine [2282]

R=$(CH_3O)_2$: $C_6H_3 \cdot$ CO–
(XLIV) Veratroylpseudoaconine [2284]

2279b) Jacobs, W. A. *et al*: *J. Biol. Chem.*, **128**, 439 (1939).

2280) Marion, L. *et al*: *Can. J. Chem.*, **41**, 1634 (1963); *Index. Chem.*, **10**, 30719 (1963).

2281) Schneider, W. *et al*: *Ann.*, **628**, 114 (1960).   2282) Marion, L. *et al*: *Tetr. Letters*, **1962**, 923.

2283) Marion, L. *et al*: *Pure Appl. Chem.*, **6**, 621 (1963); *C.A.*, **59**, 7571 b (1963).

2284) Marion, L. *et al*: *Can. J. Chem.*, **41**, 1485 (1963); *Index Chem.*, **10**, 30702 (1963).

2285) Marion, L. *et al*: *Proc. Chem. Soc.*, **1959**, 192.

2285a) Marion, L. *et al*: *Can. J. Chem.*, **34**, 1315 (1956); *C.A.*, **51**, 1996 g (1956).

2285b) Trevett, M.E. *et al*: *Chem. & Ind.*, **1956**, 276; *C.A.*, **50**, 15552 g (1956).

2286) Marion, L. *et al*: *Can. J. Chem.*, **34**, 158 (1956).

2286a) Suginome, H. *et al*: *Bull. Soc. Japan*, **32**, 819 (1959).

2287) Suginome, H. *et al*: *ibid.*, **32**, 354 (1959).

2288) Amiya, T: *ibid.*, **30**, 677 (1957), **33**, 644, 1133 (1960).

2289) Suzuki, A. *et al*: *ibid.*, **34**, 455 (1961).   2290) Marion, L. *et al*: *Tetrahedron*, **4**, 157 (1958).

2290a) Marion, L. *et al*: *J.A.C.S.*, **80**, 4434 (1958).   2291) Valenta, Z. *et al*: *Tetrahedron*, **9**, 284 (1960).

2291a) Manske, R.H.F: The Alkaloids, Chemistry and Physiology, **IV** p.275~333 (Academic Press, 1954).

2292) Edwards, O.E. *et al*: *Can. J. Chem.*, **41**, 477 (1963); *C.A.*, **58**, 7989 c (1963).

R=H, R′=Me (XLV) Delsoline [2290) 2290a)]
R=R′=H (XLV-a) Delcosine [2289) 2290a)]

(XLVI) Browniine [2292)]

(XLVII) Lycoctonine [2286)]

(XLVII) Isolycoctonine [2285)]

(a) [2285)]   or   (b) [2291)]

(XLVIII) Hydroxylycoctonine

(XLIX) Hydroxylycoctonine anhalonium iodide[2285)]

(a) Aconitine (XL)[2280) ~ 2282) 2279a)]

≪Occurrence≫ *Aconitum napellus, A. fauriei, A. grossedentatum, A. hakusanense, A. ibiukiense, A.majimai, A. sanyoense* var. *typicum* etc., mp. 205°, $[\alpha]_D$ +19°, $C_{34}H_{47}O_{11}N$

(b) Mesaconitine (XL-a)[2291a)]

≪Occurrence≫ *Aconitum napellus, A. fischeri* and other *Aconitum* spp.

mp. 209°, $[\alpha]_D$ +26°, $C_{33}H_{45}O_{11}N$

Aconitine (XL) ⟶   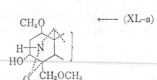   ⟵ (XL-a)

Oxonitine [2279b)]

(c) **Indaconitine** (XLI)[2282) 2283)]

    ≪Occurrence≫   *Aconitum chasmanthum* STAPF,   mp. 203°, $[\alpha]_D$ +18°, $C_{34}H_{47}O_{10}N$

(d) **Pseudoaconitine** (XLII)[2282) 2283)]

    ≪Occurrence≫   *Aconitum balfourii* STAPF., *A. deinorrhizum* STAPF.   mp. 213°, $[\alpha]_D$ +17°, $C_{36}H_{51}O_{12}N$

(e) **Delphinine** (XLIII)[2282)]

    ≪Occurrence≫   *Delphinium staphisagria*,   mp. 200°, $[\alpha]_D$ +25°, $C_{33}H_{45}O_9N$

(f) **Veratroylpseudoaconine** (XLIV)[2284)]

    mp. 199°, $[\alpha]_D$ −38°, $C_{34}H_{49}O_{11}N$

(g) **Delsoline** (XLV)[2290)]

    ≪Occurrence≫   *Delphinium consolida* L.,   mp. 216°, $[\alpha]_D$ +52°, pK 6.4, $C_{25}H_{41}O_7N$

IR ν_{max} cm^{-1}: 3442, 3300

Monoacetyldelsoline. $HClO_4$ mp. 212∼214°

IR ν_{max} cm^{-1}: 3430, 3380 (OH), 3140 $\left(\!\!>\!\!N\overset{\oplus}{H}\right)$, 1762, 1220 (OAc)

    Delcosine (XLV–a) $\xrightarrow{\text{NaH+CH}_3\text{I}}$ (XLV) [2290a)]

(h) **Delcosine** (Lucaconine)[2287) 2289)] (XLV–a)[2290a)]

    ≪Occurrence≫   *Delphinium consolida* L., *Aconitum lucidusculum* NAKAI[2288)]

        mp. 200°, $[\alpha]_D$ +57°, $C_{24}H_{39}O_7N$

NMR in pyridine, 40 Mc cps: 184 s (2) $\left(2\text{ HO–}\overset{|}{\underset{|}{C}}\text{–H}\right)$, 225 d (19) (piperidine ring $-CH_2-$, $-N-CH_2-CH_3$,

$-\overset{|}{\underset{|}{N}}-\overset{|}{\underset{|}{C}}-H$, OCH_3, CH_3O-CH_2-, $CH_3O-\overset{|}{\underset{|}{C}}-H$), 253 d (4) (4 $-\overset{|}{\underset{|}{C}}-H$), 264 br.m (8) (ring $-CH_2-$), 295 t

(3) (N–CH$_2$–CH$_3$)[2289)]

(XLV–a)⟶Diacetate $\xrightarrow{\text{KMnO}_4}$ Diacetyloxodelcosine (XLV–b)⟶Oxodelcosine (XLV–c) $\xrightarrow{\text{Na}_2\text{Cr}_2\text{O}_7}$
Didehydrooxodelcosine (XLV–d)

R=Ac   (XLV–b)
R=H   (XLV–c)

(VLV–d)

(XLV–b) mp. 103∼105°, $C_{28}H_{41}O_{10}N$, IR λ_{max} cm^{-1}: 1644 (lactam C=O)

(XLV–c) mp. 245∼246°, $C_{24}H_{37}O_8N$, IR ν_{max} cm^{-1}: 1649, 1622 (split lactam band)

(XLV–d) mp. 211∼212°, $C_{24}H_{33}O_8N$, UV λ_{max} mμ (log ε): 297 (2.07), IR ν_{max} cm^{-1}: 3346 (OH),
1757 (cyclopentanone), 1720 (cyclohexanone), 1653 (lactam C=O)

(i) **Browniine** (XLVI)[2292)]

    ≪Occurrence≫   *Delphinium brownii* RYDB.

(j) **Lycoctonine** (XLVII)[2285) 2286) 2286a) 2291a)]

    ≪Occurrence≫[2291c)]   *Delphinium consolida* and other spp.—Anthra–noyl–(XLVII), *D. ajacis*—*N*–
        Acetylanthranoyl–(XLVII) (Ajacine), *Aconitum lycoctonum*, *N*–Succinylanthranoyl–(XLVII)
        (Lycaconitine)

mp. 143°, $[\alpha]_D$ +53°, $C_{25}H_{41}O_7N$

Anthranoyllycoctonine   mp. 135°/165°, $[\alpha]_D$ +54°, $C_{32}H_{46}O_8N_2$

Ajacine   mp. 154°, $[\alpha]_D$ +53°, $C_{34}H_{48}O_9N_2$

Lycaconitine $[\alpha]_D$ $+42°$, $C_{36}H_{48}O_{10}N_2$

(XLVII) $\xrightarrow{\text{Pb(OAc)}_4 \text{ in AcOH}}$ Hydroxylycoctonine (XLVIII)

$\xrightarrow[\text{Na}_2\text{CO}_3]{\text{HI}}$ —— anhydronium-iodide (XLIX)[2285a)b)]

Hydroxylycoctonine anhalonium-iodide (XLIX)[2285a)]

mp. 189° (decomp.)

UV λ_{max} mμ (log ε): 330 (1.85),   IR ν_{max} cm^{-1}: 1720 (C=O), 1670 ($>C=\overset{+}{N}<$)

(k) Sachaconitine (L)[2261)]

≪Occurrence≫   *Aconitum miyabei* NAKAI, $C_{23}H_{37}O_4N$

(l) Denudatine and Denudatidine[2273)]

≪Occurrence≫   *Delphinium denudatum* WALL.

Denudatine (LI)   mp. 248~249°, $[\alpha]_D$ $+0.15°$ (EtOH), $C_{21}H_{33}O_2N$

UV $\lambda_{max}^{\text{EtOH}}$ mμ (ε): 210 (4925),

IR ν_{max}^{nujol} cm^{-1}: 3270 (OH), 1094, 1079 (C–O), 2995 (C=CH$_2$), 1650 (C=C), 902 (C–H), 1460, 1392

($>$C–Me)

Denudatidine (LII)   mp. 273°, $[\alpha]_D$ $+31.56°$ (EtOH), $C_{23}H_{35}O_5N$

[N]   Miscellaneous Alkaloids, undetermined

(1) Cephalotaxine (I)[2293)]

(I) Cephalotaxine

≪Occurrence≫   *Cephalotaxus drupacea, C. fortunei*, $C_{18}H_{21}O_4N$

(2) UV and IR

UV Colchicine, substance F[2240)]

IR Colchicine, substance F[2241)],

(3) Colchicine and related Alkaloids[2240)~2249)]

(I) Colchicine

(II) Colchiceine

(a) Colchicine (I)

≪Occurrence≫   *Colchicum autumnale* L. and other spp.

mp. 155°, $[\alpha]_D$ $-119.9°$ (CHCl$_3$), pK 1.8, 7.2, 10.3, $C_{22}H_{25}O_6N$

dl-**Colchicine** (Synthetic)[2246]

mp. 277~279°(decomp.), $C_{22}H_{25}O_6N$

UV λ_{max}^{EtOH} mμ (log ε): 234 sh (4.46), 246 (4.49), 357 (4.22)

IR $\nu_{max}^{CHCl_3}$ cm^{-1}: 3460, 3285 (NH), 1674, 1620, 1597 (C=O)

UV and IR of *dl*-desacetylisocolchicine (III) and *dl*-desacetylcolchicine (IV)

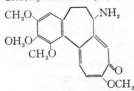

(III) Desacetylisocolchicine
mp. 164°

(IV) Desacetylcolchicine
mp. 147°

UV (III) λ_{max}^{EtOH} mμ (log ε): 346 (4.26), 246 (4.45), 227 sh (4.38)

(IV) λ_{max}^{EtOH} mμ (log ε): 355 (4.16), 242 (4.41), 232 s (4.40)

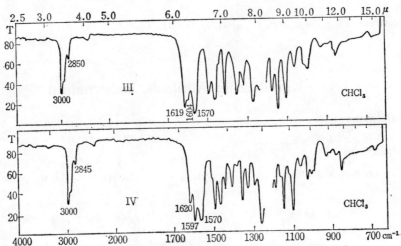

Fig. 398-a IR Specta of *dl*-Desacetylisocolchicine (III) and
dl-Desacetylcolchicine (IV) in CHCl$_3$
(Schreiber *et al.*)[2246]

Synthesis[2240]~[2246][2242a] and Biosynthesis[2248]

(b) Colchiceine (II)

[Production] Colchicine (I) $\xrightarrow[\text{w. dil. HCl}]{\text{Reflux 2 hrs.}}$ (II)

mp. 178~179°, $[\alpha]_D$ −252.5° (CHCl$_3$), $C_{21}H_{23}O_6N$

2240) Ueno, Y: *Y. Z.*, **73**, 1238 (1953). 2241) Ueno, Y: *ibid.*, **73**, 1235 (1953).
2242a) Loewenthal, H.J.E.: *Soc.*, **1961**, 1421.
2243) Van Tamelen, E. E. *et al*: *J.A.C.S.*, **81**, 6341 (1959).
2244) Van Tamelen, E. E. *et al*: *Tetrahedron*, **14**, 8 (1961).
2245) Hardegger, E. *et al*: *Helv. Chim. Acta*, **38**, 2030 (1955).
2246) Schreiber, J. *et al*: *ibid.*, **44**, 540 (1961).
2247) Nakamura, T: *Chem. Pharm. Bull.*, **10**, 281, 291, 299 (1962).
2248) Battersby, A. R. *et al*: *Proc. Chem. Soc.*, **1964**, 86.
2249) Wiesner, K. *et al*: *J.A.C.S.*, **78**, 2867 (1956).

dl-**Colchiceine** (Synthetic)[2247]   mp. 167°, $C_{21}H_{23}O_6N$

UV λ_{max}^{EtOH} mμ (log ε): 243 (4.55), 349 (4.28), 4.05 (3.32)

IR $\lambda_{max}^{CHCl_3}$ μ: 2.86, 3.04, 3.15, 3.34, 3.42, 3.54, 5.95, 6.20, 6.44, 6.71, 6.87, 6.98, 7.10, 7.39, 7.55, 7.84, 7.94, 8.10, 8.35, 8.51, 8.76, 9.11, 9.53, 9.96, 10.17, 10.50, 10.78, 10.94, 11.06, 11.64, 11.83

[O] Purine Derivatives

(1) UV and pK of Xanthines[2294]

Table. 111 UV Absorption and pK of Xanthines (Pfleiderer *et al.*)

Xanthines

Str. No.	Xanthines	pH	pK_{H_2O}	$\lambda_{max}^{H_2O}$	log ε
(I)	Xanthine	5	7.70	225, 266	3.49, 4.03
		10	11.94	241, 276	3.95, 3.97
(II)	1–Me–X	5	7.90	223, 267	3.52, 4.04
		10	12.23	242, 276	3.93, 3.98
(III)	3–Me–X	6	8.45	270	4.05
		10	11.92	274	8.43
(IV)	7–Me–X	6	8.42	268	4.01
		11	>13	231, 287	3.67, 3.93
(V)	9–Me–X	3	6.12	234, 265	3.92, 4.01
		10	>13	245, 277	3.98, 3.97
(VI)	1,3–Di–Me–X	6	8.68	270	4.02
		11		274	4.09
(VII)	1,7–Di–Me–X	6	8.65	268	4.02
		11		233, 288	3.70, 3,94
(VIII)	1,9–Di–Me–X	3	5.99	238, 263	3.82, 3.98
		9		248, 276	3.94, 3.95
(IX)	3,7–Di–Me–X	7	10	271	4.01
		13		234, 273	3.85, 4.01
(X)	3,9–Di–Me–X	7	10.14	235, 268	3.91, 3.99
		13		(240), 269	(3.82), 4.01
(XI)	1,3,7–Tri–Me–X	6		272	4.02
(XII)	1,3,9–Tri–Me–X	6		237, 268	3.99, 4.00

2293) Paudler, W.M. *et al*: *J. Org. Chem.*, **28**, 2194 (1963).
2294) Pfleiderer, W. *et al*: *Ann.*, **647**, 155 (1961).

(2)   **Triacanthine** (XIII)[2295) 2296) 2296a) b)]

≪Occurrence≫   *Gleditschia triacanthos*, *G. horrida* MAKINO (サイカ
チ),   mp. 228∼229°, $C_{10}H_{13}N_5$

UV λ_{max}^{EtOH} mμ (ε): 273 (12500); $\lambda_{max}^{pH=1-EtOH}$ mμ: 239

IR ν_{max} cm^{-1}: 3400, 3240 (NH), 1682, 1630, 1557 (arom.)

NMR   of *N*–Benzyl–(XIII) in CDCl$_3$, τ: 2.77 s (5) (arom. H of benzyl),
8.14 (2Me of dimethylallyl), 2.05 and 2.23 (2H, attached directly
to the purine nucleus), 4.49 t∼5.01 d, (J=7.6) (=CH–CH$_2$), 5.62

Mass   M 203 ($C_{10}H_{13}N_5^+$), 188 ($C_9H_{10}N_5^+$), 135 ($C_5H_5N_5^+$)
Activity: Hypotensive and antispasmodic[2296a) b)]

(3)   **Herbipoline** (XIV)[2296c)]

≪Occurrence≫   *Geodia gigas* (Riesen Kieselschwamm)

(4)   **Biogenesis of purines**[2297)]

2295)   Leonard, N. J. *et al*: *J.A.C.S.*, **84**, 2148 (1962), **82**, 6202 (1960).
2296)   Morimoto, H. *et al*: *Chem. Pharm. Bull.*, **11**, 1320 (1963).
2296a)   Ignatéva, M.A. *et al*: *C.A.*, **51**, 10765 (1957).
2296b)   Goutarel, R. *et al*: *Compt. rend. Biol.*, **155**, 470 (1961).
2296c)   Pfleiderer, W: *Ann.*, **647**, 167 (1961).
2297)   Buchanan, J.M. *et al*: *J. Biol. Chem.*, **237**, 485, 491 (1962).

Index of Plant and Animal Names

Index of Compounds